MW00974346

Bioluminescence and Chemiluminescence

E. Newton Harvey

This volume is dedicated to the memory of
EDMUND NEWTON HARVEY 1887–1959
Pioneer in the study of bioluminescence and
chemiluminescence

Bioluminescence and Chemiluminescence

Molecular Reporting with Photons

Proceedings of the 9th International Symposium on Bioluminescence and Chemiluminescence held at Woods Hole, Massachusetts, October 1996

Edited by

J W Hastings, L J Kricka, P E Stanley

JOHN WILEY & SONS
Chichester • New York • Weinheim • Brisbane • Singapore • Toronto

National 01243 779777
International (+44) 1243 779777
e-mail (for orders and customer service enquiries): cs-books@wiley.co.uk
Visit our Home Page on http://www.wiley.co.uk
or http://www.wiley.com

Other Wiley Editorial Offices

John Wiley and Sons, Inc., 605 Third Avenue,
New York, NY 10158-0012, USA

VCH Verlagsgesellschaft mbH, Pappelallee 3,
D-69469 Weinheim, Germany

Brisbane Singapore Toronto

Library of Congress Cataloging-in-Publication Data

International Symposium on Bioluminescence and Chemiluminescence (9th : 1996 : Woods Hole, Mass,)
Bioluminescence and chemiluminescence : molecular reporting from photons : proceedings of the 9th International Symposium on Bioluminescence and Chemiluminescence at Woods Hole, Massachusetts, October 1996 / edited by J. W. Hastings, L. J. Kricka, P. Stanley.
p. cm.
Includes bibliographical references and index.
ISBN 0–471–97502–8 (alk. paper)
1. Bioluminescence asssay—Congresses. 2. Chemiluminescence assay–Congresses. 3. Bioluminescence—Congresses. 4. Chemiluminescence–Congresses I. Hastings, J. Woodland (John Woodland), 1927–
II. Kricka, Larry J. 1947- . III. Stanley, P. E. (Phillip E.)
IV. Title.
QP519.9.B55158 1996
572′.4358—dc21 96–40502
CIP

British Library Cataloguing in Publication Data:

A catalogue record for this book is available from the British Library

ISBN 0 471 97502 8

Produced from camera-ready material supplied by the contributorsPrinted and bound in Great Britain by Bookcraft (Bath) Ltd. This book is printed on acid-free paper responsibly manufactured from sustainable forestation, for which at least two trees are planted for each one used for paper production.

Table Of Contents

Part 4: BIOLUMINESCENCE IN NATURE: PHYSIOLOGY, FUNCTION AND EVOLUTION

Part 5: LUMINESCENCE IN SCIENCE EDUCATION

Part 9: GREEN FLUORESCENT PROTEIN, AEQUORIN AND OBELIN

PREFACE

This volume contains the Proceedings of the *9th International Symposium on Bioluminescence and Chemiluminescence* held at the Marine Biological Laboratory, Woods Hole, Massachussets, USA between 4 and 8 October 1996 . The Abstracts for the Symposium were published in *J Biolumin Chemilumin* 11 (4+5) 1996 and late abstracts in volume 12 1997.

Previous Symposia in this series have been held in Brussels (1978), San Diego (1980), Birmingham (1984), Freiburg (1986), Florence (1988), Cambridge (1990), Banff (1993) and Cambridge (1994).

It is planned to have the next Symposium at Bologna, Italy in 1998.

Marlene DeLuca Memorial Prize
The Prize, generously sponsored by EG&G Berthold, is awarded at these Symposia for outstanding oral or poster presentations made by young workers. The Memorial Prize Committee considered a number of excellent presentations and awarded the Prize as follows:

Dr CY Wang (University of Utah)
Dr Liming Li (Harvard University)
Dr Patrizia Pasini (University of Bologna)

We wish to express our thanks to:
The International Scientific Programme committee:
JW Hastings (USA) Chair, R Allen (USA), T Baldwin USA), J-M Bassot (France), A Berthold (Germany), AK Campbell (UK), J Case (USA), A Egorov (Russia), J Gitelson (Russia), P Grant (UK), LJ Kricka (USA), M Knight (UK), A Lundin (Sweden), F McCapra (UK), E Meighen (Canada), J-C Nicolas (France), M-T Nicolas (France), D O'Kane (USA), M Pazzagli (Italy), A Roda (Italy), E Schram (Belgium), AP Schaap (USA), P Stanley (UK), G Stewart (UK), A Szalay (USA), A Tsuji (Japan), N Ugarova (Russia), S Ulitzur (Israel), K Wood (USA).

Local Organizing Committee:
JW Hastings (Chair), T Baldwin, I Bronstein, P Dunlap, Y Kishi, D Prasher, G Reynolds, O Shimomura, M Ziegler.

To the exhibitors and sponsors:
US Office of Naval Research

BMG LabTechnologies, Inc., (Durham, NC, USA)
Boehringer Mannheim Biochemicals, (Indianapolis, IN, USA)
Clontech Laboratories, Inc., (Palo Alto, CA, USA)
Dynex Technologies, Inc., (Chantilly, VA, USA)

EG & G Berthold, c/o Wallac Inc.
Hamamatsu Photonic Systems, (Bridgewater, NJ, USA)
John Wiley & Sons, Publishers (Chichester, UK)
Lab Systems-Denley, (Needham Heights, MA, USA)
Laboratory Technologies, Inc., (Roselle, IL, USA)
Lumigen, Inc., (Southfield, MI, USA)
MGM Instruments, Inc., (Hamden, CT, USA)
Packard Instrument Company, Inc., (Meriden, CT, USA)
PharMingen, (San Diego, CA, USA)
Photek Limited, (St Leonards on Sea, UK)
Princeton Instruments, Inc., (Trenton, NJ, USA)
Promega, (Madison, WI, USA)
STRATEC Electronic GmbH, (Birkenfeld, Germany)
Tropix, Inc., (Bedford, MA, USA)
Turner Designs, Inc., Sunnyvale, CA, USA)
Universal Imaging Corporation, (West Chester, PA, USA)
Wallac Inc., an EG&G Company, (Gaithersburg, MD, USA)

We also thank Dr James Deeny of John Wiley and Sons and the *Journal of Bioluminescence and Chemiluminescence* for their help not only with this publication, but also for publishing the Symposium abstracts.

Editors note: These proceedings were compiled from camera ready manuscripts, which were not subjected to peer review. The Editors take no responsibility for the scientific or priority content of the papers. In the interest of rapid and timely publication of these Proceedings the editors have only made minor changes to the contributions to minimise inconsistencies in style.

JW Hastings, LJ Kricka, PE Stanley Editors

The Symposium

INTRODUCTORY REMARKS

J Woodland Hastings
The Biological Laboratories, Harvard University
16 Divinity Avenue, Cambridge, Massachusetts 02138, USA

This 9th International Symposium on Bioluminescence and Chemiluminescence was held at the Marine Biological Laboratory in Woods Hole, Massachusetts, 4-8 October 1996. The MBL, founded in 1888, is the oldest private marine station in the United States and provides specimens along with laboratory and library facilities to investigators coming from all over the world. Its facilities for the collection and maintenance of live animals, which members of the Symposium were able to see, are outstanding. Many important discoveries in modern biology can be traced to MBL, including the nature of the nerve action potential (squid giant axon) and cyclin (cell cycle in surf clam eggs). Numerous Nobel prize winners have carried out research at Woods Hole, including Albert Szent-Györgyi, Otto Loewi, George Wald and Keffer Hartline. Many investigators have studied bioluminescence here, including E. Newton Harvey and Bill McElroy.

It was thus highly appropriate that the Symposium marking the first meeting of the newly established International Society for Bioluminescence and Chemiluminescence should take place at Woods Hole, and we are grateful to the Director and staff of the laboratory for the accommodations and wonderful hospitality they have afforded us. I also want to thank the many persons who helped in the planning and organization of the meeting, especially Karen Christians and the members of the local committee.

At this Symposium we honored Professor Eric Schram of the Vrije Universiteit Brussel, Chairman of the first of these successful and now very important meetings, held in Brussels in September 1978. Together with Dr. Philip Stanley, who has continued to serve as an Editor of most of the volumes, Eric recognized the then emerging applications of bioluminescence and chemiluminescence, and the possibility for enhancing such developments by bringing together basic and applied scientists with those prepared to go forward with commercializing the applications. We salute him for this, and for his continuing contributions to these symposia. His presence and after-dinner comments at the banquet graced the meeting.

The continuing vigorous development of applications of bioluminescence and chemiluminescence was evident from the strong commercial presence at this symposium, with more than 25 companies represented, and 18 exhibiting. Of the 170 abstracts submitted, all of which are published in the Journal of Bioluminescence and Chemiluminescence, a large fraction were concerned with direct applications.

The oral presentations placed considerable emphasis on the basic science, on the continuing search for understanding of mechanisms and new insights into the fundamental nature of the phenomenon of luminescence. As repeatedly pointed out and widely accepted, new applications are virtually completely dependent upon basic research - undirected investigations that seek answers to questions that often seem to have little or no importance and no apparent applications.

As an example, the green fluorescent protein (GFP), studies and applications of which figured so prominently at this symposium (Ward, Phillips, Cormack and others), was first discovered over 25 years ago with my graduate student Jim Morin, in photocytes of the colonial hydroid *Obelia geniculata* collected and studied right here at the MBL in Woods Hole. Curiosity was the main impetus for that research. The front cover illustrates the recently elucidated unusual structural features of this protein.

Likewise, some of the early studies leading to the discovery of autoinducer in luminous bacteria were also carried out in Woods Hole, these by Kenneth Nealson and myself; this is also an example of a finding that lay dormant without any apparent broader significance for over 20 years. Presentations by Greenberg, Bassler, Ulitzur, Winkler, Meighen and Escher focused on the present status of this rapidly developing field, where it has been found that its function in "quorum sensing" is important in diverse species and different ecological situations. Presentations also included studies of systems where the findings and their possible significance are still to come. Delegates to the Symposium heard about the intriguing luminescent squid *Euprymnes* and its bacterial symbiont, where the details of the two way conversation are being studied (McFall-Ngai, Ruby, Nishiguchi). We were also treated to a marvellous account of the circadian (daily) clock in the prokaryote *Synechococcus* using the bacterial lux gene as a reporter (Johnson).

For bioluminescence and chemiluminescence, my mentor E. Newton Harvey stands out as the dominant figure in the field during the first half of the century. Over a period of nearly 50 years he observed and experimented with luminous organisms in every major group and concerned himself with every aspect of the subject, ranging from structure and taxonomy, ecology and evolution to physiology and biochemistry. Long before the advent of molecular evolution, he recognized from still very scant biochemical data the independent evolutionary origins of the many different groups of luminous species and their sporadic phylogenetic distribution, "cropping up....here and there as if a handful of damp sand has been cast over the names of various groups written on a blackboard, with luminous species appearing wherever a mass of sand struck".

Harvey worked at the Marine Biological Laboratory during summers throughout his long career, and it is here that much of the research and writing was done for his many scientific articles and four monographs (The Nature of Animal Light, 1920; Living Light, 1940; Bioluminescence, 1952; The History of Luminescence, 1957). The latter two continue to serve as invaluable reference books, and anyone who sees a copy of either for sale should snap it up. It is with great respect and admiration for his pioneering contributions that we dedicate this volume of the Proceedings of the 9th Symposium to Edmund Newton Harvey

PART 1

Chemiluminescence and Enzymology of Light Emitting Reactions

MECHANISMS IN CHEMILUMINESCENCE AND BIOLUMINESCENCE - UNFINISHED BUSINESS

Frank McCapra
School of Chemistry and Molecular Sciences, University of Sussex, Brighton BN1 9QJ, England

Introduction

The study of mechanism in chemiluminescence and bioluminescence has reached an interesting, but frustrating, stage. It is over thirty years since the first suggestion was made that the formation of dioxetanes would constitute a major route to strong visible chemiluminescence and bioluminescence (1,2). Based on this prediction, experiments on the chemical mechanism of light emission in the crustacean *Cypridina* (now *Vargula*), and the firefly, established dioxetanones as the only credible route to *efficient* chemiluminescence and by extension, bioluminescence. A similar mechanism for coelenterate luciferin (coelenterazine) was also established. The intervening years have only served to confirm their place in the mechanism of the chemical generation of light (3,4).

Parallel work on electrogenerated chemiluminescence, during which radical anions and cations are generated and annihilated, and the earlier work on the reaction of tetralin hydroperoxide with zinc tetraphenylporphin (5), gave rise to the theoretical possibility that the luminescent reactions of peroxides could involve the formation of a radical anion/cation pair, and that the transfer of an electron from anion to cation would result in the formation of an excited state. In spite of the acceptance of this idea, and its extensive and ingenious development (6), for *efficient* chemiluminescence it only remains an attractive *possibility*. It is a remarkable fact that of the many so called CIEEL (Chemically Initiated Electron Exchange Luminescence) systems so skilfully studied, in only two cases is the actual light yield given, and in one of these a subsequent correction indicates a quantum yield of $2x10^{-5}$. There may well be no grounds for accepting the CIEEL mechanism in efficient (say, $Ø_{overall}$ >1%) chemiluminescence (7). The case for its involvement in bioluminescence is even less well supported by the evidence, as will be explained in the sequel. The persistence of the CIEEL description is a tribute to the power of an acronym , and to the lack of a coherent alternative.

Discussion

Although no truly efficient chemiluminescence has ever been discovered using the concept of CIEEL as a guide, several previously existing bright chemiluminescent reactions have been interpreted in this way. The first of these is the oxalate system, in which active oxalate esters react with hydrogen peroxide, to form derivatives of the peroxide, followed by interaction with a fluorescer, preferably a polycyclic aromatic hydrocarbon, to give carbon dioxide and the excited singlet state of the fluorescer as the main products. The logical and theoretically most effective peroxide derivative, dioxetanedione, has never been properly identified, and many other peroxides have been suggested as taking part in the reaction, which is clearly more complex than indicated by the reaction shown (Scheme 1) as illustration (8,9). We are so far from a precise structure for the actual reactant, that we are justified in generalising, and concentrating on the principles involved, in

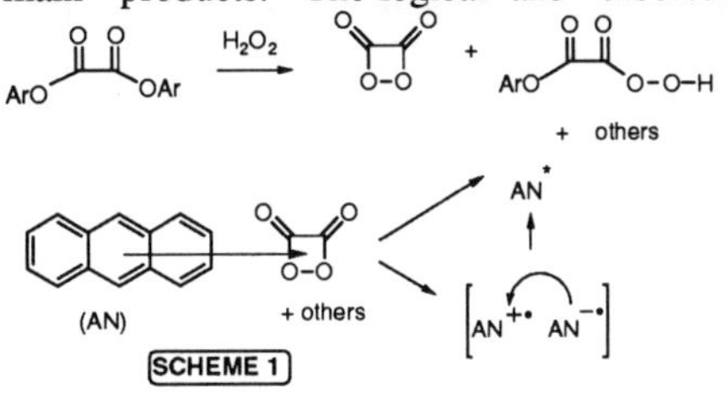

SCHEME 1

determining the nature of the excitation step.

The excited state could arise directly from the ground state charge transfer complex or full electron transfer could yield the radical ion pair, with subsequent annihilation by electron transfer. There is an ideal experimental method available (10) for the study of this notoriously difficult distinction. Irradiation by fast electrons, or by positrons, of biphenyl or naphthalene, creates geminate radical ion pairs (Figure 1).

$$\text{Solv.} \xrightarrow{\text{Radiation}} \text{Solv.}^{+\bullet} + e$$

$$\text{N} + e \longrightarrow \text{N}^{-\bullet}$$

$$\text{Solv.}^{+\bullet} + \text{N} \longrightarrow \text{N}^{+\bullet}$$

$$\text{N}^{+\bullet} + \text{N}^{-\bullet} \longrightarrow \text{N} + \text{N}^{*}$$

[Solv. = solvent; N = naphthalene]

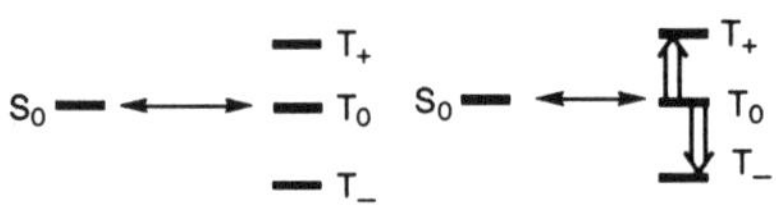

T_+, T_0, T_-, S_0

(a) Singlet-triplet mixing in a high magnetic field (b) Additional triplet population by absorption of microwaves

FIGURE 1

Radical pairs are formed in the singlet state since the preceding solvent molecule had paired electrons. The hyperfine interaction can populate the triplet level To the extent of 50%. In a high magnetic field the triplet levels are as shown schematically, and microwave radiation can induce resonance between them. This depletes the singlet state still further. Since the generation of the singlet excited state that emits light depends absolutely on the annihilation of the singlet radical ions, the depletion is marked by a reduction in the intensity of the light emitted.

An ESR spectrometer has most of the hardware necessary to carry out this experiment, with an enormous advantage over a simple magnetic field. The diminution in light yield as a result of the population of triplets at the expense of singlets is now a resonance phenomenon, and as the field is swept in the ESR machine, light emission is modulated in exact synchrony i.e. the emission of light and the ESR signal are "locked". All "lock in" experiments of this sort give excellent signal to noise ratios, and are as a consequence, extraordinarily sensitive. The Russian workers were able to observe the presence of as few as 20 radical ion pairs per sample tube! Now CIEEL is a process just made for this technique, and we have attempted to carry out this experiment on the oxalates and other candidates for the mechanism, using an aromatic hydrocarbon expected to form a particularly stable radical cation - with negative results (11). However, I must emphasise that we did not have the equipment, nor to be honest, the expertise, to make a satisfactory experiment. I hope that someone will repeat this work, since it seems to be the best means on offer of detecting radical ion pairs formed as a result of chemical reaction, in a quantitative fashion.

CIEEL in Perspective. Those reactions in which the evidence for CIEEL is best substantiated, do not number among the most efficient. Our mechanisms for the luciferins, now fully accepted, were devised using very efficient model compounds. Although we have considered electron transfer as an explanation, for the last 15 years we have been pointing out that charge transfer is a better idea, but harder to verify. We believe that in the majority of efficient luminescent reactions there is no discreet formation of separate radical ion entities, but excitation is achieved by the bond making and breaking merging with a charge transfer excited state. This charge transfer state may well be initiated in the ground state of the reactants (see the representation in the oxalate case above), but at the present time it is not clear whether this is obligatory or not, although it would seem to be in intermolecular reactions. This argument has also been touched on by Wilson (12), who has pointed out that many of the criteria used in determining mechanism are shared by both the charge transfer and discreet electron transfer mechanisms.

The excited states formed from the dioxetanes in the luciferins and models such as the acridans are unequivocally of the *singlet* π,π^* variety, and not the *triplet* n,π^* states invariably found in simple dioxetanes. When a correlation diagram is constructed (and by extension, one of the many methods of MO calculation) which includes the oxygen non-bonding orbitals, then stretching the O-O bond leads to a crossing point for the potential energy surfaces with the characteristics of a diradical *and* an n,π* excited state. In these simple dioxetanes, almost all the calculations and experiments have shown that the C-C bond only breaks when the O-O bond is fully dissociated. The rate of decomposition of the simple dioxetanes is, as expected, unaffected by substitution on the C-C bond, *whereas the visibly efficient ones are remarkably destabilised by electron donating substitution.*

If we construct an orbital correlation diagram that actually reflects the undoubted π,π* character of the event, this π-system-only diagram has two ***repulsive, charge transfer***, antisymmetric surfaces - the lower of which crosses the ground state surface. It is possible that the lowering of the energy implied by the curvature imposed by the repulsive nature of the two states, and the changed configuration about the crossing point, increases the yield of excited state. Many factors affect efficiency in non-adiabatic reactions such as chemiluminescence, for example the relative slopes at the crossing point, and whether the crossing occurs before or after the ground state curve maximum. Any explanation of efficient chemiluminescence must confront the fact that the population of the singlet excited state occurs to the extent of 90% in the firefly. This virtually demands a fully concerted, intramolecular reaction, with no opportunity for potentially wasteful radical side reactions and spin inversion.

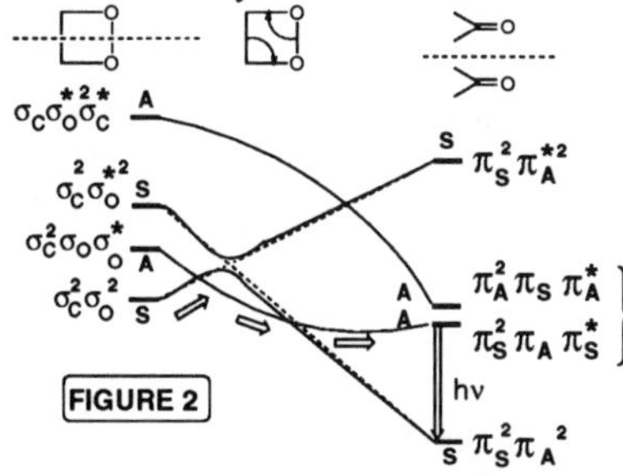

FIGURE 2

I indicated earlier that the distinction between discrete electron transfer and charge transfer is difficult to substantiate experimentally because both processes have so many criteria in common. However there is one system that seems to indicate that charge transfer and not electron transfer is occurring. The para-substituted dioxetane actually decomposes faster (about 4,000 times) than the meta-substituted (both as oxyanions) yet has a quantum yield 140 times lower. In the first place, the decomposition is actually initiated by the creation of the oxyanion so that either electron or charge transfer must be the source of the reaction. Secondly, it is very unlikely that the oxidation potentials of the phenolates (Scheme 2) are very different (they are only isomers, not differently substituted) and so should not show this large rate difference if electron transfer were responsible. Explanations for the difference in terms of charge transfer have already been advanced(13,14,15).

OCH_3 OR OCH_3 O⊖ → Products + Light

SCHEME 2

Bacterial and Flavin Luminescence. With the current information available for bacterial bioluminescence, it is extremely unlikely that dioxetanes play a part in the mechanism of light emission. Indeed, the character of this reaction is a paradigm of the search for a viable non-dioxetane process.

More excellent work has been carried out on this system than on any other bioluminescence. Many suggestions have been made over the years, but the distillation of all the considerations, and the application of the principles of CIEEL, has led to the mechanism in Scheme 3. The strongest support for the mechanism is provided by the observations resulting from the substitution of the benzenoid ring of the flavin (16). Electron transfer will be inhibited by electron withdrawing groups (X in Scheme 3) and enhanced by electron donating groups. If this step is at least partially rate determining, then the rate of reaction should increase and decrease respectively. Such a trend was indeed discovered. The major competitor for the most reasonable fit to the facts is the Baeyer-Villiger reaction shown in **E**. The opposite trend should have been observed in this case.

However, there is a serious flaw in the reasoning concerning the reaction kinetic scheme. The reaction as presented (16) does not include the equilibrium between the 4a-flavin hydroperoxide [**A**], and the hemiperoxyacetal [**B**]. This equilibrium is one of the most mobile and easily set up in organic chemistry, as has been amply demonstrated in this very reaction, using the luciferase (17). It is also very sensitive to the acidity of the nucleophile (the peroxide). Electron withdrawing groups will shift the equilibrium to the left, increasing the value of k_{-1} and reducing the size of K_{eq}. The rate will therefore decrease for this reason, and the very unlikely electron transfer need not be invoked as an explanation. An additional objection to such an electron transfer is that peroxides such as [**F**] have a half life of seven days at room temperature. This peroxide is particularly pertinent, in that steric compression would be expected to stretch the O-O bond, lowering the energy for the transfer of an electron. It could be argued that the enzyme creates conditions for electron transfer not seen in model compounds. I have already discounted this argument since all luciferins with an accepted mechanism for light generation have effective models with fully comparable light yields. In any case, the evidence for the apparently exclusive support of CIEEL is more supportive of other mechanisms, and specifically in favour of the Baeyer-Villiger reaction.

SCHEME 3

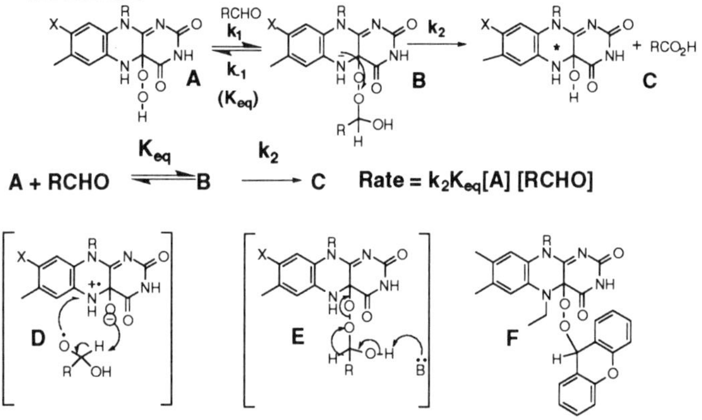

$$A + RCHO \underset{}{\overset{K_{eq}}{\rightleftharpoons}} B \xrightarrow{k_2} C \qquad \text{Rate} = k_2K_{eq}[A]\,[RCHO]$$

Extensive work on the oxidative reactions of the 4a-flavin hydroperoxide has shown that electron density is withdrawn from the peroxide bond, inducing heterolytic cleavage and consequent oxidative reactions. Eberhard and Hastings were the first to suggest a Baeyer-Villiger reaction, and the work designed to show that CIEEL is operating is fully consistent with just such a reaction. Most significantly, there is now evidence to show that bacterial luciferase actually catalyses the Baeyer-Villiger reaction, with a variety of aliphatic ketones, without light emission. For example, oxidation of 2-tridecanone causes migration mainly of the methyl group, to form dodecanoic acid, presumably by hydrolysis of the intermediate methyl dodecanoate (18). Any defence of the status quo must thus deal with the following observations.

(1) Bacterial luciferase is a "mainstream" flavin mono-oxygenase, performing Baeyer-Villiger reactions.
(2) The peroxide bond in the intermediate peroxyhemiacetal and peroxyhemiketal for aldehyde and ketone reactions respectively, will have virtually the same electron

affinity. The electron transfer is equally likely, and if it occurred would not allow the observed Baeyer-Villiger.

(3) There is excellent evidence of an isotope effect for light emission using deuterated aldehyde. The scheme involving CIEEL cannot be made to account for this.

It is clear that we have returned to a much earlier position, in which a Baeyer-Villiger reaction fits the evidence best. Yet all workers in this field will be dissatisfied with this conclusion, since there is no reason why a linear peroxide decomposition should generate an excited state. The difference between the dark linear and high quantum yield cyclic routes in the reaction between the acridinium esters and hydrogen peroxide come to mind in this context (Scheme 4). We have looked at several Baeyer-Villiger reactions involving aldehydes and fluorescent compounds, and have not seen any light. There are various configurations possible for such reactions, and there is little doubt that the range of possibilities should be examined more thoroughly, in view of the very strong evidence in favour of this reaction type. Other mechanisms are dependent on some considerable ingenuity in assessing the very large body of evidence available, and of course we must include electron transfer mechanisms as a possibility. However, *at present*, there is no evidence of any sort in favour of a CIEEL mechanism.

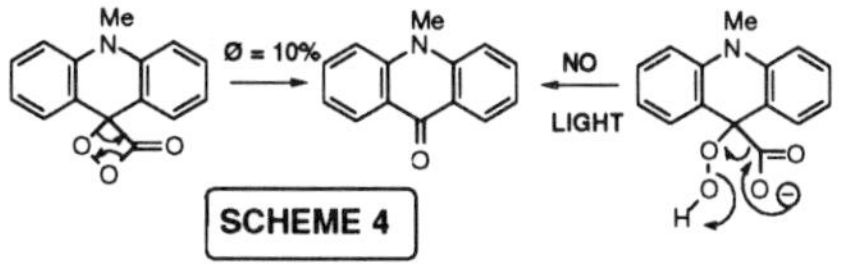

Ø = 10% NO LIGHT

SCHEME 4

The reactions of peroxides with fluorescent compounds almost always leads to light emission at some level, often proceeding by reactions between the fluorescer and peroxide to give a multitude of products. One reaction in particular sends a warning against reliance on seductive theory, and a plea for the re-examination of the wide range of peroxide reactions known to be chemiluminescent. The late Karl Gundermann pointed out several times when confronted by enthusiasts for CIEEL that the reaction between phthaloyl peroxide and aromatic hydrocarbon fluorescers was visibly (and of course, quantitatively) much brighter than any of the CIEEL reactions, with their attractive theoretical rationale. It is in fact some 20 times more efficient than the flagship of CIEEL - the reaction of diphenoyl peroxide with the same fluorescers. The diphenoyl peroxide differs most strikingly from the phthaloyl peroxide in having an acceptable mechanism for the return of the electron to the radical cation, thus generating the excited state. This is an important requirement for any reaction involving radical cations, since without a rapid route to electron return, other reactions of the highly reactive radical anion will seriously reduce the light yield. Phthaloyl peroxide differs in this and many other properties, and no neat mechanism for the excitation step comes readily to mind.

Colour in Beetle Luminescence. The various species of Coleoptera show a wide range of colour of light emission, often within a single individual, as in the click beetles. Recently, cloning of the luciferase, and manipulations of the DNA have given rise to a large number of clones with an even wider range of colour. It has been known for a long time that a variety of denaturing agents and environmental changes affect the luciferase. An almost invariable response to agents such as heavy metals, heat, alkylating agents, urea and lower pH is the appearance of a red (maximum around 600nm) luminescence in place of, or in addition to, the typical yellow (560nm) light. In these experiments, intermediate colours are not seen, and when both maxima are present, there is a characteristic double curve spectrum. This "either - or" colour change was satisfactorily explained for many years by the observation, in purely chemical experiments to generate light from active luciferin esters, of red fluorescence in relatively low strength base or yellow light with increasing base strength. Repetition of this work by the same research group (19) has shown that this observation was erroneous, and the explanation for the in vivo colour change of proton removal from the first formed excited state became untenable. Concern over explanations for this dual colour change have been overtaken by the fact that we now need to explain why every colour from green to red can be found in the reactions of the luciferase clones. These many spectra do not show shoulders or any other evidence of being a mixture of curves, if a single luciferase is considered. The range is large, and I think we can take it that two and only two discreet chemical entities are quite insufficient to cover that range, especially as all the spectra recorded show smooth monotonic curves. It is assumed that the various emissions are from the enzyme bound oxyluciferins.

Our first thoughts on this puzzle were that the enzyme binding sites were sufficiently changed by aminoacid substitution so as to provide a large variety of environments created by conformational change. Many experiments involving a variety of very different media, using 5,5-dimethyloxyluciferin as the emitter failed to deliver more than a few nanometers of spectral shift. It does remain a possibility, however, that the heteroatoms in the oxyluciferin especially the oxygen and nitrogen atoms, are specifically hydrogen bonded, and that this pattern is conformationally different for each clone. The spectrum of the emission could well be sensitive to this and other perturbations such as the proximity of charged aminoacid residues.

Nevertheless, the 5,5-dimethyloxyluciferin seems very resistant to the bulk effects of all these influences, and we turned to a more specific possibility. By analogy with biphenyl and binaphthyl, varying degrees of rotation about the single bond linking the benzothiazole with the thiazolone ring would give rise to a related range of energies and hence colours in the emission spectra of the various enzyme bound oxyluciferin emitters. Biphenyl (20) **[2]**, Scheme 5, unrestrained by solvent takes up a dihedral angle (Ø) of 42 ± 2°, as a result of the interference between the H-atoms at the ortho-positions. On excitation the rings become planar (Ø = 0°), and if biphenyl is "frozen" into its ground state dihedral angle by solution in a glassy solid at low temperature there is a considerable blue shift in the absorption spectrum. 1,1'-Binaphthyl (21) **[3]**, has an absorption spectrum almost identical to that of the single naphthalene molecule, since the two rings are so hindered about the single bond that they are nearly orthogonal. Freezing of this conformation results in a fluorescence maximum that is strongly blue shifted when compared to the spectrum in free solution, where rotation is allowed.

The hindrance to free rotation in both luciferin and oxyluciferin is considerably less, since there are no H-atoms in the ortho positions (we calculate the barrier to rotation about the linking bond to be as low as 7-11 $kJ.mol^{-1}$), and X-ray studies of crystalline luciferin show that the benzothiazole and thiazoline (and by extension, the thiazolone) rings are almost co-planar. In the ground state of these two molecules there is charge transfer interaction, resulting in a significant degree of double bond character between the rings. However, luciferin is perforce enzyme bound, and the various binding interactions, notably H-bonding to the N-atoms, may impose a degree of twist in the linking C(2)-C(2') bond. The case can also be considered as analogous to that of binaphthyl, although the amount of twisting in the luciferins is at present unknown. Promotion of an electron to the excited state manifold, will, by the Franck-Condon principle, result in the population of a high vibrational level of the excited state, with the geometry of the ground state. If this state is twisted, as described above, it will relax to the ground vibrational level of the excited state, in equilibrium with its surroundings. It is from this level that fluorescence occurs, determining the energy of the photon emitted and the colour of the light emitted. (Figure 3) *The equilibrium rotamer is not that of a molecule in the gas phase, but one which is immersed in the binding interactions of the luciferase and its clones.* The resulting geometries i.e. degree of twist, and the related vibrational level in the excited state equilibrium, will depend on the conformation of the enzyme. In terms of the exact fit of the oxyluciferin and its precursor luciferin, a single amino acid substitution remote from the active site could well alter the binding to give the variation in energy level and colour of light seen in beetle bioluminescence.

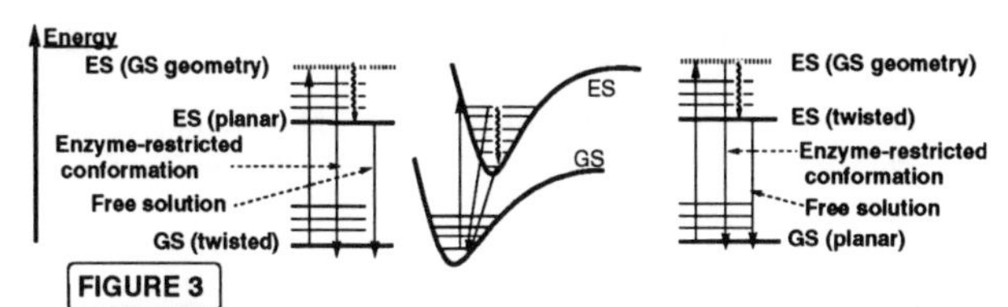

FIGURE 3

The formation of the excited state is represented in terms of fluorescence, but the same considerations apply in chemiluminescence. Either a twisted ground state forms a twisted ES which relaxes only to the extent allowed by the luciferase, or a near planar GS forms a twisted ES, also held in a unique conformation

We decided to learn more about the effect of twisting on the spectra of both ground and excited states by calculation, using available physical data as an aid to parameterisation.

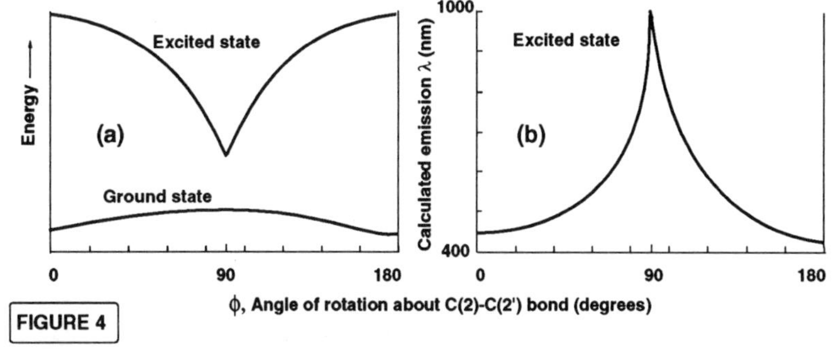

FIGURE 4

(a) The rotation about the C(2)-C(2') bond surprisingly leads to a sharp minimum at the fully twisted position in the excited state. Red light would be emitted from this conformer. The ground state surface appears as expected.
(b) The rotation also provides a continuum of colours, depending on the angle imposed by the luciferase.

The results, using AM1 calculations, surprised us. The oxyluciferin anion ground state showed a normal and expected energy profile on rotation around the C(2)-C(2') bond, (Figure 4) and again as expected the barrier to rotation had a higher maximum than that of the protonated compound, both at Ø = 90°. However, the potential energy curve of the excited state of the oxyluciferin anion had a minimum at 92° with rapid rises to maxima in energy at around 0° and 180°. The calculation of the emission wavelength with rotation followed the pattern of the energy curve, indicating that the twisted state with Ø = 90° absorbed in the red, and that deviations from 90° led to progressively shorter wavelength emissions. The shortest wavelength of emission should be similar to that of the 6-hydroxybenzothiazole anion, 421nm, and the calculated value of 440nm is actually in very good agreement. Such a short wavelength is not likely to be seen in vivo, since the steepness of the energy curve means that the enzyme induced rotation will not proceed much beyond the values leading to the observed blue shift (lowest observed wavelength about 540nm).The degree of twist giving the best fit at the active site need not be large in either the ground or excited state to produce the range of colours. Although we do not know the energy of interaction between the protein and luciferin/oxyluciferin, it is likely to be within the range required. Calculations of excited state properties are notoriously unreliable, the more so for complex, anionic heterocycles! However a twisted charge transfer excited state is not unknown in fluorescence (22), and since it is formed in this case from a precursor which differs in structure from the oxyluciferin, Franck-Condon factors do not prohibit its population, and curve crossing occurs readily in biaryls generally.

It may be significant that the decomposition of the dioxetanone precursor has charge transfer character, as argued in this paper. In any event, we commend the further exploration of these ideas to our successors.

References.

1. McCapra F, Richardson DG. The mechanism of chemiluminescence. Tetrahedron Lett. 1964; 43:3167-72.
2. McCapra F, Richardson DG, Chang YC. Chemiluminescence involving peroxide decomposition. Photochem Photobiol. 1965;4:1111-21.
3 Mayer A, Neuenhofer S. Luminescent labels - more than just an alternative to radioisotopes?. Angew. Chem. Int. Ed. Engl. 1994;33;1044-72.
4. McCapra F. Chemical mechanisms in bioluminescence. Acc Chem Res. 1976;9:201-8.
5 Linschitz H. Chemiluminescence in porphyrin-catalyzed decomposition of peroxides. In: McElroy WD, Glass B, editors. Light and Life. Baltimore, The Johns Hopkins Press. 1961:173-82.
6 Schuster GB, Chemiluminescence of organic peroxides. Acc Chem Res 1979;12:366-73
7. Wilson T. Mechanisms of peroxide chemiluminescence. In: Frimer AA, editor. Singlet Oxygen Boca Raton FL: CRC, 1985 Vol II, 37-57.
8. Rauhut MM, Chemiluminescence from concerted peroxide decompositions. Acc Chem Res 1969; 2:451-60.
9. Alvarez FJ, Parekh NJ, Matuszewski B, Givens RS, Higuchi T, Schowen RL. Multiple intermediates generate fluorophore-derived light in the oxalate/peroxide chemiluminescence system. J Am Chem Soc.1986;108:6435-7.
10. Molin YN, Anisimov OA, Grigoryants VM, Molchanov VK Salikhov KM. Optical detection of ESR spectra of short-lived ion-radical pairs produced in solution by ionising radiation. J Phys Chem 1980;84:1853-6.
11 McCapra F, Perring K, Hart RJ, Hann RA. Photochemistry without light - reaction of oxalate esters with anthracenophanes. Tetrahedron Lett 1981;22:5087-90.
12. Catalani LH, Wilson T. Electron transfer and chemiluminescence. Two inefficient systems: 1,4-dimethoxy-9,10-diphenylanthracene peroxide and diphenoyl peroxide. J Am Chem Soc 1989;111:2633-9.
13. Edwards B, Sparks A, Voyta JC, Bronstein I. Unusual properties of odd- and even -substituted naphthyl-derivatised dioxetanes. J Biolum Chemilum 1990;5:1-4.
14. McCapra F. Charge transfer dioxetanes - a simple rationalisation. Tetrahedron Lett 1993;34:6941-4.
15. Schaap AP, Chen TS, Handley RS, DeSilva R, Giri BP. Chemical and enzymatic triggering of 1,2-dioxetanes. Tetrahedron Lett 1987;28:1155-8.

16. Eckstein JW, Hastings JW, Ghisla S. Mechanism of bacterial bioluminescence: 4a,5-dihydroflavin analogs as models for luciferase hydroperoxide intermediates and the effect of substituents at the 8-position of flavin on luciferase kinetics . Biochem 1993;32:404-1.
17. Shannon P, Presswood RB, Spencer R, Becvar JE, Hastings JW, Walsh C. A study of deuterium isotope effects on the bacterial bioluminescence reaction. In: Singer TP, Ondarza RN, editors. Mechanism of oxidising enzymes. Amsterdam: Elsevier, 1978:69-78.
18. Villa R, Willetts A. Oxidations by microbial NADH plus FMN-dependent luciferases from Photobacterium phosphoreum and Vibrio fischeri. J Molecular Catalysis: B Enzymatic, in press. I am very grateful to Dr. Willetts for a copy of this paper, received before publication.
19. White EH, Roswell DF. Analogs and derivatives of firefly oxyluciferin. The light emitter and firefly bioluminescence. Photochem Photobiol 1991;53;131-6.
20. Hoong-Sun I, Bernstein ER. Geometry and torsional motion of biphenyl in the ground state and the first singlet excited state. J Chem Phys 1988;88:7337-47.
21. Hochstrasser R. The effect of intramolecular twisting on the emission spectra of hindered aromatic molecules. Can J Chem 1961;39:459-70.
22. Lin CT, Guan HW, McCoy RK, Spangler CW. Dual fluorescence of p,p'-disubstituted 1,6-diphenyl-1,3,5-hexatrienes: evidence of a twisted intramolecular charge transfer state. . J Phys Chem 1989;93;39-43.

ON THE MECHANISM OF THE PEROXYOXALATE REACTION: SYNTHESIS AND CHEMILUMINESCENCE CHARACTERISTICS OF AN INTERMEDIATE

WJ Baader, DF Lima and CV Stevani
Instituto de Química - Universidade de São Paulo,
C.P. 26077, 05599-970 - São Paulo, S.P., Brazil

Introduction

The peroxyoxalate reaction is one of the most efficient chemiluminescence systems known, and is believed to occur by an intermolecular "Chemically Induced Electron Exchange Luminescence" (CIEEL) mechanism (1). This system consists of a base catalyzed reaction of activated oxalic phenylesters with hydrogen peroxide in the presence of aromatic hydrocarbons with low oxidation potentials as chemiluminescence activators (ACT). Although this complex reaction is frequently utilized in analytical applications, its mechanism is still not well understood, especially with respect to the excitation step, the elementary step in which the electronically excited species are formed. Rauhut *et al.* (2) postulated 1,2-dioxetanedione (**I**) as the high energy intermediate which leads to excited state formation upon interaction with the chemiluminescence activator (ACT). Subsequent kinetic studies, using various oxalic esters and a variety of experimental conditions led to the proposal of other possible reactive intermediate structures (**II** - **IV**) (3-7).

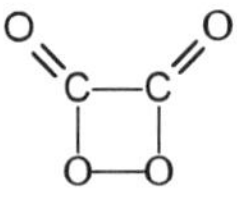

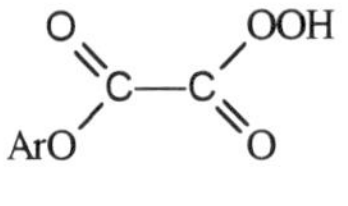

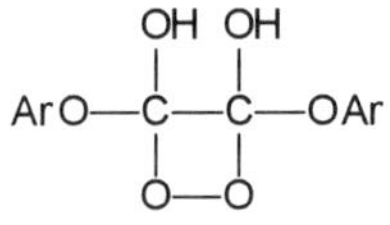

Materials and Methods

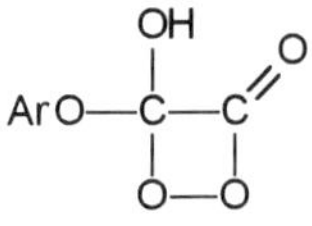

The kinetic studies on the peroxyoxalate reaction were performed in anhydrous ethyl acetate at 25 °C on a SPEX-Fluorolog 1681 spectrofluorimeter as previously described (8). Standard conditions: bis(2,4,6-trichlorophenyl) oxalate (TCPO): 0.10 mmol/L, imidazole: 1.0 mmol/L, 9,10-diphenylanthracene (DPA): 1.0 mmol/L, H_2O_2: 10 mmol/L. In the experiments with delayed H_2O_2 addition, the other reagents were pre-incubated and 35 μL of the H_2O_2 stock solution added after a delay time of 10 min.

4-Chlorophenyl O,O-hydrogen monoperoxyoxalate (**1**) was prepared as previously described (9). Kinetic studies with **1** were performed at 25 °C on a SPEX-Fluorolog 1681 spectrofluorimeter, using DPA (1.0 mmol/L) as activator. The experiments with imidazole and 1,8-bis(dimethylamino)

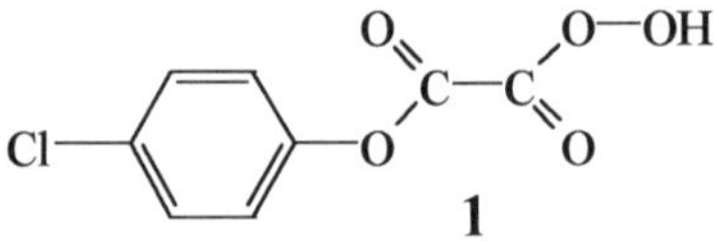

naphthalene were performed in ethyl acetate, whereas anhydrous tetrahydrofurane was used as solvent in the experiments with potassium *tert*-butoxide and potassium *p*-chlorophenolate. The studies with different activators were performed using **1** (1.0 mmol/L) and imidazole (1.0 mmol/L) as catalyst in ethyl acetate at 25 °C. The half-peak oxidation potentials ($E_{p/2}^{ox}$) of the activators were determined by cyclic voltametry in acetonitrile, on an EG&G Princeton Applied Research Model 175 potentiostate-galvanostate, using a Pt working electrode, a Ag/Ag+ reference electrode, and tetraethylammonium perchlorate (0.1 mol/L) as electrolyte.

Results and Discussion

In a recent kinetic study (8) of the system bis(trichlorophenyl) oxalate (TCPO), hydrogen peroxide, imidazole (IMI-H, as catalyst) and 9,10-diphenylanthracene (DPA, as ACT) in anhydrous ethyl acetate, we propose a mechanistic reaction scheme and attribute rate constants to two elementary steps (Scheme 1).

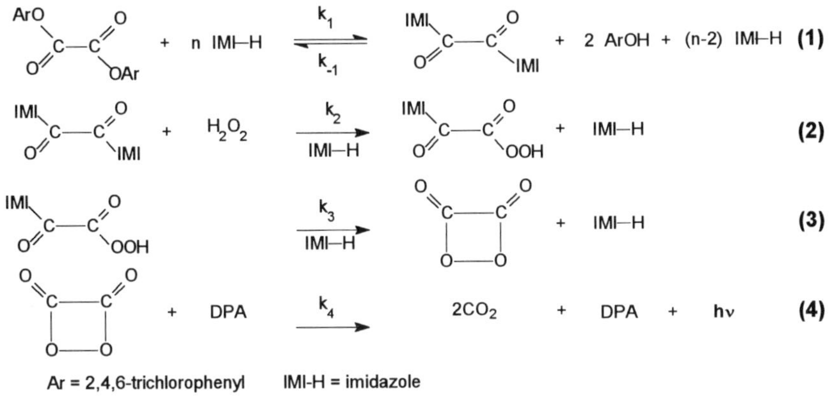

ArO–C(=O)–C(=O)–OAr + n IMI–H ⇌ (k_1, k_{-1}) IMI–C(=O)–C(=O)–IMI + 2 ArOH + (n-2) IMI–H **(1)**

IMI–C(=O)–C(=O)–IMI + H_2O_2 → (k_2, IMI–H) IMI–C(=O)–C(=O)–OOH + IMI–H **(2)**

IMI–C(=O)–C(=O)–OOH → (k_3, IMI–H) 1,2-dioxetanedione (O=C–C=O, O–O ring) + IMI–H **(3)**

1,2-dioxetanedione + DPA → (k_4) $2CO_2$ + DPA + hν **(4)**

Ar = 2,4,6-trichlorophenyl IMI-H = imidazole

Scheme 1: Kinetic scheme for the main pathway of the peroxyoxalate chemiluminescence.

A further study on the dependence of the decay rate constant (k_{obs1}) and the rise rate constant k_{obs2} of the light emission kinetic curve on the [H_2O_2] shows k_{obs1} to be independent of the [H_2O_2] at low concentrations ([H_2O_2] $\leq$ 10 mmol/L) and linearly dependent at high concentrations, with a bimolecular rate constant of k_1 = 0.232 ± 0.007 L/mol s. However, k_{obs2} shows a linear dependence on [H_2O_2] up to 15 mmol/L and a bimolecular rate constant of k_2 = 22.2 ± 0.5 L/mol s (Fig. 1A). At higher [H_2O_2], a saturation curve is obtained with a maximum rate constant of k_{max} = 0.40 ± 0.04 1/s (Fig. 1A). This behavior indicates a change in the rate limiting step in these conditions and this rate constant is assigned to the cyclization step 3 (Scheme 1). Preliminary experiments show k_{max} to be dependent on the [IMI-H] and a bimolecular rate constant of k_3 = 400 ± 40 L/mol s is obtained for this step.

On the other hand, step 2 (Scheme 1) can be observed directly, when H_2O_2 is added in delay to the otherwise complete system (Fig. 1B). From the dependence of the decay rate constant in these conditions (k_{obs2}') on the [H_2O_2], a bimolecular rate constant of k_2' = 9.1 ± 0.6 L/mol s is obtained, of the same order of magnitude as k_2, determined from the kinetics of the complete system. These results contribute to confirm the validity of the kinetic scheme proposed earlier (8); furthermore, a rate constant for the cyclization of the peracid derivative to a cyclic peroxide (step 3, Scheme 1) can be determined. However, as in former kinetic studies (3-8), no direct information on the excitation step and the nature of the reactive intermediate could be obtained. Therefore, we decided to synthesize a derivative of the proposed peracid intermediate **II**.

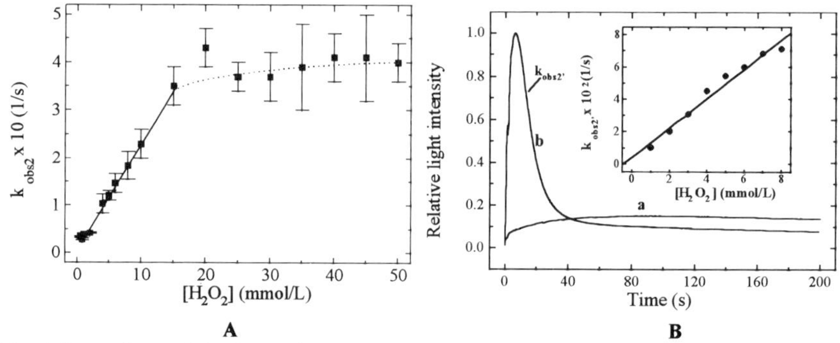

Fig. 1: Dependence of the observed rate constants on the [H_2O_2]. **A**: rise rate constant (k_{obs2}) in the complete system; **B**: kinetics of the complete system (a) and upon 10 min delayed H_2O_2 addition (b); Inset: observed decay rate constant (k_{obs2}') upon delayed hydrogen peroxide addition.

4-Chlorophenyl O,O-hydrogen monoperoxyoxalate (**1**) was prepared by the reaction of the corresponding acid chloride with H_2O_2 in the presence of pyridine at -20 °C and characterized by ^{13}C NMR spectroscopy (9). In the presence of activators alone, the peracid derivative **1** does not show any chemiluminescence and can therefore be excluded as the ultimate reactive intermediate structure. However, in the presence of several bases [potassium hydroxide, potassium *tert*-butoxide, potassium *p*-chlorophenolate, imidazole and 1,8-bis(dimethylamino)-naphthalene] and an ACT, the decomposition of **1** is accompanied by a bright light emission. Based on a detailed kinetic study, a general mechanistic scheme for the chemiluminescent base catalyzed decomposition of **1** can be proposed, where the key step is the cyclization of deprotonated **1** to the cyclic peroxide 1,2-dioxetanedione, which interacts with the ACT leading to the formation of the ACT's excited state (Scheme 2). The main conclusion obtained from these studies is the fact that a slow chemical transformation of **1** - most likely the cyclization to 1,2-dioxetanedione - must occur prior to the interaction of the reactive intermediate with the ACT, which leads to excited state formation. Although these studies lead to a more direct insight to the structure of the reactive intermediate, the interaction of this intermediate with the ACT - the excitation step - can not be observed kinetically.

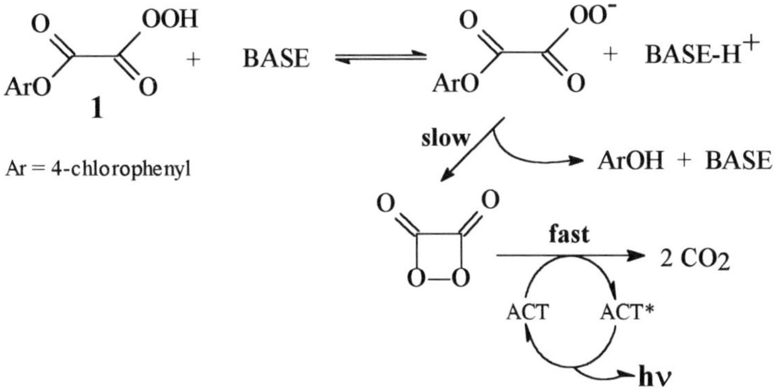

Ar = 4-chlorophenyl

ArO–CO–CO–OOH (1) + BASE ⇌ ArO–CO–CO–OO⁻ + BASE-H⁺

slow → ArOH + BASE

fast → 2 CO_2; ACT → ACT* → hν

Scheme 2: General scheme for the base catalyzed chemiluminescent decomposition of **1** in the presence of a chemiluminescence activator (ATC).

As the excitation step is supposed to proceed by the CIEEL mechanism, which involves an electron transfer in the rate limiting step, we studied the chemiluminescence properties of the imidazole catalyzed decomposition of **1** in the presence of various ACT with different oxidation potentials ($E_{p/2}^{ox}$). The double reciprocal plots of the chemiluminescence quantum yields (Φ_{CL}) versus the [ACT] show the expected linear correlation (Fig. 2), from which the quantum yields at infinite [ACT] (Φ_{∞}), as well as the ratio of the rate constant for the interaction of the reactive intermediate with the ACT (k_{et}) and the decay rate constant of this intermediate (k_D) can be obtained (Scheme 3). This quotient constitutes a relative measure for the rate constant k_{et}, as the decomposition rate constant k_D does not depend on the nature of the ACT. This parameter, as well as the Φ_{∞}, show to be dependent on the oxidation potential of the ACT (Table 1).

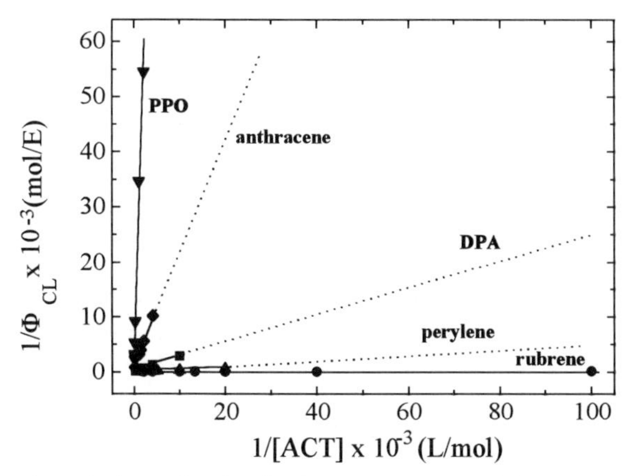

Fig. 2: Double reciprocal plot of the chemiluminescence quantum yields (Φ) *versus* the activator concentration for different activators (PPO: 2,5-diphenyloxazole).

Table 1: Quantum yields (Φ_∞), electron transfer rate constants (k_{et}/k_D) and half-peak oxidation potentials ($E_{p/2}^{ox}$) for different activators.

ACT	Φ_∞ 10^2 (E/mol)	k_{et}/k_D 10^{-2}(M)	$E_{p/2}^{ox}$ (V)
Rubrene	16 ± 1	74 ± 2	0.61
Perylene	6.1 ± 0.2	15.6 ± 0.6	0.91
DPA	1.2 ± 0.3	3.0 ± 0.8	1.06
Anthracene	0.14 ± 0.03	3.0 ± 0.6	1.18
PPO[a]	0.046 ± 0.003	0.67 ± 0.04	1.46

[a] PPO: 2,5-diphenyloxazole

The linear dependence of ln k_{et}/k_D on the $E_{p/2}^{ox}$ of the activator (Fig. 3) is consistent with the involvement of an electron transfer from the activator to the reactive intermediate in the rate limiting step. However, the dependence of the quantum yields at infinite [ACT] on the nature of the ACT is not expected, if these yields are only determined by the competition between k_{et} and k_D. Therefore, the correlation of Φ_∞ with $E_{p/2}^{ox}$ (Table 1) shows that the quantum yields are determined also by the efficiency of the back electron transfer from the radical anion of CO_2 to the radical cation of the ACT (Scheme 3). This competition between k_{et}' and k_D' should, of course, also depend on the properties of the ACT.

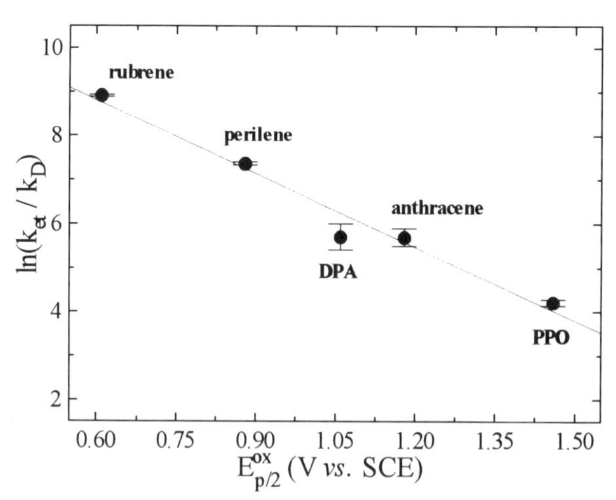

Fig. 3: Correlation of ln k_{et}/k_D with the oxidation potential of the activator ($E_{p/2}^{ox}$). $\ln(k_{et}/k_D) = \ln A + \alpha B - (\alpha /RT) E_{p/2}^{ox}$, $\alpha = 0.14 \pm 0.01$ ($r = 0.99$).

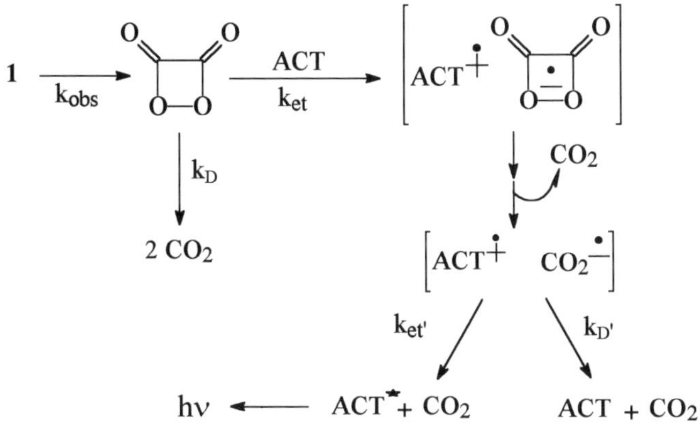

$$\frac{1}{\Phi_{CL}} = \frac{1}{\Phi_{\infty}} + \left(\frac{k_D}{k_{et}\ \Phi_{\infty}}\right) \times \frac{1}{[ACT]}$$

Scheme 3: Interaction of the reactive intermediate with the chemiluminescence activator leading to the formation of the ACT's singlet excited state.
The equation shown above is derived by applying the steady state approximation on 1,2-dioxetanedione.

Conclusions

The kinetic studies on the peroxyoxalate chemiluminescence reported in this communication lead to a value for the cyclization rate constant of the peracid intermediate to the reactive intermediate. Furthermore, the independent preparation of a peracid derivative analogous to **II**, 4-chlorophenyl O,O-hydrogen monoperoxyoxalate (**1**), is described. Kinetic studies on this isolated intermediate show that this peracid derivative is ***not the reactive intermediate*** in the peroxyoxalate chemiluminescence. The peracid **1** undergoes a slow chemical transformation, probably a cyclization to *1,2-dioxetanedione*, which is observed in the kinetic measurements. The interaction of 1,2-dioxetanedion with the chemiluminescence activator, which ultimately leads to excited state formation, occurs in a fast step, which can not be observed directly by kinetic measurements. However, on using different activators, relative values for the rate constant of this excitation step are obtained. The observed dependence of this relative rate constants on the ACT's oxidation potential suggests the involvement of a rate limiting electron transfer in the excitation step. This fact is consistent with the occurrence of a CIEEL mechanism in the peroxyoxalate chemiluminescence.

Acknowledgements

We thank FAPESP, CNPq, CAPES and FINEP Foundations for financial support and Dr. Thérèse Wilson for many helpful discussions.

References

1. Schuster GB, Schmidt SP. Chemiluminescence of organic compounds. Adv Phys Org Chem 1982;18: 187-238.
2. Rauhut MM. Chemiluminescence from concerted peroxide decomposition reactions. Acc Chem Res 1969;2:80-7.
3. Catherall CLR, Palmer TF, Cundall RB. Chemiluminescence from the reaction of bis(pentachlorophenyl) oxalate, hydrogen peroxide and fluorescent compounds. kinetics and mechanism. J Chem Soc Faraday Trans 2 1984;80: 823-36
4. Catherall CLR, Palmer TF, Cundall RB. Chemiluminescence from the reaction of bis(pentachlorophenyl) oxalate, hydrogen peroxide and fluorescent compounds. role of the fluor and nature of chemielectronic prosess(es). J Chem Soc Faraday Trans 2 1984;80: 837-49.
5. Alvarez FJ, Parekh NJ, Matuszewski B, Givens RS, Higuchi T, Schowen RL. Multiple intermediates generate fluorophore-derived light in the oxalate/peroxide chemiluminescence system. J Am Chem Soc 1986;108: 6435-7.
6. Orlovic M, Schowen RL, Givens RS, Alvarez F, Matuszewski B, Parekh N. A simplified model for the dynamics of chemiluminescence in the oxalate-hydrogen peroxide system: towards a reaction mechanism. J Org Chem 1989;54: 3606-10.
7. Givens RS, Jencen DA, Riley CM, Stobaugh J F, Chokshi H, Hanaoka N. Chemiluminescence: a sensitive and versatile method for analyte detection. J Pharmac Biomed Anal 1990;8: 477-91.
8. Stevani CV, Lima DF, Toscano VG, Baader WJ. Kinetic studies on the peroxyoxalate chemiluminescent reaction: imidazole as a nucleophilic catalyst. J Chem Soc Perkin Trans 2 1996: 989-95.
9. Stevani CV, Campos IP de A, Baader WJ. Synthesis and characterisation of an intermediate in the peroxyoxalate chemiluminescence: 4-chlorophenyl O,O-hydrogen monoperoxyoxalate. J Chem Soc Perkin Trans 2 1996: 1645-8.

SINGLET EXCITED STATE FORMATION BY INTRAMOLECULAR ELECTRON TRANSFER CATALYZED 1,2-DIOXETANE DECOMPOSITION

LH Catalani, ALP Nery and WJ Baader
Instituto de Química - Universidade de São Paulo
CP: 26077, São Paulo, 05599-970, Brazil

Introduction

The Chemically Induced Electron Exchange Luminescence (CIEEL) is a widely quoted mechanism for singlet excited state formation in the decomposition of several peroxides. This mechanism consists in a rate limiting electron transfer from an appropriate chemiluminescence activator, generally a polycyclic aromatic hydrocarbon with low oxidation potential, to the peroxide (1,2-dioxetanones, diphenoyl peroxide and related compounds) and results in the formation of the activator's singlet excited state (1). On the other hand, 1,2-dioxetanes, containing only simple alkyl or aryl substituents, do not show decomposition by an *intermolecular* CIEEL mechanism in the presence of external activators. However, several examples of an *intramolecular* CIEEL decomposition of 1,2-dioxetanes, in which the electron donor is directly attached to the peroxide ring, have been described (2-5). More recently, some interesting examples of appropriately substituted 1,2-dioxetanes, which suffer chemiluminescent decomposition by an *intramolecular* CIEEL mechanism have been reported (6-8). The decomposition of silyloxyphenyl substituted dioxetanes, for example, can be triggered by the addition of fluoride, which leads to the formation of the corresponding phenolate, an excellent electron donor. The triggered decomposition of these thermally stable dioxetanes leads to the preferential formation of singlet excited products and emission efficiencies of up to 0.25 E/mol (7, 8). This kind of compounds has already been used in several analytical applications (9).

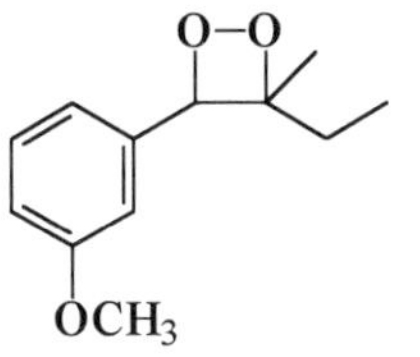

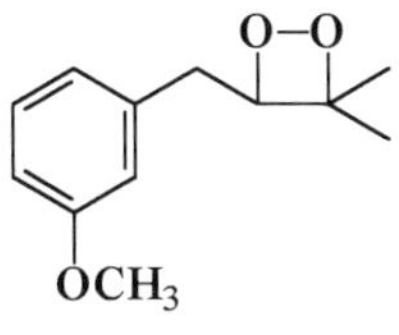

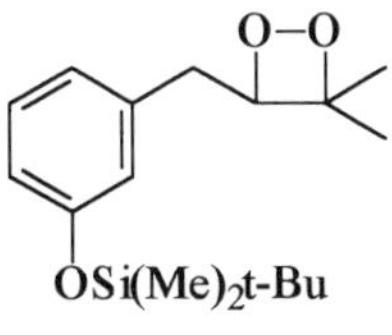

In this communication, we report the synthesis and chemiluminescence properties of the 1,2-dioxetanes **I** - **IV**, where the electron donor, after fluoride catalyzed deprotection, is either directly attached to the dioxetane ring **(II)** or separated from it by a methylene bridge **(IV)**. Therefore, these compounds serve as models to distinguish between a resonance electron transfer and a 'through space' electron transfer from the generated phenolate to the dioxetane moiety.

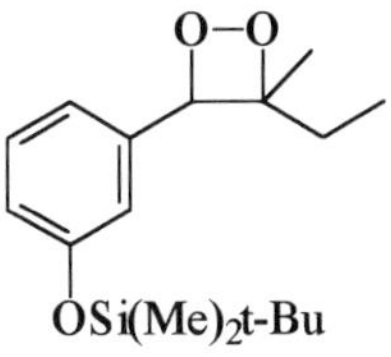

Materials and Methods

The dioxetanes **I** - **IV** were prepared from the corresponding olefins, using the classical method of Kopecky (10), purified by low temperature column chromatography and identified by ^{1}H and ^{13}C NMR spectroscopy, as well as its decomposition products. The activation parameters were obtained from the temperature dependence of the decomposition rate constants and the parameters for the chemiluminescent pathway from the temperature dependence of the emission intensities (for a review on the methodology see (11)). The chemiexcitation quantum yields were determined using 9,10-diphenylanthracene (DPA) and 9,10-dibromoanthracene (DBA) for the quantification of singlet and triplet excited states, respectively (11).

Results and Discussion

Thermal decomposition of **I** to **IV** leads exclusively to the formation of the expected carbonyl cleavage products. Direct emission, with a maximum at 412 nm, is observed for **I** and **II**, whereas in the case of **III** and **IV** excited state formation can only be detected with sensitizers, DPA and DBA (11). The activation parameters for the unimolecular decomposition of the compounds are within the range expected for trisubstituted 1,2-dioxetanes (Table 1) (11). The comparison of the values for the pairs **I/II** and **III/IV** shows the similarity in the thermal stability of the silyloxy substituted derivatives with the methoxy substituted model compounds. The fact that the $\Delta H^{\ddagger}_{chl}$ values for **I** and **II** are lower than the values determined by isothermal kinetics is most probably caused by a negative activation energy of the fluorescence from the excited products. These later activation parameters were determined by direct emission measurements, whereas the values for **III** and **IV** were obtained by DPA sensitized emission measurements.

Table 1: Activation parameters and quantum yields for the unimolecular and fluoride catalyzed decomposition of dioxetanes **I-IV**^{a}

	$\Delta H^{\ddagger}$	$\Delta S^{\ddagger}$	$\Delta G^{\ddagger}$	$\Delta H^{\ddagger}_{chl}$	ϕ^T x10^2	ϕ^S x10^2
I^b	23.3 (0.5)	-4.5 (1.4)	24.6 (0.6)	22.2 (0.3)	9.8^f(2.5)	0.26^f(0.08)
II^b	25.3 (0.4)	0.6 (0.2)	25.1 (0.1)	21.3 (0.5)	7.4^f(2.0)	0.40^f(0.15)
II+F^{-c}	20.6 (0.2)	5.0 (1.0)	19.1 (0.4)	16.2 (0.4	-	'high'^{g}
III^d	22.4 (0.7)	-6 (2)	24.2 (0.9)	24.4 (2.1)	4.9^f(1.4)	0.016^f(0.004)
IV^d	22.6 (0.9)	-5 (2)	24.1 (0.9)	24.5 (1.0)	6.4^f(0.8)	0.027^f(0.011)
IV+F^{-e}	14.6 (0.9)	-15 (3)	19.1 (0.9)	11.6 (0.9)	-	'high'^{g}

a. $\Delta H^{\ddagger}$, $\Delta H^{\ddagger}_{chl}$ and $\Delta G^{\ddagger}$ (25°C) in kcal/mol; $\Delta S^{\ddagger}$ in cal/mol K, standard deviations in brackets; **b.** [**I**]= 67μmol/L, [**II**]=0.10mmol/L, toluene, 412nm; at 65 to 85°C; **c**. [**II**]=20μmol/L, [F$^-$]=32mmol/L, THF, at 0 to 25°C, 560nm, **d.** [**III**]=83μmol/L, [**IV**]=21μmol/L, [DPA]=10mmol/L, toluene, 438nm, at 65 to 85°C; **e.** [**IV**]=41μmol/L, [F$^-$]=32mmol/L, THF, at 0 to 25°C, 550nm, **f.** [**I**]=67μmol/L, [**II**]=40μmol/L, [**III**]=83μmol/L, [**IV**]=21μmol/L, ϕ^T and ϕ^S in E/mol, DBA/DPA method in toluene, at 80°C **g.** easily visible; exact values were not obtained yet.

The chemiexitation quantum yields are the expected for trisubstituted dioxetanes and show the preferential formation of triplet excited states in the unimolecular decomposition and show to be similar for the methoxy and the silyloxy substituted derivatives.

In the presence of tetrabutylammonium fluoride, which causes the deprotection of the silyloxy substituted phenyl moiety to the corresponding phenolate, the decomposition rates of **II** and **IV** are increased drastically. The deprotected structure decomposes about 10^4 times faster than the protected ones, and a strong direct emission with a maximum at around 550 nm is observed. At low fluoride concentrations, the observed rate constants (k_{obs}) show a linear dependence on $[F^-]$, indicating the deprotection (k_1) as the rate limiting step. At high concentrations ($[F^-] > 0.01$ M for **II** and $[F^-] > 0.02$ M for **IV**), the k_{obs} values are independent of the catalyst concentration. Under these conditions k_{obs} corresponds to the decomposition (k_{et}) of the phenolate substituted dioxetanes (Scheme 1).

The activation parameters for the catalyzed decomposition of **II** and **IV** are considerably lower than for the unimolecular decomposition. Although the $\Delta H^{\ddagger}$ and $\Delta S^{\ddagger}$ values for **II** and **IV** are quite different, the $\Delta G^{\ddagger}$ values (at 25 °C) proved to be comparable (Table 1). The lower $\Delta H^{\ddagger}_{chl}$ values as compared to $\Delta H^{\ddagger}$ are, also in these cases, probably due to a negative activation energy of the fluorescence of the cleavage products. The observation of a strong direct emission from both dioxetanes indicates the efficient formation of singlet excited states in the catalyzed decomposition, although it was not yet possible to obtain exact values for the quantum yields, due to experimental difficulties.

Conclusions

In the uncatalyzed decomposition (protected forms), **I-IV** show activation parameters and excited state yields expected for trisubstituted 1,2-dioxetanes; relatively high thermal stability ($\Delta G^{\ddagger} = 25 \pm 1$ kcal/mol, 25 °C) and preferential formation of triplet excited states are observed. Fluoride catalyzes the decomposition of **II** and **IV**, the rate constants are four orders of magnitude higher than in the uncatalyzed reaction and the activation parameters considerably lower. The catalyzed decomposition leads to the preferential formation of singlet excited states as judged by the strong direct emission observed at around 550 nm. This long emission wavelength may indicate the involvement of an intramolecular exciplex as the emitting species.

The decomposition of the deprotected dioxetanes is supposed to be initiated by an electron transfer from the phenolate substituent to the peroxide ring, in analogy to the CIEEL sequence (Scheme 1). Interestingly, the activation entropy is positive for **II** and highly negative for **IV**. The positive value of $\Delta S^{\ddagger}$ in the case of **II**

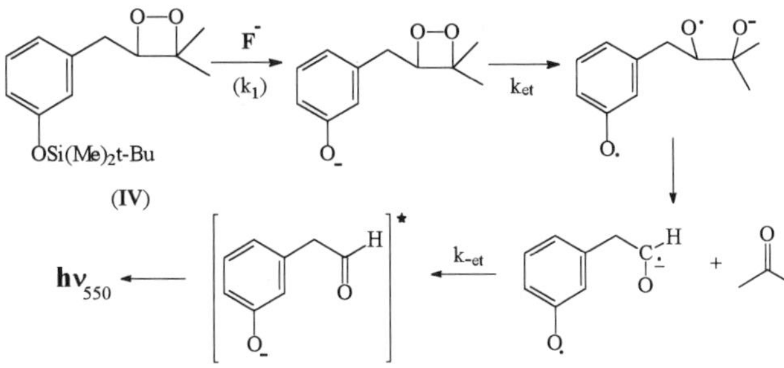

Scheme 1: Proposed intramolecular CIEEL mechanism for the fluoride catalyzed decomposition of dioxetane **IV**

may be due to a higher charge distribution in the transition state of the rate limiting electron transfer step, causing a loss in solvatation. Whereas, the negative values of $\Delta S^{\ddagger}$ for **IV** may be related to the necessity of a specific conformation for the occurrence of the electron transfer from the phenolate not directly linked to the dioxetane ring. The fluoride catalyzed decomposition of the dioxetane **IV** constitutes the first example of an intramolecular CIEEL mechanism in 1,2-dioxetanes, initiated by an electron transfer from a donor *which is not directly bonded to the peroxide ring* ('through space electron transfer').

Acknowledgments
We thank FAPESP, CNPq, CAPES and FINEP Foundations for financial support and Prof. Thérèse Wilson for many helpful discussions.

References

1. Schuster GB, Schmidt SP. Chemiluminescence of organic compounds. Adv Phys Org Chem 1982;18:187-238.
2. Zaklika KA, Kissel T, Thayer LA, Bruns PA, Schaap AP. Mechanisms of 1,2-dioxetane decomposition: the role of electron transfer. Photochem Photobiol 1979;30:35-44.
3. Nakamura H, Goto T. Studies on aminodioxetanes as model of bioluminescence intermediates. 1-(1-methyl-3-indolyl)-6-phenyl-2,5,7,8-tetraoxabicyclo[4.2.0]octane, an aminodioxetane resulting in efficient ultraviolet and exciplex chemiluminescence. Photochem Photobiol 1979; 30:27-33.
4. Lee S, Singer AL. Structural effects on the intramolecular electron transfer induced decomposition of a series of 1,2-dioxetanes derived from 9-alkylidene-10-methylacridans. J Am Chem Soc 1980;102:3823-9.
5. McCapra F, Beheshi I, Burford A, Hann RA, Zaklika KA. Singlet excited states from dioxetane decomposition. J Chem Soc Chem Commun 1977: 944-6.
6. Schaap AP, Gagon SD. Chemiluminescence of a phenoxide-substituted 1,2-dioxetane: a model for firefly bioluminescence. J Am Chem Soc 1982;104: 3504-6.
7. Schaap AP, Handley RS, Giri BP. Chemical and enzymatic triggering of 1,2-dioxetanes. 1. Aryl esterase-catalyzed chemiluminescence from a naphthyl acetate substituted dioxetane. Tetrahedron Lett 1987;28: 935-8.
8. Schaap AP, Chen T-S, Handley RS, De Silva R, Giri, BP. Chemical and enzymatic triggering of 1,2-dioxetanes. 2. Fluoride-induced chemiluminescence from tert-butyldimethylsilyloxy-substituted dioxetanes. Tetrahedron Lett 1987;28: 1155-8.
9. Beck S, Köster H. Applications of dioxetane chemiluminescent probes to molecular biology. Anal Chem, 1990;62: 2258-70.
10. Kopecky KR, Filby JE, Mumford C, Lockwood PA, Ding J-Y. Preparation and thermolysis of some 1,2-dioxetanes. Can J Chem 1975;53:1103-22.
11. Adam W, Cilento G, editors. Chemical and Biological Generation of Excited States, New York, Academic Press, 1982.

STUDIES ON THE STABILITY AND REACTIVITY OF OXYGENATED FLAVIN INTERMEDIATES OF BACTERIAL LUCIFERASE BY CHEMICAL MODELS AND MUTAGENESIS

SC Tu, H Li, S Huang, X Xin, L Xi[1] *and HIX Mager*
Dept of Biochemical and Biophysical Sciences, University of Houston, Houston, TX 77204-5934, USA

Introduction

Bacterial luciferase, classified as a flavin-dependent external monooxygenase, has long been the target for extensive enzymological studies. The bioluminescence reaction catalyzed by this luciferase involves the oxidation of reduced FMN ($FMNH_2$) and a long-chain aldehyde by molecular oxygen to yield a fatty acid, FMN, H_2O and light (λ_{max} ~490 nm). In comparison with other flavo-monooxygenases, luciferase offers special challenges and advantages for mechanistic studies. First, luciferase is the only flavo-monooxygenase known to catalyze a light-emitting reaction. The chemical mechanism of the reaction steps that efficiently couple the oxidation energy to the generation of an excited state by luciferase still remains an intriguing but largely unanswered question. Second, the luciferase turnover rate is extremely slow (~ 3 to 20 min^{-1} at 23°C depending on the aldehyde substrate used). Consequently, a number of oxygenated flavin intermediates of luciferase can be detected and/or isolated for detailed characterizations. This report focuses on our studies, mainly by the approaches of chemical modeling and site-directed mutagenesis, aimed at elucidating the properties and reactivities of certain luciferase oxygenated flavin intermediates.

Materials and Methods

Native and mutated *Vibrio harveyi* luciferases were generated and overexpressed in *Escherichia coli*, and purified following the principles and methods described previously (1). Luciferase activities were determined in phosphate buffer, pH 7.0, by the dithionite assay or the Cu(I) assay (1). Bioluminescence intensity (quantum/s) was measured using either a calibrated home-made luminometer or a Turner TD-20e Luminometer (Sunnyvale, CA USA). Uncorrected fluorescence spectra were measured using a Perkin-Elmer MPF-44A Fluorescence Spectrophotometer (Norwalk, CT USA). Rapid mixing kinetic data were collected using a Dionex Stopped-flow spectrophotometer (Sunnyvale, CA USA).

Results and Discussion

A chemically initiated electron exchange luminescence (CIEEL) mechanism was first formulated for the luciferase reaction in 1984 (2, 3). With some modifications, the basic steps of this mechanism can be summarized as the following. Luciferase first catalyzes the addition of molecular oxygen to the bound $FMNH^-$ to generate the 4a-peroxyFMNH intermediate II (HFMN-4a-OO^- or HFMN-4a-OOH; Eqs. 1a,b). Intermediate II subsequently reacts with a long-chain aliphatic aldehyde (RCHO) to yield the 4a-peroxyhemiacetalFMNH intermediate III (HFMN-4a-OOCH(OH)R; Eq. 2). A fission of the O-O bond of intermediate III together with electron reshuffling generate a caged radical pair of $RC^{\bullet}(OH)O^-$ and the novel 4a-hydroxyFMNH radical $(HFMN\text{-}4a\text{-}OH)^{+\bullet}$ (Eq. 3). A back electron transfer from the C-centered radical $RC^{\bullet}(OH)O^-$ to the N5-centered radical $(HFMN\text{-}4a\text{-}OH)^{+\bullet}$ results in the formation of a fatty acid and an excited singlet 4a-hydroxyFMNH (HFMN-4a-OH*) as the emitter (Eq. 4). Radiative relaxation of HFMN-4a-OH* to the ground state leads to

---

[1]Present address: Energy BioSystems Corporation, 4200 Research Forest Dr., The Woodlands, TX 77381.

luminescence (Eq. 5). Finally, HFMN-4a-OH decays to generate H_2O and FMN (Eq. 6) thus completing the overall chemical pathway.

$$FMNH^- + O_2 \rightarrow HFMN\text{-}4a\text{-}OO^- \quad (1a)$$
$$HFMN\text{-}4a\text{-}OO^- + H^+ \rightleftharpoons HFMN\text{-}4a\text{-}OOH \quad (1b)$$
$$HFMN\text{-}4a\text{-}OO^- + RCHO + H^+ \rightarrow HFMN\text{-}4a\text{-}OOCH(OH)R \quad (2)$$
$$HFMN\text{-}4a\text{-}OOCH(OH)R \rightarrow RC^{\bullet}(OH)O^- + (HFMN\text{-}4a\text{-}OH)^{+\bullet} \quad (3)$$
$$RC^{\bullet}(OH)O^- + (HFMN\text{-}4a\text{-}OH)^{+\bullet} \rightarrow RCOOH + HFMN\text{-}4a\text{-}OH^* \quad (4)$$
$$HFMN\text{-}4a\text{-}OH^* \rightarrow HFMN\text{-}4a\text{-}OH + Light \quad (5)$$
$$HFMN\text{-}4a\text{-}OH \rightarrow FMN + H_2O \quad (6)$$

Although a full electron separation or transfer is depicted above for Eqs. 3 and 4, these steps have later been modified to allow charge transfer as an alternative in an electron/charge transfer (ECT) mechanism (4, 5).

Chemical Models: A critical aspect of this mechanism is the hypothesized novel 4a-hydroxyFMNH radical as a key intermediate. The feasibility for the formation of a flavin radical of this type was first examined by using 5-ethyl-4a-hydroxy(or methoxy)-3-methyl-4a,5-dihydrolumiflavin (5-EtF-4a-X; X = OH, OMe) as models. One-electron removal from such model compounds has indeed been shown to yield the expected $(5\text{-EtF-4a-X})^{+\bullet}$ flavin radicals by electrochemical oxidation, chemical disproportionation, and pulse radiolysis (4, 5 and references therein).

The successful formation of $(5\text{-EtF-4a-X})^{+\bullet}$ flavin radicals has important consequences. First, the one-electron oxidation potentials were measured to be 1.29 and 1.23 V for 5-EtF-4a-OH in acetonitrile and water, respectively (6, 7), and 1.3 V for 4a-hydroxy-4a,5-dihydro-3-methyl-tetraacetylriboflavin in water (7). These measurements led to the estimate of 1.3 V for the one-electron oxidation of HFMN-4a-OH (7). This, in turn, allows the calculation of free energies in water for the fission of the O-O bond (<16 kcal/mol) and that of the C(4a)-O bond (21 kcal/mol) in flavin-4a-OOH (8). Moreover, the O-O bond dissociation enthalpies (in kcal/mol) are <26 for flavin-4a-hydroperoxide, 38 for $CH_3O\text{-}OCH_3$, 45 for $CH_3O\text{-}OH$, and 51 for HO-OH (8). Therefore, the O-O bond in flavin-4a-hydroperoxide appears to be highly activated for fission. This is in good accord with the ECT mechanism, for the light-emitting pathway requires an efficient homolytic O-O bond fission whereas the heterolytic scission of the flavin-4a-hydroperoxide C(4a)-O bond would lead to H_2O_2 formation in a dark pathway. The relationship of the H_2O_2-forming dark decay of intermediate II to the luciferase luminescence quantum yield will be addressed later.

Second, the step for the formation of the primary excited state in bioluminescence is proposed as shown in Eq. 4. Thus, it is critical to examine the energetics of this reaction in relation to the energy requirement for the excitation of HFMN-4a-OH. The energy of the radical pair of $R\text{-}C^{\bullet}(OH)O^-$ and $(HFMN\text{-}4a\text{-}OH)^{+\bullet}$ was found to be about 90 kcal/mol above the ground state HF-4a-OH and carboxylic acid (7). This is 23 kcal/mol in excess of what is required to generate HFMN-4a-OH*. Therefore, the reaction as shown in Eq. 4 is energetically sufficient to form the singlet excited emitter.

Stability and Reactivity of Luciferase Intermediate II: Site-directed mutagenesis was employed to investigate the functional consequences of structural changes in luciferase. A substantial body of evidence indicates that the α subunit of the $\alpha\beta$ dimeric luciferase participates directly in catalysis. Consequently, only the α subunit was subjected to mutations in this work. An alignment of primary sequences of different species of luciferase α subunit reveals five conserved histidyl residues at positions 44, 45, 82, 224, and 285 (based on the *V. harveyi* luciferase α sequence). Each of these histidine residue was mutated to alanine, aspartate, or lysine to obtain ten mutants: αH44D, αH44A, αH44K, αH45D, αH45A, αH45K, αH82D, αH82K, αH224A, and αH285A (1). In nonturnover assays, these mutated luciferases showed light decay rates mostly similar to that of the native luciferase using octanal, decanal, or dodecanal as a substrate. However, both the peak emission intensities and the total quantum outputs of these luciferase variants were significantly reduced from a few-fold to 7 orders of magnitude. The essentiality of αHis44 and αHis45 was indicated by the remarkable 4-7 orders of magnitude of activity reductions detected with αH44D, αH44A, αH45D, αH45A, and αH45K. The αH44K mutant was more active than the other αH44- or αH45-mutated luciferases but still retained only 2% of the activity. The drastically reduced emissions of these luciferase variants correlated well with their very low or lack of activities in consuming the aldehyde substrate. The αHis45 was chosen for additional mutagenesis studies in order to probe further the essential role of this residue in regulating the luciferase activity. In addition to the original αH45D, αH45A, and αH45K mutants, this residue was mutated to generate αH45Q, αH45S, αH45R, αH45P, αH45M, αH45C, αH45W, αH45G, αH45Y, αH45L, and αH45F. All the new luciferase variants were similar to the original three αHis45 mutants in exhibiting little changes in the light decay rates but were associated with 3 to 7 orders of magnitude reductions in emission intensities or total quantum outputs. The three-dimensional structure of luciferase has recently been solved and the location of the active site has been proposed (9). The essential roles of αHis44 and αHis45 as shown above are in good accord with the crystal structure of luciferase which suggests that αHis44 is well within the range for direct interaction with the bound flavin and αHis45 is close to the proposed active site and involved in intersubunit hydrogen bonding.

In addition to reacting with aldehyde in the light-emitting monooxygenation pathway, intermediate II also undergoes an alternative dark decay to yield FMN and H_2O_2. While the dark decay is associated with the heterolytic fission of the C(4a)-O bond in flavin 4a-hydroperoxide, the bioluminescence (and aldehyde hydroxylation) activity is coupled with the homolytic scission of the O-O bond of intermediate III. Therefore, a more efficient C(4a)-O bond fission of intermediate II is expected to result in less efficient bioluminescence. The dark decay rate of intermediate II (k_{II}) can be determined by first reacting luciferase with $FMNH_2$ and oxygen in the absence of aldehyde to generate the flavin 4a-hydroperoxide. Subsequently, aliquots of the preformed intermediate II were withdrawn after different times and added to an aldehyde solution to initiate the light emission. The k_{II} can then be determined from the time course of the decrease of the bioluminescence capacity of intermediate II as shown by the delayed-aldehyde addition method. Using such an assay, values of k_{II} were determined at 23°C for the native luciferase and all the luciferase variants described above and compared with their bioluminescence quantum yields. Interestingly, with the exception of αH285A, the degrees of bioluminescence activity reduction showed a remarkable correlation with the rates of the dark decay of intermediate II (Fig. 1).

This correlation is quite consistent with our model studies which show that the O-O bond of flavin 4a-hydroperoxide is highly activated thus favoring the light-emitting monooxygenation pathway upon reacting with aldehyde. This is apparently the case with the native luciferase. However, for all the α-mutated luciferases shown in Fig. 1, structural changes result in perturbations of the chemical properties of the intermediate II in such a way that whenever the dark decay rate of intermediate II is enhanced, a corresponding reduction of the bioluminescence activity occurs.

For a multiple-step pathway (Eqs. 1-6), the light emitting activity can be affected by the yields and reactivities of any or any combination of the reaction intermediates. Moreover, the bioluminescence quantum yield is also intimately regulated by the intrinsic emission efficiency of the exited emitter. Structure perturbations of luciferase by mutation could conceivably affect any number of these properties. Therefore, it is not surprising that a few-fold changes in k_{II} are associated with over several orders of magnitude reductions in activity. Apparently, the mutations studied in this work not only affected the partition between the O-O bond and the C(4a)-O bond fission of the oxygenated flavin intermediates but also had other functional consequences.

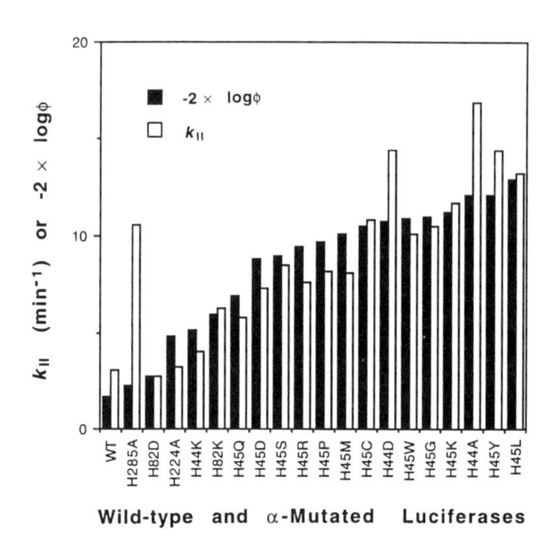

Figure 1. Bioluminescence quantum yield (ϕ) and the intermediate II dark decay rate constant (k_{II}) for native and α subunit-mutated luciferases. The ϕ is transformed to -2 × logϕ for presentation in order to use the same ordinate scale.

Although αHis44 and αHis45 are both super-critical to the expression of bioluminescence activity, their mutations resulted in two distinct patterns of effect on the yield of intermediate II. Luciferase variants with the αHis45 mutated to alanine, aspartate, lysine, glycine, and tryptophan were used for the preparation and isolation of intermediate II at 4°C using dodecanol as a stabilization agent as described previously (1). No intermediate II was obtained for any of these mutants in amounts detectable by absorption measurements. This could be due to either an extremely low yield or a very fast decay of intermediate II. To distinguish these two possibilities, αH45G and αH45W were further tested by stopped-flow experiments. Both intermediate II and FMN have much intense absorption than $FMNH_2$ in the 370-500 nm range. Moreover, intermediate II and FMN share an isosbestic point at 382 nm whereas only FMN has a pronounced absorption at 445 nm. When $FMNH_2$ was reacted with oxygen in the presence of excess αH45W, the formation of a significant amount of intermediate II would be accompanied by a sizable initial rise in A_{382} with very little increase in A_{445}. The subsequent decay of intermediate II to FMN would be associated with a pronounced increase in A_{445} with little or no change in A_{382}. On the other hand, the direct oxidation of $FMNH_2$ to FMN without involving flavin 4a-hydroperoxide as an intermediate would be characterized by the same time course monitored at either 382 or 445 nm. The results obtained with αH45W show the same kinetic patterns at 382 and 445 nm (Fig. 2). Similar results were also observed with the αH45G mutant. The apparent first-order rate constants of flavin oxidation at 0.65 mmol/L O_2 in the presence of αH45W (11 s^{-1}) and αH45G (10 s^{-1}) were slightly but significantly slower than that of the autoxidation of $FMNH_2$ (15 s^{-1}) under identical conditions, indicating that the former processes were enzyme mediated. Therefore, neither αH45W nor αH45G was capable of generating any significant amount of intermediate II detectable by the stopped-flow time scale. Such low yields of intermediate II can at most be attributed in part to a weak flavin binding; molar ratios of ~0.4 $FMNH_2$ bound per luciferase were detected for these two luciferase variants by equilibrium binding.

In contrast, intermediate II at yields of 14, 20 and 45% for the αH44D, αH44K, and αH44A (Fig. 3A) mutants were obtained following the same isolation procedures. These yields of intermediate II greatly exceeded the light emission activities of the corresponding luciferase variants. When the decay rate of αH44A intermediate II was examined at 4°C, it was found that a faster decay was detected by A_{460} changes (0.04 min^{-1}) than that by following the decreases in bioluminescence capacity (0.02 min^{-1}) (Fig. 3B). A similar experiment gave 0.02 min^{-1} for the rate of decrease in bioluminescence capacity in comparison with 0.3 min^{-1} for the A_{460} changes for the αH44D intermediate II. Under similar conditions, the native enzyme intermediate II showed the same decay rate by either of these two detection methods. These findings are consistent with the interpretation that two forms of intermediate II were generated by each of αH44A and αH44D. The major and bioluminescence-inactive species was detected by absorption whereas the minor species was detectable by its light emission capacity but was too low in yield to be observed by absorption.

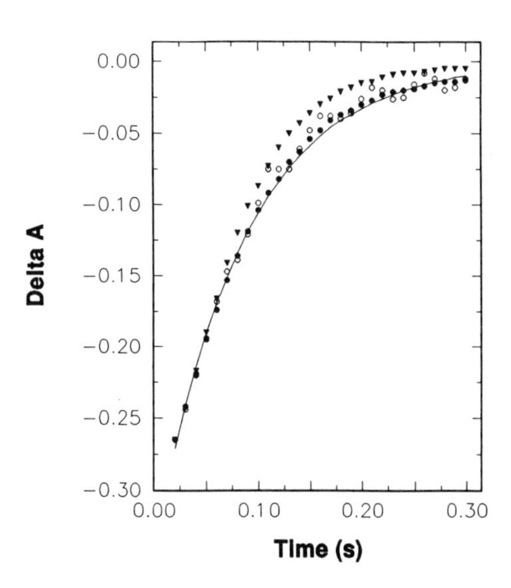

Figure 2. Kinetics of autoxidation and αH45W-catalyzed oxidation of $FMNH_2$. $FMNH_2$ at 30 μmol/L in anaerobic 0.2 mol/L phosphate, pH 7.0, containing 10 mmol/L EDTA and zero (▼) or 0.25 mmol/L αH45W (○, ●) was mixed at 23°C with an equal volume of the same buffer saturated with O_2. ΔA_{382} (× 2.02; ○) and ΔA_{445} (▼, ●) changes were followed as a function of time. The solid line is a single exponential fit for the αH45W-catalyzed oxidations.

Fluorescence of Intermediate II and Effects of KI: Flavin 4a-OH model compounds have very low fluorescence quantum yields, even in hydrophobic media, at <3 x 10^{-5} (10) but could be greatly enhanced by freezing. On the other hand, the bioluminescence quantum yield of luciferase is about 0.14. Therefore, most probably the HFMN-4a-OH* emitter is tightly and rigidly bound at the luciferase active site and is well shielded from the solvent. In the luciferase system, the HFMN-4a-OOH intermediate II is more stable than HFMN-4a-OH but the two species have similar fluorescence quantum yields and spectra. Therefore, the fluorescence properties of intermediate II, rather than HFMN-4a-OH, was subjected to further investigation. The freshly isolated intermediate II sample is known to show a weak fluorescence peaking at 520 nm, probably due to FMN contamination, and can be transformed into a highly fluorescent form (λ_{max} 490 nm) by exposure to excitation light (11). Both the weakly and highly fluorescent species have the same absorption spectra and are about equally active in bioluminescence. The light-induced transformation was repeated using

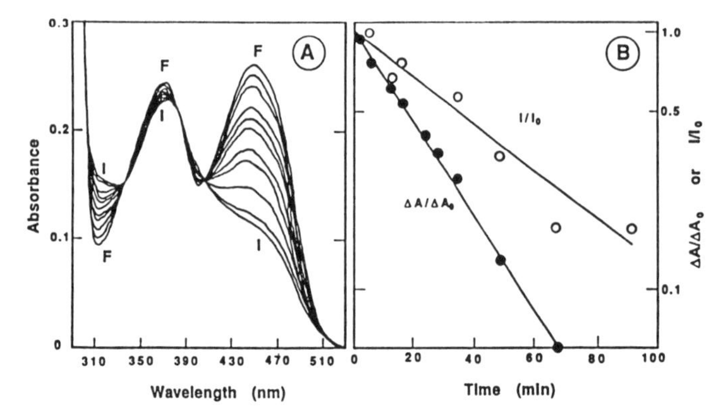

Figure 3. Decays of αH44A intermediate II at 4°C as detected by absorption and bioluminescence capacity. A: Absorption spectra of the intermediate II formed with αH44A immediately after isolation (I) using dodecanol as a stabilizing agent and the final FMN product (F) are shown along with transient spectra during the decay at 0.2, 5.7, 11.6, 15.8, 23.2, 28.0, 34.4, 49.0, and 66.3 min after isolation. B: The decay of intermediate II was determined by following the changes in A_{460} (●) and the decreases in bioluminescence capacity upon reacting with aldehyde (○). The ordinate for B is in a logarithmic scale. (Reprinted with permission from Xin X, Xi L, Tu SC. Biochemistry 1991;30:11255-62. Copyright (1991) American Chemical Society.)

intermediate II derived from the native luciferase (Fig. 4, left). At 4°C, maximal fluorescence (λ_{max} 490 nm) was obtained by exposing the intermediate II to 370-nm excitation light for 88 min. Decay of intermediate II to FMN became apparent upon further standing under excitation light or in the dark. The fluorescence of the sample under 178 min of exposure to the 370-nm light showed a major component of the FMN emission. The intermediate II derived from αH44A was subjected to the same treatment and results are shown in Fig. 4 right panel. This intermediate II species

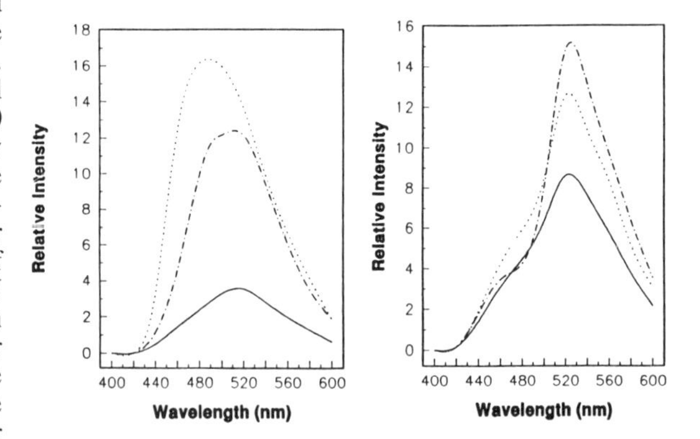

Figure 4. Fluorescence spectra and intensities of intermediate II upon standing under excitation light. Left: Intermediate II of the native luciferase was isolated under red light at 4°C in 50 mmol/L phosphate, pH 7.0, using 0.1 mmol/L dodecanol as a stabilizing agent. Fluorescence emissions were recorded for the sample immediately placed (——), or after 88 (····) or 178 (–·–) min under excitation at 370 nm at 4°C. Right: Intermediate II was isolated similarly using αH44A in place of the native luciferase. Fluorescence emission spectra were recorded for the sample immediately placed (——), or after 6 (····) or 17 (–·–) min under the excitation. For both panels, fluorescence intensities were normalized to 5 μmol/L total flavin in the samples.

showed a pronounced peak of fluorescence at 520 nm with an apparent shoulder at 480 nm immediately after isolation. Very little increase of 480-nm fluorescence occurred under the 370-nm excitation, reaching the maximal intensity at 480 nm after 6 min of exposure. Longer standing under the excitation resulted in decreases in 480-nm fluorescence and pronounced increases at 520 nm indicating a fast break down of intermediate II to FMN.

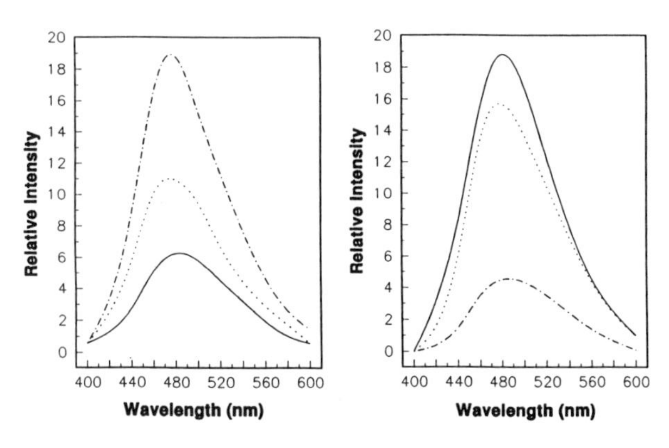

Figure 5. Effects of KI on the fluorescence spectra and intensities of intermediate II upon standing under excitation light. Left: Intermediate II of the native luciferase was isolated at 2°C as described for Fig. 4 and KI was immediately added to 0.2 mol/L. Fluorescence emission spectra were recorded at 2°C for the sample immediately placed (—), or after 8 (····) or 58 (-·-) min under excitation at 370 nm. Right: Intermediate II was isolated similarly using αH44A, and KI was added to 0.2 mol/L. Fluorescence emission spectra were recorded for the sample immediately placed (—), or after 3 (····) or 18 (-·-) min under the excitation. For both panels, fluorescence intensities were normalized to 5 μmol/L total flavin in the samples.

The same experiments described for Fig. 4 were repeated with the addition of 0.2 mol/L KI in the intermediate II samples. This concentration of KI is sufficient to quench nearly all the fluorescence of free FMN. For the intermediate II derived from the native luciferase (Fig. 5, left), only the fluorescence emission with λ_{max} at 480 nm was observed for the sample immediately placed under the 370-nm excitation. Exposure to the 370-nm light again increased the emission intensities, reaching a maximum intensity after 58 min at a rate faster than that in the absence of KI. Further standing under the excitation light resulted in a gradual decrease of the fluorescence (not shown). In all cases, the peak of fluorescence emissions remained at 480 nm with no visible shoulder at 520 nm, even during the decay phase of intermediate II. KI, at 0.2 mol/L, was apparently effective in quenching the 520-nm fluorescence from free FMN but not the 480-nm emission from HFMN-4a-OOH. Therefore, HFMN-4a-OOH at the luciferase active site must be well shielded from the quencher. Moreover, the maximal intensity at 480 nm was reached faster upon exposure to 370-nm light in the presence of KI and was actually slightly higher than that observed in the absence of KI. Such findings suggest that KI may bind to luciferase and affect its conformation, hence resulting in these effects. The same experiments described for Fig. 5, left panel were repeated for the intermediate II derived from αH44A (Fig. 5, right). Surprisingly, an intense 480-nm fluorescence was observed immediately after the addition of 0.2 mol/L KI to the freshly isolated mutant intermediate II. Further standing under the excitation light resulted in a gradual fluorescence decrease. Again, no 520-nm fluorescence was observed during the above described process. Since KI alone did not enhance the fluorescence of the native intermediate II without the exposure to the excitation light (data not shown), we believe that the marked increase in the 480-nm fluorescence of the αH44A intermediate II in the presence of KI was also induced by the exposure to the excitation light during the emission measurement. These findings, taken together, indicate that the microenvironment surrounding the bound HFMN-4a-OOH in the native enzyme active site must be distinct from that of the αH44A enzyme, as evident from the faster decay and much weaker excitation-induced fluorescence at 480 nm observed with the mutant enzyme intermediate II. Such a conclusion is also supported by the crystal structure of luciferase which suggests that the αHis44 is at the flavin site (9). Furthermore, KI apparently binds to and affects the conformation (and hence properties) of luciferase. This is evident from both the faster rates of the excitation-induced fluorescence increase for the native and mutant intermediate II and the remarkable 480-nm fluorescence enhancement of the αH44A intermediate II observed in the presence of KI. The effects of KI in increasing the light decay and decreasing the bioluminescence intensity of luciferase (12) are also consistent with the view that luciferase binds KI at specific site(s).

Acknowledgements

This work was supported by grants GM25953 from N.I.H. and E-1030 from The Robert A. Welch Foundation.

References

1. Xin X, Xi L, Tu SC. Functional consequences of site-directed mutation of conserved histidyl residues of the bacterial luciferase α subunit. Biochemistry 1991;30:11255-62.
2. Mager HIX, Addink R. On the role of some flavin adducts as one-electron donors. In: Bray RC, Engel PC, Mayhew SG, editors. Flavins and Flavoproteins. Berlin: Water de Gruyter, 1984:37-40.
3. Macheroux P, Ghisla S, Kurfürst M, Hastings JW. Studies on the bacterial luciferase reaction: Isotope effects on the light emission. Is a 'CIEEL' mechanism involved? In: Bray RC, Engel PC, Mayhew SG, editors. Flavins and Flavoproteins. Berlin: Water de Gruyter, 1984:669-72.
4. Mager HIX, Tu SC. Chemical aspects of bioluminescence. Photochem Photobiol 1995;62:607-14.
5. Tu SC, Mager HIX. Biochemistry of bacterial bioluminescence. Photochem Photobiol 1995;62:615-24.
6. Mager HIX, Sazou D, Liu YH, Tu SC, Kadish KM. Reversible one-electron generation of 4a,5-substituted flavin radical cations: Models for a postulated key intermediate in bacterial bioluminescence. J Am Chem Soc 1988;110:3759-62.
7. Merényi G, Lind J, Mager HIX, Tu SC. Properties of 4a-hydroxy-4a,5-dihydroflavin radicals in relation to bacterial luminescence. J. Phys. Chem. 1992;96:10528-33.
8. Merényi G, Lind J. Chemistry of peroxidic tetrahedral intermediates of flavin. J Am Chem Soc 1991;113:3146-53.
9. Fisher AJ, Raushel FM, Baldwin TO, Rayment I. Three-dimensional structure of bacterial luciferase from *Vibrio harveyi* at 2.4 Å resolution. Biochemistry 1995;34:6581-6.
10. Kaaret TW, Bruice TC. Electrochemical luminescence with N(5)-ethyl-4a-hydroxy-3-methyl-4a,5-dihydrolumiflavin. The mechanism of bacterial luciferase. Photochem Photobiol 1990;51:629-33.
11. Balny C, Hastings JW. Fluorescence and bioluminescence of bacterial luciferase intermediates. Biochemistry 1975;14:4719-23.
12. Sirokmán G, Wilson T, Hastings JW. A bacterial luciferase reaction with a negative temperature coefficient attributable to protein-protein interaction. Biochemistry 1995;34:13074-81.

THE INTERACTION OF THE FLUORESCENT (ANTENNA) PROTEINS WITH BACTERIAL LUCIFERASE REACTION INTERMEDIATES

V. N. Petushkov and J. Lee
Department of Biochemistry and Molecular Biology, University of Georgia, Athens, GA 30602, USA.

Introduction

Emission spectra from bioluminescent organisms are observed to range over almost the whole visible region. Even within a single species, various colors of bioluminescence may be found. The mechanism of these color variations can be explained as the result of a direct or indirect process. Among species of firefly for example, a direct mechanism is responsible for the spectral variation. The chemical structure of the emitter, the electronically excited product molecule, oxyluciferin, is the same, but each species of firefly luciferase provides a different active site environment, which shifts the energy level of the oxyluciferin. This is analogous to a solvent effect on the fluorescence spectrum of a molecule in homogeneous solution. An indirect or sensitizer mechanism of bioluminescence modulation, appears to be more common, at least there are more identified examples of it. This is where a fluorescent species not involved in the chemical process is the ultimate light emitter. In *Renilla* and *Aequorea* bioluminescence this second molecule is the famous Green Fluorescence Protein, many applications of which are described in this present volume.

A different protein modulates the bioluminescence of many species of luminous bacteria. Most species of *Photobacterium* contain a highly fluorescent 21-kDa protein called Lumazine Protein, named for its highly fluorescent ligand, 6,7-dimethyl-8-ribityllumazine. Lumazine protein is the origin of these blue bioluminescences, with emission maxima in the 470-490 nm range (1). There also exists a yellow bioluminescence strain Y1 of *Vibrio fischeri*. The responsible emitter here is the Yellow Fluorescence Protein (YFP), related to lumazine protein but having FMN or riboflavin as a ligand (2).

Our interest is in the biophysical mechanism of the bioluminescence energy transduction into the electronic excited states of lumazine protein or YFP. The inclusion of only micromolar concentrations of these fluorescent proteins in the luciferase reaction mixture, is sufficient to cause a bioluminescence spectral shift. This implies that the luciferase and the fluorescent protein must be associated in a protein-protein complex of some sort. It has been shown that lumazine protein does form a stable complex with the reaction intermediates luciferase peroxyflavin and luciferase hydroxyflavins, but not with the unreacted luciferase (3). It can be rationalized that, if the native luciferase were to preform a protein-protein complex, the lumazine protein would cover the active site and therefore hinder the binding of the $FMNH_2$ substrate to the luciferase active site.

From measurement of fluorescence anisotropy decay, a model has been proposed that the initially formed excited state of luciferase hydroxyflavin, is able to transfer its excitation to the lumazine group by the weak dipole-dipole coupled mechanism (4). On the simple model for this process to be efficient, the flavin and the lumazine groups must lie within about 15 Å.

Experimental

The source of the bacteria, growth conditions and extraction and purification of proteins, are detailed elsewhere (3, 5). Fluorescence dynamics measurements use a Nd-YAG sync-pumped, dye laser which is cavity-dumped and frequency doubled. Fluorescence is detected by a micro-channel plate photomultiplier and single photon counting electronics. Data is analyzed with the aid of the "Globals" software (Laboratory for Fluorescence Dynamics, University of Illinois).

Results and Discussion

Bacterial luciferase is a 77-kDa protein without any fluorescence above 400 nm. Interaction of lumazine protein with luciferase was first recognized by the observation that the rotational correlation time of lumazine protein was increased in the presence of luciferase (6). This correlation time can be recovered from the decay of emission anisotropy of the lumazine ligand since the luciferase contributes no fluorescence in the 470-nm region. From the simple formula for the rotational correlation time:

$$\phi = M_r \eta(v + h)/RT$$

where v is the partial specific volume and h the hydration of the protein rotator, it turns out that at the temperature of measurement, 2 °C, ϕ in units of ns is the same as M_r in kDa. For lumazine protein alone, $\phi = 20$ ns, but with luciferase this is increased towards 100 ns, indicating the presence of the 21 + 77 kDa complex. However, when these interactions with the native luciferase were analyzed, the bioluminescence effect of lumazine protein could not be accounted for quantitatively. In the case of *Photobacterium leiognathi* luciferase for example, no interaction could be observed at all (5)! Therefore it was considered that the functional interaction must occur with one of the luciferase-flavin reaction intermediates, one preceding the chemical excitation step.

Luciferase hydroxyflavin, the fluorescent transient (FT), can be prepared in relatively stable condition. Measurement of its decay of emission anisotropy yields the expected value of ϕ around 70 ns. Instead of the expected increase in ϕ, on addition of lumazine protein it decreased substantially in a concentration dependent manner down to about 2.5 ns! This is because fluorescence anisotropy can be lost by an energy transfer pathway in addition to rotational diffusion. In the protein-protein complex of luciferase peroxyflavin and lumazine protein, if the two fluorophores, the peroxyflavin and the lumazine are proximate, they could exert a weak dipole-dipole interaction. The above 2.5 ns value corresponds to a 0.4 ns^{-1} rate of energy transfer within the two fluorescent transitions (4).

Figure 1 shows the experimental data. The top decay is for a 14 μM concentration of FT and the fluorescence anisotropy decay, r(t), can be accurately fitted by a monoexponential function, yielding $\phi = 70$ ns. Lumazine protein itself also shows a monoexponential decay of anisotropy with $\phi = 20$ ns (not shown). In mixtures with lumazine protein the decay curves need to be fitted with a biexponential function:

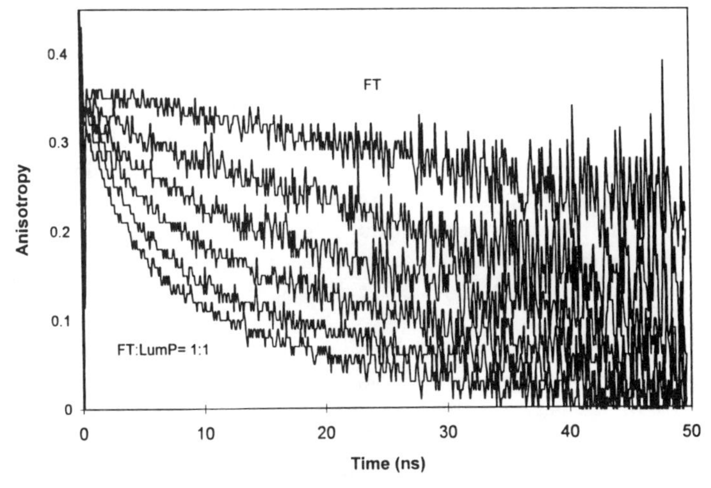

Figure 1. Fluorescence anisotropy decay of luciferase hydroxyflavin (FT, 14 μM, top), and with increasing concentrations of lumazine protein until it is equimolar. Ex 375, em 460 nm, 2 °C.

$$r(t) = \beta_1 \exp(-t/\phi_1) + \beta_2 \exp(-t/\phi_2)$$

with $\phi_1 = 70$ ns and $\phi_2 = 2.5$ ns. The lowest curve is almost monoexponential and is the one where the FT and lumazine protein are equimolar. If higher concentrations of lumazine protein are added, the curves rise again towards the one for lumazine protein alone. This can be explained by a protein-protein equilibrium:

$$FT + LumP \Leftrightarrow FT.LumP$$

We can equate the relative amplitudes (β's) with the concentrations of free FT and that bound to LumP. It is concluded that the proteins form a 1:1 complex with K_d in the micromolar range. The complex is sufficiently stable to be identified by rapid chromatography (3).

Energy transfer is responsible for the $\phi_2 = 2.5$ ns component and from the Förster equation for the rate of energy transfer, k_T, in a weakly coupled dipole-dipole electronic system:

$$k_T = AQ_F\kappa^2\tau_D^{-1} R^6 J$$

The separation *R*, between the donor, the hydroxyflavin intermediate on the luciferase, and the acceptor, the lumazine ligand on its protein, can be calculated as about 15 Å. The other symbols in this equation are Q_F, the fluorescence quantum yield of the donor and τ_D its fluorescence lifetime, κ is an orientation factor allowing for the angles between the transition dipoles of the donor and acceptor, *J* is the spectral overlap between the donor emission and acceptor absorbance, and *A* is a constant. The unknown factor in this equation is κ, usually for molecules randomly oriented in free solution a value $\kappa^2 = 0.67$ is correct. However, it is known that both donor and acceptor are rigidly fixed on their respective proteins but since in Figure 1, it is apparent that the there is no residual anisotropy after 50 ns, it assumed therefore that the transition dipoles must on the average have a "magic angle" respective orientation, justifying the random orientation value of κ^2 to be assumed. Nevertheless, this remains an important uncertainty in the estimate of R.

The other parameters in the Förster equation are all measurable. In the FTLumP case, the value of J is unusually low compared to the many homogeneous donor-acceptor systems that have received detailed study. However, τ_D = 10 ns for the FT donor is rather a slow decay so Nature makes up for the lack of sufficient spectral overlap by placing the molecules close together thus increasing the coupling. In other words, by their proximity, the probability of emission from the lumazine excited state is much greater than from the hydroxyflavin.

The dependence on spectral overlap suggests an experiment to test the weak-coupling hypothesis of the lumazine protein bioluminescence modulation. Lumazine apo-protein is a stable species and can be recharged with analogs of the native ligand, obvious candidates being riboflavin and FMN. Their spectral overlaps would be 8 times greater, they should effectively shift bioluminescence to longer wavelengths to match the flavin fluorescence (max 535 nm) with an accompanying rapid decay of anisotropy reflecting the excitation transfer. Both flavoproteins fail to affect either the bioluminescence or fluorescence property. The reason is that they do not form the necessary protein-protein complex, demonstrated by direct chromatography methods as well as fluorescence dynamics. It is concluded that the ligand itself must also participate in the protein-protein interaction. However, no conformational change in the protein can be observed by far UV circular dichroism on binding different ligands. If there is a ligand induced conformational difference it must be quite subtle (3).

The related YFP does bind FMN and is responsible for the yellow bioluminescence emission of *V. fischeri* Y1. This protein has strong sequence similarity to lumazine protein so it is expected that the bioluminescence mechanism involved will be the same as for the lumazine protein case. An additional advantage of the YFP system is that the donor and acceptor emissions are spectrally well separated. This allows the observation of change in the fluorescence lifetime of the donor to measure the rate of energy transfer, an experiment allowing greater precision and is more tractable for analysis than is the interpretation of anisotropy decay.

Figure 2 shows the fluorescence intensity decay of the FT from Y1 luciferase to be approximately monoexponential with τ_D = 10 ns. Addition of YFP makes the decay more complex and a major rapid decay rate is seen with τ_D = 0.25 ns. This rapid decay can be directly equated to the inverse of the energy transfer rate, k_T = 4 ns^{-1}. This value is ten times that calculated for the lumazine protein-FT complex to be compared with the ratio = 8 of spectral overlaps. It is concluded that the bioluminescence mechanisms of these two proteins is the same and that the topology of the two complexes must be very similar (5).

As well as the spectral effects described above, the inclusion of the fluorescent proteins in the reaction alters the bioluminescence kinetics, i.e., a change on the seconds time scale in contrast to nanoseconds. This has led to a consideration of these fluorescent proteins participating in the chemical process thereby influencing some rate determining step. The largest effects are reported for YFP and *V. fischeri* luciferase, up to a ten-times

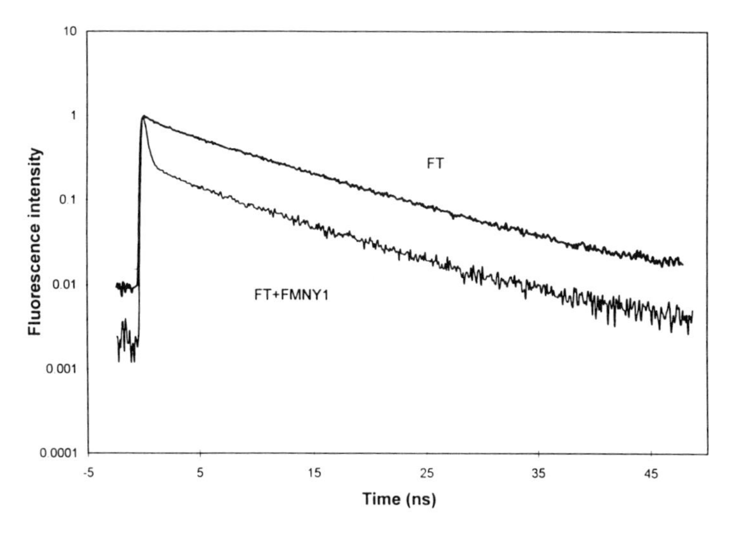

Figure 2. Fluorescence decay from luciferase hydroxyflavin (FT, 9 μM, top) and after the addition of Yellow Fluorescence Protein (FMNY1, 13 μM).

Table 1. Inhibition of steady-state bioluminescence by fluorescent proteins.

	decanal	dodecanal	tetradecanal
LumP	40	10	8
FMN-Y1protein	1.5	0.8	0.3
Rf-Y1protein			1

Concentration of added protein (μM) required for 20% inhibition (0 °C). Bioluminescence with the NADH oxido-reductase system with added dodecanol to stabilize the luciferase peroxyflavin intermediate. The Y1 protein has either FMN or riboflavin as ligand and Y1 luciferase. The LumP reaction is with *P. leiognathi* luciferase.

increase in the initial light intensity and its rate of decay. Lumazine protein produces up to three times increase in the kinetic rates with *P. phosphoreum* luciferase but the effect is much less in the *P. leiognathi* luciferase reaction.

In contrast to enhancing the kinetic rates, under certain reaction conditions inhibition of the bioluminescence can also be brought about by the fluorescent proteins. Table 1 shows some results for the approximate "steady-state" condition achievable by generating the $FMNH_2$ via NADH and oxido-reductase, with dodecanol also included to stabilize the luciferase peroxyflavin. The inhibition effect is much stronger for longer chain length aldehydes and also depends on the type of fluorescent protein. These properties parallel those conditions where the added fluorescent protein stimulates the bioluminescence reaction kinetics (7)

All these kinetic effects can be explained by a modification of the recently elaborated reaction sequence model (8). This scheme is described in detail elsewhere but can be discussed by reference to Figure 3, a picture of the protein-protein complex which attempts to incorporate all information known at this time. The shape of luciferase is traced from the recently published three-dimensional structure (9). It is proposed that the aliphatic component is essential for the association to occur, acting as a "glue" between the two proteins. Lumazine protein (or YFP), a spherical protein of Stokes radius 23 Å (10), is placed so that the ligand is 15 Å from the flavin site on the luciferase. This site cannot be resolved yet in the luciferase structure but is inferred from the structure of the analogous "non-fluorescent protein" (11). Also, by analogy a second site may exist in the interface region responsible for the ligand specificity observed in the protein-protein association.

If the luciferase peroxyflavin is stabilized by the addition of dodecanol, it is

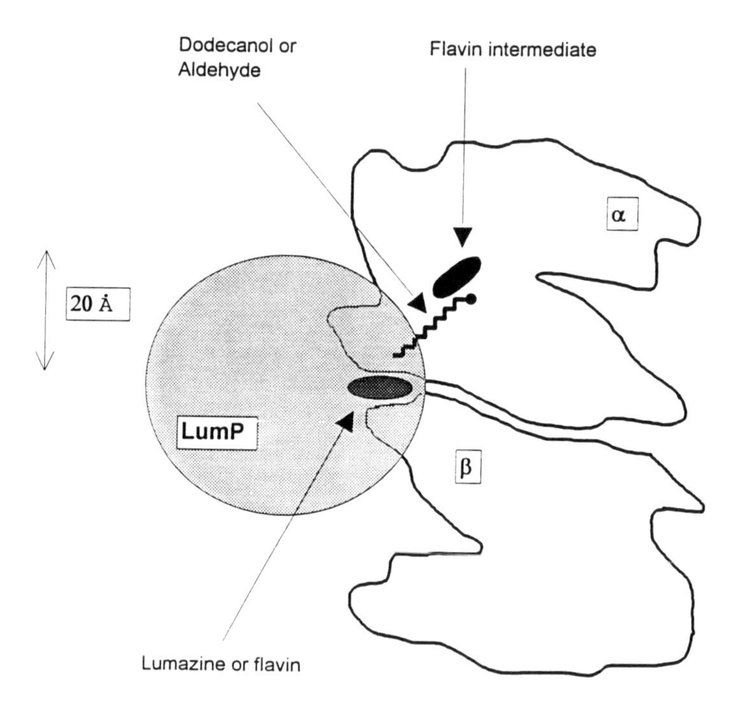

Figure 3. Model of the antenna protein-luciferase intermediate drawn to scale.

reasonable to propose that the dodecanol binds to the same site as the aldehyde and that the associated fluorescent "antenna" protein would interfere with the aldehyde-alcohol exchange, resulting in the inhibition effect. The same argument would imply that the bound antenna protein would favor the bound state of the aldehyde in that reaction where dodecanol is not used. If the concentration of this precursor species to the excitation step is increased, the rate of light reaction will therefore be stimulated.

In order to proceed further in research on the biophysical aspects of bacterial bioluminescence, two experimentally formidable tasks present themselves. First a detailed knowledge of the kinetics of all the species in the light reaction pathway must be established. Some progress in this regard has appeared recently (8). Second, the three-dimensional structure of the lumazine protein-luciferase complex is needed. In itself this is beyond the range of present technology, the luciferase peroxyflavin has a half life of only a few hours at 0 °C. Although it is probably indefinitely stable at -80 °C, to produce X-ray quality crystals under cryoscopic conditions is challenging! Lumazine protein itself fails to form crystals but its molecular weight is in the range where the solution structure can be solved by NMR methods. Progress is being made in this regard and with the recent solution of the structure of luciferase, it may soon be possible using advanced calculation methods, to produce a plausible model for the protein-protein complex.

References

1. Lee J. Lumazine protein and the excitation mechanism in bacterial bioluminescence. Biophys Chem. 1993;48:149-58.
2. Petushkov VN, Gibson BG, Lee J. The yellow bioluminescence bacterium, *Vibrio fischeri* Y1, contains a bioluminescence active riboflavin protein in addition to the yellow fluorescence FMN protein. Biochem Biophys Res Commun. 1995;211:774-9.
3. Petushkov VN, Gibson BG, Lee J. Properties of recombinant fluorescent proteins from *Photobacterium leiognathi* and their interaction with luciferase intermediates. Biochemistry. 1995;34:3300-9.
4. Lee J, Wang YY, Gibson BG. Electronic excitation transfer in the complex of lumazine protein with bacterial bioluminescence intermediates. Biochemistry. 1991;30:6825-35, erratum 10818.
5. Petushkov VN, Gibson BG, Lee J. Direct measurement of excitation transfer in the protein complex of bacterial luciferase hydroxyflavin and the associated yellow fluorescent proteins from *Vibrio fischeri* Y-1. Biochemistry. 1996;35:8413-8.
6. Visser AJ, Lee J. Association between lumazine protein and bacterial luciferase: direct demonstration from the decay of the lumazine emission anisotropy. Biochemistry. 1982;21:2218-26.
7. Petushkov VN, Gibson BG, Lee J. Interaction of *Photobacterium leiognathi* and *Vibrio fischeri* Y-1 luciferases with fluorescent (antenna) proteins: Bioluminescence effects of the aliphatic aldehyde. Biochemistry. 1996;35:Sept 25 .
8. Abu-Soud H, Mullins LS, Baldwin TO, Raushel FM. Stopped flow kinetic analysis of the bacterial luciferase reaction. Biochemistry. 1992;31:3807-13.
9. Fisher AJ, Raushel FM, Baldwin TO, Rayment I. Three dimensional structure of bacterial luciferase from *Vibrio harveyi* at 2.4 Å resolution. Biochemistry. 1995;34:6581-6.
10. O'Kane DJ, Lee J. Physical characterization of lumazine proteins from *Photobacterium*. Biochemistry. 1985;24:1484-8.
11. Moore SA, James MN. Structural refinement of the non fluorescent flavoprotein from *Photobacterium leiognathi* at 1.60 A resolution. J Mol Biol. 1995;249:195-214.

CRYSTAL STRUCTURE OF FLAVIN REDUCTASE P - A DIMERIC ENZYME THAT PRODUCES REDUCED FLAVIN IN *VIBRIO HARVEYI*

KL Krause, MD Miller, JJ Tanner
Department of Biochemical and Biophysical Sciences
University of Houston, Houston, TX 77204-5634, USA

Introduction

Bioluminescence is a prominent characteristic of several *Vibrio* species of which the most heavily studied is *V. harveyi*. In the most common light producing reaction bacterial luciferase catalyzes the oxidation of reduced flavin and aldehyde in the presence of molecular oxygen resulting in the production of water and light (1). Because of its role as a biological catalyst in bacterial bioluminescence, the luciferase enzyme has been extensively researched. Considering its role as an important and well studied flavoprotein it is somewhat paradoxical that bacterial luciferase does not contain its own flavin co-factor and is, therefore, dependent on externally supplied reduced flavin for its light production.

Reduced flavin in luminous bacteria is primarily supplied by three distinct flavin reductases which are identified as FRP, FRD and FRG. This identification has been proposed to differentiate between flavin reductases that display a preference for NADPH in their catalytic mechanism as is the case for FRP, or for NADH as is the case for FRD. FRG uses NADH and NADPH with similar efficiencies. Since FRP displays a kinetic shift from a ping-pong bisubstrate-biproduct mechanism to a sequential mechanism in the presence of luciferase, a direct transfer of reduced flavin between luciferase and FRP has been proposed (2).

$$\text{FMN} + \text{NAD(P)H} + \text{H}^+ \xrightarrow{\text{Flavin Reductase}} \text{FMNH}_2 + \text{NAD(P)}^+ \quad (1)$$

$$\text{FMNH}_2 + \text{RCHO} + \text{O}_2 \xrightarrow{\text{Bacterial Luciferase}} \text{FMN} + \text{RCOOH} + \text{H}_2\text{O} + \text{light} \quad (2)$$

Recently our group has determined the 1.8Å structure of FRP from *V. harveyi*, which is a flavin reductase that is thought to provide reduced flavin to luciferase for use in bioluminescence (3). The structure of this flavoprotein, which was determined using multiple isomorphous replacement, is a unique homodimer and is representative of a new class of flavoprotein. Analysis of this structure has allowed us to understand the enzyme's preference for FMN as a cofactor and a co-substrate in the flavin reductase reaction. From the structure of the active site we have deduced the probable geometry of hydride transfer from NADPH to FMN.

Materials and Methods

Purification and Crystallization: A detailed description of cloning and purification of FRP reductase which was accomplished in the laboratory of

Professor S. -C. Tu is available (4). Crystallization was performed by incubation of purified protein in sitting drops versus 30% polyethylene glycol 6000 and 0.1 mol/L Hepes at pH 7.0. After one to two weeks large crystals of greater than 0.2 x 0.4 x 0.6 mm in size appear that occupy space group $P2_1$ with lattice constants of a = 51.2 Å, b = 85.9 Å and c = 58.1Å. The monoclinic angle β was 109.3° with the other cell angles equal to 90°.

Instrumentation and Software: The initial crystallization trials were carried out using Crystal Screen™ and Grid Screen™ reagents both from Hampton Research. Cryschem™ plates from Charles Supper Company were used for the sitting drop crystallization setups. Heavy atom reagents were purchased from Johnson Matthey *Alfa* chemical company and Aldrich Chemical Company. Data collection at the University of Houston was performed on an Enraf Nonius *FAST* area detector mounted on a Rigaku RU-200 rotating anode and data collection at Rice University was performed on a Rigaku R-axis II imaging plate detector mounted on a Siemens rotating anode. Data collection at Rice University was performed courtesy of Professor G. N. Phillips.

Computing and structural construction was done on a Vaxstation 4000/90, a Cray YMP/EL, an Evans & Sutherland PS390 Graphics device and a Silicon Graphics Indigo/ELAN workstation. On the FAST data collection was performed using MADNES software and data were merged with XSCALE. Rigaku data collection software was used on the R-axis II. Heavy atom refinement was performed with the PHASES package supplemented with SQUASH. Refinement and fitting were performed with X-PLOR and O respectively. Complete references to these methods are available in the full structural report (3).

Data Collection and Structure Solution: Although flavoproteins from several different structural families such as flavodoxins and oxidases have been solved crystallographically (5-8), a search of sequence databases revealed no significant homology with any flavoprotein of known structure (4). Significant homology was, however, discovered between NADH oxidase of *Thermus thermophilus* (NOX) and FRP flavin reductase, but at the time of our search NOX had not been studied crystallographically. The results of our search indicated that a traditional x-ray structure determination would have to be done with the rate limiting step being the preparation of suitable heavy atom derivatives.

Prior to heavy atom preparation we collected native data sets on two area detectors for use in building and refinement. On our home area detector, which is a *FAST* detector from Enraf Nonius we collected a 2.6 Å data set containing 27, 508 observations of 13,441 reflections. This data set was used primarily in combination with the heavy atom derivative data sets. In addition we collected a native data set on an R-axis II imaging plate detector to 1.8 Å that contained 117,092 observations of 36,163 reflections. Statistics from these native data sets are shown in Table I along with statistics from the two heavy atom data sets . It is clear that for space groups possessing low symmetry like $P2_1$ the larger solid angle of the R-axis II allows for rapid collection of a large number of independent reflections at high resolution.

	Native R-AXIS	Native FAST	Me_3PbAc FAST	K_2PtCl_4 FAST
Concentration (mM)			10.0	0.75
Soaking Time (days)			12	0.7
Number of Observations	117092	27508	39833	19252
Unique Reflections	36163	13441	13013	13117
R-sym (%)	4.3	2.3	3.5	2.0
Completeness 2.6 Å (%)	95.3	88.4	85.6	86.4
2.0 Å (%)	91.6			
1.8 Å (%)	81.9			
outer 0.1 Å shell (%)	40.7	51.8	47.0	50.4
R-merge (%)			12.6	9.0
Number of Sites			2	7
Phasing Power to 2.6 Å			2.5 (2.7^d)	1.7

Table 1. Statistics from native and derivative data sets collected for FRP reductase

Soaking of native FRP crystals in trimethylead acetate yielded our first heavy atom derivative. In the process of locating heavy atom derivatives a complete data set is collected using crystals soaked in a solution containing heavy atom material. Then data collected from these soaked crystals is compared on a reflection by reflection basis with data collected from unsoaked "native" crystals. Differences in intensity between the reflections are then used to calculate a difference Patterson map. In favorable cases Patterson maps can be solved to locate the positions of the heavy atoms in the unit cell. At least two heavy atom derivatives, but usually more, are needed for a structure solution and success in performing heavy atom soaks is often a matter of protracted trial and error. Obtaining a clear Patterson map with strong distinct sites allows for a more rapid structural determination and we were fortunate to have an initial derivative that produced a excellent Patterson map (Fig. 1)

With one strong derivative in hand that also contained excellent anomalous information, we were able to readily identify a platinum tetrachloride derivative. After data collection on this second derivative we began multiple isomorphous replacement (MIR) phase refinement. In this method all of the differences in intensity between heavy atom derivative and native data sets are interpreted with the knowledge of the location of the heavy atom sites and used to provide an initial solution to the protein structure. This solution comes in the form of an electron density "shell" into which the protein sequence is fit and subsequently refined. Noncrystallographic two-fold averaging prior to fitting was also performed to allow a more clear interpretation of the electron density prior to fitting. From analysis of the initial electron density it was clear that the enzyme was a homodimer composed of two identical monomers and that one dimer was present in each asymmetric unit of the crystallographic unit cell.

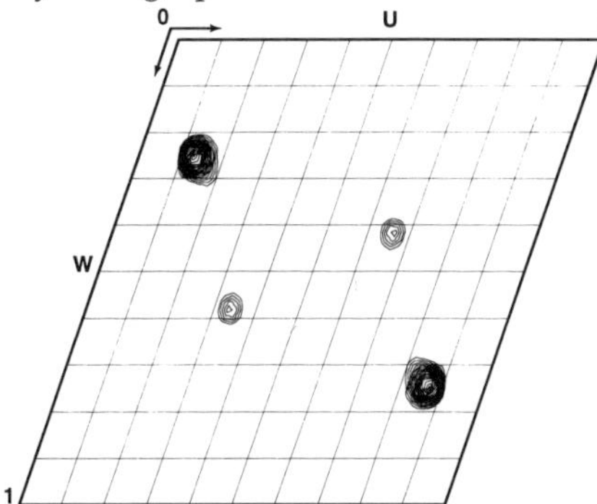

Figure 1. Harker section of the two-site difference Patterson map obtained from the trimethylead derivative of FRP native crystals. Note that the peaks are strong, quite distinct and well-shaped and that a two-fold axis of symmetry relates the two sets of peaks to each other.

Refinement and Quality of Structure

The final FRP structure was fit into density using the program O and refined with X-PLOR. The current model includes 3564 atoms comprising 460 amino acids, 2 FMN molecules, 2 inorganic PO4 molecules and 107 water molecules. Residue 1 and residues 201-209 of both monomers were omitted because of weak density, but the remainder of the structure is well fit to density. No restraints were employed between the two monomers of the homodimer during refinement.

R-factor	17.5
Free R-factor	20.7
Resolution (Å)	28-1.8
Number of reflections	32,144
Completeness 1.8 Å (%)	82
Completeness of outer shell (%)	41
Rmsd bonds (Å)	0.013
Rmsd angles (°)	1.5
Number of atoms	3,564
Number of protein residues	460
HETATM residues	2 FMN
HETATM residues	2 PO_4
Number of water molecules	107
Average B-factor (Å^2)	19.5
NCS constraint	none

Table 2. Refinement Statistics

With our current model the crystallographic R-factor is 17.5% with a real-space correlation coefficient of 0.95. The R_{free} on 3540 randomly chosen reflections left out of refinement to avoid bias is 20.7% representing a difference between crystallographic R-factor and R_{free} of about three per cent. The average B-factor for all atoms is 19.5 Å2 and 23.1 Å2 for modeled water molecules. Final refinement statistics are shown in Table 2.

Results and Discussion

Nature of the Structure: The FRP structure is a homodimer composed of two monomers that interlock extensively involving almost 9000Å of surface area (Fig 2a). In a fashion resembling the two components of a baseball cover each monomer interacts with the other (Fig. 2b). There are two FMN molecules clearly visible in the structure that identify the location of the two active sites of the enzyme. Each flavin primarily interacts with only one monomer but as the active site is located at the dimer interface some interaction between the FMN co-factor and the opposite monomer does take place. The noncrystallographic two-fold operation that relates the monomers is nearly correct with a root-mean-square-difference between the monomers following superposition of 0.18 Å for the main chain atoms.

Each monomer is composed of two domains, one major and one minor domain (Fig 2a). The major domain (residues 15-161, 226 - 240) contains most of the first half of the monomer and a portion of the carboxy-terminus. It consists of a four stranded antiparallel sheet with several helices stacked on either side. The minor domain (residues 2-14, 162-225) extends across the surface of the protein into the region of the major domain of the opposite monomer. In fact a fifth beta strand is contributed by this monomer to the central sheet of the major domain. The remaining portion of the minor domain is composed of extended structure and helix that is stacked against the opposite monomer. By inspection this fold is representative of a new class of flavoprotein resembling neither the α/β fold of most flavodoxins (5, 6, 8) or the common β barrel as found in glycolate oxidase (7). It most closely resembles the recently reported structure of NADH oxidase which is also a homodimer but which is significantly smaller in size (9). This close relationship is not surprising since from another perspective a flavin reductase reaction is, in fact, an NAD(P)H oxidation.

Figure 2. (A) A ribbon diagram of the FRP dimer viewed down the twofold axis relating the monomers. (B) A CPK depiction of the FRP dimer that emphasizes its interlocking nature .

Dimer Interface: The dimer interface of FRP reductase contains several unique features and, therefore, deserves special mention. The amount of buried surface area (9352 Å) is unusually large for an isologous homodimer suggesting a very strong dimer interaction (10). The shape of this interaction also deviates sharply from the more common circular, flat interface found in most dimers (10). To illustrate these differences we have compared the buried surface area and the dimer energetics of the FRP dimer with that of a more typical isologous dimer, the extracellular nuclease from *Serratia marcescens*. These calculations were performed with X-PLOR using the refined structure of FRP and the 1smn entry for nuclease from the Protein Data Bank. Of course, since these calculations lack an entropy component they do not represent free energies and cannot be used to predict the overall stability of the interface. In fact, since the large surface area ties up so much of the monomer surface it is possible that the large loss in conformational entropy associated with this interface may destabilize the dimer interaction. Of the buried surface area hydrophobic residues contribute 51% while polar and charged residues contribute 31% and 18% respectively. Residues in the dimer interface are involved in 57 hydrogen bonds and 5 intersubunit salt links.

Table 3. Dimer Interface Comparison: FRP and nuclease

	FRP	nuclease
Buried Surface Area	9835 $Å^2$	2016 $Å^2$
% of Surface Buried	39.7%	10.4%
van der Waals interactions	-637 kcal/mol	-124 kcal/mol
electrostatic interactions	-326 kcal/mol	-124 kcal/mol

FMN Binding Site and Mechanism: The active site region is located in a crevice between the two monomers with each FMN co-factor being bound by portions of both subunits (Fig.3). An active site which is located between monomers of an oligomer or at an interface between protein domains is so common as to be considered routine (11). In some cases only some of the active sites are active because substrate binding to one part of the enzyme interferes with action elsewhere, but there is no evidence for "half of the sites reactivity" in FRP reductase. In the case of FRP the isoalloxazine ring is wedged between strand 3 of one subunit and a loop containing 4 adjacent Ser residues from the other subunit. Both hydrogen bonds and salt links dominate the FMN binding interactions although some hydrophobic contacts do occur.

From consideration of the space filling model of the dimeric enzyme it is clear that the entire *si* face of the flavin co-factor is buried indicating that the approach of NADPH and later FMN must occur toward the *re* face of FMN. Although the static structure would not allow for the approach of either of these molecules to within the interaction distance of FMN cofactor, preliminary solvent accessible calculations (3) suggest that N1 and N5 are the strongest candidates to receive a hydride from NADPH. Additionally within the active site there is a free phosphate bound and it is hope that this may represent a possible location of the phosphate moieties of the two substrates FMN and NADPH. From steric considerations it is also clear that FAD could not be accomodated in this dimeric enzyme as a co-factor.

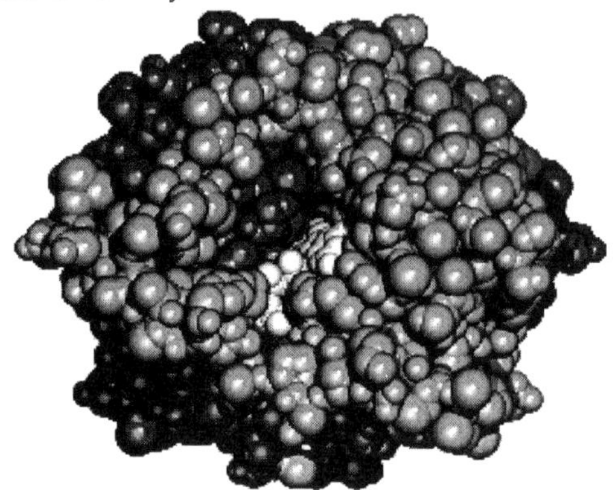

Figure 3. A space filling depiction of the FRP dimer in which the FMN molecule (white) can be seen interacting with both monomers leaving only a portion of the *re* face exposed.

Since it is presently not known how FMN substrate and NADPH substrate bind within FRP it is difficult to speculate on the geometry of their approach to FMN in the catalytic mechanism. From the current structure it seems probable that the *re* face of the FMN co-factor must be involved in the mechanism and that the N1 and N5 atoms are involved in accepting the hydride. One important missing piece of the puzzle is how reduced flavin is made available to luciferase for light generation. Simple discharge of the $FMNH_2$ into the intracellular compartment is inelegant and if deposited in this way it would likely have a brief life span (12). If, on the other hand, FMN were transferred directly into the active site of luciferase, the process would be less haphazard. This direct transfer mechanism is supported by a change in kinetic scheme from ping-pong to sequential when FRP is assayed in the presence of luciferase (2). However a sequential mechanism implies that a ternary E:FMN:NADPH complex exists. From the structure it is clear that both substrates and the FMN cofactor cannot be simulaneously accommodated within the active site. One way around this dilemma would be for FRP to transfer its FMN co-factor to luciferase. It is known that FRP enzyme binds FMN significantly more tightly than $FMNH_2$. However even a reduced flavin molecule could still form as many as 15 stabilizing hydrogen bonds within the flavin co-factor site. Perhaps, if luciferase were to bind to FRP and interact directly with the flavin cofactor or to interfere with FRP:$FMNH_2$ interactions such a transfer might more readily take place.

The FRP structure reveals an exciting new structural class of flavoprotein, but most questions about the structural basis of its activity remain unanswered. The speculations of the previous paragraph leave us concluding that there is a need for more structural work aimed at elucidating the geometry of the binding of the two substrates, a need for further biochemical and isotopic studies of the enzyme's mechanism and finally with a goal of identifying mutants that in the absence of luciferase shift the kinetic scheme.

Acknowledgements

This work was supported by the State of Texas and the W. M. Keck Foundation. We thank George Phillips for graciously allowing us to use his R-axis area detector at Rice University. Finally, we are grateful to our friend and colleague Professor Shiao-Chun Tu without whose support this work would not have been possible.

References

1. Hastings JW, Nealson KH. Bacterial bioluminescence. Ann Rev Microbiol 1977;31: 549-95.
2. Lei B, Tu S-C. Characterization of the *Vibrio harveyi* FMN:NADPH oxidoreductase expressed in Escherichia coli. In: Yagi K, editor. Flavins and Flavoproteins. Berlin: Walter de Gruyter, 1994:847-50.
3. Tanner JJ, Lei B, Tu S-C, Krause KL. Flavin Reductase P: Structure of a dimeric enzyme that reduces flavin. Biochemistry 1996;in press.
4. Tanner J, Lei B, Liu M, Tu S-C, Krause KL. Crystallization and preliminary crystallographic analysis of NADPH:FMN oxidoreductase from *Vibrio harveyi*. J Mol Biol 1994;241:183-187.
5. Watenpaugh KD, Sieker LC, Jensen LH. The binding of riboflavin-5'-phosphate in a flavoprotein: Flavodoxin at 2.0 Å resolution. Proc Nat Acad Sci USA 1973;70:3857-60.
6. Burnett RM, Darling GD, Kendall GS, et al. The structure of the oxidized form of *Clostridial* flavodoxin at 1.9 Å resolution. Description of the flavin mononucleotide binding site. J Biol Chem 1974;249:4383-92.
7. Lindqvist Y. Refined structure of spinach glycolate oxidase at 2 Å resolution. J Mol Biol 1989;209:151-66.
8. Rao ST, Shaffie F, Yu C, Satyshur KA, Stockman BJ, Markley JL, et al. Structure of the oxidized long-chain flavodoxin from *Anabaena* 7120 at 2 Å resolution. Prot Sci 1992;1:1413-27.
9. Hecht HJ, Erdmann H, Park HJ, Sprinzl M, Schmid RD. Crystal structure of NADH oxidase from *Thermus thermophilus*. Nature Struct Biol 1995;2:1109-1114.
10. Jones S, Thorton JM. Protein-protein interactions: A review of protein dimer structures. Prog Biophys Molec Biol 1995;63:31-65.
11. Creighton TE. Proteins: Structures and molecular properties. (2nd ed.) New York: W. H. Freeman and Co., 1993
12. Gibson QH, Hastings JW. The oxidation of reduced flavin mononucleotide by molecular oxygen. Biochem J 1962;83:368 - 77.

A CHARGE RELAY SYSTEM IS INVOLVED IN THE MECHANISM OF THE LUXD THIOESTERASE

Jun Li and Edward A. Meighen
Department of Biochemistry, McGill University,
Montreal, Quebec, Canada, H3G 1Y6

Introduction

Thioesterases are essential for many biochemical processes in the cell. One of the most important roles for thioesterases involves chain termination and release of fatty acids from the phosphopantetheine group that carries the acyl chain during fatty acid biosynthesis (1,2). In luminescent bacteria such as *Vibrio harveyi*, a lux-specific thioesterase (LuxD), which cleaves myristoyl-CoA and myristoyl-ACP, is responsible for specifically diverting myristic acid from fatty acid biosynthesis to the luminescent system (3). This enzyme is part of a fatty acid reductase multienzyme complex, which includes acyl-protein synthetase and acyl-CoA reductase subunits responsible for the ATP-dependent activation and NADPH-dependent reduction respectively, of the myristic acid provided by the thioesterase (4). The recent elucidation of the crystal structure of the lux-specific thioesterase has shown that this enzyme contains a proteinase-like catalytic triad with Ser^{114} in close proximity to His^{241} and Asp^{211} (5). However the proposed active site nucleophilic Ser^{114} is not part of a GxSxG consensus pentapeptide sequence common to esterases, lipases and other mammalian thioesterases (6,7,8). The lux-specific thioesterase also differs from the other thioesterases as it has a much slower turnover rate with deacylation rather than acylation being the rate-limiting step. In this report, the biochemical evidence supporting a Ser-His-Asp charge relay system similar to other thioesterases is described for LuxD.

Materials and Methods

Materials: [^{3}H]Myristic acid (14 Ci/mmol) (Amersham Corp.) was purified by thin-layer chromatography. [^{3}H]Myristoyl-CoA (14 Ci/mmol) was prepared from the radioactive fatty acid. *p*-nitrophenyl myristate (Sigma Chemical Co.). En^{3}Hance (Dupont). Hyamine hydroxide and CytoScint (ICN).

Site Directed Mutagenesis: A 1.6 kbp SacI-BamHI restriction fragment containing the entire *V. harveyi* LuxD gene encoding the thioesterase was inserted into the SacI and BamHI restriction sites of the M13 (mp19) sequencing vector. Site directed mutagenesis was performed according to the method of Kunkel using the M13 In Vitro Mutagenesis Kit from Bio-Rad. Oligonucleotide primers were obtained from the Sheldon Biotechnology Center, McGill University.

Protein expression and purification: The recombinant wild type and mutant thioesterases were expressed and purified as previously described (9,10).

Protein Assay: The protein concentration was determined from the absorbance at 280 nm using an extinction coefficient of 32,000 $M^{-1}cm^{-1}$ or by the Bradford method.

Enzyme activity: The enzyme activity of the wild type and mutant LuxD were determined from the rate of cleavage of *p*-nitrophenyl myristate which was measured at pH 8.0 in 50 mM phosphate containing 0.05% Triton X-100, 0.7 mM ß-mercaptoethanol, 1.4% glycerol and 100 μM *p*-nitrophenyl myristate by following the change in absorbance at 405 nm using an extinction coefficient of 16,800 $M^{-1}cm^{-1}$ for *p*-nitrophenol at pH 8.0. All assays were conducted at 23°C.

Protein Acylation: To detect the acylated enzyme intermediates, the purified wild type LuxD and its mutants were incubated in 50 mM phosphate, pH 7.5, 0.1 mM ß-mercaptoethanol, 0.2% glycerol with [^{3}H]myristoyl-CoA (14 Ci/mmol) at room temperature. The acylation reaction was stopped by mixing with an equal volume of 1:1 SDS sample buffer.

Gel Electrophoresis and Autoradiography: SDS-PAGE was performed by using 12% polyacrylamide resolving gels and 5% stacking gels. Gels were stained in Coomassie brilliant blue R-250, destained, soaked in En^{3}Hance, dried under vacuum and exposed to Kodak X-OMAT film overnight to 3 days at -80°C. The dried protein bands were cut from the gels, dissolved in 90% hyamine hydroxide at 37°C overnight, CytoScint added and the radioactivity counted.

Results

The ability to measure the levels of acylated enzyme intermediate during enzyme catalysis indicates that the lux-specific thioesterase has a very slow turnover rate. A pre-steady state burst of *p*-nitrophenol in assays with *p*-nitrophenyl myristate and the stimulation of the rate of cleavage of myristoyl-CoA and *p*-nitrophenyl myristate by glycerol and other compounds with hydroxyl or thiol acceptors indicates that deacylation is the rate limiting step for the turnover of the enzyme with both thioester and ester substrates (10).

As show by X-ray crystallographic data (5), Ser^{114} along with His^{241} and Asp^{211} form a typical chymotrypsin-like catalytic triad. However the putative nucleophile Ser^{114} is not within a consensus GxSxG pentapeptide but is the central serine in the sequence of AxSxS. In order to investigate the role of Ser^{114} during enzyme catalysis, Ser^{114} has been converted to Cys^{114} by site-specific mutagenesis. The results of kinetic studies of the S114C and the wild-type enzyme show that residue Ser^{114} is required for high enzyme activity (10). The acyl group is readily removed by treatment of the acylated S114C with neutral hyamine hydroxide whereas the acyl group covalently linked to the wild-type enzyme is stable. These results have clearly demonstrated that Ser^{114} is indeed the active site nucleophile.

Previous work also implicated His^{241} in the catalytic mechanism (11). In this investigation, the S114C:H241N double mutant has been constructed, expressed and purified to homogeneity. Experiments show that both H241N and S114C:H241N have a low level of activity in terms of cleavage of *p*-nitrophenyl myristate. Fig. 1 is an autoradiograph of wild type, S114C, H241N and S114C:H241N LuxD after acylation and SDS gel electrophoresis. Although the H241N mutant can not be acylated, it is possible to incorporate acyl groups into the S114C:H241N double mutant at a level somewhat lower than that of the wild type and S114C mutant. These results indicate that residue His^{241} is indispensable for acylation of the Ser^{114} nucleophile and is required for efficient acylation of the Cys^{114} nucleophile. The low level of acylation in S114C:H241N but not in H241N may reflect the higher nucleophilicity of the -SH group of cysteine over the -OH group of serine.

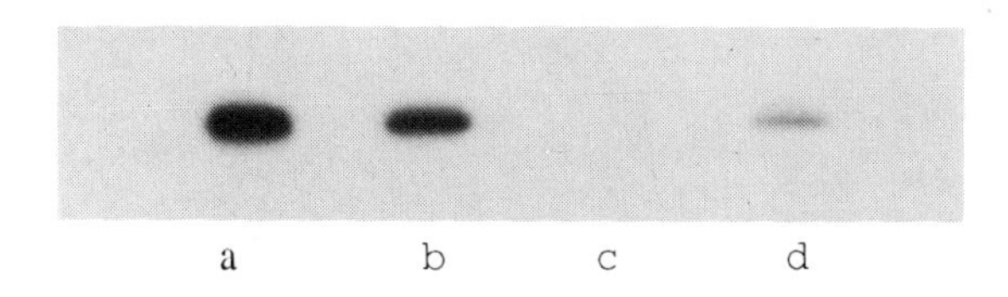

Figure 1. Comparison of the levels of acylation of wild-type LuxD thioesterase and its mutants (a)wild type, (b)S114C, (c)H241N, and (d)S114C:H241N.

Preliminary studies of the role of the third residue Asp211 in the catalytic triad have just been initiated. Mutants D211N and S114C:D211N have been constructed and expressed in extracts. Table 1 compares the esterase activities of the mutants in cleaving *p*-nitrophenyl myristate and their acylation level in extracts to that of the wild type. Both D211N and S114C:D211N show a very low levels of acylation and esterase activity which are comparable to the mutant S114C:H241N. These results indicate that Asp211 is an essential residue for esterase activity and is required to assist the Ser114 and H241 for efficient acylation of the enzyme. Purification and characterization of mutants D211N and S114C:D211N are currently in progress.

Table 1. Comparison of the esterase activities and the levels of acylation of S114C, H241N, S114C:H241N, D211N and S114C:D211N to the wild-type LuxD in extracts

	Wild type	S114C	H241N	S114C:H241N	D211N	S114C:D211N
Esterase activity	100%	8%	<2%	5%	5%	6%
Acylation level	100%	50%	0%	10%	10%	10%

Discussion

A catalytic mechanism for the lux-specific thioesterase derived from current biochemical and structural data and composed of a Ser, His, and Asp triad is given in Fig. 2. The mechanism is similar to that proposed for most mammalian thioesterases,

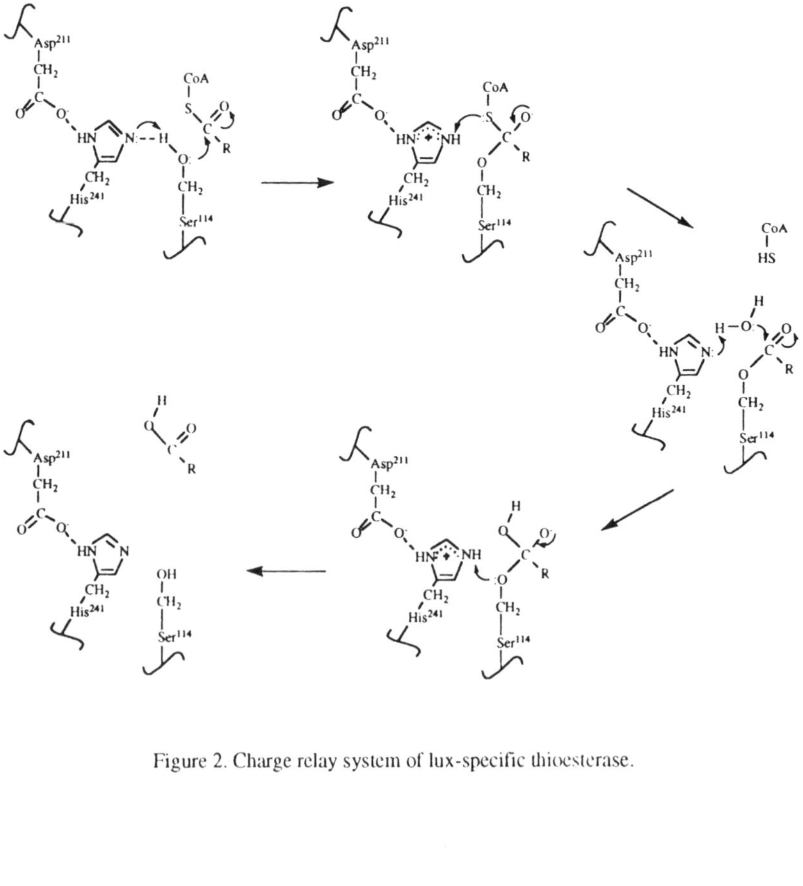

Figure 2. Charge relay system of lux-specific thioesterase.

esterases and lipases. For lux-specific thioesterase, His^{241} acts as a general base accepting a proton from the hydroxyl group of Ser^{114} thus facilitating its nucleophilic attack on the substrate. The tetrahedral intermediate then collapses with the formation of the acylated enzyme. Deacylation occurs by a nucleophilic attack of a water molecule, assisted by general base catalysis by His^{241}, yielding a second tetrahedral intermediate which breaks down to yield the product and regenerate the enzyme. Asp^{211} acts as the third component of the charge relay system by stabilizing the imidazolium ion generated during the formation of tetrahedral intermediates.

The turnover rate of lux-specific thioesterase is lower than that of mammalian thioesterases with deacylation instead of acylation being the rate limiting step. This raises the possibility that the primary function of this enzyme is to collect myristoyl groups and function as a fatty-acyl carrier until appropriate physiological conditions or interactions occur *in vivo* resulting in the release or transfer of fatty acyl group. The slow deacylation rate and turnover rate could reflect the relatively unique sequence flanking the active centre nucleophile located in the sequence $AxS^{114}xS$. Recently, a more general consensus motif for the nucleophilic serine in thioesterases has been proposed which encompasses the sequence flanking the nucleophile in LuxD(10).

Acknowledgments
The Medical Research Council of Canada (MT4314) for research support and FRSQ-FCAR, Fonds pour la Formation de Chercheurs et l'Aide à la Recherche for a scholarship (to JL) are gratefully acknowledged.

References

1. Libertini LJ, Smith S. Purification and properties of a thioesterase from lactating rat mammary gland which modifies the product specificity of fatty acid synthetase. J Biol Chem 1978;253:1393-401.
2. De Renobales M, Rogers L, Kolattukudy P. Involvement of a thioesterase in the production of short-chain fatty acids in the uropygial glands of mallard ducks (Anas platyrhynchos). Arch Biochem Biophys 1980;205:464-77.
3. Ferri SR, Meighen EA. A lux-specific myristoyl transferase in luminescent bacteria related to eukaryotic serine esterases. J Biol Chem 1991;266:12852-57.
4. Meighen EA. Molecular biology of bacterial bioluminescence. Microbiol Rev 1991; 55:123-42.
5. Lawson DM, Derewenda U, Serre L, Ferri SR, Szittner R, Wei Y, et al. Structure of a myristoyl-ACP-specific thioesterase from Vibrio harveyi. Biochemistry 1994;33:9382-88.
6. Brenner S. The molecular evolution of genes and proteins: A tale of two serines. Nature 1988;334:528-30.
7. Tai MH, Chirala SS, Wakil SJ. Roles of Ser^{101}, Asp^{236}, and His^{237} in catalysis of thioesterase II and of the C-terminal region of the enzyme in its interaction with fatty acid synthase. Proc Natl Acad Sci USA 1991;90:1852-6.
8. Witkowski A, Witkowska HE, Smith S. Reengineering the specificity of a serine active-site enzyme. J Biol Chem 1994;269:379-83.
9. Byers DM, Meighen EA. Purification and characterization of a bioluminescence-related fatty acyl esterase from Vibrio harveyi. J Biol Chem 1985;260:6938-44.
10. Li J, Szittner R, Derewenda ZS, Meighen EA. Conversion of serine-114 to cysteine-114 and the role of the active site nucleophile in acyl transfer by myristoyl-ACP thioesterase from Vibrio harveyi. Biochemistry 1996;in press.
11. Ferri SR, Meighen EA. An essential histidine residue required for fatty acylation and transfer by myristoyltransferase from luminescent bacteria. J Biol Chem 1994;269:6683-8.

ISOLATION AND CHARACTERIZATION OF A FLUORESCENT PROTEIN FROM A MARINE LUMINOUS BACTERIA

H Karatani[1], *T Konaka*[1], *Y Kitao*[2] *and S Hirayama*[2]
[1]*Department of Polymer Science and Engineering and* [2]*Laboratory of Chemistry, Kyoto Institute of Technology, Kyoto 606, Japan*

Introduction

Photobacterium phosphoreum and *P. leiognathi* produce the lumazine protein to cause the blue-shifted bioluminescence with a spectral maximum at around 475 nm (1). By contrast, *Vibrio fischeri* strain Y1 emits the striking yellow bioluminescence peaking at around 540 nm, which is attributable to the participation of the accessory yellow fluorescent protein in the luciferase reaction (2). Moreover those three species have been known to produce highly fluorescent proteins besides the two. Because of a diversity of bacterial fluorescent proteins, their characterization is interesting and important to explain the mechanism of bacterial bioluminescence (3-7).

Recently we have investigated the distribution of luminous bacteria as a function of depth in the Pacific Ocean. Almost the entire bacterial isolate from the depths of the sea emitted the blue-shifted bioluminescence *in vivo* with a spectral maximum at around 475 nm. Preliminary study of *in vitro* bioluminescence of several isolates from the deep-sea water showed that the luminous bacteria emitting the blue-shifted bioluminescence produced a fluorescent protein with two maxima at 488 and 517 nm other than a blue fluorescent protein ($\lambda_{em,max}$ ≈475 nm, probably the lumazine protein). The fluorescent protein emitting the bimodal fluorescence is referred to as the bimodal fluorescent protein in this report. In the present study, we isolated and characterized the bimodal fluorescent protein from a bacterial isolate, identified as *P. phosphoreum*.

Materials and Methods

Chemicals: The chromatographic resins were obtained from Pharmacia Biotech (Uppsala, Sweden). The standard proteins for gel filtration and sodium dodecyl sulfate polyacrylamide gel electrophoresis(SDS/PAGE) were obtained from Pharmacia and BioRad (Hercules, CA, U.S.A.), respectively. All other reagents were of the best commercial grade. All buffer solutions were prepared with deionized-distilled water.

Seawater sampling and bacterial isolation: The vertical sampling of seawater was performed by the use of the sterile Niskin butterfly sampler. The seawater sample (10 to 500 mL) was soon filtered on the sterile Nucleopore filter (0.2 μm in pore size and 47 mm in diameter, Costar Scientific Co., Cambridge, MA, U.S.A.). The bacterial isolation was attained according to the reported procedures (8). Identification of the bacterial isolates was accomplished in terms of the restriction fragment length polymorphism pattern, the growth temperature and the gas formation from D-glucose.

Cell growth: The cells of a fresh bright single colony of *P. phosphoreum*, captured at the depth of 500 m at 24° 25'N, 127° 16'E, were grown in the liquid seawater complete mainly at about 15 °C. The cells were harvested by centrifugation at near the end of the logarithmic phase recorded by monitoring *in vivo* bioluminescence intensity. The cells were also cultured in the same medium at 5 and 30 °C .

Protein purification: All subsequent procedures were conducted at about 4 °C. Cells grown at 15 °C (ca. 40 g, wet weight) were osmotically lysed with the 10 mmol/L Na/K phosphate buffer (pH 7.0) containing 10 mmol/L ethylenediaminetetraacetic acid(EDTA) and 1mmol/L dithiothreitol (DTT) (5 mL/g of cells). The centrifuged supernatant was subjected to the ammonium sulfate fractionation and the protein

precipitates (30-80 % of saturation) were collected by centrifugation. The precipitates were resuspended in 10 mmol/L Na/K phosphate buffer (pH 7.0) containing 0.1 mmol/L EDTA and 1mmol/L DTT and dialyzed against three changes of the same buffer solution. The dialyzate was loaded onto the DEAE Sepharose CL-6B column (2.2 × 25 cm) and eluted with a linear phosphate gradient from 10 to 300 mmol/L. The fractions rich in the bimodal fluorescent protein were pooled and then the proteins were precipitated by the addition of ammonium sulfate (80% of saturation). The centrifuged precipitates were dialyzed in the same way mentioned above and the dialyzate was loaded onto a calibrated Sephadex G100 column (2.0 × 72 cm). The crude luciferase eluted from the DEAE column was further purified by the aminohexyl Sepharose 4B chromatography according to the reported procedures (9). In the purification steps, luciferase activity was measured by the single turnover assay (9). The protein concentration was estimated by the Bradford method.
Luminescence measurement: Fluorescence spectra were taken with a RF540 spectrofluorophotometer with a thermoregulated cuvette holder (Shimadzu Co. Kyoto, Japan). Bioluminescence was measured by the same spectrofluorophotometer with no excitation light. The bioluminescence reaction with the purified luciferase was initiated by the electroreduction of riboflavin 5'-phosphate (FMN) in 50 mmol/L Na/K phosphate buffer (pH 7.0) (10).

Results and Discussion

Figure 1 shows the fluorescent elution pattern of the dialyzed protein precipitates on the DEAE Sepharose column. The bimodal fluorescent protein was eluted first, followed by the blue fluorescent protein ($\lambda_{em,max} \approx$ 475 nm). The ratio of the fluorescence intensity at 480 nm to that at 520 nm was nearly the same in different fractions. The bimodal fluorescent protein was purified to homogeneity by the subsequent Sephadex G100 column. In the preparation, about 20 mg of the bimodal fluorescent protein was obtained from 40 g of *P. phosphoreum* cells grown at 15 °C. Luciferase was eluted almost together with the blue fluorescent protein from the DEAE column. For the reaction with the further purified luciferase (4 μmol/L), $FMNH_2$ (~100 μmol/L) and tetradecanal (40 μmol/L), the pseudo-first-order rate constants of light emission decay at 4 and 25 °C were 0.03 and 0.15 s^{-1}, respectively. The fluorescence emission spectra of the bimodal fluorescent protein are shown in Fig. 2. The fluorescence excitation spectral maxima were observed at 339, 418, 441 and 470 nm when monitored at 485 and 530 nm. With the excitation at 339 nm, the bimodal fluorescent protein was excited at the highest efficiency. The absorption maxima were observed at 270, 386 and 440 nm.

As a result of gel filtration on the G100 column, the apparent molecular weight of the bimodal fluorescent protein was estimated to be 63.5 kDa. On the other hand, the SDS/PAGE (13 % separating gel) analysis showed a single band at the position of ca. 28 kDa, which was roughly half the value obtained by the gel filtration. These results might indicate that the bimodal fluorescent protein exists as a dimer, of which two components are similar in size.

The Stern-Volmer analysis of the bimodal fluorescent protein (2 μmol/L) in the presence of varying concentrations of KI up to 0.9 mol/L showed that the Stern-Volmer constant (K_d) for the fluorescence emission peaking at 517 nm at 29 °C was about two times greater than that at 5 °C ($K_{d,517nm,\,5^{\circ}C}$= 0.031 and $K_{d,517nm,\,29^{\circ}C}$= 0.077), whereas for the emission peaking at 488 nm the K_d was less sensitive to the change in temperature ($K_{d,488nm,\,5^{\circ}C}$ = 0.043 and $K_{d,488nm,\,29^{\circ}C}$ = 0.058). Assuming that two kinds of chromophores bind to the supposed dimer and that each polypeptide chain bears a chromophore, one polypeptide responsible for the fluorescence peaking at 517 nm appears to be susceptible to the thermal denaturation as compared to the other.

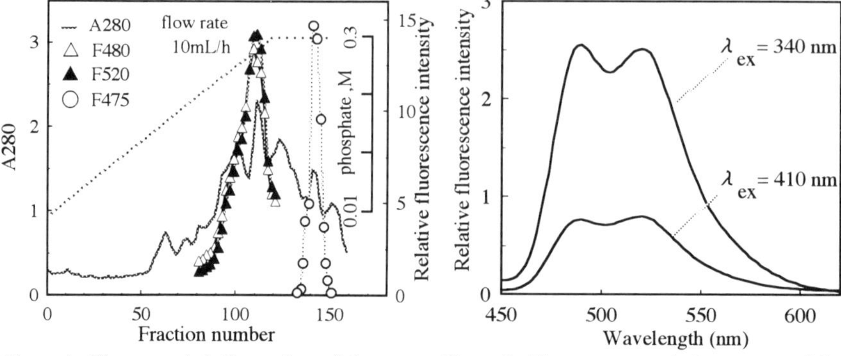

Figure 1. Fluorescent elution pattern of the dialyzed protein precipitates on DEAE column. λ_{ex}, 410 nm. Fluorescence intensity at 475 nm is reduced to a scale of 1/20.

Figure 2. Fluorescence emission spectra of the purified bimodal fluorescent protein with two different excitation wavelengths at 5 °C.

The bimodal fluorescent protein added into the *in vitro* luciferase reaction mixture with the blue fluorescent protein from *P. phosphoreum* cells modulated the blue-shifted bioluminescence, resulting in the slight red shift in the wavelength distribution.

It is of interest to examine whether temperature affects the production of the bimodal fluorescent protein because *P. phosphoreum* is the species suited to the condition of low temperature (11). To examine this, the supernatant of centrifuged lysate of *P. phosphoreum* cells cultured at 5 and 30 °C (referred to as the cell extracts) was chromatographed on the DEAE column and the elution patterns are shown in Fig. 3.

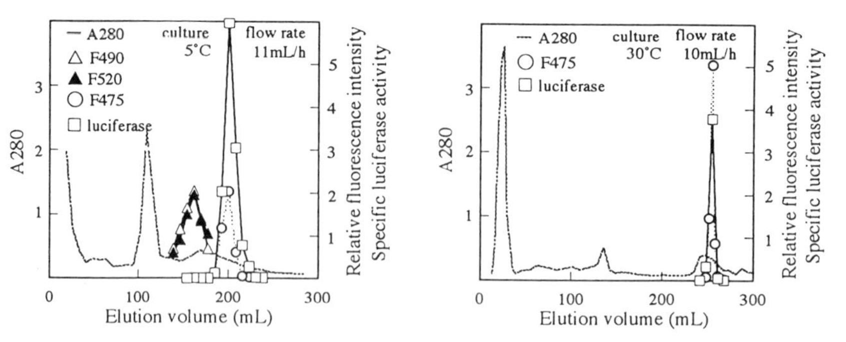

Figure 3. Elution patterns on DEAE Sepharose column (1×27cm) for the supernatant of centrifuged cell lysate of *P. phosphoreum* grown at different temperatures. Elution, phosphate gradient(10-300 mmol/L). Specific luciferase activity, bioluminescence intensity /A_{280}. A_{280} and specific luciferase activity of supernatant of cell lysate; 54.2 and 1.13 for 5 °C cell culture; 22.0 and 1.06 for 30 °C cell culture. Volume loaded , ca. 5 mL. Fluorescence intensity at 475 nm is reduced to a scale of 1/20.

The elution pattern for the cell extracts from the 30 °C cell culture showed the fractions rich in the blue fluorescent protein and luciferase at nearly the same position. This was similar to the pattern for the cell extracts from the 5 °C cell culture. A significant difference between two patterns was that the fractions rich in the bimodal fluorescent protein were not obtained in the case of the cell extracts from the 30 °C cell culture. This observation may suggest that low temperature is suited to production of the bimodal fluorescent protein in the cell. The elution pattern does not necessarily reflect the exact contents of the fluorescent proteins in the cell, because of the release of chromophores during the osmotic lysis (1). Indeed, the fluorescent fractions were obtained during the column washing. However, the observed relationship between temperature and the production of the bimodal fluorescent protein should be noted.

In conclusion, *P. phosphoreum* isolated from the depths of the sea produces the protein emitting bimodal fluorescence with maxima at 488 and 517 nm. The bimodal fluorescent protein seems to exist as a dimer and to weaken the interaction between the blue fluorescent protein and the luciferase intermediate. The low temperature condition appears to be favorable to the production of the bimodal fluorescent protein.

Acknowledgements
H.K. thanks the scientific group and crews of R/V Hakuho-Maru for their assistance during oceanic water sampling (KH95-2 research cruise, chief scientist, Prof. Kouichi Ohwada). We are also grateful to Prof. Ohwada and Dr. Kumiko Kita-Tsukamoto of University of Tokyo for the taxonomic identification of bacterial isolates.

References

1. O'Kane DJ, Lee J. Purification and properties of lumazine proteins from *Photobacterium* strains. Methods Enzymol 1986;133:149-72.
2. Macheroux P, Schmidt KU, Steinerstauch P, Ghisla S, Colepicolo P, Buntic R, et al. Purification of the yellow fluorescent protein from *Vibrio fischeri* and identity of the flavin chromophore. Biochem Biophys Res Commun 1987;146:101-6.
3. Karatani H, Wilson T, Hastings JW. A blue fluorescent protein from a yellow-emitting luminous bacterium. Photochem Photobiol 1992;55:293-9.
4. Karatani H, Hastings JW. Two active forms of the accessory yellow fluorescence protein of the luminous bacterium *Vibrio fischeri* strain Y1. J Photochem Photobiol B Biol 1993;18:227-32.
5. Sirokmán G, Wilson T, Hastings JW. A bacterial luciferase reaction with a negative temperature coefficient attributable to protein-protein interaction. Biochemistry 1995;34:13074-81.
6. Petushkov VN, Gibson BG, Lee J. Properties of recombinant fluorescent proteins from *Photobacterium leiognathi* and their interaction with luciferase intermediates. Biochemistry 1995;34:3300-9.
7. Petushkov VN, Gibson BG, Lee J. The yellow bioluminescence bacterium, *Vibrio fischeri* Y1, contains a bioluminescence active riboflavin protein in addition to the yellow fluorescence FMN protein. Biochem Biophys Res Commun 1995;211:774-9.
8. Ruby EG, Greenberg EP, Hastings JW. Planktonic marine luminous bacteria: species distribution in the water column. Appl Environ Microbiol 1980;39:302-6.
9. Hastings JW, Baldwin TO and Nicoli MZ. Bacterial luciferase: assay, purification, and properties. Methods Enzymol 1978;57,135-52.
10. Karatani H, Shizuki T, Rozana H, Nakayama E. A method for the electrochemical initiation of the *in vitro* bacterial luciferase reaction. Photochem Photobiol 1995;61:422-8.
11. Ruby EG, Morin JG. Specificity of symbiosis between deep-sea fishes and psychrotophic luminous bacteria. Deep-Sea Research 1978;25;161-7.

A THEORETICAL APPROACH TO ELUCIDATE A MECHANISM OF O_2 ADDITION TO INTERMEDIATE I IN BACTERIAL BIOLUMINESCENCE

N Wada[1], T Sugimoto[2], H Watanabe[3], S-C Tu[4] and HIX Mager[4]

[1]Dept General Edu Physics, Toyo Univ, Kawagoe 350, Japan

[2]Biophys Lab, Coll of Engin, Kanto Gakuin Univ, Matsuurra, Kanazawa-ku, Yokahama 236, Japan

[3]Hecchst Marion Roussel, Osaka 530, Japan

[4]Depts of Biochemical and Biophysical Sci and Chem, Univ of Houston TX 77204-5934, USA

Introduction

Bacterial luciferase (E) catalyzes the following reaction (1):

$$FMNH_2 + O_2 + RCHO \xrightarrow{E} FMN + RCOOH + H_2O + h\nu, \qquad (1\text{-}1)$$

where a molecular oxygen (O_2) reacts first with intermediate I (E-$FMNH_2$) to form intermediate II (E-FMNH-OOH). Subsequently, II reacts with a long chain aldehyde (RCHO) to generate intermediate III (E-FMNH-OO-CHOH-R). Bacterial luciferase is an enzyme classified as a monooxigenase. Specific interactions of luciferase with $FMNH_2$ and/or O_2 may be crucial in regulating the O_2 addition to intermediate I.

Although Baldwin et al. have recently studied the 3D-structure of *Vibrio harveyi* luciferase by X-ray diffraction spectroscopy (2) and speculated on the spatial configuration of amino acid residues surrounding its active site, the nature and effects of specific interactions between $FMNH_2$ and active site amino acids have not been elucidated. In this paper, a molecular mechanism of O_2 addition to intermediate I is studied theoretically on the basis of a "three-step model" in which a general base (B^-) residue and a general acid (AH^+) residue function as proton acceptor and donor, respectively:

$$1,5H_2\text{-FMN}(Pi^{2-}) + B^- \text{----->} 5H\text{-FMN}^-(Pi^{2-}) + BH, \qquad (2\text{-}1)$$

$$5H\text{-FMN}^-(Pi^{2-}) + O_2 \text{----->} 5H\text{-FMN-4aOO}^-(Pi^{2-}), \qquad (2\text{-}2)$$

$$5H\text{-FMN-4aOO}^-(Pi^{2-}) + AH^+ \text{----->} 5H\text{-FMN-4aOOH}(Pi^{2-}) + A, \qquad (2\text{-}3)$$

where Pi^{2-} is a diionized phosphate group in $FMNH_2$. In order to analyze these reaction steps by molecular orbital calculations, we have chosen an Asp^- and a $Lys\text{-}H^+$ on luciferase as representatives for the catalytic base and acid, respectively.

Methods

Geometrical structure and electronic states of each molecule in (2-1) to (2-3) mentioned above were optimized with respect to total energy of those reaction systems by the MNDO-PM3 molecular orbital method (3,4). In order to obtain the potential energy curve for each reaction (2-1) to (2-3), the total energy of a system consisted of two molecules interacting with each other was minimized, using the MNDO-PM3 method, by varying all geometrical parameters except the molecular distance R. The potential energy curve is shown as total-energy difference ΔE defined by $\Delta E = E(R) - E_0$, where $E(R)$ and E_0 are total energies at distance R and reference distance R_0, respectively. As shown in (2-1) to (2-3), the phosphate group is assumed to be a dianion (Pi^{2-}). It was found that the obtained potential energy curve is a unique solution for each reaction path within the framework of our model. Several different initial conditions for molecular geometry give almost the same potential energy curve.

Results and Discussion

Figure 1 shows the calculated potential energy curve for the reaction (2-1). As Asp^- approaches 1,5H_2-FMN(Pi^{2-}), the potential energy increases slightly, and a proton is transferred from the N(1) in 1,5H_2-FMN(Pi^{2-}) to the Asp^- carboxylate oxygen when the distance between the proton donor and acceptor group reaches R=2.44 Å. A stable system composed of 5H-FMN^-(Pi^{2-}) and Asp-H is produced, and then the two components separate from each other spontaneously. In this reaction, E_0 is -7658.31 eV at R_0=4 Å. The addition of O_2 at C_{4a} in 5H-FMN^-

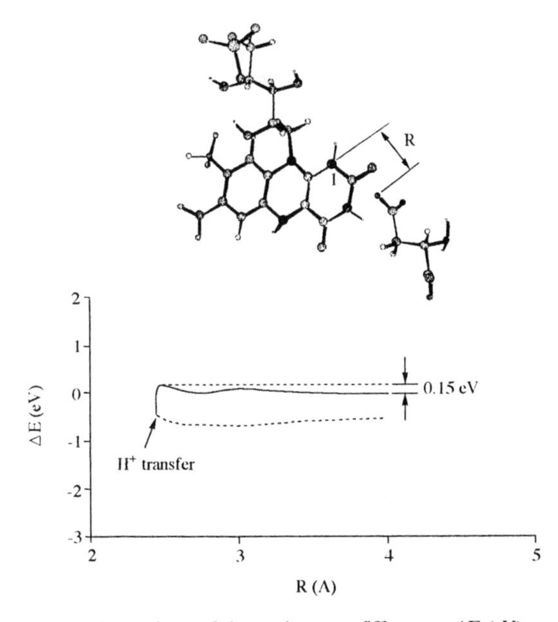

Figure 1. Dependence of the total-energy difference ΔE (eV) on the distance R (Å) between N(1) in 1,5H_2-FMN(Pi^{2-}) and O of COO^- in aspartic acid anion (Asp^-), calculated on the [1,5H_2-FMN(Pi^{2-}) + Asp^-] system. Dashed curve denotes the case where aspartic acid leaves from flavin ring after H^+-transfer. The origin of ΔE corresponds to -7658.31 eV at R=4 Å.

(Pi^{2-}) is realized along the potential curve with the barrier of 0.32 eV depicted in Fig. 2, where E_0 = -6320.87 eV at R_0=4 Å. Thus it is clearly shown that the hydroperoxide anion 5H-FMN-4aOO⁻(Pi^{2-}) exists stably at R=1.42 Å. The change in electron density of each O atom in O_2 along with the potential energy curve in Fig. 2 was also calculated. Results show that the polarization of electrons in the O-O bond is maximal on C_{4a}-OO bond forming and that the electron density on each O

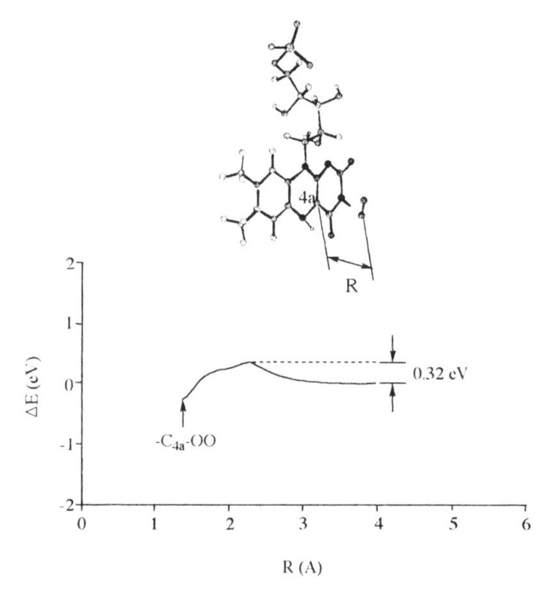

Figure 2. Dependence of ΔE (eV) on the distance R (Å) between C_{4a} and O in the [5H-FMN⁻(Pi^{2-}) + O_2] system. The origin of ΔE corresponds to -6320.87 eV at R=4 Å.

atom is nearly equivalent to result in charge depolarization after proton transfer. The potential energy for the final process in our model is depicted in Fig. 3 where 5H-FMN-4aOO⁻(Pi^{2-}) reacts with a Lys-H^+ in luciferase to form 5H-FMN-4aOOH(Pi^{2-}). This process occurs without potential barrier. It is accompanied by a proton transfer from Lys-H^+ to 5H-FMN-4aOO⁻(Pi^{2-}) at R=2.57 Å where R is the distance between the terminal O of C_{4a}-OO and N of the NH_3^+ in Lys-H^+. The two products then separate from each other with ease.

Even though O_2 addition to free 1,5H_2-FMN(Pi^{2-}) can occur to produce 5H-FMN-4aOOH(Pi^{2-}) in buffer solutions, the specific interactions between 1,5H_2-FMN(Pi^{2-}) and amino acids surrounding the active site of luciferase are necessary to form the stable flavin-hydroperoxide intermediate II with high yields. Therefore, a "three-step model" is proposed to depict the reaction path of O_2 addition to intermediate I. Thus far, the isolated luciferase intermediate II has never been identified to be a flavin-4a-OO⁻, or flavin-4a-OOH, or a mixture of both species. In this connection, our calculations show that the total energy of [5H-FMN-4aOOH(Pi^{2-}) + Lys] is 1.15 eV lower than that of [5H-FMN-4aOO⁻(Pi^{2-}) + Lys-H^+]. It should be noted, however, that flavin-4a-peroxy anion is expected to be the primary species in reacting with aldehyde for the formation of intermediate III.

The optimized structure of 5H-FMN-4aOOH(Pi^{2-}) thus calculated indicates that O_2 is bound to the C_{4a} position of flavin with the direction of the O-O bond approximately perpendicular to the plane of the flavin ring.

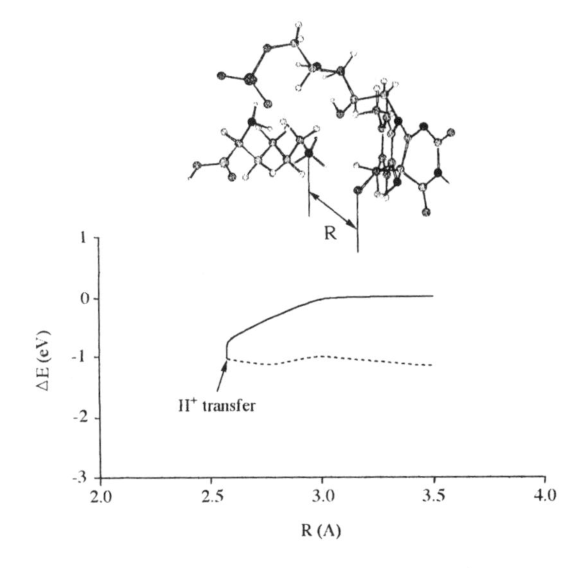

Figure 3. Dependence of ΔE (eV) on the distance R (Å) between the terminal O of C_{4a}-OO in 5H-FMN-4aOO⁻(Pi^{2-}) and N of NH_3^+ in lysine cation (Lys-H⁺), calculated on the [5H-FMN-4aOO⁻(Pi^{2-}) + Lys-H⁺] system. Dashed curve denotes the case where lysine residue leaves from flavin ring after H⁺-transfer. The origin of ΔE corresponds to -8179.39 eV at R=3.5 Å.

Acknowledgements

S,C.T. acknowledges the support by NIH grant GM25953 and The Robert A. Welch Foundation grant E-1030.

References

1. Hastings JW, Potrikus CJ, Gupta SC, Kürfurst M, Makemson JC. Biochemistry and physiology of bioluminescent bacteria. Adv Microb Physiol. 1985;26:235-91.
2. Baldwin TO, Christopher JA, Raushel FM *et al.* Structure of bacterial luciferase. Curr Opin Struct Biol. 1995;5:798-809.
3. Dewar MJS, Thiel W. Ground states of molecules. 38. The MNDO method. Approximations and parameters. J Am Chem Soc. 1977;99:4899-907.
4. Stewart JJP. Optimization of parameters for semiempirical methods. II. Applications. J Comput Chem. 1989;10:221-64.

FUNCTIONAL MAMMALIAN EXPRESSION OF CHIMERIC PROTEIN A-*VARGULA HILGENDORFII* LUCIFERASE BOTH IN FULL AND TRUNCATED FORMS

Y Maeda[1], H Ueda[1], J Kazami[2], G Kawano[2], E Suzuki[1] and T Nagamune[1]
[1].Department of Chemistry and Biotechnology, Graduate School of Engineering, The University of Tokyo, 7-3-1 Hongo, Bunkyo-ku, Tokyo 113, Japan,
[2].Toray Medical Devices & Diagnostics Research Laboratory, 3-3-3 Sonoyama, Ohtsu, Shiga 520, Japan

Introduction

The high sensitivity and the high rate at which bioluminescent assay could be performed has prompted us in utilizing the light emitting properties of luciferase in detection and quantitation of biological substances. The complete nucleotide sequence of the structural gene for stable luciferase of the marine ostracod crustacean *Vargula hilgendorfii* (1,2) has been determined recently (3), which has enabled us to use this gene as an attractive candidate for obtaining a bifunctional immunoreagent. We have employed this luciferase due to : 1. the enzyme is infinitely stable at room temperature in dried state. 2. the light-emitting reaction is very simple, consisting of the oxidation of luciferin by molecular oxygen in presence of *Vargula* luciferase and does not involve energy rich compounds like ATP (1).

As the other moiety for conjugation, we have used the well-documented Protein A of *Staphylococcus aureus* (4), which binds to the Fc region of various mammalian IgGs. Since chemical conjugation often results in loss of enzyme activity and in decrease of yield, we have conjugated the two moieties at the gene level thus obtaining a stoichiometrically controlled product. Such chimeric proteins are supposed to retain bifunctional properties and hence useful for sensitive immunoassay. Here, we elucidate how we successfully obtained a chimeric protein having both the antibody binding and the luminescing properties, by the introduction of a peptide link between the two moieties. Also, since significant amino acid sequence homology in the two regions of *Vargula* luciferase (residues 97-154 and 353-411) to the same amino-acid region of apoaequorin (residues 82-144) is reported (3), we examined the regions between 81-312 and 321-540 of *Vargula* luciferase and found that they were 19.3% amino acid homologous. In order to prove the possibility that one of the homologous regions is sufficient for light emission we produced a truncated *Vargula* luciferase having only the N-terminal homologous region (Pro28-Cys312) in mammalian COS-1 cells.

Materials and Methods

The construction of the expression plasmids were performed with standard techniques (5). *Escherichia coli* used for the propagation of DNA was XL1-Blue obtained from Stratagene.

Protein A-*Vargula* luciferase fusion vector (pRSVPAcluc) encoding a mouse V_{NP} signal sequence(6) and a single Fc binding D domain (SpA-D) from *Staphylococcus aureus* fused upstream of the mature form of *Vargula* luciferase (Pro29-Gln555) for expression in mammalian cells was constructed(7) using appropriate primers. pRSVPALcluc with linker of $(Gly)_4Ser$ was also constructed similarly (Fig. 1). For expression of the truncated luciferase, vector pRSVPALclucT was constructed(8). The N-terminal fragment of luciferase containing Pro28-Cys312 was utilized. The DNA Thermal Cycler used was from Perkin Elmer Cetus (Takara, Kyoto). All enzymes were from Takara (Kyoto, Japan) or New England BioLabs Inc. (Massachusetts USA).

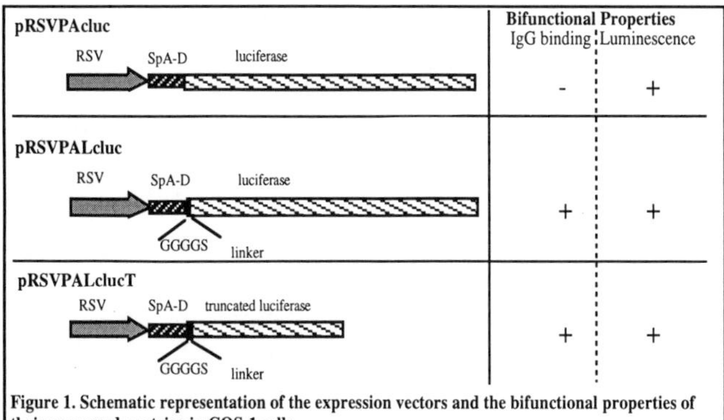

Figure 1. Schematic representation of the expression vectors and the bifunctional properties of their expressed proteins in COS-1 cells

Simian COS-1 cells (Riken Cell Bank, Tokyo) were transfected transiently with plasmids, pRSVPAcluc, pRSVPALcluc and pRSVPALclucT, by DEAE-dextran method. The cells were cultured using Dulbecco's Modified Eagle Medium (Nissui Pharmaceuticals, Tokyo) containing 10% calf serum (Gibco BRL, Grand Island, NY) in a humidified 5% CO_2 incubator at 37°C. After 3-4 days of incubation, the culture supernatant and adherent cells were collected for characterization of the expressed protein. The luminescence measurements were taken with ATP photometer (Sai Technology, CA, USA), immediately after the addition of luciferin.

Results and Discussion

Bioluminescent analysis of the chimeric proteins indicated that there was significant luciferase activity when compared to that of mock transfected cells, as shown in Table 1 (1st row). The results indicate that the expression of the chimeric proteins in COS-1 cells was successful. Next the chimeric proteins were assayed for their IgG binding ability, by incubating with IgG-Sepharose beads. Measurement of the Sepharose bound chimeric protein, showed

TABLE 1. Bioluminescence measurement and the comparison of the specific activity
Transfected COS-1 cells were sonicated in 300 µLPBS and 50 µL was taken for measurement. The band intensity of the Western blot in Fig. 2B was numerically represented for comparison of the specific bioluminescence of the truncated and the mature form of luciferase .

	Mock	pRSVPAcluc	pRSVPALcluc	pRSVPALclucT
Total cytosolic activity (cps)	1000.5	12060	18580.8	24913.7
Band intensity of protein	0	ND	8894	31022
% Specific activity	-	-	100	38.5

that the chimeric protein without the linker had no luminescence (Fig.1), while IgG binding ability of the chimeric proteins with the linker was retained. This indicates that the introduction of $(Gly)_4Ser$ between SpA-D and luciferase could enable the Fc binding domain to be restored to the wild type configuration. Western blot analysis of the cytosolic extract was performed as shown in Fig.2. Briefly, SDS-polyacrylamide gel electrophoresed proteins were transferred to a nitrocellulose membrane, and incubated serially with rabbit anti-luciferase antiserum raised against recombinant *Vargula* luciferase and peroxidase conjugated goat anti-rabbit antibody (Tago, Burlingame, CA). The blots were visualized by enhanced chemiluminescent detection system (Amersham, Buckinghamshire, England) for peroxidase labelled antibodies. A main band of truncated luciferase fused to SpA-D was observed, as shown in Fig. 2A (lane 3) of about 45 kDa. Species of that of the mature form of *Vargula* luciferase fused to SpA-D (lane 2), whose molecular weight is about 77 kDa, was also observed, as previously reported (6). In order to prove the feasibility of the chimeric protein in immunoassay, a sandwich assay for the quantitation of anti-TNP antibodies was performed (Fig 3). It can be deduced from the luciferase activity that the chimeric protein was specifically bound to antibody molecules.

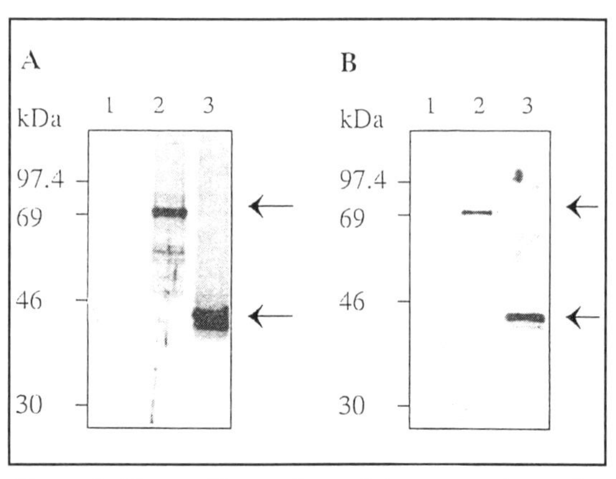

Figure 2. Western blot analyses of the expressed proteins. SDS-polyacrylamide electrophoresed proteins were detected with A. anti-Vargula luciferase and B IgG-peroxidase. Lanes 1, Mock-; 2,pRSVPALcluc; and 3, pRSVPALclucT-transfected.

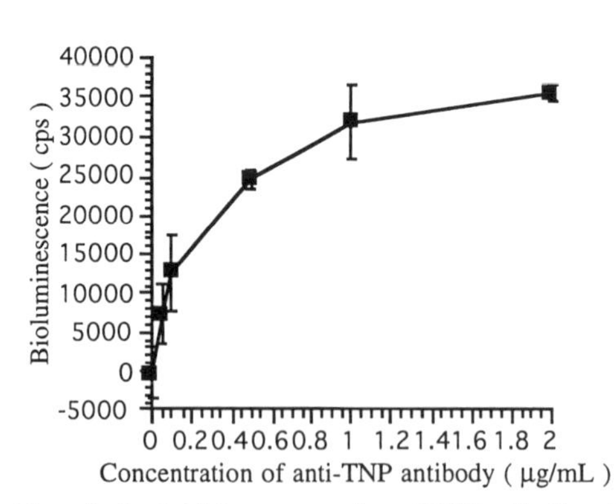

Figure 3. Sandwich immunoassay for anti-TNP antibodies . TNP immobilized Sepharose was taken and increasing amounts of mouse anti-TNP were added. Subsequently, 1 μg/mL of rabbit anti-mouse IgG was added, followed by the binding reaction between chimeric protein and the Fc region of rabbit IgG.

Comparison of the specific luminescence of the truncated luciferase to that of the mature form of luciferase was carried out by measuring the luminescence and the respective band intensity of IgG-peroxidase bound species (Fig.2B, Table 1). 38.5% of the activity of the mature form of luciferase is found to be retained by the truncated form, which proves an interesting fact that the N-terminal homologous region alone may be sufficient for light emission.

The truncated luciferase was also utilized for the quantitation of IgG. Fig.4 shows that there is a linear relationship between the amount of human IgG covalently immobilized on Sepharose and bioluminescence.

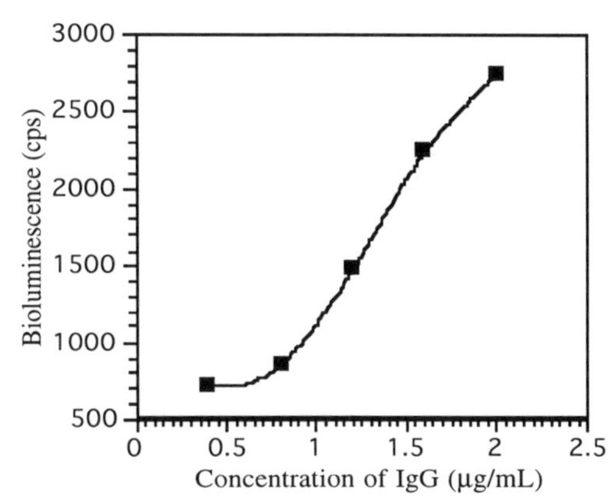

Figure 4. Dependence of luminescence on IgG concentration bound on Sepharose. Increasing amount of human IgG-Sepharose was taken, and the luciferase activity of IgG-bound chimeric protein was measured.

In conclusion, the insertion of a linker of $(Gly)_4$Ser between the two moieties of the chimeric protein resulted in retention of both luciferase and the antibody binding activity, and hence chimeric protein A-*Vargula* luciferase could be used as a useful immunological tool. In addition, it was observed that the N-terminal homologous region of *Vargula* luciferase may be sufficient for luminescence. The fact that the truncated luciferase retained some activity, may have important implications for further development of engineered luciferases.

References

1. Shimomura O, Johnson FH, and Masugi T. Cypridina bioluminescence: Light emitting oxyluciferin-luciferase complex. Science 1969;164:1299-1300.
2. Harvey EN. Bioluminescence. New York, Academic Press,1952:297-331.
3. Thompson EM, Nagata E, and Tsuji FI. Cloning and expression of cDNA for the luciferase from the marine ostracod *Vargula hilgendorfii*. Proc Natl Acad Sci USA 1989;86: 6567-71.
4. Uhlen M, Guss B, Nilsson B, Gatenbeck S, Philipson L, and Linberg M. Complete sequence of the Staphylococcal gene encoding protein A. J Biol Chem 1984;259:1695-1702.
5. Sambrook J, Fritsch EF, and Maniatis T. Molecular Cloning: A Laboratory Manual, 2nd ed. Cold Spring Harbor Laboratory Press, Cold Spring Harbor, NY.1989.
6. Ueda H, Kikuchi M, Yagi S, and Nishimura H. Antigen responsive antibody-receptor kinase chimera. Bio/Technol. 1992;10: 430-33.
7. Maeda Y, Ueda H, Hara T, Kazami J, Kawano G, Suzuki E et al. Expression of a bifunctional chimeric protein A-*Vargula hilgendorfii* luciferase in mammalian cells. BioTechniques 1996;20; 1: 116-21.
8. Maeda Y, Ueda H, Kazami J, Kawano G, Suzuki E and Nagamune T. Truncation of *Vargula* luciferase still results in retention of luminescence. J Biochem 1996;119:601-3.

COUPLING MECHANISM OF *VIBRIO HARVEYI* LUCIFERASE AND NADPH:FMN OXIDOREDUCTASE

B Lei[1] *and SC Tu*[1,2]
Departments of [1]*Biochemical & Biophysical Sciences and* [2]*Chemistry*
University of Houston, Houston, TX 77204, USA

Introduction

NAD(P)H:flavin oxidoreductases (flavin reductases, FR) catalyze the reduction of flavin at the expense of NAD(P)H and are believed to provide reduced flavin mononucleotide ($FMNH_2$) *in vivo* as a substrate for the luciferase-catalyzed bioluminescent reaction in luminous bacteria. Free reduced flavin is quite unstable because of its rapid autoxidation. Therefore, free diffusion is unlikely an efficient mechanism for inter-enzyme $FMNH_2$ transfer. Previous studies suggest a direct coupling of the reductases to luciferase (1, 2). However, in general, little is known about how $FMNH_2$ produced by FR-catalyzed reactions is efficiently utilized by luciferase. *Vibrio harveyi* NADPH-specific flavin reductase (FRP) binds FMN as a cofactor (3) to mediate the flavin reduction. In this work, we present evidences for the preferential utilization of the reduced flavin cofactor, rather than the $FMNH_2$ product, of FRP by luciferase for light emission.

Methods

Light emission in the FR-luciferase coupled assay was measured using a calibrated photometer (4) and FR activities were determined as steady emission intensities expressed in light units with 1 unit = 4.8×10^8 quantum/s. The coupled reaction was initiated by adding NAD(P)H into 1 mL of 50 mmol/L Pi containing luciferase, FR, decanal, and FMN with or without bovine serum albumin in a 12 x 75 mm glass tube.

Results and Discussion

Steady-state kinetic analyses were carried out for results of the luciferase-coupled assays of FRP and *Photobacterium fischeri* major flavin reductase (FRG) in which FRP and FRG provide $FMNH_2$ as a substrate for the luciferase reaction. The steady light intensities were measured as FR activities using different NAD(P)H concentrations at constant levels of FMN in the presence of excess decanal and luciferase. Double reciprocal plots of light intensity versus NADPH concentration at constant FMN levels display a family of converging lines for the FRP reaction (Fig. 1A). This pattern indicates a sequential mechanism with a $K_{m,FMN}$ of 0.3 μmol/L and a $K_{m,NADPH}$ of 0.02 μmol/L. Meanwhile, the same plot generates a set of parallel lines for the FRG reaction (Fig. 1B), suggesting a Ping-Pong mechanism with a $K_{m,FMN}$ of 4 μmol/L and a $K_{m,NADH}$ of 9 μmol/L. Both FRP (3) and FRG (5) tightly bind an FMN and follow the Ping-Pong mechanism in the single-enzyme spectrophotometric assay monitoring the decrease in A_{340} (2, 6). FRP has a $K_{m,FMN}$ of 8 μmol/L and a $K_{m,NADPH}$ of 20 μmol/L, and $K_{m,FMN}$ and $K_{m,NADH}$ of FRG are 220 and 120 μmol/L, respectively, in the spectrophotometric assay. Strikingly, the kinetic mechanism of

FRP changed from the Ping-Pong in the single enzyme assay to the sequential pattern in the luciferase-coupled assay.

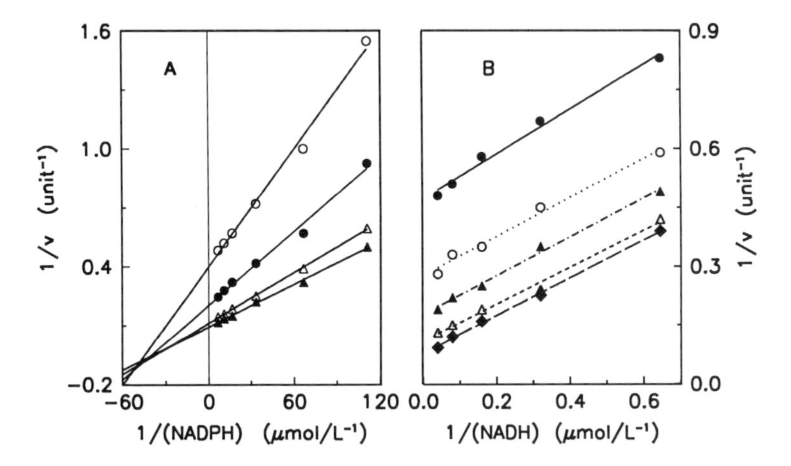

Figure 1. Steady-state kinetic analyses of FRP and FRG in the luciferase-coupled assay. The reactions were initiated by adding 10 μL of NAD(P)H into 1 mL of 50 mmol/L Pi, pH 7.0, containing 0.5 nmol/L FRP or 1.5 nmol/L FRG, 1.2 μmol/L luciferase, 10 μL of 0.02% decanal in ethanol at constant levels of FMN. Steady-state light intensities (v) versus NAD(P)H concentration are presented as double reciprocal plots. FMN concentrations are, from the top line downward, 0.1, 0.2, 0.4, and 0.6 μmol/L in panel A for FRP and 0.5, 1.0, 2.0, 4.0, 6.0, and 8.0 μmol/L in panel B for FRG.

FMN titration in the FRP luciferase-coupled reaction was conducted. Light intensity (v) increased rapidly from 0.2 to 1 μmol/L FMN and reach the maximum at about 2 μmol/L FMN. This increase in v was apparently from the sequential pathway since this range of FMN concentration was around the 0.3 μmol/L $K_{m,FMN}$ for the coupled assay. Surprisingly, v dramatically decreased with the increase of FMN concentration from 2 to 20 μmol/L which were from a few fold lower to higher than the $K_{m,FMN}$ in the Ping-Pong pathway. FMN concentration from 0.2 to 1 μmol/L had a small enhancing effect on the total quanta (Q_t) of light emission. However, Q_t decreased with further increase of FMN concentration from 2 to 20 μmol/L in a pattern similar to that for the FMN effect on light intensity. High concentrations of FMN thus cause inhibition in both v and Q_t. However, FMN has no inhibition on the rate of NADPH oxidation in the spectrophotometric assay in the presence of luciferase and decanal. These observations indicate that there are two NADPH-consuming pathways in the coupled assay. One of them is productive, and the other is substantially less active and, hence, apparently inhibitory in light emission. FMN was determined to be mixed and competitive inhibitor against NADPH and luciferase, respectively.

A minimum reaction scheme is formulated to account for the above results (Scheme 1). In the Ping-Pong pathway, NADPH (NH) first binds to FRP (E-F) and reduces the cofactor to $FMNH_2$ ($E\text{-}FH_2$). After the release of $NADP^+$ (N), an exogenously added FMN binds to $E\text{-}FH_2$ to yield product $FMNH_2$ (FH_2) and regenerate the oxidized

FRP. In the luciferase-coupled reaction, we propose that luciferase (Lu) forms a transient complex with E-FH_2 and a direct transfer of $FMNH_2$ from the latter to luciferase occurs leaving the reductase in the apoenzyme form (E). Subsequently, the exogenous FMN binds to the apoenzyme to regenerate the holoenzyme E-F. When viewed with the apoenzyme as the starting point, it is clear that the coupled assay would follow an apparent sequential kinetic pattern. Luciferase and FRP can not form stable complex and exogenously added FMN can react with E-FH_2 under the conditions used. The $FMNH_2$ product of the Ping-Pong pathway is utilized much less efficiently than the reduced FMN cofactor by luciferase. Therefore, the Ping-Pong pathway competes with luciferase for E-FH_2, resulting in an apparent inhibition of the luciferase-coupled reaction at higher FMN concentrations.

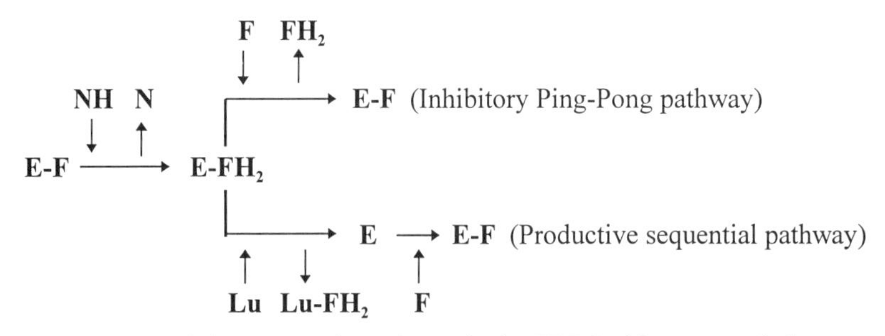

Scheme 1: Minimum reaction scheme in the FRP-luciferase coupled assay

A reconstituted active FRP derivative (FRP_S) containing 2-thioFMN as the cofactor was used in the luciferase-coupled assay to further test the above scheme. Luciferase can use both $FMNH_2$ and reduced 2-thioFMN as a substrate for bioluminescence but with much lower quantum yield for the latter flavin. The coupling efficiencies of FRP and FRP_S using FMN as substrate at about the same NADPH oxidation rate would allow us to demonstrate the source of reduced flavin for the luciferase reaction in the coupled assay. If the $FMNH_2$ product is transferred to luciferase, about same light intensities are expected for both FRP and FRP_S. In contrast, much lower light intensity is expected for FRP_S if reduced flavin cofactor of flavin reductase is preferentially used by luciferase. NADPH oxidation and light emission were compared in the reaction mixture containing FRP or FRP_S, FMN, NADPH, luciferase and decanal. As shown in Fig. 2A, FRP_S had a slightly higher rate of NADPH oxidation than FRP did under the conditions used. The light intensity increased to the maximum in a few seconds after initiation of the reaction with FRP (Fig. 2B, dotted line). However, the intensity was initially quite low and then gradually increased by 10 times to a plateau during a period of 3 min (Fig. 2B, solid line) for the reaction using FRP_S. These results demonstrate that luciferase did not efficiently use the $FMNH_2$ product of flavin reductase for light emission. The slow increase in light emission for FRP_S was due to the replacement of 2-thioFMN cofactor by exogenous FMN. These results again indicate that reduced flavin cofactor rather than $FMNH_2$ product is efficiently utilized by luciferase.

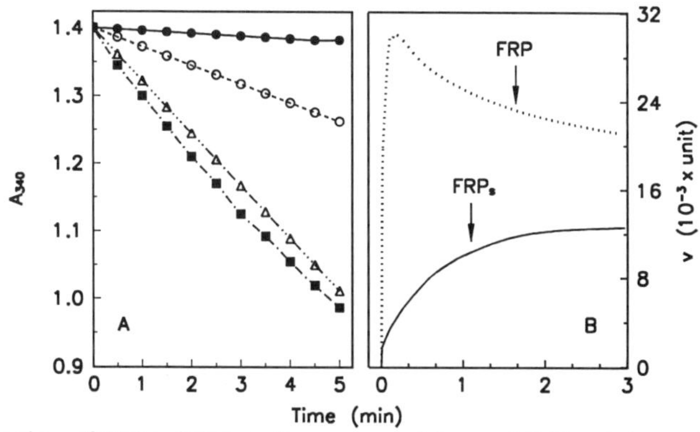

Figure 2. Comparison of FRP and FRP_S in time courses of A_{340} and light emission under identical conditions. A, A_{340} was monitored after 20 μL of 8 μmol/L FRP or FRP_S was added into 2 mL of 50 mmol/L Pi containing 2.4 μmol/L luciferase, 20 μL of 0.02% decanal in ethanol, 225 μmol/L NADPH, and 0 or 3.5 μmol/L FMN in a 12 x 75 mm glass culture tube. Symbols: ●, FRP_S/no FMN; O, FRP/no FMN; Δ, FRP/FMN; ■, FRP_S/FMN. B, the reactions using FRP/FMN (dotted line) or FRP_S/FMN (solid line) were repeated under identical conditions for the measurement of light emissions. Time was set at zero when data collection started.

The cofactor FMN of FRP can be reduced by NADPH in the absence of exogenously added FMN and the resulting $FMNH_2$ still binds to the protein (3). The autoxidation rate of the reduced FRP was estimated to be about 3% of that of autoxidation of free $FMNH_2$ at air-saturated condition. Our results demonstrate that the bound $FMNH_2$ cofactor is less accessible to autoxidation, and is more efficiently utilized than free $FMNH_2$ by luciferase.

Acknowledgment

This work was supported by grants GM 25953 from NIH and E-1030 from The Robert A. Weltch Foundation.

References

1. Duane W, Hastings JW. Flavin mononucleotide reductase of luminous bacteria. Mol Cell Biochem 1975;6:53-64.
2. Jablonski E, DeLuca M. Studies of the control of luminescence in *Beneckea harveyi*: properties of the NADH and NADPH:FMN oxidoreductases. Biochemistry 1978;17:672-8.
3. Lei B, Liu M, Huang S, Tu SC. *Vibrio harveyi* NADPH-flavin oxidoreductase: cloning, sequencing and overexpression of the gene and purification and characterization of the cloned enzyme. J Bacteriol 1994;176:3552-8.
4. Hastings JW, Weber G. Total quantum flux of isotopic sources. J Opt Soc Am 1963;53:1410-5.
5. Inouye S. NAD(P)H-flavin oxidoreductase from the bioluminescent bacterium, *Vibrio fischeri* ATCC 7744, is a flavoprotein. FEBS Letters 1994;347:163-8.
6. Tu S-C, Becvar JE, Hastings JW. Kinetic studies on the mechanism of bacterial NAD(P)H:flavin oxidoreductases. Arch Biochem Biophys 1979;193:110-6.

UPPER ELECTRON-EXCITED STATES IN BACTERIAL BIOLUMINESCENCE

NS Kudryasheva[1], YP Mechalkin[2] and DN Shigorin
[1]*Institute of Biophysics, Siberian Branch of Russian Academy of Sciences, Krasnoyarsk, 660036, Russia*
[2]*Novosibirsk Electrotechnical University, Novosibirsk, 630092, Russia*

Introduction

Generation of the electron excited states in bioluminescent systems is of a great interest. Physics approach to this problem should be based on the consideration of the emitting compound. Bacterial bioluminescent emitter supposed to be 4A-hydroxyflavine (1). According to the classification of Prof. D.N.Shigorin (2) geteroaromatic highly fluorescent molecules belong to the 5-th spectral luminescent group of compounds. In the simpliest case they are characterised by the sequence of the energy levels shown on Fig.1 with taken into account the multiplicity and the orbital nature ($n\pi^*$- or $\pi\pi^*$-type) of the electron excited states. The sequence is determined by lowest excited states of $\pi\pi^*$-type and upper excited states of $n\pi^*$-type. The first ones show the excitation delocalized over the π- system of the molecule, while the second ones show the excitation localized on a heteroatom or a carbonyl group of the molecule.

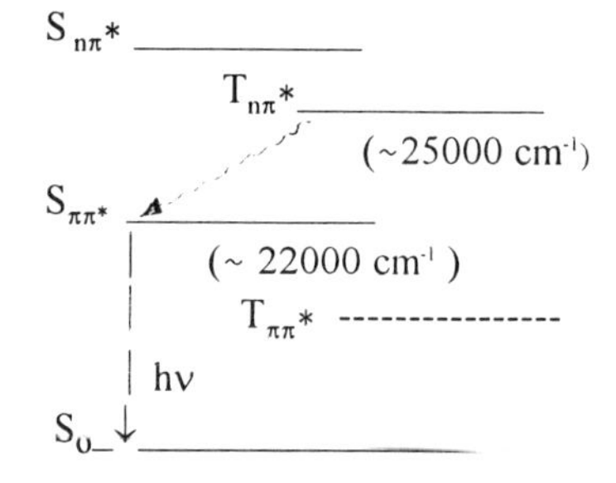

Figure 1. Electron excited states of bacterial bioluminescent emitter.

Formation of a carbonyl group is the result of the oxidation of organic compounds in the process of breathing of biological systems. There exist the possibility of generation of excitation in this process. In the paper of R.F.Vasil'ev the formation of $n\pi^*$-states of carbonyl compounds in the decomposition of organic peroxides was discussed (3). Excited and nonexcited carbonyl groups are shown on the Scheme 1:

$>C=O:$	$>\dot{C}-O\cdot$
nonexcited carbonyl group	$T_{n\pi^*}$ excited carbonyl group

Scheme 1.

In the $n\pi^*$-excitation process one uncoupled n-electron becomes localized at the non-binding n-orbital and the other one is transferred to the delocalised π^*-orbital of the molecule.

In the bacterial bioluminescence case the primary excitation can be localised in the position C(4A) (4) with the break of the aromatic cycle (Scheme 2a) and subsequent cycle transformation. The other suggestion on the localisation of $n\pi^*$-excitation is C2 -position of the isoalloxazine (Scheme 2b) (5).

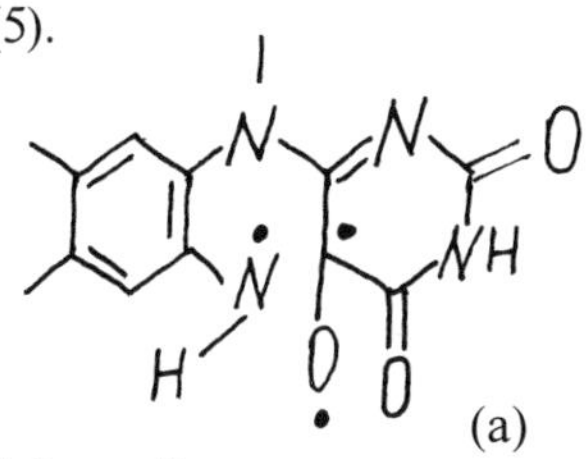

(a) 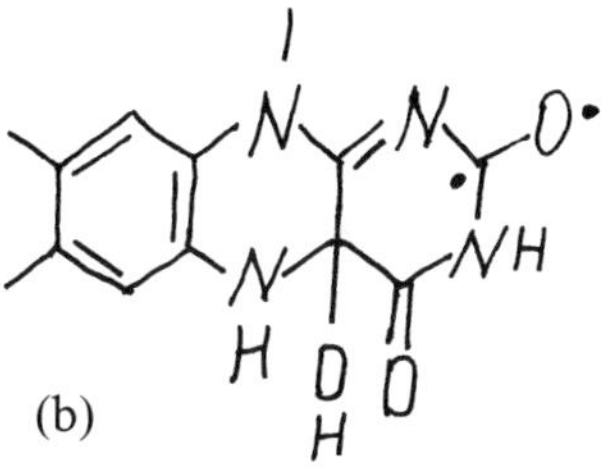 (b)

Scheme 2.

In the case of bioluminescence after formation of the $n\pi^*$-state of the bioluminescent emitter the occupation of the electron excited states of low energy is to occur. The intramolecular nonradiating energy transfer $T_{n\pi^*} \longrightarrow S_{\pi\pi^*}$ (the transfer into fluorescent $S_{\pi\pi^*}$-state) (Fig. 1) is the permitted process according to the El-Sayed rule as the transfer between the levels of different orbital nature and multiplicity (6). The rate constants of such transfers are 10^{10}-10^{11} c^{-1} (7). As a result of this process the excitation localized at the carbonyl group ($n\pi^*$-excitation) of the emitter becomes delocalized throughout the molecule ($\pi\pi^*$-excitation). The last stage of the process of the intramolecular energy migration is the light emission ($S_{\pi\pi^*} \longrightarrow S_o$, Fig.1). In 1991 for the first time the participation of the upper electron excited states in bacterial bioluminescence was proposed (5,8).

The purpose of the present work was to investigate the upper electron excited states participation in the bioluminescent process using the bacterial bioluminescent system as an example.

Materials and Methods

The coupled BL system NADH:FMN-oxidoreductase - luciferase was used. Lyophilised preparations of luciferase and NADH:FMN-oxidoreductase from *Photobacterium leiognathi* were supplied by the biotechnology sector of the Institute of Biophysics (Krasnoyarsk). Reaction mixture was described in (5,9).

Measurements were done using a BLM 8801 Bioluminometer (Special Design Bureau "Nauka", Krasnoyarsk). The effects of molecular quenchers on bioluminescence intensity were studied. Inhibition constants (K) were calculated using formular: $I/Imax = e^{-KC}$, where I is the intensity of the bioluminescent signal in the presence of a quencher of the concentration C, Imax is the intensity of the bioluminescent signal in the absence of quenchers.

Bioluminescent spectra were recorded using the spectrometer designed in Novosibirsk Electrotechnical University (Novosibirsk).

Results and Discussion

The molecules of different nature and properties were classified according to the energy of their lower triplet electron excited states (T) (Fig.2). It could be seen from the Fig.2 that the organic compounds with the energy of T less than of ~300 kJ (25000 cm^{-1}) quench bioluminescence. Compounds with the energy of T grater than ~300 kJ (25000 cm^{-1}) do not have that effect.

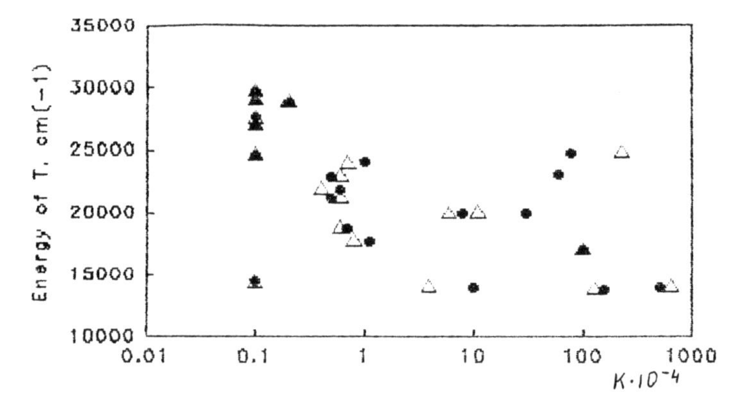

Figure 2. Inhibition constants (K) of molecules with different lowest triplet states (T) energies for a coupled BL system in the presence of decanal (o) and tetradecanal (Δ). List of molecules: rhodamine 6G, eosin, 9,10-phenanthrenquinone, 1,4-benzoquinone, 1,4-anthraquinone, 1,4-naphtoquinone, 9,10-anthraquinone, benzaldehyde, anthrone, 1,4-diphenylbutadiene, naphthalene, hydroquinone, benzophenone, anthracene, biphenyl, benzoic acid, indole, toluene, benzene.

This energy barier may be determined by the interactions (energy, electron transfers) of the upper triplet excited state of bioluminescent system and the excited states of the molecular quenchers. The data (Fig.2) can be attributed to the triplet precursor of the emitter with the energy being around 25000 cm^{-1} (≈300kJ) (see Fig.1). The effects of inorganic compounds on the bacterial bioluminescence were discussed in (10) from the point of view of interactions of electron excited states (uncluding upper) of bioluminescent emitter and cations of metals.

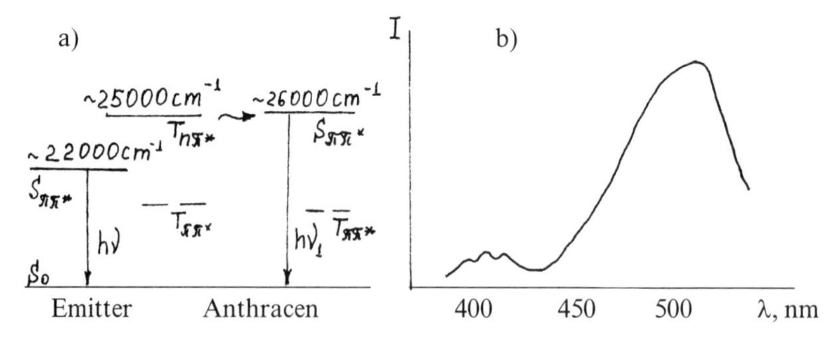

Figure 3. (a) The energy levels of anthracen and bioluminescent emiter. (b) The emission spectrum of the bioluminescent system in the presence of anthracen.

The spectra of the bacterial bioluminescent system in the presence of anthracen were recorded. The orbital nature of the anthracen's fluorescent state is $\pi\pi^*$. It's energy exceeds that of the bioluminescent emitter (Fig.3a). The absorbtion spectrum of anthracen and the luminescent spectrum of the bioluminescent system do not overlap, hence the trivial absorption was excluded. The emission spectrum of the bacterial bioluminecent system in the presence of anthracen has the low intensity peaks in the region of anthracen fluorescence (sensitized fluorescence) (Fig 3b). This result can be explained by the intermolecular energy transfer with upper excited state of the bioluminescent emitter participation as a donor. Studies to increase the yield of the sensitized luminescence of anthracene are currently in progress.

Thus, our results show that the upper electron excited states of the emitter molecule can participate in the bioluminescent process. Most likely these states are triplet and have the orbital nature of $n\pi^*$-type.

Acknowledgements - This work was supported by the grant from Russian Federal Research Program 08.05 "Newest methods of Bioengineering", subprogram Engineering in Enzymology" and the grant from Krasnoyarsk Scientific Foundation 5F0086-C.

References

1. Hastings JW, Potricus CI, Gupta SC, Kurfurst M, and Makemson IC.Biochemistry and Physiology of Bioluminescent Bacteria. In: Advances in Microbial Physiology. London: Academic Press, 1985;26:235-91.
2. Shigorin DN, edited by Kolotyrkin YM. Electron excited states of polyatomic molecules. Moscow, Nauka, 1993.
3. Vasil'ev RF. Ways of chemiluminescence excitation in organic compounds. In: Biochemiluminescence. Moscow: Nauka, 1983:31-45.
4. Kemal C, Bruice TC. Chemiluminescence accompanying the decomposition of 4A-flavin alhyl peroxide model studies of bacterial luciferase. J Am Chem Soc 1977;99:7064-9.
5. Kudryasheva NS, Belobrov PI, Kratasyuk VA, Shigorin DN. Studyingof bioluminescence mechanism by means of molecular quenchers. Krasnoyarsk: Inst.of Physics, 1991.
6. El-Sayed J. Spin-orbital coupling and radiationless process in nitrogen heterocycles. J Chem Phys 1963;38:2834-38.
7. Plotnikov VG. Theoretical basis of spectral-luminescent systematic of molecules. Uspekhi Khimii 1980;49:327-61.
8. Kudryasheva NS, Belobrov PI, Kratasyuk VA, Shigorin DN. Electron-excited states in bioluminescence. Dokl AN SSSR 1991; 321: 837-41.
9. Kudryasheva NS, Kratasyuk VA, Belobrov PI. Bioluminescent analysis. The action of toxicants: Physical-chemical regularities of the toxicants effects. Anal.Lett 1994;27:2931-48.
10. Kudryasheva NS, Zuzikova EV, Gutnyk TV. Mechanism of metallic salts action on bacterial bioluminescence in vitro. Biofizika 1996:6. In press

THE SINGLE POLYPEPTIDE CHAIN OF *GONYAULAX* LUCIFERASE HAS THREE ENZYMATICALLY ACTIVE REPEAT UNITS

Liming Li, Robert Hong and J. Woodland Hastings
Department of Molecular and Cellular Biology
Harvard University, 16 Divinity Avenue, Cambridge, Massachusetts 02138, USA

Introduction

It is well established that bioluminescence originated and evolved independently in the many different phyla where it occurs. The substrates (luciferins), enzymes (luciferases) and genes involved are different and diverse (1). The dinoflagellate luminescent system provides an excellent and interesting example of this.

Dinoflagellate luminescence is emitted following mechanical stimulation as brief (~100 msec) bright (~10^9 photons) flashes from unique small (~0.4 microns) cytoplasmic organelles, scintillons (flashing units) (2, Figure 1). In *G. polyedra*, two proteins immediately involved in the reaction are localized therein. One of these, the luciferin binding protein (LBP), sequesters the luciferin substrate, a unique open-chain tetrapyrolle (3), and releases it due to a transient pH change (Figure 2), thus giving rise to a flash. This pH change is postulated to occur as the result of a conducted action potential in the tonoplast (vacuolar membrane), which opens voltage-gated proton channels and allows H^+ to move from the acidic vacuole into the scintillon.

The enzyme in the reaction, dinoflagellate luciferase (LCF), is also unique; it catalyzes the oxidation of the tetrapyrolle by molecular oxygen leading to the chemical population of the excited state, presumably an intermediate or product tetrapyrolle in the reaction, with light emission peaking at ~474 nm (4). *Gonyaulax* luciferase is a large single chain polypeptide (Mr ~134,000). Its activity shows a strong dependence on pH, peaking at about pH 6.3 and declining 10 to 20 fold at pH 8 (5).

The genes coding for both proteins have been cloned and sequenced, and when expressed *in vitro* the proteins produced exhibit activities characteristic of the native molecules. In keeping with the generalization above, both proteins are unique; they bear no homologies to any other protein (including other luciferases) so far entered into the data base (6,7). The present communication concerns repeat structures in the full length luciferase molecule and the activities of individual peptides spanning those regions.

Materials and Methods

Isolation of full-length *lcf* cDNA: A 1.1 kb *lcf* cDNA fragment was isolated from plasmid pMM8 (8), labeled with fluorescein dUTP and used as a probe for screening a *G. polyedra* cDNA library (7) according to the manufacturer's procotol (DuPont, NEN). Positive phage plaques were identified and the pBluescript phagemids carrying the inserts were excised. The plasmid carrying the largest insert (4.0 kb), named pLL10, was used for further analysis.

Plasmid constructs for GST-luciferase expression: The cDNA fragment containing the entire region of the first repeat region corresponding to aa 1 to 566 (N to C) was obtained by digesting pLL10 with *Sma*I and *Xmn*I. The second, corresponding to aa 516 to 881 (N to C) was obtained by digesting pLL10 with *Nco*I and *Sph*I. The third, corresponding to aa 921 to 1240 (N to C) was obtained by digesting pLL10 with *Xho*I and *Fsp*I. These fragments were blunt ended and cloned into the *SmaI* site of either pGEX:2T or pGEX:3X, gluthathione S-transferase (GST) expression vectors for fusion proteins (Pharmacia). The DNA fragment containing the complete open reading frame of luciferase was isolated after codigestion of pLL10 with *Xho*I and *Fsp*I and the isolated DNA frament was blunt ended and ligated to the *Sma*I site of pGEX:3X. Plasmid preparation, digestion, DNA fragment isolation and ligation procedures were

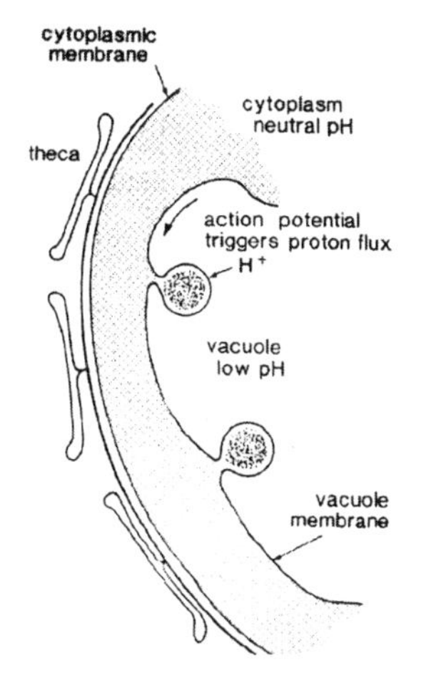

Figure 1: Schematic diagram of *G. polyedra* scintillons. Scintillons are unique organelles in bioluminescent dinoflagellates formed as invaginations of the vacuolar membrane. Action potentials generated in response to mechanical agitation trigger a proton flux from the vacuole to the scintillon, causing a pH change inside the scintillon, thereby resulting in light emission.

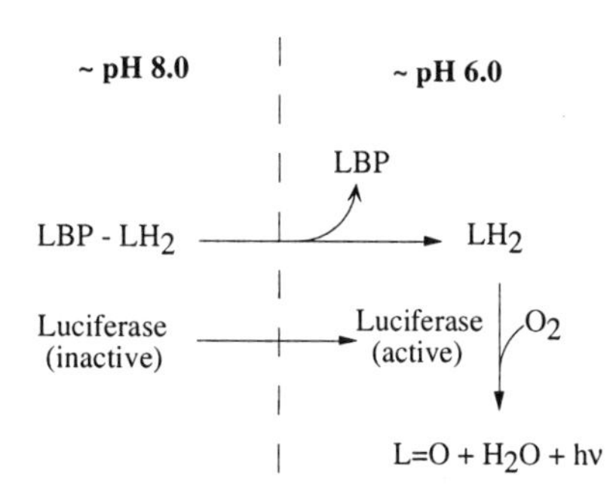

Figure 2. Postulated mechanism of the bioluminescence reaction in *G. polyedra*. At pH 8 luciferase is in an inactive form and luciferin binding protein (LBP) tightly binds the luciferin, making it inaccessible to luciferase. A change of pH from 8 to 6 causes a conformational change of both LBP and luciferase, resulting in the activation of luciferase and the release of luciferin from LBP.
LH_2: reduced luciferin (substrate), L=O: oxidized luciferin (product).

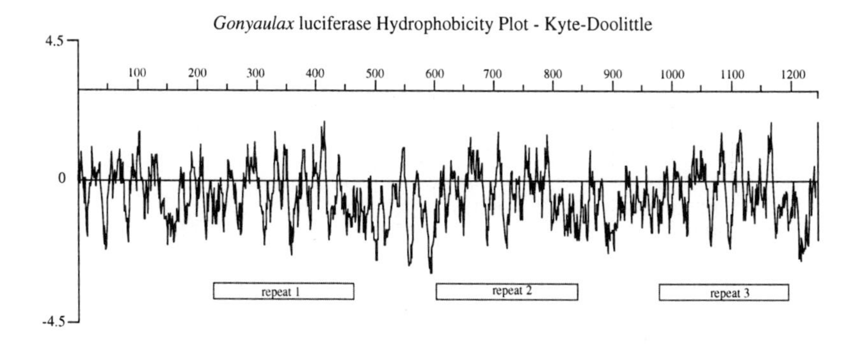

Figure 3. Hydropathy profile of *G. polyedra* luciferase. The deduced amino acid sequence from the luciferase ORF in pLL10 was used to predict the local hydropathy (Protean 1.16, DNASTAR, Inc). The average hydropathy value of a moving window of nine amino acids is plotted at the midpoint of the window.

performed by following standard protocols (9).

GST-luciferase expression and luciferase activity assay: Plasmids containing GST-luciferase fusion genes were transformed into *E. coli* cells, strain JM109, and were grown overnight in 2YT-G in the presence of 100.µg/mL ampicillin at room temperature. One part of the overnight culture was then diluted into 9 parts of fresh 2YT-G with 100 µg/mL ampicillin and grown another 3 h before the addition of IPTG to a final concentration of 0.1 mmol/L. After growing additional 3 h, the cells were harvested and lysed followed by preparation of protein extracts and purification of the GST-luciferase proteins according to the manufacturer's protocols.

The luciferase assay was done by rapidly mixing the *E. coli* extract containing GST-luciferase fusion protein and luciferin with 1 ml of 0.2 mol/L citrate buffer, pH 6.1 in a scintilation vial and recording the time course of light emission..

Results and Discussion

A partial sequence of the luciferase gene starting at its 3' end had revealed two repeat regions in which the DNA sequences are very similar (7). In that study, a peptide expressed *in vitro* from a clone embracing only one of the repeat regions was found to have luciferase activity, so the possibility that the repeat region represents the active site of the enzyme, duplicated for some unknown reason, was considered.

Northern blots indicated that the full length mRNA was about 4.1 kb, sufficient to code a protein of ~130 kDa or more, the size estimated for luciferase from SDS PAGE gels of cell extracts. Isolation of longer cDNAs, and of ones including the 5' end of the molecule, resulted in the determination of the nucleotide sequence of the full length molecule and its deduced amino acid sequence (Figure 3). This led to the present discovery of a third repeated region, with a sequence similarity to the first two of about 70% at the amino acid level, slightly less than was reported between the first two (~80%).

In order to determine if these regions were individually active in catalyzing the *Gonyaulax* luciferase reaction, subclones embracing the three repeat regions shown in Figure 3 were constructed and expressed *in vitro* as fusion proteins. The cell extracts were assayed by mixing with dinoflagellate luciferin. Activity was obtained for all three; possible differences in specific activity and kinetic parameters are under investigation.

It had been previously shown that proteolysis of native *Gonyaulax* luciferase with subtilisin resulted in a ~35 kDa fraction (called B-luciferase) with good activity (9). At the time it was assumed that this represented a single region of the full length molecule, but it is now evident that this could have been a mixture of more than one of the repeat regions. Whether or not this is so has not yet been determined.

Irrespective, an interesting feature of B-luciferase was its altered pH-activity profile. Native luciferase exhibits an asymmetric curve, with a very sharp cutoff in activity over the region of pH 6.5 to 7.5 whereas B-luciferase maintains over 50% of its maximum activity up to pH 9. It is thus of interest to determine if the peptides expressed from the cloned repeat regions exhibit pH-activity profiles similar to native or B-luciferase. Preliminary results, measured as the ratio of activities at pH 8 to 6.3, indicate the latter. For example, the ratio for the full length luciferase from the cloned gene was determined to be 0.15, while that for the polypeptide containing repeat unit #2 was 0.55. How the additional structure of the full length molecule results in the truncation of activity at alkaline pHs remains a matter of considerable interest.

When translated into protein, the repeat regions themselves correspond to proteins with molecular masses of about 26 kDa, somewhat smaller than the active proteolytic fragments of native luciferase. However, the minimum structural requirement for luciferase activity has not yet been determined for any of the repeat regions. Analysis and experimental manipulation of the regions conserved in all three regions can be expected to yield information concerning the active site of the enzyme.

There are only a few examples of the occurrence of more than one catalytic site for the same activity within a single polypeptide, the receptor-like tyrosine phosphatase HPTPalpha being an example (10). However, a functional importance is not clearly established in any of those cases, and it might be different for different enzymes. For dinoflagellate luciferase there are a number of possibilities. One requires consideration of the colligative constraints in packaging the system within a small organelle that is already packed with protein. The scintillon is only about 0.4 microns in diameter and has large amounts of the luciferin binding protein (LBP), about 100 times more than luciferase. By having three catalytic sites on a single luciferase the system can obtain additional enzyme activity with less protein in a situation where the protein osmotic strength must already be high.

Another but much less likely possibility is that the different sites have different functions. For example, one site might bind a luciferin that serves as the emitter for a reaction on an adjacent site. These and other possibilities are under consideration in the experimental studies underway.

Acknowledgements

This research was supported in part by a grant from the National Science Foundation (MCB 93-06879). We thank James Comolli for providing the dinoflagellate luciferin and for his assistance with the luciferase assay.

References

1. Hastings JW. Biological diversity, chemical mechanisms and evolutionary origins of bioluminescent systems. J Mol Evol 1983;19: 309-21.
2. Nicolas MT, Nicolas G, Johnson CH, Bassot JM, Hastings JW. Characterization of the bioluminescent organelles in *Gonyaulax polyedra* (dinoflagellates) after fast-freeze freeze fixation and antiluciferase immunogold staining. J Cell Biol 1987;105: 723-35.
3. Nakamura H, Kishi Y, Shimomura O, Morse D, Hastings JW. Structures of dinoflagellate luciferin and its enzymatic and non-enzymatic air-oxidation products. J Am Chem Soc 1989;111: 7607-11.
4. Dunlap J, Hastings JW, Shimomura O. Dinoflagellate luciferin is structurally related to chlorophyll. FEBS Letters 1981;135: 273-6.
5. Krieger N, Njus D, Hastings JW. An active proteolytic fragment of *Gonyaulax* luciferase. Biochemistry 1974;13: 2871-7.
6. Lee DH, Mittag M, Sczekan S, Morse D, Hastings JW. Molecular cloning and genomic organization of a gene for luciferin-binding protein from the dinoflagellate *Gonyaulax polyedra*. J Biol Chem 1993;268: 8842-50.
7. Bae YM, Hastings JW. Cloning, sequencing and expression of dinoflagellate luciferase DNA from a marine alga, *Gonyaulax polyedra*. Biochim Biophys Acta 1994;1219: 449-56
8. Mittag M, Lee DH, Hastings JW. Circadian expression of the luciferin-binding protein correlates with the binding of a protein to the 3' untranslated region of its mRNA. Proc Natl Acad Sci USA 1994;91: 5257-61
9. Sambrook J, Fritsch EF, Maniatis T. Molecular Cloning: A Laboratory Manual, 2nd Edn., Cold Spring Harbor Laboratory, Cold Spring Harbor,1989.
10. Wang Y, Pallen CJ. The receptor-like protein tyrosine phosphatase HPTPa has two active catalytic domains with distinct substrate specificities. EMBO J 1991;10:3231-37

PART 2

Quorum Sensing and Regulatory Elements Controlling Bacterial Bioluminescence

H-NS PROTEIN SILENCES TRANSCRIPTION OF *lux* SYSTEM OF *Vibrio fischeri* AND OTHER LUMINOUS BACTERIA CLONED IN *Escherichia coli*

S Ulitzur
Faculty of Food Engineering and Biotechnology, Technion Israel Institute of Technology, Technion City, Haifa, 32000, Israel

Introduction

Several host factors are involved in the expression of the bacterial *lux* system (1, 2). The *LuxR* autoinducer complex is the ultimate transcriptional initiator of the *lux* system of *V. fischeri*, while several regulatory factors, including cAMP, and CRP as well as GroESL are associated with the formation or activity of the LuxR protein. The involvement of RpoS and H-NS in the regulation of bacterial bioluminescence has not been studied yet. RpoS is the principle regulator of numerous stationary-phase genes in *E. coli.* Out of the many factors that are involved in the regulatory activity of RpoS, special attention has been given to H-NS (3). H-NS protein is a neutral, homodimeric protein comprised of 138 amino acid that shows high affinity for double-stranded DNA, with preference to a curved DNA. *hns* mutants show pleiotropic phenotypes and many genes, including: *drdX, bglY, osmZ, pilG* and *virR* were shown to be allelic to *hns* (3).

The *lux* genes of all luminous bacteria show a typical base composition that characterizes a curved DNA. The upstream regions of *luxR* of *V. fischeri* and the *luxC* genes of all species contain DNA fragments showing over 80% AT stretches, while other structural genes are also highly enriched with A tracts. This study shows that the *lux* systems of all the luminous bacterial species are fully and constitutively expressed in *E. coli hns* mutant and that neither LuxR nor the *V. fischeri* autoinducer are required for the formation of luminescence in *V. fischeri*

Materials and Methods

<u>Bacterial strains</u>: *E. coli* strain MC4100 and its Tn10 insertion-inactivated *hns* mutant (*hns*::kan) and *rpoS* mutant (*rpoS*::tet) as well as the double mutant MC4100 *hns*::kan *rpoS*::tet were constructed by the bacteriophage P1 transduction using *E. coli* AMS6 Tn10 hns::kan or *E.coli* AMS6 Tn10 *rpoS*::tet as donor strains, (both strains were kindly provided by A. Matin, Stanford University, Stanford, CA. USA).
<u>Plasmids:</u> pChv1 (4) contains the entire *V. fischeri lux* system cloned into the pACYC-184 plasmid. pW21A is a deleted *lluxI* derivative of pChv1(4). pHK555, kindly provided by Dr. P. Greenberg (Iowa University, Iowa, USA), is a derivative of a pChv1-like plasmid with a large deletion in the DNA binding domain of LuxR. pW21A and pHK555 clones in *E. coli* are very dim. The following plasmids were kindly provided from Dr. AE Meighen (McGill University, Montreal, Canada). For more details on the origin of these plasmids see reference 2. The clones of *P. leiognathi* , *X. luminescens* , and *V. harveyi* were ligated into the 2. 4 Kb pT7 plasmid, The clone of *V. fischeri* (pVfB1 plasmid that contains the entire *lux* system of ATCC 7744 as well as of the clone of *P. phosphoreum* system were ligated into pBR322. All of the *lux* constructs include the AT rich regions upstream of *luxC* gene. The plasmid ptv1073 (5) kindly provided by DuPont company, Wilmington, DE. USA, contains the *V. fischeri luxCDABE* genes under *lac* promoter control.
<u>Bacterial growth and bioluminescence determination.</u> *E. coli* cells were grown with shaking (200 rpm) in Luria Bertani medium (LB-10g tryptone (Difco), 5g yeast extract (Difco) and 5g NaCl in 1L distilled water). The LB medium was supplemented with chloramphenicol (cm-30μg/ml) for cultivation of pChv1 and pHK555 carrying *E. coli* strains, ampicillin (amp-50μg/ml) for pT7 and of pChv1 with a deletion in the *luxI* gene (1, 4). The pBR332 derivatives carrying *E. coli* strains, tetracycline (tet-10μg/ml) was added to *rpoS*::tet carrying *E. coli* strains and

kanamycin (kan-25µg/ml) was added to *hns*::kan carrying *E. coli* strains. In vivo and in vitro luciferase activities were determined as previously described, using decyl aldehyde (4). Light levels are given in specified light units (LU).

Results

MC4100 *rpoS*::tet carrying the *V. fischeri lux* systems (pChv1, pfVB1) growing in shaken cultures in LB at 28°C, developed significantly lower in vivo luminescence than the clones of these plasmids in the wild type strain. Externally added autoinducer (40 ng/ml) to the LB medium resulted in rapid restoration of luminescence (not shown). In growth temperatures below 24°C, the effect of *rpoS* mutation was much less pronounced. These plasmids showed thousand folds higher luminescence when transformed to MC4100 *hns* cells (data not shown) or to the double mutant MC4100 *hns*::kan *rpoS*::tet strain (figure-1). The luminescence of *V. fischeri* clone in MC4100 rpoS *hns* mutants showed a constitutive mode. The development of luminescence in this cells occurs promptly upon dilution of overnight grown cultures to a fresh LB medium. The MC4100 hns mutants and were growing very slowly, on the other hand the double mutant MC4100 hns rpoS cells showed 50% higher growth rate.

To determine whether the autoinducer or the LuxR protein of *V. fischeri* is involved in the development of light in *hns* mutants, two deletion mutants of the pChv1 plasmid were transformed into MC4100 cells. The plasmids carried the whole *lux* system of *V. fischeri* with either a deletion in the *luxI* gene (pW21A), or in the *luxR* gene (pHK555). Fig 2 shows that MC4100/pW21A cells emitted low light compared to pChv1, while MC4100/pHK555 cells were dark throughout the growth cycle. On the other hand, *hns* and *hns rpoS* mutants of MC4100/pW21A and MC4100/pHK555 strains developed high luminescence levels. All the *E. coli hns* mutants were very unstable and grew very slowly. Since MC4100 *rpoS hns* double mutant cells are more stable and showed 50% better growth rate than MC4100 hns strain, the double mutant was used for most of the experiments.

The homoserine-lactone (HSL) acts as an inducer of RpoS formation in *E. coli* (6). To exclude the possibility that MC4100 or its mutants could synthesize and excrete the HSL we determined the autoinducer activity of conditioned medium in which different *E. coli* mutant strains were grown overnight. Undetectable activity of the autoinducer (≤10pg/ml), was found in the supernatant of MC4100 cultures or in that of its respective mutants. Using the MC4100/pW21A cells for the autoinducer bioassay, we determined the accumulation of the autoinducer in the growth medium of MC4100 *hns rpoS*/pHK555 cells in comparison with that of wild-type cells (MC4100/pChv1). We found that MC4100 *hns rpoS* pHK555 cells produced high concentrations of the autoinducer at very early stages of growth while wild type cells made more autoinducer at the end of the exponential phase of growth. The maximal yield of autoinducer in both strains was comparable (data not shown). As a working hypothesis we assumed, that in the absence of H-NS, the *luxR-luxI* region undergoes topological changes that bring about LuxR-autoinducer-independent transcription of the right operon, including the *luxI* gene.

To address the question of whether the *luxR* and *luxI* genes are the only sites affected by H-NS, we used the plasmid ptv1073, which carries *luxCDAB* under control of the *lac* promoter. ptv1073 was transformed into the MC4100 wild-type strain and its *hns* and *rpoS* mutants as well as the MC4100 hns *rpoS* double mutant. The *rpoS* strain showed less than 0. 1% of the maximum light level of the wild-type, while the double mutant strain showed 500-1000 fold higher luminescence than the *rpoS* mutant cells. Addition of decyl aldehyde to the growth medium did not affect the light in either case. These results indicate the presence of other H-NS binding sites in the *luxCDABE* genes, the transcription of which is repressed by H-NS and relieved by RpoS. An alternative way to arrest H-NS protein formation is by addition of DNA synthesis inhibitors (3). Indeed, Figure 6 shows that the gyrase inhibitors novobiocin, nalidixic acid and tarivid increased the luminescence of MC4100 *rpoS*/ptv1073 by more than 100-folds after incubation for 3 h.

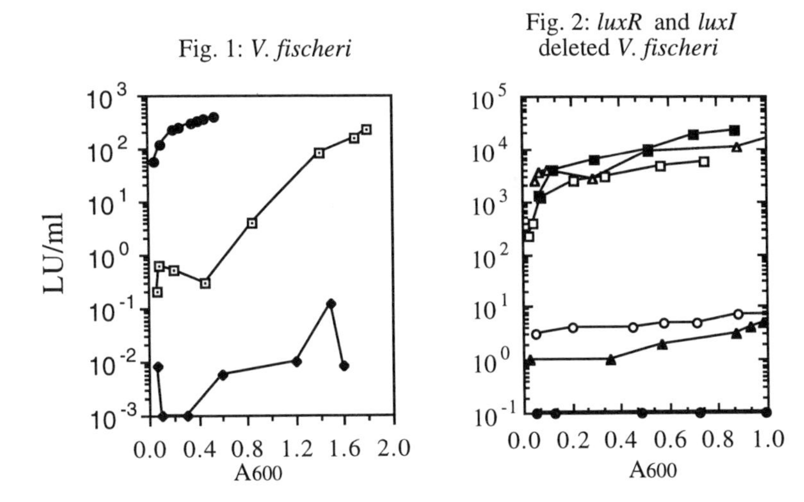

Figures 1 and 2: Development of luminescence in different strains of MC4100/pChv1. Overnight grown cultures at 37°C were diluted 50 folds in a fresh LB medium with the appropriate antibiotics. The cultures were shaken at 29°C and the optical density and the in vivo luminescence were recorded at different times. The figures show the recorded luminescence (LU/ml) as a function of cell density (A_{600}). Fig-1: ❑, MC4100/pChv1; ◆, MC4100*rpoS*/pChv1; ●, MC4100 *rpoS hns*/pChv1. 1 LU = 5. 10^9 q/s. Fig-2: ●, MC4100/pHK555; ○, MC4100/pW21A; ▲, MC4100 *rpoS*/pW21A; ❑, MC4100 hns/pW21A; Δ, MC4100 *rpoS hns*/pW21A; ■, MC4100*hns rpoS*/pHK555. 1 LU = 5. 10^7 q/s.

To test the possibility that H-NS also silences the expression of the *lux* systems of other luminous bacteria, the *lux* systems of *P. leiognathi, P. phosphoreum, V. harveyi* and *X. luminescens* were transformed to MC4100 *hns rpoS,* and MC4100 *rpoS* mutants as well as to MC4100 wild type strain, (figures 3-5). The MC4100 *rpoS hns* harboring the *lux* systems of the five species of luminous bacteria (the data for *P. phosphoreum* are not shown), exhibited very high, (1500-4000 q/s/cell), and constitutive mode of luminescence. The in vivo and in vitro luciferase activities were comparable for all the clones in MC4100 *rpoS hns* cells. The in vitro luciferase activity in MC4100 and its *rpoS* mutant was 4-7 times higher than the in vivo values (not shown). The *V. fischeri* clones pfVB1 and to a lesser extent pChv1, are very dim in MC4100 *rpoS* strain. Addition of autoinducer to the growth medium resulted in appearance of wild type level of luminescence in both clones. On the other hand, the other clones of luminous species showed comparable values of in vivo luminescence in MC4100 *rpoS* and wild type strain in liquid medium.

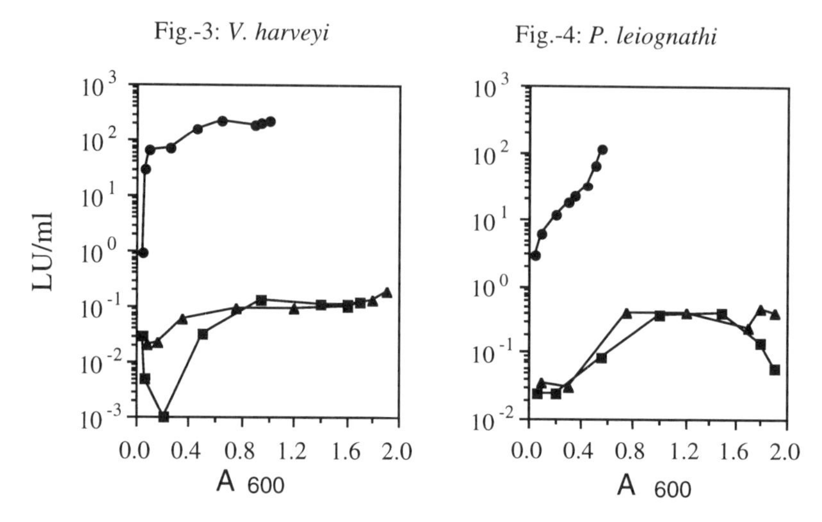

Fig. 3-5: The development of luminescence of different *lux* systems in MC4100 and its *rpoS* and *rpoS hns* mutants. The experiments were carried out as described in the legend for figures 1 and 2. ●, MC4100 *rpoS hns;* ▲, *MC4100 rpoS;* ■, *MC4100.* 1LU = 5. 10^9 q/s.

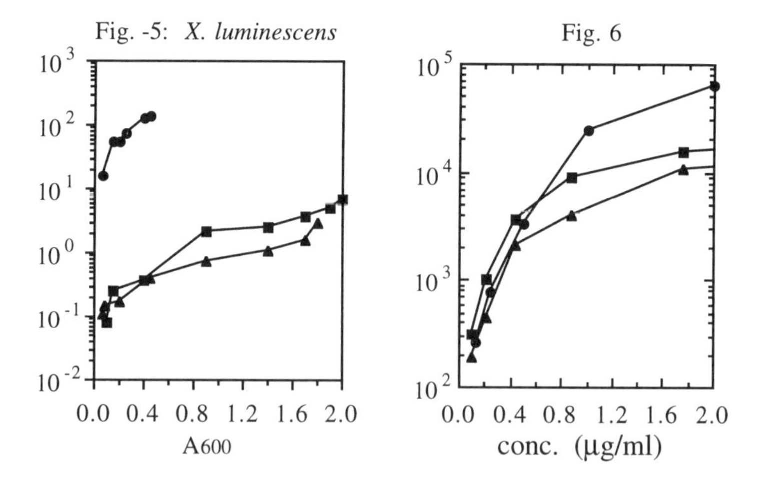

Fig-6: Effect of novobiocin, nalidixic-acid and tarivid (ofloxacin), on the luminescence of MC4100 *rpoS* /ptv1073. Overnight grown cells were diluted into LB containing serial dilutions of ●, novobiocin; ▲, nalidixic acid; or ■, tarivid. Luminescence was recorded after 3 h of incubation at 24°C. 1LU = 5. 10^6 q/s

Fig. -7: DNA sequencing of luxRI"C" of *V. fischeri and luxC* and up stream DNA of *Vibrio harveyi, Photobacterium phosphoreum, Photobacterium leiognathi* and *Xenorabdus luminescens.*

Vibrio fischeri (ATCC 7744), *luxR luxI "luxC*". Data Bank Accession-M19039.

aagcTTTacttacgtacttaaTTTTTAAAgtatgggcaatcaattgctcctgttAAAattgcTTTagAAAtacTTTggcagcggTTTgttgtattgagTTTcaTTTgc
gcattggttAAAtggAAAgtgacagtacgctcactgcagcctaataTTTttgAAAtatcccaagagcTTTTTccttcgcatgcccacgctAAAcattcTTTTTctcTTT
TggttAAAtcgttgTTTgaTTTattaTTTgctataTTTaTTTTTcgataattatcaactagagaaggaacaattaatggtatgttcatacacgcatgtAAAAAtAAActat
ctatatagttgtcTTTTTctgaatgtgcAAAActaagcattccgaagccattgttagccgtatgaatagggAAActAAAcccagtgataagacctgatgTTTTcgcttcTTTa
attacaTTTggagaTTTTTTaTTTacagcattgTTTTcAAAtatattccaattaattggtgaatgattggagttagaataatctactataggatcataTTTattAAAttagcgtc
atcataatattgcctccaTTTTTTagggtaattatctagaattgAAAtatcagaTTTaaccatagaatgaggatAAAtgatcgcgagtgttAAAtaatattcacaatgtaccaTT
TTagtcatatcagataagcattgattaatatcatatgctctacaagcTTTaaTTTTattaattattctgtatgtgtcgtcggcaTTTatgTTTTTcatacccatctcTTTatccttacct
attgTTTgtcgcaagTTTTgcgtgttatatatcattAAAAcggtaatggattgacaTTTgattctaatAAAttggaTTTTTgtcacactattgtatcgctgggaatacaattactt
aacataagcacctgtaggatcgtacaggTTTagcgaagAAAatggTTTgttatagtcgaatAAAcgcaagggaggttggatgactataatgatAAAAAAatcggaTTTT
TTggcaattccatcggaggagtatAAAggtattctaagtcttcgttatcaagtgTTTaagcAAAgacttgagtgggacttagttgtagAAAAtaaccttgaatcagatgagtatgat
aactcAAAtgcagaatataTTTatgcttgtgatgatactgAAAAtgtaagtggatgctgcgTTTatnacctacaacaggtgattatatgctgAAAAgtgTTTTTcctgaattg
cttggtcaacagagtgctcccAAAgatcctaatatagtcgaattaagtcgTTTTgctgtaggtAAAAAtagctcAAAgatAAAtactctgctagtgAAAttacaatgAAA
ctaTTTgaagctatatatAAAcacgctgttagtcaaggtattacagaatatgtaacagtaacatcaacagcaatagagcgaTTTTTAAAgcgtattAAAgttccttgtcatcgtatt
ggagacAAAgAAAttcatgtattaggtgatactAAAtcggttgtattgtctatgcctattaatgaacagTTTAAAAAAgcagtcttAAAttaatgttgttAAAtcattaaTT
TaTTTtAAAtactaagtatattatagggAAAAtaatgaatAAAtgtattccaatgataattaatggaatgattcaagaTTTTgataattatgcatatAAAgaagttAAActAA
AtaatgataatagagtAAAAttatctgtcattactgAAAgttcagTTTcAAAAacattAAAtatcAAAgatagaattaatctAAATTTAAAtcagattgtgaaTTTT
TTatataccgttggtcaacgatggAAAAgtgaagaatataatcggcgacgaacctat

Vibrio harveyi lux C. + 158 b.p. of upstream DNA. Data Bank Accession-X07084..

TTTTccAAAATTTAAAAgagaagctcttgatatggAAAAAAcacttaccTTTaatagtAAAtggacAAAttaTTTctactgaagAAAAtcgaTTTgagat
cagTTTTgaagAAAAAAAAgttaagattgattccTTTaataatcttcaTTTaacccagatggtgaatcatgattaTTTAAAtgatctAAAtattaataacatcatcaaTT
TTcTTTatacaacggggcagcgttggaagagcgaagagtattcaagaagaagggcatatattcggtctcttattacttaTTTggggtattcaccacAAAtggcAAAActagagg
cAAAttggattgcaatgatccTTTgctctaagagtgcgctctacgacattattgataccgaacttggctctacgcatatacaagatgaatggctaccgcagggtgagtgttatgtgagggc
TTTtcctAAggacgcacgatgcatttgcttgcggggaatgttcctctctccggtgtgacctcaatactacgaggcatactgacgagAAAtcaatgtattgtgagaatgtcagcatcggat
ccTTTTactgcccacgcgctagcgatgagcTTTattgacgtcgatccgaatcatccaaTTTctcgttccatctccgtattgtattggcctcatgcatcggatacgacactcgctgaagag
ttactcagtcatatggatgcagtggttgcttggggggggcgggatgccattgattgggcggttaagcattctccttcacatatcgatgTTTTgaagTTTggtccAAAgaagagTTTT
accgtgttagaccatccagccgatctagaagaagccgcctcgggtgttgcccatgataTTTgcTTtatgaccAAAAtgcctgcTTTTctactcagaataTTTacTTTTctgga
gataagtatgaagaaTTTAAAttAAAActtgttgAAAAActgaatctctatcaagaagTTTTaccAAAatcAAAAcAAAgTTTTgatgatgaagcTTTaTT
TTctatgactcgtcttgagtgtcaaTTTTctggTTTgAAAgttatatcagaaccggAAAAtaactggatgatcatcgagtcagagcccggggttgaatataaccatccattaagtc
gttgcgTTTatgtccacAAAAtAAAtaaggttgatgatgttgttcaatatatagAAAAAcatcAAAcacAAAcgaTTTcTTTTTatccatgggaatcttccaagAA
AtatcgagatgcattcgccgAAAAggggtagAAAgaatcgttgaatctgggatgaataatataTTTagagctggtggcgcacatgatgcaatgcgccctcttcaacgTTTagttc
gaTTTgTTTctcatgAAAgaccatataacttcaccactaaggatgtatctgtcgAAAtag

Photobacterium leiognathi luxC + 745 b.p of upstream DNA. Data Bank Accession-M63594.

tctagacTTTgtataTTTTcaaTTTaataccAAAcgatAAAtctaagaggaTTTTagTTTTTTAAAgatAAAAtacAAAAtacggacAAATTTTTTT
TattatcatcacAAAtaTTTTaaTTTAAAAacatAAAgaacaatcttgtaTTTattctAAAAcaccaataatattctggtgggagatagtaatatcattaattacataccatat
atagacAAAcactacctagcgataatAAAtcccatcatacccacaccAAAAAAAAAtAAAAaccacccaatataacAAAgacataactacaaccagAAAgaattA
AAAtAAATTTcaccattcaTTTaagttatgTTTattccatcaataatAAAActatAAAAataccattcaacctaaTTTAAAAattccggagtAAAttAAAAat
AAAgatAAAAagtAAAgccaatAAAtAAAAtcacAAAAAAcatAAAAAAcgaattaataatAAAAAATTTTgtaatataTTTcacttataccatcatct
atcatgAAAAatgAAAAatAAAAAAAtcagAAAgattAAAtatAAAAatattAAAttAAAAactgaTTTTattaatagtgtgaccAAAAtctcgaatag
AAAAAAgAAAAAAtAAAttagttatcacatccataTTTcTTTatctaatctgattcagctagctcatgcggcatagTTTatacgAAActacatactcagcatgtgcgaat
accAAAggagattacatgattaagaagatcccaatgattattgggggtgtagttcAAAAcacgtctggatatggcatgcgtgaactaacgctcaacaataatAAAgtgaatatccctat
catcacccAAAgtgatgttgaagctattcaatcactAAAtatagAAAAcAAAttgactatAAAtcagatagttaaTTTcttatatacagtgggacAAAAatggaagagcg
AAActtacagccgacgactcacttatattcgagatcttattaagttcctcggttactcacaagagatggcAAAActtgaagctaactggatctcaatgattctgtgtagcAAAAgtgcgt
tgtacgatattgttgagaatgatcttagctcacggcatattattgatgagtggatcccccaaggtgaatgttatgtcAAAgcgctcccAAAAggAAAAtctgtacacctattagctggta
acgtaccactatctggtgtgacttctattcttcgtgcgaTTTTgaccAAAAacgagtgcatcatAAAAacgtcatcagctgatccTTTTacagctactgcgctagttaatagTTT
TatcgatgtagatgcagaacacccgatcacacgttcaatctcagttatgtattggtcacatagcgaggatcttgctattccAAAAcAAAtaatgagctgtgctgatgtggttattgcatggg
gtggtgatgatgcaattAAAtgggctacagaacatgcaccatcacacgcagatattctAAAATTTggtcccAAAAagagtatatccattgttgacaacccaacagatattaaggct
gctgctatcggtgtagcacatgatatctgTTTTTacgatcagcaagcatgTTTctccacccaagataTTTattatattggcgatagcatagacataTTTTTTgatgaattagctcag
caattAAAtAAAtatAAAgacatattgcctAAAggtgagcggaaTTTTgatgAAAAagcagcTTTTTcTTTaacggAAAgagaatgTTTgTTTgccAA
AtatAAAgttcAAAAaggtgAAAgccaatcttggttattaacgcaatcacctgcgggatcaTTTggtaatcagccgttatcacgctcggcttatattcatcaagtAAAtgacaT
TTcagaagtcattccattcgtgcataaggcggtaacgcAAAccgtcgcaatagcgccgtgggagtcgtcTTTcAAAtatagagatatattagcagaacatggtgcagaacgaattat
agaagccggaatgaataatataTTTcgagtaggtggcgcccatgatgggatgcgtccccttcaacggcttgttaactatatatcacatgAAAggccgtcaacatataccactAAAgat
gtctcggtgAAAAtcgaacagactc

Photobacterium phoshoreum luxC + 619 b.p. upstream (*lumP*). Data Bank Accession-X54690.

cataataatctccTTTgtagacAAAcataTTTAAAAtgAAAAtattcaTTTggAAAttagtctaatggtAAAAAAAtAAAAttcaataccAAAtaataTTTT
TTgaaTTTaatctgTTTTTTataattaacTTTgagttgtaaTTTctaTTTaagatgaTTTTTTaaggTTTaaTTTacTTTgaatagataTTTtagttgaggTTTa
ggtcggtaagttgatgaTTTgAAAtggtaTTTaatAAActaagtaattaattacagatacTTTaatattaatcTTTtatgctaTTTaaTTTtaattgTTTataTTTtattgttg
ttgatattgcaaTTTaTTTAAAAtcaagtatatgcatcaggttcttattgTTTtatAAAAAtataattcTTTAAAttaatatgtagttAAAtgctcttaTTTataTTTAA
AggTTTTTTAAAAtagacatgaattaatAAAAAAcAAAAATTTcccttacatacctaatgAAATTTagttagtctctagtcatacctatgcagcagggttgtatgct
gattgagtatgtgcaatgcggaaTTTAAAggagattgtatgatAAAgAAAAtcccaatgattattggtggcgcagagagggatacttcagaacatgaatatcgtgaactcacact
AAAtagctatAAAgttagtatacctatcattaatcaagatgatgttgaggcgattAAAtcacAAAAtgtggAAAAtaatctAAAtatcaatcagatagtgaaTTTcttataca
ctgttggtcagAAAtggAAAAgtgagaattattctcgtcgactaacctatattcgtgaTTTggtcagaTTTctcggatattctcctgAAAtggcAAAActagaagctaactgg
atctcaatgatcttgagctcAAAAAgtgccttatatgatattgttgAAAcagagttaggttctcgtcatattgtagatgaatggttacctcagggtgattgttatgttaaggccatgccAAA
AggAAAAtctgttcaTTTgctagcaggtaatgtgcctctatctggtgttacttctattattagagcaattctgactAAAAAtgaatgtatcattAAAAcatcatcagcagatccaTT
TacggcaatagcattagcttcaagTTTattgatacagatgaacaccatccaattagccgctcaatgtcggtaatgtattggtcgcataacgaagatattgcaatcccacaacAAAttatgaa
ttgtgctgatgttgttgttagttggggtggatatgatgcaattAAAtgggcaacagaacatacgccggtAAAcgtcgacatattAAAATTTgggccgaagAAAAgtattgcgat
tgttgataatcctgtagatattacagcttctgctattggtgtggctcatgataTTTgTTTtatgatcagcaggcctgTTTtcaactcaagatatctattatataggcgataacattgatgcgTT
TTTTgatgagcttgtagaacaattAAAtctatatatggatatattgccAAAAggcgatcAAAcaTTTgatgAAAAggcatcaTTTcattaattgAAAAAgagtgtcaa
TTTgcAAAAtatAAAgttgagAAAggtgataatcaatcttggttacttgttAAAtcaccgc

Xenorhadbus luminescens luxC + 459 b.p. of upstream DNA. Data Bank Accession-M90093.

gaattcttcTTTagAAAtctgccggtAAAAAttagattgctattcaatctaTTTctatcggtaTTTgtgAAAtaatactcaggataataaTTTacatAAAtattatcacgcatt
agagaagagcatgacTTTTTTaaTTTAAAcTTTTcattaacAAAtcttgttgatatgAAAATTTTccTTTgctaTTTTaacagatattAAAAcgggaatatggc
gttatattgacgatccattcagttagattAAAAAccttgagcagAAAATTTatattattatcataattatgacgAAAgttacaggccaggaaccacgtagtcagaatctgaTTTT
ctatataTTTgttaTTTacatcgtcataacacAAAAAtataagaagcaagtgttggtacgaccagttcgcaagatagttAAAcagcaacttaagttgAAAttacccccattAAA
tggatggcAAAtatgactAAAAAAATTTcattcattattaacggccaggttgAAAtcTTTcccgAAAgtgatgaTTTagtgcaatccattaaTTTTggtgataatagt
gTTTacctgccaatattgaatgactctcatgtAAAAAAcattattgattgtaatggAAAtaacgaattacggttgcataacattgtcaaTTTTctctatacggtagaatatatgggata
ttcagaagAAAtggctaagctagaggccaattggatatctatgaTTTTatgttctAAAggcggccTTTatgatgttgtagAAAAtgaacttggttctcgccatatcatggatgaat
ggctacctcaggatgAAAgttatgttcgggcTTTTccgAAAggtAAAtctgtacatctgttggcaggtaatgttccattatctgggatcatgtctatattacgcgcaaTTTTaacta
agaatcagtgtattatAAAAAcatcgtcaaccgatccTTTTaccgctaatgcattagcgttaagTTTTattgatgtagaccctaatcatccgataacgcgctcTTTatctgttatatat
tggccccaccaaggtgatacatcactcgcAAAAgAAAttatgcgacatgcggatgttattgtcgcttggggagggccagatgcgattaattgggcggtagagcatgcgccatcttatg
ctgatgtgattAAATTTggttctAAAAAgagtcTTTgcattatcgataatcctgttgaTTTgacgtccgcagcgacaggtgcggctcatgatgTTTgTTTTTacgatcagc
gagcttgTTTTTctgcccAAAAcatatattacatgggAAAtcattatgaggaaTTTaagttagcgttgatagAAAAActtaatctatatgcgcatatattaccgaatgccAA
AAAAgaTTTTgatgAAAAggcggcctattcTTTagttcAAAAAgAAAgcttgTTTgctggattAAAgtagaggtggatattcatcaacgttggatgattattgagtc
AAAtgcaggtgtggaaTTTaatcaaccacttggcagatgtgtgtaccttcatcacgtcgataatattgagcAAAtattgccttatgttcAAAAAAAtaagacgcAAAccatatc
taTTTTTccttgggagtcatcaTTTAAAtatcgagatgcgttagcattAAAAggtgcggAAAggattgtagaagcaggaatgaataacataTTTcgagttggtggatctcat
gacggaatgagaccgttgcacgattagtgacatataTTTctcatgAAAggccatctaactatacggctaaggatgttgcggttgAAAtagaacagactcgattcctggagaagataag
ttccttgtat.

Discussion

The present study shows that H-NS controls the expression of *lux* clones of luminous bacteria in *E. coli.* The exact DNA binding site of H-NS that results in transcriptional silencing of the right operon of *V. fischeri* clone in *E. coli* has not been characterized in this study. Figure-7 shows the DNA sequence of *luxR* and *luxI* of *V. fischeri* as well as the sequence of *luxC* and of all the studied species of luminous bacteria, (the data was taken from the DNA data bank and the accession number is indicated). The *lux* genes of all luminous bacteria show a typical base composition that characterizes a curved DNA. The upstream regions of *luxR* of *V. fischeri*, and of *luxC* gene for all the other species, contain over 75% AT while the other structural genes are also highly enriched with A tracts. It has been proposed that H-NS might exert its effect on gene expression directly by binding to the DNA either as a silencer for extended chromosomal regions, or as a repressor by binding to specific promoters. Alternatively, H-NS might effect expression of some genes indirectly through its promoting effect on RpoS formation (3). Since the *rpoS hns* mutants showed a similar level of luminescence as the *hns* mutants, it seems unlikely that H-NS acts through activation of *rpoS* gene. The possibility that H-NS acts as a repressor of the *lux* system is supported by the studies of Choi and Greenberg (7). These authors showed that deletion of the fragment between 31 to 60 bp from the first codon of the *luxR* gene resulted in dramatically increased transcription of both operons of the *V. fischeri lux* system. As the 5' end of *luxR* is highly enriched with poly-d(A) stretches characteristic of the H-NS binding site, it is likely that H-NS exerts its inhibitory effect by binding to this region. A second DNA binding site of HNS might occur in the upstream of *luxAB* or *luxC* regions. Forsberg et al. (8) have recently showed that 200 nucleotides upstream of *V. fischeri luxAB* coding sequence contain tracts of d(A) that interact with H-NS.

The requirement for RpoS for the expression of the *lux* systems has been shown for *V. fischeri* clones, while other *lux* systems were less dependent on RpoS functions. However, one should remember that the clones of *V. fischeri* were the only ones that showed high luminescence in wild type strains of *E. coli..* Of the many genes that are transcriptionally repressed by H-NS and relieved by RpoS, the *lux* genes resemble most the *csgBA* genes. The *csgA* gene encodes the curlin subunit protein. The transcriptional start site of *csgA* is located downstream of an AT-rich activating sequence (UAS) (9). Similar to the *lux* system, the repressed transcription of the *csgBA* proteins in MC4100 *rpoS* mutants is relieved in MC4100 *hns rpoS* strains. Arnqvist et al. (9) suggested that the repression mediated by the H-NS protein is relieved due to the accumulation of RpoS as cells approach the stationary phase of growth. It has been already shown that H-NS increases the cellular content of *RpoS* by more than ten fold, while *RpoS* relieves H-NS repression of the promoters of as many as 20 genes (3. 10).

Although the importance of RpoS and H-NS in the regulation of the *V. fischeri lux* system is evident, one should remember that the right operon of *V. fischeri* is fully expressed in MC4100 *rpoS* cells upon addition of the autoinducer to the growth medium. Similarly, the other studied clones of the *lux* systems showed high luminescence when T7 polymerase coding gene was cloned in trans to the plasmids applied in this study. It thus appears that more than one factor controls the induction of the *lux* systems. In growth tempertures above 28°C, when the LuxR is partially inactivated (1), RpoS and LuxR-autoinducer complex are required to initiate the transcription of *V. fischeri lux* system, while in the absence of H-NS, none of these factors are required. Many commonly used *E. coli* K-12 strains carry an amber mutation in the *rpoS* gene (11). Expression of the functional *RpoS* in these cells depends on the presence of amber suppressers. Strain MC4100, which is *rpoS*+ and *supo*, emits significantly higher luminescence than *lux* carrying *E. coli rpoS* strains like W3110 or HB101 cells, (unpublished observation). The appearance of *rpoS* missense mutants is often observed in old stationary phase *E. coli* cultures (12). Similar type of mutants seems to develop in old cultures of freshly isolated marine luminous bacteria. These very dim pleiotropic mutants (K-variants), regain high luminescence in the gyrase inhibitors inhibitors and DNA intercalating agents (13. Thus, it is possible that the K-variants are *rpoS*-mutants of luminous bacteria.

Another agent that acts as an H-NS reliever is intracellular potassium ions at high concentrations (14). *E. coli* cells accumulate potassium ions when grown in the presence of high concentrations of salts or sugars. We have shown (14), that the presence of salts (0. 3-0. 4M) or sugars (0. 5-0. 7M) in LB medium resulted in rapid induction of luminescence in recombinant *E. coli* harboring the whole *lux* system of*V. fischeri.* Moreover, these cultures showed high luminescence at 37°C (a temperature in which wild-type cells are dark). Interesting enough, Watanaba et al. (17) have also showed the enhanced effect of nalidixic acid and NaCl on *P. phosphoreum* luminescence and concluded that these agents act on the expression of the *lux* system by changing the DNA supercoiling

The rate of H-NS synthesis during log and stationary phases of growth is a controversial issue. Recent studies (16) showed that the H-NS:DNA ratio is constant through all stages of growth and the level of *hns* mRNA declines at the onset of the stationary phase. This may explain why the *lux* system undergoes an intense period of induction only during the mid to late log phase of growth.

LuxR is a well studied member of the UhpA/FixJ family of prokaryotic transcriptional activators that share homology throughout their entire amino acid sequence. (1). It is not surprising, therefore, that like *luxR*, these genes are also enriched with stretches of d(A) and d(T). These genes are induced, in most cases, in late log phase upon accumulation of homoserine lactone derivatives in the growth medium. It is possible that the role of LuxR proteins is to relieve H-NS repression and thus induce the formation of proteins essential for cell survival at the stationary phase.

Acknowledgments

Special thanks to E.A. Meighen for providing the clones of the *lux* systems and for N. Ulitzur for critical reading of the manuscript. The assistance of C.D. Fraley, Dina Sachar and Michael Roseff and S. Yannai is also appreciated. Part of this work was performed at Stanford University where it was supported by Microbics Corporation and by National Science Foundation grant DCB 920701 to A. Matin.

References

1. Ulitzur S, Dunlap P. Regulatory circuitry controlling luminescence autoinduction in *Vibrio fischeri*. Photobiol Photochem 1995;62:625-32.
2. Meighen EA, Dunlap PV. Physiology biochemical and genetic control of bacterial bioluminescence. Adv Microbiol Physiol 1993:34;1-67.
3. Loewen CP, Hengge-Aronis R. The role of the RpoS (KatF) in bacterial global regulation Annu Rev Microbiol 1994;48:53-8.
4 Adar Y, Siman M, Ulitzur S. Formation of the LuxR protein in the *Vibrio fischeri lux* system is controlled by HtpR through the GroESL proteins. J Bacteriol 1992;174:71338-42.
5. VanDyk TK, Majarian WM, Konstantinov KB, Young RM, Dhurjati PS, LaRossa RA. Rapid and sensitive pollutant detection by induction of heat shock gene-bioluminescence gene fusions. Appl Environ Microbiol 1994;60:1414-20
6. Huisman GW, Kolter R. Sensing starvation: a homoserine lactone-dependent signaling pathway in *Escherichia coli* Science 1994;265:537-39.
7. Choi SM, and Greenberg EP. The C terminal region of *V. fischeri* LuxR protein contains an inducer-independent *lux* gene activating domain. Proc Natl Acad Sci USA 1991;88:11115-19.
8. Forsberg AJ, Pavitt GD, Higgins C. Use of transcriptional fusions to monitor gene expression: a cautionary tale. J Bacteriol 1994;176:2128-32.
9. Arnqvist A, Olsen A, Normark S. RpoS dependent growth-phase induction of csgBA promoter in *Escherichia coli* can be achieved in vivo by sigma-70 in the absence of the nucleoid-associated protein H-NS. Mol Microbiol 1994;13:1021-32.
10. Olsen A, Arnqvist A, Hammer M, Sukupolvi S, Normark S. The RpoS sigma factor relieves H-NS mediated transcriptional repression of csgA the subunit gene of fibonectin-binding curli in *Escherichia coli*. Mol Microbiol 1993; 94:523-36.
11. Kaasen I, Falkenberg P, Styrvold DB, Strom AR. Molecular cloning and physical mapping of the otsBA genes which encode the osmoregulatory trehalose pathway of *Escherichia coli*: evidence that transcription is activated by KatF (AppR). J Bacteriol 1992;174:889-98.
12. Zambrano MM, Siegele D, Almiron M, Tormo A, Kolter R. Microbial competition: *Escherichia coli* mutants that take over stationary phase cultures. Science 1993;259:1757-60.
13. Weiser I, Ulitzur S, Yannai S. DNA damaging agents and DNA synthesis inhibitors induce luminescence in dark variants of luminous bacteria. Mut Res 1981;91:443-503.
14. Mechthild B, Marschall C, Muffler A, Fischer D, Hengge-Aronis R. Role for the histone-like H-NS in growth phase-dependent and osmotic regulation of sigma S and many sigma-S dependent genes in *Escherichia coli*. J Bacteriol 1995;177:3455-64.
15. Ulitzur S, Osman M. The expression of the whole *lux* system of *Vibrio fischeri* in recombinant *Escherichia coli* at 37°C requires high ionic strength and the presence of either GroESL proteins or externally added inducer. In: Campbell AK, Kricka LJ, Stanley PE. editors. Bioluminescence and Chemiluminescence-Fundamentals and Applied Aspects;. Chichester, J. Wiley & Sons, 1994:515-20.
16. Free A, and Dorman C. Coupling of *Escherichia coli hns* mRNA levels to DNA synthesis by autoregulation: implications for growth phase control. Mol Microbiol 1995;18:101-13.
17. Watanaba H, Inaba H, Hastings JW. Osmoregulation of bioluminescence expression in Photobacterium phosphoreum is related to gyrase activity. In: Stanley PE, Kricka LJ, editors, Bioluminescence and chemiluminescence current status, Great Britiain: John Wiley & Sons, 1991:43-46.

THE GLUCOSE EFFECT ON BACTERIAL BIOLUMINESCENCE SEEMS TO BE PARTIALLY DUE TO INHIBITION OF AUTOINDUCER SYNTHASE BY PROTEIN EIIA^{Glc}

J Sicher, S Pöhling and UK Winkler
Dept of Biology, Ruhr-University,
D-44780 Bochum, Germany

Introduction

In batch cultures of *Photobacterium fischeri* glucose causes both transient and catabolite repression of bioluminescence and synthesis of luciferase as well (1). Glucose repression of bacterial functions is usually linked to the vectorial phosphorylation and uptake of the sugar by the PEP-sugar-phosphotransferase system (PTS) (2). Protein EIIA^{Glc} coded by *crr* participates in the sequential phosphotransfer to glucose and additionally has regulatory functions: In the presence of glucose, EIIA^{Glc} is dephosphorylated; it then deactivates the adenylcyclase thereby decreasing the intracellular cAMP level (*catabolite repression*) and moreover inhibits some permeases specific for non-PTS-sugars (*inducer exclusion*). Since the glucose repression of bioluminescence not fully responds to exogenous cAMP we propose the hypothesis that the cAMP-refractory repression might be due to a mechanism similar to inducer exclusion. The goal of this study was to test this hypothesis. Our data indicate that dephosphorylated EIIA^{Glc} probably inhibits the autoinducer (AI) synthase (LuxI).

Materials and Methods

Bacteria, plasmids: *Escherichia coli* strains TP2503 and C600 are wildtype with respect to PTS; TP2811Δ*ptsIptsHcrr* (3); TP2862Δ*crr* (3); JLV86*crr*-1zfb::Tn10 (J. Lengeler, unpubl.). Plasmids: Both pJE202 (4) and pW21A (5) harbor the *lux* regulon of *P. fischeri* but the latter has *lux*I deleted. pJRC contains the *E. coli crr* gene inserted into pJRD215 (6). pAR34 (this study) contains the *P. fischeri* MJ-1 *lux*I gene inserted into pBTac1 (Boehringer, Mannheim/Germany). pUHS*4i-Z1 constitutively expressing the lac repressor was obtained from Dr. R. Lutz (Heidelberg).

Procedures: *E. coli* was grown in Luria broth (LB) with or without 2 g/L of an appropriate sugar at 30°C. Autoinducer was bio-assayed using *E. coli* C600(pW21A) as indicator (5); a concentration of 1 nmol/L corresponded to about 700 relative light units. cAMP concentrations were determined with the cAMP EIA kit (Amersham, Braunschweig/Germany). Genetic methods and luciferase assays as well were performed according to standard procedures.

Instrumentation: Bacterial light-emission was measured using a LUMIStox instrument (Dr. B. Lange, Berlin/Germany).

Results and Discussion

Only PTS sugars repress bioluminescence: In batch cultures of *E. coli* TP2503(pJE202) bioluminescence was repressed only by PTS sugars such as glucose (Fig. 1), fructose, N-acetylglucosamine and mannitol. Glucose also inhibited the biosynthesis of autoinducer (Fig. 1) and luciferase. Non-PTS sugars, e.g. lactose, maltose, sucrose, galactose and cellobiose were unable to repress bioluminescence. Exogenous cAMP (4 mmol/L) only partially abolished glucose repression.

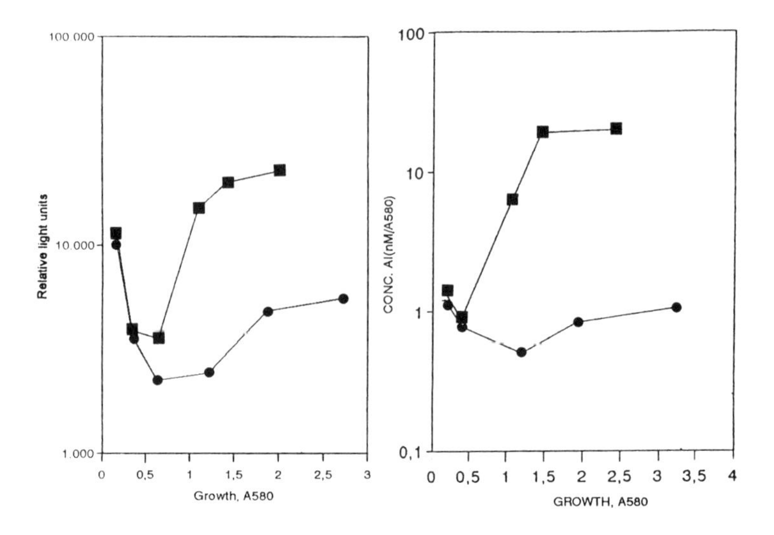

Figure 1. Bioluminescence (left) and concentration of autoinducer (AI) in the growth medium (right) of *E. coli* TP2503(pJE202) as a function of cell density. [■] LB; [●] LB containing 2 g/L glucose.

Bioluminescence of PTS mutants harboring pJE202: Bioluminescence of *E. coli* TP2862(Δ*crr*) and TP2811(Δ*ptsIptsH crr*) was not repressed by glucose or other PTS sugars though TP2862 could take up glucose by the mannose PTS. When TP2811(pJE202) was provided with *crr* from *E. coli* in *trans* the bioluminescence decreased by a factor of 100 (Fig.2). The expression of *crr* (inserted in pJRD215) was demonstrated using $EIIA^{Glc}$ specific antibodies. Protein $EIIA^{Glc}$ probably existed in its dephosphorylated form because the general PTS enzymes EI and HPr were missing in TP2811. Consequently, it seems that dephosphorylated $EIIA^{Glc}$ acted inhibitory on bioluminescence even in the absence of glucose. Since the intracellular cAMP level (0.8-1.5 μmol/L) was not significantly influenced by pJRC one might conclude that the very low bioluminescence of TP2811(pJE202,pJRC) was rather due to dephosphorylated $EIIA^{Glc}$ itself than to a shortage of cAMP. On the other hand, we observed that the bioluminescence of the *crr* mutants TP2811, JLV86 and TP2862 all harboring pJE202 and grown in the absence of glucose could be greatly enhanced by exogenous cAMP (4 mmol/L). The bioluminescence of *E. coli* TP2503(pJE202), however, did not respond to exogenous cAMP.

Protein LuxI possesses a putative $EIIA^{Glc}$ binding site: Proteins which are allosterically inhibited by dephosphorylated $EIIA^{Glc}$ share a sequence motif (VGANXSL) probably functioning as binding site for $EIIA^{Glc}$ (7). The amino acids at the positions Nos. 107 to 113 of LuxI of *P. fischeri* ATCC7744 (8) are VGKNSSK, i.e. this sequence is very similar to the consensus. None of all the other Lux proteins (ABCDEGR) of *P. fischeri* (8) contains a stretch of seven amino acids showing a similarity to the consensus as high as that of LuxI.

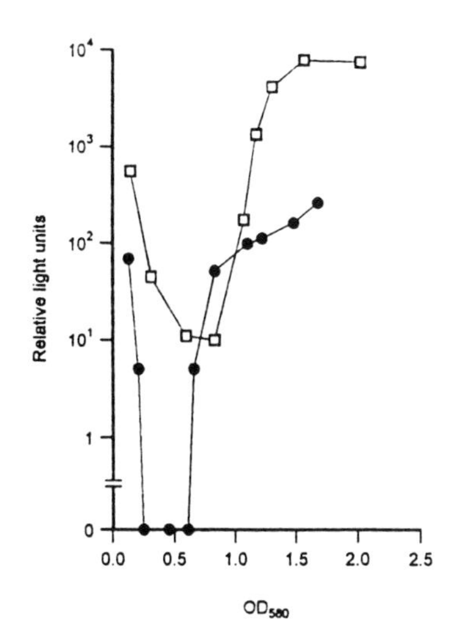

Figure 2. Bioluminescence of *E. coli* TP2811(pJE202) in the absence [□] and presence [●] of pJRC containing the wildtype allele of *crr* from *E. coli*.

Influence of $EIIA^{Glc}$ upon autoinducer synthase: Using pJE202 as source the coding region of *lux*I was amplified by PCR simultaneously generating new restriction sites at both ends. The PCR product inserted in pBTac1 was now under regulatory control of the cAMP-independent *Tac* promotor. The resulting plasmid pAR34 was transformed into *E. coli* TP2503 as well as TP2862 (Δ*crr*). Both strains also contained pUHS*4i-Z1 for improving the control of gene expression by IPTG. *E. coli* TP2503 induced with 25 μmol/L IPTG produced AI much less in the presence of glucose than in its absence (Fig. 3). The synthesis of AI by mutant TP2862, however, did not respond to glucose (Fig. 3). For yet unknown reason the mutant yielded much less AI than the wildtype. In conclusion, our results indicate that $EIIA^{Glc}$ is involved in glucose repression of bioluminescence beyond its function as a regulator of cAMP synthesis. An allosteric inhibition of AI synthase by dephosphorylated $EIIA^{Glc}$ we consider to be most likely. Since *P. fischeri* NRRL-B11177 contains a 20 kDa protein homologous to $EIIA^{Glc}{}_{E.c.}$ (S. Pöhling, unpubl.) we assume that our above conclusion is true not only for *E. coli* but also for *P. fischeri*. A critical test of our hypothesis requires the mutational exchange of single amino acids within the putative $EIIA^{Glc}$ binding site of LuxI.

Acknowledgements
We thank Dr K Müller-Breitkreutz for initiating the project and helpful discussions. We also thank Drs A Danchin, W Hengstenberg, J Lengeler, R Lutz, M Silverman, and S Ulitzur for supplying us with bacterial strains, plasmids and antiserum. We gratefully aknowledge the help of Dr U Strych who performed a data bank search for us.

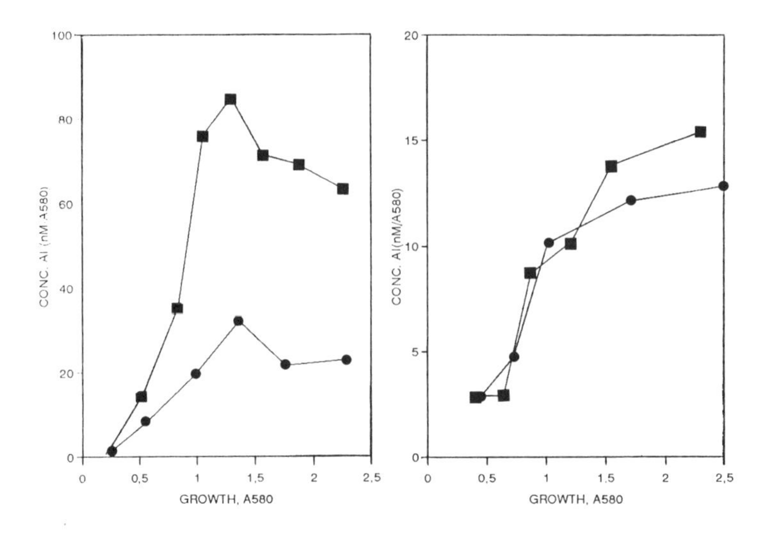

Figure 3. Concentration of autoinducer (AI) in the growth medium of *E. coli* TP2503 (left) and TP2862 (right) both harboring pAR34 and pUHS*4i-Z1. Symbols as in figure 1.

References

1. Nealson KH, Eberhard A, Hastings JW. Catabolite repression of bacterial bioluminescence: Functional implications. Proc Nat Acad Sci USA 1972; 69:1073-6.
2. Saier MH Jr. Protein phosphorylation and allosteric control of inducer exclusion and catabolite repression by the bacterial phosphoenolpyruvate: sugar phosphotransferase system. Microbiol Rev 1989; 53:109-20.
3. Levy S, Zeng GQ, Danchin A. Cyclic AMP synthesis in *Escherichia* coli strains bearing known deletions in the *pts* phosphotransferase operon. Gene 1990;86:27-33.
4. Engebrecht J, Nealson KH, Silverman M. Bacterial bioluminescence: Isolation and genetic analysis of functions from *Vibrio fischeri*. Cell 1983;32:773-81.
5. Adar YY, Simaan M, Ulitzur S. Formation of the LuxR protein in the *Vibrio fischeri lux* system is controlled by HtpR through the GroESL proteins. J Bacteriol 1992;174:7138-43.
6. Davison J, Heusterspreute M, Brunel F. Restriction site bank vectors for cloning in gram-negative bacteria and yeast. Meth Enzymol 1987;153:34-54.
7. Titgemeier F, Mason RE, Saier MH Jr. Regulation of the raffinose permease of *Escherichia coli* by the glucose-specific enzyme IIA of the phosphoenolpyruvate: sugar phosphotransferase system. J Bacteriol 1994;176:543-46.
8. Baldwin TO, Devine JH, Heckel RC, Lin JW, Shadel GS. The complete nucleotide sequence of the *lux* regulon of *Vibrio fischeri* and the *lux*ABN region of *Photobacterium leiognathi* and the mechanism of control of bacterial bioluminescence. J Biolumin Chemilumin 1989;4:326-41.

ANAEROBIC EXPRESSION OF THE *PHOTOBACTERIUM FISCHERI* LUX REGULON REQUIRES THE FNR PROTEIN WHICH ACTS UPON THE LEFT OPERON

K Jekosch and UK Winkler
Dept of Biology, Ruhr-University,
Bochum D-44780, Germany

Introduction

In *Photobacterium fischeri* all *lux* genes reside in a single regulon, consisting of two divergently transcribed operons. The expression of the *lux* regulon is controlled by several cellular and environmental factors. Even though the light-emitting reaction catalyzed by luciferase requires molecular oxygen, micro- or even anaerobic conditions are favorable for the luciferase-synthesis (1). This latter phenomenon is due to the fact that the expression of the *lux* regulon is likely to depend on the transcriptional regulatory protein FNR (2). This protein is able to sense different oxygen levels and, at low oxygen tension, to initiate the transcription of many bacterial genes of anaerobic metabolism (3). Within the *lux* regulon of *P. fischeri* MJ-1 a putative FNR binding site has been recognized upstream the *lux*R transcription start site (2). Chromosomal DNA of *P. fischeri* contains a gene coding for a FNR-like protein (4). The objective of this study was to further analyze the FNR requirement of the anaerobic expression of the *lux* regulon. In our first FNR-study (2) we used a missense mutant, but its residual FNR activity sometimes obscured the FNR requirement of bioluminescence. Therefore, here we applied a *fnr* transposon mutant.

Material and Methods

Bacteria, plasmids: *E. coli* M182 and M182*fnr*::Tn10 (5). Plasmid pJE202 (*lux* regulon from *P. fischeri* MJ-1 (6), pHV200 (*lux* regulon from *P. fischeri* ES114) (7) pJE413 (*lux*C-*lac*Z) and pJE455 (*lux*R-*lac*Z) (6).

Procedures: The bacteria were grown either aerobically or anaerobically, mostly at 30^{0}C. The medium used was Luria broth (LB) supplemented with 2 g/L maltose. All genetic methods applied followed standard schemes, luciferase, ß-galactosidase and some FNR-dependent enzymes (fumarate-, nitrate-, molybdate-reductase) were assayed by standard techniques.

Instrumentation: Bacterial light-emission was measured by using a LUMIStox Instrument (Dr. B. Lange, Berlin/Germany). Specific bioluminescence was defined as relative light units per OD_{580} of the respective bacterial culture.

Results and Discussion

Anaerobic expression of the lux regulon is FNR-dependent: In anaerobic cultures of *E. coli* M182*fnr*::Tn10, neither fumarate- nor nitrate- or molybdate-reductase activities could be detected, i.e. the mutant did not show any FNR activity. The mutant and its parental strain M182 were both transformed with pJE202. Under anaerobic (fermentation) conditions and in the absence of FNR, the *lux* regulon could be induced only very little (Fig. 1). The specific luciferase activity of the *fnr* mutant was only 2 % of that of the parental strain. The anaerobic expression of the *lux* regulon could be fully restored by providing mutant M182*fnr*::Tn10(pJE202) with $fnr^{+}_{E.c.}$ in *trans*.

Furthermore, the FNR deficiency of the mutant could be partially compensated by growing it in the absence of oxygen but at high concentrations of cAMP (5 mmol/L) or autoinducer (30 μg/ml).

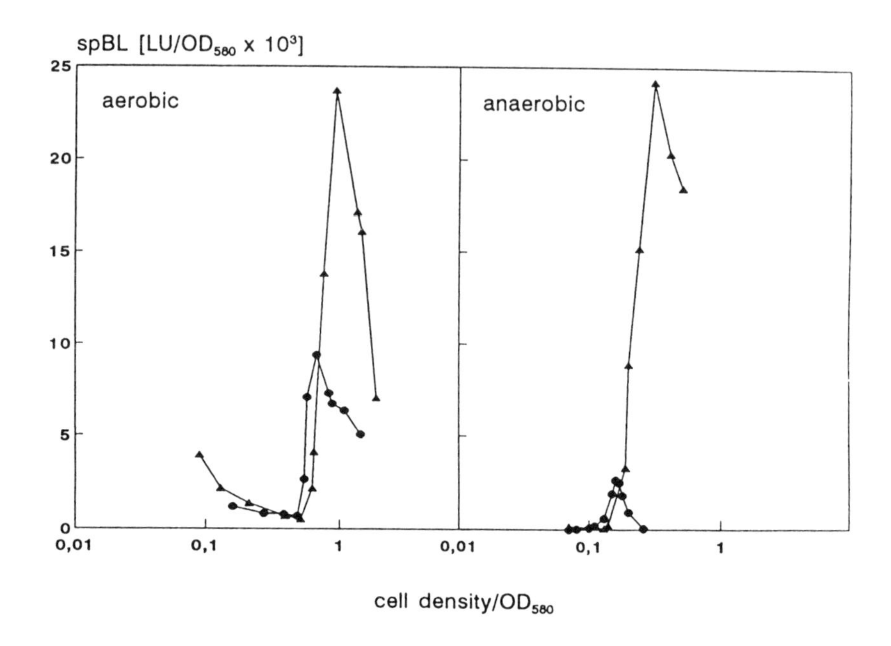

Figure 1. Specific bioluminescence of *E. coli* M182(pJE202)[▲] and M182*fnr*::TN10(pJE202) [●] as a function of cell density.

<u>FNR activates the transcription of *lux*R:</u> Theoretically, FNR might be required for activating the transcription either of the left (*lux*R) or the right (*lux*CDABE) operon. In order to decide between these alternatives, the *lac*Z gene was used as reporter, fused either to *lux*R (pJE455) or to *lux*C (pJE413). *E. coli* M182 was transformed with these plasmids. Both the aerobic and anaerobic culture of *E. coli*(pJE413) showed the same maximum ß-galactosidase activity (Fig. 2). The corresponding enzyme activity of *E. coli*(pJE455), however, was 3 to 4 fold higher under anaerobic than aerobic growth conditions (Fig. 2). This indicated that FNR is primarily required for the transcription of the left operon. Experiments performed with *E. coli* M182*fnr*::Tn10(pJE455) confirmed this conclusion: in an anaerobic culture the ß-galactosidase activity dropped to 2 % of that of the fnr^+ control whereas aerobic conditions caused a decrease to 42 % only (data not shown). Providing the mutant strain carrying pJE455 with $fnr^+_{E.c.}$ in *trans* led to a ß-galactosidase activity similar to that of the wildtype. The results presented here agree well with the fact that the putative FNR binding site within the *lux* regulon is located about 35 bp upstream of the *lux*R transcription start site (2).

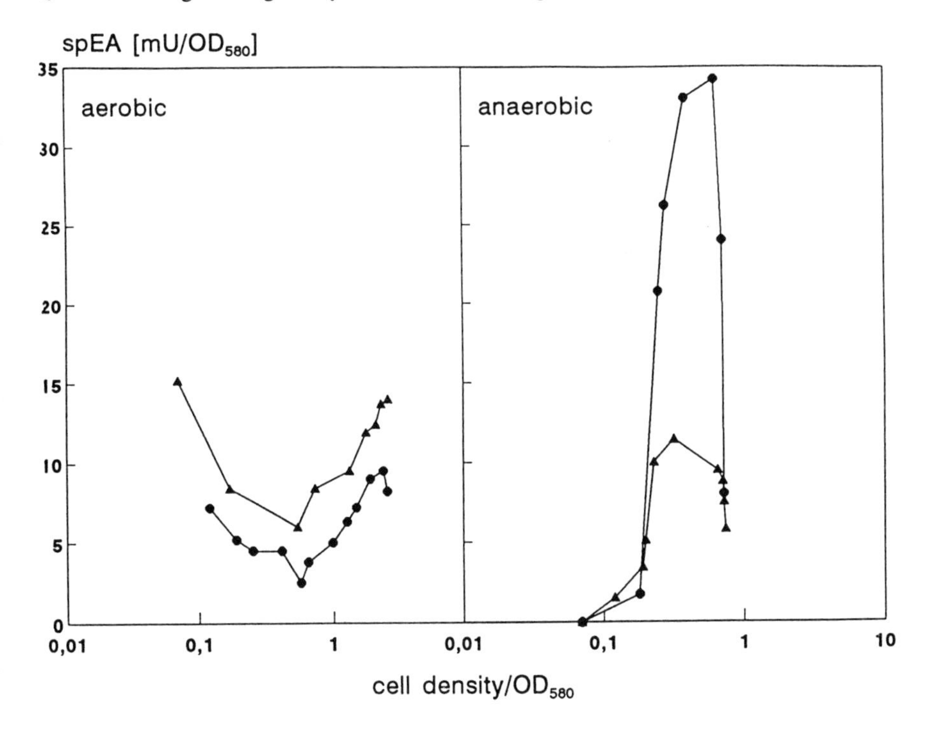

Figure 2. Specific ß-galactosidase activity of *E. coli* M182 (pJE413) [▲] and M182(pJE455) [●] as a function of cell density.

Induction of bioluminescence in *P. fischeri* ES114: On the chromosome of *P. fischeri* MJ-1 and ES114, the *lux* genes are arranged in the same way, but both strains greatly differ with respect to the base sequence of their intergenic region (7). We therefore compared the expression of the *lux* regulons derived from MJ-1 (→pJE202) and ES114 (→pHV200). In aerobically grown *E. coli* wildtype, both *lux* regulons were expressed, but in anaerobic cultures only the bacteria harboring pJE202 showed expression (Fig. 3). The negative result obtained with *E. coli*(pHV200) matches the fact that a FNR consensus sequence is missing within the intergenic region of the *lux* regulon of strain ES114: Within a typical 22 bp-long palindromic FNR consensus sequence the base positions Nos. 5-9 (TTGAT) and 14-18 (ATCAA) are highly conserved (8). These sequences are also found within the putative FNR site of the lux_{MJ-1} regulon (2) except position No 16 (C →A). The corresponding sequence of ES114, however, deviates at four bp positions from the consensus probably preventing the binding of the FNR homodimer (6). It would be worth-while to test the effect of anaerobic regulatory proteins other than FNR upon the expression of the *lux* regulon, e.g. NarL which is involved in nitrate control (3). The promoter of *lux*R from strain ES114 contains several putative NarL binding sites.

Acknowledgements
We thank Drs M Silverman, EP Greenberg, A Bell, EG Ruby and JR Guest for providing us with bacterial strains and plasmids. We aknowledge Dr U Strych for helpful suggestions.

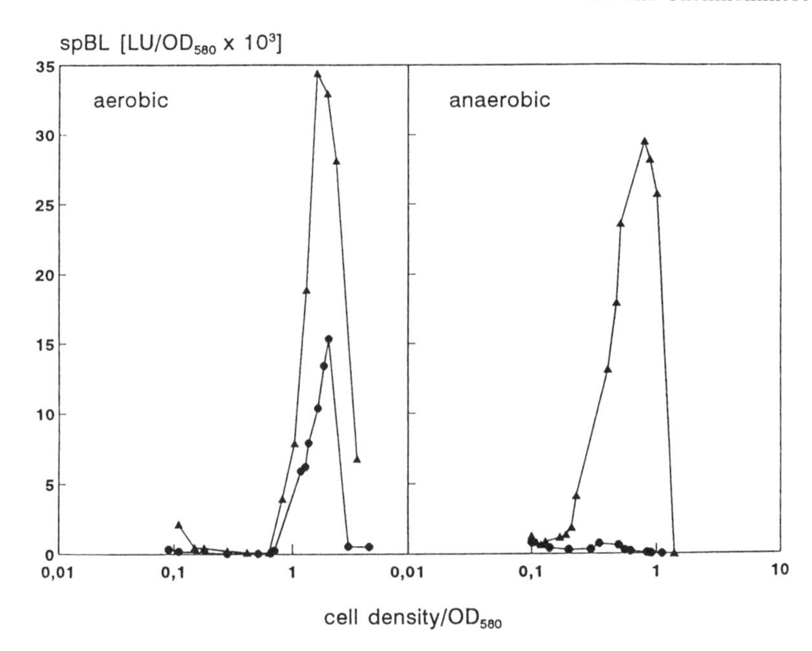

Figure 3. Specific bioluminescence of *E. coli* M182(pJE202) [▲] and M182(pHV200) [●] as a function of cell density.

References

1. Nealson KH, Hastings JW. Low oxygen is optimal for luciferase synthesis in some bacteria. Arch Microbiol 1977;112:9-16.
2. Müller-Breitkreutz K, Winkler UK. Anaerobic expression of the *Vibrio fischeri lux* regulon in *E. coli* is FNR-dependent. In: Szalay AA, Kricka LJ, Stanley P (eds.) Bioluminescence and Chemiluminescence. Status Report. Chichester: John Wiley & Sons, 1993:142-6.
3. Unden G, Becker S, Bongaerts J, Schirawski J, Six S. Oxygen regulated gene expression in facultatively anaerobic bacteria. Antonie Leeuwenhoek 1994; 66:3-22.
4. Tolksdorf P, Sommer I, Pöhling S, Winkler UK. Genetic control of bioluminescence in *Vibrio fischeri* by FNR and PTS. 17th Int. Congress of Genetics, Birmingham UK, 1993.
5. Jayaraman PS, Gaston KL, Cole JA, Busby SJW. The *nir*B promoter of *Escherichia coli*: Location of nucleotide sequences essential for the regulation by oxygen, the FNR protein and nitrite. Mol Microbiol 1988;2:527-30.
6. Engebrecht J, Nealson K, Silverman M. Bacterial bioluminescence: Isolation and genetic analysis of functions from *Vibrio fischeri*. Cell 1983;32:773-81.
7. Gray KM, Greenberg EP. Sequencing and analysis of *lux*R and *lux*I, the luminescence regulatory genes from the squid light organ symbiont *Vibrio fischeri* ES114. Mol Mar Biol Biotech 1992;1:414-9.
8. Bell AI, Gaston KL, Cole JA, Busby SJW. Cloning of binding sequences for the *Escherichia coli* transcription activators, FNR and CRP: location of bases involved in discrimination between FNR and CRP. Nucl Acids Res 1989;17:3865-74.

INSECT PATHOGENIC *XENORHABDUS NEMATOPHILUS* (ENTEROBACTERIACEAE) MAY HAVE AN AUTOINDUCER REGULATORY SYSTEM SIMILAR TO *VIBRIO HARVEYI*

GB Dunphy[1], CM Miyamoto[2] and EA Meighen[2]
[1]Dept of Natural Resources Sciences & [2]Dept of Biochemistry, McGill University, Montreal, Quebec, H3G1Y6, Canada.

Introduction

The luminescence systems of *Vibrio harveyi* and *Photobacterium fischeri* are both controlled by N-acyl homoserine lactone autoinducers(1). In spite of this common feature, most of the lux regulatory proteins in these marine bacteria appear to be quite different. In *P. fischeri*, *lux*I and *lux*R genes code for proteins involved in autoinducer synthesis and reception(2). In *V. harveyi*, a set of proteins (LuxL-R) have been implicated in the regulation of luminescence including synthesis and reception of the β-hydroxybutanoyl homoserine autoinducer (LuxM, LuxN) and a second as yet undefined inducer (2). However, these proteins are completely unrelated to *P. fischeri* LuxR-LuxI not only in sequence but also in gene organization and relationship to the primary lux operon containing the structural genes. Recently, however, proteins implicated in the synthesis-reception of a second homoserine lactone inducer in *P. fischeri* have been found to be related in sequence to LuxM-LuxN of *V. harveyi* (3).

Protein homologs of the prototype *P. fischeri* LuxI-LuxR regulatory system involving a N-acyl-homoserine lactone inducer have been discovered controlling a number of different functions implicated in secondary metabolism in a wide range of nonluminescent bacteria (4). In contrast, a homoserine lactone regulatory system comparable to that controlling the *V. harveyi* lux system has not yet been identified in another bacterial species. The present paper provides evidence that the insect pathogenic bacteria *Xenorhabdus nematophilus* may contain a system analogous to the β-hydroxybutanoyl homoserine lactone regulatory system of *V. harveyi*.

Materials and Methods

Strains and Growth Conditions: *X. nematophilus* was grown in LB media at 27°C. Luminescent bacteria were produced by transconjugation of the pKT plasmid containing the *V. harveyi lux* operon with the help of the pRK plasmid at 27°C. Cells were grown in LB media containing ampillicin, kanamycin and streptomycin. In some cases, the media were conditioned by prior growth of the appropriate strain followed by passing through 0.22 μm sterile filters.

Luminescent Measurements: Luminescence was measured in light units(LU) using a phototube, where 1 LU corresponds to 5 x 10^9 quanta/sec.

Results

On transfer of the *V. harveyi lux* operon into *X. nematophilus*, a luminescent phenotype was generated whose level of light emission was approximately 5-20% of that for *V. harveyi*. This result was surprising since transfer of the same plasmid into *Escherichia coli* gave much lower light levels (<0.1%);high light intensity required cotransformation of the *V. harveyi* regulator, LuxR.

Fig. 1 shows a plot of the light intensity per unit cell growth(LU/OD_{660}) for *V. harveyi* and the transconjugated *X. nematophilus*. The luminescence per unit of cell growth decreases at the early stages of cell growth and then increases at higher cell density not only for *V. harveyi* cells but also for the transconjugant, *X. nematophilus*. This result is reflective of the autoinducer regulatory system for *V. harveyi* and suggests that a similar system may be present in *X. nematophilus* that allows expression of the *V. harveyi lux* operon in a density-dependent manner indicative of a possible autoinducer.

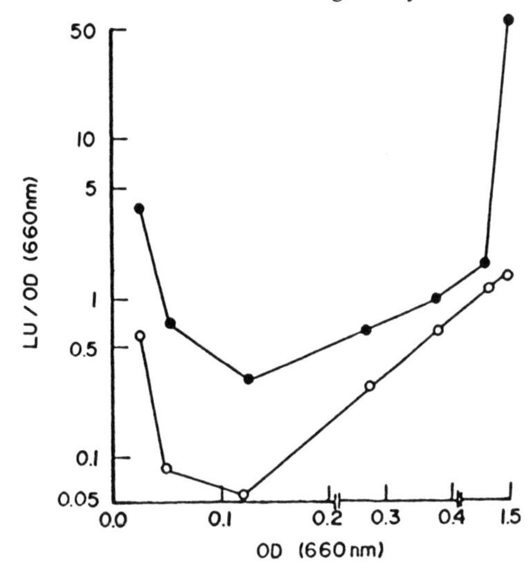

Figure 1. Dependence of luminescence (for 1 ml) of *V. harveyi* (closed circles) and X. nematophilus transconjugated with the lux operon (open circles) on growth in LB media.

During growth of *V. harveyi* cells, the β-hydroxybutanoyl homoserine lactone autoinducer is excreted into the media. After sterilization of the media, reinoculation of the conditioned media with *V. harveyi* will result in an increase in luminescence at a lower cell density(OD_{660}) than in nonconditioned media as shown in Fig.2. When medium was conditioned by *X. nematophilus*, luminescence of *V. harveyi* cells was also stimulated over cells grown in non-conditioned medium although the light intensity was far less at high cell density than that of cells grown in *V. harveyi*-conditioned medium. Excretion of a lower level of autoinducer or related compound by *X.nematophilus* compared to the amount of hydroxybutanoyl homoserine lactone excreted by *V. harveyi* could account for this result. It should be noted that conditioning of media by *X. nematophilus* is quite complex as a number of compounds that function as antibiotics and inhibitors of growth are excreted particularly at higher cell density (5) that may impinge upon the

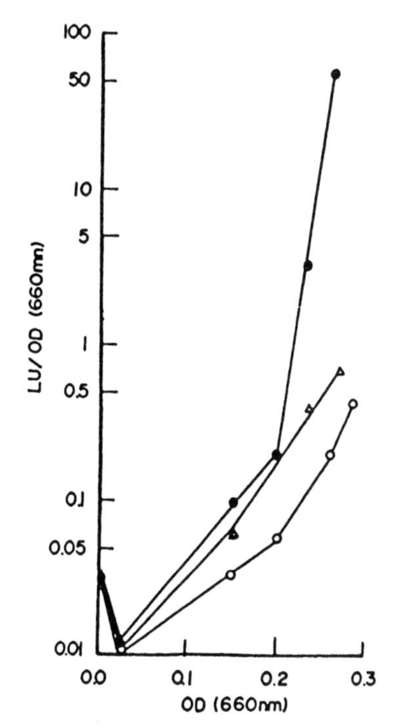

Fig. 2. Effect of conditioned LB media on luminescence of *V. harveyi* cells. *V. harveyi* conditioned media (closed circles); X. nematophilus conditioned media (triangles); nonconditioned media (open circles).

induction of luminescence by putative excreted autoinducers.

Similarly, conditioning of media with *V. harveyi* or *X. nematophilus* cells resulted in an increase in the light intensity emitted by transconjugated *X. nematophilus* cells compared to the emission of light by the same cells in untreated media as well as a shift of the luminescence induction to lower cell density(data not shown). This induction of luminescence in *X. nematophilus* occurred without acceleration of the development of the stationary phase. These results suggest that *X. nematophilus* may produce an autoinducer identical or similar to the β-hydroxybutanoyl homoserine lactone of *V. harveyi* and raises the possibility that it may be implicated in the control of specific secondary metabolism processes as cells enter the stationary phase.

Discussion

The emission of light by *X. nematophilus* cells transconjugated with the *V. harveyi luxCDABEGH* operon coupled with the increase in luminescence at higher cell density was quite surprising since the *V. harveyi lux* operon can generally not be expressed in other bacterial systems without a strong external promoter or the LuxR regulatory protein. This result raised the possibility that a regulatory system may exist in *X. nematophilus* that allowed expression of the *lux* operon in an analogous manner to its expression in V. harveyi. The stimulation of luminescence at lower cell densities by conditioned media provides indirect evidence that a lux stimulatory factor is excreted by the cells and that this factor can stimulate the *V. harveyi lux* operon. These results are similar to earlier reports that the luminescence of *V. harveyi* was enhanced by some nonluminescent bacteria including *V. parahaemolytica, V. anguillara* and *V. cholerae*(6,7). Recent results suggest that the compound excreted by *X. nematophilus* may indeed be identical to the autoinducer of *V. harveyi*.

The function of a quorum sensing mechanism in *X. nematophilus* is unknown. However, it is interesting to note that the luminescent insect pathogen *Photorhabdus* luminescens, does not emit light in a cell density-dependent fashion (8). This result supports the proposal that clear biochemical differences must exist between these two closely related pathogens in their regulatory systems (8,9).

Acknowledgment

Supported by a grant from the Medical Research Council of Canada (MT 7672).

References

1. Meighen E. Genetics of bacterial bioluminescence. Annu Rev Genet 1994;28:117-39.
2. Bassler BL, Wright LM, Silverman MR. Multiple signalling systems controlling expression of luminescence in *Vibrio harveyi*: sequence and function of genes encoding a second sensory pathway. Mol Microbiol 1994;13:273-86.
3. Gilson L, Kuo A, Dunlap PV. AinS and a new family of autoinducer synthesis proteins. J Bacteriol 1995;177:6946-51.
4. Fuqua WC, Winans SC, Greenberg EP. Minireview: quorum sensing in bacteria: the LuxR-LuxI family of cell density-responsive transcriptional regulators. J Bacteriol 1994; 176:269-75.
5. Maxwell PW, Chen G, Webster JM, Dunphy GB. Stability and activities of antibiotics produced during infection of the insect *Galleria mellonella* by two isolates of *Xenorhabdus nematophilus*. Appl Environ Microbiol 1994;60:715-21.
6. Greenberg EP, Hastings JW, Ulitzur S. Induction of luciferase synthesis in *Beneckea harveyi* in other marine bacteria. Arch Microbiol 1979;120:87-91.
7. Hada HS, Stemmler J, Grossbard ML, West PA, Potrikus CJ, Hastings JW et al. Characterization of non-O1 serovar *Vibrio cholerae*(*Vibrio albensis*). System Appl Microbiol 1985;6:203-9.
8. Forst S, Nealson K. Molecular biology of the symbiotic-pathogenic bacteria *Xenorhabdus* spp. and *Photorhabdus* spp. Microbiol Rev 1996;60:21-43.
9. Dunphy GB. Physicochemical properties and surface components of *Photorhabdus luminescens* influencing bacterial interaction with non-self response systems of non-immune *Galleria mellonella* larvae. J Invertebr Pathol 1995;65:25-34.

GENETIC STUDY OF CHAPERONIN-BACTERIAL LUCIFERASE INTERACTIONS

A.P. Escher, G. Bagi, and A.A. Szalay
Center for Molecular Biology and Gene therapy,
Department of Microbiology and Molecular Genetics,
School of Medicine
Loma Linda University, Loma Linda, CA 92350, USA

Introduction

Molecular chaperonins are essential proteins required for the productive folding of many prokaryotic and eukaryotic proteins *in vivo*. The most studied and best characterized chaperonins are the GroEL (MW:57kDa) and GroES (MW:10kDa) chaperonins of *Escherichia coli*. GroEL chaperonin is functional as a 14mer and directly interacts with a folding polypeptide. GroES chaperonin is functional as a 7mer and interacts with GroEL, coordinating the ATPase activity and thus the conformational changes of the GroEL 14mer during mediation of protein folding. The mechanism of GroESL-mediated protein folding is not known.

GroEL chaperonin from *E. coli* interacts with the LuxAB bacterial luciferase from *Vibrio harveyi* MAV (B392) *in vitro* and *in vivo*. GroEL chaperonin interacts with the molten globule state of separately synthesized LuxA and LuxB subunits *in vitro*, but not with native LuxAB luciferase (1). Interaction between MAV LuxAB luciferase and GroEL can also be observed in *E. coli* at 42°C, a temperature at which this luciferase requires both GroEL and GroES to fold (2). In contrast, the LuxAB luciferase from thermotolerant *V. harveyi* CTP5, designated $LuxA_cB_c$, shows decreased interaction with GroEL when compared to LuxAB, and does not require GroE chaperonins to fold at that temperature (2).

Work done *in vivo* with $LuxA_cB$ and $LuxAB_c$ hybrid luciferases indicates that at 42°C in *E.coli* the $LuxA_cB$ hybrid luciferase shows a bioluminescence activity similar to that of the LuxAB luciferase, and that the $LuxAB_c$ luciferase shows a bioluminescence activity similar to that of the $LuxA_cB_c$ luciferase (3). These results suggest that the LuxB subunit may be responsible for the temperature sensitive folding of LuxAB luciferase at 42°C, and thus for the requirement that this luciferase has for GroE mediated folding at that temperature.

Here we report that the amino acid residues that differentiate the $LuxB_c$ subunit from the LuxB subunit can suppress the requirement of LuxAB luciferase for GroE-mediated folding at 42°C *in vivo* when introduced individually in the LuxB subunit. In addition, these amino acid substitutions cause reduced interaction of LuxAB luciferase with GroEL, albeit to different extent.

Materials and Methods

Plasmids construction: A 1.1kb EcoRI-PvuII DNA fragment containing the *luxB* gene from *V. harveyi* MAV was cloned in vector M13mp18. Site-specific mutagenesis was then performed on this fragment according to the method of Kunkel (4). Five oligonucleotides were designed to introduce separately the following amino acid substitutions from $LuxB_c$ CTP5 into the LuxB MAV subunit polypeptide, thereby generating five different LuxB mutant subunits: Ala102 to Val (LuxB-1), Phe163 to Tyr (LuxB-2), Arg193 to Lys (LuxB-3), Ser256 to Pro (LuxB-4), and Gly265 to Val (LuxB-5). In order to reconstruct *luxAB* dicistrons, the genes encoding these different subunits were cloned together with the wild type MAV *luxA* gene into the BamHI site of vector pACYC184 under the transcriptional control of the tetracycline promoter, yielding plasmids pLX403ab-1 to -5.

Bioluminescence assay: *E.coli* HB101 cells were transformed with the five plasmids described above, and with plasmids pLX403ab and $pLX403a_cb_c$ carrying respectively the wild type *luxAB* and $luxA_cB_c$ genes. These cells were also cotransformed with vector pT7/T3-19 to titer endogenous GroESL chaperonins through over expression of the beta-lactamase gene. This step is necessary to show that a luciferase requires GroESL chaperonins to be functional. Cells were grown at 42°C in liquid LB medium under antibiotics selection (30 ug/mL chloramphenicol and 100 ug/mL ampicillin), and bioluminescence was measured *in vivo* as a function of cell culture density using a luminometer (Turner TD-20e) using 50 uL cell culture aliquot as previously described (2).

Immunoprecipitation of GroEL: *E.coli* HB101 cells transformed with plasmids pLX403ab and pLX403ab-1 to -5 were grown at 42°C as liquid cultures to an $O.D._{600}$ of 1.0, pelleted, and sonicated. GroEL protein present in the lysates was then immunoprecipitated using an anti-GroEL polyclonal antibody (from Epicentre Technologies) and an immunoprecipitation kit (Protein A kit from Boehringer-Mannheim).

Gel separation of proteins and immunoblot analysis of luciferase proteins: GroEL-immunoprecipitated proteins as well as total cell extracts were separated on a gradient SDS-polyacrylamide gel (7.5%-20%). Proteins were then transferred onto a PVDF membrane. The membrane was reacted first with polyclonal anti-LuxAB primary antibodies, and then with a monoclonal anti-rabbit anti-IgG peroxidase conjugate as secondary antibody. Chemiluminescence was used to detect presence of luciferase proteins on an x-ray film.

Results and Discussion

Amino acid substitutions in the LuxB subunit affect the requirement of LuxAB luciferase for GroE-mediated folding: It has been shown previously that an *E.coli* cell culture density vs. bioluminescence activity curve with slope zero or less indicates the requirement of a luciferase for GroE-mediated folding (2). Figure 1 shows bioluminescence activities of *E.coli* cells grown at 42°C and expressing wild type *luxAB* and $luxA_cB_c$ genes, as well as the five mutant luciferase gene constructs *luxAB-1* to *-5*. The data indicate that only LuxAB luciferase requires GroESL chaperonins for folding, and that luciferases $LuxA_cB_c$, and LuxAB-1 to -5 do not.

Thus each of the five amino acid residues that characterizes the $LuxB_c$ subunit of CTP5 luciferase alleviates the requirement of LuxAB MAV luciferase for GroE-mediated folding at 42°C when substituted into the LuxB MAV subunit. The significance of such a redundancy is not clear at the present time. However, since each of the five amino acid substitutions also causes decreased interactions with GroEL chaperonin

(Figure 2), it is possible that in thermotolerant *V. harveyi* there is a selective advantage

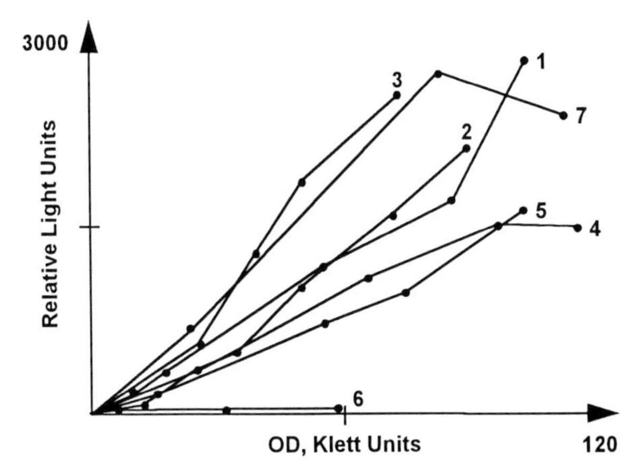

Figure 1. Bioluminescence activity as a function of cell culture density of *E.coli* cells carrying different luciferase gene constructs. Cells were grown at 42°C and were expressing the following luciferase genes: 1, luxAB-1, 2, luxAB-2, 3, luxAB-3, 4, luxAB-4, 5, luxAB-5, 6, luxAB, 7, luxA$_c$B$_c$.

to the loss of interaction between luciferase and chaperonins at elevated temperatures. This advantage could be a more rapid folding of luciferase, or an increase in the levels of chaperonins available for other proteins.

Amino acid substitutions in the LuxB subunit affect interactions between LuxAB luciferase and GroEL chaperonin: Figure 2A shows that LuxAB luciferase has the strongest interaction with GroEL (lane 2). Each amino acid residue characteristic of the LuxB$_c$ CTP5 subunit causes decreased interaction of LuxAB MAV luciferase with GroEL chaperonin at 42°C when introduced into the LuxB MAV subunit (Fig.2A, lanes 3 to 7).

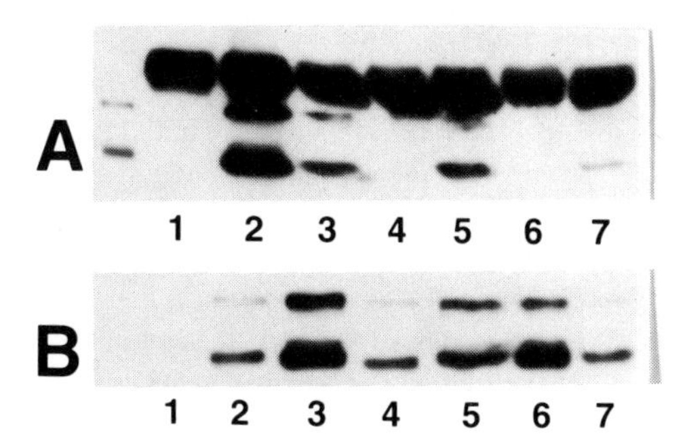

Figure 2. Luciferase immunoblot analysis of GroEL immunoprecipitates (A) and total extracts (B) from lysates of *E.coli* cell expressing different luciferase constructs at 42°C. Cells were transformed with: lane 1, vector only; lane 2, pLX403ab; lane 3, pLX403ab-1; lane 4, pLX403ab-2, lane 5, pLX403ab-3; lane 6, pLX403ab-4; lane 7, pLX403ab-5. The unmarked lane in A contains isolated LuxAB luciferase as a control.

These reduced interactions are not due to lower amounts of synthesized mutant luciferases (Fig. 2B), and thus reflect decreased binding affinities between luciferase enzymes and GroEL protein.

Proportional decreased amounts of the LuxA subunit are coimmunoprecipitated with the different LuxB subunits in each case (Fig.2A). These results indicate that GroEL interacts with already assembled LuxAB luciferase *in vivo*, and also indicate that GroEL does not show significant interactions with free LuxA subunit *in vivo*.

There are significant differences in the amounts of luciferase bound by GroEL, depending on the amino acid substitution introduced in the LuxB subunit. Three substitutions in particular cause a drastic decrease in GroEL-bound luciferase: Phe163 to Tyr, Ser256 to Pro, and Gly265 to Val (Fig2A, lane 4, 6, and 7, respectively). There is no evidence presented here to suggest that these positions are within binding sites for GroEL chaperonin. Since these substitutions also suppress the requirement for GroE-mediated luciferase folding, it is possible that the decreased binding of the luciferase mutants to GroEL is as a result of GroEL not recognizing some feature of a misfolding luciferase present in another domain. However, data obtained with a LuxA-LuxB luciferase fusion , referred to as Fab9, suggest that positions 163 and 265 in the LuxB subunit may be within binding sites for GroEL. Fab9 luciferase exhibits temperature sensitive folding at 37°C (5), and requires GroESL chaperonins to fold *in vivo* at that temperature (2). When the amino acid substitutions described here are introduced individually into the LuxB moiety of Fab9, substituting positions 528 and 630 (equivalent to positions 163 and 265 in the LuxB subunit) prevent GroE-mediated folding of Fab9 luciferase (Escher and Szalay, unpublished data). These results suggest a direct correlation between decreased binding of GroEL to luciferase, and the decreased ability of GroEL to mediate luciferase folding.

In the future, identification of specific positions within luciferase polypeptides that affect productive binding by GroEL will allow us to use site-specific mutagenesis to vary GroEL-luciferase binding affinity, and to study the effects of these changes on productive luciferase folding directly *in vivo*.

References

1. Flynn GC, Beckers CJM, Baase WA, Dahlquist FW. Individual subunits of bacterial luciferase are molten globules and interact with molecular chaperones. Proc. Natl. Acad. Sci. USA 1993; 90:10826-30.
2. Escher A, Szalay AA. GroE mediated folding of bacterial luciferases *in vivo*. Mol. Gen. Genet. 1993; 238:65-78.
3. Escher A, O'Kane DJ, Lee J, Szalay AA. The beta subunit polypeptide of *Vibrio harveyi* luciferase determines light emission at 42°C. Mol. Gen. Genet. 1991; 230:385-93.
4. Kunkel TA. Rapid and efficient site-specific mutagenesis without phenotypic function. Proc. Natl. Acad. Sci. USA 1985; 82: 488-92.
5. Escher A, O'Kane DJ, Lee J, Szalay AA. Bacterial luciferase alpha-beta fusion protein is fully active as a monomer and highly sensitive in vivo to elevated temperature. Proc. Natl. Acad. Sci. USA 1989; 86:6528-32.

MUTATION OF THE PROXIMAL LUXR BINDING SITE ON THE *VIBRIO HARVEYI* LUX OPERON

YH Lin and EA Meighen
Dept of Biochemistry, McGill University, Montreal, Quebec, H3G 1Y6, Canada

Introduction

The LuxR regulatory protein from *Vibrio harveyi* has been implicated directly in the activation of the promoter for the *luxCDABEGH* operon although its exact role is not yet known (1,2). Gene complementation experiments have shown that *luxR* is required for expression of luminescence in *V. harveyi* and to enable high light production in *Escherichia coli* containing the *lux* operon under control of its own promoter. Previous primer extension experiments on the *lux* operon promoter in *E.coli* showed that the transcription start site switched from 123 bp to 26 bp upstream of the ATG initiation start codon for *luxC* on the addition of LuxR (3). In addition, the *lux* mRNA level increased 250-fold suggesting that the -26 bp site is a strong promoter while the -123 bp site is relatively weak.

Footprint analyses of the binding of LuxR to the lux operator/promoter demonstrated that the LuxR protein bound to two sites centered about 120 bp (proximal B site) and 250 bp (distal A site) upstream of the translational start site of *luxC* (4). Partial deletion of the DNA upstream of the *lux* operon which was linked to the chloramphenicol actyltransferase gene (cat) followed by transconjugation into *V.* harveyi showed that the proximal B site is critical for maximum activation of the *lux* promoter by LuxR (5).

To map the bases involved in the control of the *lux* operon promoter, random mutagenesis of the proximal B site was performed by PCR (6). Mutants were selected by nucleotide sequencing followed by measurement of luminescence in *E.coli* complemented with *luxR* to determine the effects of specific base changes on the activity of the promoter.

Materials and Methods

Random PCR Mutagenesis: Mutations within the *lux*C promoter were introduced by PCR (Polymerase Chain Reaction). The template was plasmid SSpII322B which carries a wild type *lux*C promoter in front of the reporter luxAB gene, with a unique BstBI site installed between the A and B binding sites (- 200 bp) by site directed mutagenesis. The reaction mixture contained 10 ng of template plasmid DNA, 50 pmol each of primers flanking the translation start site (5'GTCCATTTACTATTAAAGGTAAGTGTTTTTC 3') and the BstBI site (5' CCGCTAGTGTTTAATAGCGCTGAT 3) ', 25 mmol/L TAPS (pH 9.3), 2 $MgCl_2$ mmol/L, 50 mmol/L KCl, 1 mmol/L DTT, 0.5 mg/mL bovine serum albumin, 0.2 mM of each dNTP and 1 unit of Taq DNA polymerase. Thirty cycles of amplification were performed (94 °C for 1 min, 55 °C for 1 min, 72 °C for 30 sec). In order to increase the frequency of miscorporated bases, PCR amplifications were performed by limiting one of the of the four dNTPs to 14 μmol/L, Mg^{2+} was increased to 5 mmol/L and 200 μmol/L dITP was added (6).

Subcloning and selection of mutated PCR fragments: Amplified fragments were digested at PacI and BstBI restriction sites upstream and downstream of the LuxR B binding site. The fragments (120 bp) were then ligated into the original plasmid, SSPII322B, previously cut by the same enzymes. The resulting ligation mixtures were then cotransformed with plasmid pMGM100Rcat (which expresses the luxR gene) into *E.coli* MM294 and individual colonies were analyzed by nucleotide sequencing of the region between the PacI and BstBI sites. The luminescence of the colonies containing the mutated plasmids was followed during growth.

Results and Discussion:

Random mutagenesis of the 120 bp region encompassing the proximal B binding site for LuxR was accomplished by incorporation of the wrong basesduring PCR by limiting one of the dNTPs and adding dITP as well as increasing the Mg^{2+} concentration. Among the 120 clones sequenced, 40 contained at least one substitution in the 120 bp fragment encompassing the LuxR proximal binding site. Eighteen of these clones were mutated in the LuxR binding site as depicted in Fig 1. Most of the mutants were A to G substitutions. Fifteen of the eighteen mutants were single point substitutions while three of the mutants (A167G, A137G, and T133C) also had an additional substitution outside the LuxR proximal binding site but within the mutated 120 bp fragment between the BstBI and PacI sites.

Expression of the *luxC* promoter with the mutated LuxR binding site was accomplished by using the *luxAB* gene as a reporter and measuring the light intensity on addition of decanal to *E.coli* containing the mutated construct and LuxR. Average light intensity during the exponential phase of growth (OD_{660}= 0.5 to 1.0) was then compared to that for the wild type promoter in the same construct. Sixteen of the eighteen mutants including the three double mutants

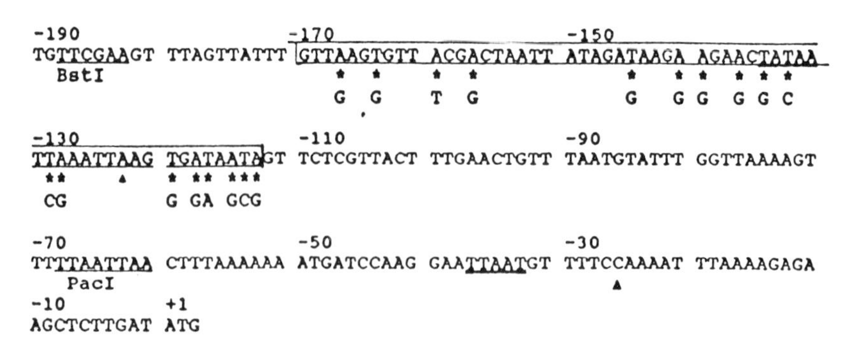

Fig 1. Location of the mutations in the LuxR proximal binding site (boxed). The base changes are shown under the sequence. - 26(△) bp and -123 bp (▲)transcriptional start sites are indicated, with the Pribnow boxes bold-underlined. The first nucleotide of the *luxC* translational start site is designated as +1 and the PacI and BstI restriction sites are underlined.

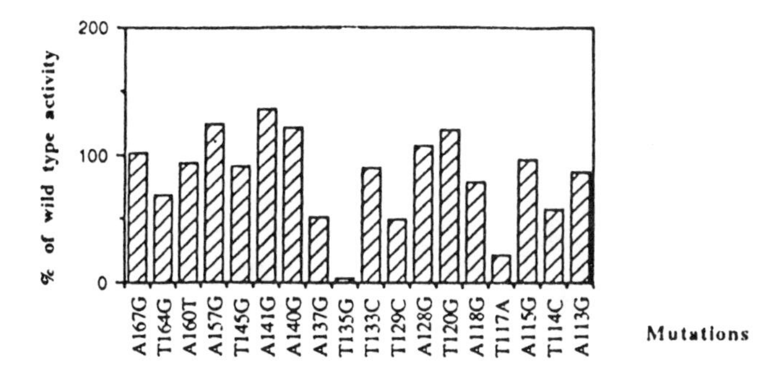

Fig 2. Luminescence intensities of mutants in the proximal LuxR binding site. Luminescence was measured after growing *E.coli*(*luxR*) with the mutant constructs to $O.D._{660}$ of 0.5 and 1.0. The ratio of the luminescence activities of the mutant to that of the wild type were averaged and are shown in the vertical bar graph. Position of the mutation as well as base changes (e.g. A137G = change from A to G at position -137) are indicated at the bottom.

exhibited luminescence intensities comparable to that of the wild type (>40% activity) (Fig 2.). However, prelimary results have indicated that some of these mutants active in *E.coli* showed much lower activities than the wild type construct on transconjugation into luciferase-deficient dark *V. harveyi*. One possible explanation for this result is that the high level of LuxR synthesized in *E.coli* due to the multiple copies of the *luxR* expression vector might obscure differences in expression between the mutant and wild type constructs.

Two mutants, T117A and T135G are expressed at much lower levels than the wild type construct even in *E.coli* (+*luxR*). The construct with mutation T117A has about 20% the activity of the wild type construct while mutation T135C results in only 5% of the luminescence intensity of the wild type construct. Interestingly, the mutant with the lowest expression was mutated in the Pribnow box of the weak promoter site (-123 bp site) used by RNA polymerase (RNAP) in the absence of LuxR (Fig 3a), changing it from a TATAA to a GATAA sequence (Fig 2). One possibility to explain this result is that RNAP may still interact at the -123 bp site but can not initiate transcription in the presence of LuxR and would then scan downstream to the strong -26 bp site before transcription could be initiated as shown in the model in Fig 3b. Mutation of position 135 would prevent binding of RNAP to this site and block the scanning process to initiate at the -26 bp start site. Alternatively, binding of LuxR to the proximal B site could activate RNAP in the downstream strong promoter site (e.g. by protein-protein interaction) as shown in Fig 3c. Mutation at the -135 bp position may prevent binding of LuxR in an effective manner to activate initiation by RNAP at the strong -26 bp promoter. The loss of activity on mutation of position 117 could also occurr by either preventing entry of RNAP or binding of LuxR.

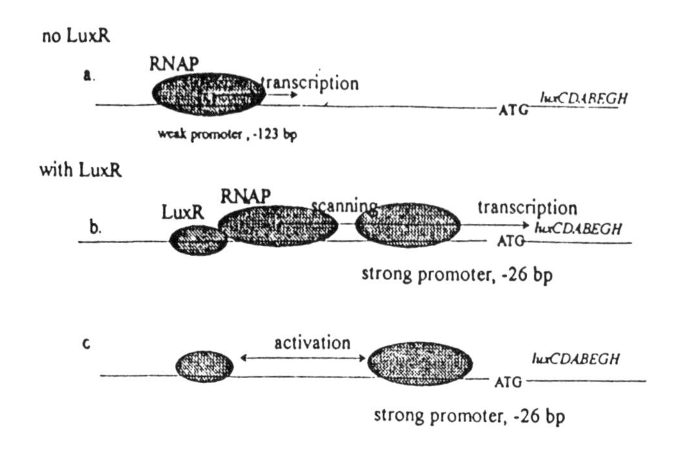

Fig 3. Models illustrating the possible mechanisms of LuxR and RNA polymerase(RNAP) interaction in the *lux* promoter region.(a) In the absence of LuxR, RNAP uses the -123 bp weak promoter site to start transcription. Two models are suggested in the presence of LuxR; (b) RNAP interacts at the -123 bp site and scans downstream to initiate at the -26 bp site. (c) LuxR blocks the binding of RNAP in the -123 bp site and the -26 bp site is then chosen as the transcription site.

Acknowledgments

Supported by a grant (MT 7672) from the Medical Research Council of Canada.

References

1. Martin M, Showalter RE, Silverman M. Identification of a locus controlling expression of luminescence genes in *Vibrio harveyi*. J Bacteriol 1989;171: 2406-14.
2. Showalter RE, Martin M, Silverman M. Cloning and nucleotide sequence of luxR, a regulatory gene controlling bioluminescence in *Vibrio harveyi*. J Bacteriol 1990;172: 2946-54.
3. Swarztman, E, Silverman, M, Meighen, EA. The luxR gene product of *Vibrio harveyi* as a transcriptional activator of the lux promoter. J Bacteriol 1992; 174: 7490-3.
4. Swarztman E, Meighen EA. Purification and characterization of a poly (dA- dT) lux-specific DNA binding protein from *Vibrio harveyi* and identification of LuxR. J Biol Chem 1993;268: 16706-16.
5. Miyamoto CM, Smith EE, Swarztman E, Cao JG, Graham AF, Meighen EA. Proximal and distal sites bind LuxR independently and activate expression of the *Vibrio harveyi* lux operon. Mol Microbiol 1994;14: 255-62.
6. Spee JH, de Vos WM, Kuipers O. Efficient random mutagenesis method with adjustable mutation frequency by use of PCR and dITP. Nucl Acids Res 1993; 21: 777-8.

LUX AUTOINDUCER RESPONSE IN *E. COLI* AND *VIBRIO FISCHERI*

MD Thomas,[1] *JB Schineller*[2] *and TO Baldwin*[1]
[1]*Department of Biochemistry and Biophysics and Center for Macromolecular Design, Texas A&M University, College Station, Texas 77843-2128 and* [2]*Department of Chemistry, Humboldt State University, Arcata CA 95521, U. S. A.*

Introduction

N-acyl homoserine lactones (*N*-acyl HSLs) function as intracellular signalling molecules to control various population-dependent functions in a number of bacteria (1). However, *N*-acyl HSLs can also act as intercellular signalling molecules. Multiple *N*-acyl HSLs have been found in cells of *Vibrio fischeri* (2), *Pseudomonas aeruginosa* (3), and *Agrobacterium tumefaciens* (4). Activity of the primary autoinducer of *V. fischeri*, *N*-(3-oxohexanoyl) HSL, is inhibited by a second autoinducer, *N*-octanoyl HSL, which by itself activates the *lux* genes to a limited extent (5, 6). The molecules can also affect other bacteria within a microbial community. The *N*-acyl HSL produced by *Rhizobium leguminosarum* acts as a bacteriocin to inhibit the growth of closely related bacteria (7). The mechanism of autoinducer recognition and specificity is crucial to our understanding of bacterial cell-cell communication.

Autoinduction has been best studied in *Vibrio fischeri*. The autoinducer is produced by the *luxI* gene product and interacts with the N-terminal domain of the *trans*-acting factor LuxR to autoinduce *luxI* and *luxR*, along with inducing genes for bioluminescence. However, neither the molecular interaction nor the mechanism for recognition specificity of the *N*-acyl HSL autoinducer molecule with its cognate *trans*-acting factor have yet been fully elucidated. *Escherichia coli* has been used as a model to investigate genetic control of the bioluminescence *lux* regulon of *V. fischeri* as expression of these genes was thought to be identical in the two different bacteria. Sitnikov *et al.* (8) reported that *N*-(3-hydroxybutanoyl) HSL was without effect on induction of bioluminescence in *V. fischeri* but activated transcription in *E. coli*, and that *N*-decanoyl HSL, which inhibited autoinduction in *V. fischeri*, activated transcription in *E. coli*. Consequently, they proposed that there may be a receptor molecule that mediates the interaction between the autoinducer and the transcription factor LuxR. Schaefer *et al.* (9) did not confirm these findings and suggested that, while there were subtle differences in autoinducer responses in the two organisms, the original model of a direct autoinducer-LuxR interaction was more reasonable. In an effort to clarify this issue, we investigated autoinducer responses in a naturally occurring dim mutant of *V. fischeri* (B61) and a *luxI* reporter (pJHD500) in *E. coli.*

Materials and Methods

<u>Bacterial culture</u> *Vibrio fischeri* B61 and *E. coli* IGC1 [LE392 containing pJHD500, a pBR322 derivative with *luxR*•*lux* control region•truncated *luxI* from *V. fischeri* and *luxAB* from *V. harveyi* downstream of *luxI* (10)] were used. *Vibrio fischeri* was grown at 22°C in NaCl complete medium (11). IGC1 was grown at 37°C (reduced to 27°C when monitoring luciferase expression) in complete broth (10 g peptone, 5 g yeast extract, 3g K_2HPO_4 per L) with 50 µg/mL carbenicillin. *N*-3-oxohexanoyl-L-HSL and *N*-3-oxohexanoyl-DL-HSL were obtained from Sigma (St. Louis MO), *N*-3-oxohexanoyl-D-HSL was provided by A. Eberhard. *N*-(3-hydroxybutanoyl) HSL, *N*-nonanoyl HSL, and *N*-decanoyl HSL were provided by J. Schineller.
<u>Procedures</u> The bioluminescence assay of Eberhard *et al.* (5, 12) was used with slight

modifications. *N*-acyl HSLs were solublized in ethanol, added to microtitre plates, and dried under vacuum. Bacterial suspensions were added to the wells of a 96 well microtitre plate containing appropriate treatments. In some experiments the microtitre plates were incubated with shaking and 5 μL aliquots of cells were transferred at various times to a new microtitre plate for measurement of luminescence on a ML3000 Microtiter® Plate Luminometer (Dynex Laboratories Inc., Chantilly, VA, USA) using integrate flash mode and read for 4 s following a 1 s delay after injection of decanal (final concentration 0.05% in 200 μL). In other experiments the microtitre plates were incubated in the ML3000 with shaking and light was monitored continuously, with each well being measured using the cycle mode for 0.04 s/min.

Results and Discussion

The response to the *N*-(3-oxohexanoyl) HSL autoinducer was stereospecific in both *V. fischeri* and *E. coli*. The L isomer of *N*-(3-oxohexanoyl) HSL was approximately twice as active as the LD preparation. In order to further examine the activity of the D isomer, we obtained a preparation of *N*-(3-oxohexanoyl)-D-HSL. This preparation was estimated to be 90% D isomer and 10% L isomer (A. Eberhard, personal communication). In *V. fischeri* B61, as little as 1 nmol/L L isomer or 10 nmol/L D isomer gave a detectable response (both induced a 12-fold increase in expression after 7 h). Maximal induction was observed at 100 nmol/L L isomer and 1 μmol/L D isomer at which there was a 10,000-fold increase in luciferase expression. Strain IGC1 of *E. coli* was slightly less sensitive to the autoinducer: 1.7 nmol/L (L isomer) and 50 nmol/L (D isomer) gave an increase in expression, while maximum expression was observed at 170 nmol/L (L isomer) and 1.7 μmol/L (D isomer). The slopes of the response at intermediate concentrations on a log plot were similar for both isomers in each organism. *N*-(3-oxohexanoyl)-D-HSL did not appear to inhibit or compete with the activity of the L isomer in either *V. fischeri* or *E. coli* as the slopes of the response were equal at subsaturating autoinducer concentrations and luciferase expression levels were the same at greater concentrations of either the L or D isomer in each organism.

N-(3-hydroxybutanoyl) HSL is the autoinducer produced by *Vibrio harveyi*. We found that the response to *N*-(3-hydroxybutanoyl) HSL was similar in both *V. fischeri* and *E. coli*, i.e., it did not induce expression of the *lux* operon over a concentration range of 200 nmol/L to 2 mM, while it inhibited induction mediated by the *V. fischeri* autoinducer (Figure 1). Our findings do not confirm the observations of Sitnikov *et al.* (8) with respect to *N*-(3-hydroxybutanoyl) HSL inducing *lux* gene expression in *E. coli*. Other studies did not examine the influence of *N*-(3-hydroxybutanoyl) HSL on either *V. fischeri* or *E. coli* (5, 9).

In previous studies, *N*-decanoyl HSL was found to *i*) inhibit the binding of *N*-(3-oxohexanoyl) HSL to *E. coli* cells containing LuxR, *ii*) not activate *V. fischeri lux* genes at concentrations as high as 200 nM, and *iii*) inhibit *N*-(3-oxohexanoyl) HSL-mediated luminescence in *E. coli* (9). There was less inhibition of *N*-(3-oxohexanoyl) HSL-mediated luminescence by *N*-decanoyl HSL in *E. coli* than was observed in *V. fischeri* (5, 9). Our results confirm these findings, and we observed no activation of the *lux* genes in *E. coli* at concentrations as high as 20 μmol/L *N*-decanoyl HSL. At higher concentrations, we observed an induction of bioluminescence, thus confirming the observations of Sitnikov *et al.* (8). However, the maximum level of expression that was observed with *N*-decanoyl HSL was only 7% that mediated by *N*-(3-oxohexanoyl) HSL (Figure 1D). Direct comparison of our work with previous studies is difficult as different plasmids were used and assay conditions influence the magnitude of the response. *N*-decanoyl HSL activation of the *lux* genes may not be biologically significant as the required concentrations are unusually high; however, it may be

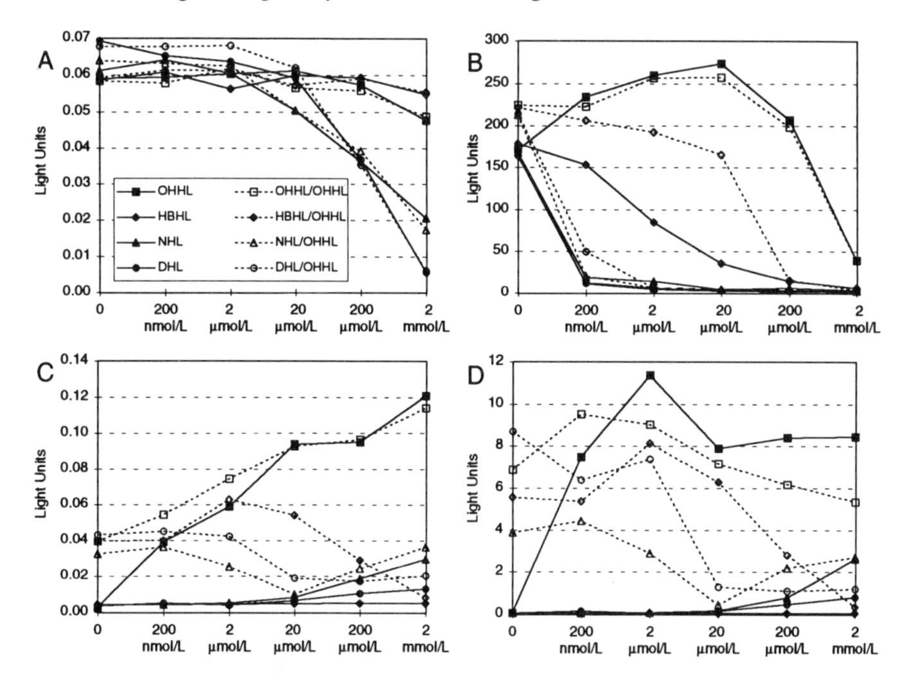

Figure 1. Response of *V. fischeri* and *E. coli* to various *N*-acyl HSLs. A. *V. fischeri* B61 at 5 min after addition of *N*-acyl HSLs. B. Cultures in A after 6 h. C. *E. coli* IGC1 at 1 h after addition of *N*-acyl HSLs. D. Cultures in C after 6 h. OHHL = *N*-(3-oxohexanoyl) HSL; HBHL = *N*-(3-hydroxybutanoyl) HSL; NHL = *N*-nonanoyl HSL; DHL = *N*-decanoyl HSL. Data points are the average of 2 replications; the experiment was repeated 4 times with similar results.

mechanistically significant if *N*-decanoyl HSL interacts differently with the *lux* system in *E. coli* than in *V. fischeri*. Further work is required to verify that the observed response is not due to minor impurities contaminating the *N*-decanol HSL or due to a general cellular response unrelated to the *lux* system that is mediated by the unusually high concentrations.

N-nonanoyl HSL was more effective at activating the *lux* genes in *E. coli* than *N*-decanoyl HSL. We observed no activation of the *lux* genes in *V. fischeri* with either compound, and they were potent inhibitors of *N*-(3-oxohexanoyl) HSL-mediated bioluminescence (Figure 1B). In preliminary experiments, it appeared that *N*-nonanoyl HSL and *N*-decanoyl HSL not only inhibited induction of the *lux* genes in *V. fischeri*, but they also blocked basal expression. The concentration of B61 cells in the experiment reported in Figure 1 was adjusted to yield a measurable amount of light before addition of the *N*-acyl HSLs. Normally, a lower concentration of cells is used in order to minimize light produced due to basal expression. Both *N*-nonanoyl HSL and *N*-decanoyl HSL inhibited basal expression of the *lux* genes (Figure 1B); however, they also inhibited bioluminescence due to preformed *lux* gene products (Figure 1A). There was a decrease in bioluminescence within 45 s upon addition of the cells of B61 to wells containing *N*-nonanoyl HSL or *N*-decanoyl HSL. Bioluminescence in the untreated samples had stabilized by 5 min after the addition of the *N*-acyl HSLs, and at that time *N*-nonanoyl HSL and *N*-decanoyl HSL inhibited bioluminescence in a dosage-dependent manner with at least 60% inhibition at 2 mM. While this immediate inhibition of bioluminescence may suggest interruption of general cellular functions, no growth inhibition was observed. It is possible that longer chain *N*-acyl HSLs compete with the aldehyde substrate of luciferase.

The differences in the response of *V. fischeri* and *E. coli* to *N*-acyl HSLs suggest that the interaction between the autoinducer and the response-regulator *trans*-acting factor is more complex that a simple binding phenomenon. This conjecture is also supported by the lack of observed binding between *N*-(3-oxohexanoyl) HSL and LuxR. However, we have no evidence of a intermediary receptor protein or other factors to date.

Acknowledgments
We wish to thank A. Eberhard for supplying *N*-(3-oxohexanoyl)-D-HSL. This research was supported by the Texas Advanced Technology Program and Amgen, Inc.

References
1. Salmond GP, Bycroft BW, Stewart GS, Williams P. The bacterial 'enigma': cracking the code of cell-cell communication. Mol Microbiol 1995;16:615-24.
2. Kuo A, Blough NV, Dunlap PV. Multiple N-acyl-L-homoserine lactone autoinducers of luminescence in the marine symbiotic bacterium Vibrio fischeri. J Bacteriol 1994;176:7558-65.
3. Pearson JP, Passador L, Iglewski BH, Greenberg EP. A second N-acylhomoserine lactone signal produced by Pseudomonas aeruginosa. Proc Natl Acad Sci USA 1995;92:1490-4.
4. Zhang L, Murphy PJ, Kerr A, Tate ME. Agrobacterium conjugation and gene regulation by N-acyl-L-homoserine lactones. Nature 1993;362:446-8.
5. Eberhard A, Widrig CA, McBath P, Schineller JB. Analogs of the autoinducer of bioluminescence in Vibrio fischeri. Arch Microbiol 1986;146:35-40.
6. Kuo A, Callahan SM, Dunlap PV. Modulation of luminescence operon expression by N-octanoyl-L-homoserine lactone in ainS mutants of Vibrio fischeri. J Bacteriol 1996;178:971-6.
7 Schripsema J, de Rudder KE, van Vliet TB, Lankhorst PP, de Vroom E, Kijne JW, van Brussel AA. Bacteriocin small of Rhizobium leguminosarum belongs to the class of N-acyl-L-homoserine lactone molecules, known as autoinducers and as quorum sensing co-transcription factors. J Bacteriol 1996;178:366-71.
8. Sitnikov DM, Schineller JB, Baldwin TO. Transcriptional regulation of bioluminescence gene from Vibrio fischeri. Mol Microbiol 1995;17:801-12.
9. Schaefer AL, Hanzelka BL, Eberhard A, Greenberg EP. Quorum sensing in Vibrio fischeri: probing autoinducer-LuxR interactions with autoinducer analogs. J Bacteriol 1996;178:2897-901.
10. Devine JH, Shadel GS, Baldwin TO. Identification of the operator of the lux regulon from the Vibrio fischeri strain ATCC7744. Proc Natl Acad Sci USA 1989;86:5688-92.
11. Hastings JW, Baldwin TO, Nicoli MZ. Bacterial luciferase: assay, purification, and properties. Meth Enzymol 1978;57:135-52.
12. Eberhard A, Burlingame AL, Eberhard C, Kenyon GL, Nealson KH, Oppenheimer. Structural identification of autoinducer of Photobacterium fischeri. Biochemistry 1981;20:2444-9.

TRANSCRIPT INITIATION IN THE *LUX* OPERONS OF *VIBRIO FISCHERI*

AB van Tilburg, MD Thomas, and TO Baldwin
Center for Macromolecular Design and the Department of Biochemistry and Biophysics, Texas A&M University, College Station, Texas 77843-2128, U. S. A.

Introduction

Genetic regulation of the *Vibrio fischeri lux* regulon has been extensively studied in a number of laboratories. It is known that the divergently-transcribed operons share a common control region containing cAMP-cyclic AMP response protein binding motifs and the *lux* box which interacts with LuxR, the response-regulator protein. Although mutational analysis has identified domains on LuxR that interact with either the autoinducer (3–oxohexanoyl homoserine lactone) or the *lux* box, neither direct binding of the autoinducer to LuxR nor binding of the autoinducer-LuxR complex to *cis*-elements within the control region has been demonstrated.

Earlier work (1, 2) revealed that transcription of both the left and right *lux* operons can be initiated from several different sites. This may be due to the unusual structure of the control region. There are a number of motifs within this region that approximate the consensus sequences of the -10 and -35 signal elements, but only with respect to the right operon are two of these hexamers appropriately positioned. Furthermore, the *lux* box is immediately upstream of the putative -35 hexamer for the right operon and shares one nucleotide with it (2). The lack of definitive signal elements for the left operon and positioning of the *lux* box with respect to the right operon may lead to some imprecision in the binding of RNA polymerase, resulting in the multiple transcription start sites observed.

In the previous studies, the investigators isolated RNA from one to three strains of *E. coli* containing various plasmids carrying *lux* genes. In an attempt to better understand the three-dimensional topology of the transcriptional complexes, we have expanded on the earlier work and reevaluated transcript initiation of the right and left *lux* operons by isolating RNA from four strains of *E. coli* transformed with one of nine recombinant plasmids and from three strains of *V. fischeri*.

Materials and Methods

Bacterial strains and culture The following strains were used: *V. fischeri* MJ-1, B61, and ATCC7744; *E. coli* IGC1 [LE392 + pJHD500, a pBR322 derivative containing *luxR•lux* control region•truncated *luxI* from *V. fischeri* and *luxAB* from *V. harveyi* downstream of *luxI* (3)]; IGC50 [LE392 + pGS138, which contains *V. fischeri luxR•lux* control region•truncated *luxI* with *V. harveyi luxAB* downstream of *luxR* (2)]; IGC60 (LE392 + pVFC, which contains the entire ATCC7744 *lux* regulon), IGC87 (TB1/pJHD500), IGC88 (*ΔcyaA854*/pJHD500), IGC89 (*ΔcyaA854 Δcrp-45*/pJHD500). Other plasmids used are described in Table 1.

Cultures of *V. fischeri* were incubated at 22°C in salt medium (4). The *E. coli* strains were grown at 37°C (reduced to 27°C when monitoring luciferase expression) in complete broth (CB, 10 g peptone, 5 g yeast extract, 3g K_2HPO_4 per L) supplemented with 50 µg/mL carbenicillin. *N*-3-oxohexanoyl-L-homoserine lactone (autoinducer) was obtained from Sigma (St. Louis MO).

Determination of luminescence Expression studies utilized overnight cultures diluted 100-fold into fresh medium and allowed to grow for 2 h. After another 100-fold dilution, 200 µL of the bacterial suspension was added to the wells of a 96-well microtitre plate. The microtitre plate was shaken for 2 h, at which time autoinducer was added to appropriate wells at a final concentration of 100 nmol/L. After a further 2 h of shaking 10 µL from each well was transferred to a clean microtitre plate for measurement of luminescence on a ML3000 Microtiter® Plate Luminometer (Dynex Laboratories Inc., Chantilly, VA, USA). Light was measured for 4 s after a 1 s delay following injection of 90 µL CB and 100 µL 0.1% decanal.

RNA isolation Overnight cultures were centrifuged, resuspended in a small volume of fresh medium, and used to inoculate fresh medium to a final OD_{600} of 1. When appropriate, cAMP (10 mmol/L) was added and after 2 h incubation autoinducer (100 nmol/L) was added. The cultures were shaken an additional 2 h and up to 10 ml of a culture with an OD_{600} of 10 was then centrifuged (7 min, 3000 *g*). The cells were resuspended in 1:1:2 phenol:$CHCl_3$:7 mol/L guanidine HCl. The cell lysate was centrifuged (10 min, 100,000 *g*), the aqueous phase removed to a new tube, and the RNA precipitated in an equal volume of isopropanol. After an additional 7 mol/L guanidine HCl treatment and isopropanol precipitation, the RNA was dissolved in a minimal volume of H_2O and quantitated spectrophotometrically and electrophoretically.

Determination of transcription start site SuperScript™ II Reverse Transcriptase (GibcoBRL, Bethesda MD) was used to prepare first strand cDNA from 2.5 µg total RNA and the following oligonucleotide primers for *luxI* and *luxR*, respectively: CGAAGACTTAGAATACC, CTGTAAGTGTCGTCGGC. The cDNA was labelled with ^{35}S dATP and ^{35}S dCTP (DuPont NEN, Boston MA) incorporated during primer extension. The labelled cDNA was run on a 7 mol/L urea/8% polyacrylamide gel with a sequencing ladder produced using the appropriate primer by the CircumVent™ Thermal Cycle DNA Sequencing Kit (New England Biolabs, Beverly MA).

Results and Discussion

Primer extension using RNA isolated from *V. fischeri* B61, ATCC7744, and MJ1 and from *E. coli* containing various *lux* plasmids revealed multiple transcriptional start sites for the left and right *lux* operons under basal and induced expression (Fig. 1). The large number of distinct transcriptional start sites seems unusual as genes more commonly appear to be transcribed from one or only a few start sites. For example, *luxR* from *V. harveyi* has been shown to have a single transcription start site (5). Very short transcripts appear to be produced by both operons in addition to longer ones. Although there are a number of sequences within the control region that approximate the consensus sequences for -35 and -10 signals, few of these sequences are appropriately positioned. In the right operon, all of the transcriptional start sites are downstream of the putative Pribnow box (959-964) and -35 sequence (937-942) as denoted by Shadel and Baldwin (2). Sequence analysis does not reveal any appropriately positioned -10 and -35 signal sequences in the left operon.

Engebrecht and Silverman (1) used S1 nuclease protection to map the transcription start sites of both the left and right operons, observing a number of start sites in *E. coli* containing *V. fischeri* MJ-1 sequences with and without added autoinducer. However, they attributed the shorter fragments to S1 nuclease digestion of the DNA probes. Shadel and Baldwin (2) also found multiple start sites for the left operon (*luxR*). These transcriptional start sites were grouped in three localized areas, two of which were cAMP-CRP dependent. Our results for the left operon generally confirm these results. In preliminary studies we found that 10 mmol/L cAMP resulted in maximal expression of *cyaA* strains. Consequently, this concentration of cAMP was used in cultures for RNA isolation. We found two groups of start sites that were dependent on cAMP-CRP, and one group (comprising the longer transcripts) that did not appear to be cAMP-CRP dependent. The shortest transcripts that we observed were not detected in the previous study (2), possibly because the primers that were used were closer to the control region than the ones used here, including one that overlapped the region 779 to 796. Start site differences between that study and ours may have resulted from the use of different sequencing templates for the marker ladder.

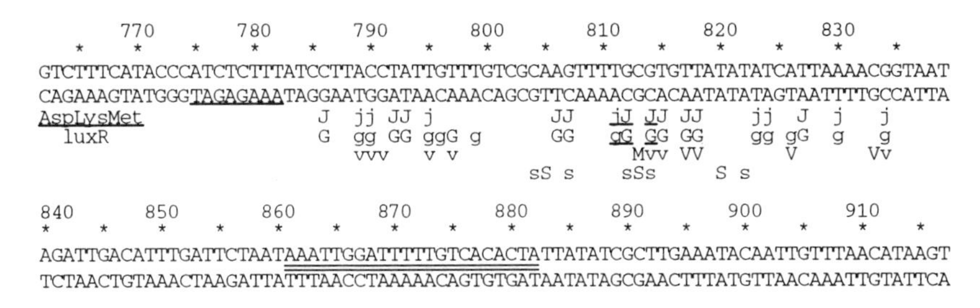

Fig. 1. Location of transcriptional start sites in the *lux* control region. Nucleotide numbering is that of GenBank accession Y00509. Major start sites are upper case letters, minor are lower case, underlined start sites were cAMP-CRP dependent (*luxR*) or LuxR:autoinducer dependent (*luxI*); J = pJHD500, G = pGS138, V = *V. fischeri* B61, MJ-1, ATCC7744, and plasmid pVFC in *E. coli*, M = those reported for MJ-1 (1), S = those reported by Shadel and Baldwin (2). Shine-Dalgarno sequences (6) are underlined, cAMP-CRP binding site identified in (2) is double underlined on one strand, the *lux* box is double underlined on both strands.

The function of the short transcripts is not clear. A number of the major start sites are either within the Shine-Dalgarno sequence or are sufficiently close to it to suggest that these transcripts are not translated. It is conceivable that the short transcripts participate in transcriptional regulation. Multiple start sites in the *E. coli* gene *pyrC* contribute to transcriptional regulation as short transcripts form a stable hairpin and are consequently poorly translated (7). However, we have not been able to identify any sequences within the *lux* control region that would be likely to form a similar hairpin structure. Alternatively, the short transcripts may result from torsional forces due to the simultaneous *versus* sequential transition from closed to open transcriptional complex formation occurring when both operons are being transcribed. To test this possibility, a series of *luxR* deletion mutants was examined. The basal expression levels of these mutants (Table 1) are similar to those observed by Shadel (8). Transcriptional start sites for the right operon were identical among these *luxR* mutants: removal of all or part of the left operon did not alter the start sites observed for *luxI*.

Plasmid	Description	Light Units	Expression
pJHD500	$luxRluxI^*LuxAB_{Vh}$	4.8	1
pJHD505	Δ15-917	150.1	31
pGS580	Δ716-917	4.4	0.9
pGS582	Δ190-917	157.6	33
pGS584	Δ496-593	8.0	1.7
pGS586	Δ496-917	90.6	19
pGS591	Δ190-716	29.0	6.1

Table 1. Expression of *luxR* deletion mutants. Expression is relative to pJHD500; deletions are identified by their approximate nucleotide numbers, numbered as in Fig.1.

We observed fewer transcriptional start sites for the left and right operons in the intact *lux* regulon than in either pJHD500 or pGS138. These start sites did not vary significantly with respect to the cellular background (three different strains of *V. fischeri* or *E. coli*). The start site differences between the intact regulon and pJHD500 or pGS138 suggest that selection of transcription initiation sites for the *lux* operons is influenced more by DNA sequences than by RNA polymerase or *trans-*

acting factors as the DNA sequences were the only factor varied. In addition, *lux* start site selection appears to be influenced by areas outside of the control region as the control region is identical in pVFC, pJHD500 and pGS138. This conjecture is further supported by the observation that removal of all or part of the left operon had no effect upon start site selection for the right operon even though it had a profound effect on basal expression levels (Table 1).

Transcription of the left and right *lux* operons appears to be interrelated and controlled in part by the physical structure of the DNA. The binding of cAMP-CRP bends DNA, and many cAMP-CRP dependent promoters are sensitive to DNA bending, i.e., in-frame poly-A tracts can substitute for binding of cAMP-CRP. cAMP-CRP binding has been shown to induce transcription of one gene while repressing expression from a divergent promoter adjacent to it (9). This phenomenon may be acting in the *lux* control region, resulting in the cAMP-CRP-mediated 5-fold induction of *luxR* and 5-fold repression of *luxI* (2, 10). It is not yet clear in what way the LuxR-autoinducer complex modifies the cAMP-CRP activation to yield induction of both operons under certain conditions and activation of the right operon and repression of the left operon under other conditions (11).

Acknowledgments

Support for this work was provided by the Texas Advanced Technology Program and Amgen, Inc., and is gratefully acknowledged.

References

1. Engebrecht J, Silverman M. Nucleotide sequence of the regulatory locus controlling expression of bacterial genes for bioluminescence. Nuc Acids Res 1987;15:10455-67.
2. Shadel GS, Baldwin TO. Positive autoregulation of the Vibrio fischeri luxR gene. J Biol Chem 1992;267:7696-702.
3. Devine JH, Shadel GS, Baldwin TO. Identification of the operator of the lux regulon from the Vibrio fischeri strain ATCC7744. Proc Natl Acad Sci USA 1989;86:5688-92.
4. Hastings JW, Baldwin TO, Nicoli MZ. Bacterial luciferase: assay, purification, and properties. Meth Enzymol 1978;57:135-52.
5. Miyamoto CM, Chatterjee J, Swartzman E, Szittner R, Meighen EA. The role of the lux autoinducer in regulating luminescence in Vibrio harveyi; control of luxR expression. Mol Microbiol 1996;19:767-75.
6. Devine JH, Countryman C, Baldwin TO. Nucleotide sequence of the luxR and luxI genes and structure of the primary regulatory region of the lux regulon of Vibrio fischeri ATCC 7744. Biochemistry 1988;27:837-42.
7. Liu J, Turnbough CL. Effects of transcriptional start site sequence and position on nucleotide-sensitive selection of alternative start sites at the pyrC promoter in Escherichia coli. J Bacteriol 1994;176:2938-45.
8. Shadel GS. 1991. Control of Vibrio fischeri bioluminescence: genetic analysis of the LuxR protein and the mechanism of transcriptional autoregulation of the luxR gene. Ph.D. dissertation. Texas A&M University, College Station, TX.
9. Kolb A, Busby S, Buc H, Garges S, Adhya S. Transcriptional regulation by cAMP and its receptor protein. Annu Rev Biochem 1993;62:749-95.
10. Dunlap PV, Greenberg EP. Control of Vibrio fischeri lux gene transcription by a cyclic AMP receptor protein-luxR protein regulatory circuit. J Bacteriol 1988;170:4040-46.
11. Shadel GS, Baldwin, TO. The Vibrio fischeri LuxR protein is capable of bidirectional stimulation of transcription and both positive and negative regulation of the luxR gene. J Bacteriol 1991;173:568-74.

PART 3

Symbioses of Luminous Bacteria With Higher Organisms

NEW GENETIC TOOLS FOR USE IN THE MARINE BIOLUMINESCENT BACTERIUM VIBRIO FISCHERI

KL Visick and EG Ruby
Pacific Biomedical Research Center, University of Hawaii, Honolulu, HI 96822 USA

Introduction

In recent years, tools for the genetic analysis of marine *Vibrio* species have become increasingly established (1). In particular, transposon mutagenesis has been applied as a powerful technique. However, not all of the techniques that have been developed in one species of *Vibrio* can be successfully applied to another. For example, phage P1 is an effective delivery system for transposon mutagenesis of *V. harveyi* (1), but not for *V. fischeri* (Ruby, unpublished data), a species whose symbiotic association with the squid *Euprymna scolopes* has become an emergent model for the study of symbiosis genes (2). The most efficient method for the introduction of DNA into *V. fischeri* is through conjugative matings with *Escherichia coli*, which result in a high efficiency of plating both for the transfer of plasmids (3) and for transposition (4).

To date, two different transposons have been reported for the study of *V. fischeri* genes in the squid symbiont strain ES114, and its derivatives (2). Mutagenesis with the temperature sensitive MudI1681 (5) transposon yielded a pool of mutants with random chromosomal insertions, a subset of which were flagella mutants (4). Although insertions with this transposon were readily obtained, there was a tendency for the delivery vector to be maintained, resulting in multiple insertions under conditions where a low temperature was not strictly maintained. Mini-Tn*5* Cm (6), a derivative of Tn*5* in which the transposase is external to the transposon, has also been useful in generating a transposon-insertion library of *V. fischeri* (2). This transposon is one of a set of Tn*5*-derived transposons that contain many useful features, including a multiple cloning site internal to the transposon and the suicide origin, oriR6K, in the delivery vector. We have had some difficulty, however, in obtaining expression from Tn*5* derivatives carrying either promoterless *lac* or *lux* constructs introduced into *V. fischeri* (unpublished data).

A new transposon system (7) that has become available is based on Tn*10* and has some of the same useful features as the mini-Tn*5* system--a transposase gene external to the transposon, a multiple cloning site internal to the transposon, an origin of replication that is suicidal in *V. fischeri*, and a conjugation-based delivery system. In this paper we describe two derivatives of this Tn*10*-derived transposon that we have constructed, as well as other genetic tools that we expect will be useful for the genetic analysis of *V. fischeri*.

Materials and Methods

Reagents: Restriction enzymes and T4 DNA ligase were obtained from either Promega (Madison, WI, USA) or New England Biolabs (Beverly, MA, USA). X-gal (5-bromo-4-chloro-3-indoyl-β-D galactopyranoside) was purchased from 5 Prime 3 Prime (Boulder, CO, USA)
Procedures: Standard cloning techniques (8) were used to construct the plasmids described in this paper: parent plasmids were digested with appropriate restriction enzymes, the DNA was agarose gel purified, and the correct fragments were extracted from the agarose using the GeneClean kit (Bio101). The fragments were ligated together and the resulting mix was used to transform $CaCl_2$ competent *E. coli* cells (either strain DH5α or CC118λ*pir*). For conjugations, plasmids were transferred into S17-1λ*pir E. coli* cells by electroporation (8).

Results and Discussion

Mini-Tn*10* derivatives. Mini-Tn*10*-carrying plasmids pKV32 and pKV34 were derived from pBSL181 (7). Plasmid pKV32 (Fig. 1A) was constructed by the insertion of the 7-kb BamHI fragment carrying the promoterless *lacZYA* genes from pMC903 (9) into pBSL181 digested with BamHI. S17-1λ*pir* cells carrying pKV32 were conjugated with *V. fischeri* strain ESR1 (4), a rifampicin-resistant derivative of ES114 (2). The cells were spread on LBS plates containing rifampicin, chloramphenicol, and X-gal. After incubation at 28°C, a small number of blue and white colonies arose at approximately equal frequencies. These data suggested to us that the transposon was functional in *V. fischeri* and that promoterless genes internal to the transposon could be expressed by promoters upstream of the inserted DNA.

Plasmid pKV34 (Fig. 1B) carries the promoterless *luxAB* genes from *V. harveyi* inside the Tn*10* transposon. It was constructed by the ligation of a 3-kb *Bam*HI-*Pvu*II (*luxAB*) fragment from pHV100 with pBSL181 digested with *Bam*HI and *Sma*I. A number of colonies, appearing at a frequency similar to that obtained above, arose when S17-1λ*pir* cells carrying plasmid pKV34 were conjugated with KV150, a *luxA* mutant of ESR1 (10).

A subset of the colonies obtained were examined for luminescence in culture. Specific luminescence levels varied from just above background (the *luxA*-deletion parent emits no luminescence) to greater than the level observed for *lux*+ strain ESR1. These data suggest that the promoterless *luxAB* genes are located downstream (and are being transcribed) from promoters of varying strengths in the chromosome of *V. fischeri*. The mutants were checked by Southern analysis to determine the number and location of the transposon insertions. All strains checked, except one that apparently maintained the suicide vector, had a single transposon insertion. The location of the insertions was different for each strain, suggesting that the transposition of Tn*10* in *V. fischeri* may be random. Thus, both pKV32 and pKV34 are useful constructs both for the introduction of gene fusions into the chromosome of *V. fischeri*, and for the study of gene expression. Similar constructs can also be made that encode sequences for protein fusions.

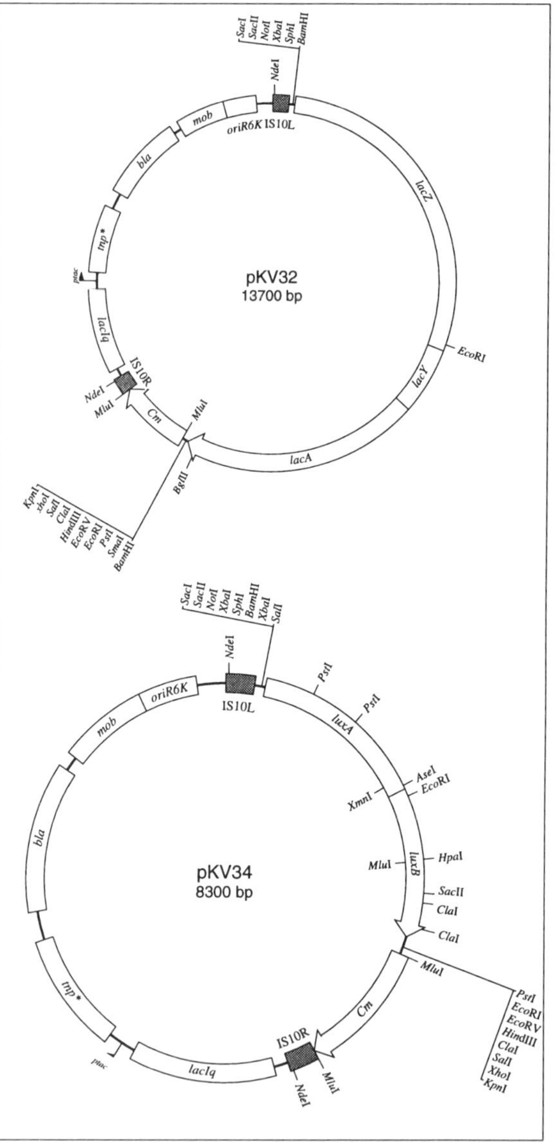

Figure 1. (A) Plasmid map of pKV32. (B) Plasmid map of pKV34.

<u>Tn7</u>. Transposon Tn*7* in *E. coli* inserts into a single chromosomal site (11). We investigated the ability of Tn*7* to insert into the chromosome of *V. fischeri*. Plasmid pSUP2017 is a mobilizable vector that encodes chloramphenicol resistance and carries Tn*7* (12). Tn*7* itself encodes resistance to trimethoprim, an antibiotic to which *V. fischeri* is sensitive. When this plasmid is introduced into ESR1, both chloramphenicol-resistant and trimethoprim-resistant colonies can be obtained. Strains that became chloramphenicol sensitive but remained trimethoprim resistant were isolated, suggesting that the Tn*7* transposon was able to transpose into native sequences from the vector, which was subsequently lost. Preliminary Southern analysis (unpublished data) indicated that the Tn*7* transposon targeted the same site in each of 6 isolates. These data support the conclusion that Tn*7* inserts into a single site in *V. fischeri*; we predict that Tn*7* will be useful as a vector for introducing genes in single copy in the chromosome.

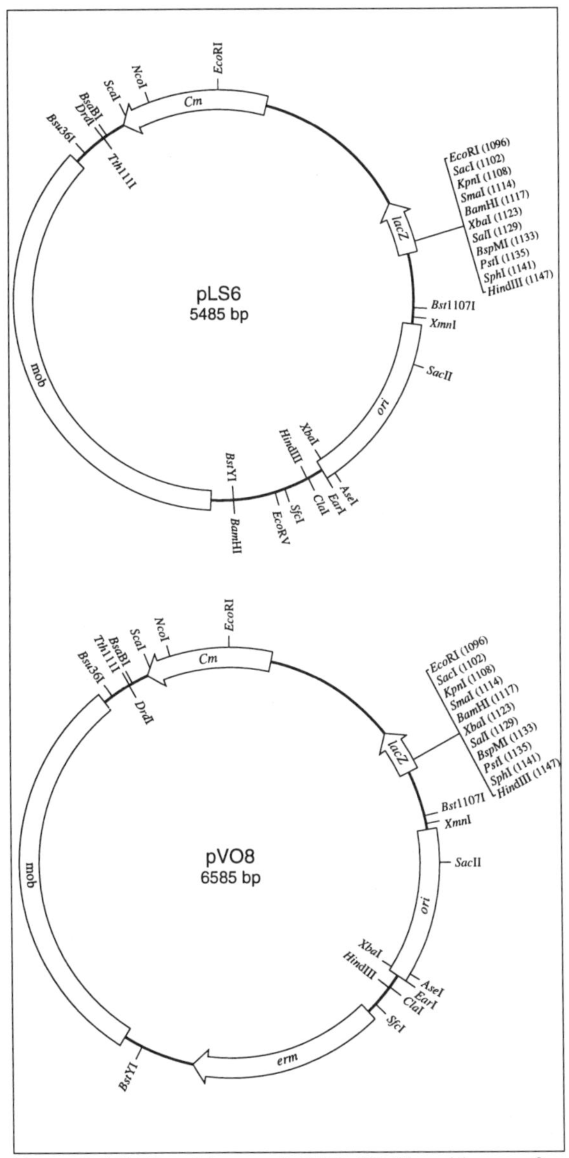

Figure 2 (A) Plasmid map of pLS6. (B) Plasmid map of pVO8.

<u>Useful vectors</u>. In addition to being sensitive to the antibiotics chloramphenicol and trimethoprim, *V. fischeri* strain ES114 is also sensitive to erythromycin, tetracycline, gentamycin, and kanamycin. Of these, only chloramphenicol, trimethoprim and erythromycin have proven useful for our studies. The tetracycline resistance gene encoded by pACYC184 is unable to confer resistance to *V. fischeri*, and the same is true for the single gentamycin resistance gene that we have checked (unpublished data). While kanamycin can and has been used as an antibiotic, there is a high background of spontaneously resistant colonies regardless of the concentration of kanamycin used. *V. fischeri* cells are also resistant to ampicillin, making most of the commonly available vectors of no

practical use. In addition, the high-copy-number vectors such as pBluescript (Stratagene, Inc.) that replicate in *E. coli* are relatively unstable in *V. fischeri*.

Derivatives of pACYC184 (8) (or any vector that contains the p15A origin of replication) appear to replicate fairly well in *V. fischeri*, and thus are useful for complementation or overexpression studies. For this reason, we have constructed modified vectors that can be used in *V. fischeri*. Plasmid pLS6 (Fig. 2A) was derived in two steps from pSUP102, a mobilizable version of pACYC184. The first construction step abolished the tetracycline resistance gene of pSUP102 by a deletion of a 1.6-kb BstYI restriction fragment. The second step resulted in an insertion of the multiple cloning site and *lacZ* cassette from pUC19 (8), thereby increasing the ease of cloning genes of interest into this vector. Plasmid pLS6 was further modified to contain the gene for erythromycin resistance. This vector, pVO8 (Fig. 2B), will be useful for complementing the chloramphenicol-resistant mutants obtained by transposon mutagenesis.

Acknowledgements
We would like to thank the following individuals for their contributions to this work: Ann Stevens, for pSUP2017 and helpful information regarding antibiotic resistances; Lornie Phillips for his work with Tn7; Lisa Shiroishi for her work in constructing pLS6; Vivian Orlando for her work in constructing pVO8; and Doug Bartlett for sending us pMC903. This work was supported by NSF grant (IBN-9220482) to EGR and an NRSA (7F32GM17424-02) award to KLV.

References

1. Silverman M, Showalter R, McCarter L. Genetic analysis in *Vibrio*. Methods Enzymol 1991; 204:515-36.
2. Ruby, E G. Lessons from a cooperative, bacterial-animal association: the *Vibrio fischeri-Euprymna scolopes* light organ symbiosis. Annu Rev Microbiol 50:591-624.
3. Dunlap PV, Kuo A. Cell-density modulation of the *Vibrio fischeri* luminescence system in the absence of autoinducer and LuxR protein. J Bacteriol 1992; 174:2440-8.
4. Graf J, Dunlap PV, Ruby EG. Effect of transposon-induced motility mutations on colonization of the host light organ by *Vibrio fischeri*. J Bacteriol 1994; 176:6986-91.
5. Castilho BA, Olfson P, Casadaban MJ. Plasmid insertion mutagenesis and *lac* gene fusion with mini-Mu bacteriophage transposons. J Bacteriol 1984; 158:488-95.
6. de Lorenzo V, Herrero M, Jakubzik U, Timmis KN. Mini-Tn*5* transposon derivatives for insertion mutagenesis, promoter probing, and chromosomal insertion of cloned DNA in gram-negative eubacteria. J Bacteriol 1990; 172:6568-72.
7. Alexeyev MF, Shokolenko IN. Mini-Tn*10* transposon derivatives for insertion mutagenesis and gene delivery into the chromosome of Gram-negative bacteria. Gene 1995; 160:59-62.
8. Sambrook J, Fritsch EF, Maniatis T. 1989. Molecular cloning: a laboratory manual. Cold Spring Harbor Laboratory, Cold Spring Harbor, N.Y.
9. Casadaban MJ, Chou J, Cohen SN. *In vitro* gene fusions that join an enzymatically active β-galactosidase segment to amino-terminal fragments of exogenous proteins: *Escherichia coli* plasmid vectors for the detection and cloning of translational initiation signals. J Bacteriol 1980; 143:971-80.
10. Visick KL, Ruby EG. Construction and symbiotic competence of a *luxA*-deletion mutant of *Vibrio fischeri*. Gene 1996; 175:89-94.
11. Craig, NL. Transposon Tn7. In: Howe, MM., editor. Mobile DNA. Washington, DC: Am Soc Microbiol, 1989:211-26.
12. Simon, R. Priefer, U, Puhler, A. 1983. A broad host range mobilization system for *in vivo* genetic engineering: transposon mutagenesis in gram negative bacteria. Bio/Technology 1:784-91.

PHENOTYPIC BIOLUMINESCENCE AS AN INDICATOR OF COMPETITIVE DOMINANCE IN THE *EUPRYMNA-VIBRIO* SYMBIOSIS

M.K. Nishiguchi, E.G. Ruby, and M.J. McFall-Ngai
Department of Biological Sciences, University of Southern California, Los Angeles, CA 90089-0371, USA and
Pacific Biomedical Research Center, University of Hawai'i
41 Ahui St., Honolulu, HI 96813, USA

Introduction

The study of coevolutionary relationships among symbiotic associations has been an important avenue for understanding the establishment and development of two independent, but closely integrated organisms (1-3). Most studies investigating the evolutionary relatedness amongst host-symbiont pairs have alluded to the parallelisms that occur with the onset of a particular host speciation event that eventually leads to the isolation of the symbiont population (4). Hence, symbionts from an "ancestral" host species may have similar biochemical, physiological, or molecular characteristics that group them within the same "species". Yet, the delineation between strains or biovars within a species is much harder to define; strain characteristics that would normally define a particular bacterial population may be consistent throughout the entire group of symbionts. Although sequencing hypervariable regions of specific loci is a technique that can genetically differentiate between the strains (5), there are few ways that one can phenotypically distinguish similar strains from one other.

The sepiolid squid-bioluminescent bacterial association offers several advantages as a model system to study the coordinated influence of luminous bacteria on the coevolution and speciation between partners (6). Both the bacteria and the host squid can be cultured separately, allowing the specific comparison between different symbiotic strains during colonization of a particular host squid light organ. Initiation, colonization, and persistence of each strain can then be monitored individually, or, in a competition experiment, where two strains compete for dominance in a particular species of host squid (7). Previous research has clearly indicated that symbiotic bacterial strains can be genetically differentiated from each other, either through restriction fragment length polymorphisms from specific DNA fragments (Lee and Ruby, unpubl. data), or through direct sequencing of loci that are variable between strains (8). Although molecular probes designed from these types of experiments can be used to distinguish and identify various strains in each individual light organ, this method is laborious and is limited by the number of colony blots one utilizes per infection assay (most infection/competition experiments utilize approximately 100 juveniles) (9). It has been recently discovered that variation in luminescence intensity of individual strains of symbiotic bacteria can be utilized to phenotypically distinguish strains from each other in an infected juvenile sepiolid light organ (8, Figure 1). This phenotypic variation allows the direct physiological comparison between different strains of symbiotic bacteria in a particular host squid, and can be compared to other types of data for determining the evolutionary relatedness of different host species and the symbiotic bacteria that have coevolved with them.

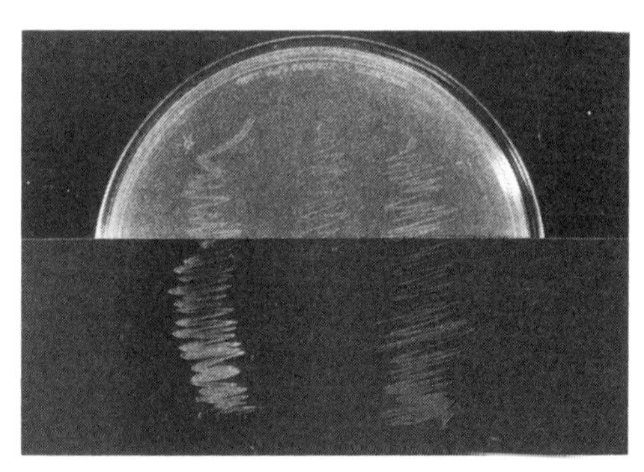

Figure 1. Light organ symbionts (left to right: EM17, ES114, and ET101) from three species of sepiolid squid (*Euprymna morsei*, *E. scolopes*, and *E. tasmanica*, respectively) have similar growth characteristics on agar plates. However, when observed in the dark, it can be seen that the light output per bacterial cell differs significantly between the strains.

Methods and Materials

For colonization assays, luminescent symbiotic bacteria were isolated from frozen or freshly collected light organs of several species of squid: *Euprymna scolopes* (Hawaii), *E. morsei* (Japan), *E. tasmanica* (Australia), *Sepiola affinis* (France), and *Loliolus noctiluca* (Australia) (10, Table 1).

Table 1. Sepiolid and loliginid squids and their respective light organ symbionts

Species	Representative light organ symbiont strain	Concentration in *E. scolopes* light organ	Luminescence on seawater nutrient agar medium
Euprymna scolopes	ES114*	2.6×10^6	non-visible
Euprymna morsei	EM17*	5×10^6	bright
Euprymna tasmanica	ET101*	3×10^6	dim
Sepiola affinis	SA1*	4×10^6	moderately bright
Loliolus noctiluca	LN101†	0	bright

* = *Vibrio sp.*
† = *Photobacterium leiognathi*

All strains isolated from the sepiolid squids were identified as *Vibrio* species; the bacterium isolated from the loliginid squid was identified as *Photobacterium leiognathi*. For experiments, cell suspensions consisting of approximately equal concentrations [about 10^3 colony forming units (CFU's)] were used as the bacterial inocula. Using isolates of symbiotically competent bacteria from *Euprymna*, *Sepiola*, and *Loliolus* species, infection experiments were completed to determine whether non-native strains could infect juvniles from a closely related sepiolid host, *E. scolopes* . The number of symbionts of each strain that had colonized a light organ 48 hours after infection was determined by simply observing the percentage of brightly, dimly, or non-visibly luminous CFU's arising from platings of light organ homogenates on seawater tryptone (SWT) agar medium (Table 1). The relative luminescence of the colonies was easily distinguished by viewing them in a dark room.

Results and Discussion

When presented individually, all strains of sepiolid symbionts, whether isolated from *Euprymna* or *Sepiola,* were capable of initiating and maintaining a symbiotic population in *E. scolopes* juvenile light organs. The only strain that did not infect *E. scolopes* juveniles was the light organ symbiont of *L. noctiluca,* which belongs to the species *Photobacterium leiognathi*. Although concentrations of bacteria homogenized from light organs were similar between the symbiotic strains, luminescence from these isolates on agar plates varied between visibly luminous and non-visibly luminous phenotypes (Table 1). This indicates that a symbiont associated with a particular *Euprymna* host may vary in the luminescence it produces outside the light organ; once colonization has occurred luminescence in the squid has little variation. Thus, once the symbiosis is established it can be deduced that the light organ has a direct effect on the production of light and luminescence of each bacterial strain, regardless of which host it has evolved. Previous results of luminescence variation have shown direct correlation with the amount of autoinducer present in the light organ (11), irrespective of the evolutionary or selective agents that differentiate luminous symbiotic bacteria.

The evidence of bacterial specificity and phenotypic variability between symbiotically competent strains that are "genetically" the same species (<2% 16s sequence identity) provides new insights into the speciation and evolutionary events that have occurred after the origin of the association (1, 8). Bacterial speciation within symbiotic populations is difficult to monitor; unless the molecular clock of either host or symbiont can be traced, there is little evidence that alludes to the predisposition of the individual partners before the onset of the symbioses (12). Present day models of symbiosis can only reveal the biological mechanisms that control the interactions, and unless there are distinct methods for defining characteristics of similar strains or biovars, evolutionary tracking is obscure. Thus, the use of bioluminescence as a phenotypic character in the sepiolid squid-bioluminescent bacteria symbioses is a unique and informative quality that enables one to distinguish the competency of individual strains, and provides evidence that speciation and diversification have occurred after the evolutionary origin of the symbioses (8).

Acknowledgements

We would like to thank S.v. Boletzky, M. Norman, T. Okutani, and R. Young for help collecting and obtaining specimens. This research was supported by NSF #OCE-9321645 to MKN, ONR #NOOO17-91-1347 to MMN, and NSF #IBN 9220492 to MMN and EGR.

References

1. Doyle JJ. Phylogeny of the Legume Family: An approach to understanding the origins of nodulation. Annu Rev Ecol Syst 1994;25: 325-49.
2. Haygood MG, Distel DL. Polymerase chain reaction and 16s rRNA gene sequences from the luminous bacterial symbionts of two deep sea anglerfishes. Nature 1993;363: 154-56.
3. Hinkle G, Wetterer JK, Schultz TR, Sogin ML. Phylogeny of the Attine ant fungi based on analysis of small subunit rRNA gene sequences. Science 1994;266: 1695-97.
4. Baumann P, Lai C-Y, Baumann L, Rougbakhsh D, Moran NA, Clarke MA. Mutualistic associations of aphids and prokaryotes: Biology of the genus *Buchnera.* Appl Environ Microbiol 1995;61: 1-6.
5. Cary SC, Warren W, Anderson E, Giovannoni SJ. Identification and localization of bacterial endosymbionts in hydrothermal vent taxa with symbiont-specific polymerase chain reaction amplification and *in situ* hybridization techniques. Mar Mol Biol Biotech 1993;2: 51-62.
6. McFall-Ngai MJ, Ruby EG. Symbiont recognition and subsequent morphogenesis as early events in an animal-bacterial mutualism. Science 1991;254: 1491-94.
7. Lee K-H, Ruby EG. Competition between *Vibrio fischeri* strains during initiation and maintenance of a light organ symbiosis. J Bacteriol 1994;176: 1985-91.
8. Nishiguchi MK, Ruby EG, McFall-Ngai MJ. Competitive dominance during colonization is an indicator of coevolution in an animal-bacterial symbioses. Submitted.
9. Lee K-H, Ruby EG. Detection of the light organ symbiont, *Vibrio fischeri* in Hawaiian seawater by using *lux* gene probes. Appl Environ Microbiol 1992;58: 942-47.
10. Boettcher KJ, Ruby EG. Depressed light emission by symbiotic *Vibrio fischeri* of the sepiolid squid *Euprymna scolopes.* J Bacteriol 1990;172: 3701-06.
11. Nealson KH. Autoinduction of bacterial luciferase. Arch Microbiol 1977;112: 73-9.
12. Moran NA, Baumann P, Ishikawa H. A molecular clock in endosymbiotic bacteria is calibrated using the insect host. Proc R Soc Lond B. 1993;253: 167-71.

FREE -LIVING AND ASSOCIATED LUMINOUS BACTERIA

GA Vydryakova, AM Kuznetsov
Institute of Biophysics, Siberian Branch of Russian Academy of Sciences,
Krasnoyarsk, 660036, Russia

Introduction

The investigation of the World Ocean bioluminescence showed the luminous bacteria to be among the most numerous unicellular inhabitants of sea waters and gastroenteric tract of marine animals. Luminous bacteria associated with marine fauna are represented by symbionts isolated from light organs, and protocooperative luminous bacteria inhabiting the gastrointestinal tract. Among the luminous bacteria isolated from various natural sourcses, the most frequently found are the following: *Photobacterium phosphoreum, Photobacterium leiognathi, Vibrio fischeri* and *Vibrio harveyi*. Not all species of luminous bacteria are among symbionts. So *V. harveyi* has not been isolated from light organs of marine animals. There are some hypotheses how an animal obtain and maintain the proper symbiotic species. We propose one more. However, direct evidence of it still remains to be produced.

Materials and methods

Samples of sea water and marine animals were taken from 64 stations during the 17-th trip of Scientific Investigation Ship "Vityaz" in the western part of the Indian Ocean in winter period (Figure 1). For isolation of luminous bacteria from samples, Photobacterium agar ("Difco") was used. Cultures were grown at 20-24° C. Physiological and biochemical properties of luminous bacteria were used for their identification.

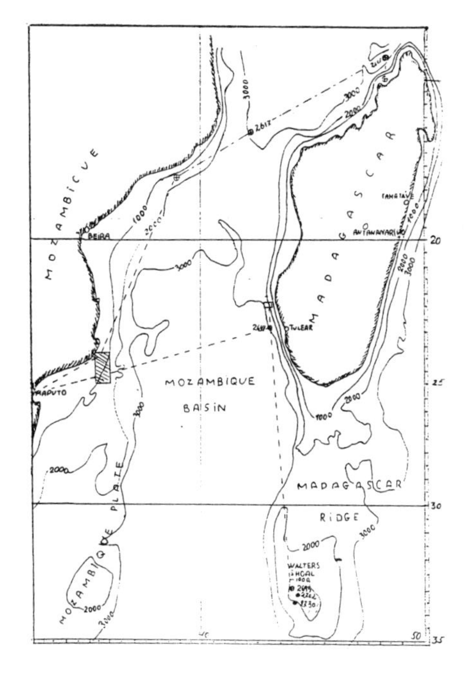

Figure 1. Map of same microbiological stations

⊕ - station, ▭ - ground - - - - way of ship.

Results and discussion

120 fishes belonging to 24 families and several representatives of cephalopods, prawns, and euphausiids were investigated for the presence of luminescent bacteria. Species identification of more than 500 of isolated free-living and associated luminescent bacteria was performed. The frequence and the ratio of luminous bacteria occurence in the gastrointestinal microflora of marine animals were determined. Luminescent bacteria occured in 23-65% of fishes, depending on the habitat depth, and their ratio varied from 8 to 60% of the total gastrointestinal microflora of fish. The free-living luminescent bacteria were found in 50% of the seawater samples from depths down to 1000 m, and their ratio contained from 0 to 50% of water microflora. The luminescent bacterium *P. phosphoreum* was dominant among the isolated cultures. The strains of *V. harveyi* were isolated from warm coastal waters and were found in a composition of shrimps microflora from depths from 40 to 400 m. Cultures *P. leiognathi* were isolated from water samples and samples of marine animals from depth to 400 m (Table 1). Bacteria *V. fischeri* were found only in several cases in the alimentary canal of fishes of fam. *Macrouridae* and fam. *Alepocephalidae* from depths 600-1000 m. One nonfermentative culture *Alteromonas hanedae* was allocated from the stomach of fish *Cynoglossus zauzibareusis*.

The composition of symbiotic luminous bacteria, symbionts of the marine animals, has been studied. All isolates from light organs were luminous. The strains of the species *P. phosphoreum* have been found in symbiotic association with deep-water fishes and fishes of medium depth (fam. *Macrouridae*, fam. *Moridae*, fam *Clorophtalmidae*, fam. *Opistoproctidae*, fam. *Acropomatidae*). Bacteria *P. leiognathi* were found in light organs of *Loligo sibogae* (fam. *Loliginidae*) from depths to 70 m.

Depth, m	Luminous bacteria from water samples	from gastro-enteric tract	from light organs
0-100	*V. harveyi* *P. leiognathi*	*P. leiognathi* *V. harveyi*	*P. leiognathi*
100-400	*P. leiognathi* *P. phosphoreum*	*P. phosphoreum* *P. leiognathi* *V. harveyi**	*P. phosphoreum*
400-1000	*P. phosphoreum*	*P. phosphoreum* *V. harveyi** *V. fischeri**	*P. phosphoreum*

* ocasional isolation.

Table 1. Distribution of luminous bacteria in depth dependent ecological groups.

It is known that the luminous bacteria of light organs of the marine animals are always pure culture of one strain. Besides it is known that cells of luminescent bacteria that have lost ability to emit light, are eliminated from light organs of the animals. Apparently, there is a uniform machanism of regulating the selection of symbiont's cells. This mechanism can be based on antigene - antybody interaction of owner's cells and symbiotic microorganism cells. Probably, lectines are such specific selective factors on cell-surface. It can be assumed that several lectines should be responsible for selection of symbiont's cells by the organism of the owner. Such lectines should identify type specifity of cells of symbionts and the presence of luciferase in them. Proceeding from the given assumption, it is possible to conclude that the luciferase is connected to a cellular membrane rather than located in cytoplasm of the cell.

Protocooperation of the luminous bacteria from gastrointestinal tract with sea animals turns into nearly mutualistic relationship in the light organs of them. Apparently, luminous bacteria eliminated from the light organs and excreted by animals form a group of free-living luminous bacteria. The areal and seasonal fluctuations of luminous bacteria in the World Ocean most likely determined by temperature.

PART 4

Bioluminescence in Nature: Physiology, Function and Evolution

BIOLUMINESCENT SIGNALS AND SYSTEMS: VARIETY IS THE SPICE OF LIGHT

P J Herring
Southampton Oceanography Centre, Empress Dock, Southampton SO14 3ZH, UK

Introduction

A signal transmits information from a source to a receiver through an intervening medium. The "quality" of the information emitted by the source is not the same as that arriving at the receiver, nor can the receiver necessarily decipher all the information that does reach it. The information content of the message is affected, first, by the transmission characteristics of the intervening medium (as manifested through the path length) and, second, by the transduction characteristics ("sensitivity") of the receiver. All signalling systems are subject to these factors. The optimum signal range, duration and specificity determines which system is used

Bioluminescence is the ultimate "Molecular signaling with photons" in that the signalling is (usually) between individuals (1). It is only one of several signalling systems employed by living organisms. The other two main ones involve the use of chemical messages, and sound or pressure waves. Each has its benefits and its disadvantages. Electromagnetic waves suffer the disadvantage of wavelength-dependent absorption and scattering, but in the visible wavelengths offer the potential for single photon sensitivity at a low metabolic cost. Biological materials can easily be employed to achieve high precision directionality and filtering of these wavelengths (by reflection and absorption) with minimal associated signal loss.

A comparison of the metabolic cost demonstrates how attractive an option bioluminescence can be for limited range communication. For a fish to receive a threshold acoustic signal of 1 dyne cm^{-2} at a range of 10 m from a source (with a frequency of 200-500 Hz) requires an output power of about 4μW. A threshold signal for the human eye from a 480nm source at the same range requires a light output in clear water of only about 10^{-3} μW.

Information in the light signal can be achieved through simultaneous encoding in flash frequency and modulation, in spectral and spatial distribution, and potentially in polarization. In a well-illuminated environment rapid visual signal patterns are provided instead by body movements which display specific reflective areas. Fast changes in the reflective areas themselves are a feature of shallow water cephalopods and some fish, which can change both the reflective tissues and their associated chromatophores. The equivalent bioluminescent signals in a dark environment can be expressed either as a spatial pattern of multiple light sources on an animal's body or by characteristic movements of the whole luminescing individual or its appendages.

Given that photon signalling provides an economical and highly flexible communication system in the sea, what bioluminescent variety can be recognized and how has it been achieved? Three categories are immediately apparent, namely biochemical, physiological and morphological variety. The combination of these three in a behavioural context provides functional variety.

Biochemical variety

One of the most surprising results of recent years has been the very limited biochemical variety that has been recognized among marine luciferins. The widespread occurrence of coelenterazine across the phyla, and the apparent absence of imidazolopyrazine systems other than in the marine environment, suggests that this particular system was probably a very early attribute of marine organisms. It was not the first luciferin to be characterized from the sea, that honour going to its imidazolopyrazine sibling in *Vargula*. Nor was it the first to be investigated: Dubois' studies of *Pholas* preceded study of coelenterazine systems by several decades.

It is, of course, much easier to extend the distribution of known systems than it is to characterize new ones. The extent to which coelenterazine and vargulin are employed will gradually become established and the tetrapyrrollic systems of euphausiids and dinoflagellates are now sufficiently characterized for an effective search to be made for them elsewhere in the living world. Bacterial luminescence biochemistry remains steadfastly unique. There are strong indications of other marine chemistries to be found at least in the polychaetes (*Chaetopterus*) molluscs (*Pholas*) and the echinoderms (*Amphiura*), but these are still very few in comparison with the great variety of luminous species in the sea. Land-based bioluminescence chemistries are dominated by the single type of system identified in fireflies and other arthropods, but there are additional types known in the molluscs (*Latia*), earthworms, and fungi, despitc the limited number of species involved. Hastings' (2) attempt to assess the number of separate evolutionary appearances of bioluminescence is solidly founded on fundamental differences of biochemistry. This must be the appropriate criterion for estimating the minimum number of times the capability has arisen but the identification of parallel evolution of similar chemistries is a more difficult task. Molecular studies of luciferase relationships may hold the key to this Pandora's box.

Physiological variety

Under this general heading come the many variables recognized in the bioluminescent responses of different organisms, including control and kinetics, the mechanics of external production and the spectral features.

The diversity of control and kinetics is bewildering. At the simplest level the light is continuous and control is metabolically determined by the expression of particular genes. The analysis of this condition in bacteria has given great insight into the role of the *lux* operon which has become a model system against which others can be evaluated. Continuous light emission is a rare phenomenon in other organisms and is best exemplified by the fungi, although some myriapods, springtails and larval glow-worms approach this condition. Continuous light emission is not known in the sea other than from bacteria. The ability to glow for long periods has, however, developed with the extensive employment of counterillumination camouflage systems. In some of these the observed glow is the visible integration of a large number of scintillating microsources (e.g. some decapod crustaceans). A few other marine animals have responses lasting for tens of seconds (e.g.*Pyrosoma, Spirula*) and may modify the intensity. Unless it is specifically tailored to camouflage, steady bioluminescence will make an organism easy to detect, particularly in the clear environment of the open ocean. Most marine bioluminescent signals are therefore transient events. On land, firefly signals probably involved the refining of lengthy signalling glows, such as that of female *Lampyris*, into shorter pulses of light with more specific information content and less risk from predators (3). In the sea the pressure to achieve this is acute.

Animals with luminous bacterial symbionts do have the capability of continuous bioluminescence but have usually developed elaborate systems for controlling the light. Anglerfish can and do utilize the continous emission of their symbionts to provide a steady light source for their lures, but this can be controlled at will, although the mechanism is still not clear. The only case in which it is known that bacteria are extinguished is that of the intracellular symbionts of *Pyrosoma*. It is not known how this is achieved. The muscular and reflective mechanisms whereby the emission from symbiont-associated light organs are controlled are known in the flashlight and pony fishes, but are still not understood in e.g. the squid *Heteroteuthis*.

If longterm steady light emission is a rare phenomenon among marine bioluminescent systems, the variations on short term responses are legion. Very rapid flashes, with durations of only a few hundred milliseconds, have been recorded from dinoflagellates, cnidarians, copepods, echinoderms and fish. Response kinetics are not well known, but already it seems that fast flashes are often part of a multiple response rather than isolated events. The repetition of flashes in the form of a train seems intuitively designed to reinforce their effectiveness to an observer but I am not aware of any research that has been carried out which demonstrates the general applicability of this hypothesis to bioluminescent signals. The efficacy of the repetition rate will be determined by the integration time of the observers eye (flicker fusion frequency). In experimental circumstances many bioluminescent marine animals will produce instrumentally distinguishable flashes in response to stimulus trains of up to 30 Hz, but this does not necessarily mean that similar frequencies are produced in natural encounters or that even if they were they could be distinguished by the observer. In many species which produce multiple flash trains there is also the capability of propagating the response to other parts of the body (or colony). Cnidarians, ophiuroids, holothurians and polychaete worms are typical examples; this provides the potential for spatial coding of the signal if the pattern of propagation is consistent. Some of the more visually spectacular patterns arise from the generation of multiple propagating waves from a single stimulus. The complexity of this pattern will only provide information if the observer can resolve the spatial detail. It seems likely that the rapid and complex patterns generated by e.g. *Renilla* and *Atolla* may not be fully perceived;. Seaslugs grazing on *Renilla* colonies seem unaffected by the wheeling patterns of light they induce.The resolution of the firefly eye is clearly adequate to recognize the specific subtleties of its respective signals, including the spatial elements (3). The same cannot yet be said of the eyes of many marine species yet must be the case for the species of the ostracod *Vargula* whose sexual signals involve complex spatial patterns (4).

Many of the more dramatic bioluminescent responses are produced by animals which themselves have little or no visual capability; they can only have evolved as general responses to an interspecific encounter. The signaller responding with a bioluminescent flash to a potentially dangerous chance encounter cannot anticipate the identity of the observer, let alone their sensory abilities. Fast and slow flash responses conform to no convenient taxonomic pattern; each species has evolved the control of its bioluminescence in response to the particular selective pressures it has faced. Evolutionary trends are not recognizable among the diversity of flash kinetics. Indeed many species have more than one type of emission; dinoflagellates can manifest a fast flash or a slow glow and some shrimp combine squirted luminescence with a counterilluminating glow. Lanternfish exhibit another example, namely a steady ventral glow from the photophores and a rapid train of flashes from the tail photophore or other luminous tissue. There is no experimental information on how these kinetic differences are achieved at the cellular level.

Secretory luminescence is a very widespread capability, being found in most marine groups. The expulsion of a cloud of luminescence presents little potential difficulty for a large animal such as a fish or squid. Thus the squid *Heteroteuthis* can eject ink and luminescent material simultaneously by contraction of the mantle musculature. What seems more surprising is that this tactic is not more widely employed among both fish and squid. We find instead that secretory luminescence is much commoner in the cnidarians, worms and crustaceans, with a few examples also among the molluscs. The omni-directional light of secretions is a valuable feature of the sexual bioluminescence of males of species of *Vargula* and provides the equivalent of sky writing for the receptive benthic females, just as it does in the fireworm *Odontosyllis*, though here the females do the skywriting to attract the males. In these animals, as in many others, the secretion is not so much ejected as swum away from. To a small aquatic animal the high kinematic viscosity of water makes the projection of luminescent material beyond its boundary layer a severe hydrodynamic problem, which becomes increasingly acute as the animal gets smaller and the Reynolds number decreases. The secretory bioluminescence of copepods, in particular, attaches to the pore of the gland until it is wafted away on a respiratory or swimming current or flicked off by rapid movements. It does not diffuse away but remains as a drop of light.

Animals living in a tube and capable of producing a rapid water current (such as *Pholas* or *Chaetopterus*) have no difficulty in expelling a luminescent secretion. Non-tube dwellers can achieve the same result if the secretion is inserted into a respiratory current (which will itself have been developed to escape from the producers space). In the mysid *Gnathophausia* this is achieved by siting the luminous gland on the maxilla, in decapods by ejecting material from the mouth into the two exhalant currents. Secretory luminescence is common among the cnidarians, whether or not accompanied by separate flash responses. The material is shed from particular regions of the body surface located close to the swimming system and is propelled by the swimming pulse or flow. In ctenophores this means injcction between the comb plates or at the downstream end of the comb row and in medusae such as *Periphylla* from the lower portion of the bell. Some medusae (*Pelagia* and *Poralia*), produce a luminous mucus from all over the bell and this response may be more to "mark" a contact than to act as a diversionary cloud. It is not always clear whether such exudates are natural secretions or the consequence of local contact damage. In the pelagic holothurian *Enypniastes* the latter may be the case, but the effect on a potential predator will not differ. In the case of *Periphylla* and many ctenophores each particle within the discharge scintillates separately. It will entirely depend upon the specific visual capabilities of a predator as to whether this is particularly disorienting or distracting. Searsiid fish secrete groups of cells which similarly flash independently in the water. Mixing the luminescence with a mucus carrier can maintain its spatial coherence, reducing local diffusion and dispersion. The best example of this ploy is the secretion of *Heteroteuthis*, which is so agglutinated with the mucoid ink that it may have a spatial persistence of several minutes.

Rhythms present a special case of control of bioluminescence kinetics. The internal clock setting the luminescent rhythm of autotrophic dinoflagellates is amenable to molecular analysis and its effect has been traced to mRNA transcription. The role of light in entraining the rhythm, and the absence of the rhythm in most heterotrophic species (e.g. *Noctiluca*), offers great opportunities to dissect the system still further. The only other confirmed rhythm is in the control of the luminous symbionts of the squid *Euprymna*, though light certainly inhibits the luminescence of some ctenophores. Rhythms have been invoked for the erratic bioluminescence of. the land snail *Dyakia*. Environmental control of bioluminescence expression (which reaches its peak in the entrainment of a rhythm) can appear as seasonal features. The specific daily period of a firefly's or an ostracod's sexual signal is presumably initiated through the eye but how is its seasonality determined? Developmental changes in the luminescence capability in marine animals might be induced by a changing depth environment, cued by pressure or temperature change during an ontogenetic migration. Environmental conditions may determine the sex ratio of protandric hermaphrodites and thus any associated sexually dimorphic luminescent signals.

The spectral characteristics of many species of marine organism have now been characterized. There is a natural division between the terrestrial and marine species in that the former have emission spectra clustering in the green-yellow region while the latter cluster in the blue-green. The differences reflect the environmental transmission and the visual sensitivities of the respective fauna. Within the confines of the clustering there remain consistent variations. Some emission spectra are of relatively narrow bandwidth. This almost invariably results from the secondary involvement of cellular fluors (e.g. many cnidarians) or the addition of narrow bandwidth filters (e.g. the hatchetfish *Argyropelecus* (5)). Most bioluminescent emission spectra have a broad bandwidth. This should ensure visibility, given a reasonable overlap with the observer's spectral sensitivity. How good does this have to be? How important is it that the spectral emission be fine tuned? In the midwater shrimp *Systellaspis debilis* a shift of 20nm in the spectral sensitivity peak produces a change in photon capture of only 2% (6). If the bioluminescence is a general defensive signal in encounters with unpredictable partners how can the organism maximize the effectiveness of its emission spectrum? Are the spectral variations that have been reported, between, say, the secretion of a decapod and that of a copepod, the result of particular selection pressures or are they equally adequate and therefore not subject to further evolutionary refining? Are dinoflagellate spectra geared to the eyes of grazing copepods or to the next level predator attracted by the burglar alarm effect (7)? Perhaps selection for high quantum yield outweighs any minor resulting spectral differences. The loss of light (up to 75%) resulting from the use of accessory filters (5) puts a high premium on quantum efficiency in those cases. It is very likely that close matching occurs between the spectral characteristics of bioluminescence used for intraspecific communication and the emitter's visual sensitivity. This kind of congruence has been elegantly demonstrated in fireflies (8) but not yet in most marine species. There are exceptions: fish which emit both blue and red light have red-sensitive visual pigments in addition to the blue-sensitive ones present in other fish. However, some decapods with photophores have additional short wavelength-sensitive visual pigments not matched by equivalent bioluminescence emission spectra (6). The yellow bioluminescence of the pelagic worm *Tomopteris* has no apparent visual counterpart, nor does the yellow-green luminescence of the fish *Benthalbella*. Perhaps the function of their bioluminescence is not intraspecific.

What of specific or individual variety? There has been no determined study of the potential for spectral variation in any one species. It has been one of the paradigms of bioluminescence that each species has its own characteristic emission spectrum; any apparent measured variability is usually ascribed to instrument limitations. Many factors can affect the emission spectra *in vitro* so there is no *a priori* reason why they should be inviolate *in vivo.* Indeed if evolutionary pressures act to refine the emission spectra it must be assumed there is some natural variation on which they operate. The molecular approach has also shown the spectral shifts that can be brought about in fireflies by a single aminoacid substitution (i.e. a point mutation) in the luciferase molecule and in aequorin by peptide substitutions. Perhaps this is the basis of the population and individual differences in the spectral emission of the coelenterates *Stachyptilus* and *Parazoanthus* (9).

If the spectral diversity has any consistent pattern it is hard to recognize (9,10). The chemistry is not the prime determinant, for coelenterazine systems have emission maxima ranging from 445nm (Radiolaria) to at least 475nm (Copepoda), excluding those in which fluors have further effects. Euphausiids and dinoflagellates have similar tetrapyrrolic chemistries but different emission spectra. Habitat seems to play little part; even within the echinoderms there are benthic asteroids, holothurians and ophiuroids with emission spectra ranging from blue to green. Are these targetting predators with different visual sensitivities?

Morphological variety

If physiological variety presents a picture of great variation so too does morphological variety. The latter can be examined and analysed in many more species than is ever likely to be possible for the equivalent physiological variety. In theory the larger data set should provide more scope for rationalization of the diversity.

Diversity of both site and structure are immediately apparent. There is probably no region of the body that one animal or another has not used to site a luminous structure. In bacteria and dinoflagellates isotropic emission is inevitable but in multicellular organisms the obvious choice of location for the photocytes is the surface of the body, particularly for secretions but also to minimize the optical losses of intracellular systems through self-absorption or scattering. In a few species photocytes are randomly distributed over most of the surface, thus occurring in very large numbers, and are frequently difficult to identify. Examples include some cnidarians (*Periphylla, Hippopodius*), echinoderms (*Benthogone*), polychaetes (*Chaetopterus*), and molluscs such as *Phyllirrhoe*. More often the photocytes are restricted to a limited number of sites but are otherwise "simple". These occur in most animal groups, and are the only luminescent structures in cnidarians, annelids, echinoderms and some molluscs, as well as in the springtails and myriapods. Accessory optical structures appear in the amphipod *Parapronoe*, decapod and euphausiid shrimp, cephalopods, fish, and most insects. These can reach quite astonishing degrees of complexity, involving structures for increasing the efficiency and directionality of emission (pigment cups, lenses, reflectors and light guides) and for modifying its intensity and/or spectral emission (chromatophores, shutters, rotatory systems, interference or absorption filters and monochromatic reflectors). One of the features of morphological variety, exemplified by these elaborate adaptations, is the degree of parallel evolution that has occurred in the development of similar systems in different groups of animals and the use of different materials to achieve the same functional result. Lenses may be derived from chitin (decapods), muscle (sepiolid squid) or protein similar to eye lens proteins (*Porichthys*). Collimating devices of different construction are found in squid, fish and euphausiid photophores. Pigments in the absorption filters in photophores range from carotenoproteins (shrimp) to porphyrins (squid and fish) and cytochromes (fish). The reflector materials and designs range from diffuse granular layers (shrimp, searsiid fish) to multilayer interference reflectors of chitin in euphausiids, of protein or collagen in squid and of guanine or connective tissue in fish.

The parallels extend to the distribution of photophores in different groups of animals. The common function of counterillumination determines a ventral directionality of photophores to obscure the silhouette. It is nevertheless remarkable that three quite separate families of decapod shrimp should have sited their photophores almost identically on the eyes, thorax, limbs and abdomen (*Systellaspis* and *Oplophorus* (Oplophoridae), *Sergia* (Sergestidae) and *Gennadas* (Penaeidae)). Similarly the modification of liver tubules to form counterilluminating photophores has evolved separately in the shrimps *Sergestes, Stylopandalus* and *Thalassocaris*.

The elaborate lures of anglerfish, with a bacterial light at the end of a fin ray, have a possible parallel in the elongate dorsal fin ray of the stomiiform fish *Chauliodus* and its (unconfirmed) associated light organ. Internal photophores (other than the hepatic ones noted above) are particularly common in squid and fish. In squid the transparency of the mantle musculature has permitted the extensive development of internal photophores, often sited on the ink sac (e.g. *Heteroteuthis*, *Ctenopteryx*, *Megalocranchia*, *Onychoteuthis*) while in the opaque bodies of fish internal photophores have required the associated development of elaborate means of transmitting the light to the exterior (leiognathids, macrurids, *Opisthoproctus* and *Coccorella*). The gut-associated photophores of many fishes owe their general origin to the ready availability of symbiotic luminous bacteria in the gut lumen. They occur in five different orders but related species have light organs that are morphologically very similar but which lack bacteria (e.g. *Howella*, *Coccorella*). Indeed in the family Apogonidae *Siphamia* has a bacterial photophore whereas other genera have intrinsic ones. We cannot judge whether the one has evolved from the other or whether they are independent responses to the same selection pressures. Perhaps the most remarkable case is that of the anglerfish genus *Linophryne* which has developed its own intrinsic photophores (on the barbel) in addition to the typical bacterial lure (on the elongate fin ray). Bioluminescence has been developed on two separate occasions.

This example raises the question of whether there are other recognizable cases of multiple evolution of bioluminescence. One approach to this is to examine other animals which have a range of morphologically different photophores and to consider whether the differences are fundamental.. Although there are very many photophores on sharks they nevertheless are of one type only, and the variety described on some decapods is referrable to different assemblies of a common subunit. Squid and teleost fishes present the greatest variety of photophore types. In squid the complexity of the photophores can usually be ascribed to complexity in the accessory optical structures which are each associated with the same type of photocyte. However, in some cases the differences in photocyte appearance at different sites suggest that different photophores may be the result of separate evolutionary events (e.g. *Octopoteuthis*).

The greatest scope for identifying multiple bioluminescent events in the evolution of a single species is to be found in the fishes. Two or three radically different kinds of photophore are often present, including glandular and non-glandular in the same species (e.g. *Searsia*, *Gonostoma*, most stomiiforms). No complete study of the different types on any one species has yet been conducted and the case for multiple evolution remains unproven. In the case of the myctophid fishes wide superficial differences in the photophores at different locations disguise a common photocyte architecture. Molecular methods of analysis should soon make it possible to recognize whether there are fundamental differences of biochemistry (and therefore of origin) between individual photophores in a single species.

One feature of the subcellular structure of the photocytes in several different groups of animals (shrimp, squid, fish and fireflies) is the presence of complex paracrystalline granules. Thus this particular type of packaging has been adopted for at least three different chemical systems. Yet there are other species in which the photocytes are structurally hardly distinguishable from general epithelial cells (e.g. many cnidarians). In a few cases an existing organ or tissue has subsequently been modified for light production. This has been recognized in the fish *Benthalbella* (and perhaps the octopod *Japetella*), in which the light organ is formed from muscle, in the midge *Arachnocampa* in which the Malpighian tubule is utilized, and in the liver light organs of the shrimp *Sergestes*. The morphological diversity that is apparent at all levels does not encourage the view that bioluminescence is an ancient feature of many groups and derives from their evolutionary history. Instead it supports the biochemical data for a multiplicity of later origins.

One consequence of life in the sea is that optical signals are much more limited by the transmission characteristics of seawater than are those in air. A glowworm or firefly may still be visible (to the human eye) at a range of 100m, the limit determined largely by inverse square law attenuation. In the *clearest* of ocean water a blue light (475nm) of similar intensity would be invisible by 40m, the reduction being largely due to absorption. Most seawater contains scattering particles and in practice these further greatly reduce the range of visibility throughout most of the ocean. There is little difference in the transmission of different visible wavelengths through air - the visibility of a blue, yellow or red light is determined primarily by the receiver. In the sea this is not so; red light is much more rapidly absorbed than blue light. A red light (700nm) of intensity equivalent to that of the hypothetical firefly would be visible at only 8-9m. Clearly blue bioluminescence achieves maximum range for minimum power in clear seawater. Those fish using red and blue lights can only use the red at short range *unless* the power output is increased exponentially. Since up to 75% of the light produced by the red light organ of *Malacosteus* is absorbed by the filter (to achieve the observed narrow bandwidth) the cost-benefit of increasing the power is of very limited value. In evolutionary terms this has been carefully balanced against the value of a signal invisible to other organisms. It could theoretically be a means of range-finding; if the transmission characteristics of the water change little and if the relative power output of the red and blue light organs remains constant then one fish could judge the range of another by the relative brightness of the two lights.

When the signal reaches the receiver (the eye) much of the potential information may not be accessible. Detection of the bioluminescence will be determined by the sensitivity of the eye, itself a consequence of light-gathering power (aperture size and optics) and the receptor characteristics (absorption by the visual pigment and degree of neural integration). Structural analysis suggests that most deep-sea animals have very sensitive eyes; the shrimp *Oplophorus* has an eye almost 100 x as sensitive as the dark-adapted human eye. The resolving power of the eye of this shrimp, however, is less than 1% that of the dark-adapted human eye. A receiver with high sensitivity will probably not be able to access all the spatial or temporal information potentially available in the signal. The ability to recognize complex patterns of luminescence, such as those of many fish photophores, will be very limited. Temporal resolution (flicker fusion frequency) determines the ability to distinguish the information that may be present in a bioluminescence flash train and it reduces with temperature. It is unlikely that most deep-sea animals with high sensitivity eyes can perceive flashes at a frequency higher than about 5-10Hz, despite the production of higher flash rates under experimental stimulation. This implies that there will be little selection pressure for high flash rates; this contrasts with fireflies in which females respond preferentially to those males with a flash rate higher than that of the population mean (over a range of 2.5-4Hz at 23.3°C) (11).

In most applications the information content in "molecular signaling with photons" is primarily limited to the binary "on" or "off" and instrument development is concerned with sensitivity. In bioluminescence the signal information content is very much greater and the detector development (eye evolution) is constrained by the relative importance in the life of the individual organism of the different information elements in the signal. Our interpretation of the functions of bioluminescent signals requires a better understanding of the "message"; this can only be fully resolved by analysis of the receiver's capabilities. The behavioural response which follows integrates that message with those from other sensory systems and determines whether the individual eats, mates, lives or dies.

REFERENCES

1. Herring PJ. Bioluminescent communication in the sea. In: Herring PJ, Campbell AK, Whitfield M, Maddock L editors. Light and life in the sea. Cambridge: Cambridge University Press, 1990:245-64.
2. Hastings JW. Biological diversity, chemical mechanisms and the evolutionary origins of bioluminescent systems. J Molec Evol 1983;19:309-21.
3. Case JF. Vision in mating behavior of fireflies. In: Lewis T. editor. Insect communication. London:Academic Press, 1984:195-222.
4. Morin JG. "Firefleas" of the sea: luminescent signaling in marine ostracode crustaceans. Fla Entomol 1986;69:105-21.
5. Denton EJ, Herring PJ, Widder EA, Latz MI, Case JF. The roles of filters in the photophores of oceanic animals and their relation to vision in the marine environment. Proc R Soc Lond (B) 1983;225:63-97.
6. Frank T, Case JF. Visual spectral sensitivities of bioluminescent deep-sea crustaceans. Biol Bull 1988;175:261-73.
7. Mensinger AF, Case JF. Dinoflagellate luminescence increases susceptibility of zooplankton to teleost predation. Mar Biol 1992;112:207-10.
8. Lall AB, Seliger HH, Biggley WH, Lloyd JE. Ecology of colors of firefly bioluminescence. Science 1980;210:560-2.
9. Widder EA, Latz MI, Case JF. Marine bioluminescence spectra measured with an optical multi-channel detection system. Biol Bull 1983;165:791-810.
10. Herring PJ. The spectral characteristics of luminous marine organisms. Proc R Soc Lond B 1983;220:183-217.
11. Branham MA, Greenfield MD. Flashing males win mate success. Nature Lond 1996;381:745-6.

SYNCHRONIES AND MECHANISMS IN THE NORTH AMERICAN FIREFLY *PHOTINUS CAROLINUS:* THE LINE-OF-SIGHT HYPOTHESIS

A Moiseff[1] *and J Copeland*[2]

[1]*Dept of Physiology and Neurobiology, Univ CT, Storrs CT 06269-3042 USA*

[2]*Dept of Biology, Georgia Southern Univ, Statesboro, GA 30460-8042 USA*

Introduction

At its simplest, firefly flash communication involves an exchange of flashes between a single flying male and a stationary female (1), but more complex flash behavior can be observed as well. For example, large numbers of males flash in synchrony in certain Southeast Asian firefly species (2). The emergence of synchrony at the level of the population comes about through the flash entrainment, and this is based (at least in some Southeast Asian fireflies) on the phase-advance and phase delay of one male's flash by another male's pacer flash (3).

We have been studying *Photinus carolinus* a North American species of firefly that also exhibits synchronous flashing. *P. carolinus* males produce a species specific pattern that consists of 5-8 pulses at 2 Hz (4). Flying males repeat these bursts every 10-12 secs. When two or more males are present, *P. carolinus* produces unison synchrony.

Our study site in Great Smoky Mountains National Park is characterized by the juxtaposition of a steep hillside and a relatively flat area. An observer viewing the flat area would characterize the fireflies flashing there as producing a unison synchrony. Viewing the hillside from the flat area, the same observer would describe the hillside flashing as wave synchrony.

We documented these observations using videographic techniques and confirmed the existence of these two forms of synchrony at the study site. We then compared the flash parameters of individuals that made up the populations exhibiting these synchronies. Our data revealed a similarity in the flash parameters of individual fireflies that was consistent with a single neural mechanism being responsible for both synchronies. A field experiment was performed to test this hypothesis. Then, lab experiments were carried out to suggest possible mechanisms that control the synchronies.

Materials and Methods

Free-flying males were videographed using a tripod mounted NVS Model 600 image intensifier (NVS Model 600, Night Vision Systems, Emmaus, PA, USA) coupled to a CCD camera (Panasonic Model WV-BL202). Recordings were made with a Hi-8 video recorder (SONY EV-C100). Frame-by-frame analysis, photometric analysis, and stimulation with green LEDs are described elsewhere (4,5). For the hillside field experiment, a string of white miniature incandescent lights (NOMA Moonlights, NOMA International, Inc., Forest Park, IL, USA) was used.

Results

Unison and Wave Synchrony The flashing (and videography) started 20-30 minutes after dusk and lasted for 1.5-2 hours. It took about 30 minutes for the synchrony to emerge across the entire field.

When the fireflies were observed within a discrete section of the flat area, they flashed in unison synchrony (Fig. 1A). Between flash bursts, flashing rarely

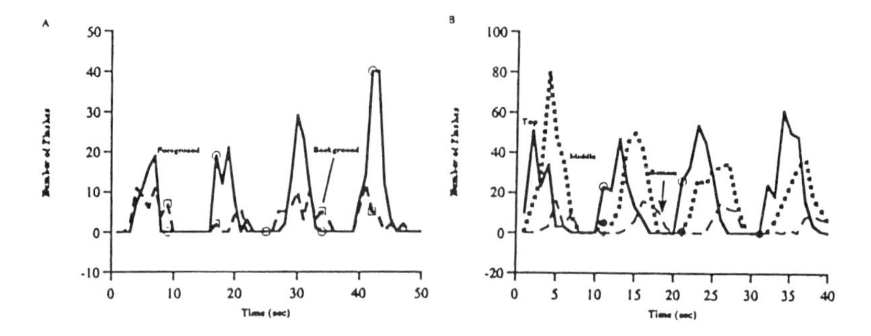

Fig. 1: Frame-by-frame analysis of unison (A) and wave (B) synchrony. Initiation of flashing and peak number of flashes are either concurrent (A) in unison synchrony or sequential (B) in wave synchrony. T = 18°C (A), 21.5°C (B).

occurred. The flashing on the hillside, when observed from the flat area, exhibited wave synchrony (Fig. 1B). The direction of the hillside wave varied during the evening, sometimes travelling in horizontal bands up the hillside (early in the evening), down the hillside (later in the evening), and sometimes even diagonally down the hillside. From cycle-to-cycle, the location of the individual flashes varied resulting in an ever changing pattern of the flashes that comprised the wave. It took 10-20 sec for a wave to travel 52 m (21 °C).

This wave synchrony was readily observed on the hillside. Wave synchrony was also observed in the flat area but it was substantially more subtle. After 8-10 sec of complete darkness on the forest floor, a bout of synchronous flashing spread from one discrete section of the flat area to another. The start of flashing took 10-20 sec to travel 52 m (21 °C). The propagation of flashing across the flat area differed from cycle to cycle, but it usually spread from west to east. When viewed close up, the synchrony within each area (node) was always unison synchrony.

Frame-by-frame analysis of the flash behavior of individual fireflies participating in unison or wave synchrony revealed little difference in interflash interval, number of consecutive flashes in a burst, average interburst interval, and average flash duration (t-test, $p > 0.05$) (Fig. 2). To the unaided eye, there was no difference between the flash behavior of fireflies participating in unison synchrony, wave synchrony, and those few fireflies that flashed in isolation along a path or the meadow.

The similarity of flash parameters describing the flashes of individuals on the hillside and the flat area suggested that the mechanisms underlying unison and wave synchrony were the same. We developed the hypothesis that the lines-of-sight among fireflies on the hillside were more restricted than the lines-of-sight among the flat area fireflies, and that this was the basis of the different synchronies. This hypothesis was tested by artificially providing the equivalent of increased lines-of-sight to the hillside fireflies and observing whether unison synchrony would result.

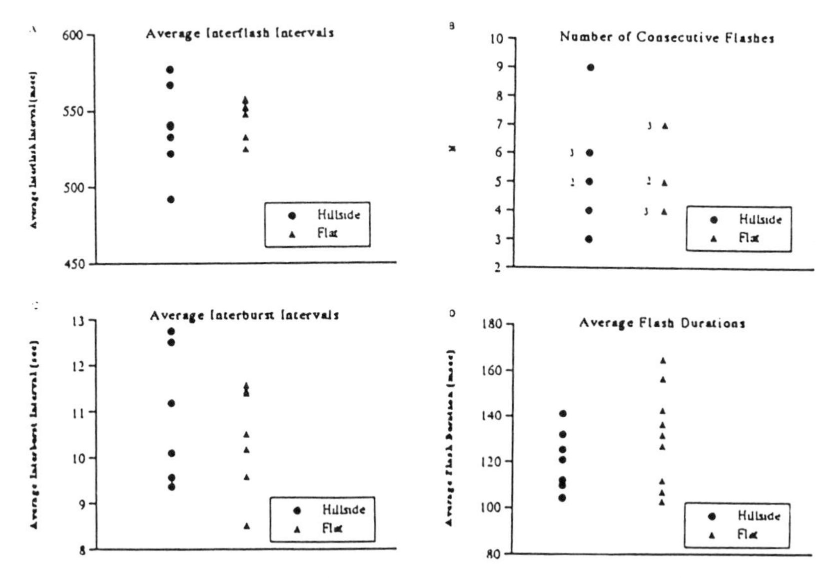

Fig. 2: Frame-by-frame analysis of flash parameters produced by individual fireflies. Average interflash intervals (A), number of consecutive flashes (B), average interburst interval (C), and average flash durations (D) are shown for individual fireflies participating in wave (hillisde) and unison (flat area) synchrony. There was no difference between hillside responses and responses from the flat area. T = 21° Hillside, 20.2°C Flat.

To attempt this conversion, a string of incandescent lights was hung from a small tree located 46 m up the hillside. After 5 minutes of spontaneous wave synchrony, computer_triggered species specific flashes were presented. Experiments were performed using flash patterns with different numbers of stimulus flashes (8, 4, 4 with 2nd flash suppressed, 2, and 1). In the majority of trials, wave synchrony was convertable to unison synchrony (Table 1). When the conversion occurred, the delay between the last stimulus flash and the massed firefly response was about 450 msec (Table 2), similar to the firefly's spontaneous interflash interval (IFI) or some multiple of it (4,5). This effect lasted from 10-20 seconds following the stimulus, and then wave synchrony resumed on the hillside.

Lines-of-sight could be affected by either local ground cover (plants, bushes, and trees) or variation in local topography. To estimate the effect of ground cover, photographs were made of a target held 3 and 9 m from a camera and 0.6 and 1.2 m off the ground. No differences were found in comparisons between hillside, flat area, and connecting meadow. Both hillside and flat area seemed similar in micro-topography. Boulders, ridges, valleys, and depressions created localized areas (> 9 m diameter) (nodes) that appeared to be visually separated from the adjacent areas. The hillside differed in that the bottom 2-3 m was covered with rhodadendron and mountain laurel, thus obstructing part of the view of the hillside from the forest floor. Additionally, when viewed from the forest floor, the hillside slope appeared to be smooth. When climbed and trod on, the hillside slope actually consisted of small flat areas (deer paths and small plateaus) that were separated by areas of steep incline. Using a USGS topographical map, we calculated the average hillside incline to be 67°. The steepness of the slope might restrict lines-of-sight for a firefly on the hillside.

TABLE 1: Conversion from Wave Synchrony to Unison Synchrony

STIMULUS	UNISON (%)	INHIB (%)	NR (%)	SCINT (%)	N
8	.	100	.	.	4
4	80	14.7	2.9	1.5	68
1,3 4	64.2	28.6	.	7.1	14
2	91.7	8.3	.	.	12
1	33.3	50	13.3	3.3	30

N= number of stimulus presentations

TABLE 2 Delay in Conversion from Wave to Unison

STIMULUS	DELAY (msec) X±S D	N	n
8	inhibition	4	1
4	468.9±67.6 63	6	
	915.0±17.3 4	1	
1,3 4	486.0±104.8	5	1
	990.0± 42.4	2	1
	1350	1	1
	1897.5	1	1
	2430	1	1
2	429.0 ±126.5	10	1
1	452.0±78.1 15	1	
	915.0±113.6 4	1	

N = Number of stimulus presentations n = Number of stimulus bouts

Mechanisms Because paired or four flashes were so effective at converting the hillside from wave to unison synchrony (Table 1), two types of experiments were done using single caged males: 1) The relationship between stimulus IFI, spontaneous IFI, flash delay (the time between the most immediate stimulus flash and the first firefly triggered flash), and stimulus flash number was determined by taking photometric recordings of single isolated males that were stimulated with a computer_synthesized flash 2) To test for the importance of stimulus location and the presence of pairs of stimulus flashes, videographic recordings were made from 8 individually caged males that were stimulated simultaneously with a computer_synthesized flash. Stimulation with 1 or 2 LEDs (separated by 2 m) produced a series of 2-6 flashes from the caged individuals, and these were called triggered flashes (4).

Using photometry and single isolated males, the spontaneous IFI was shown to be independent of the stimulus IFI over a stimulus range of 250 - 500 msec (Fig. 3A). While spontaneous IFI was affected by stimulus IFI, it was not influenced to the extent that a direct correlation could be made between them (Fig. 3B). The flash delay (time from the stimulus flash to the first triggered flash) was independent of the spontaneous IFI at all stimulus IFIs used (300, 400, 500, 700 msec) (Fig. 3C). Furthermore, the delay did not change if the first triggered flash was concurrent with stimulus flash 1 or with stimulus flash 5 (Fig. 3D). Sometimes a given stimulus IFI would produce two triggered flash delays: one at the spontaneous IFI and one at 100-200 msec (Fig. 3C-D).

To assess the importance of paired flashes, we used videography and 1 or 2 LEDs to stimulate caged individuals. These were labelled R and L to signify their relative position. Stimuli were delivered from R and L simultaneously, sequentially, alternately, or from just one LED. The number of flashes used in each stimulus was 8, 8 with 2nd and 3rd flashes suppressed, 8 with every other flash suppressed, 2, and

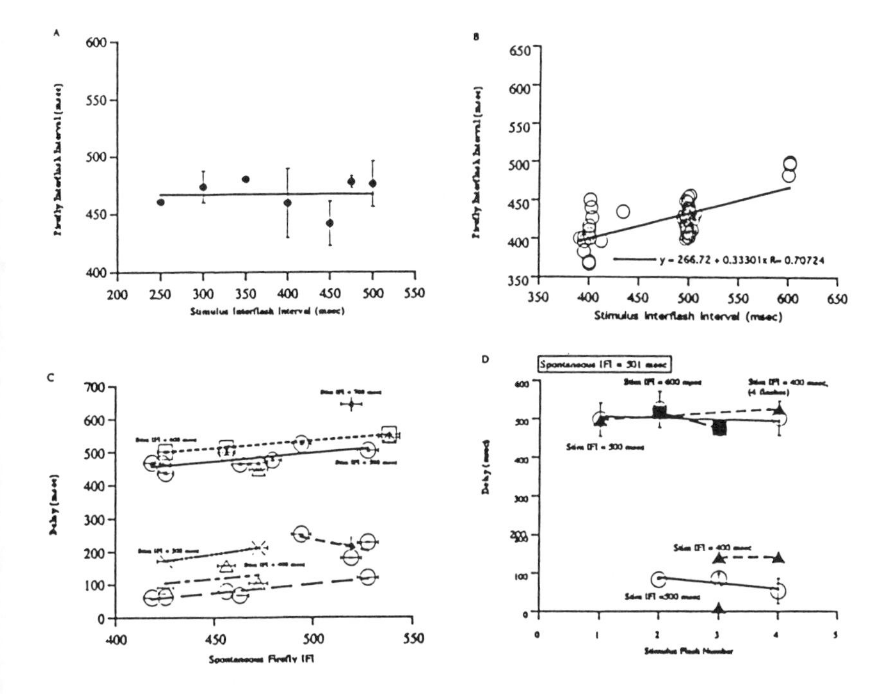

Fig. 3: Effects of stimulus interflash interval (IFI) on triggered response flash delay as measured by phtometry. A) Stimulus IFI was independent of firefly flash delay. B) The firefly's IFI was affected by the stimulus IFI but could not match it. C) The delay (time from the stimulus flash to the first flash of the firefly's response) was indepdendent of the firefly's spontaneous IFI. D) The delay was independent of the stimulus flash number. Six fireflies were used in (C). All others used 1 firefly.

1. A group of eight caged fireflies was used, each capable of seeing the stimulus flash and the flash of its neighbors. Total number of flashes were counted for each stimulus flash.

When stimulus presentations were shifted from one LED to the other, the peak number of flashes decreased but the location of the peak stayed the same (Fig. 4A). The number of flashes was slightly less when both R and L flashed simultaneously [(RL) x 8], alternately ((RL), or sequentially (RRRRLLLL) (Fig. 4A), but the location of the peak number of flashes stayed the same. However, paired flashes (RL) produced a curve that was similar to 8 flashes except that fewer flashes were produced. When the series of stimulus flashes was interrupted (R^R^R^R^, R^^^^^^^, and R^^RRRRR), the peak of each resulting curve shifted to the right (Fig. 4B), i.e., the peak was delayed. When the second and third flashes were deleted from the 8 flash stimulus, flashes even occurred after stimulus flash 8 (when stimulus flash 9 would have occurred) (Fig. 4B, arrow), highlighting the fact that the previous stimulus flash did not cause the firefly's flash, i.e., triggering seemed likely. Thus, the occurrence of two flashes seemed to be critical for evoking flashing in this group of fireflies.

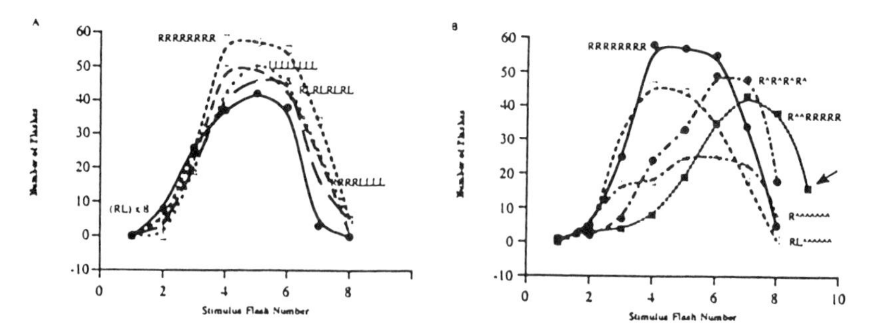

Fig. 4: Effects of spatial separation (A) and patterning of flashes (B) on triggered responses. Eight fireflies were used, each in a clear plastic cage. They were arrayed in a straight line with the 2 LEDs above them.

Discussion

Synchrony or Synchronies? For a single species of firefly to show two types of synchrony is not without precedent. Wave synchrony has been described in the Southeast Asian unison synchronizer *Pteroptyx tener* (7). *P. tener* is found in the trees and shrubs that line the banks of brackish rivers in Southeast Asia. The suggestion was made that the wave was an illusion caused by separate unison nodes that flashed sequentially. For a single species of firefly to produce two context specific (topographic specific) synchronies in *P. carolinus* would be remarkable.

P. carolinus appears to exhibit both unison synchrony or wave synchrony (Fig. 1). However, individual *P. carolinus* flashed similarly whether they were participating in unison or wave synchrony (Fig. 2). Since an individual firefly flashed the same way regardless of the synchrony in which it was observed to participate, it became necessary to experimentally separate what the fireflies were doing from what we were seeing. This was done two ways: *via* the hillside conversion experiment that tested free-flying males and *via* stimulation of caged single males.

If wave synchrony resulted from sequential activation of spatially separate nodes of fireflies (each node flashing in unison), it should have been possible to convert wave synchrony to unison synchrony by providing a sufficiently bright stimulus to trigger all nodes, thus overcoming the restriction of lines-of-sight. This was true in most cases (Table I). In a small number of cases, the stimulus produced either an inhibition of flashing on the hillside or the seemingly random flashing that has been termed scintillations (Table I) (4). Occasionally, there was no response to stimulation (Table I). Thus, a stimulus that acted like a counterfeit unison synchrony could consistently convert the hillside wave synchrony to unison synchrony. This is in keeping with the hypothesis that a single neuronal mechanism controls unison and wave synchrony.

A type of wave was also seen in the flat area. Here, discrete areas (nodes) flashed in a unison synchrony. Initiation of flashing spread in a wave, and this appeared to the observer on the forest floor as the spread of a line of light. If the 2 dimensional plane of the forest floor were stood on end at a steep angle, it is likely that the line of light seen on the forest floor would appear as a 2-dimensional wave. Thus, whether the flashing appears as a unison synchrony or a wave synchrony might depend on the viewing position of the observer and the type of flashing being viewed.

The kaleidescopic-like variability of the flashes from burst to burst can also be explained. The fireflies were flying and flashing and moving from place to place on both hillside and flat area. This might be one source of variability. Since the CNS excitability of the fireflies probably changes over time, it is likely that different fireflies (perhaps even in different nodes) could initiate a localized unison flashing that spread. This would predict both the stability and the variability which was seen in wave and unison synchronies.

Mechanisms underlying Synchronies Stimuli that included paired flashes were effective in triggering a firefly response (Fig. 4, Table I)(4). Paired flashes were so important that when the second and third flashes of an 8 flash stimulus were suppressed, the peak number of flashes was delayed, as predicted. (However, single flashes could trigger the same flashes (Fig. 4, Table I)). Thus, pairs of flashes seem important in triggering a species typical response.

The spontaneous IFI is in *P. carolinus* is invarient (Fig. 3A) (4,5). The triggered flash delay is similarly invariant and is similar to the spontaneous IFI (Fig. 3C, Table II). No matter when the delay occurs with respect to the stimulus, the triggered flash delay is the same (Fig. 3D).

It may be that the neural mechanisms that control the spontaneous IFI and those that control the delay are the same. The response viewed, because it continues even when the stimulus has been withdrawn, could be considered a fixed action pattern (8). The concept of the fixed action patterns has proven useful in the study of the neural control of insect behavior (8) and might be key to our neural analysis of synchrony.

At a set stimulus interflash, delays of two classes occurred: a long delay (similar to the firefly's spontaneous IFI) and a short delay (100-200 msec) (Fig. 3C-D). Some short delay flashes are likely to have occurred when the spontaneous IFI was 100-200 msec longer than the stimulus IFI, e.g., stimulus IFI = 400 msec and spontaneous IFI = 500 msec. Some short delays could represent the 90 - 200 msec reflex flashes reported for some fireflies (9).

The Light Show We suspect that the ever changing wave synchrony, the "light show" of the Elkmont, TN fireflies (10), comes about through a mechanism that, at the level of the individual firefly, is the same: if you see two flashes (or if you see an especially "good" single flash, "good" meaning a really well timed flash *vis-a-vis* your own trigger circuit excitability), fire your trigger circuit. (The trigger circuit and the IFI circuit might even be the same.) In a simplified case, one individual could trigger several individuals and a unison synchrony would result. Wave synchrony could be produced by a delay in the triggering (for example, a firefly that only saw the leader firefly's 7th and 8th flash) or if most of the flashing in one node not being seen by fireflies in an adjacent node). Thus, synchrony in *P. carolinus* could be different from some Southeast Asian synchronies. In the latter, phase-shifting occurred and in some cases the interflash interval also changed (2, 3). In *P. carolinus*, because the IFI and delay do not vary, (Fig. 3D, ref 4), the opportunity for flash plasticity to occur appears to reside in the neural mechanisms that control the interburst interval oscillator.

Thus, to the flying *P. carolinus* male, all synchrony is unison synchrony. To the observer standing at a distance from the fireflies and with a field of view that is probably different from that of the firefly, a wave of light can be perceived that travels across the observer's field of view. When the field of view is the flat area, nearby unison synchrony can be seen, as can a wave of initiations of unison synchronies across the flat area. When the field of view happens to be the hillside, the spectacular wave synchrony is seen that makes up the "Light Show".

We suspect that, although the wave synchrony is spectacular to us, it does not have any mating (communicative) significance to the firefly. If the wave were of communicative significance, we would expect the flashes within the wave to be more stable. Thus, it is likely that the hillside fireflies only perceive that they are engaged in a local unison synchrony. We make the assumption that the unison synchrony of *P.*

carolinus promotes courtship and mating in these fireflies, but the behavioral (evolutionary) function of virtually any synchrony is still undiscovered (2).

Acknowledgements
Supported by National Geographic Society grant 5262-94 and a grant from Georgia Southern University. We thank the National Park Service (Great Smoky Mountain National Park) and especially Keith Langdon for permission to work at the Elkmont site and numerous kindnesses; Lynn Faust, Elaine Burt, Fabiana Kubke, Jason Smith-Metcalfe , and Paul Talierco for help with the field work; and Jason Smith-Metcalfe and Ursula Sterling for commenting on the manuscript.

References

1. Carlson AD, Copeland J. Flash communication in fireflies. Q Rev Bio 1985; 60:415-36.
2. Buck JB. Synchrony in fireflies II. Q Rev Bio 1988; 63:263-81.
3. Hanson FE. Comparative study of firefly pacemakers. Fed Proc 1978; 37:2158-64.
4. Copeland J, Moiseff A. The occurrence of synchrony in the North American firefly Photinus carolinus (Coleoptera: Lampyridae). J Insect Behav 1995; 8:381-94.
5. Moiseff A, Copeland J. Mechanisms underlying synchrony in the North American firefly Photinus carolinus (Coleoptera: Lampyridae). J Insect Beh 1994; 8:395-407.
6. Gatlinburg Quadrangle, Tennessee-Sevier County, 7.5 minute series (topographic) 157-NE. Geological Survey. 1979.
7. Case, J. Vision in the mating behavior of fireflies. In: Lewis T, editor. Insect Communication. London: Academic Press, 1984:195-222.
8. Camhi, J. Neuroethology. Sunderland: Sinaurer Press, 1984.
9. Case, JF, Buck, J. Control of flashing in the firefly II. Role of the CNS. Biol Bull 1963; 125:234-50.
10. Billman K, Mills, D, Landry, Bill. The Light Show. WBIR-tv, Knoxville. 1994.

THE MICROSCOPICAL STRUCTURE OF THE BIOLUMINESCENCE SYSTEM IN THE MEDUSA *PERIPHYLLA PERIPHYLLA*

PR. Flood[1], J-M Bassot[2] and P J Herring[3]
[1]Dept. of Zoology, University of Bergen, 5007 Bergen, Norway,
[2]NAM, Université Paris Sud, 91405 Orsay, France and
[3]Southampton Oceanography Centre, Southampton SO14 3ZH, United Kingdom

Introduction

Bioluminescence is present in many marine species and the chemical identity of this property is often known (1). However, the cellular basis for how such bioluminescence is produced within the animal is known only in a few exceptional cases (2) and Cnidaria are no exception to this rule. Direct evidence which relates the light itself, or the chemical compounds needed for its production, to specific cells or organelles within these fragile animals is technically difficult to obtain (3). In most publications only circumstantial evidence pointing to likely sources of bioluminescence has been presented (e.g. 4, 5, 6). In the older literature, epithelial cells with cytoplasmic granules have been pointed out as the likely photocytes in several scyphozoans (7, 8).

Material and methods

In Lurefjorden, Western Norway, large specimens of the scyphozoan *Periphylla periphylla* are present in large numbers (9). During night in the late winter months these may be collected directly in buckets from the surface water. We have studied such undamaged material in a multidisiplinary way. In addition to imaging and recording of bioluminescent displays in intact and experimentally altered medusae (Herring, Bassot and Flood, this volume) and chemical studies (Shimomura and Flood, work in progress), we have processed dome and lappet exumbrellar epithelium and ripening oocytes from the gonads for several microscopical techniques. These include vital-, epifluorescence- and classical bright field light microscopy, and transmission and scanning electron microscopy.

Bioluminescence and autofluorescence were studied through a Nikon Labophot microscope equipped with a 100W mercury lamp and an epifluorescence unit with UV (330-380nm) and blue (420-485nm) excitation filters, 400 and 510nm low pass dichroic mirrors and 420 and 520nm barrier filters. Images were recorded on sVHS tape through a Dage-MTI GenIIsys image intensifier and a Dage CCD-72 B&W video camera, or on 35mm colour slide film through a Nikon photoautomat.A JEOL 100CX transmission electron microscope and a JEOL 6400 scanning electron microscope were also used.

Results

Bioluminescence was associated with two distinct sources in these medusae. One was found in females only. Here, minute and irregularly shaped granules in the cortical layer of maturing eggs emitted light when exposed to distilled water (Fig.1a), 1% Triton X-100 in seawater (Fig 1b), or when the gonad was rubbed mechanically. These granules also revealed yellow-green autofluorescence when excited by both UV and blue light (Fig. 1c). In toluidine blue stained Historesin sections they corresponded to irregular and dense, μm-sized granules gathered around much larger, pale and perfectly spherical bodies in the cortical ooplasm (Fig. 1d). The ultrastructure of these luminescent granules has yet to be studied.

The second source of bioluminescence was found in the exumbrellar epithelium of all individuals of both sexes. This bioluminescence was punctate and scattered throughout the dome and lappet exumbrella with less than 1 mm spacing (Fig. 2a). Each bioluminescent point corresponded to a highly specialized photocyte characterized by a single, hyaline cylinder, surrounded by 20 to 40 ovoid to crescent-shaped and highly autofluorescent bodies (Fig. 2b). In UV light this autofluorescence was initially clearly pinkish. However, this colour faded rapidly away and left behind a dull blue. In blue excitation light the autofluorescence similarly started out as strong yellow-green and ended up as dull green. Nematocysts in the epidermis and granulated cells in the underlying mesogloea were also autofluorescent, but this was unrelated to luminescence.

The rod-like assemblies generally stood perpendicular to the epithelial surface. However, oblique or curved rods were also found, particularly near the lappet rim. The cell nucleus was located lateral to the rod assembly and the entire structure usually embedded in a large vacuole traversing most of the thickness of the epithelium (Fig. 2c). Near their basis, most of the photocytes appeared to be contacted by one or several nerve fibres. In fact, the photocytes were usually found near the crossing points of these fibres (Figs. 2c,d)

Under exposure to 1% Triton-X-100 solutions each photocyte lit up for 1 to 2 min.The highest intensity of light then emanated from the interface between the central hyaline rod and its surrounding crescent-shaped and autofluorescent bodies (Figs. 2e-h). If the cell was exposed briefly to UV or blue light during this period the bioluminescence dissapeared temporarily before it slowly reappeared and reached its previous intensity. This pronounced photoinhibition was observed with fixed image intensifier gain.

By transmission electron microscopy each crescent-shaped body was clearly seen to be made up of closely packed membrane bound, 0.3μm wide organelles with electron dense and finely granular matrix (Fig. 3).

In response to strong swimming contractions and gentle handling numerous flashing particles were expelled from the exumbrellar epidermis, particularly at the lappet rims. Such particles were recovered by filtration of the water and verifed to be photocyte rods. In medusae caught by nets such loss of photocyte rods made the histological demonstration of these in the animal extremely difficult.

Discussion

Our results indicate that the bioluminescence systems of maturing eggs and exumbrellar epithelial photocytes of *Periphylla periphylla* differ from each other with respect to autofluorescence properties. This corresponds to differences in the chemical organization of the two systems (Flood, unpubl. res.).

Further, our results demonstrate for the first time the indisputable presence of highly specialized epithelial photocytes in the exumbrellar epidermis of a scyphozoan. These photocytes also appear to be heavily innervated by the diffuse epithelial nerve net,

---

Figure legends

Fig. 1. Maturing eggs of *Periphylla periphylla.* **a** and **b**) Image intensifier video prints of bioluminescence stimulated by distilled water. **c**) UV-epifluorescence photomicrograph of vital cortical ooplasm. **d**) Bright field photomicrograph of toluidine blue stained 0.5μm Epon section through cortical ooplasm. The dark, small grains (arrows) represent the bioluminescent sources.

Fig. 2. Exumbrellar epidermis of *Periphylla periphylla.* **a**) Bioluminescence imprint of lappet epidermis mounted on 50ASA slide film covered by Saran wrap. Electrical stimulation. **b**) UV-epifluorescence photomicrographs of epidermal photocytes in the lappet region. **c** and **d**) Photomicrographs of toluidine blue stained 1μm Historesin sections perpendicular (**d**) and tangential (**e**) to the epidermal surface. Arrows point to nerves. **e-h**) Image intensified video print pairs of photocytes seen by UV-autofluorescence (**e**, **g**) and bioluminescence (**f**, **h**) respectively.

Fig. 3. Transmission electron micrograph of cross-sectioned epidermal photocyte rod of *Periphylla.* Note the presence of numerous membrane-bound organelles within each crescent-shaped body.

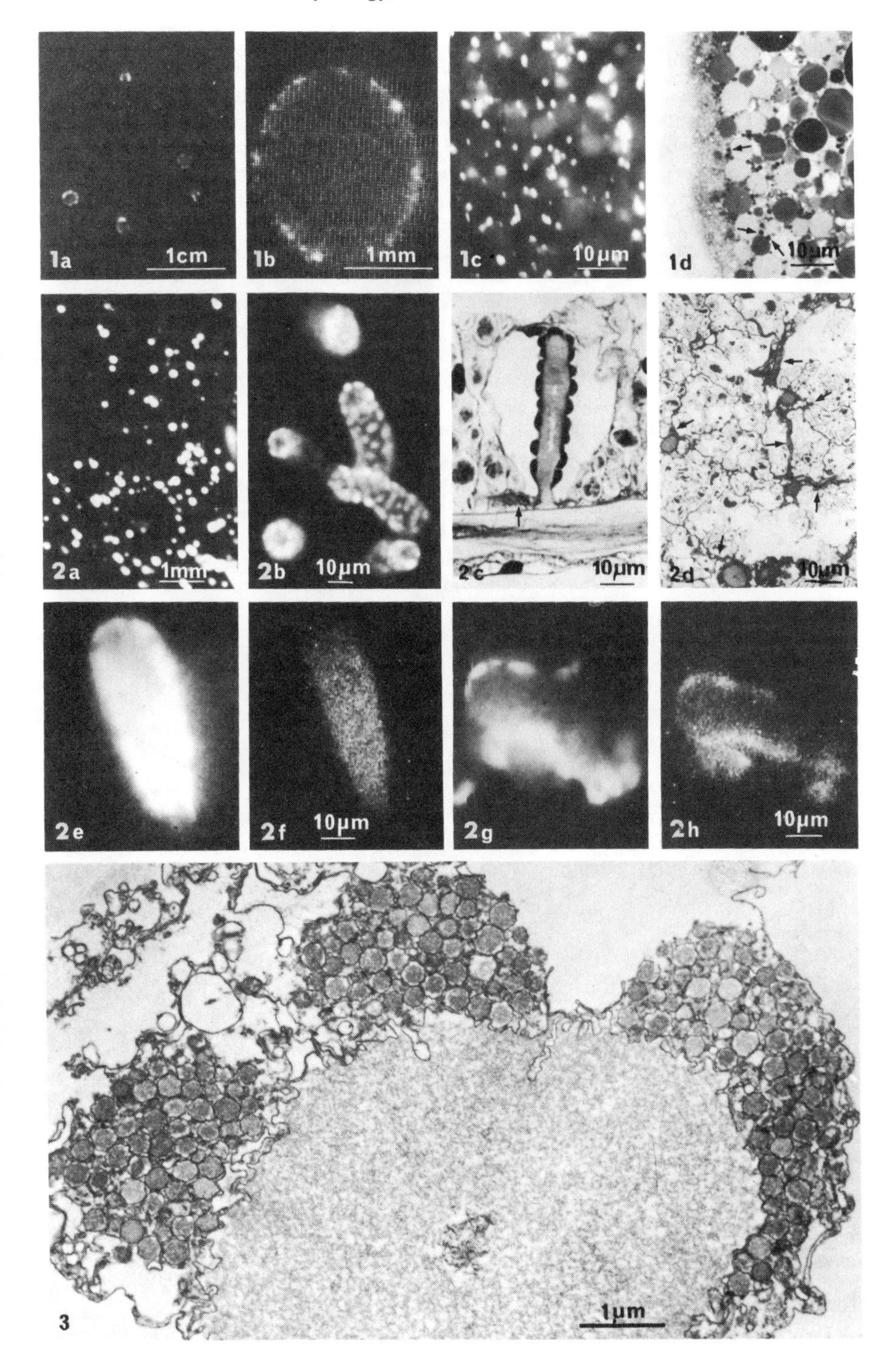
1a
1cm
1b
1mm
1c
10µm
1d
10µm
2a
1mm
2b
10µm
2c
10µm
2d
10µm
2e
2f
10µm
2g
2h
10µm
3
1µm

designated as the "small multipolar cell net" (10). In the dome region of the medusae, where at least muscle fibres are lacking and other effector and sensory cells appear to be few, the numerous photocytes may even represent the prime effector cell for this nerve net. Propagated bioluminescent responses associated with the exumbrellar epithelium are known to occur in at least two other scyphozoans: *Pelagia noctiluca* (11) and *Atolla wyvillei* (12). In *Pelagia* the bioluminescence is supposed to be associated with mucus secreted from granulated cells (8). *In Atolla* (12) bioluminescence could not be related to any particular epidermal cell type. We believe these difficulties in proving the identity of epithelial photocytes are related to the great ease by which photocytes may be expelled from the animal to the surrounding water (or mucus) during the collecting phase. Such loss of photocytes may also explain previous reports about extracellular bioluminescence among scyphozoans (5, 8). We are surprised by how few photocyte rods remain in the epithelium of *Periphylla* caught by nets, even when the heavily pigmented exumbrellar epithelium remains seemingly intact.

Although Triton X-100 represents an unphysiological stimulant of bioluminescence, it is interesting to note that the highest intensity of light was recorded from the interphase between the hyaline cylinder and the surrounding crescent-shaped and autofluorescent clusters rather than from the interior of these clusters themselves. This spatial separation of light and autofluorescence may indicate that the luciferin is stored separate from its luciferase and that diffusion of one of them is involved. This seems not to be in contradiction to the rapid time course of individual flashes in the living medusa (duration *~0.1s)* since we are dealing with diffusion distances in the sub-micrometer range.

Epithelial photocytes, with cytoplasmic granules, but otherwise with a structure distinct from *Periphylla* photocytes, are probably present in the narcomedusa *Aegina citrea* (Flood, unpubl. obs.) and in the siphonophore *Nanomia cara* (3).

Acknowledgement
We wish to express our gratitude to Mrs. Anne Nyhaug and Mrs. Teresa Cieplinska for technical assistance. The University of Bergen provided a three months guest professorship for J.M.B. and research time onboard R/V Håkon Mosby. The Norwegian Research Council is acknowledged for financial support.

References.

1. Herring PJ, Campbell AK, Whitefield M, Maddock L, editors. Light and life in the sea. Cambridge; Cambridge Univ Press, 1990.
2. Sweeney BM. Intracellular sources of bioluminescence. Int Rev Cytol 1980;68:173-95.
3. Freeman, G. Localization of bioluminescence in the siphonophore *Nanomia cara*. Mar Biol 1987;93:535-41.
4. Germain G, and Anctil M. Luminescent activity and ultrastructural characterization of photocytes dissociated from the coelenterate *Renilla kollikeri*. Tissue Cell 1988;20:701-20.
5. Morin JG. Coelenterate bioluminescence. In Muscatine L, Lenhoff HM, editors. Coelenterate biology: Review and new perspectives. New York: Academic Press, 1974;397-38.
6. Spurlock BO, Cormier M.J. A fine structure study of the anthocodium in *Renilla mülleri*. Evidence for the existence of a bioluminescent organelle, the luminelle. J Cell Biol 1975;64:15-28.

7. Dahlgren U. The production of light by animals. Porifera and coelenterates. J Franklin Inst 1916;181:243-61.
8. Nicol JAC. Observation on bioluminescence in pelagic animals. J mar biol Ass UK, 1958;37:705-52.
9. Fosså JH. Mass occurrence of *Periphylla periphylla* (Scyphozoa, Coronatae) in a Norwegian fjord. Sarsia 1992;77:237-51.
10. Bullock TH, Horridge BA. Structure and function of the nervous system of invertebrates. San Francisco: Freeman & Co. 1965; vol 1
11. Panceri P. Études sur la phosphorescence des animaux marins. Annls Sci Nat (sér. b: Zool.) 1872;16 (8):1-67.
12. Herring PJ. Bioluminescent responses of the deep-sea scyphozoan *Atolla wyvillei*. Mar Biol 1990;106:413-17.

BIOLUMINESCENT RESPONSES OF THE SCYPHOZOAN *PERIPHYLLA PERIPHYLLA* FROM A NORWEGIAN FJORD

[1]P J Herring, [2]J-M Bassot & [3]P R Flood
[1] Southampton Oceanography Centre, Empress Dock, Southampton SO14 3ZH, UK,
[2] NAM, Université Paris Sud, Paris, France,
[3] Institute of Zoology, University of Bergen, Bergen, Norway

Introduction

Species of the luminescent scyphozoan genera *Periphylla* and *Atolla* are common throughout the world's oceans, the former usually being found at depths of 400-1500m. The luminescent responses of net-caught *Atolla wyvillei* have been reported (1) but undamaged (submersible-caught) small animals of both genera are much more responsive (Herring & Widder, unpublished). Here we report work on specimens from a population of large *Periphylla* in the Lurefjorden (2), which at night migrate right to the surface under certain hydrographic conditions.

Methods

Animals at the surface were individually collected, with as little mechanical damage as possible, in buckets from an inflatable. Most had a coronal groove diameter of 100-140mm. They were brought back aboard R/V Håkon Mosby and stored on deck under thick black plastic sheet until required. Animals were stimulated electrically (usually with 20v pulses of 5ms duration) through platinum electrodes using a Grass square wave stimulator. Video recordings of the luminous responses were made with a Dage-MTI GenIIsys image intensifier system and CCD videocamera. Recordings were analyzed on a frame-by-frame basis.

Flash kinetics were analyzed with a 1mm fibre optic probe mounted over an isolated strip of epidermis and coupled to a photomultiplier. The flash responses to electrical stimuli were displayed on an oscilloscope and recorded photographically.

Results

1. Whole animal responses

A single stimulus delivered to any point on the aboral surface of the animal initiates a non-polarized wave of luminescence over the entire surface. The visual appearance is two-fold: a wave front of uniform intensity spreads over the surface while just behind it is a second wave of brighter point sources persisting for up to 1s. Over most of the animal a small minority of points persists for a few seconds as independently scintillating sources, whose flash frequency rapidly declines. The appearance at the distal lappet margins is different: all but the first few stimuli induce a very bright aggregation of scintillating sources persisting for many seconds. These continue to scintillate in the wake of a swimming animal.

A train of stimuli at about 0.5 Hz results in increasingly intense responses to the first few stimuli, though propagation usually extends over the whole animal from the start. Many tens, even hundreds, of responses can be elicited from a single animal. The first flashes propagate at a rate of some 200mm s^{-1} at a temperature of 6-7 °C; this rate falls to as low as 15-20 mm s^{-1} at the end of long flash sequences although the rate of decline is not always a smooth one (Fig. 1). Temperature has a marked effect on propagation rate.The initial rate increased from about 225 mm s^{-1} at 10.8 °C to some 500 mm s^{-1} at 19.7 °C.

The uniform passage of the early waves across the surface of the animal gives way during later ones to more erratic and disjunct propagation. The later stimuli in a long sequence may illuminate only the dome or only the lappet region (according to the site of stimulation) with the coronal groove acting as a visible barrier to the passage of weak bioluminescent responses. Early stimuli in a sequence only induce single waves. Later ones may induce multiple responses. Luminous waves propagating in a circular pathway round an animal from a single stimulus may have different speeds in the clockwise and anticlockwise directions but light production ceases where they meet.

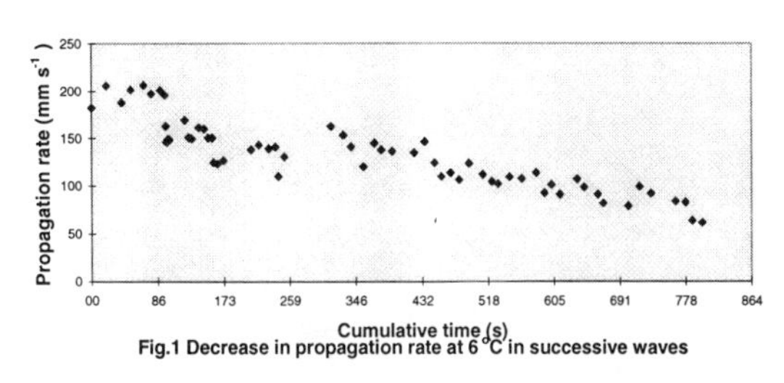

Fig.1 Decrease in propagation rate at 6 °C in successive waves

The propagation pathways were explored further by making shallow cuts in the epidermal layer of the dome. Any damage to the surface blocked the conduction of the wave of luminescence. A narrow epithelial bridge between two otherwise separated areas allowed the wave to pass through and expand beyond as a new wavefront. A rectangle or triangle of incisions prevented any penetration of an external wave into the enclosed area, and stimuli within the area did not spread beyond its margins. Complex maze-like pathways could be prepared which the wavefront would follow providing there was tissue continuity.

2. Light and chemical inhibition

Exposure to high light levels (e.g.daylight for several tens of minutes) greatly reduced the intensity of subsequent responses. A dome preparation was shielded with black plastic pierced by an aperture; 30 minutes fibre-optic microscope illumination through such an aperture blocked the conduction of luminous waves. A shorter exposure (10 minutes) irreversibly blocked the light production but not the conduction: the response crossed the light-affected (dark) area, with the bioluminescence reappearing beyond it. The luminescence was reversibly blocked by isotonic $MgCl_2$. Octanol in seawater did not block conduction but it did initiate some scintillation.

3. Flash recordings

Flash recordings from ~1mm^2 of stimulated tissue show a range of flash characteristics, though most are ca 150ms with fast rise times. Flash responses follow stimulus frequencies of up to 3Hz for short periods but synchrony soon breaks down, resulting in a flash rate much closer to 0.5-1Hz. In some cases there is a fast initial spike followed by a slower second flash. Both show facilitation during successive stimuli. The second flash may have complex kinetics and be followed by rhythmic scintillation (Fig. 2). Scintillation of lappet material is different; sources remain at the same intensity and have an initial rate of about 8Hz, declining to 1Hz. During a short train of stimuli the increasing intensity of flashes is due both to the facilitatory effect on single sources and to recruitment of additional sources during the train (Fig. 3).

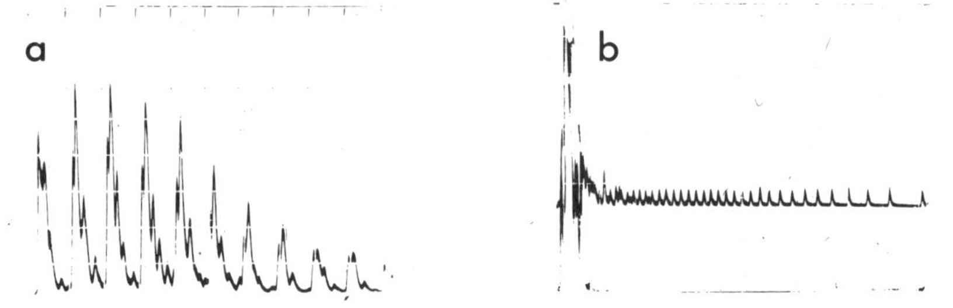

Fig. 2 Flash records showing examples of (a) facilitation (2 Hz stimuli, top)and (b) post-stimulus scintillation (centre; recorded by a 5mm fibre optic probe;same time scale as a)

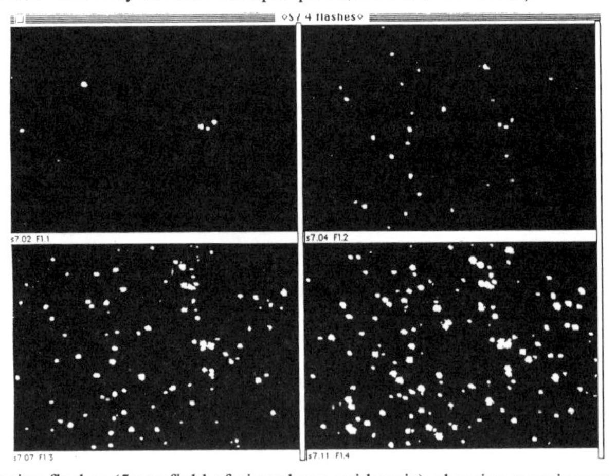

Fig.3 Four successive flashes (5mm field of view:dome epidermis), showing recruitment and facilitation

Discussion

Scyphozoa have at least a giant fibre nerve net (GFNN) and a diffuse nerve net (DNN) (3), while other Cnidaria may have multiple conducting systems. The diffuse propagation in *Periphylla* could be through the DNN or directly through epithelial conduction. It is not easy to distinguish between these two possibilities from the responses, and examples in other Cnidaria show a wide range of propagation rates in both pathways (4,5). Nerve nets are reversibly blocked by $MgCl_2$ while epithelial conduction is usually blocked by octanol. Our results, in conjunction with the fatigue-related changes (5), suggest that a diffuse ectodermal nerve net (DNN) is the most likely propagation pathway.

Each photocyte appears to contain a large rod-like structure with groups of associated granules (Flood, Bassot and Herring, this volume). The appearance of rhythmic scintillation on the epidermis may reflect the occasional release of these luminous structures which then flash spontaneously. The release of similar structures at the lappet margin is more likely to be a physiologically controlled defensive response designed to deliver them into the wake of a swimming individual..

Both facilitation and recruitment contribute to the increased intensity resulting from stimulus trains. This is typical of other luminescent systems (e.g. the siphonophore *Hippopodius hippopus*, (6)) and the recruitment presumably reflects facilitation at the nerve - photocyte junction. The slowing of the propagation rate seems to be the result of fatigue in conduction (5) rather than in the photocyte transduction system, because there is no apparent increase in luminous latency close to the point of stimulation. The fatigue-related spontaneous waves and erratic propagation suggests that the control of the system begins to break down, as in the multiple responses and autoexcitation of luminescent waves in colonies of the sea pansy *Renilla* (7). Conduction velocities in this species (40-70mm s^{-1} at 20-25°C) represent epithelial conduction between zooids.

Light inhibition seems to operate at two different sites. Short exposures only block photocyte luminescence and may inactivate the luciferin (or the photoprotein as occurs in *Mnemiopsis* (8)). The long light exposure damage to the conduction pathways may be more fundamental; it is possible that the large quantities of free protoporphyrin in the tissues acts as a photosensitizer (9).

Acknowledgement: We are most grateful for the support from the University of Bergen which enabled PJH and J-MB to participate in this research programme.

REFERENCES

1. Herring PJ. Bioluminescent responses of the deep-sea scyphozoan *Atolla wyvillei*. Mar Biol 1990;106:413-7.
2. Fosså JH. Mass occurrence of *Periphylla periphylla* in a norwegian fjord. Sarsia 1992;77:237-51.
3. Horridge GA. The nervous system of the ephyra larva of *Aurellia aurita*. Q J Microscop Sci 1956;97:59-74.
4. Josephson RK. Cnidarian neurobiology. In: Muscatine L, Lenhoff HM editors. Coelenterate biology:reviews and new perspectives. New York: Academic Press, 1974:245-78.
5. Passano LM. Variability in the initiaition of diffuse nerve-net impulses in the mangrove jellyfish *Cassiopea xamachana* (Coelenterata:Scyphozoa). Comp Biochem Physiol 1988;91C:273-9.
6. Bassot J-M, Bilbaut A, Mackie GO, Passano LM, Pavans de Ceccatty M. Bioluminescence and other responses spread by epithelial conduction in the siphonophore *Hippopodius*. Biol Bull 1978;155:473-98.
7. Buck J. Bioluminescent behavior in *Renilla*. I. Colonial responses. Biol Bull 1973;144:19-42.
8. Anctil M, Shimomura O. Mechanism of photoinactivation and re-activation in the bioluminescence system of the ctenophore *Mnemiopsis*. Biochem J 1984; 221:269-72.
9. Herring PJ. Porphyrin pigmentation in some deep-sea medusae. Nature 1972; 238:276-7.

PHOLAS DACTYLUS, THE REMARKABLE MOLLUSC
A FILM

Robert Knight and Jan Knight
Knight Scientific Limited, 18 Western College Road, Plymouth, PL4 7AG, UK

The famous, bioluminescent, marine, rock-boring, bivalve mollusc *Pholas dactylus*, the Common Piddock, was once widely distributed along the Atlantic coasts of Europe and the Mediterranean basin (1,2). However, by the year 1978 it was considered in France virtually extinct.(1). Across the Channel in England the situation was much the same with many large colonies barren and others nearly denuded of living specimens (2). And the potential loss of a species was made even more acute by the many applications of the luciférine of *Pholas dacylus*, which we call Pholasin®. We therefore embarked on a mission to cultivate the mollusc in a land-based system to obtain commercial quantities of *P. dactylus* and to recolonise barren areas.

We celebrate our success in the conservation and perhaps even resurrection of a remarkable mollusc with this film of the life history of *Pholas dactylus* which was filmed on S-VHS format. The tape lasting some 19 minutes was edited from 200 hours of film taken over a 2 year period. Most of the filming was through stereo as well as high power microscopes. The film follows the liberation of eggs and sperm by the separate sexes and the fertilization and development of the egg from a rapidly dividing ball of cells into a typical shelled D-larva. The D-larvae are filmed as they transform into exquisitely beautiful free-swimming veligers. From there we trace the changes that occur at metamorphosis when the pedal veliger gradually becomes a unique miniature adult. Along the way we made new discoveries, for example that the gills develop while the velum still functions, that the siphons form as the velum recedes and, most remarkably, that the sperm become activated by light into frenzied motion. Other new observations include: the short-lived locating knob and depression on the juvenile shell at the early stages of drilling, the mechanism for dealing with rock spoil, and that the young piddock appears to have a left-handed thread! And some puzzles of the adult's anatomy are understood immediately. The fast paced film, which also includes the discovery and naming of a new genus and species of ciliated protozoan (3,4) unique to the gills of *P. dactylus* is narrated, choreographed and set to music and designed to appeal to specialist and non-specialist alike. This is the story of *Pholas dactylus*, the not so common, Common Piddock.

References

1. Michelson AM. Purification and properties of *Pholas dactylus* luciférin and luciférase Meth Enzymol 1978; 57:385-406.
2. Knight J. Studies on the biology and biochemistry of *Pholas dactylus* L. PhD thesis University of London, 1984.
3. Knight R, Thorne (=Knight) J. *Syncilancistrumina elegantissima* (Scuticociliatida: Thigmotrichina), a new genus and species of ciliated protozoon from *Pholas dactylus* (Mollusca: Bivalvia) the Common Piddock. Prostistologica. 1982;18:53-66.
4. Knight J, Knight R.The blood vascular system of the gills of *Pholas dactylus* L. (Mollusc, Bivalvia, Eulamellibranchia). Phil Trans. R Soc Lond; 1986;B3 13:509-23.

IN SITU VIDEO RECORDINGS OF BIOLUMINESCENCE IN THE OSTRACOD, *CONCHOECIA ELEGANS*, AND CO-OCCURRING BIOLUMINESCENT PLANKTON IN THE GULF OF MAINE

EA Widder
Harbor Branch Oceanographic Institution, 5600 US 1 N, Fort Pierce 34946, USA

Introduction

An *in situ* video transect technique has been used to identify and map bioluminescent zooplankton based on the spatial and temporal patterns of stimulated bioluminescent displays (1-4). The utility of this mapping technique depends on having an adequate data base of identified displays. Since displays from captured specimens are often radically different from those recorded *in situ* (1,2), definitive identification of a particular organism with a particular bioluminescent display needs to be confirmed by *in situ* recordings. For larger, slow moving organisms, like the gelatinous zooplankton, correlation of display with organism is simply a matter of turning on the submersible's lights to reveal the source of the display and then capturing the specimen for definitive taxonomic identification (1,2). For smaller, more motile plankton, a less direct approach is required. For example, bioluminescent displays of the euphausiid *Meganyctiphanes norvegica* were identified by correlating the abundance of a particular display type with the abundance of euphausiids as quantified concurrently with high-resolution SONAR (2,5). In this investigation a similar correlation analysis technique was used to identify *in situ* display parameters of the pelagic ostracod *Conchoecia elegans.*

Of the three marine genera of ostracods which are known to be bioluminescent (*Vargula*, *Cypridina* and *Conchoecia*), *in situ* observations have been documented only in *Vargula* and *Cypridina* (6-8). In these coastal genera bioluminescence is produced as a luminous secretion from glands in the labrum. *In situ* observations have revealed that these genera release luminescent chemicals into the water to produce extrinsic displays (7,8). *Vargula* produces complex courtship displays by releasing glowing pulses of light into the water column in species specific spatial patterns, while antipredatory displays involve the release of a brilliant cloud of luminescence which glows for many seconds (7). Species of *Cypridina* also emit long lasting (3-4 sec) luminous clouds into the water when photically stimulated with a flashlight and even longer lasting luminous excretions in response to mechanical stimuli (8). In the cosmopolitan pelagic genus *Conchoecia*, luminescent glands are located on the carapace margin and laboratory investigations of net captured specimens have demonstrated that bioluminescence may be excreted or retained within the glands (6,9). Although there are no reports on the nature of bioluminescent displays produced by *Conchoecia in situ,* recorded displays from captured specimens have been described as "multi-apex impulses" (10) and "fast, repetitive flashes" (MI Latz, personal communication). Whether or not the luminescence is excreted or retained within the glands during these displays has not been reported.

Materials and Methods

During a series of *Johnson-Sea-Link* (JSL) submersible dives in the Gulf of Maine (Wilkinson Basin 42° N: 69° W) in August 1992, video recordings of stimulated bioluminescence were made with an intensified video camera (ISIT 66, Dage-MTI, Michigan City, IN), mounted inside the observation sphere of the submersible. A 1 m diameter hoop with 1800 μm mesh Nitex® nylon stretched across it was mounted in front of the sphere and served to stimulate bioluminescence in the plane of focus of the ISIT video camera during horizontal transects. Each transect lasted 4 min at a forward speed of 0.6 knots. During each transect a quantitative plankton sample was collected by a suction pump with an in-line flow meter and 64 μm mesh screen at the outflow of the collection buckets. A minimum of 300 L was pumped for each transect and transects were carried out at discrete depths between the bottom (247-274 m) and 140 m during daytime dives and between the bottom and 15 m during night-time dives. Collections of fragile gelatinous zooplankton and euphausiids were made using the submersible's detritus samplers (2). Identification of organisms responsible for unidentified displays was based on correlation of the number of such displays per m^3 with the abundance of identified bioluminescent organisms in the quantitative plankton samples. Computer image analysis (1024XM, Megavision, Santa Barabara, CA) of the video-recordings was used to assess the kinetics of emission of different display types.

Results and Discussion

Analysis of the intensified video transects revealed an as yet unidentified luminescent display, which consisted of a series of rapid, repetitive flashes (RRFs). Possible sources of bioluminescence in the water column identified from pumped samples and detritus sampler collections were the dinoflagellate *Protoperidinium depressum*, the ostracod *Conchoecia elegans*, the copepods *Metridia lucens* and *M. longa*, the euphausiids *Meganyctiphanes norvegica*, *Thysanoessa gregaria*, *T. raschi*, and *T. inermis*, the polychaete *Tomopteris* sp., the siphonophore *Nanomia cara*, the ctenophores *Euplokamis* sp., *Beroe cucumis* and *Bolinopsis infundibulum* and the anthomedusa *Leuckartiara octona.* Initially displays were classified based on whether the bioluminescence appeared to be intrinsic or extrinsic and whether or not the displays were spatially distinct. Examples of these different displays are shown in Fig 1 with representative non spatially distinct intrinsic displays shown in white outlines. The kinetics of such displays, shown in Fig 2, were their primary identifying characteristics. In Fig 1A several of the large extrinsic displays characteristic of the cydippid ctenophore *Euplokamis* sp. (2) are shown as well as 2 examples of non spatially distinct intrinsic displays from other organisms. The display shown in the square outline is an RRF. The display shown in the triangle in the middle and bottom frames consisted of a single flash and may have originated from an RRF source which passed through the screen (see below). Fig 1B was from a transect at 30 m where the copepod *Metridia lucens* was found in high abundance. The extrinsic displays associated with these bioluminescent copepods were easily distinguished from those of *Euplokamis* sp. as they appeared smaller, less persistent and diffuse. The intrinsic display shown in the box in Fig 1B is characteristic of euphausiids which are too large to pass through the 1800 μm mesh of the transect screen and which therefore produce displays which are more persistent (Fig. 2A) than those of smaller organisms, like dinoflagellates, which pass through the screen. Fig 1C shows a transect from 60 m where the dinoflagellate *Protoperidinium depressum* was abundant. These non spatially distinct intrinsic displays, one of which is shown in the square, consisted of a single brief

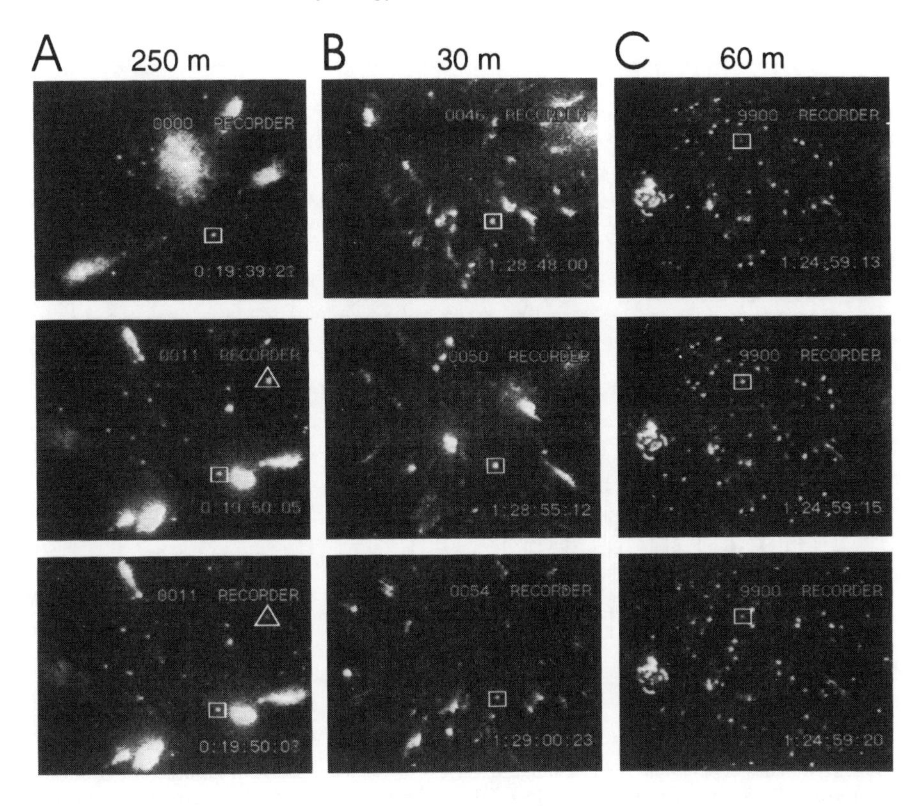

Fig 1 A series of video frames showing bioluminescent displays taken from transects made at three different depths during one night dive (Dive 2440). The field of view shown from each frame is 78 X 95 cm. The time stamp at the bottom of each frame represents hours, minutes, seconds and frames (30 frames/s). In several frames the vertical support bar of the transect screen is dimly visible in the middle of the field. A. Sequence from a transect run at 250 m showing extrinsic release of bioluminescent particles by the ctenophore *Euplokamis* sp. The intrinsic display indicated in the square produced a rapid repetitive flash (RRF) (see Fig. 2B). The display indicated in the triangle in the middle and bottom frames was an intrinsic display which consisted of a single flash and may be from specimens of *Conchoecia elegans* which passed through, rather than stuck to the transect screen (see Fig. 2D). B. Sequence from transect at 30 m where the bioluminescent copepod *Metridia lucens*, was abundant (63 m^3) and probably responsible for the extrinsic displays which appeared as small diffuse clouds of luminescence that passed through the screen. The intrinsic display indicated in the square produced the prolonged glow characteristic of euphausiids (see Fig. 2A). C. Sequence from transect at 60 m where the bioluminescent dinoflagellate, *Protoperidinium depressum*, was abundant (7600 m^3) and was responsible for intrinsic displays like the one indicated in the square (see Fig. 2B). The bright intrinsic display on the left is the "figure-8" pattern characteristic of the lobate ctenophore *Bolinopsis infundibulum*.

flash (Fig. 2B) apparently stimulated by the shear stress of flow through the screen. Also in this transect, on the left, is the spatially distinct "figure 8" display characteristic of the lobate ctenophore *Bolinopsis infundibulum* (Fig 1C) (3).

The unknown displays designated as RRFs (Fig 1A, Fig 2C) consisted of from 2 to 120 flashes per display. Maximal flash frequencies as determined from the video, which

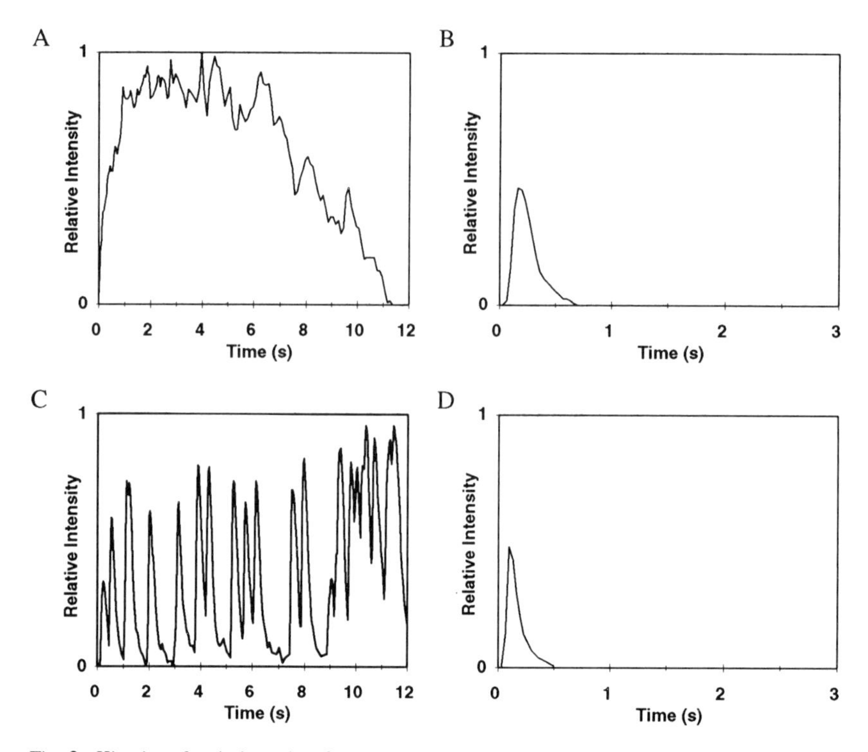

Fig. 2 Kinetics of emission taken from image anaylsis of video recordings of intrinsic displays which lacked any distinct spatial pattern. The prolonged emission shown in A was characterisitc of euphausiids while the brief flash in B was characteristic of dinoflagellates. The emissions shown in C and D were two previously unidentified displays here attributed to *Conchoecia elegans*. C was the RRF of a specimen which apparently stuck to the transect screen while D was a single flash, thought to be due to a specimen passing through the screen. All four emissions are plotted on the same relative intensity scale.

records 30 fps, were 7 flashes per second. Most RRFs consisted of a brief burst of flashes as the source appeared to stick momentarily to the screen. Occasionally a burst ended as a source appeared to actively swim away from the original point of contact while still flashing. The determination of active swimming was based on a trajectory that was not in line with other particles flowing through the screen. Flash bursts were sometimes separated by periods of no visible light emission. With prolonged flash bursts, flash intensity gradually decreased, as did flash frequency. Glowing between flash bursts or following a prolonged burst was not observed.

Initially, the observation that RRFs originated from organisms that did not pass through the transect screen made the possibility that these displays originated from *Conchoecia elegans* seem unlikely. Although a few of the largest ostracods from the pumped samples had carapace lengths of more than 1800 μm, all specimens readily passed through the screen when the carapace was oriented on axis with the flow. However, an analysis of the abundance of RRF displays, as compared to the abundance of luminescent organisms in pumped samples collected during each transect, revealed a significant positive correlation ($r=0.86$; $P<<0.001$) with the distribution of *Conchoecia elegans* (Fig 3).

The discrepancy between the small size of the ostracods and their apparent retention by the transect screen may be accounted for by the very large number of gelatinous organisms, (primarily the ctenophores *Bolinopsis infundibulum* and *Euplokamis* sp. and the siphonophore *Nanomia cara*) which stuck to the screen during transects. This hypothesis, that the RRFs occurred only when bioluminescent ostracods adhered to gelatinous material on the screen, would account for the fact that the concentration of RRFs was less than 1% of the concentration of *Conchoecia elegans* collected in the pumped samples. This would also account for the other unidentified display (shown in the triangle in Fig 1A) which consisted of a single brief flash (Fig 2D) which may have been a single emission from a specimen of *Conchoecia elegans* that passed through the screen These displays were not included in the correlation analysis with *Conchoecia elegans* because, at present, abundance estimates of individual display types are made by visually scanning the video transects and it proved difficult to distinguish this display from that of dinoflagellates. However, computer image analysis of such displays from transects, where *Conchoecia elegans* was most abundant and where dinoflagellates were generally absent, indicated that these flashes (Fig 2D) had a more rapid decay phase than dinoflagellate flashes (Fig 2B). Therefore, with the image recognition algorithms which are currently under development (3,4) it may be possible to distinguish these displays in future analyses.

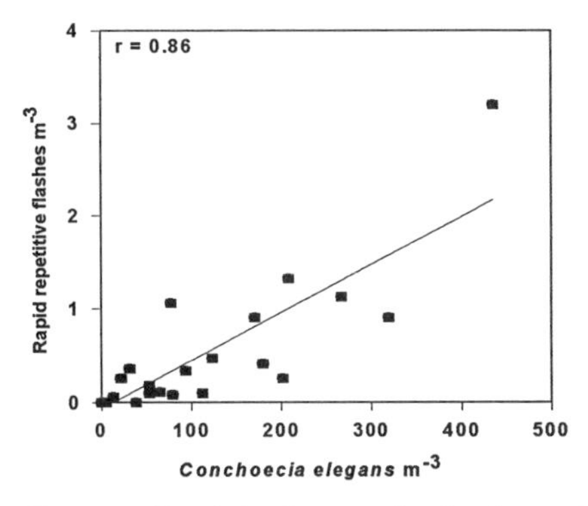

Figure 3. Correlation between the abundance of RRFs in the video transects and the abundance of *Conchoecia elegans* found in the pumped samples collected during each video transect. Line represents the calculated regression, given by the equation $Y = 0.005 X - 0.066$ ($n = 29$).

Rapid repetitive flashes are more common in intrinsic light producing systems that are under direct neural control, than in secretory systems (12). The appearance of the RRFs in the video transects would suggest that *Conchoecia elegans* either retains its bioluminescent emissions within the carapace or, if there is some release of material, the kinetics of light emission are so rapid that there is no apparent separation of the luminescence from the source. In either case the kinetics of emission for this pelagic ostracod are very different from the luminous secretions characteristic of the coastal genera *Vargula* and *Cypridina.* It is interesting that similar fast repetitive flashes have been recorded from the small poecilostomatoid copepod *Oncaea conifera,* which is unique for producing an intrinsic display while all other luminescent copepods produce secretory emissions. It has been suggested that the intraglandular mode of luminescence in *O. conifera* is an adaptation to its small size, which prohibits a rapid escape from secretions entrained in its boundary layer (11). Although *Conchoecia elegans* is generally smaller than *Vargula,* which can be as large as 3 mm (7), it is comparable in size to *Cypridina* (8). Whether or not the rapid repetitive flashes of *Conchoecia elegans* are a hydrodynamically driven adaptation, it is probable that this display serves an antipredator function given that it is initiated by prolonged mechanical stimulation. We still have much to learn about the visual ecology of the marine environment before we can begin to interpret the adaptive significance of these complex bioluminescent displays.

Acknowledgements

This research was supported by grant N00014-90-J-1819 from the Office of Naval Research and by grant UCAP-92-21 from NOAA/NURP. For assistance at sea I thank the captain and crew of the RV *Seward Johnson* and the crew of the *Johnson Sea Link* submersible as well as T. Frank, S. Bernstein and S. Haddock. I thank D. Smith for assistance with computer image analysis and P. Herring for providing the English translation of reference 10.

References

1. Widder EA, Bernstein SA, Bracher DF, Case JF, Reisenbichler KR, Torres JJ et al. Bioluminescence in Monterey Submarine Canyon: image analysis of video recordings from a midwater submersible. Mar Biol 1989;100:541-51.
2. Widder EA, Greene CH, Youngbluth MJ. Bioluminescence of sound-scattering layers in the Gulf of Maine. J. Plank. Res. 1992;14:1607-24.
3. Widder EA. 3-D bioluminescence mapping. U. S. GLOBEC News 1993:6.
4. Kocak DM, Lobo NdV, Widder EA. Tracking and mapping underwater bioluminescent displays using snakes. Oceans '96:in press.
5. Greene CH, Widder EA, Youngbluth MJ, Tamse A, Johnson GE. The migration behavior, fine structure and bioluminescent activity of krill sound-scattering layers. Limnol Oceanogr 1992;37:650-8.
6. Herring PJ. Bioluminescence in the crustacea. J Crus Biol 1985;5(4):557-73.
7. Morin JG. "Firefleas" of the sea: Luminescent signaling in marine ostracode crustaceans. Florida Entomologist 1986;69:105-21.
8. Tsuji FI, Lynch RVI, Haneda Y. Studies on the bioluminescence of the marine ostracod crustacean *Cypridina serrata.*. Biol Bull 1970;139:386-401.
9. Angel MV. Bioluminescence in planktonic halocyprid ostracods. J mar biol Ass UK 1968;48:255-7.
10. Rudjakov JA. The study of the luminescence of pelagic ostracoda. Bioenergetics and Biological Spectrophotometry, Moscow: Nauka 1967;52-62.
11. Herring PJ, Latz MI, Bannister NJ, Widder EA. Bioluminescence of the poecilostomatoid copepod *Oncaea conifera*. Mar Ecol Prog Ser 1993;94:297-309.
12. Case JF, Strause LG. Neurally controlled luminescenct systems. In: Herring PJ, editors. Bioluminescence in action. London, New York, San Francisco: Academic Press, 1978.

THE ESTIMATION OF SEA BIOLUMINESCENCE BY MESOZOOPLANKTON BIOMASS DISTRIBUTION IN THE UPPER LAYER OF THE CENTRAL ATLANTIC

II Gitelson[1], *LA Levin*[1], *JA Rudjakov*[2] *and RN Utyushev* [1]
[1]*Institute of Biophysics, Akademgorodok, Krasnoyarsk 660036, Russia*
[2]*P.P.Shirshov Institute of Oceanology, 23 Krasikov St., Moscow 117218, Russia*

Introduction

The observations of sea bioluminescence which were carried out in the last decades demonstrated the wide distribution of this phenomenon which came to the attention of oceanographers, navigators, and military specialists. It unequivocally says about necessity to learn how to predict bioluminescence using other biooceanographical data. The inverse problem is no less important since the remote sensing easily applicable to phytoplankton and bioluminescence studies cannot supply an information on distribution of organisms representing higher links of the food web.

The potential bioluminosity of sea water depends on abundance of luminescent phyto- and zooplankton (1, 2). As to its dependence on total mesozooplankton biomass, their cause and effect relation is less straightforward, though some statistically significant correlation was stated at meso- and microscale (3, 4). This paper is devoted to the inquiry if this relation exists over larger space-temporal scales, comparable to the scales of tropical part of the Atlantic ocean.

Material and Methods

The bioluminescence intensitywas measured at 17 stations during 37 trip of RV "Akademic Kurchatov", at the cross section along 22° N in August-October, 1983. The station sites are given in Fig. 1.

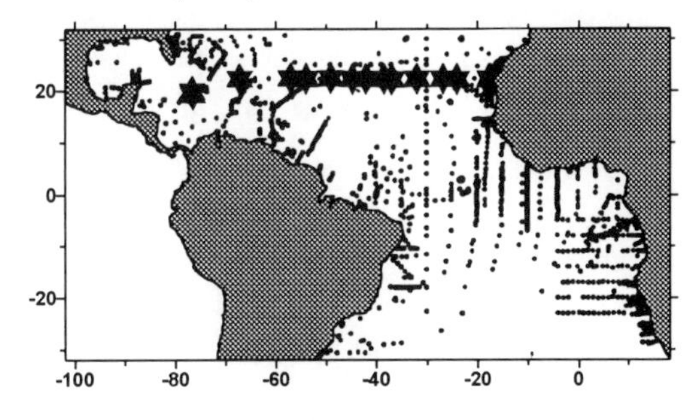

Figure 1. Station sites. Asterisks - bioluminescence measurements, dots - zooplankton biomass determinations

The changes of bioluminescence with depth were studied using the device "Daisy-3" (5, 6). The construction of the device greatly reduces the ratio of inside and outside illumination (by 140 DB). This furnishes an opportunity to measure the bioluminescence intensity regardless of outside illumination.

To calibrate the bioluminescence sensor, the radioactive standard light source was used, the light flux F due to bioluminescence being equal to (7):

$$F = A\,D_b/D_e \qquad (1)$$

where D_b and D_e are light fluxes emitted by luminous organisms and by the standard source respectively and A is a constant depending on the device features and on the spectral characteristics of both the photomultiplier and the standard source.

At every station, 3-5 bioluminescence profiles down to the depth 100-150 m were taken at the device velocity 1.2-1.3 m sec^{-1}. The preliminary treatment of the data included the calculation of the average bioluminescence intensity of 5-meter layers. Then, on the basis of the data on diurnal bioluminescence rhythmicity (3, 8), the values were adjusted to dark time of the day when bioluminescence intensity keeps at the level close to stationary. After that, the average bioluminescence intensity in the layer 0-100 m was calculated for every vertical profile and, finally, - the same for every station.

To compare the field of bioluminescence with the same of zooplankton biomass, the biomass data gathered on 1506 oceanographic stations between 35°N and 35°S (Fig. 1) were used. In order to use to the maximum degree the data collected with nets of different mesh sizes, from different layers, in different seasons, the same set of conversion factors was used as in the paper devoted to mesoplankton biomass mapping (9). In the result of the conversion all measurements were expressed in units of wet weight of animals of the size range 0.2-20 mm per meter cube in the layer 0-100 m. These values were adjusted to the mean annual level about which the seasonal oscillations run at the specified latitude. Then, for every 5-degree square of the ocean area, the arithmetic mean of all station values was calculated. Finally, the "Surfer 6.01" Golden Software program was applied to calculate the grid values, which were computed using the "minimum curvature" method. The grid was smoothed by the matrix 3x3 with the weight of matrix center equals 2. At the next step of the data analysis, a vertical slice was taken through the biomass grid surface along the line of the station sites where bioluminescence was measured. And finally, interpolated in this way biomass and bioluminescence intensity values (Table 1) were compared using the conventional methods of correlation and regression analysis.

St	D	t	Lat	Lon	log(B)	log(BL)
3790	08-02	17-37	22.47	48.98	1.663	2.261
3793	08-06	17-01	22.33	67.18	1.590	2.651
3794	08-22	00-13	19.92	76.65	1.664	2.654
3797	08-26	00-00	19.55	76.63	1.663	2.619
3800	08-28	15-23	19.28	76.52	1.658	2.504
3822	09-10	22-43	22.40	57.00	1.683	2.353
3823	09-11	21-02	22.37	53.92	1.636	2.427
3827	09-14	01-26	22.28	44.68	1.666	2.590
3829	09-14	21-36	22.32	41.98	1.704	2.564
3831	09-15	21-00	22.33	38.52	1.810	2.695
3832	09-18	00-15	22.18	36.97	1.830	2.651
3834	09-21	21-32	22.32	32.05	1.768	2.547
3836	09-23	02-49	22.33	26.85	1.783	2.474
3841	09-25	21-59	22.33	17.63	2.058	3.268
3842	09-26	01-01	22.33	17.32	2.071	3.148
3843	09-26	03-45	22.17	17.12	2.083	3.332
3879	10-11	22-42	22.25	24.12	1.843	2.838

Table 1. Stations (St), date (D, m-d, 1983), and zone time (t, h-m), of measurements, coordinates (Lat, latitude; Lon, longitude west, both in degrees), logarithms (to the base 10) of mean annual mesoplankton biomass in the layer 0-100 m (B, mg m^{-3}), and of bioluminescence intensity in the same layer (BL, 10^{-6} $\mu W\ cm^{-2}$)

Results and Discussion

The dependence of bioluminescence intensity on mesozooplankton biomass is shown in Fig. 2. These data were fitted to the linear regression equation:

$$BL = -0.327 + 1.695\ B, \qquad (2)$$

where BL - logarithm (to the base 10) of bioluminescence intensity (10^{-6} $\mu W\ cm^{-2}$) and B - logarithm of mesozooplankton biomass (mg m^{-3}). Standard error of regression coefficient is 0.233 and, hence, the regression is significant at $p<0.000003$. Correlation coefficient between logarithms of biomass and bioluminescence intensity proved to be as high as 0.883. It means that 77.9% of variability of bioluminescence intensity is explained by biomass variability.

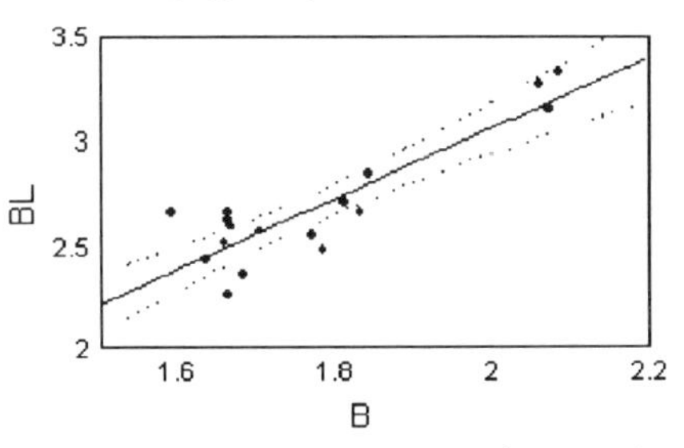

Figure 2. Bioluminescence (BL, 10^{-6} $\mu W\ cm^{-2}$) versus zooplankton biomass (B, mg m^{-3}) and .95 confidence bands for means

In spite of the fact that the regression above was gained using the limited bioluminescence data taken along 22°N, we believe that it is valid and applicable to much more wide area between 35°N and 35°S which includes two central and one equatorial pelagic communities comprising very similar sets of neritic and pelagic species in their near-shore and off-shore parts respectively (10). To predict values of bioluminescence intensity, each grid biomass value was converted into bioluminescence one using the equation (2). The resulting grid was visualized by the "Surfer" program as the contour map given in Fig. 3.

The derived map of potential bioluminescence can be compared with two maps produced in earlier studies, one of which (11) is prepared using the results of direct bioluminescence measurements, and the other (3) is constructed using the correlation between zooplankton biomass and bioluminescence intensity. In general, it can be said that all three maps equally reflect the most general tendencies of potential bioluminosity pattern. As to details, the differences can be attributed to data limitation and/or, to differences in methods of biomass evaluation, the disparate results of which were used earlier in bioluminescence intensity calculations. Much more important and troublesome are the prominent differences in absolute bioluminescence values. This dissimilarity is too great to be explained by differences in plankton collection methods and/or plankton samples variability and definitely says about the necessity to carry out the intercalibration of bioluminescence measurement methods used by the scientists from different oceanographic institutions.

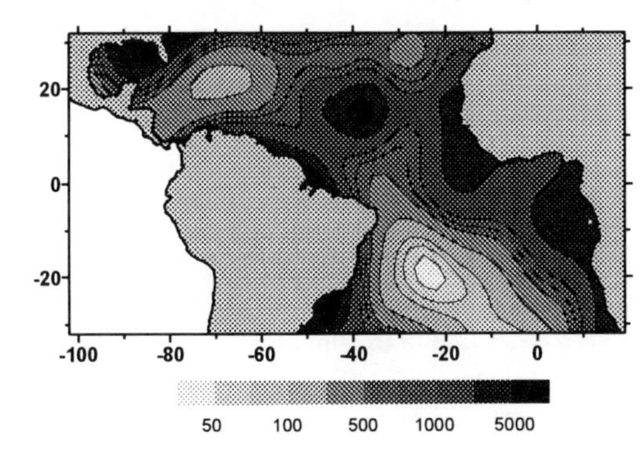

Figure 3. Potential bioluminosity of the Central Atlantic (10^{-6} $\mu W\ cm^{-2}$)

A special consideration is provoked by the high correlation (0.88) between macroscale changes in mesozooplankton biomass distribution and bioluminescence intensity which was found in this study. The correlation coefficients like this (0.75 - 0.88) were derived also between these variables measured simultaneously at lesser space-temporal scales, at different depths of the same station or at different stations of the same trip (3). It demonstrates that micro-, meso- and macrocomponents of the stimulated bioluminescence variability are approximately equal and comparatively easy gained measurements of bioluminescence can be used for the express estimation of temporal and space variability of the ocean productivity over wide range of space-temporal scales.

References

1. Rudjakov JA, Voronina NM. Plankton and bioluminescence in the Red Sea and the Gulf of Aden. Okeanologya, 1967; 7:1076-88.
2. Rudjakov JA. Bioluminescence potential and its relation to concentration of luminescent plankton. Okeanologya, 1968; 8:888-94.
3. Gitelson II, Levin LA, Utyushev RN, Cherepanov OA, Chugunov JV Bioluminescence in the Ocean. Moscow, Nauka, 1992.
4. Levin LA, Utyushev RN, Artemkin AS. Distribution of bioluminescence field intensity in the Equatorial Pacific. Trans Inst Oceanology, USSR Acad Sci, 1975; 102:94-101.
5. Gitelson II, Levin LA, Shevirnogov AP, Utyushev RN, Artemkin AS. Bathiphotometric profiling in the oceanic pelagial and possibilities of its use in the investigations of biocenosis space structure. In: Vinogradov ME, editor. Functioning of pelagic communities in the tropical regions of the Ocean. Moscow, Nauka, 1971:50-63.
6. Levin LA, Chugunov JV, Ermakov VV. Automated complex of biophysical devices for the analysis of space structure of pelagic communities. Congress of Soviet Oceanologists, Yalta. Sebastopol, 1982; 6:5-6.
7. Utyushev RN, Levin LA. Energy estimation of bathyphotometric measurement results. In: Gitelson II, editor. Bioluminescence in the Pacific. Krasnoyarsk, p. 1982:111-22.
8. Utyushev RN, Levin LA, Cherepanov OA. Experience on estimation of bioluminescence field structure using data derived in daylight hours. Okeanologya, 1984; 24:701-5.
9. Rudjakov JA, Tseitlin VB, Kitain VJ. The mean annual zooplankton biomass distribution in the upper layer of the Pacific. Okeanologya, 1996; 36 (in press).
10. Beklemishev CW. Ecology and biogeography of the open ocean. Moscow, Nauka, 1969.
11. Piontkovski SA, Tokarev YN, Bitukov EP, Williams R, Kiefer D. The bioluminescent field of the Atlantic Ocean. 1997.(this volume)

THE BIOLUMINESCENCE FIELD AS AN INDICATOR OF THE SPATIAL STRUCTURE OF THE PLANKTONIC COMMUNITY OF THE MEDITERRANEAN SEA BASIN

EP Bitukov[1], YuN Tokarev[1], SA Piontkovski[1], VI Vasilenko[1], R Williams[2], BG Sokolov[1]

[1] *Institute of Biology of the Southern Seas 335011 Sevastopol, Ukraine,*

[2] *Plymouth Marine Laboratory, Prospect Place, Plymouth PL1 3DH, UK*

Introduction

The modern methodology of systematic analysis of the structural-functional characteristics of the plankton community is characterized by the combination of classic and express methods. The latter ones are based on measurements of the amplitude and the frequency parameters of the bioluminescent, fluorescent and acoustic fields of the ocean and the assessments of their relationships with physical and chemical fields (1,2).

The Mediterranean Sea has been a region where bioluminescence studies have been carried out on the level of single organisms (3) and the bioluminescent field as a whole (4). The last 25 years of field research carried out by the Institute of Biology of the Southern Seas has enabled a comprehensive survey of the bioluminescence in the Mediterranean basin to be executed. This paper deals with some major trends evaluated in the spatial-temporal variability of bioluminescence in the micro- and macroscales not previously published.

Materials and Methods

The instrumental measurements of bioluminescence in the Mediterranean basin were started by the Laboratory of Biophysical Ecology, IBSS in 1965 (5,6).

The submersible bathyphotometer, built in the laboratory was used (7). This gear has been updated during the last few years, although its linear size and the basic principles of parameter measurements are still the same. A photodiode was used to measure the light flux works in a linear mode within the range up to 57 dB. Radio-luminescent light sources were applied to calibrate the bioluminescence sensor. A light-protecting unit was used to restrict the volume of the scanned space (the chamber), to reduce the astronomic light value and to initiate the mechanical stimulation of bioluminescence from plankton organisms. It was designed in a form a rotor impeller in the latest version of the submersible module **SALPA**. This module carries sensors for the measurement of temperature, conductivity, and bioluminescence; operating simultaneously. An inter-calibration procedure has been applied to obtain transfer coefficients when the submersible modules have been upgraded over the years.

Normally, the field measurements were carried out 1 hour after the sunset in the majority of expeditions. This enables us to consider the whole data set as having been collected within the same period of the day, which is important in terms of data comparison. The data bank contains 3500 vertical casts of bioluminescent potential obtained at 375 oceanographic stations carried out during 21 expeditions to the Mediterranean Sea basin (1968-1993).

Results and Discussion

The well developed vertical stratification of the bioluminescent field can be noted as a typical feature within the upper 100m. The minimal thickness of layers consist from 3 to 7 m. The degree of development of such a stratification increases in waters with enhanced biological productivity.

However, almost in all cases, there is an upper low-bioluminescent layer (8-15 m thick), a quasi-stationary layer of enhanced bioluminescence in the thermocline, an adjacent layer with depressed bioluminescence, 15-30m thick, beneath the thermocline, and, finally, a well developed bioluminescent layer at depths of 60-130 m. Bioluminescence in the deep layers below is normally formed by the single flashes (i.e. single organisms which do not form a dense planktonic field).

A general trend of increase in bioluminescence, from the Aegean Sea towards the west, was evaluated on the scale of the whole basin. Maximal numbers of bioluminescence were noted within zones where the Atlantic and Mediterranean water masses interact, within cyclonic eddies and the divergent zones. For instance, bioluminescence recorded within these local dynamic zones exceed the average value in the Ionian Sea by 30 times.

Bioluminescence in the central part of the Black Sea is 3 times weaker than that in the Alboran Sea. However, it is one order of magnitude higher than that observed in the central part of the Mediterranean Sea.

Seasonal changes are also well developed and mostly observed in the Black Sea. Two intensive periods of bioluminescence were recorded, one in May-June and the other one, more intensive, in October-November. The bioluminescent potential was 1.4 10-2 microwatt cm^{-2} l^{-1}, which exceeded the minimal numbers observed in February by 500 times.

Seasonal cycles are weakly developed in the oligotrophic regions of the Mediterranean Sea. Differences between summer and winter periods were only 3.5 times. In the center of the Sardinian Sea, where the average value of bioluminescence suggests that it is a mesotrophic region, seasonal changes are also weakly developed. Thus, differences in bioluminescence in September and December rarely exceeded 20%.

Macroscale trends of the change in bioluminescence within the Mediterranean basin are comparable with that of the plankton spatial distribution (8). Zones of the intensive bioluminescence are related to the position of enhanced plankton biomass zones. This tendency is partly mirrored in the statistical relationship between abundance of Dinoflagellates and the bioluminescent potential found for non-polluted waters of the basin (Fig. 1).

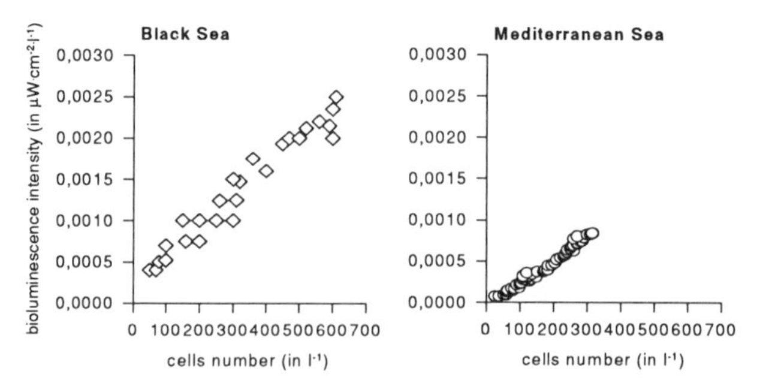

Fig.1. Regressions between bioluminescence intensity and numbers of dinophyte algae in the upper 100-m.

This type of relationship has been proposed earlier, on the basis of experimental studies of bioluminescence of single phytoplankton species (9). On the other hand, a strong correlation between Dinophyta algae and bioluminescence, revealed for the other regions of the Mediterranean Sea, was quite unexpected, because such strong bioluminescent organisms as copepods, numerous in the plankton, should contribute considerably to the total bioluminescent potential (10). This means that the models where Dinoflagellata play the dominant role in the formation of bioluminescent field (11,12) should be more applicable for the Mediterranean Sea basin.

Acknowledgments

This work was part of an International cooperation program between IBSS and Plymouth Marine Laboratory and has been funded by ONR grant No. N00014-95-1-0089

References

1. Gitelson II, Gladyshev MI, Degermendzhy AG, Levin LA, Sidko FYa. The ecological biophysics and its part in investigation of aquatic ecosystems. Biophysics (Biofizika) 1993;38:1069-78 (in Russian).
2. Tokarev YuN, Bityukov EP, Sokolov. BG Fine scale spatial structure of the acoustic field of the upper productivity layer of the Atlantic ocean. Hydrobiological J (Hidrobiologicheskii Jurnal), 1995;39:79-86.
3. Giesbrecht W. Mittheilungen uber Copepoden. 8 Uber das Leuchten der pelagischen Copepoden und das tierische. Leuchten im Allgemeinen. Mitt Zool Sta Neapel, 1895;11:631-94.
4. Clarke GL., Breslau LR. Measurements of bioluminescence off Monaco and Northern Corsica. Bull Inst Oceanogr 1959;1147:1-29.
5. Bityukov EP, Rybasov VP, Shaida VG. Annual variations of the bioluminescence field intensity in the neritic zone of the Black sea. Oceanology (Okeanologiya), 1967;7:1089-99 (in Russian).
6. Bityukov EP. Brief characteristic of the bioluminescent field at the Mediterranean sea and at the North-Western Atlantic. In: G.A.Kolesnikov (ed). Investigations of the interdepartmental cruise at the North-Western Atlantic. Sevastopol, MGI, 1969:166-69.
7. Bityukov EP, Vasilenko VI, Tokarev YuN, Shaida VG. Bathyphotometer with distance-switched sensitivity for estimating intensity of bioluminescent field. Hydrobiological J (Hidrobiologicheskii Jurnal) 1969;5:82-6 (in Russian).
8. Greze VN. Pelagial of the Mediterranean Sea as an ecological system. Kiev, Naukova Dumka, 1989.
9. Bityukov EP, Evstigneev PV, Tokarev YuN. Luminous Dinoflagellata of the Black Sea under anthropogenic impact. Hydrobiol J (Hidrobiologicheskii Jurnal) 1993;29:27-34 (in Russian).
10. Evstigneev PV, Bityukov EP. Bioluminescence of the marine Copepoda. Kiev, Naukova Dumka, 1990.
11. Kelly MG. The occurrence of dinoflagellate luminescence at Woods Hole. Biol Bull 1968:135:279-95.
12. Tett PB. The relation between dinoflagellates and the bioluminescence of sea water. J Mar Biol Ass UK, 1971;51:183-206.

PART 5

Luminescence in Science Education

POLYMERASE CHAIN REACTION-TRANSFORMATION EXPERIMENT USING BIOLUMINESCENCE GENES OF *VIBRIO FISCHERI* AS A TEACHING TOOL

JA Slock, LA Jablouski, CM Wozniak and JM Agati
Dept of Biology, King's College, Wilkes-Barre, PA 18711, USA

Introduction

Polymerase Chain Reaction (PCR)-transformation experiments are an important tool in recombinant DNA technology. The addition of new genes to a recipient cell introduces a heritable modification in the recipient cell's phenotype. Most laboratory teaching exercises involve the transformation of plasmid DNA into *E. coli* followed by selection of transformants containing an antibiotic resistance marker. While these experiments are successful and instructive, they lack the wonderment that is the essence of genetic engineering.

Research on the regulation of bioluminescence in the marine bacterium *Vibrio fischeri* has provided us with exciting teaching materials that can be used in a series of molecular biology experiments, including transformation. There are seven genes required to produce bioluminescence in *E. coli*. These seven *lux* genes from *Vibrio fischeri* have been cloned into two plasmids. The plasmid pHK724 contains the *lux R* gene whose gene product is a transcriptional regulatory protein. The plasmid pHK555 (*lux ICDABE*) contains the structural genes required to make light. When pHK724 is transformed into *E. coli* containing pHK555, the resultant colonies that grow on selective media bioluminesce(1).

Another recombinant DNA technique that utilizes transformation is PCR. The application of PCR is far reaching and laboratory experiments are needed to teach this technology to students. The combination of the *lux* system of bioluminescence and PCR produces an exciting teaching experiment. The plasmid pHK724 is used as the template for *lux* R gene amplification. The amplified *lux* R is ligated into a plasmid vector and the student generated chimeric molecule is transformed into *E. coli* /pHK555.

Bioluminescence transformation experiments show students the excitement and power of recombinant DNA technology.

Materials and Methods

Preparation of Competent Cells: Preparation of competent cells of *E. coli*/pHK555 was previously described (1). Commercial solutions can be used to make cells competent (EZ-Comp; 5 Prime 3 Prime, Inc., 1-800-533-5703). The bacterial strain *E. coli* /pHK555 may be obtained by writing the author.

Plasmid Preparation: Purified plasmid pHK724 may be obtained by writing the author. The plasmid pHK724 can then be transformed into *E. coli* . There are numerous techniques for extracting plasmids from *E. coli* (1).

Transformation: The transformation of *E. coli* /pHK555 with the plasmid pHK555 is accomplished by heat shock (1).

Amplification of *lux R* Gene: The plasmid pHK724 was used for the *lux* R gene template. The amplification program consisted of: 94° C, 15s; 55° C, 15 s; 72° C, 60 s; 25 cycles.

Primers for Amplification of *lux* R Gene: Upstream primer; 5' - TCA CAC AGG AAA CAG GGT ATG-3'. Downstream primer; 5' - CGT ACT TAA TTT TTA AAG TAT GGG CA-3'.

Vector System: Any of the commercially available plasmid expression vectors.

DNA Polymerase: Any of the commercially available polymerases.

Selective Medium: Lauria agar containing 80 μg/mL ampicillin and 50 μg/mL

chloramphenicol.
Incubation of Transformants: Selective media plates are incubated at 30° C for 48-72 h. The bioluminescent colonies must be viewed in a dark room.

Results and Discussion

The *lux* system of bioluminescence offers an exciting approach to conducting laboratory experiments in recombinant DNA technology. These teaching experiments are an outgrowth from research on the regulation of bioluminescence in the marine bacterium, *Vibrio fischeri*. This system can be used to teach students gene amplification, primer design, complementarity of nucleotides, regulation of gene expression, plasmid mini prep preparation, agarose gel electrophoresis, deletion mutagenesis, site-directed mutagenesis, cloning and transformation. The cost of the transformation experiment is relatively inexpensive. The cost of the instrumentation and reagents for PCR is high but can be justified because PCR is so fundamental to our exploration of molecular biology.

References

1. Slock JA. Transformation Experiment Using Bioluminescence Genes of *Vibrio fischeri*. Am Biol Teacher 1995; 57:225-7.

BACTERIAL BIOLUMINESCENCE IN ECOLOGICAL EDUCATION

VA Kratasyuk, AM Kuznetsov, JI Gitelson,
Institute of Biophysics, Russian Academy of Sciences,
Krasnoyarsk, 660036, Russia

Introduction

Investigation of bioluminescence and its employment in our teaching work at the University and at high school convinced us that this phenomenon (a gift of nature) is a mighty and pluripotent instrument for science educational and enlightenment activities (1-2). The aim of this presentation is to describe our investigations in teaching ecology through bioluminescent visualization.

Ecological practical course are based on inhibition of luminescence of marine luminous bacteria and bacterial luciferases by toxic substances and media (3). The proposed practical course can be realized both in classrooms and in station camps during field practicals for schoolchildren, students and teacher ecological education.

Materials and Methods

The lyophilized bacteria, NADH:FMN-oxidoreductase and luciferase were obtained from the Institute of Biophysics, Krasnoyarsk (4,5). Luciferase activity and the activity of the two-enzyme coupled system were measured with luminometer as described previously (6,7). The lyophilized bacteria were used as described previously (7).

Results and Discussion

We usually begin ecological course with the description of the phenomenon of bioluminescence, a brief overview of the major luminous organisms with the special attention to luminous bacteria.

Then we talk about the opportunities of bioluminescence in teaching. We'll try to outline this field by systematizing the areas of educational-scientific application of bioluminescence and illustrating by individual examples.

Then we consider application of bioluminescence in the main fields in ecology: Biophysics of populations; Evolution of organic life; Man and biosphere; Toxicology and environmental monitoring. There are some examples for demonstrations of bioluminescence in these fields:

1. Collection strains of luminous bacteria, associated with marine fauna, are represented by symbiontes isolated from light organs, and commensals inhabiting the gastrointestinal tract. The photobacteria, inhabiting the light organs of fishes, are the object for demonstration of the fundamental biological phenomenon of co-adaptation of metabolism of symbiont and host, as well as an example of controlled cultivation for engineering microbiology.
2. Taxonomic criterion of a species taught by way of example of luminous bacteria. Identification of particular species of bacteria by kinetics of their luciferases.
3. Mutagenic effects of different substances were determined by reversed mutants.

4. Variability of organism for evolution can be easily observed by heterogeneity of colonies in population of luminous bacteria.
5. Results of artificial selection by way of example of dark mutants of luminous bacteria and superproducing strains. The collection of luminous bacteria can serve as a basis for researching the strain-superproducers of luciferases, NADH:FMN-oxidoreductase, decarboxylase, endonuclease of restriction, chitinase producers.
6. The classes considering the topic "Man and biosphere" can be made interesting by demonstrating the possibility of unlimited or limited growth of luminous bacteria in the absence of growth-limiting factors and in their presence (nutrition, oxygen, xenobiotics, etc.);
7. The modeling a man-made biosphere (co-existence of algae and luminous bacteria in closed space).

The teacher and students usually participated in these experiments, but with schoolchildren we only demonstrated this experiments.

The main part of Ecological practical course are the experiments, presenting basic methods of research, such as assay of pollutants in water, air and food. It is the following experiments: bioluminescence kinetics varying in different concentration of pollutants; the effect of compounds with various degree of toxicity on bioluminescence *in vivo* and *in vitro*; effect of different classes of compounds on kinetics of the bioluminescent reactions; the mechanisms of pollutants effects on bioluminescence; the additive and non-additive effects of compounds on bacterial bioluminescence; effective dose of pollutants for man; assays of quality of potable and natural waters (river, lake, etc.); evaluation of air pollutions and toxicity of food.

In ecological research the bioluminescent tests *in vivo* and *in vitro* were used simultaneously. The comparison of results of biotests *in vivo* and *in vitro* gives the additional information on mechanisms of pollutant's effect. Preliminary bioluminescent tests were calibrated with standards of nontoxic water. In the presence of the pollutants the light emission can be decreased or increased.

We have a lot of results on water ecological monitoring after student's ecological experiments and sent scientific papers to the real scientific journals. Every year we organized Krasnoyarsk university's scientific students conference, where our students present their results. The example of this investigations are the ecological monitoring of portable and natural water of 7 Altai's regions. The results of bioluminescent biotests are confirmed by chemical data(7). It was shown that several Altai's region had unsatisfactory quality of water. The similar results were obtained during shipboard expedition on river Yenisei. The most of water samples had pollutants but the levels of contamination were different and depended on the concentrations and characteristics of pollutants. Bioluminescent biotests shows the integral quality of water and time of one assay is less then 1 minutes. The expressivity of bioluminescent tests permit us to investigate a great number of samples and to choose the samples with toxicity for chemical analysis to understand the nature or mechanisms of toxicity.

At school the ecological practical course is the same but more simple. We try to conduct it without bioluminometer. In the dark room every student make experiments with a test-tubes containing luminous bacteria, which light emission can be decreased after injection of pollutant 's solution.

At the end of our course we have the general discussion. Our students and teachers told us about their expressions about the properties of bioluminescence which make it so promising for education:

1. First of all it is vividness. Developed in the course of evolution extremely efficient mechanism that transforms the energy of substrate oxidation into light emission combined with extremely high ocular sensitivity makes possible to observe bioluminescence directly. In this connection it should be borne in mind that for man more than 80% of all information is provided by vision. This makes visualization the most direct way to obtain knowledge. Primary school children who have not yet access to instrumental techniques get immediately involved into demonstration of new phenomena. Many experiments at lessons can be made much cheaper.
2. Close correlation of bioluminescence with basic vital processes. The basis of bioluminescence is enzymatic oxidation of special substrates - luciferins - by specific enzymes - luciferases employing nonspecific co-factors: ATP, NADH, FMN. Energy branched for bioluminescence from the common energy pot of the cell opens opportunities to develop bioluminescent probes for hundreds of substances.
3. Occurrence of bioluminescence. The number of different species of animals emitting light is small but they are abundant and at sea they are almost ubiquitous.
4. Easily isolated bioluminescent systems of different species. They can be preserved for years, and easily activated, if necessary - for example, diluting lyophilized organisms by water at lectures, lessons, seminars, practical work.
5. Extensive use is made in research and in practice of limit-sensitive meters for emission of optical range - up to individual quanta. They are relatively simple and cheap. This makes possible to measure bioluminescence in quality, using it as a measure of numerous physiological, biochemical and biophysical processes. The advantages of the method result from: the sensitivity of photodetectors to the point level of quantum counting, a high quantum efficiency of the reaction, a wide range for the linear response, the specificity of enzymatic reactions, the simplicity the signal can be registered and transformed with and the development of the measuring techniques required.
6. Finally, emotional attractiveness is the importance of this feature of bioluminescence for the teaching process can hardly be overestimated. How does it attract? The first is a novelty, absence of correlation between light and heat. The second is certain mystery (which is especially important for the junior it can be observed in darkness only). And the next is the beauty of its manifestation both in the ocean and in a Petri dish. But all this can hardly provide explanation to the passion for bioluminescence. I can only say that the authors after many years of work with this object have not lost their admiration for it. After all these years it remains aesthetically attractive. Especially strong is their first impression of students meeting light emitted by life. This is a lasting memory, for some people - for their entire life; it brings into the memory phenomena which are demonstrated and comprehended by bioluminescence.

7. Achievements of last decades in bioluminescent research are tremendous. Its biochemical and biophysical mechanisms have been deciphered. Genetic and physiological regulation has been studied. The insight into ecological role of luminescence has not progressed as markedly, but here the achievements are also considerable: dramatically sophisticated methods of its use in evolution have been discovered.

Acknowledgements
This work has been funded by Russian Humanitarian Scientific Foundation Krasnoyarsk Foundation "Education" and Krasnoyarsk Scientific Foundation .

References

1. Kratasyuk VA, Gitelson JI. Life is a language of life. Krasnoyarsk: Institute of Biophysics, 1993.
2. Kratasyuk VA, Gitelson JI. Bioluminescent visualization of vital processes and its application in biophysical education. Progress in Biophysics and Molecular Biology; 1996; 65, suppl 1: 24.
3. Kratasyuk VA. Principle of luciferase biotesting. In: Jezowska-Trzebiatowska B, Kochel B, Slawinski J, Strek W, editors. Biological Luminescence. Singapore: World Scientific, 1990:550-8.
4. Kuznetsov AM, Primakova GA, Fish AM. Lyophilized luminous bacteria as a toxicity biotest. In: Jezowska-Trzebiatowska B, Kochel B, Slawinski J, Strek W, editors. Biological Luminescence. Singapore: World Scientific, 1990:559-563.
5. Tyulkova NA Purification of bacterial luciferase from Photobacterium leiognathi with the use of AN FILCH-system. In: Jezowska-Trzebiatowska B, Kochel B, Slawinski J, Strek W, editors. Biological Luminescence. Singapore: World Scientific, 1990:369-374
6. Kratasyuk VA, Abakumova VV, Kim NB. A gel model for the functioning of luciferase in the cell. Biochemistry (Moscow); 1994;59:761-5.
7. Kratasyuk VA, Kuznetsov AM, Rodicheva EK, Egorova OI, Abakumova VV, Gribovska IV. Problems and prospects of bioluminescent biotests in ecological monitoring. Siberian ecol J; 1996; 5, in press

REAL SCIENTIST: AN EDUCATIONAL KIT USING BIOLUMINESCENT BACTERIA AND CD-ROM

PE Andreotti[1], *T Berthold*[1], *PE Stanley*[2] *and F Berthold*[1]
[1] *MicroLab Systems Corporation, Boca Raton, FL 33428, USA*
[2] *Cambridge Research & Technology Transfer Ltd, Cambridge CB1 2HF, UK*

Introduction

The growth of a culture of bioluminescent bacteria *Photobacterium phosphoreum* is readily observed by the dark-adapted human eye. Such a culture is ideal for teaching about bioluminescence and microbiology. In an effort to use bioluminescence as a teaching tool and provide a fun and educational product for children ages 10 - 14, we have incorporated *Photobacterium phosphoreum* into a series of activities and experiments in a stand-alone science education kit along with CD-ROM software for both IBM and Macintosh computers. The kit also includes a guide with background information on bioluminescence, thought-provoking questions to encourage further exploration, and complete instructions for the hands-on activities, experiments, and software. The REAL SCIENTIST kit was developed with teachers and parents to actively engage children in learning about science, particularly biology, ecology and computers, while having fun.

Materials

In addition to the CD-ROM and instruction guide, the kit includes absorbent paper, 50 ml plastic beaker, powdered culture medium, plastic medium bottle, sterile plastic culture flask, sterile loop, 9 sterile pipettes, 2 plastic petri dishes, 2 filter paper disks, 2 sterile cotton tip applicators, 2 wood spatulas, 2 glass vials, and 1 glass vial with chlorinated pollution sample. A vial of *P. phosphoreum* is provided through the mail after providing a registration number for the software

Results and Discussion

The hands-on activities and experiments in REAL SCIENTIST include three components. The first activity is to grow *Photobacterium phosphoreum* in liquid culture in a tissue culture flask and record observations every twelve hours for the amount of bioluminescence given off and the amount of growth over a period of four days. A simple, subjective scale of 0 - 3 is used to record observations for bioluminescence and growth. This activity is coupled with didactic, written information and questions for how bacteria grow, and how bioluminescence is produced.

The second hands-on component is a petri dish based experiment comparing the growth of *Photobacterium phosphoreum* on filter paper disks coated with water or growth medium. Scientific method including the testing of a hypothesis and the use of controls in a well designed experiment are practically introduced. This teaching is coupled with the fun of painting with the *Photobacterium phosphoreum* on the filter paper disks.

The third hands-on component is to have the child design an experiment and test for solid or liquid pollutants in the environment using the *Photobacterium phosphoreum* as an indicator organism. This experiment is based on the hypothesis that a toxic pollutant will extinguish the bioluminescence from *Photobacterium phosphoreum*. A positive control "pollution sample" of chlorine granules is included.

The interactive hybrid CD-ROM includes two programs: LIVING LIGHT which primarily involves the subject of bioluminescence and is designed for use with the hands-on activities and experiments, and ROBOT COMMAND which is a one or two player maze game that teaches about computer programming fundamentals.

LIVING LIGHT is divided into four modules:

- What Is Bioluminescence
- Show Me Some Examples!
- How Do I Grow
- Graph Results

The What Is Bioluminescence module teaches about "living light" with a basic introduction to organisms, species, cells, and the firefly luciferase reaction which teaches about a chemical reaction using cartoon graphics.

The Show Me Some Examples module includes photographs with written descriptions for *Lampyris noctiluca, Obelia geniculata, Pholas dactylus, Photinus pyralis, Vibrio fischeri, Vibrio harveyi, Mycena, Omphalotus, Panellus stypticus, and Renilla.* Images were kindly provided by Drs. Anthony Campbell, Keith Wood, Dennis O'Kane, and Larry Kricka.

The How Do I Grow module is designed for use with the hands-on activity for growing *P. phosphoreum*. It first describes the process of binary fission for bacterial replication and exponential growth using graphics and text, and then it provides an interactive game-like graphics screen which allows the user to experiment with growing bioluminescence *P. phosphoreum* using computer simulation.

The Graph Results module is also designed for use with the hands-on activity for growing *P. Phosphoreum*. Using text and graphics, it first describes what a X-Y line graph is and how it might be used to picture experimental results, and then it provides an interactive graphics screen which allows the user to graph their observations for *P. phosphoreum* bioluminescence and growth which they recorded using the subjective 0 - 3 scale.

The ROBOT COMMAND program, while unrelated to bioluminescence or the hands-on activities and experiments, is an entertaining game that teaches the logic and principles of computer programming. Players "write a program" to choose the steps to navigate a robot through a maze, then immediately see the results as the computer plays back their program.

Through the use of *P. phosphoreum* as both a fun and educational tool, children are able to discover the principles of scientific procedures, learn to use laboratory materials and techniques, explore how common substances may pollute their environment, learn about how bacteria grow and how to chart experimental laboratory results using a computer, while at the same time learning about bioluminescence and species from around the world that glow in the dark. Additional details can be found on the internet home page[1].

References

1. http://www.microlabsystems.com

A BIOLUMINESCENCE/CHEMILUMINESCENCE (BL/CL) BIBLIOGRAPHIC DATABASE FOR RESEARCH AND EDUCATION

DJ O'Kane
Dept. of Laboratory Medicine and Pathology, Mayo Foundation for Medical Education and Research, Rochester, Minnesota, USA 55905

Introduction

The amount of information and the number of published manuscripts in science and medicine is increasing rapidly. The number of citations retrieved from searching the Medline clinical science database doubles approximately every five years in the specialty area of bioluminescence and chemiluminescence (BL/CL) (Fig. 1). Nearly 2750 references pertaining to BL/CL will be published in the year 2000 if this rate of growth is maintained. How to manage this expanding body of works is a fundamental concern. Recent advances in importation technologies facilitate capturing references from on-line databases, bibliographic reference research services, and compact disks. This minimizes the resources required to build and maintain a specialty bibliographic database and to manage the increasing amounts of information pertaining to BL/CL. More than 20 single-user bibliographic database management software programs are available commercially (Hoke, 1994). Once created through one of these programs, the database can be searched by author, title, year, journal, keyword(s) or combinations of the forcgoing linked through Boolean operators (OR, AND, NOT) with great ease. The goals of the present study are:

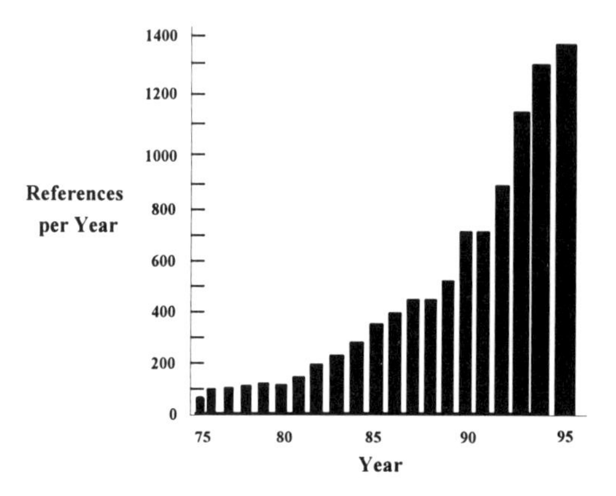

Figure 1. BL/CL citations in Medline

i. to develop a bibliographic specialty database for BL/CL;
ii. to assess limitations that may impact bibliographic database utilization.

Materials and Methods

Search and Retrieval: The majority of references were incorporated into the BL/CL bibliographic database by searching primary reference databases using OVID software maintained on an Institutional UNIX server. The following primary databases were provided under license from CD Plus Technologies (New York, NY): Current Contents (CC), Medline, Embase, Health, Aidsline, Cancerlit, and PsychInfo. Search strategies were tailored for the individual primary databases. The searches were performed through telnet sessions using "smart button" macro commands to speed the operations. Default searches were performed in the most recently updated CC database. The search strategy was comprised of: 10 lines of code searching for 66 textwords, including variants, in abstracts, title, keywords, and key phrases using the "OR" Boolean operator; 6 lines of code excluding references containing confounding textwords with the "NOT" Boolean operator; compiling the retrieval reference set; and finally limiting the reference retrieval set to the current database update. Medline and Embase required a more complex search strategy in which all the textwords had to be searched explicitly as "Textwords" in addition to the abstract, title, and keyword searches performed in CC. This increased the number of lines of code dealing with searches and exclusions and compilation to 80 with a final line limiting the retrieval set to the most recent database update. The limited reference sets were saved as text files using the OVID reprint format. Patents were retrieved from the MicroPatent database either from compact disks provided by the company or by download through the World Wide Web at "http://www.micropat.com/" using keyword searches.

BL/CL Database: The specialty BL/CL database was created using Reference Manager (RM) for Windows (Research Information Systems, Carlsbad, CA) Version 5 that has been upgraded to Version 7. RM has been described previously (2). The saved OVID text files from CC searches required global word processing prior to importation into RM to eliminate problems with keyword format. All text files required global word processing to permit proper incorporation of abstracts, institution addresses of the corresponding author, unique identifiers, ISSN/ISBN numbers, and paper type and language of publication. These editing changes were incorporated utilizing Word for Windows. The edited text files were captured into temporary unique databases and edited in RM to eliminate certain keywords. The temporary databases were rebuilt by a RM subroutine prior to merging with the BL/CL database to eliminate errors that might corrupt the main BL/CL database. Patents were entered manually in the correct RM format, as were a minority of other BL/CL references not obtained through OVID searches.

Results and Discussion

BL/CL Database Description: The BL/CL bibliographic database assembled during the past 3 years facilitates organizing and utilizing information pertaining to the ever increasing number of applications of BL and CL in diverse disciplines with an emphasis on clinical research and experimental pathology. An updated BL/CL database is by definition a work-in progress that will continue to grow in size and utility over time. As of the update on September 3, 1996, the database contained 18,888 references. Retrieved references are assigned a database ID number sequentially during the merging process with the BL/CL database. The reference type composition of the database is presented in Table 1. In addition to BL/CL, the few references pertaining to selected related luminescent phenomena (*i.e.* sonoluminescence, triboluminescence, *etc.*) are incorporated as well. References pertaining solely to fluorescence and phosphorescence are excluded from the database. Periodical names and abbreviations are

Table 1. Reference Type Composition of the Retrieved BL/CL Database

Reference Type	Number	Reference Type	Number
Books	11	Journal Article	18050
Book Chapters	303	Comments	12
Book Reviews	1	Communications	2
Conference Proceedings	2	Editorials	11
Magazine Articles	1	Errata	54
Meeting Abstracts	403	Letters	58
Patents	220	News	5
		Notes	121
		Reviews	516

not consistent in the several primary bibliographic databases searched. Accordingly, the periodical names selected for use with the BL/CL database are those utilized by the National Library of Medicine (3). Periodical name variants utilized in the primary databases are assigned as synonyms of the NLM designations to minimize the incorporation of duplicate references in the BL/CL database. Errata reported in the primary databases are collected and filed in the BL/CL database with the same ID number as the parent reference, but ending with a "C" to indicate a correction. The language of publication is also stored with the reference material if other than English. Non-English references are designated by the translated title demarcated with square brackets and are retrievable using keywords in the form "[Language]". Twenty-two different non-English languages are represented in the references stored in the BL/CL database. Several types of reference materials dealing with BL and CL are not incorporated in the database at present. These include slides, photographs, motion pictures, and videos dealing with BL/ CL. More importantly, commercial literature pertaining to BL/CL products and applications has not as yet been incorporated into the database despite the increased popularity of these products. These data sources will be introduced into the database in the near future. Personal reprints are not maintained as part of the database, but instead are indexed by the database ID numbers. At present, nearly 8500 personal reprints (45% of total references) are archived by the sequential database ID number.

Database Utility: The utility of a bibliographic database resides not in its archival features, but rather in the capability to find information and retrieve pertinent references. The BL/CL database may be searched for information using a variety of parameters or combinations of terms. Searches can be based on authors' names (37,465), editors (62), keywords (25,751), document types (*e.g.* book, chapter, article, patent, etc.), journals (2553), reprint status (*e.g.* On Request, In File, Not In File), year, title, abstracts, miscellaneous fields, or combinations of terms using Boolean operators (AND, OR, NOT). A search strategy frequently utilized to obtain information stored in the database takes the form of "Keyword" AND "Year range", with additional "Keyword(s)" added to focus the search. Another strategy is used to retrieve a particular reference by a particular author ["Author Name" AND "Keyword" AND ("Journal"/"Year Range")].

The original purpose of RM was to organize archived references and generate bibliography lists for publication in the reference format for selected journals produced by RM-generated filters. Although the BL/CL database is used for this purpose, other functions have been added by generating user-defined reference format outputs. Additional roles for the BL/CL database include the generation of reprint request cards, requests for photocopy and inter-library loan services, generation of labels to index reprints to the database, and WWW intranet output (Fig. 2). Each of these additional roles requires additional, sometimes considerable,

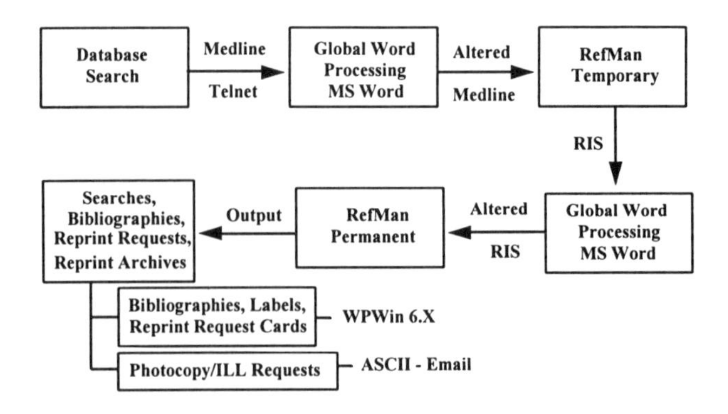

Figure 2. Flow diagram of BL/CL database and output functions.

intermediate word processing steps. While all these features are utilized, it is clear that they are not integrated in an intuitive, "user-friendly", package.

Research and Education Usage: The BL/CL database has been used extensively in research activities and as a resource for writing and reviewing manuscripts for publication. Research groups from several departments, including Biochemistry and Molecular Biology, Pharmacology, Gastroenterology, Laboratory Medicine and Pathology, and Neuroimmunology have utilized the database for bibliographic searches. In addition, the database is used to prepare lectures for Clinical Chemistry, Experimental Pathology, and a Clinical Chemiluminescence Training Program. Residents and Training fellows are provided additional training in the use of the BL/CL database as a model for organizing personal databases in their areas of interest.

Database Problems: While RM facilitates performing bibliography management, three major problems have been noted that limit the scope of this commercial database manager. First, references are organized in a few, very large, common files. Information in RM is stored in 6 files. The largest file occupies over 30 Mbytes of disk space in the BL/CL database out of 34 Mbytes total. The first problem attributable to large file structure that restricts database utility is the elimination of errors from the files. The RM program utilizes a repair function to detect and restore data in corrupted files. However, the time to rebuild the database using this procedure is approximately 1 hour per Mbyte of database using a 100 MHz Pentium computer. A second difficulty attributable to large file size is that sharing the database on a common drive is cumbersome. The amount of intermediate word processing required to obtain all the desired database features and to implement all the output options is a final difficulty.

Future Prospects: The resources necessary to manage the increasing numbers of references on BL/CL will rely heavily on a relational database format accessible through a network (4,5). This is in contrast to present reliance on personal computer strategies. A prototype bibliographic relational database written in Sybase is undergoing development at Mayo. The advantages of this approach are that the relational database is platform-independent and can be accessed simultaneously by multiple users. The file structures for references are small, approximately 1.5 kbytes per reference, allowing corrupted files to be repaired rapidly. The database can be searched using Boolean operators and multiple outputs may be selected from an integrated package. The relational database approach will streamline the maintenance of the BL/CL database and permit more flexible utilization in future research and education activities.

Acknowledgments. Support for database development was received from the Research and Development Committee, DLMP, Mayo Foundation for Medical Education and Research.

References

1. Hoke F. Making the on-line connection with bibliographic-database software. The Scientist 1994 (June 27):18-19.
2. Matus N, Beutler EB. Reference Update and Reference Manager: personal computer programs for locating and managing references. BioTechniques 1989;7:636-639.
3. Anon. List of Serials Indexed for Online Users. National Library of Medicine, Bethesda, MD, 1994.
4. Mosley LG, Mead DM. Good relations: the use of a relational database for large-scale data analysis. J Adv Nursing 1993;18:1795-1805.
5. Banhart F, Klaren H. A graphical query generator for clinical research databases. Meth Inf Med 1995;34:328-339.

APPLYING BIOLUMINESCENCE TO GENERAL SCIENCE EDUCATION: SCIENCE WITHOUT WALLS TELECOURSE

JD Andrade
Center for Integrated Science Education, 2480 MEB
University of Utah, Salt Lake City, UT 84112, USA
Joe.Andrade@m.cc.utah.edu

Introduction

Bioluminescence is a nearly ideal subject with which to experience the scientific process and critical science concepts and themes.

We have developed bioluminescent dinoflagellate cultures which enable upper elementary and junior high teachers and students (1) to readily experience bioluminescence, closed ecosystems, circadian rhythms, protozoa and optics. Much of the experience is conducted in the dark. Science in the Dark has been an effective way to reduce science anxieties and fears and to encourage teachers to develop a fresh, positive and instructive attitude towards hands-on science in their classrooms.

These materials have now been included in a television based distance learning course: *Science Without Walls: Science in Your World* (2). We utilize bioluminescence as an effective way of imparting the scientific experience and method to television viewers throughout Utah.

Science Without Walls Telecourse

We have previously reported (1) our experience with a ten hour hands on inservice course for teachers titled Integrated Science Concepts and Themes. This course extensively utilized bioluminescence, particularly the dinoflagellate Pyrocystis lunula, as a unique experimental tool with which to develop scientific observation skills and provide the opportunity to formulate many different and specific scientific hypotheses. With such observational skills and hypotheses in hand, the students can move forward to design, conduct, and analyze simple experiments using only the Pyrocystis lunula cultures.

Bioluminescence readily connects to practically all of the basic concepts and themes developed in Project 2061 (3) (Figure 1) and used in Science Without Walls (2), (Figure 2). The course connects science with the arts and with the humanities and relies heavily on integrated science concepts and themes, philosophies which came out of the American Association for the Advancement of Science Project 2061 Report: *Science for All Americans.* (3) (Figure 1).

Figure 1. Basic Concepts & Themes, Traditional Disciplines, and Bioluminescence

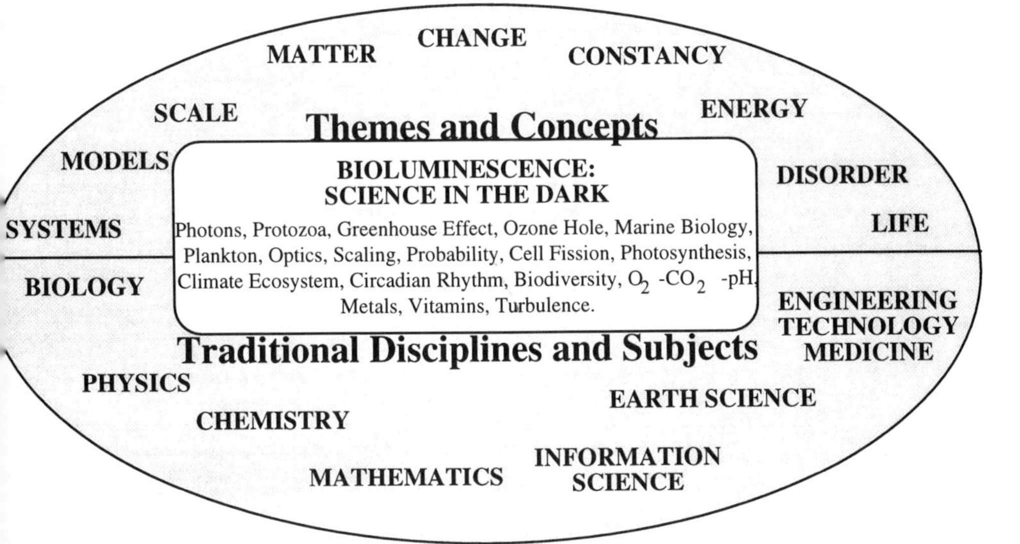

Figure 2. General Design of the Science Without Walls Telecourse

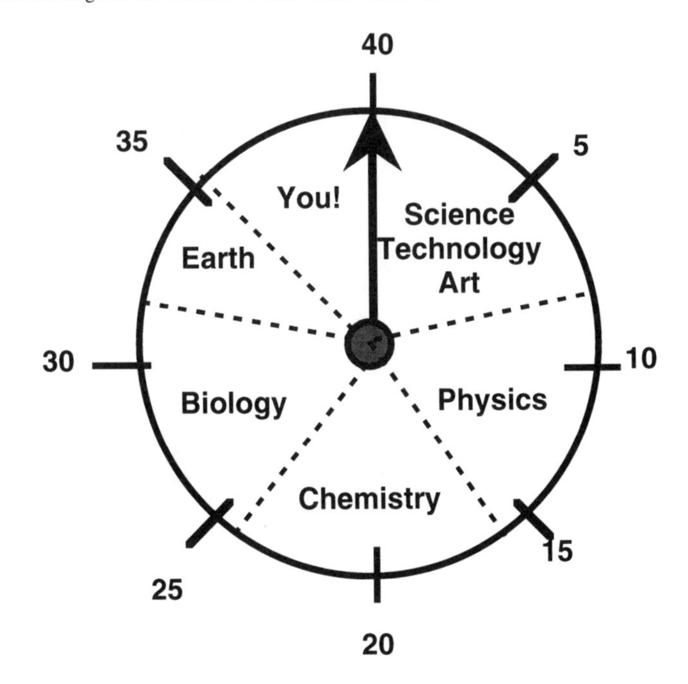

In order to minimize the teaching of science as a virtual subject rather than as a real subject, the course also involves a unique Labless Lab, basically a science kit which all students are required to obtain and use to conduct weekly, semi-quantitative experiments. (4). This course took about 15 months to write and produce, and was launched October 1, 1996 as a telecourse for the University of Utah's Division of Continuing Education, aired through their local KULC Channel 9, a special television channel for Continuing Education/Distance Learning courses.

The course consists of forty half hour programs covering all of science in an integrated coherent fashion. (Table 1) It is designed primarily for non-science majors and for the general public and has as one of its major goals the empowerment of citizens and residents to become involved in public and political issues which may have a science or technological component.

Table 1. The forty half hour programs/topics in Science Without Walls: Bioluminescence is used mainly in Program 32 and parts of 27 and 28.

1: The World of Science-The World of Art
2: Observing And Perceiving: The Senses
3: Patterns And Numbers
4: Extending Your Senses
5: Integrated Concepts And Themes: Systems and Models
6: Scale
7: Constancy, Change, & Matter
8: Energy, Disorder & Life
9: Physicists In The Wild
10: Inertia, Gravity, & Senator Garn
11: Energy, Efficiency, Entropy
12: Interstate Physics
13: Action At A Distance
14: From Magnets To Electricity
15: From Electrons To Light
16: From Newton To Quanta
17: Chemists In the Wild
18 Your Personal Periodic Table
19: From Atoms to Molecules
20: From Metals To Water
21: From Water To Solutions
22: Molecular Alchemy
23: Very Personal Chemistry
24: Guns And Bombs
25: Biologists In The Wild
26: What is Life?--Diversity And Extinction
27: What Is Life?--The Very Early Days
28: What Is Life?--From Bacteria to You
29: Energy In: Fuel & Light
30: Energy Out: Biomass and Work
31: Information In: The Senses
32: Information Out: Language
33: Your Brain And Consciousness
34: Is There Intelligent Life on Earth?
35: Planetary Medicine: Gaia
36: Your Stuff: Cars And Transportation
37: Luck And Risk: Personal Statistics
38: Medicine & Health--Yours
39: Creativity--Yours
40: Where Do We Go From Here?

The Future

We are developing programs (going beyond Program 38) where we discuss health care, encouraging the student to be interested not only in clean and healthy living styles, but also in functioning as their physician's assistant, to help serve as eyes and ears, as an information gathering source, to aid health care practitioners in their efforts in diagnosing and treating the student's ailments.

Acknowledgments

Most of our work on the development of bioluminescence for science education has been funded by Protein Solutions, Inc., Salt Lake City, UT, USA. Our courses for teachers on Integrated Science Concepts and Themes based on bioluminescence were funded by the U.S. Department of Education, Eisenhower Grant Funds, administered by the Utah State Board of Regents.

Bioluminescence footage was provided by Edi Widder, James Morin, and Protein Solutions, Inc., Salt Lake City. The overall telecourse project was funded by the State of Utah Higher Education Technology Initiative and by the University of Utah, including its Center for Integrated Science Education.

We thank all of those who provided help, assistance, advice, footage, visuals, ideas, and support.

References

1. Andrade JD, Lisonbee M, Min D. Using Novel Biological Phenomena to Enhance Integrated Science Education. Stanley PE, Campbell AK, Kricka LJ, eds. New York, Wiley 1994: 373-8.
2. Andrade JD. Science Without Walls: Science in Your World, Center for Integrated Science Education and Division of Continuing Education, University of Utah, Salt Lake City, Utah, 84112, USA, 1996.
3. Rutherford FJ, Ahlgren A, Science for all Americans (The Project 2061 Report), Oxford University Press, 1990.
4. Scheer RJ, Andrade JD, Applying Intelligent Materials to Materials Education, J. Intelligent Material Systems 1995; 6; 13-21.
5. Taylor FJR, ed., The Biology of Dinoflagellates, Blackwell Sci., 1987.
6. Campbell AK, Chemiluminescence VCH Publ, 1988.
7. Tinnin R, and Buskey E , Bioluminescence: Living light of the sea, Currents 1994; 12: 10-3

GFP GENE AS A MODEL GENE EXPRESSION MARKER FOR rDNA LABORATORY COURSES

DA Yernool[1], PV Reddy[2], DF Davis[1] and WW Ward[1]
[1]Dept. of Biochemistry and Microbiology, [2]Dept. of Plant Pathology, Cook College, Rutgers University, New Brunswick, NJ 08903, USA

Introduction

The green fluorescent protein (GFP) derived from the jellyfish *Aequorea victoria* is a 238 amino acid peptide, which absorbs blue light at 395 nm and emits green light at 510 nm. The gene for the protein has been cloned, sequenced and expressed in a variety of heterologous cell types (1). The gfp gene has been extensively used in cell and developmental biology as a fluorescent marker for gene expression and intracellular sorting/targeting of proteins. The detection of the protein is facile requiring only a standard long-wave UV light source. Mutant forms of GFP, including red- and blue-shifted variants have been developed, enabling simultaneous labeling of different subcellular structures in cells (2).

This paper describes the use of the gfp gene as a model gene expression marker for an rDNA course at Rutgers University. rDNA manipulations such as expression cloning by PCR, restriction mapping of DNA and Southern blotting are carried out on the gfp gene. Site-directed mutagenesis is done to convert the green fluorescent protein to a blue fluorescent protein and both mutants and wild-type are sequenced to establish structure/ function relationships. The use of the gfp gene has a dramatic effect on the perception of the students of the power of rDNA techniques. For this reason GFP is used as an instructional tool in biochemistry laboratory courses at Rutgers University (3).

Materials and methods.

Construction of gfp expression system: the open reading frame from pGFP (Clontech Inc., Palo Alto, CA) was PCR amplified using 5' GFP and 3' GFP primers as shown in table 1. The resulting PCR product was cut with XbaI and EcoRI and ligated in-frame with the lacZ alpha-peptide of pUC119. The recombinant molecules are transformed and plated on LB-ampicillin-isopropyl thio-galctoside (IPTG) plates and cloning efficiency is determined by counting the fluorescent and non-fluorescent colonies.
Restriction mapping and southern blotting: The pGFP DNA is cut with the restriction enzymes EcoRI, NcoI, NdeI and XbaI (Life Technologies, Gaithersburg, USA) and electrophoresed on a 1% agarose gel and photographed. The gene fragments are blotted onto a nylon membrane by capillary blotting. Hybridization is carried out using a biotin labelled gfp-specific probe and detected using the Photogene detection system (Life Technologies, Gaithersburg, USA).
Site-directed mutagenesis: PCR site directed mutagenesis is carried out using the Quick Change site-directed mutagenesis kit (Stratagene, La Jolla, USA) with the plasmid TU58 (1) as a template. The primers for mutagenesis are outlined in table 1 and the amino acid tyrosine at position sixty six is changed to histidine. The mutagenesis efficiency is calculated after transforming mutagenised DNA into BL21(DE3) *E. coli* cells by counting the number of blue (mutant) and green (wild-type) colonies on LB-ampicillin-IPTG plates.
DNA sequencing: Plasmid DNA is prepared from selected wild-type (green) and mutant (blue) colonies. Sequencing reactions are carried out using the Cy5 Autocycle

sequencing kit (Pharmacia Biotech, Piscataway, USA) with M13 reverse primer for wild-type and T-7 promoter primer for the mutant DNA molecules. Sequencing and data analysis are done on the ALFexpress DNA sequencer (Pharmacia Biotech, Piscataway, USA).

Results and Discussion

The PCR is carried out using primers annealing to regions as shown in fig 1. The resulting 781 bp PCR product is cloned into pUC119 and screened for clones

Table 1. Primers used for expression PCR and Site-directed mutagenesis.

PRIMER NAME	PRIMER SEQUENCE
GFP 5' Primer	GCTATGACCATGATTACGCCAAGC [1]
GFP 3' Primer	TCATGAATTCAATGTGTAATCCCAGCA [1]
BM5' (Y66H)	GTCACTACTTTCTCT**CAT**GGTGTTCAATGC [2]
BM3' (Y66H)	GCATTGAACACC**ATG**AGAGAAAGTAGTGAC [2]

[1] Primers used in expression PCR.
[2] Primers used in Site-directed mutagenesis, mutant codons are in bold.

exhibiting green fluorescence. Cloning efficiencies ranging from 70-98% were obtained by various classroom laboratory groups. In addition the PCR product was labeled by random primer labelling with biotin and used as a probe to detect and demonstrate the technique of Southern blotting. The restriction pattern obtained after digestion of the gfp gene was analyzed and a map was prepared and compared to a preexisting maps of the gfp gene. This exercise gives students hands on experience in creating restriction maps of DNA molecules.

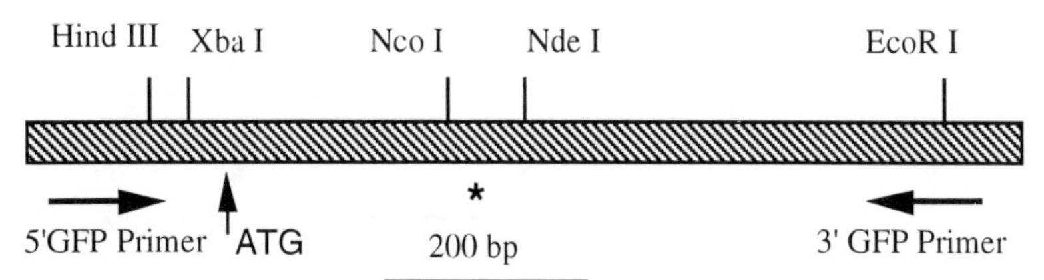

Figure 1. Restriction map of gfp gene showing PCR primer binding sites. The start of the open reading frame is indicated by ATG and * indicates the position for site-directed mutagenesis.

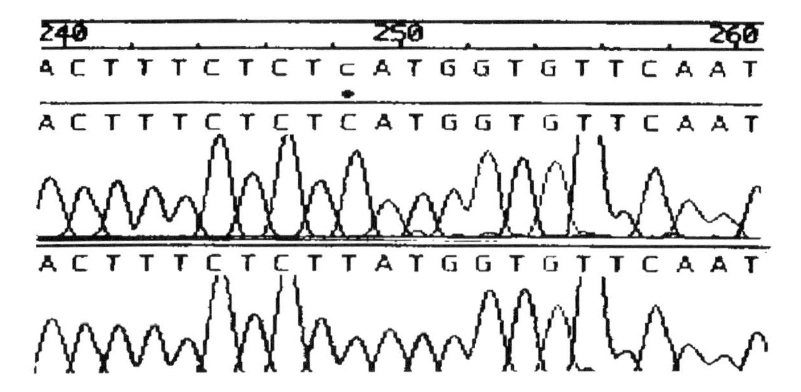

Figure 2. Comparison of wild-type (lower panel) and blue mutant (upper panel) sequences, showing change from TAT to CAT (tyrosine to histidine)

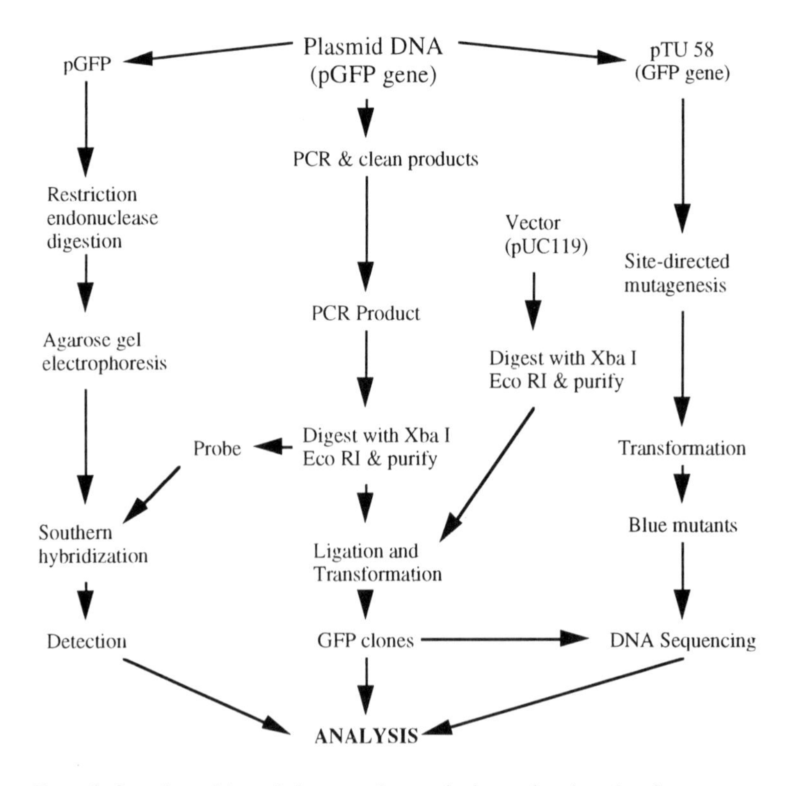

Figure 3. Overview of the techniques used to manipulate and analyse the gfp gene.

In order to demonstrate site-directed mutagenesis the wild type (green) clone was changed to a mutant (blue) form. Mutagenesis efficiencies ranging from 50-73% were obtained by various classroom laboratory groups. The changes in the phenotype from green to blue fluorescence were linked to changes in the primary structure via DNA sequencing. The change in the tyrosine codon TAT to CAT (the histidine codon) at position 66 is shown in fig. 2. The change in color from green to blue has dramatic effect on the perception of the students of the power of rDNA techniques in manipulating macromolecules.

The overview of the techniques used is presented in fig 3. All manipulations are carried out on the gfp gene. This experimental format is currently followed in the five day rDNA short course offered by the Cook College Office of Continuing Professional Education, Rutgers University. The course has been designed in a modular form and can be expanded easily to form a full semester undergraduate rDNA laboratory course.

Conclusion

The gfp gene has proved to be an excellent instructional tool for gene expression analysis in rDNA courses. The students can directly visualize the effects of various manipulations on this gene which include changing the fluorescence of the protein from green to blue. To date twenty four people have been trained and all agree that gfp gene makes a fascinating instructional tool.

Acknowledgments

NSF grant # DUE-9602356 to WWW and DFD. Modular approach to biotechnology laboratory instruction based on a novel green fluorescent protein.

References

1. Chalfie M, Tu Y, Euskirchen G, Ward W, Prasher D. Green fluorescent protein as a marker for gene expression. Science 1994; 263: 802.
2. Heim R, Prasher D, Tsein R. Wavelength mutations and posttranslational auto oxidation of green fluorescent protein. Proc Natl Acad Sci USA 1994; 91: 12501.
3. Cutler M, Davies D, Ward W. GFP as an instructional tool for Biochemistry laboratory courses. Campbell A, Kricka L, and Stanley P editors: Bioluminescence and Chemiluminescence. Fundamentals and Applied Aspects. John Wiley and Sons Chichester, England. 1994: 383.

NEW FASCINATING DEMONSTRATIONS WITH CHEMILUMINESCENCE

H Brandl[1], S Albrecht[2], T Zimmermann[3]
[1]Gymnasium Kaltenkirchen, D-24568 Kaltenkirchen, Germany
[2]Dept of Gynecology and Obstetrics and [3]Dept of Surgery,
Technical University Dresden, Fetscherstrasse 74, D-01307 Dresden, Germany

Introduction

Reactions that emit "cold light" visible to the human eye have drawn much attention from the scientific community. Simple bio- and chemiluminescence demonstrations are useful for scientific motivation in education. In this communication, we report about 4 new chemiluminescent demonstration experiments with fascinating show effects.

Material and Methods

Experiment 1: *Oscillating Chemiluminescence* [1-3]

Part a) Demonstration of the oscillating Luminol Chemiluminescence

Chemicals:
- Solution A: 10 ml H_2O_2 (30 p.c.) are mixed with 90 ml of dest. water.
- Solution B: 1,46 g KSCN are dissolved in 100 ml of dest. water.
- Solution C: 0,4 g NaOH (solid) are dissolved together with 0,07 g luminol in 100 ml dest. water.
- Solution D: 0,01 g $CuSO_4 \cdot 5H_2O$ are dissolved in 100 ml dest. water.

Procedure: 10 ml of solution A, B and D are mixed in a 100 ml glas tube resulting in a yellow color. After adding 10 ml of solution C in the darkness a blue weak chemiluminescence is visible. The mixture is placed in a hot water bath (50-60 °C) and stirred with a magnetic stirrer. The oscillations begin after a short incubation time. The weak blue CL will be stronger and after a short time the whole mixture gives strong blue light for 3-6 seconds. The periods of strong light emission come back after 20-30 seconds and will be longer in duration. The color of light emission is variable by adding a trace of useable fluorescers (Rhodamine, Fluorescein, Thiazolycllow).

To achieve oscillations relating to space the solutions A-D (2 ml) are mixed in a glass tube and heated for short time (water bath) up to 60 °C. Resulting blue chemiluminescent rings are moving up to the ground of the glass.

Part b) Chemiluminescent chemical waves in the Belousov-Zhabotinski-reaction

Chemicals:
- Solution A: 2 ml H_2SO_4 (conc.) are added to 50 ml dest. water. In this solution are dissolved 5 g $KBrO_3$.
- Solution B: 1 g NaBr is dissolved in 10 ml dest. water.
- Solution C: 1 g malonic acid is dissolved in 10 ml dest. water.
- Wöhler's Siloxene [3]: 5 g Calciumdisilicide are added to 50 ml conc. HCl (violent reaction). Then 25 ml conc. HCl are added by boiling for 10 min. After adding of 150 ml dest. water the yellow-green suspension is filtered and washed with water, ethanol and diethylether.

Procedure: 6 ml solution A, 1/2 ml solution B and 2 ml of solution C are mixed in glass tube. After 5 min 4 drops of $Ru(bipy)_3Cl_2$-solution (10 p.c. in water) are added. The mixture is placed in a petridish. 0,1 mg Ce(IV)sulfate and 0,07 g Wöhler's Siloxene are dissolved in 1,5 ml dest. water. This solution is added quickly to the petridish. In the darkness a good visible yellow-red chemiluminescence is obtained. After some minutes chemiluminescent ring structures and waves are generated.

Experiment 2: *Red chemiluminescence of Tea* [4,5]
Chemicals: ethylacetate, H_2O_2 (30 p.c.), bis(2,4-dinitrophenyl)oxalate (DNPO)
Procedure: 100 ml ethylacetate are placed in a tea glass. Then 4-5 ml of H_2O_2 and 0,5 g DNPO are suspended in the ethylacetate. By addition of a tea bag and stirring, the solution emits a good visible red light causes by peroxyoxalate chemiluminescence in presence of porphyrin derivatives.

Experiment 3: *Double chemiluminescence of singlet oxygen* [6]
Chemicals: chlorine (generated from $KMnO_4$/HCl in an erlenmeyer flask), NaOH (10 p.c.), H_2O_2 (30 p.c.), luminol, lucigenin
Procedure: Alkaline H_2O_2 solution (5 ml NaOH + 1 ml H_2O_2) is placed in a fermentation tube. By bubbling chlorine through the tube (generated in an erlenmeyer flask for example) red $^1O_2$ chemiluminescence is visible. By adding a trace of luminol or lucigenin to the fermentation tube a blue or green chemiluminescence is generated beside the red $^1O_2$ luminescence.

Experiment 4: *Chemiluminescent smoke* [1]
Chemicals: hexamethylentetramine, triethylaluminium, tetrahydrofuran, tetrakis-(dimethylamino)ethylene
Procedure: 10 g of hexamethylentetramine and 14 g triethylaluminium are mixed in a two-neck-flask under anaerobic conditions. After a few minutes 1 ml THF and 16 ml tetrakis(dimethylamino)ethylen are added. By adding 20 ml of water to the mixture a big green chemiluminescent smoke cloud is generated (up to 6 m in high).

Results and Discussion

Experiment 1

Orban's oscillating system is based by the Cu^{2+}-catalized reaction of H_2O_2 and KSCN in alkaline solution. Luminol reacts by light emission with H_2O_2 in the presence of the oscillating Cu^{2+}/Cu^{1+}-NH_3-complex-ions. The reaction is useful to explain basic conditions for oscillating chemiluminescence in aqueous solution. Hypothetical reaction mechanisms are discussed in the original literature. Because of the simple procedure the experiment is also practicable for school teaching.
Part b) of the experiment can be discussed in the same way. The well known BZ-reaction results in light emissions in presence of chemiluminescent siloxene.

Experiment 2

Special porphyrins including chlorophyll (except complexes of metals with several stable valence states, e. g. iron and cobalt) are known as excellent sensitizers in peroxyoxalate-chemiluminescence-systems. This experiment demonstrates impressive the presence of chlorophyll in commercial tea products:

$$2\ \text{DNPO} + H_2O_2 + F \rightarrow 2\ \text{Dinitrophenol} + 2\ CO_2 + F^*$$

$$F^* \rightarrow F + h\nu \qquad (F = \text{Chlorophyll})$$

Experiment 3

This strong chemiluminescent reaction causes on the cooxydation of luminol or lucigenin by H_2O_2 and OCl^- in alkaline solution. Beside singlet oxygen is generated, which can also react with luminol or lucigenin. The emitting light consists of two parts: red light ($^1O_2$) and blue/green light (luminol, lucigenin).

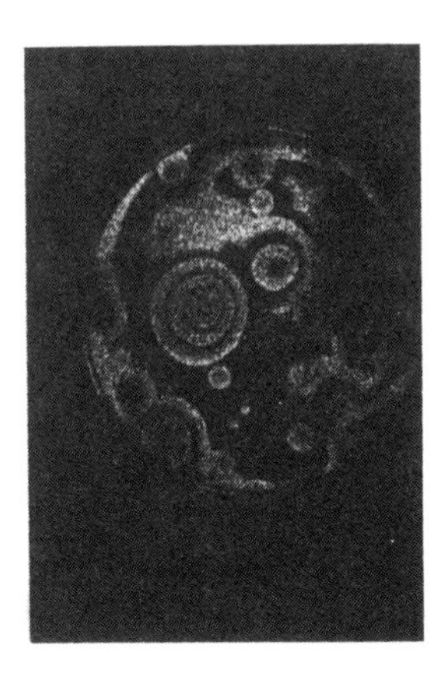

Fig.1: Chemiluminescent chemical waves in a petridish

Fig.2: Generation of luminol-$^1O_2$-double chemiluminescence in a fermentation tube

Experiment 4

This experiment is predestinated for demonstrations in lecture halls. The well known chemiluminescence of tetrakis(dimethylamino)ethylene (TDE) is combined with the generation of $Al(OH)_3$-aerosols. The $Al(OH)_3$-particles are loaded with TDE and by intensive air contact green chemiluminescence occurs. For details of the reaction mechanism see the original literature.

References

1. Brandl H, Albrecht S, Haufe M. Neue Experimente zur Chemolumineszenz. Chem unserer Zeit 1993;27:303.
2. Orban M. Chemical oscillations. J Am Chem Soc 1986;108:6893.
3. Brandl H. Oszillierende Chemolumineszenz im BZ-Siloxen-System. PdN Chem 1988;37-1:32.
4. Brandl H. Zur Chemolumineszenz von Chlorophyll. Chem unserer Zeit 1986;20:63
5. Albrecht S, Brandl H, Köstler E. Ein neuer Suchtest zur Porphyrie-Diagnostik. Z Klin Med 1989;44:2071.
6. Brandl H. Häm-Biosynthese und Porphyrine. PdN Chem 1993; 42-5:35.

PART 6

Firefly Luminescence and Applications

CHAPERONE DnaK AND ATP PARTICIPATE IN THE *in vivo* FOLDING OF FIREFLY LUCIFERASE SYNTHESIZED BY *E.COLI* CELLS

OV Leontieva and NN Ugarova
Faculty of Chemistry,Lomonosov Moscow State University,
119899, Moscow, Russia

Introduction

We showed previously (1), that in *E. coli* strain CA, transformed by a plasmid with the *luc* gene and termoinducible λP_R promoter, after 3 h of thermoinduction (42^0C) practically inactive firefly luciferase is synthesized; the luciferase adopts the native catalytically active conformation during subsequent incubation of the cell culture at 21^0C. The intracellular ATP concentration, intra- and extracellular pH increase in parallel with enzyme activity growth. It was supposed that *E. coli* heat shock proteins (chaperones) take part in the process of luciferase activation. Earlier it was shown, that in solution it is possible to renaturate thermally or chemically denaturated firefly luciferase with the help of some DnaK homologs, members of the Hsp70 chaperone family (2). The presence in the reaction mixture DnaK, DnaJ and GrpE chaperones results 70% renaturation of *Photinus pyralis* luciferase activity. ATP is essential for all these renaturing processes (3). Other literature data confirm the participation of chaperones from yeast (Ydjlp, Ssapl) anf TRIC-chaperone in this process (4). Summarizing the data mentioned above, it is possible to say that the greatest extend of luciferase renaturation *in vitro* is observed in case of using *E.coli* chaperones.

The goal of this work was to elucidate the roles of ATP and DnaK and GroEL chaperones in the *in vivo* folding of firefly lucuferase synthesized in *E.coli* cells on a plasmid with termoinducible λP_R promoter. We used the DnaK-deficient *E. coli* strain Ω238 and the GroEL-deficient *E. coli* strain B178groE7 and corresponding control strains.

Materials and Methods

The materials and methods used were described previously (1). *E. coli* strains Ω237 (F^-, sup^- , galK2, thr::Tn10), Ω238(F^-, sup^-, galK2, thr::Tn10, dnaK7), W3110 (F^-, sup^- ,trpA), B178groE7 (F^-, sup^- , galE, groE7) were kindly provided by the Institute of Molecular Genetics of the Russian Academy of Science. *E. coli* strain CA (Hfr H, thi - 1, rel -1, lacI22, lacZ13, supC70, cI_{857}, χis1, Rec A^-).was also used. Plasmid pJGλ with the *luc* gene was described earlier [1]. Polymyxin (Sigma) was dissolved in Hepes, pH 7.2, its final concentration was $2x10^5$ units/mL. ATP solution (0.5 mol/L) was adjusted to pH 6.5 with concentrated NaOH.

The DnaK-deficient *E. coli* strain Ω238, the GroEL-deficient *E. coli* strain B178groE7, and control strains Ω237, W3110, and CA, were transformed with pJGλ plasmid carrying the firefly *Luciola mingrelica* gene under the control of the termoinducible λP_R promoter. When the temperature scheme 28-42-21 ^{0}C was used, the cell culture was cultivated at 28^{0}C until A^{590} = 0.6, then incubated at 42^{0}C for 3 h, then incubated without shaking at 21^{0}C. When the temperature scheme 28-21^{0}C was used, the cell culture was cultivated at 28^{0}C until A^{590} = 1.6 and then incubated at 21^{0}C without shaking. During the whole period of incubation cells suspension aliquots were taken at selected time intervals, lysed, and assayed for enzyme activity by bioluminescent method as described earlier (1). The amount of luciferase protein was determined by SDS gel electrophoresis. The resulting electrophoregrams were analysed using a Shimadzu CS-9000 laser densitometer.

Results and Discussion

Time course of firefly luciferase protein accumulation according to SDS gel electrophoresis data. As shown earlier (1), SDS gels of the lysates of *E.coli* cells which do not carry *luc* gene have no band corresponding to a 64-kD protein (the molecular weight of the firefly luciferase). Thus, it is possible to calculate the luciferase protein content by scanning the electrophoregram. For both temperature schemes (28-42-21 and 28-21^{0}C) the kinetics of firefly luciferse protein accumulation were practically the same in all strains. Luciferase protein synthesis began simultaneously with the beginning of cell culture incubation and stopped 15-17 h after the end of thermoinduction and shaking. In all strains the content of luciferase protein reached approximately the same value (3.4-3.9% of total cell protein for temperature scheme 28-42-21^{0}C and 3.0-3.4% for temperature scheme 28-21^{0}C) (Fig. 1). The heat shock results in a 10-15% increase of synthesized lucuferase quantity. Thus, for all examined systems, the nature of host strain has no influence on the content of synthesized luciferase protein. The same can be said about the presence of the thermoinducible λP_R promoter on the plasmid because the cI_{857}-repressor is synthesised only in the CA strain.

Time course of recombinant luciferase activity. For *E. coli* cultures Ω237:pJGλ, W3110:pJGλ, B178groE7:pJGλ luciferase activity reached 40-80 arb.units per μL lysate after 3 h of thermoinduction. During the subsequent incubation at 21^{0}C the luciferase activity began to grow after some delay and reached the maximum (6.0-6.4)$*10^3$ arb.units per μL lysate in about 70-75 h from the beginning of incubation. It remaind at this level for the next ~40 h and then slowly decreased (Fig.2). In case of incubation according to the temperature scheme 28-21^{0}C luciferase activity appeared in 3-3.5 h from the beginning of culture incubation, then monotonously increased and reached values of (6.8-7.0)$*10^3$ arb.units per μL lysate by the late post-stationary phase. The observed phenomenon of recombinant luciferase activation in the late post-stationary phase was similar to that found for *E. coli* strain CA carrying pJGλ plasmid and synthesizing cI_{857}-repressor (1).

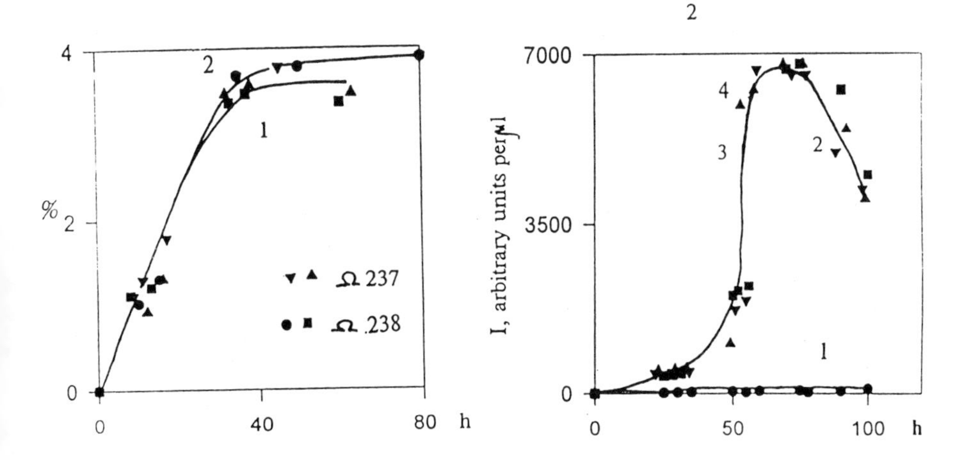

Figure 1. The content of the firefly *L. mingrelica* luciferase protein in percent to total cell proteins for *E. coli* strains Ω237 and Ω238 with temperature scheme 28-42-21°C (1) and 28-21°C (2).
Figure 2. Time-courses of the firefly *L. mingrelica* luciferase activity in lysates of *E. coli* cells strain Ω238 (1), Ω237 (2), W3110 (3), and B178groE7 (4). The cells were incubated under the temperature scheme 28-42-21 °C.

Our data demonstrate that luciferase is synthesized in *E. coli* cells in an inactive form, and for its folding into the active form the presence of DnaK chaperone is necessary. This chaperone is an ATP-dependent protein and thus its activity may depend on intracellular ATP concentration (ATP_{in}). So, we examined the influence of ATP_{in} on the time course of luciferase transformation into catalytically active conformer.

Luciferase activation is an ATP-dependent process. It was shown earlier for *E. coli* strain CA carring pJGλ plasmid that before the beginning of luciferase transformation into catalytically active form the ATP_{in} concentration equal 5-7 mmol/L began to grow in parallel with the increase in luciferase activity. It reached a maximum (50-70 mmol/L) simultaneously with the luciferase activity maximum (1).

Perhaps an artificial increase in the ATP_{in} concentration after the end of luciferase synthesis can significantly accelerate the folding of newly synthesized protein. To check this, using *E. coli* strain CA we obtained time courses of luciferase activation for different concentrations of ATP_{in}.

To introduce the additional ATP into *E. coli* cells, we treated them preliminary with polymyxin (this antibiotic increases membrane permeability for low-molecular-weight reagents (5)). At the final concentration $2x10^4$ U/mL medium, the ATP concentrations inside and outside the *E. coli* cell were equal. At the same time, the cells retained 85-90% of their proliferation ability. When the luciferase protein content reached its maximum (17 h after the end of thermoinduction), together with polymyxin, 0.5 mol/L ATP solution, pH 6.5, was added to the cell culture up to the final ATP concentrations in the medium 1, 5, 25 or 50 mmol/L. Fig. 3 shows that the protein activation is considerably accelerated with increase in ATP concentration: 1-1.5 h in comparison with 50 h in the intact cells.

When we treated the cell culture only with polymyxin without ATP or in presence of 1 mmol/L ATP, we did not observe the enzyme activation. In the presence of 5 mmol/L ATP the luciferase activity increased 20-fold, and with 25-50 mM ATP it grows 200-250-fold. Therefore, the process of luciferase protein transformation from inactive into active conformation is very ATP-dependent. In the standard experiment, without ATP and polymyxin treatment, ATP concentration of 50-70 mmol/L was reached only in the late post-stationary phase, and luciferase activation with the help of DnaK chaperone became the most effective.

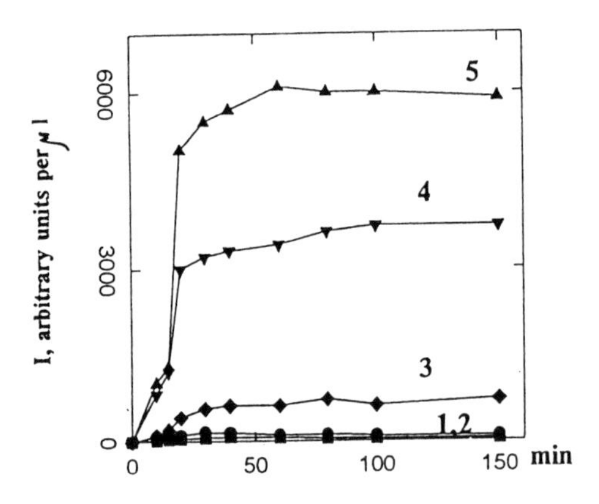

Figure.3. Luciferase activation in conditions of different ATP concentrations. Cell suspension was preliminary treated by polymyxin. ATP final concentration, mmol/L: 1- 0, 2 - 1, 3 - 5, 4 - 25, 5 -50.

Thus, the formation of catalytically active luciferase in *E. coli* cells can be described by the following scheme (Fig. 4).

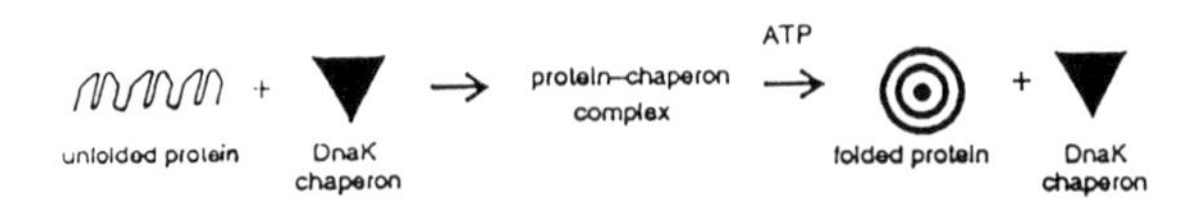

Figure.4. Scheme for recombinant luciferase activation in *E.coli* cells.

The synthesized unfolded (or incorrectly folded) luciferase protein is bound to the DnaK chaperone. Under the action of ATP this complex decomposes into catalytically active enzyme and the free DnaK molecule. Our results concerning DnaK role in the folding of newly synthesized luciferase *in vivo* coincide with the data from paper (2) where the participation of DnaK in renaturation of thermodenaturated luciferase *in vitro* is shown. For the first time we have shown experimentally that the same DnaK chaperone participates in luciferase folding *in vivo*.

Acknowledgements
This work was supported by the State Scientific Technical Program "Newest Methods of Bioengineering ", and International Science Foundation (Soros Foundation).

References

1. Kutuzova GD,Leontieva OV, Skripkin EA, Ugarova NN. Kinetics of expression of luciferase genes from the fireflies *Photinus pyralis* and *Luciola mingrelica* in *Escherichia coli.* 1. Kinetics of conversion of recombinant luciferases into enzymatically active conformers. Biochemistry (Moscow) 1993;59,102-11.
2. Schroder H, Langer T, Hartl FU, Bukau B. DnaK, DnaJ and GrpE form a cellular chaperone machinery capable of repairing heat-induced protein damage. EMBO J. 1993;12,4137-44.
3. Levy EJ, McCarty J, Bukau B, Chirico WJ. Conserved ATPase and luciferase refolding activities between bacteria and yeast Hsp70 chaperones and modulators. FEBS Lett.1995;368,435-40.
4. Frydman J, Hartl FU. Principles of chaperone-assisted protein folding: differences betwen *in vitro* and *in vivo* mechanisms. Science. 1996;272,1497-502.
5. Egorov NS. Basic knowledge on antibiotics (in russian). 1986; Moscow, Visshaya Shkola.

STRUCTURE OF THE CATALYTIC SITE OF FIREFLY LUCIFERASE AND BIOLUMINESCENCE COLOR

L Yu Brovko, EI Dementieva and NN Ugarova
Dept of Chemistry, Lomonosov Moscow State University, Moscow 119899, Russia

Introduction

Firefly luciferase (FFL) catalyses bioluminescent reaction of firefly luciferin (LH_2) oxidation by air oxygen in the presence of ATP and magnesium salt (1). The bioluminescent emitter is proved to be reaction product – oxyluciferin (LO) in the singlet electronically excited state. The amino acid sequences of the FFL of different species were shown to be highly homological (2,3). The crystal structure of *Photinus pyralis* FFL was reported lately (4). But till now the role of protein globe in the bioluminescence process, location and structure of catalytic site remains unknown.

Methods of fluorescence provide wide opportunities for investigation of protein dynamic, conformational changes and topography of protein globe. Firefly bioluminescent system has several features, that make fluorescence a powerful instrument for investigation luciferase structure. Both substrate and product of bioluminescenct reaction are intensive fluorophores. The unique tryptophan residue in *Luciola mingrelica* FFL is highly conservative in the family of firefly and click beetle luciferases. It was shown that its fluorescence is quenched and the fluorescence parameters of luciferin is not changed upon formation of the enzyme-substrate complex (5). The life-time of the excited tryptophan was not changed upon complex formation, thus indicating, that conformational changes occur in the protein during enzyme-substrate complex formation and that Trp residue is close to LH_2 binding site. On the other hand according to the 3D-structure of *P.pyralis* FFL the highly conservative Trp417 is located in the close vicinity of the cleft between two domains of the protein globe proposed as a possible ATP binding site. The computer analysis of the amino acid sequences in the superfamily of the enzymes catalysing acyladenylate formation from ATP and compound with carboxyl group showed, that this Trp residue was located in the conserved motif possibly involved in the catalytic site of these enzymes (6). Thus tryptophan fluorescence in FFL can serve as a good indicator of conformational changes and dynamic of the protein globe in the region of catalytic centre during the bioluminescence. Comparison between both steady-state and time-resolved tryptophan fluorescence for *P.pyralis* and *L.mingrelica* FFL was performed to investigate the dynamic behaviour of both proteins near the catalytic site of the enzymes.

Since singlet electronically excited state of LO was shown to be emitter, fluorescence of LO was used as a good model of the bioluminescence (7). Preliminary data concerning the fluorescence of LH_2 and LO in the complex with luciferase showed, that substrate and product binding sites in FFL were not so hydrophobic as thought before. The fluorescence spectra of the islolated enzyme-product complex and native bioluminescent spectra were similar to fluorescence in water solutions (7). Fluorescence spectra of luciferin in different organic solvents together with the spectral characteristics of native bioluminescence and protein structure were analysed in order to establish the correlation between the properties of the substrate binding site, microenvironment of the emitter and bioluminescence color.

Materials and Methods

Instrumentation: Fluorescent spectra were obtained using Perkin-Elmer LS50B spectrofluorimeter. All spectra were corrected for PMT sensitivity. Time-resolved fluorescence measurements were carried out on spectrometer described in (7).
Reagents: FFL from *P.pyralis* fireflies was purchased from Sigma. *L.mingrelica* FFL was isolated from fireflies' lanterns as in (8). All other reagents were of analytical grade.
Procedures: Luciferin solutions in organic solvents and water (10 μg/mL) were prepared just before experiment. The tryptophan fluorescence decay curves were measured in the range 310-380 nm with the step of 3 nm. Method of reconstruction of the time-resolved fluorescent spectra was used for analysing the data.

Results and Discussion

Steady-state and time-resolved tryptophan fluorescence as indicator of protein dynamic in FFL. Steady-state tryptophan fluorescence spectra of both *L.mingrelica* and *P.pyralis* FFL were identical with the maximum at 340±0.5 nm. The obtained time-resolved fluorescent spectra showed the increase of the average fluorescence life-time (τ_{av}) at the long wave part of the spectra and the corresponding red shift (from 336 to 342 nm) of the tryptophan fluorescence maximum (Table 1).

Table 1. Life times (τ_{av}) and tryptophan fluorescence intensity (I) at different wave lengths for *P.pyralis* and *L.mingrelica* FFL

FFL	λ, nm	τ_{av}, ns	I_{310}/I_{370} 1 ns after excitation	I_{310}/I_{370} 8 ns after excitation
P.pyralis	310	4.41		
	340	6.21	1.63	1.25
	370	5.87		
L.mingrelica	310	2.99		
	340	3.79	1.34	1.38
	370	5.17		

The detailed analysis of fluorescence data showed pronounced difference in dynamic properties of both FFL. For *L.mingrelica* FFL the relation of fluorescence intensities in different parts of the spectra in 1 and 8 ns after excitation remains constant, indicating that the observed red shift of the fluorescence maximum proceeds without changing in the shape of the spectrum. It means, that relaxation of the protein matrix around excited tryptophan moiety proceeds during the time less, then the life time of the excited tryptophan, provides thus indirect evidence for the rather high flexibility of the *L.mingrelica* FFL protein globe. Comparison of the intensities in different part of the fluorescence spectra of *P.pyralis* FFL at 1 and 8 ns after excitation showed, that in this case the evolution of the spectra resulted not only in the long wave shift, but also in decreasing of the spectrum's width. This indicated, that the process of relaxation of the *P.pyralis* FFL protein globe was slower, than the process of excited tryptophan deactivation. Thus, the structure of the *P.pyralis* FFL in the region of active site is probably more rigid, than *L.mingrelica* FFL. For *L.mingrelica* FFL it was shown before, that luciferin-luciferase complex formation did not affect the dynamic behaviour of the protein (5). The higher rigidity of. *P pyralis* protein globe in comparisson with *L.mingrelica* FFL coincides with the data of higher thermostability of the former enzyme.

Role of specific and nonspecific interactions of the emitter with its microenvironment in the FFL bioluminescence color. The exact location and properties of the luciferin binding site is not known yet. The problem of bioluminescence color remains one of the most intriguing. Several hypotheses were proposed to explain the changes of bioluminescence color for different beetle luciferases. Among them are the specific interactions of emitter with protein globe (1), properties of the excited LO microenvironment, when protein is being considered as a dielectric continuum (9), and the possible rotation of the excited oxyluciferin along C2-C2' bond (10). Amino acid residues 223, 247, 351-352 were proposed to be responsible for the bioluminescence color variations between species (1,11). According to the principles of fluorescent spectroscopy the factors, that can cause solvatochromic emission spectra's shifts are divided in two main groups: 1)specific interactions with solvent molecules resulted in formation of hydrogen bond or complex with fluorophore. In this case discrete changes in emission spectra are usually observed. For example interaction of LO with basic water solution results in dissociation of 6'-phenol proton and shift of the fluorescence peak from 440 to 550 nm (7). 2)Non-specific interaction of emitter with microenvironment, that depends on the polarizability of the medium according to equation 1 (12).

$$\nu_{max} = -C\Delta f, \text{ where}$$
$$\Delta f = (\varepsilon - 1) / (2\varepsilon + 1) - (n^2 - 1) / (4n^2 + 2) \qquad \text{(eqn.1)}$$

Δf is orientational polarizability, ε- dielectric constant, n - refractive index of the solvent, and ν_{max} - fluorescence maximum in cm^{-1}. In this case monotonous dependence of fluorescence maximum on dielectric properties of the medium is observed. According to (12) the spectrum's shift proportionally depends not only on the ability of relaxation of the solvent cage around the excited molecule, but also on the difference between the dipole moments of the emissive and ground states of the emitter. The more is the difference between dipole moments in the ground and excited singlet states the more is the observed emission spectra shift. For LO molecule the big changes in dipole moments upon excitation where calculated (13), that is why the significant shifts of the fluorescence spectra of LO may be expected even when minor changes in polarizability of the microenvironment occur. Analysis of the bioluminescent spectra of click beetle luciferases and its mutants showed the best correlation between the position of the emission maximum and the polarizability of the aminoacid residues, which substitution caused the bioluminescent spectra shifts (11).

Luciferin molecule was chosen as an analogue of emitter because of its high stability. To mimic the microenvironment of the emitter in native bioluminescent system we used organic solvents with different polarizabilities. All solvents contained of 3% v/v of triethylamine (except water) to provide deprotonation of the 6'-phenol group without changes of the polar properties of the solvent, because in native bioluminescence phenolate form of LO was proved to be emitter (7). The following solvents were used in the experiment: chloroform (CH_3Cl, $\Delta f = 0.254$), dichlormethan (CH_2Cl_2, $\Delta f = 0.319$), acetone ($(CH_3)_2CO$, $\Delta f = 0.375$), dimethylformamide ($(CH_3)_2NCOH$, $\Delta f = 0.377$), ethanol (C_2H_5OH, $\Delta f = 0.379$), acetonitrile (CH_3CN, $\Delta f = 0.393$), water (H_2O, $\Delta f = 0.406$). The dependence of the fluorescent maximum of LH_2 (ν_{max}) on the orientational polarizability of the solvent (Δf) is presented in Fig.1A. The data of fluorescence maximum of phenolate form of dehydroluciferol in different organic solvents reported in (14) are presented by us in the same coordinates in Fig. 1B.

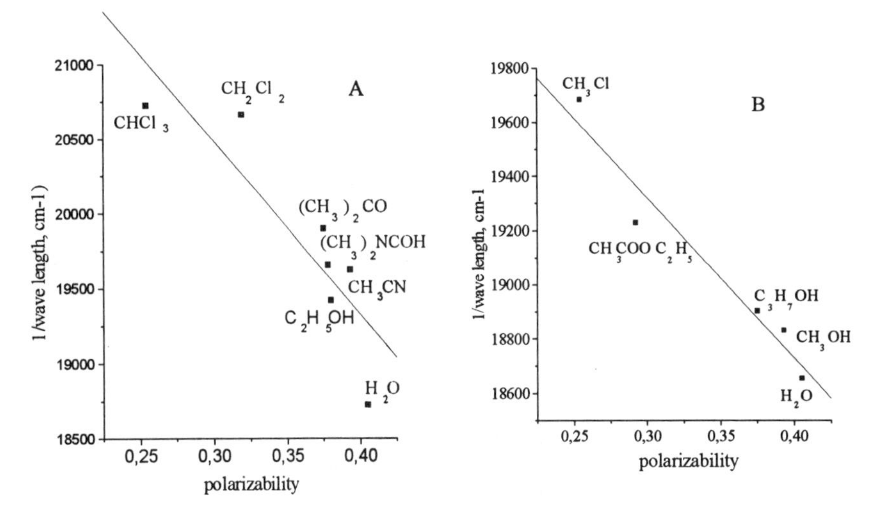

Fig. 1. Dependence of the fluorescence maxima of the phenolate forms of luciferin (A) and dehydroluciferol (B) on the polarizability of the solvent

The obtained results showed, that changes in polarizability of the solvent caused significant shifts (482 → 534 nm and 508 → 536 nm) in emission spectra of both luciferin and dehydroluciferol, respectively. The linear correlation between the fluorescence maximum and polarizability of the environment ($R = 0.875$ and $R = 0.974$ for luciferin and dehydroluciferol respectively) confirmed the proposed mechanism of nonspecific interaction of the emitter with solvent molecules as a reason of spectra shifts. According to the results obtained we may propose, that the color of bioluminescence depends both on specific and nonspecific interaction of the emitter with the protein globe. The specific interactions of LO with some basic groups in the substrate binding site of FFL should provide deprotonation of the phenolic and enolic groups of the LO molecule and formation of the dianionic form of yellow-green emitter, the only one registered in native bioluminescence. The changes in the luminescence spectra of this form of LO in rather wide range of the spectra may occur due to the changes in the effective polarizability of the substrate binding site. The observed bioluminescence spectra for different beetle luciferases lie in the region from green through yellow to orange (550-590 nm). According to the proposed hypothesis the green bioluminescent color should be observed for the protein with low polarizability of the luciferin binding site and on the opposite the orange bioluminescence should be observed for the protein, that has high polarizability of the catalytic site. On the other hand the polarizability of the environment characterize the ability of the surrounding molecules to rearrange around the formed excited dipole, so this property is connected with dynamic mobility of the environment. From this point of view the bioluminescence maxima of 562 and 570 nm for *P pyralis* and *L.mingrelica* FFL respectively, correspond to the data of higher flexibility of *L.mingrelica* FFL and support our results.

Acknowledgements.
The authors are grateful to Dr.A.Chikishev for time-resolved fluorescence measurements. The research was supported by Russian Foundation of Fundamental Science and State Scientific Technological Program "Newest Methods in Bioengineering".

References

1. Wood KV. The chemical mechanism and evolutionary development of beetle bioluminescence. Photochem Photobiol 1995; 62: 662-73.
2. Devine JH, Kutuzova GD, Green VA, Ugarova NN, Baldwin TO. Luciferase from the east European firefly *Luciola mingrelica*: cloning and nucleotide sequence of the cDNA, overexpression in *Escherichia coli* and purification of the enzyme. Biochim Biophys Acta 1993; 1173: 121-32.
3. Tatsumi H, Kajiyama N, Nakano E. Molecular cloning and expression in *Escherichia coli* of a cDNA clone encoding luciferase of a firefly *Luciola lateralis*. Biochim Biophys Acta 1992; 1131: 161-5.
4. Conti E, Franks NP, Brick P. Crystal structure of firefly luciferase throws light on a superfamily of adenylate-forming enzymes. Structure 1996; 4: 287-98.
5. Brovko LYu, Demetieva EI, Gandelman OA, Chikishev AYu, Shkurinov AP, Ugarova NN. Steady-state and time-resolved fluorescence for investigation of firefly bioluminescent system. J Fluorescence 1996; in press.
6. Morozov VM, Ugarova NN. Conserved motifs in superfamily of enzymes catalysing acyladenylate formation from ATP and compounds with carboxyl groups. Biochemistry (Moscow, in russian) 1996; 61: 1511-17.
7. Gandelman OA, Brovko LYu, Chikishev AYu, Shkurinov AP, Ugarova NN. Investigation of the interaction between firefly luciferase and oxyluciferin and its analogues by steady-state and subnanosecond time-resolved fluorescence. J Photochem Photobiol B:Biol 1994; 22: 203-9.
8. Dementieva EI, Kutuzova GD, Ugarova NN. Biochemical properties and stability of homogeneous luciferase of *Luciola mingrelica* fireflies. Moscow University Chem Bulletin (in Russian) 1989; 44: 69-73.
9. Ohmiya Y, Hirano T, Ohashi M. The structural origin of the color differences in the bioluminescence of firefly luciferase. FEBS Lett 1996; 384: 83-6.
10. McCapra F, Gilfoyl DJ, Young DW, Church NJ, Spencer P. The chemical origin of colour differences in beetle bioluminescence. In Campbell AK, Kricka LJ,

Stanley PE, editors. Bioluminescence and Chemiluminescence. Fundamentals and Applied Aspects. Chichester-NY: John Wiley&Sons 1995: 387-91.

11. Morozov VM, Ivanisenko VA, Eroshkin AM, Ugarova NN. Computer analysis of the dependence between bioluminescence spectra maximum and amino acid sequence of beetle luciferases: identification of the site responsible for color changes. Mol Biol (in Russian) 1996; 30: 1167-72.
12. Scherer T, Hielkema W, Krijnen B, Hermant RM, Eijckelhoff C, Kerkhof F et al. The synthesis and exploratory photophysical investigation of donor-bridge-acceptor systems derived from N-substituted 4-piperidones. Recl Trav Chim Pays-Bas 1993 ;112: 535-48.
13. Brovko LYu, Gandelman OA, Savich WI. Fluorescent and quantum-chemical evaluation of emitter structure in firefly bioluminescence. In Campbell AK, Kricka LJ, Stanley PE, editors. Bioluminescence and Chemiluminescence. Fundamentals and Applied Aspects. Chichester-NY: John Wiley&Sons 1995 :525-7.
14. Bowie LJ, Irwin R, Loken J, DeLuca M, Brand L. Excited state proton transfer and the mechanism of action of firefly luciferase. Biochemistry 1973 ;12 :1852-7.

CHEMICAL MODIFICATION OF FIREFLY LUCIFERASE

SR Ford, L Ye, LM Buck, SJ Pangburn, ABD Bior and FR Leach
Dept Biochem & Mol. Biol., Oklahoma State University, Stillwater OK 74078-3035, USA

Introduction

Early studies in the McElroy and DeLuca laboratories characterized the presumed active site(s) of firefly luciferase (FL, E.C. # 1.13.12.7) by chemical means (1-3). Since then the FL gene from the North American firefly, *Photinus pyralis*, has been cloned and the nucleotides sequenced (4). Surprisingly, none of the peptides identified in those early studies were present in the deduced amino acid sequence (5). Two different –SH-containing sequences have now been labeled with [^{14}C]-*N*-ethylmaleimide. They are TAC*BT and GELC*VR with the Cs being amino acid residues 216 and 391, respectively. Vellom has substituted alanines individually for each of those C residues; catalytic activity was retained (6) proving that the cysteines are not required for the reaction.

This paper reports the results of chemical modification studies of FL with the lysine-directed reactants thiourea dioxide (TUD) and fluorescein 5′-isothiocyanate (FITC).

Materials and Methods

Materials: Firefly luciferases from Sigma (St. Louis, MO, USA) and crystalline FL prepared in this laboratory were used interchangeably. Other chemicals were obtained from Sigma. Radioactive thiourea dioxide was synthesized from thiourea obtained either from Sigma or from New England Nuclear (Boston, MA, USA) using the procedure described by Barnett (7).

Methods: The inactivation of FL was performed as described by Ford et al. (8). For ^{14}C-TUD incorporation, FL (16.4 µmol/L) was incubated with 50 mmol/L ^{14}C-TUD (0.096 mCi/mmole). Enzyme activity was measured at 30 min (~ 50-60% inhibition) and 100 min (~ 95% inhibition). The samples (500 µL) were dialyzed, the protein was cleaved by incubation with trypsin, and the peptides separated by HPLC. The labeled peptides were sequenced by the Molecular Biology Facility, William K. Warren Medical Research Institute (Oklahoma City, OK, USA). For FITC labeling, one mg samples of FL were incubated in 50 mmol/L Tricine, pH 7.8, 2 % (v/v) DMSO with FITC (dissolved in DMSO and diluted 1/50 into FL solution), and with or without 10 mmol/L ATP. The samples were diluted with an equal volume of ice cold 10% $(NH_4)_2SO_4$, 1 mmol/L EDTA and concentrated in a Centricon 30 microconcentrator (Amicon, Beverly, MA, USA). After several washes the protein was cleaved with sequencing grade trypsin. The peptides were separated by HPLC and were sequenced in the Sarkeys Biotechnology DNA/Protein Core Laboratory (this department).

Results and Discussion

Survey of Lysine-reactive Reagents: Fluorosulfonylbenzoyl adenosine (FSBA) was the reagent DeLuca and colleagues used to label groups around the presumed ATP-binding site (3). We studied the inactivation of FL by FSBA. Incubation of FL with 2 mmol/L FSBA added every 20 min for two h produced an 85% decrease in enzyme activity. ATP protected the enzyme against the inactivation. Incubation with the amino-reactive reagent *o*-phthalaldehyde (200 µmol/L) rapidly inactivated luciferase ($t_{1/2}$= 7.5 min). Incubation of FL with pyridoxal phosphate (5 mmol/L) reversibly inactivated the enzyme. Dilution of the pyridoxal phosphate-treated enzyme with buffer reversed the inhibition, while sodium borohydride reduction of the incubation mixture "fixed" the enzyme in its inhibited form. ATP protection studies on the *o*-phthalaldehyde and pyridoxal phosphate inactivations were inconclusive. TUD and

FITC treatment inhibited firefly luciferase activity. In both these cases, the inhibition was prevented by inclusion of ATP in the reaction mixture. These two chemical modifiers were used for further study.

Modification of Firefly Luciferase with Thiourea Dioxide: TUD (50 mmol/L) has been used by Colanduoni and Villafranca (9) to modify a single lysine residue (the treatment also modifies the N-terminal S and oxidizes H 269; see (10)) in glutamine synthetase ($t_{1/2}$ = 15 min) with a total loss of enzymatic activity. Since glutamate protected from inactivation, the K is presumably at the glutamate binding site. Robertson et al. (11) found that CTP synthetase was >99 % inactivated by incubation with 20 mmol/L TUD for 30 min; this inactivation occurred in the absence of oxygen. Nucleotides (UTP, ATP, or CTP) protected CTP synthetase from inactivation (a combination of nucleotides was better than a single one). Three mols of TUD were incorporated per mol of enzyme monomer.

We studied TUD inhibition of FL. Figure 1 shows that TUD inactivation of FL was time- and concentration-dependent. Thiourea (100 mmol/L) was a much less effective inhibitor producing only 25 % inhibition after 60 min incubation with 100 mmol/L as compared to 68 % inhibition achieved with the same concentration of TUD. Under the same conditions sodium thiosulfate did not inhibit. When incubated with FL for 30 minutes, 50 mmol/L TUD produced 50% inhibition. Figure 2 shows that inclusion of 10 mmol/L ATP in the incubation mixture reduced the inactivation of FL . This suggests that an ATP binding site might be involved.

Radioactive TUD was prepared as described in the Materials and Methods section. Incubation of FL with 50 mmol/L TUD resulted in 76% inhibition in 30 min and 97% inhibition in 100 min. After dialysis to remove excess TUD, the inhibitions were 42 and 50%, respectively, when compared to dialyzed controls not exposed to TUD. Incorporation of ^{14}C–TUD was approximately 1 mol/mol FL at 30 min and about 2 mol/mol FL at 100 min. The inclusion of ATP in the incubation mixture reduces the inhibition of enzyme activity and the incorporation of radioactive TUD. In one experiment, 2.4 mol TUD/mol FL was incorporated in a 2 hour incubation in the absence of ATP, while only 1.4 mol TUD/mol FL was incorporated in the presence of 100 mmol/L ATP.

Tryptic peptides were produced from FL samples labeled with ^{14}C-TUD in the presence and in the absence of ATP as described in the Materials and Methods section. The peptides were separated by HPLC. Two fractions of the sample labeled in the absence of ATP contained the majority of the radioactivity; these fractions were analyzed for amino acid sequence. The sequence of the peptide with a 30 min retention time was GLTGK (residues 525-529) and the peptide with a 59 min retention time had a sequence of SGYVNNPEATNALI (residues 399-412). The sample labeled in the presence of ATP contained only one radioactive fraction, which corresponded with the 59 min peak of the other sample. In the presence of ATP, the 30 min peak was not labeled.

Modification of FL with FITC: Fluorescein 5′-isothiocyanate is a lysine-specific affinity probe with structural similarity to ATP. Pavela-Vrancie et al. (12) used FITC to selective modify tyrocidine synthetase 1. They noted that the two sequences labeled in tyrocidine synthetase 1 are found in *P. pyralis* FL. These sequences are: G416-E455 with K443 being the putative labeling location and V506-R536 with K529 being the putative labeling site. This is the K in the GLTGK peptide that was labeled with TUD only in the absence of ATP.

We modified FL with FITC to determine if the sequences identified by Pavela-Vrancie et al. (12) would be labeled. Figure 3 shows the time and concentration dependence of inactivation. With 100 μmol/L FITC the $t_{1/2}$ of FL is ~7 min. Addition of 10 mmol/L ATP to the incubation

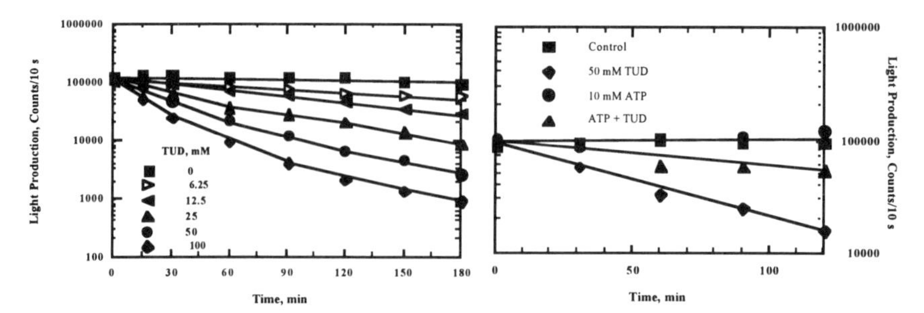

Figure 1. Effect of Thiourea Dioxide Concentration. Figure 2. Protection by ATP Against Inactivation

mixture protected FL from inactivation as is illustrated in Figure 4. Using absorbance at 495 nm to determine the number of FITC molecules reacted per molecule of FL, we found one molecule reacted in a 50% inhibited sample. The corresponding sample protected from inhibition by ATP contained 0.6 molecule FITC/FL. Tryptic peptides were prepared from samples labeled to 50% inhibition with FITC and samples labeled for the same time in the presence of ATP. Peptides from these samples were separated on HPLC. The major fluorescent peak eluted at ~64 min for both protected and unprotected samples (the fluorescence was much stronger in the unprotected sample). This fraction was subjected to protein sequencing. In the unprotected sample, two major components, GPMIMSGYVNNPEATNALI (residues 394-412) and QGYGLTETTSAILITPEGD (residues 338-356) and a minor component TIALIMNSSGSTGLPK (residues 191-206) were found. In the ATP protected sample, only the two major components were identified. The GLTGK sequence, eluting at ~30 min in the TUD labeled samples, was not labeled in either sample.

The IMNSSGSTGLPK sequence is consistent with the established AMP-binding domain signature sequence. The GLTGK sequence which was labeled with TUD was also labeled with FITC in the analogous protein tyrocidine S synthetase and is a presumed ATP binding site.

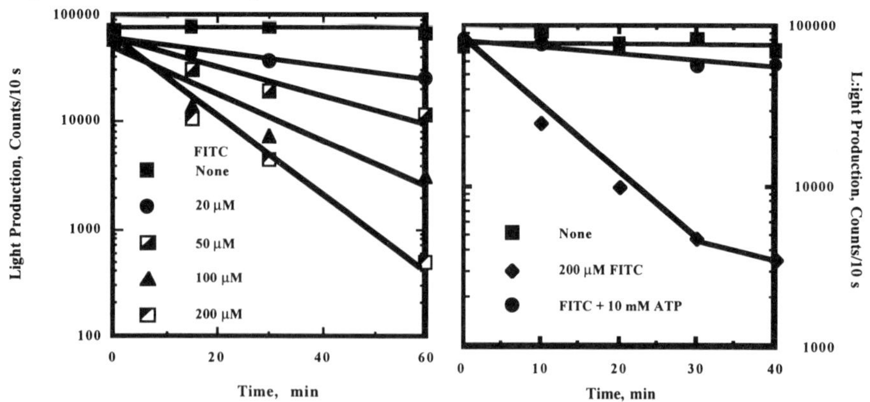

Figure 3. FITC Inactivation. Figure 4. ATP Protection Against Inactivation.

Acknowledgments
Supported in part by the Oklahoma Agricultural Experiment Station.

References

1. Travis J, McElroy WD. Isolation and sequence of an essential sulfhydryl peptide at the active site of firefly luciferase. Biochemistry 1966;5:2170-6.
2. Lee RT, McElroy WD. Isolation and partial characterization of a peptide derived from the luciferin binding site of firefly luciferase. Arch Biochem Biophys 1971;146:551-6.
3. Lee Y, Esch FS, DeLuca M. Identification of a lysine residues at a nucleotide binding site in the firefly luciferase with *p*-fluorosulfonyl[^{14}C]benzoyl-5´-adenosine. Biochemistry 1981;20:1253-6.
4. de Wet JR, Wood, KV, DeLuca M, Helinski DR, Subramani S. Firefly luciferase gene: structure and expression in mammalian cells. Mol Cell Biol 1987;7:725-7.
5. Hill F, de Wet JR, Wood KV, Helinski DR, DeLuca M. Properties of cloned firefly luciferase. Fed Proc 1986;45:1913.
6. Vellom DC. The active-site cysteines of firefly luciferase: chemical modification and mutagenesis studies. Ph. D. Thesis, UCSD, Dissertation Abstracts Order No. 9119000, 1990.
7. Barnett EdB. The action of hydrogen dioxide on thiocarbamines. J Chem Soc 1910;97:63-5.
8. Ford SR, Chenault KD, Hall MS, Pangburn SJ, Leach FR. Effect of periodate-oxidized ATP and other nucleotides on firefly luciferase. Arch Biochem Biophys 1994;314:261-7.
9. Colanduoni J, Villafranca JJ. Labeling of specific lysine residues at the active site of glutamine synthetase. J Biol Chem 1985;260:15042-50.
10. Di Ianni CL. Chemical modification of *Escherichia coli* glutamine synthetase. Ph.D Thesis, Penn State Univ, Dissertation Abstracts Order No. AAC 8826734, 1988.
11. Robertson JG, Sparvero LJ, Villafranca JJ. Inactivation and covalent modification of CTP synthetase by thiourea dioxide. Prot Sci 1992;1:1298-307.
12. Pravela-Vrancie M, Pfeider E, Schroder W, von Dohrens H, Kleinkauf H. Identification of the ATP binding site in tyrocidine synthetase 1 by selective modifcation with fluorescein 5´-isothiocyanate. J Biol Chem 1994;269:14962-6.

HIS-433 ASA A KEY RESIDUE FOR THE COLOR DIFFERENCE IN FIREFLY LUCIFERASE, *HOTARIA PARVULA*

Hiroshi Ueda[1], Hiroya Yamanouchi[1], Atsushi Kitayama[1], Kazunori Inoue[1], Takashi Hirano[2], Eiji Suzuki[1], Teruyuki Nagamune[1], and Yoshihiro Ohmiya[3]
[1]Dept. of Chem. & Biotechnol., Univ. of Tokyo, Tokyo 113, Japan
[2]Dept. of Applied Phys. & Chem., Univ. of Electro-Communications, Chofu, Tokyo 182, Japan
[3]Div. of Chem., Shizuoka Univ., Shizuoka 422, Japan

Introduction

Luciferases are the enzymes that catalyze the light-emitting reactions in bioluminescence organisms. Firefly luciferase catalyzes the oxidation of luciferin in the presence of ATP, Mg^{2+}, and molecular oxygen, yielding as products light, oxyluciferin, CO_2 and AMP . The bioluminescence reaction of firefly luciferases *in vitro* emits light ranging from green (λmax=547nm) to red (λmax=604nm) , although the firefly luciferin is a substrate common to all luminous beetles, including the true firefly. Until now, three mechanisms were proposed to explain the basis of the color changes in beetle luciferases (1). They are 1) The mechanism based on differences in the electronic conjugation states of oxyluciferin, 2) The one involving the polarity of the oxyluciferin binding site in luciferase, and 3) Recently proposed mechanism based on rotation of the C2-C2' bond of oxyluciferin.

To examine the relative relevances of these mechanisms, we attempted random mutagenesis of recently cloned *Hotaria parvula* luciferase (2) which has high homology with *Luciola mingrelica* enzyme, to obtain mutants with different color of light. Among several mutants analyzed, a mutant with marked red shifted spectre with an abnormal subpeak at shorter wavelength was selected and further characterized.

Materials and Methods

Enzymes and chemicals: Restriction enzymes were obtained from Takara Shuzo (Kyoto, Japan), and *Pfu* DNA polymerase is from Stratagene (La Jolla, CA). Isopropyl β-D-thiogalactopyranoside (IPTG), firefly D-luciferin-Na, and ATP-Na were provided by Wako Pure Chemicals (Osaka, Japan). All other chemicals were of the highest grade commercially available.

Random mutagenesis of *Hotaria parvula* luciferase: *Hotaria parvula* luciferase cDNA encoded in pB-HpL (2) was PCR amplified with primers specific for the N-terminus (HpLNcoback, 5'-CGCCATGGAAATGGAAAAGGAG-3') and M13 M3 primer (Takara) using *Pfu* DNA polymerase. The amplified fragment of 2.0 kb was digested with *Nde* I and *Hin*d III, inserted into corresponding sites of pET20b (Novagen, Madison, WI) encoding T7 promoter, and designated pET-HpL. *Escherichia coli* XL1-Red competent cell (Stratagene) was transformed with purified pET-HpL, and 20 colonies were transfered to 5 mL LB media containing 50μg/mL ampicillin. The overnight cultures at 37°C were mixed and extracted for the plasmids by alkaline/SDS method. Then 5ng of purified plasmid was used for electroporation of 100μL electrocompetent BL21 (DE3) cells per cuvette. A cuvette with 1mm electrode width was used for electroporation with Electroporator II (Invitrogen, San Diego, CA), set at 50μF, 150 ohm, and applied with 1500 V. Pulsed cells were diluted immidiately with 1mL SOC medium and plated to LB agar plates containing 50μg/mL ampicillin at 160μL per plate. The plates were incubated for 48 hrs at 26°C, splayed with 200μL per plate of 12.5mM IPTG to induce expression, and incubated for further 24 hrs at 26°C.

Colonies were transfered to Protran BA85 nitrocellulose membrane (Schleicher & Schull, Dassel, Germany) and soaked into 0.5mM D-luciferin -Na in 100mM sodium citrate pH 5.0. after 5 min soaking, colonies which emit different colors were picked with toothpicks and isolated.

Nucleotide sequence determination: Nucleotide sequences of the plasmids isolated from mutant strains were determined by dideoxy method using Texas Red-labeled synthetic primers. Primers were labeled with 5' oligonucleotide iodoacetyl Texas Red labelling kit (Amersham, UK) and the reaction was performed using Thermo Sequenase sequencing kit (Amersham). SQ-5500 DNA sequencer (Hitachi, Ltd., Tokyo, Japan) was used for the sequencing and subsequent analysis.

Purification of His-tagged luciferases: DNA encoding wild type and H433Y mutant was PCR-amplified and inserted into pET30b (Novagen) plasmid having N-terminal His-tag sequence. The resultant plasmids (pET30HpL, pET30HpLH433Y) were transformed into BL21(DE3) and cultured at 25°C. Cells in 100 mL LB containing 50μg/mL kanamycin were induced at OD_{600}=0.3, and incubated for 3 hrs at 25°C. The cells were pelleted, resuspended in 10mL 50mM sodium phosphate, 10mM Tris HCl pH 8.0, sonicated with Branson Sonifier 250, and lysate was subjected to purification with TALON histidine-binding column (Clontech, Palo Alto, CA) according to the manufacturer's instruction.

Spectral analysis of wild type and mutant enzymes: 10μL purified, concentrated protein or 50μL lysate in 10mM Tris HCl pH 7.8 was used for spectral analysis. Samples were mixed in minicuvettes with 100μL 2mM ATP-Na, 0.5mM luciferin, 5mM $MgCl_2$, 0.3mM coenzyme A-Na in 20mM Good buffers. Luminescence spectra were monitored with PMA-10 (Hamamatsu Photonics, Hamamatsu, Japan).

Results and Discussion

To obtain color mutants of *Hotaria Parvula* luciferase, a modified protocol of that of Kajiyama *et al.* (3) was employed. At first, random mutagenesis of the expression plasmid was accomplished using error-prone *E. coli* host XL1-Red, which is devoid of three DNA repair enzymes and about 5000 fold mutagenic compared with wild type strains (4). The strain was transformed with the expression plasmid pET-HpL which was pET20b inserted with *Hotaria parvula* luciferase cDNA with T7 promoter as a driving unit, and colonies were cultured overnight. The plasmid thus mutagenized was extracted and re-transformed to BL21(DE3) having IPTG inducible T7 polymerase gene on its genome. Colonies were grown at 26 °C, induced with IPTG after 48 hrs, incubated further 24 hrs, and examined for their color of light emission after transfer to nitrocellurose and soaking into luciferin solution in dark room.

From approximately 10000 colonies, 33 color mutants were selected by eye inspection. As expected from earlier study (3), most of them showed red-shifted colors except faint blueish ones. The isolated colonies were grown further and analyzed for their luminescence spectre of induced cell lysate. 50% of mutants showed red-shifted peak wavelength compared with that of wild type luciferase of λmax = 565nm (12/24 analyzed clones; not shown). Others gave mostly wild type spectra, except for three giving fainter light emissions. Fig. 1 shows some spectre patterns of arbitrarily selected mutants at pH 7.8. Although most mutants showed single peak with moderate slope, two (#5 and #30) showed spectre patterns with distinct shoulder at shorter wavelength. In addition, the peak wavelengths of these mutants were longest among obtained mutants (λmax = 610nm). So we decided to further analyze these mutants.

From the nucleotide sequencing, the both mutants were decided to have a single mutation at histidine 433, which had been changed to tyrosine (H433Y). This was interesting because the mutation was exactly the same as reported for one of red-shifted mutants previously reported for *Luciola cruciata* enzyme which was reported to have λmax of 612nm (3).

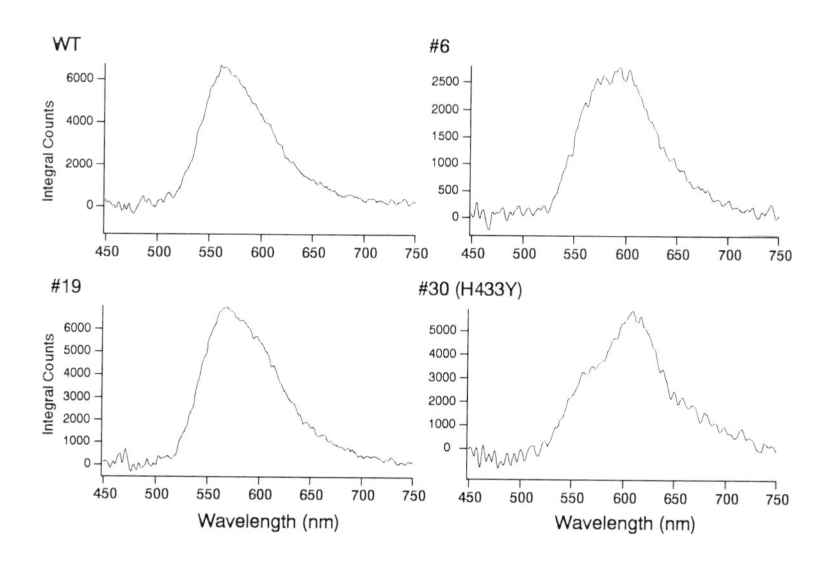

Figure 1. Luminescence spectra of wild type and mutant luciferases at pH 7.8

One specultated mechanism of color difference caused by lower pH and /or cations such as Zn^{2+} or Cd^{2+}, in firefly luminescence reaction is the involvement of a histidine residue in deprotonation reaction of decarboxy-keto-luciferin monoanion (5). It is partially because the monoanion intermediate which has been thought to emit red luminescence, and the deprotonated dianion intermediate which has been attributed to green-yellow luminescence, had pKa of ~6.5. So we tried to determine pKa of wild type and H433Y mutant enzymes. The enzymes were genetically tagged with N-terminal six histidine residues, expressed, and purified with metal affinity resin without use of imidazole, which was inhibitory in luminescent reaction, in elution procedure. The purified enzymes were mixed with buffers of different pH to measure their spectral change. Fig. 2 shows the spectra. While wild type enzyme showed distinct two peaks near pH ~6.5, similar to *Photinus pyralis* enzyme (5), the H433Y mutant showed two peaks at higher pH than 7.8. When we used lysate instead of pure enzyme, similar result was obtained (not shown). If we assume the shoulder peak as the result of partial enolization, the result shows reduced enolization efficiency of H433Y mutant even in higher pH.

The results thus obtained suggest two possibilities; i) His 433 is a residue responsible for enolization reaction. The substituted tyrosine hydroxyl group serves as an electron donor, thus enable enolization of luciferin at higher pH. The result is consistent with pKa =10 of Tyr. ii) His 433 is not directly involved in enolization reaction, but indirectly support the structure of enolization reaction center, which is disturbed by the mutation, thus decrease the reaction efficiency. Further mutagenesis of this and other histidine residues will clarify the exact role of this residue in spectre alteration.

References

1. Wood KV. The chemical mechanism and evolutionary development of beetle bioluminescence. Photochem. Photobiol. 1995;62:662-73.
2. Ohmiya Y, Ohba N, Toh H, Tsuji FI. Cloning, expression and sequence analysis of cDNA for the luciferases from the Japanese fireflies, *Pyrocoelia miyako* and *Hotaria parvula*. Photochem. Photobiol. 1995;62:309-13.
3. Kajiyama N, Nakano E. Isolation and characterization of mutants of firefly luciferase which produce different colors of light. Prot. Eng. 1991;4:691-3.
4. Stratagene. Instruction manual for Epiculian Coli XL1-Red competent cells. 1994.
5. McElroy WD, Seliger HH, White EH. Mechanism of bioluminescence, chemiluminescence and enzyme function in the oxidation of firefly luciferin. Photochem. Photobiol. 1969;10:153-70.

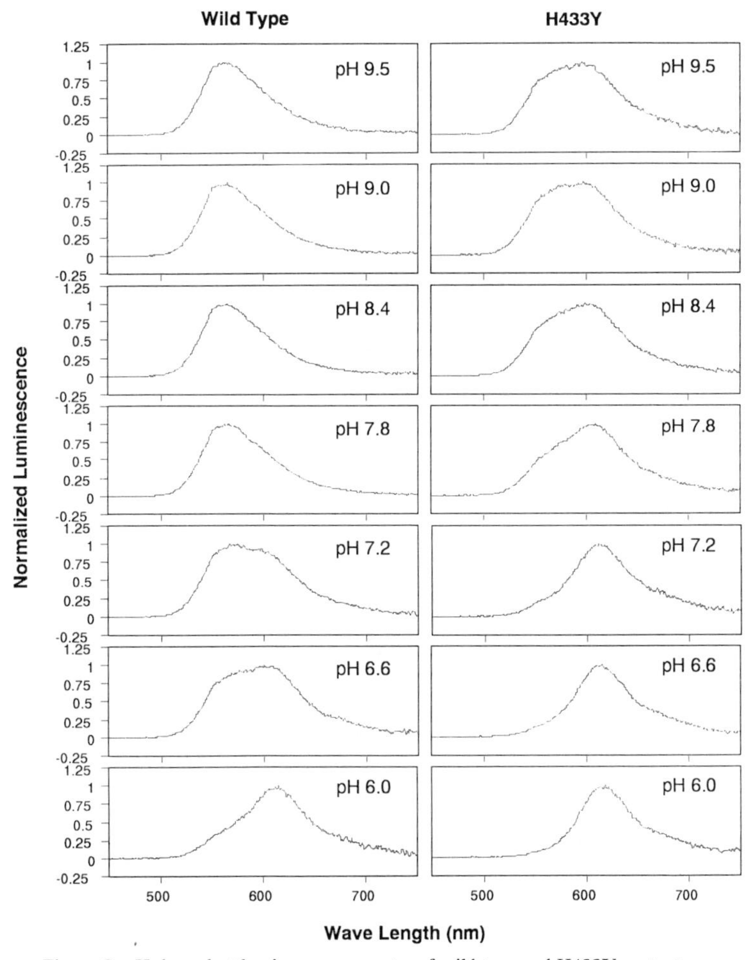

Figure 2. pH dependent luminescence spectra of wild type and H433Y mutant

GENETIC ENGINEERING OF FIREFLY LUCIFERASE TOWARDS ITS USE AS A LABEL IN GENE PROBE ASSAYS AND IMMUNOASSAYS

[1]*RL Price,* [1]*DJ Squirrell,* [1]*MJ Murphy and* [2]*PJ White*
[1]*PLSD, Porton Down, Wiltshire, SP4 0JQ, UK,*
[2]*Babraham Institute, Cambridge, CB2 4AT, UK*

Introduction

Firefly luciferase should be the ideal label for use as a reporter molecule in gene probe assays and immunoassays. It is very easy to measure and background contamination problems are non-existent. 7 femtograms or 10^{-19} moles of luciferase can be detected (1). The sensitivity obtainable from luciferase could be particularly useful for gene probe assays. These are not affected by the affinity constant restraints which, rather than reporter characteristics, are generally the limiting factor in the detection limits achievable in antibody based assays. In immunoassay, luciferase could be of particular value in evanescence-based formats where proximity coupling of emitted light would allow simpler optical geometries than those needed for current approaches with fluorescence detection (2).

Unfortunately, natural luciferase has inherent drawbacks which have meant that it has not successfully been incorporated into any usable binding assays. The first problem is its thermostability. At temperatures much exceeding 25° C it rapidly loses activity whereas for immunoassays, stability at 37° C would be desirable. To be of practical value in gene probe assays luciferase would have to be able to survive incubation at 50-60° C. Ideally it would still function after being heated to DNA melting temperatures of around 95° C. Secondly luciferase is a labile molecule which is readily inactivated by a range of reactive chemicals and this includes those required to couple it to antibodies and nucleic acid probes.

We showed previously that the inactivation caused by coupling reagents could be ameliorated by substrate protection (3). Here we present results of further efforts to allow the potential of luciferase to be realised. Random and site-directed mutagenesis were employed to increase its thermostability. Additionally a short, cysteine-containing tail was added to the C-terminus to facilitate covalent attachment to binding molecules. We report the outcome of the first attempts to couple luciferase to a nucleic acid.

Materials and Methods

Instrumentation Luminescence, in 3.5mL assay tubes, was measured using a portable "Multilite" luminometer (Biotrace Ltd., Bridgend, UK).
Reagents Recombinant luciferase was purchased from Promega Corp., Madison, Wisconsin, USA. Native luciferase, dithiothreitol (DTT), bovine serum albumin (BSA), ethylenediamine tetraacetic acid (EDTA), adenosine-5'-triphosphate (ATP), N-2(hydroxy-ethyl)piperazine-N'-ethanesulphonic acid (HEPES), tris(hydroxymethyl)aminomethane (Tris), glycerol, phenyl-methylsulphonyl fluoride (PMSF), sulphosuccinimidyl-4-(N-maleidomethyl) cyclohexane-1-carboxylate (sulpho-SMCC), and horseradish peroxidase (HRP) were purchased from Sigma Chemical Co., Poole, Dorset, UK. D-Luciferin was from Fluka Chemicals Ltd., Gillingham, Dorset, UK. Magnesium sulphate and L-cysteine were AR grade from BDH, Poole, Dorset. 5'-thiol-oligo-dG_{20} and 3'-amino-oligo-dC_{20} were supplied by Cruachem Ltd., Glasgow, UK.
Engineering of Luciferase Thermostable luciferases were prepared by a combination of random and site-directed mutatagenesis starting from pGEM-*luc* plasmid from Promega Corp. The single mutations were glutamate to lysine at position 354 (E354K) and alanine to leucine at position 215 (A215L). The double mutant incorporated both of these. The polymerase chain reaction was used to create 'tailed' double mutant luciferase with an extended C-terminus.
Luciferase Production Mutated luciferases were produced in *Escherichia coli* JM109 grown

at 37° C overnight in 400mL cultures. Cells were harvested by centrifugation and resuspended in 2mL of buffer (pH8.0, 50mmol/L Tris-HCl, 20mmol/L NaCl, 1mmol/L DTT, 1mmol/L EDTA and 1.2mmol/L PMSF protease inhibitor). Luciferase was released by freezing the cells for 10s in liquid nitrogen and then thawing on ice for 3h. The lysate was centrifuged and the supernatant retained. The luciferase was purified by ammonium sulphate precipitation and anion exchange chromatography on Q-sepharose. It was stored at -20° C in 25mmol/L pH7.5 phosphate buffer containing 0.5mmol/L DTT, 1mmol/L EDTA and 50% gycerol.

Methods Luciferase activity was measured as in (3). Thermostability determinations were made with 50μL samples (at 10ng/mL in pH7.75 HEPES with 1% BSA, 2mmol/L EDTA and 2mmol/L DTT) in luminometer tubes in a water bath. Samples were withdrawn at timed intervals and residual activity was determined. Activation of double mutant and tailed double mutant luciferases with SMCC was carried out as in (3). Activated luciferase was then reacted with thiol-tagged oligo-dG in a molar ratio of 1:1 in the presence of 0.25mmol/L EDTA, 0.5mmol/L ATP and 2.5mmol/L magnesium sulphate. The coupling was allowed to proceeed at room temperature for 30min when 10mmol/L cysteine was added to stop the reaction.

Results and Discussion

Figure 1 shows a comparison of the thermal denaturation kinetics of native luciferase, recombinant wild type, E354K and A215L single mutant, and E354K+A215L double mutant luciferases at 37° C. The double mutant combines the thermostability gains of the single mutants and maintained greater than 50% activity for over 5 hours. We thus have an enzyme able to withstand the temperatures needed for immunoassays for longer than the assays take to perform.

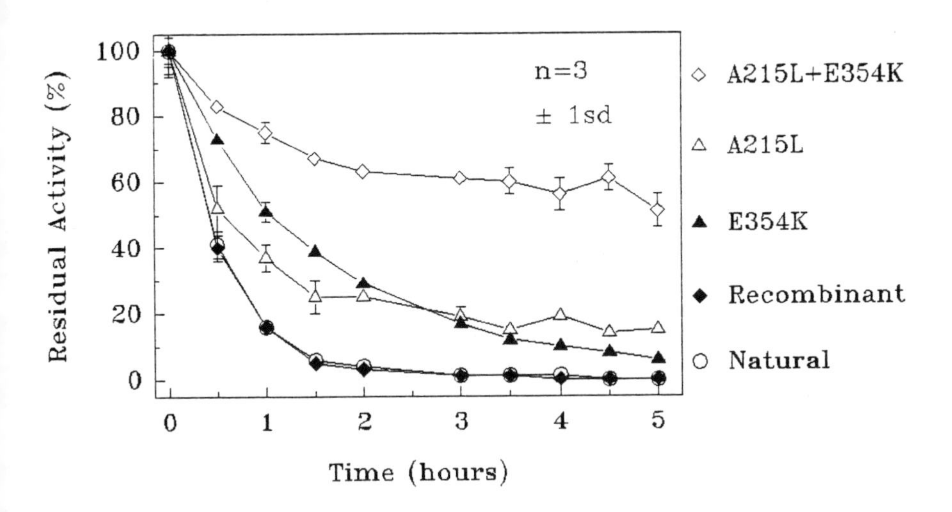

Figure 1: Stability of native, recombinant wild type, single mutant and double mutant luciferases at 37°C.

At 42° C the half life of the double mutant was reduced to 20 minutes. At 50° C the half-life was only 4 minutes, (Figure 2). This is not sufficient to be of general use in gene probe assays. Further improvement in the thermostability of the enzyme itself or, possibly of the assay media to enhance thermostability, are therefore needed. Mutated luciferase from the Japanese firefly *Luciola cruciata* has been shown (4) to be more robust than that from the North American firefly, *Photinus pyralis*, from which the thermostable luciferase used here was derived. This suggests that there is further scope for engineering the desired characeristics into the *P.pyralis* enzyme.

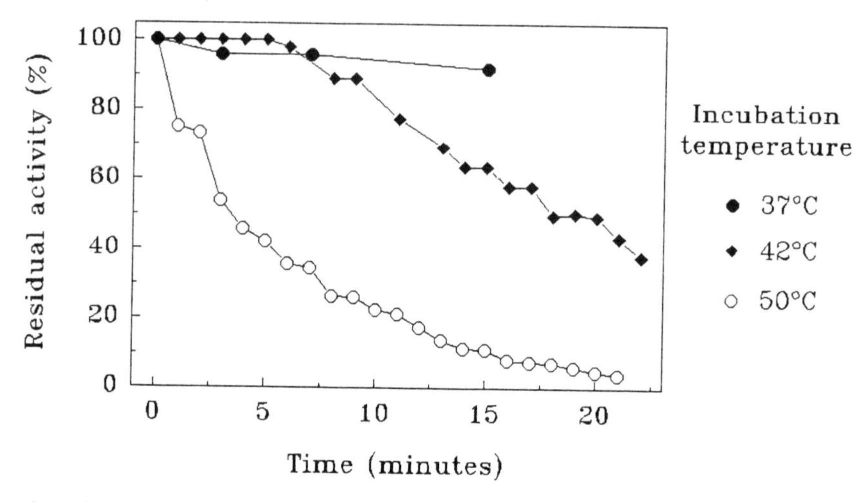

Figure 2: Stability of genetically engineered double mutant luciferase at 37°, 42° and 50°C.

With a simple model gene probe assay of oligo-dG binding to oligo-dC, elevated temperatures are unnecessary. Coupling of oligo-dG to luciferase was carried out using SMCC chemistry with MgATP for substrate protection (3). There are four cysteines in natural luciferase. A fifth thiol group was added by creating a tailed mutant with an extra -Proline-Proline-Glycine-Cysteine sequence added to the C-terminal end of the polypeptide chain of the double mutant. Expression of this mutant was found to be relatively poor. The reason for this turned out to be low thermostability. Tailed double mutant had even lower thermostability than wild type luciferase at 37° C. It was also found to be less stable than tailed wild type luciferase where addition of the extra four amino acids had little effect at 37° C, although its comparative instability was greatly increased at 42° C.

The gene probe coupling reaction was therefore performed using double mutant luciferase (without the extension) as the substrate for attachment of the oligo-dG. Activation of the luciferase with SMCC, at which point it is chemically modified mainly at the side chain amino-group of lysine residues, had no effect on enzyme activity. However, when the thiol oligo-dG was added, (Figure 3) the luciferase gradually lost activity with light emission dropping to 19% of its original level after 30 minutes, when cysteine was added to quench the reaction. After purification 97.0% of original activity had been lost.

A parallel reaction with horseradish peroxidase as a control to see whether the problems were specific to luciferase showed no inhibition of activity when the enzyme was conjugated to the gene probe, (also Figure 3). The conjugated HRP-oligo-dG product was also successfully hybridised to immobilised oligo-dC and specific colour formation demonstrated (results not shown). High concentrations of oligonucleotides free in solution reduced the activity of luciferase. Inhibition of enzyme activity by the coupling of a nucleic acid to luciferase probably resulted from competitive inhibition of the ATP binding site. This would be expected to contain polar amino acid residues to coordinate with the strong charges on the substrate's tri-phosphate group and its magnesium counter-ion. It might therefore be readily blocked by a nucleic acid bound in its immediate vicinity which would, effectively, be at a high local concentration. Alternatively the attached nucleic acid may have had an indirect effect through distortion of the protein structure.

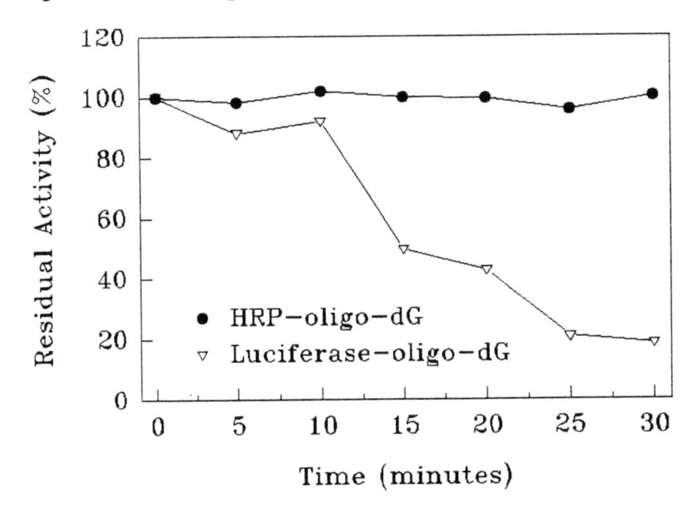

Figure 3: Comparison of the effects on activity of conjugating luciferase and horseradish peroxidase to an oligo-dG_{20} gene probe using SMCC as a coupling reagent.

Considerable further effort may be needed to overcome the problems identified here and to realise the potential of luciferase as a direct gene probe label. Indirect approaches with attachment through intermediates such as streptavidin or Protein A (5) may prove to be more successful.

The increased thermostability gained through genetic engineering together with substrate protection methods developed previously suggest that the practical application of luciferase for use in immunoassays is more realisable now than has previously been the case.

References

1. Lundin A. Optimised assay of firefly luciferase with stable light emission. In: Szalay A, Kricka L, Stanley P, editors. Bioluminescence and Chemiluminescence: Status Report. Chichester: John Wiley & Sons, 1993:291-5.
2. Graham CR, Leslie D, Squirrell DJ. Gene probe assays on a fibre-optic evanescent wave biosensor. Biosens Bioelectron 1992:7:487-93.
3. Murphy MJ and Squirrell DJ. Covalent coupling of firefly luciferase to antibodies. In: Campbell AK, Kricka L, Stanley PE, editors. Bioluminescence and Chemiluminescence: Fundamentals and Applied Aspects. Chichester: John Wiley & Sons, 1994:301-4.
4. Kajiyama N and Nakano E. Thermostabilization of firefly luciferase by a single amino acid substitution at position 217. Biochemistry 1993:32:13795- 99.
5. Kobatake E, Ikariyama Y, Aizawa M. Bioluminescent immunoassay using a genetically fused Protein-A luciferase. In: Szalay A, Kricka L, Stanley P, editors. Bioluminescence and Chemiluminescence: Status Report. Chichester: John Wiley & Sons,1993:337-41.

BIOTINYLATION OF FIREFLY LUCIFERASE IN VIVO: PURIFICATION AND IMMOBILIZATION OF BIFUNCTIONAL RECOMBINANT LUCIFERASE

CY Wang, S Hitz, JD Andrade, and RJ Stewart
Dept. of Bioengineering, University of Utah, SLC, UT 84112, USA

Introduction

Firefly luciferase is an enzyme which catalyzes a bioluminescent reaction in the presence of luciferin, adenosine 5'-triphosphate (ATP), magnesium ion and oxygen (1). The bioluminescent reaction features a high sensitivity for ATP and a high quantum yield, 0.88 (2). The methods used to obtain high purity luciferases from firefly lanterns have been reviewed (3). Long purification times and low yields makes highly purified native luciferase expensive. Distinguishing firefly habitats makes collection of fireflies ecologically inappropriate. The labile nature of firefly luciferase also makes immobilization of luciferase less successful than other enzymes (3).

The success of de Wet et al. who isolated firefly luciferase cDNA (4), broadens the applications of firefly luciferase through genetic engineering. The plasmid containing the luciferase gene (luc) was constructed and inserted as a gene reporter. Novel fusion proteins, such as protein A-luciferase and streptavidin-luciferase are now available (5, 6).

A plasmid encoding a novel fusion luciferase (BCCP-luciferase) was constructed and inserted into E. coli in our laboratory. The BCCP-luciferase contains a biotin carboxyl carrier protein (BCCP), a subunit of acetyl-CoA carboxylase (7). Post-translational modification of the BCCP domain on the fusion protein results in a biotin molecule on a specific lysine residue. The BCCP-luciferase also contains six histidine residues at the N-terminal, which bind strongly to Ni^{2+}-nitrilotriacetic acid (Ni^{2+}-NTA) agarose (8, 9). The highly purified BCCP-luciferase was obtained with a one-step purification protocol and compared with native luciferase purified from firefly lanterns (9). Both luciferases presented similar activity and stability.

Avidin, a glycoprotein with four biotin binding sites, was immobilized on Emphaze beads (see below) and served as the "receptor" for BCCP-luciferase. Via biotin-avidin bridges, the biotinylated firefly luciferase is immobilized on the avidin beads without direct modification of the luciferase domain. This novel bi-functional luciferase is easy to purify and immobilize with high bioluminescence activity.

Materials and Methods

Construction of the pRSET-BCCP-luc plasmid and culture of transformed E. coli were described elsewhere (9). The cells were harvested and disintegrated by sonication (Fisher Scientific 550 sonic dismembrator) on an ice bath with 1 mmol/L phenylmethylsulfonyl fluoride (PMSF, in isopropanol, from Sigma). The supernatant containing the recombinant luciferase was collected and applied directly to 3 mL of Ni^{2+}-NTA agarose (from Qiagen), which was packed in an Econo-column (BioRad) and washed with 20 mL Tris buffer (50 mmol/L, pH 7.8). The unbound proteins were eluted with two aliquots of 2 mL Tris buffer. Three aliquots of 2 mL imidazole solutions at 10, 50, 100, 250, and 500 mmol/L were applied to the Ni^{2+}-NTA column sequentially. The eluate was collected and the buffer changed to glycylglycine buffer (0.45 mol/L, pH 7.8) or phosphate buffer saline (PBS) with a PD-10 column (Pharmacia). Molecular weight and purity of purified BCCP-luciferase was determined by 10% sodium dodecyl sulphate-polyacrylamide gel electrophoresis (SDS-PAGE) and detected with Coomassie Blue.

Luciferase activity was determined by injecting 250 μL ATP solution (0.1 mmol/L) into the mixture of 20 μL luciferase solution and 100 μL stock solution (0.1 mmol/L luciferin and 10 mmol/L $MgSO_4$ in 25 mmol/L glycylglycine buffer, pH 7.8). The bioluminescence was recorded with a spectrofluorometer (I.S.S., Inc.) and the peak height was interpreted as the enzyme activity. The amount of total protein was determined by BioRad protein assay with a Perkin Elmer Lambda 6 UV/VIS spectrophotometer. A standard curve (A_{595} vs. protein amount) was generated for bovine serum albumin (BSA).

SoftLink soft release avidin resin (from Promega), with covalently linked monomeric avidin, was used to determine the extent of biotinylation. 1 mL BCCP-luciferase (0.4 mg/mL) was fully mixed with avidin resin and packed to the column. The eluate was collected and applied to the second column packed with fresh avidin resin. BSA was used in the same procedure to calculate the dilution factor. The amount and activity of proteins before and after binding were determined by UV absorbance at 280 nm and by ATP assay.

Avidin was mixed with dry Emphaze Biosupport Medium AB1 (a gift from 3M) to make avidin-coupled beads. The cell lysate was mixed with avidin-coupled Emphaze beads and packed into the column. Unbound luciferase was removed by washing the beads with 0.8 mL of glycylglycine buffer (25 mmol/L with 2 mmol/L dithiothreitol (DTT)). Luciferase activity was determined by applying 250 μL of the ATP solution (2 mmol/L DTT) and 100 μL of the stock solution to the column. The coupling of luciferase directly to Emphaze beads and quenched beads (no azlactone functionality) were also studied with the same method.

The beads containing immobilized luciferase were regenerated by washing the column with 0.8 mL glyculglycine buffer (with DTT) three times. The activity of luciferase on the regenerated beads were tested immediately or stored at 4 °C for later use. The work was also repeated without DTT in all solutions.

Results and Discussion

An average yield of 1.1 g wet E. coli cells were recovered from 200 mL culture. The cells were disintegrated and the supernatant of cell lysate was loaded directly on to the Ni^{2+}-NTA column. The activity of BCCP-luciferase was first detected when the concentration of imidazole increased to 100 mmol/L. The highest activity was found in the fraction eluted by 250 mmol/L imidazole. The result of SDS-PAGE showed a single band (Figure 1) suggesting that the single-step purification with the Ni^{2+}-NTA column is effective to obtain high purity BCCP-luciferase. A 200 mL cell culture yielded 9.1 mg pure luciferase with 72% total activity recovered, which is about the

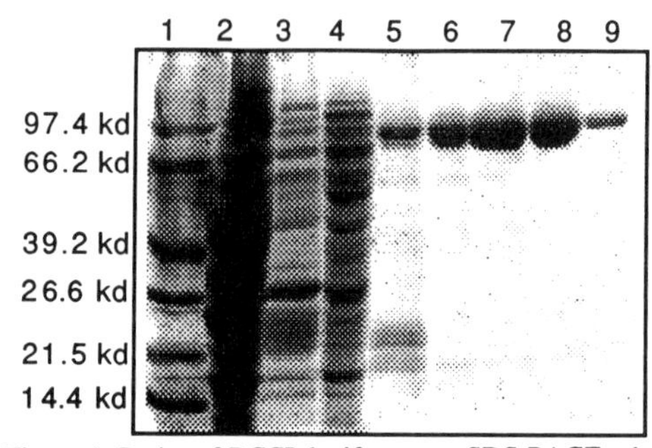

Figure 1. Purity of BCCP-luciferase on SDS-PAGE. 1: marker; 2: eluted by buffer; 3-5: eluted by 10, 50 and 100 mmol/L imidazole; 6-8: eluted by 250 mmol/L imidazole; 9: eluted by 500 mmol/L imidazole.

Table 1. Reusability of immobilized BCCP-luciferase stored at 4 °C (% activity retained).

	Storage with 2 mmol/L DTT	Storage without 2 mmol/L DTT
Fresh	100	100
After 10 uses	91	52
1 day storage	94	39
3 day storage	90	3
1 week storage	67	0

amount of luciferase purified from 1800 fireflies by the method of DeLuca & McElroy (10). The specific activity is about 6% higher than that of Sigma luciferase. The high activity recovery and purity of BCCP-luciferase suggest the Ni^{2+}-NTA column can purify BCCP-luciferase efficiently with minor protein inactivation. The molecular weight calculated from the SDS-PAGE data (Figure 1) is 77 kd, slightly greater than the sum of amino acid residues, 71 kd. The small discrepancy may be caused by the unusual amino acid sequence of BCCP domain (7).

The percentage of biotinylated luciferase was estimated to be 55% of the luciferase recovered. The solution of BCCP-luciferase after binding reaction retained 42% bioluminescence, suggesting up to 58% of BCCP-luciferase binding to avidin resin. Re-applying the non-binding BCCP-luciferase to fresh avidin resin resulted in trace bioluminescence on the avidin resin (about 0.5 % of the first avidin resin), indicating the binding is complete in the first column. The biotinylation extent was thus estimated within 55% to 58% of the total purified BCCP-luciferases.

The supernatant from the E. coli lysate was directly incubated with avidin coupled Emphaze beads. BCCP-luciferase bound to the immobilized avidin via biotin-avidin bridges. Unlike the solution assay, there was no flash of bioluminescence upon the addition of ATP and luciferin stock solution to the immobilized BCCP-luciferase. There was no luciferase activity after incubation with quenched Emphaze beads and very low Sigma luciferase activity after incubation with avidin-beads, suggesting the binding of BCCP-luciferase on avidin-beads was specific. The activity of BCCP-luciferase immobilized via biotin-avidin bridges was more than three fold of luciferases directly immobilized on Emphaze beads. Luciferase directly coupled to Emphaze beads may be inactivated during the chemical modification.

One important advantage of immobilized enzyme is its reusuability. BCCP-luciferase in buffer solution retained 98% activity after one week storage at 4 °C (9). The activity of regenerated luciferase was shown in Table 1. The results suggested that the major mechanism of luciferase inactivation was via formation of disulfide bond. Exhaustion of DTT may speed the formation of disulfide bond and the inactivation of immobilized luciferase. The stabilization effect of DTT on immobilized luciferase was consistent with other reports (3).

The one-step purification protocol using the Ni^{2+}-NTA column is efficient to obtain high purity BCCP-luciferase, which was as stable as native luciferase (9). In vivo biotinylation of BCCP-luciferase provides a specific binding capability to avidin or streptavidin. High activity of immobilized BCCP-luciferase can be obtained by directly incubating the cell lysate with avidin coupled Emphaze beads. This immobilization protocol prevents possible inactivation pathways and can be applied in the design of luciferase based biosensors.

Acknowledgement
This work was partially supported by Protein Solutions, Inc., Salt Lake City, Utah via funding provided by an NSF STTR grant. We thank Drs. J. Herron and V. Hlady for access to equipment and facilities. We thank the 3M corp. (Dr. P. Coleman) for a gift of Emphaze beads. We thank Dr. J. Cronan, Jr., University of Illinois, for helpful discussions concerning biotinylation of proteins in E. coli.

References

1. McElroy WD Seliger HH. Mechanisms of Bioluminescent reactions. In: McElroy WD, Glass B, editor. Light and Life. Baltimore: John Hopkins Press, 1961:219-57.
2. Seliger HH, McElroy WD. Spectral emission and quantum yield of firefly bioluminescence. Arch Biochem Biophys 1960;88:136-45.
3. Rajgopal S, Vijayalakshmi MA. Firefly luciferase: purification and immobilization. Enzyme Microb Technol 1984;6:482-90.
4. De Wet JR, Wood KV, Helinski DR. DeLuca M. Cloning of firefly luciferase cDNA and the expression of active luciferase in Escherichia coli. Proc. Natl Acad Sci USA 1985;82:7870-73.
5. Kobatake E, Iwai T, Ikariyama Y, Aizawa M. Bioluminescent immunoassay with a protein A-luciferase fusion protein. Anal Biochem, 1993;208:300-5.
6. Karp M, Lindqvist C, Nissinen R, Wahlbeck S, Akerman K, Oker-Blom C. Identification of biotinylated molecules using a baculovirus-expressed luciferase-streptavidin fusion protein. BioTechniques 1996;20:452-459.
7. Li S, Cronan Jr., JE. The gene encoding the biotin carboxylase subunit of Escherichia coli acetyl-CoA carboxylase. J Biol Chem 1992;267:855-63.
8. Janknecht R, de Martynoff G, Lou J, Hipskind RA, Nordheim A, Stunnenberg HG. Rapid and efficient purification of native histidine-tagged protein expressed by recombinant vaccinia virus. Proc Natl Acad Sci USA 1991;88:8972-6.
9. Wang C-Y, Hitz S, Andrade JD, Stewart R. Biotinylation of firefly luciferase in vivo: a recombinant protein with a specific immobilization site," in preparation 1996; Wang C-Y, Ph. D. Dissertation, University of Utah, Oct. 1996.
10. DeLuca M. McElroy WD. Purification and properties of firefly luciferase. Methods Enzymol 1978;57:3-15.

CO-REPORTER TECHNOLOGY INTEGRATING FIREFLY AND RENILLA LUCIFERASE ASSAYS

B Sherf, S Navarro, R Hannah and K Wood
Promega Corp. 2800 Woods Hollow Rd, Madison. WI, 53711,USA

Introduction

Co-transfecting cells with both an experimental reporter vector and a distinct control reporter vector is a commonly used method for attaining accurate, normalized measurements of the experimental reporter gene expression. Typically, expression of the primary experimental reporter is coupled to the activity of a specific regulatory element under investigation, while the control reporter is coupled to a constitutive promoter that is unresponsive to the various treatments and conditions of the experiment. Thus, by normalizing measurements of the experimental reporter to the control reporter, relative changes caused by the regulatory elemenet may be related to an internal standard through ratiometric measurements. Such ratiometric measurements minimize variables inherent to the experimental process that can undermine accurate reporter quantitation, such as plate-to-plate differences in the number, health and lysis of the treated cells, and in particular the efficiency of cell transfection (1).

The extreme speed and sensitivity of the firefly luciferase reporter assay has made firefly luciferase (*luc*) a popular experimental reporter gene for the study of genetic regulation. However, traditional reporters such as CAT, β-Gal, SEAP and GUS are inconvenient in co-reporter applications with *luc* due to the wide differences in their respective assay chemistries, handling requirements and measurement characteristics (ref. 2 for a review). The objective of this work was to develop a superior co-reporter gene and assay system that complements, and provides all the performance advantages of, firefly luciferase. This was achieved by constructing a family of co-reporter vectors expressing sea pansy (*Renilla reniformis*) luciferase, and developing a novel assay chemistry that allows rapid sequential quantitation of expressed firefly and Renilla luciferases from within a single sample of cell lysate.

Materials and Methods

Reagents: The firefly luciferase (*luc*+) expression vector pGL3-Control (3), the Renilla luciferase (R*luc*) expression vector pRL-SV40 (4) and purified recombinant firefly luciferase are available from Promega Corp. (Madison, WI). Passive Lysis Buffer, Luciferase Assay Reagent II and Stop & Glo™ Reagent are available as components of the Dual-Luciferase Reporter™ Assay System (Promega Corp., Madison, WI). Determinations for sensitivity and linearity for firefly and Renilla luciferase assays were performed using purified enzymes serially diluted in Passive Lysis Buffer supplemented with 1 mg BSA/mL. Recombinant Renilla luciferase (NovaLite™) was graciously provided by SeaLite Sciences, Inc. (Norcross, GA). The cDNA encoding luciferase from *R. reniformis*, the subject of U.S. Patent No. 5,292,658, is sublicensed from SeaLite Sciences, Inc. (Norcross, GA).

Co-transfection of cultured cells: A 1:1 pre-mix of pGL3-Control Vector and pRL-SV40 Vector was used for cell co-transfection experiments. A total of 5 μg DNA of mixed vectors was prepared as a suspension of calcium/phosphate aggregates and used to transfect $3x10^6$ CHO cells equally divided between three 15 x 60 mm culture dishes. At 30 h post-transfection cultures were rinsed once with PBS and harvested by scraping in the presence of 400 μL Passive Lysis Buffer per dish.

Dual-Luciferase Reporter (DLR) Assays: DLR assays were performed, as depicted in Figure 1, using a Turner Designs (Sunnyvale, CA) Model 20 or TD-20/20 luminometer. The firefly luciferase assay was initiated by adding 20 μL of prepared cell lysate into 100 μL Luciferase Assay Reagent II previously dispensed into 8x50 mm luminometer tubes. Firefly luciferase activity was quantified by integrating light emission over a 10 second reaction period after an initial 2 second pre-measurement delay. Upon immediate completion of the firefly luciferase assay, the firefly luminescence was quenched and Renilla luciferase luminescence activated by adding 100 μL Stop & Glo™ Reagent into the sample tube and mixing by pulse vortex. Renilla luciferase activity was quantified in an identical manner. The ability of Stop&Glo™ Reagent to quench the firefly reaction was demonstrated using the sameassay format, but using a modified Stop&Glo™ Reagent lacking coelenterazine.

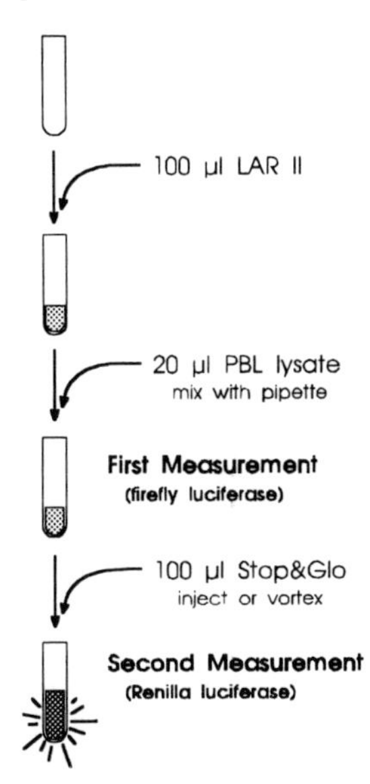

Figure 1.The Dual-Luciferase™ reporter assay

Results and Discussion

Firefly luciferase is a 61 kD monomeric protein that does not require post-translational processing for enzymatic activity. Photon emission occurs via oxidation of beetle luciferin in a reaction requiring ATP, Mg^{+2} and O_2. Renilla luciferase is a 36 kD (based on amino acid composition) monomeric protein that utilizes O_2 to catalyze the oxidation of coelenterate-luciferin (coelenterazine) with concomitant photon emission (5,6). Like firefly luciferase, Renilla luciferase does not require post-translational modification for activity. Therefore, both enzymes may function as reporters immediately following *in vivo* or *in vitro* synthesis.

Firefly and Renilla luciferases are of distinct evolutionary origin, and have dissimilar enzyme structures and substrate requirements. These differences allowed for the development of an integrated assay chemistry that selectively discriminates between the luminescent assays of these two reporter enzymes. This novel assay chemistry is now available as the Dual-Luciferase™ Reporter (DLR) Assay System (Promega Corp., Madison, WI). Using the DLR assay chemistry, the firefly luciferase reaction provides a persistent luminescence signal ($t1/2$ = 5.5 min), and the Renilla luciferase luminescence reaction decays gradually ($t1/2$ = 1.75 min) over the course of the measurement (Figure 2). Because *Renilla* and firefly luciferases exhibit glow-type reaction kinetics, performing the DLR assay does not require the use of reagent injectors dedicated to low-throughput luminometers. The sequential measurement of firefly and Renilla luciferase activities using a manual luminometer is easily performed in an elapsed time of 30 seconds or less. Plate reading luminometers must be equipped with two injectors to perform Dual-Luciferase™ assays. High-throughput DLR assays may be performed with elapsed times of 10 seconds or less per sample, thus allowing quantitation of 96 co-reporter assays within 16 minutes.

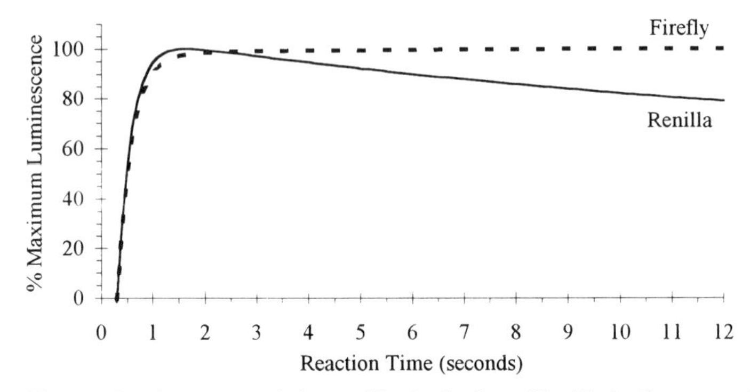

Figure 2. Luminescence emission profiles for firefly and Renilla luciferase reactions

Cell lysates prepared with Passive Lysis Buffer retain fully active firefly and Renilla luciferase reporter enzymes for at least 6 hours at room temperature, and at least 21 hours when stored on ice (4). Unlike other commonly used lysis agents, Passive Lysis Buffer contributes minimally to coelenterazine auto-luminescence (4). Hence, the luminescent reaction catalyzed by Renilla luciferase using the DLR chemistry provides assay sensitivity similar to that of firefly luciferase, with a linear range extending 7 logs (Figure 3).

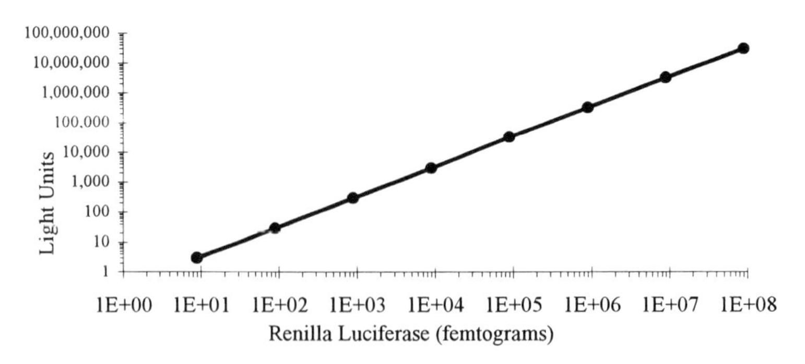

Figure 3. Sensitivity and linear range of the Renilla luciferase luminescence reaction

As seen in Table 1, Stop & Glo™ Reagent selectively and rapidly extinguishes the luminescent reaction catalyzed by the firefly luciferase. The ability of Stop&Glo™ Reagent to quench firefly luminescence by greater than 5 logs eliminates the possibility of residual firefly luminescence contributing background to the subsequent measurement of Renilla luciferase. Stop&Glo™ Reagent, which contains coelenterazine, is stable in glass vials at room temperature for at least 6 hours, or up to 2 weeks when stored frozen.

DLR Assay Reagent	RLU	Reporter Assay Measured
LARII	80,600	Firefly (Reporter #1)
LARII + "S&G" (- Cz)	0.28	Quenched Firefly
LARII + S&G (+ Cz)	116,800	Quenched Firefly, Activated Renilla (Reporter #2)

Table 1. Efficiency of Stop&Glo™ Reagent in quenching firefly luciferase activity and activating Renilla luciferase activity. A lysate of CHO cells co-transfected with pGL3-Control Vector and pRL-SV40 Vector was used to assay for firefly and Renilla luciferase activities using the DLR assay format. The luminescent activity of Reporter 1 is rapidly quenched by greater than 100,000-fold, contributing insignificant background luminescence to the subsequent measurement of Reporter 2. Luciferase Assay Reagent II, LARII; Stop&Glo™ Reagent, S&G; modified Stop&Glo™ Reagent, "S&G", lacking coelenterazine, Cz.

Conclusions

Promega's Dual-Luciferase™ Reporter Assay chemistry provides exceptional speed, sensitivity and convenience for quantifying two independent luciferase assays for co-reporter applications. Using the DLR assay, luminescence from the firefly and Renilla luciferase reactions are measured sequentially from within a single sample. After completing measurement of firefly luciferase activity (the "experimental" reporter), addition of a second reagent delivers immediate quenching of the firefly luciferase reaction and simultaneous activation of the Renilla luciferase luminescent reaction (the "control" reporter activity). Even when using a manual luminometer, the standard DLR assay may be performed in 30 seconds or less. Further, high-throughput DLR assay may be performed in significantly less time, requiring no more than 16 minutes to process 96 samples contained in a multi-well plate. In addition to genetic reporter applications, the DLR assay may greatly benefit applications using cell-free expression systems for *in vitro* translation, or coupled transcription and translation, of luciferase co-reporter molecules.

References

1. Hollon T, Yoshimura FK. Variation in enzymatic transient gene expression assays. Anal Biochem 1989; 182; 411-18.
2. Bronstein I, Fortin J, Stanley PE, Stewart GS, Kricka LJ. Chemiluminescent and bioluminescent reporter gene assays. Anal Biochem 1994; 219; 169-81.
3. Groskreutz DJ, Sherf BA, Wood KV, Schenborn ET. Increased expression and convenience with the new pGL3 luciferase reporter vectors. Promega Notes 1995; 50; 2-8.
4. Sherf BA, Navarro SL, Hannah RR, Wood KV. Dual-Luciferase™ reporter assay: an advanced co-reporter technology integrating firefly and Renilla luciferase assays. Promega Notes 1996; 57; 2-9.
5. Lorenz WW, McCann RO, Longiaru M, Cormier M. Isolation and expression of a cDNA encoding *Renilla reniformis* luciferase. Proc Natl Acad Sci USA 1991; 88; 4438-42.
6. Cormier MJ. Applications of Renilla bioluminescence. Methods Enzymol 1978; LVII; 237-44.

EXPRESSION IN *ESCHERICHIA COLI* OF BIOTINYLATED LUCIFERASES USING BIOTIN ACCEPTOR PEPTIDES

Hiroki Tatsumi, Satoshi Fukuda, Mamoru Kikuchi and Yasuji Koyama
Research and Development Division, Kikkoman Corporation, 399 Noda, Noda-city, Chiba 278, Japan

Introduction

Firefly luciferase catalyzes the oxidation of luciferin in the presence of ATP, O_2, and Mg^{2+}, producing light. The luciferase of the North American firefly *Photinus pyralis* (PpL) has been extensively characterized (1). Because of high quantum yield (0.9) firefly luciferase could be an attractive marker enzyme for enzyme immunoassay (EIA), but up to now no convenient method of preparing enzyme conjugate has been reported. PpL becomes inactivated during chemical coupling reactions (2, 3). PpL has been genetically fused with Protein A, but the fusion protein could only detect as little as 100 ng/mL of human immunoglobulin G (IgG) with EIA (4).

Recently a method to biotinylate proteins by means of genetic engineering became available. Biotin enzymes contain a biotin molecule which is covalently attached to a unique Lys *via* an amino linkage catalyzed by biotin holoenzyme synthetase (5). A highly conserved sequence was observed around the biotin-attachment site. Proteins which were genetically fused with the biotin acceptor peptides were biotinylated at the Lys *in vivo* (6-9).

Another problem about PpL is thermal stability. It retains less than 50% of the initial activity after 30 min at 40°C (10). This instability makes it difficult to use PpL for EIA. Recently Kajiyama and Nakano obtained thermostable mutant luciferases from Japanese fireflies, *Luciola cruciata* and *Luciola lateralis*, by means of genetic engineering (11,12). Among them, the mutant (Ala217Leu) of *Luciola lateralis* luciferase (LlL-217L) shows highest thermal stability and retains over 70% of the initial activity after 60 min at 50°C.

In this paper we describe the expression in *Escherichia coli* of two fusion proteins consisting of a thermostable luciferase, LlL-217L, and a biotin acceptor peptide. We demonstrated that both fusion proteins not only fully retained luciferase activity but also were efficiently biotinylated *in vivo*. Using these biotinylated luciferases, we were able to detect 2.5 pg/mL of mouse IgG_1. Our bioluminescent EIA system was found to be more sensitive than a chemiluminescent EIA system using biotinylated alkaline phosphatase.

Materials and Methods

Strains and Media: *E. coli* strain JM101 [Δ(*lac-pro*) *supE thi* (F' *traD36 proAB*+ *lacIqZ* ΔM15)] was used as a carrier for recombinant plasmids. *E. coli* cells were grown in LB broth (1% Bacto tryptone, 0.5% yeast extract, 0.5% NaCl). When necessary, 50 μg ampicillin /mL was added.

Assay of Luciferase Activities: Sample solutions were added to the wells of a Microlite 2 microtiter plate (DYNATECH). The plate was placed in a microplate luminescence reader Luminous CT-9000D (DIA-IATRON), and 50 μL of a substrate solution (40 mM ATP, 1.4 mM luciferin, 300 mM $MgSO_4$, 50 mM HEPES, pH 7.5) was injected into each well. The count emitted was integrated for 10 s.

Results and Discussion

Expression in *E. coli*, Purification and Characterization of bL203 and bL248: Recombinant plasmids pHLf203 and pHLf248, which, under the control of lactose promoter, direct the expression of fusion protein bL203 and bL248, respectively, were constructed. bL203 consists of LlL-217L with the first 4 residues removed and the 5th residue mutated to Leu, and an artificial biotin acceptor peptide #84 (7) of 23 amino acids (MAFSLRSILEAGKMELRNTPGGS, The 13th Lys is biotinylated) at the N-terminal (Figure 1). bL248 consists of LlL-217L with the last Met replaced by Gly-Ser, and the natural biotin acceptor peptide BCCP-87 (the carboxyl terminal 87 residues of a biotin carboxyl carrier protein, a subunit of acetyl-CoA carboxylase from *E. coli*) (8, 9) at the C-terminal (Figure 1). Recombinant *E. coli* cells JM101[pHLf203] and JM101[pHLf248] were each grown in 20 mL of LB broth with shaking at 30°C over night. The cells were inoculated in 1 L of LB broth containing 0.2 mM IPTG and 10 μg/mL biotin, and grown with shaking at 30°C for 5 h, then centrifuged. From the pellets, bL203 and bL248 were purified to homogeneity as reported (11).

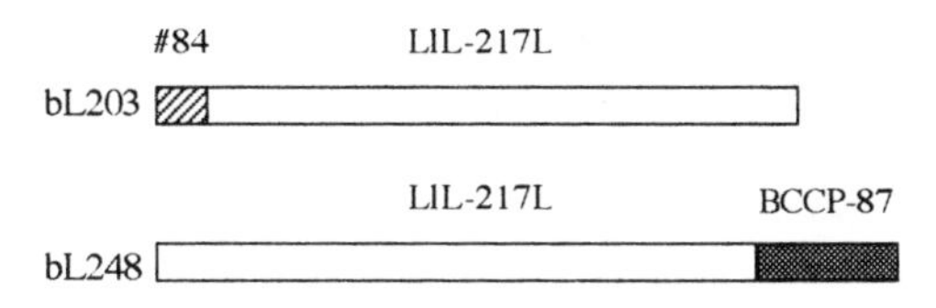

Figure 1. Structure of bL203 and bL248.

The luciferase activities per molecule of bL203 and bL248 were the same as that of LlL-217L, demonstrating that addition of a biotinylation peptide at either terminal of LlL-217L does not cause activity loss. The thermal stability at 50°C of bL203 and bL248 were found to be as stable as LlL-217L. No activity loss was observed when bL203 and bL248 were incubated at 37°C for 2 days. The pH stability (pH6-10) of bL203 and bL248 were measured and found to remain as stable as LlL-217L.

bL203 and bL248 were each mixed with approximately twice by molar ratio of streptavidin and subjected to gel filtration column chromatography. The fusion proteins which bound with streptavidin were separated from the free fusion proteins. We measured the luciferase activity of each fraction, and found that more than 95% of both fusion proteins were eluted at high molecular weight fractions. This result shows that the biotinylation ratios of them are more than 95%, suggesting that the biotin holoenzyme synthetase in the host cells (*E. coli* JM101) worked efficiently. bL203 and bL248 were each mixed with approximately ten times by molar ratio of streptavidin and were left at room temperature for 1 hr. The activity of bL203-streptavidin mixture was approximately 80% of the initial activity, while no decrease in the activity was observed with bL248-streptavidin mixture.

The novel biotinylated luciferases with full specific activity; high thermal stability; high biotinylation ratio; no or small activity decrease when they bind with streptavidin; and a homogeneous structure with one biotin attached at the specific Lys residue, should be suitable marker enzymes for highly sensitive bioluminescent EIA.

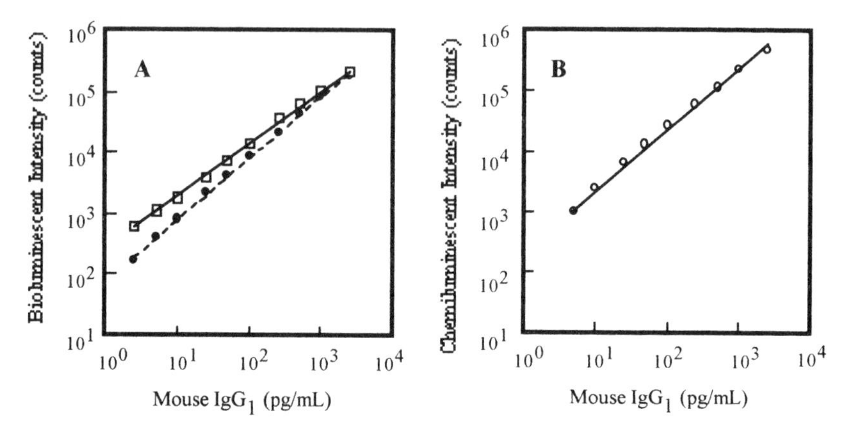

Figure 2. Sandwich EIA for mouse IgG_1. (A) Bioluminescent EIAs using biotinylated luciferases bL203 (open square) and bL248 (closed circle). (B) Chemiluminescent EIA using biotinylated alkaline phosphatase (open circle) incorporated with CSPD as a substrate.

Sandwich EIAs for Mouse IgG_1: Bioluminescent and chemiluminescent EIAs for mouse IgG_1 as a model antigen were performed. First antibodies that were bound to a solid phase were goat anti-mouse IgG Fc antibodies, and second antibodies that were biotinylated were goat $F(ab')_2$ fragments of anti-mouse IgG $F(ab')_2$ antibodies. Two biotinylated luciferases (bL203 and bL248) and biotinylated alkaline phosphatase were respectively premixed with streptavidin to prepare biotinylated enzyme-streptavidin complex. In chemiluminescent EIA, CSPD was used as alkaline phosphatase substrate with luminescence enhancer Sapphire-II. Figure 2 shows the relationship between mouse IgG_1 concentration and luminescence. When the luminescent intensities of samples greater than blank value plus 3SD (blank; without mouse IgG_1, n=6) were taken to be significant, the bioluminescent EIAs using bL203 and bL248 were able to detect 2.5 pg/mL of mouse IgG_1. In chemiluminescent EIA, the detection limit were 5 pg/mL. The correlation coefficients (r) were 0.997 for bL203 and 0.999 for bL248. The coefficient of variations (CV%, n=6) for each point were 2.2% to 7.0% (mean=3.9%) for bL203, and 0.7% to 2.7% (mean=1.9%) for bL248. In the case of the chemiluminescent EIA, the correlation coefficient (r) was 0.997 and the coefficient of variations (CV%, n=6) were 3.4% to 8.4% (mean=5.3%). In the sandwich EIA for mouse IgG_1, detection limits of the bioluminescent EIAs using the biotinylated luciferases were lower than the detection limits of chemiluminescent EIA using biotinylated alkaline phosphatase. The low detection limit of the bioluminescent EIA is probably attributable to that the CVs (2.8% for bL203 and 2.1% for bL248) of bioluminescent EIA without mouse IgG_1 were lower than CV (8.4%) of the chemiluminescent EIA without mouse IgG_1. Regression analysis showed good correlations between mouse IgG_1 concentration and bioluminescence. These results suggest that the biotinylated luciferases bL203 and bL248 will be useful for highly sensitive EIAs.

References

1. DeLuca M, McElroy WD. Purification and properties of firefly luciferase. Methods Enzymol 1978;57:3-15.
2. Arakawa H, Maeda M, Tsuji A. Bioluminescent homogeneous enzyme binding assay for biotin using luciferase as a label. Anal Letters 1992;25:1055-63.
3. Wannlund J, Azari J, Levine L, DeLuca M. A bioluminescent immunoassay for methotrexate a the subpicomole level. Biochem Biophys Res Commun 1980; 96:440-6.
4. Kobatake E, Iwai T, Ikariyama Y, Aizawa M. Bioluminescent immunoassay with a Protein A-luciferase fusion protein. Anal Biochem 1993;208:300-5.
5. Samols D, Thornton CG, Murtif VL, Kumar GK, Haase FC, Wood HG. Evolutionary conservation among biotin enzymes. J Biol Chem 1988;263:6461-4.
6. Cronan JE, Jr. Biotination of proteins in vivo. J Biol Chem 1990;265:10327-33.
7. Schatz PJ. Use of peptide libraries to map the substrate specificity of a peptide-modifying enzyme: a 13 residue consensus peptide specifies biotinylation in Escherichia coli. Bio/Technology 1993;11:1138-43.
8. Chapman-Smith A, Turner DL, Cronan JE, Jr., Morris TW, Wallace JC. Expression, biotinylation and purification of a biotin-domain peptide from the biotin carboxy carrier protein of Escherichia coli acetyl-CoA carboxylase. Biochem J 1994;302:881-7.
9. Weiss E, Chatellier J, Orfanoudakis G. In vivo biotinylated recombinant antibodies: construction, characterization, and application of a bifunctional Fab-BCCP fusion protein produced in Escherichia coli. Protein Expression and Purification 1994;5:509-17.
10. Kajiyama N, Masuda T, Tatsumi H, Nakano E. Purification and characterization of luciferases from fireflies, Luciola cruciata and Luciola lateralis. Biochim Biophys Acta 1992;1120:228-32.
11. Kajiyama N, Nakano E. Thermostabilization of firefly luciferase by a single amino acid substitution. Biochemistry 1993;32:13795-9.
12. Kajiyama N, Nakano E. Enhancement of thermostability of firefly luciferase from Luciola lateralis by a single amino acid substitution. Biosci Biotech Biochem 1994;58:1170-1.

CONSERVED MOTIFS IN SUPERFAMILY OF ENZYMES CATALYZING ACYL ADENYLATE FORMATION FROM ATP AND COMPOUNDS WITH THE CARBOXYL GROUP

VM Morozov and NN Ugarova
Faculty of Chemistry, Lomonosov Moscow State University, Moscow 119899, Russia

Introduction

The superfamily of enzymes catalyzing acyl adenylate formation from ATP and compounds with the carboxyl group includes acetyl-CoA synthetases, peptide synthetases, 4-coumaryl-CoA ligases, beetle luciferases, and other proteins (1). These enzymes form the intermediate acyl adenylate in the presence of MgATP. In particular, the firefly luciferase catalyzing bioluminescent oxidation of luciferin forms the luciferyl adenylate at the first (activation) step (2). Highly conserved motif 1 [LIVMFY]-X(2)-[STG](2)-G-[ST]-[STE]-[SG]-X-[PALIVM]-K is found in all enzymes of this superfamily (1). The other conserved motifs share the peptide synthetase and acyl-CoA ligase subfamilies (3). The functional significance of the conserved TGD-stretch was supported by mutations (4). Although the adenylate-forming reaction catalyzed by the enzymes of this superfamily is similar to one catalyzed by the aminoacyl-tRNA synthetases, no sequence or structural similarity is found between these enzyme families. We performed the computer search of conserved regions in the whole superfamily of enzymes catalyzing acyl adenylate formation from ATP and compounds with the carboxyl group.

Materials and Methods

All protein sequences with comment "Similarity to other enzymes which act via an ATP-dependent covalent binding of AMP to their substrate" were extracted from the protein database SWISS-PROT (release 30). Closely related sequences with the similarity score of greater than 150 were removed from the 66 selected sequences by the PURGE program to reduce sequence redundancy and calculation time. Then sequence fragments were removed. The final representative set of the 15 unaligned sequences was analyzed for conserved regions by the MACAW program.

At the next step we analyzed the occurrence of the detected conserved motifs in both the given protein family and other proteins. The "profile" PATMAT algorithm was used to search sequence matches with the conserved motifs along the SWISS-PROT database. The database matches with homology from 30 to 70% had been included into the profile-building process and database search was repeated. However, the third iterative search didn't' add new database matches. In addition, complete sequences of proteins of the superfamily in which any motif was not detected were analyzed by the Gibbs sampling algorithm, at that analyzed sequences were aligned with the refined sequence blocks. Secondary structure prediction was produced by the neural network PHD algorithm via InterNet.

Results and Discussion

The search for conserved regions in representative set revealed two conserved motifs: known motif 1 and new motif 2 (Figure). The screening the SWISS-PROT database with the profiles corresponding these motifs detected proteins of the given superfamily with highest matching scores. The conserved motifs repeatedly occur in the peptide synthetases because these enzymes are constructed of modules, each activating one molecule of the substrate. Only one of the motifs was detected in several proteins which are mostly fragments.

```
                  Motif 1                                          Motif 2
             # ##$$##$ #$#      ## #$$          #       # # #$      #      $  #     ##
             1   5    10        1   5     10    15    20    25    30  35     40

Consensus    ILFTSGTTGKPKG      WYRTGDLGFLDENGY----LTITGRTDDQFNLG-GERISPEEIE
                                            4-Coumaroyl-CoA ligase
             206         218    435                                      473
4CL_ORYSA    LPYS.....L...      .LH...I.YV.DDDE----IF.VD.LKEIIKYR-.FQVA.A.L.

                                              Acetate CoA ligase
             267         279    507                                      545
ACSA_ALCEU   V.Y...S......      L.LA..GSIR.KDTG---YF..M..I..VL.VS-.H.MGTM...

                                Long-chain-fatty-acid-CoA ligases
             267         279    533                                      572
LCF1_YEAST   .MY...S..E...      .FK...I.EWEA..H----.K.ID.KKNLVKTMN..Y.AL.KL.

                                               Other CoA ligases
             182         194    412                                      450
ALKK_PSEOL   LCY......N...      .FS...VATI.SD.F----M..CD.AK.IIKS.-..W..TV.L.
             177         189    396                                      434
BAIB_EUBSP   .SLSG..S..M.F      FRSV..I.YV..Q..----.YFSD.RS.MLVS.-..NVFAT.V.
             183         195    404                                      442
CAIC_ECOLI   .........SR...     .LH...T.YR..EDF----FYFVD.RCNMIKR.-..NV.CV.L.
             140         152
MENE_BACSU   LMY..........
             139         151    331                                      368
MENE_ECOLI   MTL...S..L..A      ..A.R.R.EMH-..K----...V..L.NL.FS.-..G.Q...V.

                      Luciferin-4-monooxygenases   (firefly luciferases)
             197         209    419                                      457
LUCI_LUCCR   .MNS..S..L...      .LH...I.YY..EKH----FF.VD.LKSLIKYK-.YQVP.A.L.
             197         209    419                                      457
LUCI_LUCLA   .MNS..S..L...      .LH...I.YY..EKH----FF.VD.LKSLIKYK-.YQVP.A.L.
             195         207    417                                      455
LUCI_PHOPY   .MNS..S..L...      .LHS..IAYW..DEH----FF.VD.LKSLIKYK-.YQVA.A.L.

                                              Peptide synthetases
             610         622    840                                      878
GRSB_BACBR   .IY..........      M......ARWLPD.N----IEFL..A.H.VKIR-.H..ELG...
             1651        1663   1876                                    1914
GRSB_BACBR   .MY...S......      M......ARWLPD.T----IEYL..I.Q.VKIR-.Y..E.G...
             2687        2699   2923                                    2961
GRSB_BACBR   .IY..........      L......ARWMPD.N----VEFL..N.H.VKIR-.I..ELG...
             3734        3746   3960                                    3998
GRSB_BACBR   .IY..........      M.K....AKWRSD.M----IEYV..V.E.VKVR-.Y..ELG...
             175         187    396                                      434
TYCA_BACBR   VIY..........      M......AKWLTD.T----IEFL..I.H.VKIR-.H..ELG...

                                      Other   adenylating   proteins
             63           75    263                                      304
IAAL_PSESS   .G.....S.NI.R      R..L..TAT.SMK.DKLY-..DIQ.E.MS..FM-.NL.GLGI.Q
```

```
             192         204     430                                   467
PKSJ_BACSU   L.L...S..T..A       .FE.......R-..R---.......K.AIIIN-.INYYSHA..
             187         199     410                                   448
ENTE_ECOLI   FQLSG....T..L       F.CS...ISI.PE..----I.VQ..EK..I.R.-..K.AA....
             600         612     835                                   873
ENTF_ECOLI   .I....S..R...       M.....VARWLD..A----VEYL..S...LKIR-.Q..ELG..D
             148         160     377                                   414
DLTA_BACSU   .IY...S..N...       A.....A..IQ-D.Q----IFCQ..L.F.IK.H-.Y.MEL....
             418         430     677                                   715
LYS2_YEAST   LS....SE.I...       L.......RYLP..D----CECC..A...VKIR-.F..ELG..D
             370         382     592                                   630
AAS_ECOLI    ......SE.H...       ..D...IVRF..Q.F----VQ.Q..AKRFAKIA-..MV.L.MV.
             598         610     827                                   865
ANGR_VIBAN   .IY...S..T...       ......M.CYWPD.T----.EFL..R.K.VKV.-.Y..ELG...
```

Figure. Sequence alignment of conserved motifs 1 and 2 in enzymes catalyzing acyl adenylate formation from substrates and ATP. The first column indicates sequence database identification. The top line represents the conserved feature: $ the position with conserved amino acid; # the position with the conserved property. The top alignment row is the consensus sequence representing the most frequent residue at each position. The superscript ‘f’ marks the proteins annotated as fragments in the SWISS-PROT database. The symbol ‘∇’ indicates the locations where one sequence letter is excluded for better alignment (R in PABL_STRG).

Motif 2 is located in 200-250 residues to the C-terminal of motif 1 and appropriate sequence part had been not discovered in the proteins where a second motif was not found. Thus, we suggest that conserved motifs 1 and 2 are necessary for the common function of the given enzyme superfamily, activation of the substrate carboxyl group via acyl adenylate formation using ATP.

Motif 2 is longer than motif 1. Besides the known TGD-stretch, motif 2 includes other highly conserved positions. The bulky residues (Leu, Tyr, Phe) occupy positions 1 and 2 (residue numbering according to Figure). Also, Trp and Met occur in position 1. There are the hydrophobic and bulky residues (Val, Ile, Leu, Phe, Tyr) in position 20 and 22. The residues with the propensity to hydrogen bonding via main chain carbonic oxygen (Gly, Asp) occur in position 24. There are invariant Arg and Gly in 25 and 35 position, respectively. There are the residues with the carbonic group in side chain (Asp, Asn, Glu, Gln) in position 44. The mutation experiments show that the conserved residues of these motifs are involved into adenylate-forming. The replacements Lys186→Arg,Thr (in motif 1), Asp 401 → Ser (in motif 2) in the tyrocidine synthetase 1 (TYCA_BACBR) (7), and replacements Gly1656→Asp (in motif 1), Gly 870 → Glu, Ala, Val, Arg, Trp (in motif 2) in the gramicidin S synthetase II (GRSB_BACBR) (5) significantly decreased an ATP-[^{32}P]PPi exchange by the presence of substrates, phenylalanine and proline. Also, the tyrocidine synthetase 1 bond fluorescein 5'-isothiocyanate competitively to ATP with covalent modification Lys 422 inside motif 2 (6). By photoaffinity labeling of this enzyme with 2-azidoadenosine triphosphate the triptic fragments Gly373-Lys384, Try405-Arg416, Leu483-Lys494 contained the label (7).

Predicted secondary structure of the sequence region flanked by the motifs is similar in two representative enzymes of the superfamily, the firefly luciferase and the indoleacetate-lysine ligase which have not statistically significant sequence homology. Motifs 1 and 2 are separated by approximately equal number of amino acid residues in all protein sequences. These facts indicate that these conserved motifs belong to the conserved structural unit. According to the data on firefly luciferase three-dimensional structure (8) motif 1 and the [TS]-G-D stretch of motif 2 are spatially close and hydrogen bound with the other conserved motif [YFW]-[GASW]-x-[TSA]-E (residues 340-344 in the luciferase). Motif 1 and regions 24-28 and 34-35 of motif 2 (residue numbering according to Figure) are flexible loops. The 24-28 loop connect two domains separated by the wide cleft which presumably is the substrate binding site. Glu 455 in the luciferase, corresponding to 44 position in motif 2 is located on a surface of the domain and points into the interdomain cleft.

Thus, our computer analysis of the superfamily of enzymes catalyzing acyl adelylate formation from substrate and ATP revealed new conserved motif 2. It was shown that this motif and previously known motif 1 occur in all proteins of the superfamily. The conserved residues in these motifs apparently play a key role in catalysis. The recently published data on the firefly luciferase tertiary structure support our assumption that both conserved motifs belong to a structural unit involved in acyl adelylate formation.

Acknowlegements
This work was supported by the State Scientific and Technical Program "Newest methods of bioengineering"and International Science Foundation (Soros Foundation).

References

1. Fulda M, Heinz E, Wolter FP. The fadD gene of *Escherichia coli* K12 is located close to rnd at 39.6 min of the chromosomal map and is a new member of the AMP-binding protein family. Mol Gen Genet 1994; 242: 241-9.
2. McElroy WD, DeLuca M, Travis J. Molecular uniformity in biological catalyzes. The enzymes concerned with firefly luciferin, amino acid, and fatty acid utilization are compared. Science 1967; 157: 150-60.
3. Dieckmann R, Lee YO, Van-Liempt H, Von-Dohren H, Kleinkauf H. Expression of an active adenylate-forming domain of peptide syntheteses corresponding to acyl-CoA-synthetases. FEBS lett 1995; 357: 212-6.
4. Gocht M, Marahiel MA. Analysis of core sequences in the D-Phe activating domain of multifunctional peptide synthetase TycA by site-directed mutagenesis. J Bacteriol 1994; 176: 2654-62.
5. Saito M, Hori K, Kurotsu T, Kanda M, Saito Y. Three conserved glycine residues in valine activation of gramicidin S synthetase 2 from *Bacillus brevis*. J Biochem (Tokyo) 1995; 117: 276-82.
6. Pavela-Vrancic M, Pfeifer E, Schoder W, Dohren HV, Kleinkauf H. Identification of the ATP binding site in tyrocidine synthetase 1 by selective modification with fluorescein 5'-isothiocyanate. J Biol Chem 1994; 269: 14962-5.
7. Pavela Vrancic M, Pfeifer E, Van Liempt H, Schafer HJ, Van Dohren H, Kleinkauf H. ATP binding in peptide synthetases; determination of contact sites of the adenine moiety by photoaffinity labeling of tyrocidine synthetase 1 with 2-azidoadenosine triphosphate. Biochemistry 1994; 33: 6276-83.
8. Conti E, Franks NP, Brick P. Crystal structure of firefly luciferase throws light on a superfamily of adenylate-forming enzymes. Structure 1996;4: 287-98.

CLONING AND SEQUENCING OF A FIREFLY LUCIFERASE FROM *PHOTURIS PENNSYLVANICA*

FR Leach, L Ye, HJ Schaeffer and LM Buck
Dept Biochem & Mol. Biol., Oklahoma State University, Stillwater OK 74078-3035 USA

Introduction

This paper reports the cloning and sequencing of a cDNA for firefly luciferase from *Photuris pennsylvanica*; this is the first species of the Photurinae subfamily to have a sequence determined and reported to the GenBank (U31240). *P. pennsylvanica* is a twilight-night active firefly while the common North American species, *Photinus pyralis*, is dusk-active. Before mating, *Photuris* firefly females respond to courtship signals of conspecific males, but after mating, they become "*femmes fatales*" (1).

Strause *et al.* (2) found that during *Photuris* development the larval light organ regressed and was replaced by the adult lantern. Strause and DeLuca (3) found a distinct luciferase isozyme in larval *P. pennsylanica*. Two isozymes were identified in the adult. The K_m values, pH optima, and molecular weights were similar between the adult and larval forms, but there were differences in antigenicity and pI values. We separated two firefly luciferases from adult *P. pennsylvanica* lanterns during Sephadex G-150 chromatography (4).

Since the cloning of the firefly luciferase gene of the North American firefly, *Photinus pyralis*, by DeLuca and colleagues (5), several other beetle luciferase cDNAs or genes have been cloned (see Table 1).

Table 1. Cloned Firefly Luciferases

Y	Species	Abbrev.	GenBank	PIR	# AA	% same	pI	λ_{max}	MW	Reference
87	*Photinus pyralis*	*Ppy*	M15077	A26772	550	60.4	6.42	562	60745	5
89	*Pyrophorus plagiophthalamus*									6
	Green - LucGR	*Ppl*(GR)		S29352	543	48.0	6.69	546	60691	
	Yellow green - LucYG	*Ppl*(YG)		S29353	543	48.9	6.69	560	60464	
	Yellow - LucYE	*Ppl*(YE)		S29354	543	49.3	6.39	578	60451	
	Orange - LucOR	*Ppl*(OR)		S29355	542	49.1	6.69	593	60335	
89	*Luciola cruciata*	*Lcr*	M26194	JS0181	548	55.3	7.07	562	60017	7
92	*Luciola lateralis*	*Lla*	X66919	S23437	548	55.5	6.50	552	60125	8
93	*Luciola mingrelica*	*Lmi*	S61961	S33788	548	53.7	6.24	570	60494	9
95	*Hotaria parvula*	*Hpa*	L39929		548	55.3	6.27	568	60365	10
95	*Pyrocoelia miyako*	*Pmi*	L39928		548	60.1	6.12	550	60956	10
95	*Lampyris noctiluca*	*Lno*	X89479		547	61.7	6.09	550	60365	Gene bank, 11
95	*Luciola lateralis* gene	*Lla*(g)	Z49891		548	56.2	7.08	552	60062	Gene bank
94	*Photuris sp*	*Ppe*1(J19)			552	58	7.55		60970	Patent*
	Photuris	*Ppe*1(KW)			552	55	7.55	560s	60970	Wood**
	Photuris	*Ppe*2(KW)			545	100	8.39	540s	60649	Wood**
95	*Photuris pennsylvanica*	*Ppe*2(LY)	U31240		545	-	8.40		60649	This lab, (4)

From GenBank and Protein Identification Resource at Johns Hopkins (PIR). *CAS-registry 160831-30-1. The abbreviation form was suggested by Keith Wood. (g) = gene; others from cDNA; (GR) = isozyme producing green light; 1 (adult) and 2 (larval) isoforms. The pI and MW were calculated from the data base sequences entered into the ExPASy server. The % same amino acids were calcualted using *Ppe*2(LY) as the reference. *protein sequence for JP 94303982 A2, not entered into CA until after Li Ye thesis was submitted; ** Keith Wood, personal communication.

Materials and Methods

Fireflies were collected locally with nets and stored until the evening's collection was completed. The live fireflies were taken to the laboratory, sorted according to species, and frozen in liquid nitrogen. The frozen fireflies were stored at –80 °C.

The lanterns were removed under liquid nitrogen; one g of lanterns was ground to a powder under liquid nitrogen with a mortar and pestle. Lantern RNA was isolated by using a Stratagene (La Jolla, CA, USA) RNA Isolation Kit. mRNA was then isolated using Stratagene's Quik™ mRNA Isolation Kit. A Stratagene's ZAP-cDNA™ Synthesis Kit was used for construction of the cDNA library. *In vivo* excision of pBluescript® from Uni-ZAP™ XR was done according to the instructions provided.

The pBluescript® phagemid was used for sequencing the cDNA. DNA sequencing was in the Sarkeys Biotechnology Laboratory (this Department). The sequencing was started using T3 and T7 primers and then continued in both directions using primers based on the determined sequence. The primers were synthesized by the Sarkeys Biotechnology Laboratory. The following programs and data bases (versions) were used to analyze the results: ProDom, release 28, and PROSITE, release 13.

Results and Discussion

Description of the Clone: The cDNA library was screened for light production after luciferin addition. Since the screening detected expressed bioluminescence, only functional cDNA sequences were identified. The insert size as determined after *Eco*R I and *Xho* I digestion and 1 % agarose gel electrophoresis was about 1.8 kb. Figure 1 shows the contigs that were mapped.

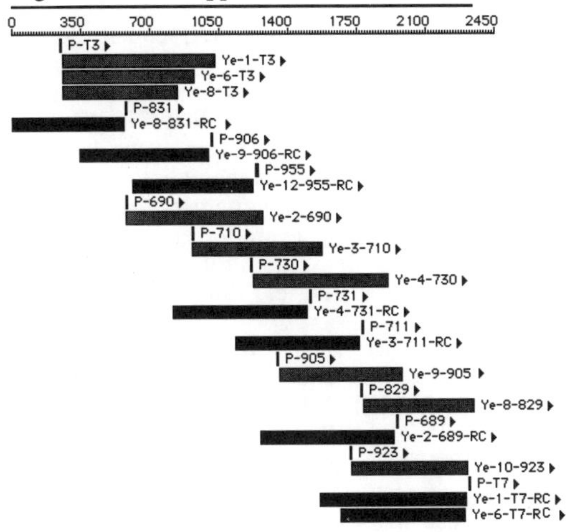

Figure 1. Contigs used to determine the nucleotide sequence of *Photuris* cDNA. The lighter bars represent sequencing from the 5′ end of the insert and the darker is from the 3′ end.

The Sequence: Figure 2 shows the deduced amino acid sequence for the largest open reading frame. From the sequence analysis, the cDNA was 1831 bp long with an open reading frame (ORF) of 1635 bp. The ORF encodes a protein of 545 amino acids. The 5′ noncoding region contains 61 bp and the 3′ noncoding region has 135 bp. The 3′ noncoding region contained a poly(A) tail of 24 nucleotides.

Similarities to Other Luciferases: The amino acid sequences deduced from the 16 cDNAs and genes sequenced for firefly luciferases have been aligned to allow determination of conserved amino acid residues and suggest possible functional portions. There are 154 residues conserved among all the luciferases (about 28 % of the total residues). In the *P. pennsylvanica* luciferase cloned in this lab (*Ppe2*(LY)), the amino acids at 276 positions are the same at corresponding positions of at least one other species (not including the corresponding sequence obtained by Wood). One hundred and fifteen amino acid residues are unique to the *Ppe*2s.

MEDKNILYGP$_{10}$ EPFHPLADGT$_{20}$ AGEQMFYALS$_{30}$ RYADISGCIA$_{40}$ LTNAHTKENV$_{50}$
LYEEFLKLSC$_{60}$ RLAESFKKYG$_{70}$ LKQNDTIAVC$_{80}$ SENGLQFFLP$_{90}$ LIASLYLGII$_{100}$
AAPVSDKYIE$_{110}$ RELIHSLGIV$_{120}$ KPRIIFCSKN$_{130}$ TFQKVLNVKS$_{140}$ KLKYVETIII$_{150}$
LDLNEDLGGY$_{160}$ QCLNNFISQN$_{170}$ SDINLDVKKF$_{180}$ KPNSFNRDDQ$_{190}$ VALVMFSSGT$_{200}$
TGVSKGVMLT$_{210}$ HKNIVARFSH$_{220}$ CKDPTFGNAI$_{230}$ NPTTAILTVI$_{240}$ PFHHGFGMTT$_{250}$
TLGYFTCGFR$_{260}$ VALMHTFEEK$_{270}$ LFLQSLQDYK$_{280}$ VESTLLVPTL$_{290}$ MAFFPKSALV$_{300}$
EKYDLSHLKE$_{310}$ IASGGAPLSK$_{320}$ EIGEMVKKRF$_{330}$ KLNFVRQGYG$_{340}$ LTETTSAVLI$_{350}$
TPDTDVRPGS$_{360}$ TGKIVPFHAV$_{370}$ KVVDPTTGKI$_{380}$ LGPNETGELY$_{390}$ FKGDMIMKSY$_{400}$
YNNEEATKAI$_{410}$ INKDGWLRSG$_{420}$ DIAYYDNDGH$_{430}$ FYIVDRLKSL$_{440}$ IKYKGYQVAP$_{450}$
AEIEGILLQH$_{460}$ PYIVDAGVTG$_{470}$ IPDEAAGELP$_{480}$ AAGVVVQTGK$_{490}$ YLNEQIVQNF$_{500}$
VSSQVSTAKW$_{510}$ LRGGVKFLDE$_{520}$ IPKGSTGKID$_{530}$ RKVLRQMFEK$_{540}$ HKSKL

Figure 2. The amino acid of Photuris luciferase deduced from the nucleotide of cDNA. See GenBank U31240. The conserved amino acids are underlined.

A domain structure map for the predicted amino acid sequence of *Photuris* firefly luciferase was developed. Figure 3 shows these results. For ProDoms starting from the N-terminus of the enzyme, ProDom 3894 and ProDom 3895 are sequences that are conserved in other firefly luciferases. ProDom 170 is found in the 4-coumarate-CoA and several peptide synthetases. The ProDom 184 segment is found in peptide synthetases and contains ProSite PS00455 which is the putative AMP-binding domain signature. ProDom 3896 is found in other firefly luciferases and ProDom 937 is from 4-coumarate CoA ligase and is also found in several other ligases. ProDoms 10748 and 10747 are found in 4-coumarate ligases. ProDom 10832 is found in 2-acylglycerolphosphoethanolamine acyltransferase. ProDom 3891 is found in other firefly luciferases.

The T-250TLGYFT-256 sequence is the AMP binding block BL00455B. PS00339 is the AA tRNA ligase II.2 sequence whose consensus sequence is [GSTALVF]-{DENQHRKP}-[GSTA]-[LIVMF]-[DE]-R-[LIVMF]-x-[LIVMSTAG]-[LIVMFY], was found as F-431YIVDRLKSL-441 (correct in 8/10 positions). The G-339YGLTRYSAVLITPDTDVRPGSTG-363 sequence is domain II that is conserved in acyladenylate-synthesizing enzymes (12). The adenylate kinase signature, PS00113, of consensus sequence [LIVMFYW](3)-D-G-[FY]-P-R-x(3)-[NQ] was tentatively found as I-412NKDGWLRSGDI-423 (correct in 7/12 positions). The G-416WLRSGD-422 sequence is domain III that is conserved in acyladenylate-synthesizing enzymes [12]. The SKL sequence at the C-terminus is the microbody-directing sequence (PS00342).

The putative AMP-binding domain signature of [LIVMFV]-x(2)-[STG](2)-G-[ST]-[STE]-[SG]-x-[PALIVM]-K (ProSite P00455) was found as V-194MFSSGTTGVSK-205 and is highly conserved among the various firefly species.

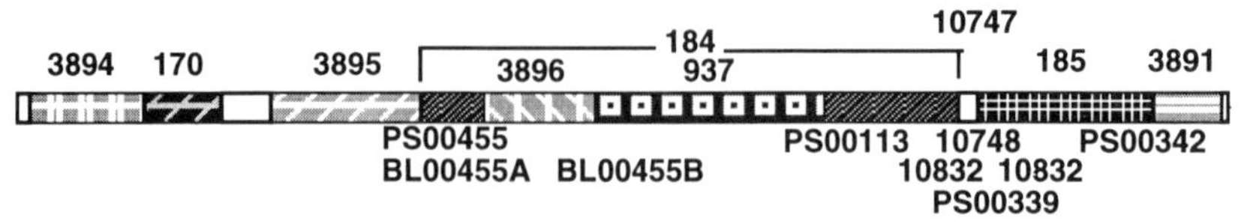

Figure 3. ProDom, ProSite, and PRINTS sequences found on *Photuris* luciferase.

Other Photuris Sequences: Wood (personal communication) has cloned and sequenced two cDNAs for luciferases from *Photuris* collected in Maryland. One cDNA is for the adult (yellow light-producing) luciferase (designated *Ppe*1) while the other was for the larval luciferase that produced green light (designated *Ppe*2). The sequence that Wood obtained for the larval cDNA is almost identical to the one we report. It is interesting that there are only two amino acid differences between these larval firefly luciferases cloned from fireflies obtained ~1300 miles apart. A Japanese group sequenced the cDNA from Wood's library and obtained the adult luciferase (*Ppe*1(J19)).

The firefly luciferase sequence presented in this paper is for the larval enzyme form (Wood, personal communication). We found (4) a 1/3 larval and 2/3 adult distribution in luciferases prepared from adult *Photuris* collected locally.

Acknowledgments

This research was supported in part by the Oklahoma Agricultural Experiment Station (Project 1806) and is published with the approval of the director. Dr. Don Arnold of the K.C. Emerson Entomology Museum, Oklahoma State University identified firefly species. Dr. Ulrich Melcher evaluated the sequencing data. Dr. Melanie Palmer introduced us to specific molecular biology procedures used with insects. Drs. Jerry Devine and Thomas Baldwin provided technical advice on screening clones for luciferase production. The sequences for *Pyrocoelia miyako* and *Hotaria parvula* were provided by Drs. Y. Ohmiya, N. Ohba, H. Toh, and F.I. Tsuji before publication. The unpublished Photuris sequences were provideed by Dr. Keith Wood of Promega who also pointed out a typographical error in our sequence.

References

1. Lloyd JE. Aggressive mimicry in Photuris: firefly femmes fatales. Science 1965;149:653-4.
2. Strause LG, DeLuca M, Case JF. Biochemical and morphological changes accompanying light organ development in the firefly *Photuris pennsylvanica*. J Insect Physiol 1979;25:339-48.
3. Strause LG, DeLuca M. Characterization of luciferases from a variety of firefly species: evidence for the presenceof luciferase isozymes. Insect Biochem 1981;11:417-22.
4. Ye L. Firefly luciferase: modification and cloning. Dissertation, Oklahoma State University, Order No. AADAA-19525455, Dissertation Abstracts, 1994
5. de Wet JR, Wood KV, DeLuca M, Helinski DR, Subramani S. Firefly luciferase gene: structure and expression in mammalian cells. Mol Cell Biol 1987;7:725-37.
6. Wood KV, Lam YA, Seliger HH, McElroy WD. Complementary DNA coding for click beetle luciferases can elicit bioluminescence of different colors. Science 1989;244:700-2.
7. Masuda T, Tatsumi M, Nakano E. Cloning and sequence analysis of cDNA for luciferase of a Japanese firefly *Luciola cruciata*. Gene 1989;77:265-70.
8. Tatsumi H, Kajiyama N, Nakano E. Molecular cloning and expression in *Escherichia coli* of a cDNA clone encoding luciferase of a firefly, *Luciola lateralis*. Biochim Biophys Acta 1992;1131:161-5.
9. Devine JH, Kutuzova GD, Green VA, Ugarova NN, Baldwin TO. Luciferase from the east European firefly, *Luciola mingrelica*, cloning and nucleotide sequence of the cDNA, overexpression in *Escherichia coli* and purification of the enzyme. Biochim Biophys Acta 1993;1173:121-32.
10. Ohmiya Y, Ohba N, Toh H, Tsuji FI. Cloning, expression and sequence analysis of cDNA for the luciferase from the Japanese fireflies, *Pyrocoelia miyako* and *Hoaria parvula*. Photochem Photobiol 1995;62:309-13.
11. Sala-Newby GB, Thomson CM, Campbell AK. Sequence and biochemical similarities between the luciferases of the glowworm *Lampyris noctiluca* and the firefly *Photinus*. Biochem J 1996;313:761-7.
12. Jackowski S, Jackson PD, Rock CO. Sequence and function of the aas gene in *Escherichia coli*. J Biol Chem 1994;269:2921-8.

CLONING AND EXPRESSION OF A *PHENGODES* LUCIFERASE

Monika G. Gruber, Galina D. Kutuzova and Keith V. Wood
Promega Corporation, Madison, WI 53711, USA

Introduction

The luciferases of luminous beetles all utilize the same substrates and presumably catalyze light emission via the same mechanism. Our knowledge of the structural and functional diversity within this enzyme group has been extended largely through the cloning of new enzymes from different beetle species. Bioluminescent beetles are found predominately in three major families: Lampyridae (fireflies), Elateridae (click beetles), and Phengodidae (glow-worms). While several luciferases from the first two families have been previously cloned and characterized, there has been very little study of phengodid luciferases. Two genera of Phengodidae are found in North America, *Zarhipis* and *Phengodes*. We report here the cloning and initial characterization of two cDNAs from the genus *Phengodes*.

The amino acid sequences encoded in the open reading frames of the two cDNAs are 55% identical with one another, and are roughly 50% identical with the sequences of other beetle luciferases. The sequences also contain the characteristic peroxisomal translocation sequences located at the C-termini of beetle luciferases. One cDNA was further expressed in *E. coli* to yield green luminescence upon addition of beetle luciferin (538-543 nm). In extracts, the reaction is ATP dependent and is similarly influenced by the addition of coenzyme A (CoA) as found with the firefly luciferase of *Photinus pyralis* (1,2). The second isolated *Phengodes* cDNA is still being characterized.

Materials and Methods

Cloning and expression: Messenger RNA was isolated using a PolyATtract® kit (Promega, Madison, WI) and then incorporated into a Uni-ZAP™XR cDNA library (Stratagene, La Jolla, CA). *In vitro* translation of 10 µg of mRNA was performed at 28 C in 120 µL of nuclease treated rabbit reticulocyte lysate (Promega). Luciferase activity in 5 µL of translation mixture was measured by adding 0.1 mL Luciferase Assay Reagent (Promega). Estimated by comparison with the translation of *P. pyralis* luciferase mRNA, the luciferase mRNA content is 0.01% of the total mRNA. The library was screened using affinity-purified antibody raised against *P. pyralis* luciferase (Promega). The same antibody was also used for Western blots. DNA sequences were determined from both strands of the cDNA clones using the Sequenase™ Version 2.0 kit (USB, Cleveland, OH). Luciferases were expressed from the pSx(*tac*) vector in *E. coli* JM109 grown at 37 C in CAA/Glc medium (3) with 10 µg/mL tetracycline and induced with 1 mmol/L final IPTG.

Evolutionary analysis: We used the Wisconsin Sequence Analysis Package Version 8 (GCG, Madison, WI). From a sequence alignment of the various beetle luciferases we calculated a distance matrix using the Kimura Protein Distance algorithm which corrects for multiple substitutions at single sites (4). A phylogenetic tree was then calculated using the neighbor-joining method (5). The tree was rooted by adding four 4-coumarate CoA ligase sequences as an outgroup to the analysis which exhibit distance values of 130 to 160 in pairwise comparisons with the beetle luciferases.

Preparation of cell extracts: Cultures of *E. coli* JM109 expressing luciferase were grown in modified LB medium (casein peptone 100 g/L, yeast extract 100 g/L, NaCl 50 g/L) with 10 µg/mL tetracycline for 1.5 h at 37 C and for 18 h at 28 C in the presence of 1 mmol/L IPTG. Cell pellets were resuspended (0.1 mmol/L KH_2PO_4 pH 7.4 / 2 mmol/L CDTA), lysed (50 mmol/L KCl / 0.2% PEI / 1 mg/mL lysozyme / 1x Cell Culture Lysis Reagent (Promega)), and luciferase activity was concentrated by resuspending the 40-65% ammonium sulfate fraction in 100 mmol/L KH_2PO_4 pH 6.9 / 2 mmol/L DTT / 0.5 mmol/L EDTA / 10% glycerol. Aliquots were stored at -70 C and used for kinetic and spectral analysis.

Kinetic analysis: Assays were performed by injecting 0.2 mL assay reagent into 10 μL of an aliquot of extract diluted 10,000-fold or 1 pg of purified firefly luciferase. Luminescence was measured as the peak light intensity using a Turner TD-20e luminometer (Turner, Sunnyvale, CA). The assay reagent contained 100 mmol/L Tricine pH 8.0 / 10 mmol/L $MgSO_4$ and either 0.5 mmol/L luciferin plus 0, 0.1, 0.3, 1, and 3 mmol/L ATP for estimating the K_m of ATP, or 1 mmol/L ATP plus 0, 0.1, 0.3, and 1 mmol/L luciferin for estimating the K_m of luciferin. The luminescence data was corrected for decay of enzyme activity over the time required for all measurements, and apparent K_m's were determined using a least-squares fit to a Michaelis-Menton equation.
Spectral analysis: Spectral measurements were performed using a Fastie-Ebert type spectrometer. The data were corrected for temporal changes of luminescent intensity and for the spectral sensitivity of the instrument. Spectral width was determined as full width at half maximum. *In vivo* spectra were taken from *E. coli* colonies grown on nitrocellulose filter strips which were soaked in 1 mmol/L luciferin / 100 mmol/L Na-Citrate pH 5.5 and taped to the outside surface of the spectrometer cuvette. *In vitro* spectra were taken by mixing 150 μL of extract with 1.1 mL reagent (100 mmol/L buffer (MES for pH 5-6, TRICINE for pH 7-9, CAPS for pH 10-11) / 0.5 mmol/L beetle luciferin / 1 mmol/L ATP / 1 mmol/L CoA / 10 mmol/L $MgSO_4$ / 2 mmol/L EDTA / 0.1 mg/mL BSA).

Results and Discussion

Phengodidae emit continuous luminescence from transverse bands and lateral spots along the worm-like body of the larvae and wingless larviform females, hence the common name "glow-worms". The adult male has a more typical adult-type beetle morphology and is not luminescent in the North American genera. The larviform *Phengodes* collected in Wisconsin varied in color from cream to brown. Luminescence spectra from specimens of both coloration extremes had intensity maxima at 544 nm and 547 nm and spectral widths of 77 nm and 83 nm, respectively. The spectral difference might be due to reflections from the body pigmentation. Crude extracts of specimens emitted light upon addition of beetle luciferin. Denaturing gel electrophoresis of the *Phengodes* extract followed by immunoblotting revealed a prominent antigenic band which comigrated with purified *P. pyralis* luciferase.

The cDNA library was created from mRNA isolated from whole specimens since the light organs were too difficult to remove. Fourteen clones were isolated by immunoscreening and the Uni-ZAP™XR lambda cloning vectors were converted into their plasmid forms for expression in *E. coli.* Two of these clones yielded a luminescence signal about 1,000-fold above the luminometer background; all other clones emitted no light. Sequence analysis of the luminescent clones revealed an identical amino acid sequence (Luc*Phg*, 546 amino acids, 61 kDa) with 48-53 percent sequence identity to other beetle luciferases (Table 1). Overall, 24 percent of the amino acids are conserved among all the beetle luciferases. This information was used to construct a phylogenetic tree (Figure 1). Surprisingly, initial sequence analysis of the remaining non-luminescent clones revealed that one partial clone contained an open reading frame of 291 amino acids which is 55% identical to the full length Luc*Phg* protein.

We attempted to improve the expression of Luc*Phg* by inserting the cDNA downstream of a *tac* promoter in a medium copy number plasmid, and properly positioning the initiation codon of the gene downstream of a ribosome binding site. Aliquots of an *E. coli* overnight culture expressing Luc*Phg* were used to perform denaturing gel electrophoresis and immunoblot analysis using antibody raised against firefly luciferase (Luc*Ppy*). The blot revealed an antigenic band comigrating with purified Luc*Ppy*. The predominant bands, however, were of smaller molecular weight, suggesting degradation. Despite efforts to optimize expression, Luc*Phg* yielded 60,000-fold lower luminescence than Luc*Ppy* expressed from the same vector measured in cells 6 hours after induction.

Table 1. Pairwise comparison of beetle luciferase protein sequences

	Lcr	Lla	Lmi	Pmi	Ppy	Lno	Ppe1	Ppe2	Phg	Ppl
Lcr	-	94	82	66	67	67	61	55	50	49
Lla	(7)	-	82	66	67	68	61	55	49	48
Lmi	(22)	(21)	-	64	67	65	60	52	48	48
Pmi	(43)	(42)	(47)	-	81	95	68	60	53	47
Ppy	(41)	(41)	(42)	(20)	-	84	70	60	53	48
Lno	(41)	(40)	(44)	(5)	(18)	-	70	61	53	48
Ppe1	(51)	(51)	(53)	(39)	(38)	(37)	-	56	51	49
Ppe2	(65)	(63)	(69)	(56)	(57)	(54)	(63)	-	51	47
Phg	(79)	(81)	(83)	(68)	(70)	(69)	(76)	(73)	-	50
Ppl	(81)	(83)	(84)	(86)	(86)	(83)	(80)	(84)	(78)	-

Percent amino acid identity (without parentheses) and evolutionary distances (in parentheses). Lcr: *Luciola cruciata* (1), Lla: *Luciola lateralis* (1), Lmi: *Luciola mingrelica* (1), Pmi: *Pyrocoelia miyako* (6), Ppy: *Photinus pyralis* (1), Lno: *Lampyris noctiluca* (1), Ppe1,2: *Photuris pennsylvanica* clone 1,2 (K. V. Wood, unpublished), Phg: *Phengodes species*, Ppl: *Pyrophorus plagiophthalamus* (ventral organ, green) (1).

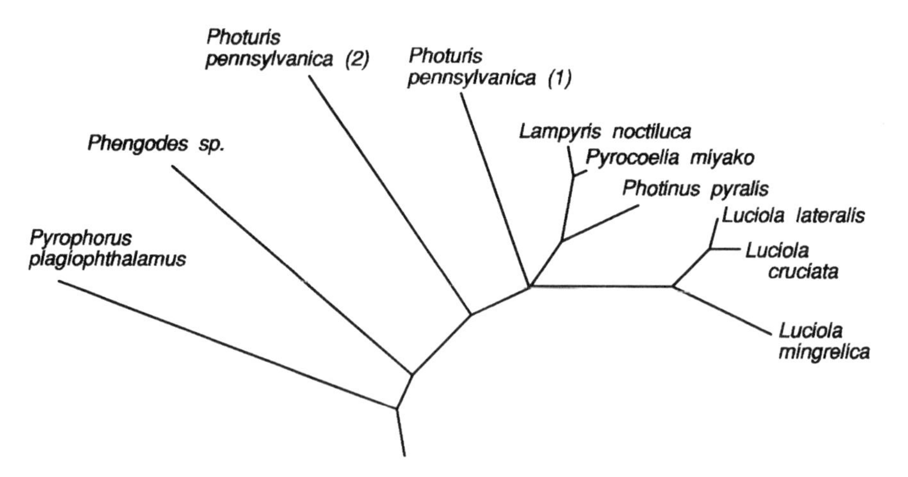

Figure 1. Phylogenetic tree of beetle luciferases. Branch lengths were calculated from the distance values in Table 1.

Table 2. Spectral parameters of Luc*Phg in vitro* luminescence at different pHs.

pH	6.0	7.0	8.0	9.0	10.0	11.0
λ_{max} (nm)	548	546	549	552	551	563
width (nm)	79	78	79	80	79	80

Spectral intensity maxima (λ_{max}) and spectral widths were calculated from the average of two independent sets of measurements, each consisting of 5 spectral scans.

The K_m values for beetle luciferin and ATP were estimated from crude extracts of Luc*Phg* expressed in *E. coli* cells. The K_m values for luciferin and ATP were 64 µmol/L and 171 µmol/L, respectively. These values are 3-4 times higher than the corresponding K_m values for purified Luc*Ppy* under identical conditions (18 µmol/L and 47 µmol/L). The assay kinetics of Luc*Phg* showed an initial slow flash of light after mixing the extract and substrates, followed by a slow decay to about 2 percent of the maximum intensity after 10 minutes. Similar to its effect on the luminescent reaction of *P. pyralis* luciferase (2), we found that CoA reduces the decay of luminescence from the reaction of Luc*Phg*. In the presence of 1 mmol/L CoA about 60 percent of the peak intensity remained after 10 minutes.

The *in vivo* bioluminescence spectra from cells of *E. coli* cells expressing Luc*Phg* revealed an intensity maximum of 538 nm and a width of 73 nm. A second measurement of the same cells after longer incubation with the luciferin solution yielded an intensity maximum of 543 nm with a secondary peak around 538 nm, suggesting that the spectrum might shift in response to changing conditions. To investigate this further we measured the spectrum of Luc*Phg* from cell extracts at pH values ranging from 6.0 to 11.0; the luminescence intensity at pH 5.0 was too low for spectral measurements.

Although there are small variations in the spectral maximum of Luc*Phg* between pH 6.0 and 10.0, we observed a clear shift towards the red only at pH 11.0 (Table 2). The luminescence intensity was highest at pH 8.0 and the spectral width was not affected significantly by the change in pH. Similar to the *in vivo* data, the spectra recorded *in vitro* between pH 6.0 and 10.0 did not show a defined single maximum but rather a varying mixture of two peaks at around 545 and 552 nm. It is not clear why the peaks *in vitro* are at longer wavelengths. Since the *in vitro* spectra were recorded in the presence of 1 mmol/L CoA to maintain better signal strength we compared *in vitro* spectra acquired at pH 8.0 in the presence and absence of CoA but did not observe any detectable effect on the spectral maximum or width.

With the cloning and expression of the first Phengodidae luciferase we were able to extend our knowledge of the structural and functional diversity of beetle luciferases to the third major family of luminescent beetles. The phylogenetic tree showing the sequence relationships between the luciferases parallels the taxonomy of the host beetle species, revealing the expected division into two beetle superfamilies, Cantharoidea (fireflies and glow-worms) and Elateroidea (click beetles), as well as the separation into the three major families of luminescent beetles. While the new *Phengodes* luciferase shares a strong effect of CoA on the luminescent reaction with the well characterized *P. pyralis* firefly luciferase, its spectral characteristics resemble more closely that of the click beetle luciferases. Specifically, it does not exhibit the large spectral shift to red luminescence in acid pH characteristic of the firefly luciferases.

Acknowledgements
This research was supported by a grant from the National Institutes of Health.

References

1. Wood KV. The Chemical Mechanism and Evolutionary Development of Beetle Bioluminescence. Photochem Photobiol 1995; 62:662-73.
2. Wood KV. Bioluminescence and Chemiluminescence: Current Status. John Wiley & Sons, Chichester. 1991. p. 11-14.
3. Gruber MG. Investigations on the Development of Multiplexed Living Cell Biosensors Using Beetle Luciferases. University of Wisconsin-Madison: Master's thesis, 1993.
4. Kimura M. The Neutral Theory of Molecular Evolution. Cambridge University Press, Cambridge, 1983.
5. Swofford DL, Olsen GJ, Waddell PJ, Hillis DM. Molecular Systematics. Ed. Hillis and Moritz, Sinauer Associates Inc. 1990; Ch.11.
6. Ohmiya Y, Ohban N, Toh H, Tsuji F. Cloning, expression and sequence analysis of cDNA for the luciferases from the Japanese fireflies, Pyrocoelia miyako and Hotaria parvula. Photochem Photobiol 1995; 62:309-13.

BIOLUMINESCENCE COLOR VARIATION AND KINETIC BEHAVIOR RELATIONSHIPS AMONG BEETLE LUCIFERASES

Galina D. Kutuzova, Rita R. Hannah and Keith V. Wood
Promega Corporation, Madison, WI 53711, USA

Introduction

There exist only two bioluminescent beetle superfamilies, Cantharoidea (includes all fireflies) and Elateroidea (click beetles). Luciferase from the North American *Photinus pyralis* fireflies is widely known because of its broad applications in industry and medicine for ATP detection, and the luciferase gene has applications in scientific research as a reporter of genetic activity in living cells. *P. pyralis* luciferase (*Ppy* luciferase) was one of the first enzymes investigated in biochemical details. For decades the kinetics and mechanism of action have been extensively studied only for two yellow-green (560-562 nm) light emitting luciferases isolated from fireflies *Photinus pyralis* and *Luciola mingrelica* (from the Black Sea Beach in the South part of Russia) (1-3). Both luciferases have been cloned and their genes have been expressed in *E. coli* thus generating a new source for large quantities of the enzymes (4,5). In this work we describe the properties of two new luciferases from the North American firefly *Photuris pennsylvanica* (*Ppe*1 and *Ppe*2 luciferases) that have been cloned recently, and of four luciferases cloned from the click beetle *Pyrophorus plagiophthalamus* (*Ppl*GR, *Ppl*YG, *Ppl*YE, and *Ppl*OR luciferases).

*Ppe*1 luciferase emits yellow-green (560 nm) light and *Ppe*2 emits green light having the shortest wavelength of any cloned beetle luciferases (538 nm). The luciferases of the bioluminescent click beetle *P. plagiophtalamus* from Jamaica have been long known to emit light of different colors ranging from green to orange (544 nm to 592 nm) (6). However, enzymological study of these various luciferases have been hampered by limited availability of the beetle species and by the need to identify and separate the different isoenzymes. Cloning and the expression of cDNA's encoding these luciferases in *E.coli* results in quantities of the enzymes suitable for the investigation of their spectral properties, kinetics and mechanism of reaction.

All beetle luciferases catalyze the same multi-step reaction of oxidation of beetle luciferin (LH_2) by atmospheric oxygen (O_2) with the involvement of adenosin-5'-triphosphate (ATP) and magnesium ions (3):

$$\mathbf{E + LH_2 + Mg{\bullet}ATP \underset{PP_i}{\rightleftharpoons} E{\bullet}Mg{\bullet}LH_2AMP \xrightarrow[AMP,\ CO_2]{O_2} EP^* \underset{hn}{\rightleftharpoons} E + P}$$

In this process, the chemical energy is converted into the light with the highest known efficiency of nearly 1. Since the substrates for all beetle luciferases are the same (7), the different colors of the emitted light must be due to the differences in the enzyme structures, which also must dictate any other differences evident in the catalytic properties.

Little has been known thus far about the enzymological and bioluminescence properties of the beetle luciferases other than firefly luciferases of *P. pyralis* and *L. mingrelica*. We present here a comparative study of the kinetic properties and bioluminescence spectra under various reaction conditions of three firefly luciferases and four click beetle luciferases that emit light of different colors. Our data show that the bioluminescent colors of these luciferases correlate with other enzymological properties.

Materials and Methods

Expression and purification of beetle luciferases: The two recombinant firefly luciferase isoenzymes of *Photuris pennsylvanica* and four recombinant click beetle luciferase isoenzymes of *Pyrophorus plagiophtalamus* were purified to homogeneity from *E. coli* strain JM109 expressing their respective cDNA clones. Expression was driven by the *tac* promoter in vectors containing both ampicillin and tetracycline resistance genes. Cells were grown in 20 liter fermentors containing LB medium with 100 and 10 mg/L of each antibiotic respectively. Purification of the recombinant enzymes was carried out on High S ion exchange (Bio-Rad, CA) and ceramic hydroxyapatite (AIC, MA) columns using a Waters 650E Advanced Protein Purification System (Waters, MA). Purified *Photinus pyralis* firefly luciferase expressed from a cDNA in *E. coli* was purchased from Promega.

Activity measurements and kinetic constants evaluation: The bioluminescence intensity of each luciferase reaction was registered on a TD-20e luminometer with injection system (Turner Designs, CA). Into a 0.2 mL reaction mixture [0.1 mol/L buffer with the appropriate pH (MES for pH 5-6, TRICINE for pH 7-9, CAPS for pH 9.5-11), 2 mmol/L EDTA, 10 mmol/L $MgSO_4$, 10 nmol/L luciferase] containing either ATP (Pharmacia, Sweden) or beetle luciferin (LH_2; Promega, WI) was injected 0.2 mL of the other substrate in the same buffer. Apparent kinetic constants (K_m and V) were calculated by the Lineweaver-Burk method from the dependencies of bioluminescence intensity maxima vs. $[S_o]$. In some experiments, coenzyme A (CoA; Pharmacia, Sweden) was also included.

Bioluminescence spectra were taken on a Fastie-Ebert type spectrometer and the data analyzed using specially designed computer software. The data were corrected for time course changes of light intensity and the spectral sensitivity of the instrument. Spectra were measured on 1 mL reactions using the same buffer as for the activity measurements with ATP and LH_2 in saturating concentrations. The bioluminescence reactions were carried out at different temperatures using a specially designed flow cuvette combined with an Isotemp Refrigerated Circulator 9100 (Fisher, PA) and the temperature was monitored by digital thermometer with the probe (Fisher, PA).

Fluorescence spectra were measured at room temperature on a Perkin-Elmer Luminescence spectrometer LS-50 in 2.0 mL of 0.1 mmol/L tris-acetate, 2 mmol/L EDTA, 10 mmol/L $MgSO_4$, pH 8.5 and 0.1 mg/mL beetle luciferase with $l_{exit.}$ 295 nm. Luciferin-luciferase binding constants (K_s) were determined from the quenching experiments and calculated using Stern-Volmer method, making the corrections for the inner filter effect.

Results and Discussion

Kinetic properties of beetle luciferases. The luminescence for all beetle luciferases studied was linearly proportional to enzyme concentration in the range of $5x10^{-15}$–$1x10^{-5}$mol/L (10^{-18}–10^{-9}mol). Protein-protein or protein-cofactor interactions are therefore unlikely in the bioluminescent reaction within this concentration range.

The $K_m(LH_2)$ values among luciferases ranged over 2 orders of magnitude at pH 8.5, and increased with pH (Fig.1A). A similar increase with pH was observed in K_m(ATP) (Fig1B), although the effect was small for *Ppy* luciferase. Over the pH range examined (6.5–9.5), the values of $K_m(LH_2)$ and K_m(ATP) each ranged over 4 orders of magnitude among the luciferases. The decreasing affinity toward luciferin and ATP with increasing pH may possibly be due to the loss of positive charges on the protein, since both substrates carry negative charges in the pH range. Both $K_m(LH_2)$ and K_m(ATP) were apparently independent of luciferase concentration.

CoA increases the activity of all luciferases in the concentration range of 0-25 mmol/L, with the highest increase (1.2–1.7 fold) at 0.5-2.0 mmol/L CoA. At higher CoA concentrations the activity slowly decreases. CoA influences the shape of the bioluminescent signals in different manners: it increases both the maximum of the light intensity and flattens the signal for *Ppy*, *Ppe*1, *Ppl*YE and *Ppl*OR luciferases but increases maximum of the light intensity only for *Ppe*2, *Ppl*GR and *Ppl*YG luciferases without the change of the luminescence decay rate.

The dependencies of V_{max} vs. pH are bell-shaped and pH optima of activity for beetle luciferases are given in the table.

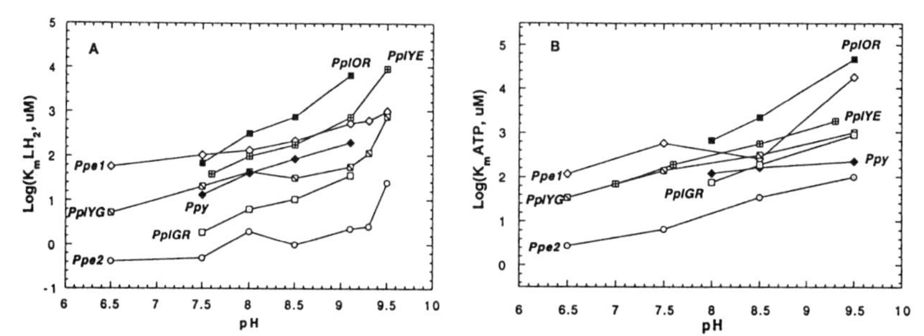

Fig. 1. pH-Dependencies of $K_m(LH_2)$ (A) and $K_m(ATP)$ (B) for firefly luciferases *Photinus pyralis* (*Ppy*), *Photuris pennsylvanica* (*Ppe*1 and *Ppe*2) and click beetle luciferases *Pyrophorus plagiophtalamus* emitting the light of green (*Ppl*GR), yellow-green (*Ppl*YG), yellow (*Ppl*YE) and orange (*Ppl*OR) colors.

Table. Properties of beetle luciferases

luciferase	***Ppy***	***Ppe*1**	***Ppe*2**	***Ppl*GR**	***Ppl*YG**	***Ppl*YE**	***Ppl*OR**
pH optimum	8.0-9.0	8.0-9.5	9.0-10.0	8.2-9.2	8.2-9.2	8.2-9.2	8.0-9.6
λ_{max}, nm (RT)	560	558	538	548	561	578	593
$K_s(LH_2)$, μM	37	37	20	23	20	30	29

Spectral properties of beetle luciferases (Fig. 2 A, B). Spectra of *Ppe*2, *Ppl*YG, *Ppl*YE and *Ppl*OR luciferases show little change over the broad pH range 5–11 (Fig.2 A); *Ppl*GR luciferase undergoes a transition to a yellow-emitting form at alkali pH (> 9.5). Both *Ppy* and *Ppe*1 firefly luciferases exist in two spectrally different conformations, one in neutral and alkali pH producing yellow-green light (560 nm) and the other in acid pH producing red light (605 - 615 nm) (Fig. 2 A). The estimated pK_a of the protein group responsible for the conformational transition is 7.3 for *Ppy* luciferase and 6.5 for *Ppe*1 luciferase. These values coincide roughly with the pK_a of the specific activity decrease at acid pH for these luciferases.

With increased temperature there is an overall trend toward increased wavelengths of the spectral maxima, the effect being most evident for luciferases having shorter spectral maxima. At temperatures above 35 C, the *Ppy* and *Ppe*1 luciferases deviated from this trend by showing a much sharper transition to a red-emitting conformation similar to that evident in acid pH. In all spectral measurements, no spectral shift was evident during the course of the bioluminescent reaction for any luciferase.

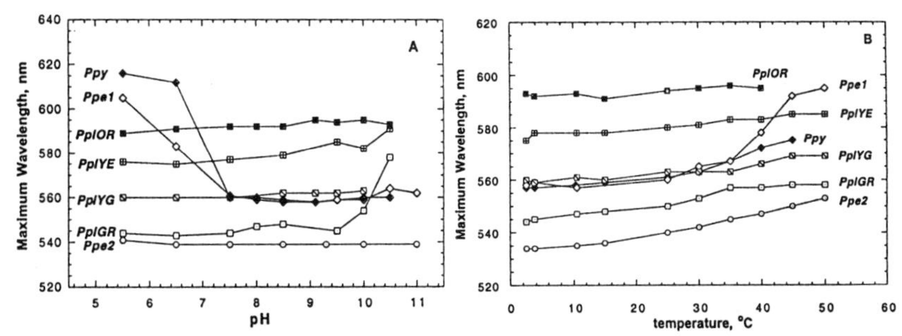

Fig. 2. Changes in bioluminescence spectral maxima with pH (A) or temperature (B) for firefly luciferases *Photinus pyralis* (*Ppy*), *Photuris pennsylvanica* (*Ppe*1 and *Ppe*2) and click beetle luciferases *Pyrophorus plagiophtalamus* emitting the light of green (*Ppl*GR), yellow-green (*Ppl*YG), yellow (*Ppl*YE) and orange (*Ppl*OR) colors.

Luciferin-luciferase binding constants. The binding constant for luciferin (K_s), measured for each luciferase using tryptophane fluorescence quenching at the pH optimum, was similar for all luciferases studied (Table). Only the K_s values for *Ppy* and *Ppl*YG luciferases are close to their $K_m(LH_2)$ values. Sequence alignment reveals that only one of the tryptophanyl residue is conserved among all the beetle luciferases, located in the C-terminal part of the protein sequence. The similarity of the K_s values may reflect the similarity of the luciferin binding site in their three dimensional structures (or similarity of this particular conservative Trp residue location).

Conclusions. The comparison of Michaelis constants for luciferin and bioluminescence spectra maxima in the pH range 7.4 - 10.0 reveals a correlation between the color of emitted light and the apparent $K_m(LH_2)$. In general, shorter wavelength of the emitted light corresponds to lower values of $K_m(LH_2)$. The difference of 55 nm in spectral maxima corresponds roughly to a difference of 300-fold in $K_m(LH_2)$. Much less correlation was observed for $K_m(ATP)$. Generally, K_m is considered the characteristic of the enzyme's substrate specificity: a higher K_m corresponds a lower specificity toward substrate. Thus, the enzyme emitting orange light (*Ppl*OR) should have the lowest substrate specificity, and the enzymes which emit green light (*Ppe*2 and *Ppl*GR) should have the highest substrate specificity. Changes in $K_m(LH_2)$ may be explained in terms of alterations in the structure of the active site. We believe the correlation between the energy of the emitted light and substrate specificity among beetle luciferases may be a consequence of the degree of compactness in the catalytic cavity which is formed during the catalytic process. The more compact this cavity is, the higher the energy of emitted light and the greater the enzyme specificity toward luciferin. This hypothesis is consistent with the "substrate flexibility" hypothesis of McCapra, *et al.* (8). Comparison of the three dimensional structures of all beetle luciferases should lead to a better understanding of the molecular mechanism giving rise to the different colors.

Acknowledgments
The authors are grateful to Dr. L.Mezei and M.Gruber for their help with bioluminescence spectra analysis. Authors are indebted to Dr. E.I.Dement'eva for the help with fluorescence spectra measurements. This work was supported by the National Institute of Health and National Science Foundation grants.

References

1. DeLuca M, McElroy WD. Purification and properties of firefly luciferase. In: Colowick SP, Kaplan NO, editors. Methods in enzymology. New York: Academic Press, 1978;57:3-15.
2. Ugarova NN. Luciferase of *L. mingrelica* fireflies. Kinetics and regulation mechanism. Biolumin Chemilumin J 1989;4:406-18.
3. Dement'eva EI, Kutuzova GD, Ugarova NN. Kinetics of firefly reaction. The influence of Cys-residue mutations on the kinetics and enzyme stability. In: Campbell AK, Kricka LJ, Stanley PE, editors. Bioluminescence and Chemiluminescence. Chichester: John Wiley and Sons, 1994:415-8.
4. DeWet JR, Wood KV, DeLuca M, Helinski DR, Subramani S. Firefly luciferase: structure and expression in mammalian cells. Mol Cell Biol 1987;7:725-37.
5. Devine, JH, Kutuzova, GD, Green VA, Ugarova, NN, Baldwin, TO. Luciferase from the east european firefly *L. mingrelica*: cloning and nucleotide sequence of the cDNA, overexpression in *Escherichia coli* and purification of the enzyme. Biochim Biophys Acta 1993;121-32.
6. Wood KV, Lam YA, Seliger HH, McElroy WD. Complementary DNA coding click beetle luciferases can elicit bioluminescence of different colors. Science 1989;244:700-2.
7. Seliger HH, McElroy WD. The colors of firefly bioluminescence: enzyme configuration and species specificity. Proc Natl Acad Sci US 1964;52:75-81.
8. McCapra F, Gilfoyle DJ, Young DW, Church NJ, Spencer P. The chemical origin of colour differences in beetle bioluminescence. In: Campbell AK, Kricka LJ, Stanley PE, editors. Bioluminescence and Chemiluminescence. Chichester: John Wiley and Sons, 1994:387-91.

SURFACTANTS AND COENZYME A AS COOPERATIVE ENHANCERS OF THE ACTIVITY OF FIREFLY LUCIFERASE

CY Wang and JD Andrade
Dept. of Bioengineering, University of Utah, SLC, UT 84112, USA

Introduction

Since protein folding depends on physical interactions with the solution, activators or inhibitors may modify protein structure via changes in the local solution environment, facilitating or inhibiting protein activity. Kricka and DeLuca found that nonionic surfactants, such as Triton X-100 and Tween 20, and some polymers, such as polyethylene glycol (PEG) and polyvinylpyrrolidone (PVP), stimulated luciferase activity (1). However, anionic and cationic surfactants failed to stimulate luciferase activity and behaved as inhibitors. They further titrated the active sites with dehydroluciferin and ATP-Mg^{2+} and showed the stimulation effect was not caused by recruitment of new active sites. Although the actual mechanism of stimulation is still unknown, they suggested that the binding of Triton X-100 or polymers to luciferase mat account for the stimulation effect. Other studies showed similar stimulation effects in some cases but not in others (2, 3). The inconsistent results may result from different preparations of firefly luciferases.

Coenzyme A (CoA) was found to reduce the decay rate of luciferase bioluminescence (4). Schroder found firefly luciferase (Photinus pyralis) was related to a plant enzyme, 4-coumarate CoA ligase, which utilized ATP and CoA (5). Since similarity between enzymes may suggest the conservation of specific functions, the existence of a CoA binding site is possible, although CoA is not a required substrate for firefly bioluminescence.

The enhancement effects of surfactants and polymers were determined with and without CoA. A significantly cooperative enhancement was observed in the presence of both surfactant/polymer and CoA. Triton X-100 was also found to be a protectant against the inhibition effect of 1,2-dioleoyl-sn-glycerol-3-[phospho-L-serine] (DOPS). Luciferase inactivated by DOPS recovered over 60% activity after incubation in Triton X-100.

Materials and Methods

Triton X-100 (avg. MW 650, density 1.059), Tween 20 (polyoxyethylene sorbitan monolaurate, MW 1228), PVP (MW 40,000) and PEG (MW 5,000) were purchased from Sigma. The stock solution for 10% (w/w%) Triton X-100 and Tween 20 and 10% (w/v%) PVP and PEG were prepared with glycylglycine buffer (0.45 mol/L, pH 7.8). The stock solutions were then diluted to 1%, and 0.1% with the same buffer.

Firefly luciferase (from Sigma, chromatographically prepared and lyophilized; showed a single band on SDS-PAGE) was used to prepare the enzyme-surfactant mixtures. 20 μL of luciferase solution (0.5 mg/mL) was mixed with 13.1 μL of glycylglycine buffer and 248.5 μL of surfactant or polymer solution. The final concentration of additives is 0.4800%, 0.0480%, and 0.0048%. In the control experiment, glycylglycine buffer was used to replace surfactant solution. 100 μL of luciferin stock solution (0.1 mmol/L luciferin and 10 mmol/L $MgSO_4$) was added to 20 μL of luciferase mixture and incubated for 3 minutes. 250 μL of ATP solution (10^{-4} mol/L or 10^{-6} mol/L) was then injected to initiate the bioluminescence reaction. The light intensity was recorded with a spectrofluorometer (I.S.S., Inc.).

The optimum concentration (the highest enhancement effect) of each individual additive was then mixed in combinations of two. The additivity of enhancements on luciferase activity was examined by assaying luciferase activity.

CoA (free acid, from Sigma) was added to ATP solution (10^{-4} mol/L and 10^{-6} mol/L) to make a 10^{-4} mol/L CoA solution. The ATP solution containing CoA was then added to the luciferase/surfactant mixtures and the enzyme activity determined as before.

For the protection experiment, 20 μL of luciferase (0.5 mg/mL) was first incubated with 248.5 μL Triton X-100 (10%) for 30 min. Preparation of 1,2-dioleoyl-sn-glycerol-3-phosphocholine (DOPC) and DOPS liposomes was described elsewhere (6). 13.1 μL of DOPS liposomes were added to the mixture and incubated for another 30 min. The final concentration of Triton X-100 was 0.48% and the molar ratio of DOPS to luciferase was 500:1. The activity of luciferase was then assayed. For the reactivation experiment, luciferase was first incubated with DOPS for 30 min. The Triton X-100 was then added and incubated for another 30 min. The activity of luciferase recoved was assayed. DOPC liposomes were used (in a molar ratio of 500 DOPC to 1 luciferase) to replace Triton X-100 and the experiments were repeated.

Results and Discussion

Some nonionic surfactants and polymers were effective in enhancing luciferase activity (1-3). In this study, all four additives were found to increase luciferase activity, which showed 28% to 118% higher initial velocity (maximum light intensity) at the 10^{-4} mol/L ATP concentration. PEG 5000 and Triton X-100 showed the most promiment enhancing effect at the peak intensity of bioluminescence. At the optimum concentration, Triton X-100 showed the slowest decay rate, which was maintained up to 326% of control at the plateau of bioluminescence, while PEG 5000 maintained the lowest bioluminescence (160%) in the same period (Table 1).

The optimum concentration for the four additives was further examined at 10^{-6} mol/L ATP. Higher stimulation effects at low ATP concentration was observed than at high ATP concentration (Table 1).

Table 1. Enhancement of additives at high and low ATP concentration. The bioluminescence intensity of pure luciferase solution is normalized as 100%.

Additives	Final concentration	Peak (%)	Plateau (%)	Peak (%)	Plateau (%)
		10^{-4} mol/L ATP		10^{-6} mol/L ATP	
Triton X-100	0.4800%	218	326	324	333
Tween 20	0.0048%	187	242	212	194
PEG 5000	0.4800%	198	160	267	140
PVP 40000	0.0480%	179	173	288	151

The surfactants and polymers were mixed in pairs to examine the additivity of the enhancement effects. The results showed that Triton X-100 mixed with Tween 20 and PEG 5000 further enhanced the peak intensity 28 to 48%. There was no significantly further enhancement for the other combinations. The presence of other additives in the Triton X-100 solution further enhanced the light intensity at plateau. However, PVP and PEG hinder the enhancing effect of Tween 20 at plateau.

Coenzyme A was effective in preventing the decay of bioluminescence. The total light output increased 6 fold over a 30 sec period when using 10^{-4} mol/L ATP and 0.1 mol/L CoA as the substrates. CoA mixed with any one of the four surfactants/ polymers resulted in a significant cooperative enhancement (Table 2). The combination of Triton X-100 and CoA had the most significant enhancement, increasing the total light output up to 23 fold for the first 30 sec (Figure 1). The mechanisms of stimulation for these two chemicals was different and cooperative. A less prominent enhancement was also observed for 10^{-6} mol/L ATP.

Table 2. Cooperative enhancing effects of CoA (final concentration: 0.068 mmol/L) and surfactants/polymers (at concentration as Table 1). Light count was integrated for first 30 sec. The count of luciferase solution without additives and CoA was used as the reference. The enhancement was calculated by dividing each output result by the reference.

	Enhancement (fold)	
Components	ATP (10^{-4} mol/L)	ATP (10^{-6} mol/L)
Luciferase and CoA	7	1.1
Luciferase, CoA and Triton X-100	23	5.4
Luciferase, CoA and Tween 20	20	4.7
Luciferase, CoA and PEG 5000	12	1.9
Luciferase, CoA and PVP 40000	19	2.3

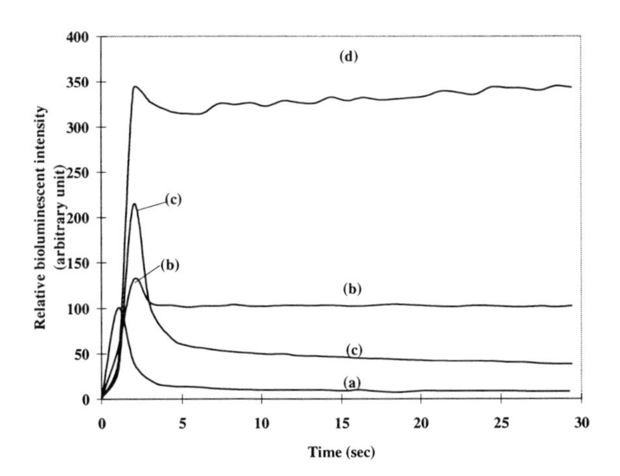

Figure 1. The Cooperative enhancement of Triton X-100 (0.4800%) and CoA on luciferase activity. (a) Luciferase; (b) luciferase with CoA; (c) luciferase with Triton X-100; (d) luciferase with CoA and Triton X-100.

Firefly luciferase was inactivated by negative charged lipids (6, 7). The presence of Triton X-100 (0.4800%) prevented inactivation by DOPS. Luciferase inactivated by DOPS can be partially reactivated by incubation with Triton X-100, the activity recovering from 17% to 64%. Although the mechanism is unknown, the recovered enzyme activity may be due to the replacement of DOPS by Triton X-100 at the same binding site; Triton X-100 may also enhance the activity of the remaining native enzyme. In the same experiment, DOPC was used to replace Triton X-100. There was no protection or reactivation effect for DOPC against DOPS.

The micelle environment was not required condition for the stimulation, since PEG and PVP do not form micelles. A large complex found by Kricka and DeLuca was proposed to be the enzyme-surfactant complex (1). The lack of an additivity effect between the combination of surfactants/polymers indicate a similarity to the enhancing mechanism. The binding of surfactants/polymers helps reactivate luciferase, which may have been inhibited by its substrates. Triton X-100 and Tween 20 may also help to modify the conformation and speed the release of oxyluciferin, enhancing enzyme turnover. Coenzyme A behaves differently with surfactants and polymers and probably binds to luciferase at a specific site. When added with Triton X-100, the bioluminescence intensity was enhanced further and maintained at a maximum level for at least 2 minutes. The high turnover rate was not inhibited by the accumulation of oxyluciferin. Other combinations with CoA produce less but still prominant total light output. The additivity effect enables a constant light output at high intensity. It is useful to develop direct reading biosensors which require those characteristics.

Acknowledgement

This work was partially supported by Protein Solutions, Inc., Salt Lake City, Utah via funding provided by an NSF STTR grant. We thank Drs. J. Herron and V. Hlady for access to equipment and facilities.

References

1. Kricka LJ, DeLuca M. Effect of solvents on the catalytic activity of firefly luciferase. Arch Biochem Biophys 1982;217:674-81.
2. Yeh PY. The properties of firefly luciferase in solution and at interfaces. Master thesis, University of Utah, 1989.
3. Simpson WJ, Hammond JRM. The effect of detergents on firefly luciferase reactions. J Biolumin Chemilumin 1991;6:97-106.
4. Airth RL, Rhodes WC, McElroy WD. The function of coenzyme A in luminescence. Biochim Biophys Acta 1958;27:519-32.
5. Schroder J, Protein sequence homology between plant 4-coumarate: CoA ligase and firefly luciferase. Nucleic Acids Res 1989;17:460-2.
6. Wang C-Y, Firefly Luciferase: Activity, Stability and Sensor Applications. Ph.D. Dissertation, University of Utah, Oct. 1996.
7. Dukhovich AF, Filippova NYu, Efimov AI, Ugarova NN, Berezin IB. Choline-containing phospholipids as specific activators and stabilizers of firefly luciferase. Dokl Akad Nauk SSSR, 1988;298:1257-60.

FACTORS AFFECTING THE SHELF LIFE OF FREEZE-DRIED FIREFLY LUCIFERASE REAGENTS

MJ Reilly and HC Jenkins
Biotrace Limited, The Science Park, Bridgend, Mid Glamorgan, CF31 3NA, UK

Introduction

Commercially produced firefly luciferase reagents have been available for more than 30 years (1). They contain all the components needed to initiate light emission, with the exception of ATP, which is usually the analyte. The reagents are commonly preserved for transit and storage by freeze-drying and supplied to users as dehydrated powders in small bottles (1). Before use, they are rehydrated, usually using a solution provided by the manufacturer. The shelf life of freeze-dried reagents is superior to that of the liquid reagents from which they are produced. But, to retain performance most suppliers of reagents recommend that the materials are stored at 4°C or less.

However, at some stage in their history, reagents may be exposed to higher temperatures. It is important from the viewpoint of reagent robustness that they can withstand the greatest possible abuse with the minimum possible loss in performance. Here, we report on the storage stability of a commercial formulation and an improved reagent.

Materials and Methods

Instrumentation: A Biotrace Unilite™ luminometer (Biotrace Limited, Bridgend, UK) was used for all measurements.

Reagents: Reagents for measurement of ATP contained luciferase (*Photinus pyralis* EC 1.13.12.7), D-luciferin, buffer salts, metal co-factors, ethylenediaminetetraacetic acid, dithiothreitol and a range of proprietary stabilizers and kinetic enhancers. After preparation, they were freeze-dried *in situ* in their containers then stored at 4°C, 25°C or 37°C for various times. Individual vials of reagents (sufficient to make 1.2 mL) were rehydrated immediately before use.

Assessment of reagent performance: Reagent performance was assessed using standard solutions of ATP at 20°C. ATP solution (100 μL; 200 pg ATP) was mixed with assay reagent (100 μL) in ATP-free disposable cuvettes. Emitted light was measured using a Biotrace Unilite™ luminometer. All measurements were made in triplicate. Luciferase activity was measured using a commercially available luciferase assay kit (Promega, Madison, USA). D-luciferin and dehydroluciferin contents were measured by uv spectrophotometry.

Results and Discussion

Dry stability of the original formulation: At 4°C the original reagent had excellent stability. No significant loss was detected over 15 weeks storage. Losses in activity occurred at higher temperatures (Figure 1). For example, after 15 weeks storage, 14% and <1% activity remained in reagents stored at 25°C and 37°C respectively.

In the case of our original formulation, the reagent could withstand 1 - 2 weeks at 25°C and still retain an acceptable level of performance. However, if the abuse temperature was 37°C, reagent performance was commercially unacceptable after only one week's storage. Comparison of fresh and aged reagents revealed a reduction in luciferase activity on storage. However, this was insufficient to completely account for the loss in response of the reagent to ATP. Analysis of luciferin contents of aged reagents revealed that only small losses occurred. We have found that inhibitors of the firefly luciferase reaction are formed in the reagents during storage. Some of these may be oxidation products of D-luciferin. For example, dehydroluciferin, which we detected in aged reagents, is a potent inhibitor of the firefly luciferase reaction (K_i = <1 μmol/L) (2).

One manifestation of the formation of inhibitors during storage is their influence on test performance. Measurements of light emission from fresh and aged reagents at different reagent concentrations, but constant ATP concentration, revealed some interesting effects (Figure 2).

In the case of the fresh reagent, dilution relative to the concentration of ATP caused a disproportionate loss in light emission. This was to be expected. By way of an example, consider the case of a two-fold dilution of the reagent. Since reaction rate

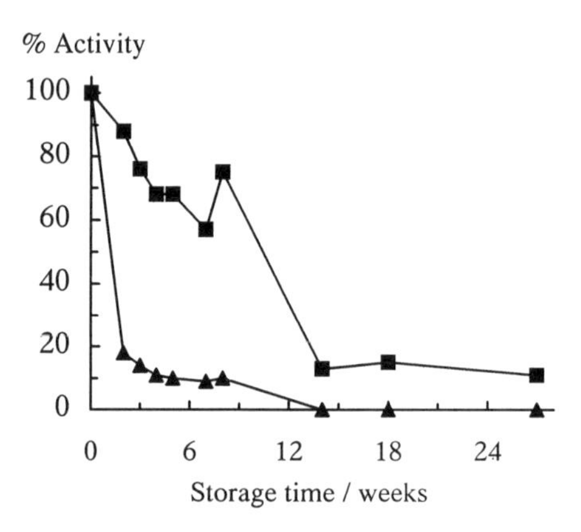

Figure 1. Stability of the original formulation at (■) 25°C and (▲) 37°C. Data shown are the means of three determinations.

is proportional to enzyme concentration, this results in a 50% reduction in light emission. A two-fold dilution of the reagent also brings about a reduction in the D-luciferin concentration of 50%. At the concentration of D-luciferin used in the formulation (somewhat below K_m), this brings about a loss in response of about 20%. The combined effect of these two factors is that the response from the reagent is reduced by a factor that exceeds that of the dilution itself.

In the case of the aged reagent the situation is more complex. Firstly, dilution brings about a loss in response caused by the reduction in luciferase and D-luciferin concentrations. However, the concentration of inhibitors is also reduced. Depending on their initial concentration, dilution can result in a disproportionate increase in reagent response as the inhibition is released (see Figure 2). The response of the fresh reagent to ATP was markedly reduced upon dilution. Although the response of undiluted reagent to ATP was less than that of the fresh reagent, the loss in response on dilution was less. The difference can be explained by the fact that inhibitors were present in the aged sample.

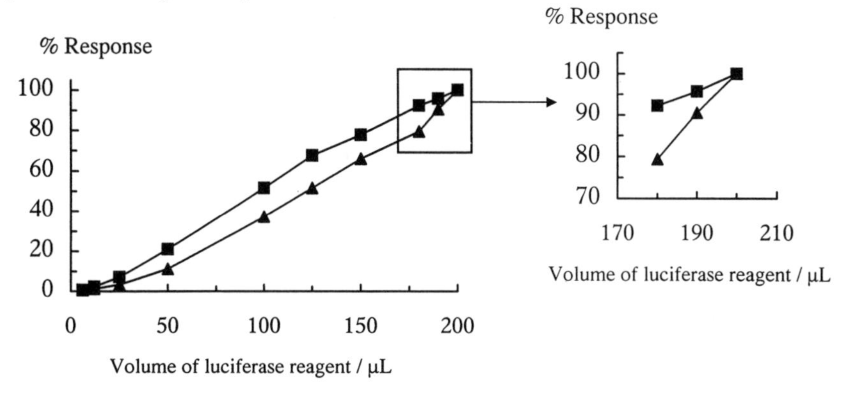

Figure 2. Performance of fresh (▲) and aged (■) reagents at a range of reagent concentrations and constant ATP concentration (5 nmol/L). Data shown are the means of triplicate determinations. Note that dilution of the aged reagent has less influence on response than in the case of the fresh reagent, especially within the upper part of the dilution curve (inset).

Strategies to improve reagent stability: Since both enzymic and non-enzymic factors determine the stability of commercially-produced firefly luciferase reagents, formulation and handling must be optimized to take account of both. Luciferase is sensitive to oxidation, so a thiol protectant, such as dithiothreitol, is usually included in commercial formulations (1). The enzyme can also be inactivated in solution by contaminants, such as fatty acids (3). Factors affecting the stability of firefly luciferase include (1, 4 - 6) (i) storage temperature; (ii) type of luciferase used in the formulation; (iii) type of buffer used in the formulation; (iv) reagent pH value prior to freeze-drying; (v) presence of substrates; (vi) presence of thiol-protecting agents and antioxidants; (vii) presence of protective proteins, such as serum albumin; (ix) presence of polyols, such as glycerol and trehalose.

Factors affecting the stability of D-luciferin include (1, 4, 5) (i) storage temperature, (ii) reagent pH value prior to freeze-drying; (iii) presence of oxygen and oxidizing agents; (iv) presence of antioxidants; (v) exposure to light.

Improvements in reagent stability: Modification of the reagent formulation, inclusion of proprietary stabilizers and optimization of the freeze-drying conditions, allowed us to improve reagent stability. The new reagent has excellent stability at 4°C, and its stability at 25°C and 37°C has also been markedly improved (Figure 3). Such reagents retain 83% and 64% activity after 15 weeks storage at 25°C and 37°C respectively.

Consequences of improved stability: These new reagents offer a greater degree of protection against damage caused by temperature abuse. Exposure to higher than anticipated temperatures may take place during transit, during storage by the distributor or the user, or immediately prior to use. The improved reagents are considerably more robust than before. While robustness is important for traditional multi-test formats, it is critical to in the case of 'single shot' test applications in which reagent performance is more difficult for the user to validate.

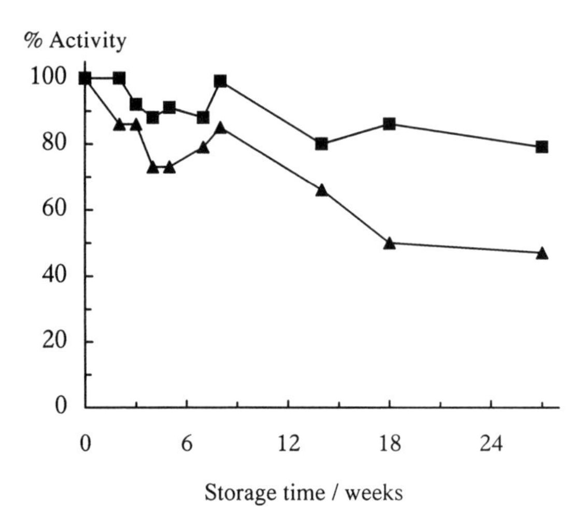

Figure 3. Stability of the improved formulation at (■) 25°C and (▲) 37°C. Data shown are the means of three determinations.

Acknowledgements
We thank the Directors of Biotrace Limited for permission to publish his work.

References

1. Leach FR, Webster JJ. Commercially available firefly luciferase reagents. Meth Enzymol 1986;133:51-70.
2. Denburg, JL, Lee RT, McElroy, WD. Substrate binding properties of firefly luciferase. Arch Biochem Biophys 1969;134,381-94.
3. Simpson WJ, Hammond JRM. The effect of detergents on firefly luciferase reactions. J Biolumin Chemilumin 1991;6:97-106.
4. Lundin A, Thore A. Analytical information obtainable by evaluation of the time course of firefly bioluminescence in the assay of ATP. Anal Biochem 1975;66:47-63.
5. Webster JJ, Leach FR. Optimization of the firefly luciferase assay for ATP. J Appl Biochem 1980;2:469-79.
6. Hall MS, Leach FR. Stability of firefly luciferase in tricine buffer and in a commercial enzyme stabilizer. J Biolumin Chemilumin 1988;2:41-4.

PART 7

Applications of Bioluminescence

NOVEL BIOLUMINESCENCE TESTS FOR THE DETERMINATION OF AOC AND BOD IN WATER

R Shomer-Ben Avraham[1], *S Belkin*[2], *A Bulich*[3] *and S Ulitzur*[1]
[1]*Faculty of Food Engineering and Biotechnology, Technion, Haifa 32000, Israel*
[2]*Environmental Microbiology, Desert Research Institute, Ben-Gurion University, Sede Boker 84990, Israel*
[3]*Microbics Co. Carlsbad, CA, USA*

Introduction

Growth of bacteria in drinking water distribution and storage systems can lead to proliferation of health hazards, to the development of undesired taste and odor, and to water turbidity (1). Regrowth of bacteria in water after disinfection is influenced mainly by the concentration of biodegradable organic carbon (BDOC) or assimilable organic carbon (AOC). Both parameters represent the fraction of the dissolved organic carbon (DOC) that can be assimilated or metabolized by bacteria in a defined period of time (2). The classical BOD test, used for the determination of bio-oxidizable organic matter in treated or untreated waste waters, is not sensitive enough for detecting the low concentrations (10-300 μg/liter) of assimilable organic matter usually found in drinking water (3). Sensitive and reliable AOC or BOD assays for water quality monitoring are therefore essential. Most of the standard or proposed AOC methodologies are hindered by their long duration. They involve the inoculation of the studied water sample with a bacterial inoculum (intrinsic population or a standard culture); after a time period varying between 5 and 30 days, the increase in viable counts and/or the decrease in DOC are determined (4).

This study describes a novel approach for the determination of AOC and BOD in water. The new bioassay utilizes starving or ATP-depleted marine luminous bacteria, the luminescence of which is directly proportional to the concentration of utilizable organic material. Using this approach, AOC data may be obtained in a matter of a few hours; in addition, the effect of potentially assimilable macromolecules can be determined after a short hydrolysis of the bio-polymers into utilizable organic compounds.

Materials and Methods

Bacterial strains and assay medium: The marine bacterium *Vibrio fischeri* strain A6 is a dim, inducer-requiring mutant isolated in the course of this study. *Photobacterium leiognathi* EGMB has been described (5). Lyophilized cultures of both bacteria were provided by Microbics Corp. (Carlsbad, CA, USA). Artificial seawater medium buffer (ASWMB) contained (in g/L): NaCl, 17.5; KCl, 0.75; $MgSO_4 \cdot 7H_2O$, 12.3; $CaCl_2 \cdot 2H_20$, 1.45; NH_4Cl, 1; $K_2HPO_4 \cdot 3H_2O$, 0.075 and MOPS (Sigma, St. Louis, USA), 20mM (pH 7.0).

Bioluminescence tests for AOC and BOD:

a) The A6 test: hydrated lyophilized cells of *V. fischeri* A6 were starved for 60 minutes by shaking at 28°C, and added to the assay medium to a final concentration of about 10^5 viable cells/mL. The first assay vial contained 1 mL of the sample and 1 mL of 200% ASWMB. This sample was double diluted 8-10 times in 1 mL of ASWMB in a set of clean new glass vials. Distilled water served as a negative control. As positive controls, different concentrations of glucose or a glucose:glutamate mix (1:1) were included in each test. The assay medium was supplemented with 20 ng/mL of *V. fischeri* synthetic autoinducer which initiates the induction of transcription of the *lux* operon. Luminescence was recorded after 120 minutes of incubation at 24°C.

b) The *P. leiognathi* EGMB test: hydrated lyophilized cells of the highly luminescent EGMB cells were depleted of ATP by incubation with 40 mM sodium arsenate at room temperature for 60 minutes. Consequently, the *in vivo* luminescence decreased 10,000 fold. The cells were then added to the tested water samples (to 10^5 cells/mL), prepared as above. The ASWMB medium was supplemented with 0.3M $NaNO_3$. The results were recorded after 45 minutes of incubation at 24°C, using the same controls as for the A6 test. In both tests the basic unit was defined as the minimal sample concentration that resulted in a two fold increase in luminescence.

Standard AOC test: the test determines changes in viable counts of a seeded inocolum of *Pseudomonas fluorescence* P17 and *Spirillum NOX* after 7 days of incubation at 15°C. The results are expressed as the acetate concentration required to obtain an equivalent growth increment under the test conditions (6). The test was conducted according to Standard Methods (7).

Standard BOD test: the standard 5 day BOD test (7) was used.

Acid hydrolysis of water samples: one mL of 6N HCl was added to a carbon-free test tube containing 9 mL of the tested water sample, to a final HCl concentration of 0.6N. Following 60 minutes in a boiling water bath, the test tubes were allowed to cool, 0.6 mL of NaOH (10N) was added, and pH was adjusted to 7.0. The "water hydrolyzate" was then serially diluted in ASWMB medium and analyzed by one of the bioluminescence tests described above.

Measurement of luminescence: A Kontron Betamatic BMI scintillation counter or an Anthos (Salzburg, Austria) Lucy1 were used. Arbitrary light units are presented.

Results and Discussion

The potential efficiency of the two test systems was examined by assaying their responses to various carbon sources at different concentrations. The luminescent responses in the A6 and the EGMB tests are presented in Figures 1 and 2, respectively. It may be observed that at certain concentration ranges, linear correlation between the concentration of tested carbon sources and the developed

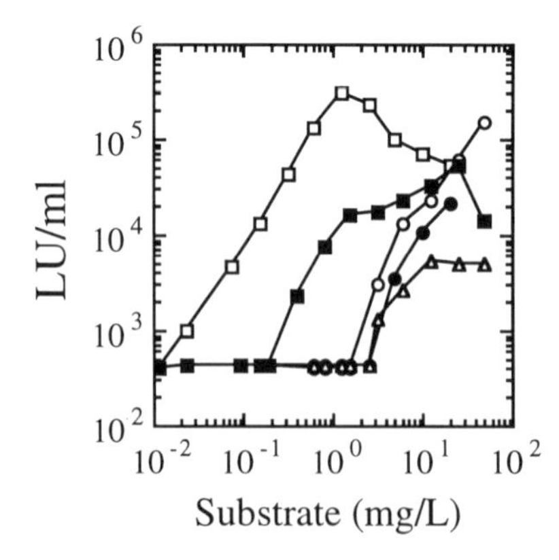

Figure 1. Determination of different organic compounds in the A6 test. ■, sodium acetate; ❑ , glucose; ●, sodium pyruvate; o, glycerol; Δ, sodium oxalate.

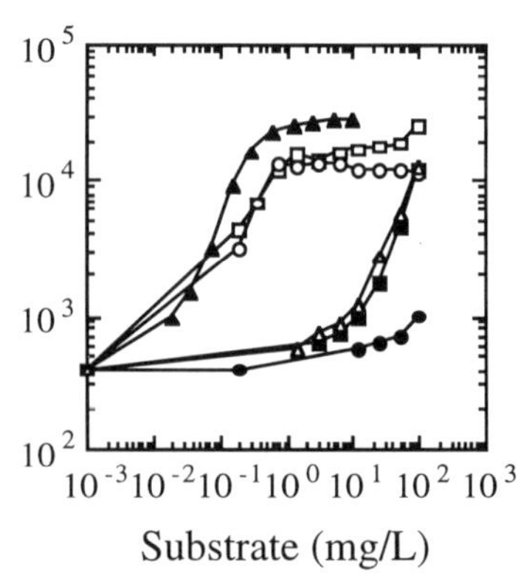

Figure 2. The *P. leiognathi* EGMB test. ▲, a glucose:glutamate mix (1:1); other symbols as in Figure 1.

luminescence were established. Substrate levels as low as 20 μg/L of glucose, and 0.5-2 mg/L of glycerol or other carbon sources were rapidly and sensitively detected.

To demonstrate the applicability of the system to "real" water samples, various drinking water samples collected at different locations along the Water Carrier system in Israel were assayed using the A6 test. Figure 3 summarizes the correlation between the standard AOC test (expressed as μg acetate/L), and the A6 test results (expressed as μg glucose/L) obtained for the same samples. A high correlation (r^2=0.864) was found between the results of these tests.

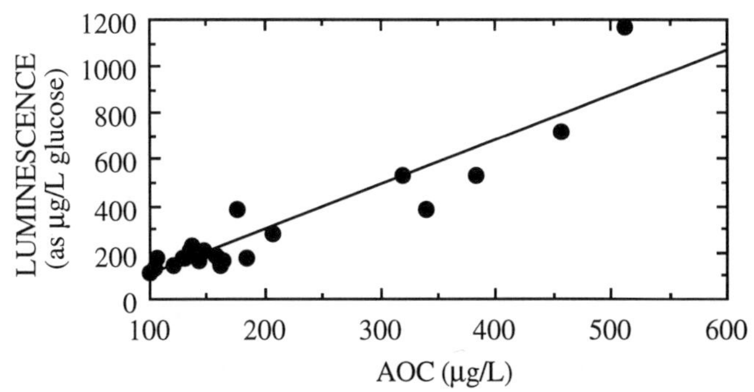

Figure 3. Correlation between the standard AOC and the A6 tests. Drinking water samples were collected from sites along the main Israeli Water Carrier, and assayed in parallel in both tests as described under Materials and Methods

Surprisingly, in contrast to drinking water samples, raw sewage as well as treated effluents exhibited low activity in both bioluminescent tests. This phenomenon may be explained by low levels of organic compounds which are readily available to the test organisms during the short assay (2 hours), even though the dissolved oragnic carbon concentrations were high. The bulk of the dissolved organic carbon, therefore, may be composed of macromolecules which need to be extracellularly hydrolyzed prior to their utilization. Indeed, pretreatment of these samples by a mild acid hydrolysis resulted in up to a 1,000 fold increase in the concentration of available organic compounds as expressed in the bioluminescence tests. Table 1 compares the results of the standard BOD test, expressed as mg O_2/L, and the A6 test, expressed in terms of mg/L of a standard BOD mix (glucose/sodium glutamate, 1:1).

exp .#	BOD (mg/L O_2)	A6 test, control mg/L BOD mix	A6 test after acid hydrolysis mg/L as BOD mix
1	4	0.055	8.8
2	7	<0.01	1.1
3	7.4	0.18	5.7
4	11.5	<0.01	14.3
5	170	0.14	73
6	243	0.16	111

Table 1. Correlation between the standard BOD test and the A6 test. Six samples were tested in the standard BOD and in the *V. fischeri* A6 tests. Results are expressed as mg/L of O_2 and mg/L of a BOD mix (glucose:glutamate, 1:1), respectively.

The correlation between these tests was very poor (r^2=0.32) for non-hydrolyzed sewage and effluent samples, but high (r^2=0.986) after acid hydrolysis. It appears, therefore, that most of the organic matter found in sewage is not in a form which is readily available to the luminous bacteria in a short-term assay. Acid hydrolysis of an aqueous suspension of starch or casein, under the conditions described under

Materials and Methods, exhibited similar luminescence levels to water samples containing equimolar concentration of glucose or amino acids, respectively (not shown).

A comparison between the two luminescent tests indicates that The *P. leiognathi* EGMB assay required a shorter incubation period and exhibited a higher sensitivity for several carbon compounds (Figs. 1, 2). This assay is based on the expression of existing luciferase enzyme molecules, and therefore does not require *de-novo* synthesis of proteins. In contrast, the development of luminescence in A6 cells involves synthesis of the *lux* system proteins as well as many other factors. This is reflected in the good correlation of the standard AOC and BOD results to those of the A6 test.

In addition to the use of the marine bacteria *V. fischeri* and *P. leiognathi*, we have also developed sensitive AOC and BOD tests based on recombinant *E. coli* strains harboring the *V. fischeri lux* system. While these tests showed comparable or better sensitivity and performance, they were not further developed in the course of this study. We believe this approach bears an excellent potential for the design and development of tailor-made water quality assays in the future, when the use of recombinant DNA technology for environmental purposes is more widely accepted.

In summary, we have presented preliminary results demonstrating the use of bioluminescent bacteria to assay two very important water quality parameters, AOC and BOD. The most important advantage of the proposed approach is in the very short time interval between sample collection and data availability. Furthermore, the bioluminescence tests described here for assimilable organic carbon may be adapted towards for the determination of any potentially limiting growth factor required for luminescence, including nitrogen and phosphorous. Such tests may therefore also serve as powerful tools in basic aquatic ecology.

Acknowledgments

The development of the AOC test was supported by a grant from BMBF, Germany, (WT 09353/01204G), administered by the Israeli Ministry of Science and Arts. The development of the bioluminescent BOD test was supported by Microbics Co. (Carlsbad, CA, USA). Special thanks to M. Peer and D. Frimerman for their assistance.

References

1. Lechevallier MW, Schultz W, Lee R. Bacterial nutrients in drinking water. Appl Environ Microbiol 1991;57:857-62.
2. Servais P, Anzil A, Ventresque C. Simple method for the determination of biodegradable dissolved organic carbon in water. Appl Environ Microbiol 1989;55:2732-34.
3. Lechevallier MW, Shaw NE, Kaplan LA, Bott TL. Development of a rapid assimilable organic carbon method for water. Appl Environ Microbiol 1993;59:1526-31.
4. Huck PM. Measurement of biodegradable organic matter and bacterial growth potential in drinking water. J AWWA 1990;7:78-86.
5. Katznelson R, Ulitzur S. Control of luciferase synthesis in a newly isolated strain of *Photobacterium leiognathi*. Arch Microbiol 1977;115:347-51.
6. Van der Kooij D, Visser A, Hijnen WAM. Determining the concentration of easily assimilable organic carbon in water. J AWWA 1982;10:540-45.
7. Biochemical oxygen demand and assimilable organic carbon. In: Rand MC, Greenberg AE, Taras MJ, editors. Standard Methods for the Examination of water and wastwater, Washington DC, American Public Health Association, 1992.

AN ATP BASED BIOASSAY SYSTEM FOR MYCOTOXIN DETECTION

MF Sanders, SA Russell and R Blasco
Central Science Laboratory, Sand Hutton, York YO4 1LZ, UK

Introduction

The desirability of challenging a living system with a potentially toxic substance to assess its biological activity has lead to the development of many different types of bioassay. The use of bioassays for detecting mycotoxins has been extensively reviewed (1) and has included bioassays based on micro-organisms, plants, crustacea, fish, birds, mammals and tissue cultures. One bioassay method listed for the presumptive detection of mycotoxins in food materials was the brine shrimp test (2). This bioassay is currently widely used, for example, for assessing the toxicity of fungi isolated from cheese (3), the toxicity of mycotoxins extracted from maize (4) and the toxin producing ability of fungi extracted from potato tubers (5). The test involves the exposure of brine shrimp larvae (*Artemia salina*) to a sample extract or mycotoxin, the toxicity being calculated from the observed as loss of motility, measured after a 20 hour incubation period. However, this test has several limitations, it is time consuming, labour intensive and the end point is open to subjective judgement. To address these problems the possibility of using an adenosine triphosphate (ATP) based test was investigated. We have previously suggested the development of an ATP based method to determine the efficiency of fumigation as a means of killing white fly scales *(Bemisia tabaci*) on poinsettia cuttings, imported into the UK. These insects are an efficient vector of a large number of economically important plant virus diseases (6), and at the scale stage of their life cycle are non motile and it is difficult to see if they have been killed. It was thought that after insect death the cellular ATP levels should fall to near zero. Preliminary experiments with other insects, killed by freezing and their ATP extracted with dimethyl sulphoxide, indicated that this was the case, and work was extended to the brine shrimp assay. A number of trial experiments were conducted with aflatoxin B_1 and acute toxins to establish the feasibility of bioluminescence technology, some of these experiments were run in parallel with the conventional brine shrimp test.

Materials and Methods

Instrumentation: Light measurements were made using a model LB953 luminometer (Berthold Instruments U.K. Ltd., St. Albans, Hertfordshire, UK.).

Reagents: ATP assay mix, containing firefly luciferase and luciferin (FL-AAM; Sigma Chemical Company Ltd., Poole, UK.). ATP assay mix dilution buffer (FL-AAB; Sigma Chemical Company Ltd.). Each vial of ATP assay mix was reconstituted with 5 mL of distilled water and diluted 1:25 with assay mix dilution buffer immediately before use.
Dilution assay buffer, 25 mmol/L HEPES pH 7.75. Aflatoxin B_1 (A 6636; Sigma Chemical Company Ltd.). Whatman polypropylene micro-centrifuge filter tubes (Whatman International Ltd., Maidstone, UK.).

Procedures: Preparation of brine shrimps: Brine shrimp eggs were incubated in aerated artificial sea water in a flask with side illumination at 30°C for 24 h. After incubation, hatched larvae that had been attracted by the light and migrated to the side of the flask, were harvested.

Bioassay: Fifty µL volumes of aflatoxin B_1 standards, to give a concentration within the 1-400 µg/mL range, dissolved in chloroform, were applied to the filter base of the upper chamber of Whatman micro-centrifuge tubes, using a positive displacement pipette, and allowed to evaporate to dryness. Other toxic compounds, such as sodium cyanide, were dissolved in water. Two hundred and fifty µL volumes of the hatched larvae (containing approx. 20-30 shrimp) were added to the upper chambers of the Whatman micro-centrifuge tubes and incubated at 30°C for 24 hours. After this time, the tubes were removed from the incubator and centrifuged at 120 *g* for 10 min to pellet the brine shrimp in the upper chamber. Each tube's upper chamber was removed and placed into a clean sterile glass test-tube for ATP extraction.

ATP extraction: A 450µL volume of dimethyl sulphoxide (DMSO) was added to the brine shrimp contained in each tube, and incubated, with intermittent mixing, for 20 min at room temperature. A 4.5mL volume of HEPES dilution assay buffer at pH 7.75 was then added to each tube and mixed thoroughly. A 50µL volume from each tube was placed in a luminometer tube and ATP measured as peak counts per second (CPS) over a 10s measuring period, after the automatic injection of 50µL of ATP assay reagents.

Results and Discussion

The relationship between ATP levels of brine shrimp exposed to aflatoxin at levels of up to 400 µg/mL and observed mortality is illustrated in Figure 1. This experiment gave a good correlation between shrimp death and decreasing ATP levels, with the light output decreasing from approx. 2.5E+04 CPS at 1 µg/mL to near zero at 400 µg/mL aflatoxin. This result was, however, not consistently repeatable.

Further experiments with similar numbers of brine shrimp showed an almost identical rise in mortality, with increasing aflatoxin concentration was shown, but the ATP levels measured as counts per second were significantly higher. In one of these

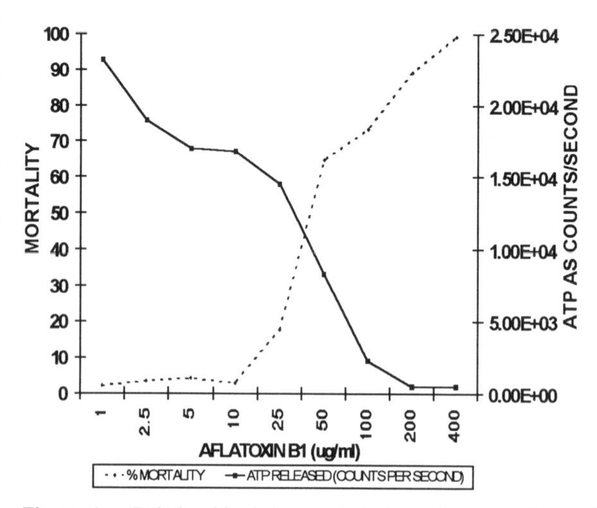

Figure 1. Relationship between *Artemia salina* mortality and extracted ATP.

experiments (Table 1.), the initial ATP levels were approximately 3.70E+05 CPS, falling to 1.58E+05 at the 400 μg/mL aflatoxin level.

Aflatoxin B_1 (μg/mL)	% Artemia salina mortality	ATP as counts/second
1	0	3.73E+05
2.5	2	4.00E+05
5	2.8	4.82E+05
10	5.9	4.27E+05
25	17.7	4.51E+05
50	46.3	2.94E+05
100	65.7	2.76E+05
200	89.1	1.93E+05
400	89.7	1.58E+05

Table 1. Relationship between *Artemia salina* mortality and extracted ATP.

Despite repeating the experiment several times it was not possible to obtain consistent results. A possible explanation for this was found by high power microscopic examination of the brine shrimp after exposure to the aflatoxin for 24 hours. For the standard brine shrimp bioassay low power microscopy is used, and shrimp death is defined as the loss of motility of the shrimp (7), however, under high power, it was possible to detect movement of internal structures within many of the shrimp exposed to the higher levels of up to 400 μg/mL aflatoxin. These shrimp were non motile, but still alive, and still contained high levels of ATP. Due to this it is not possible to use a brine shrimp/ATP bioassay for aflatoxin without extending the incubation time to allow for death to occur. Toxins differ in the way they act on living systems, some are acute and kill at the cellular level, while others like aflatoxin, seem to disrupt individual body systems and kill because the organism can no longer function in a co-ordinated manner (8). We have demonstrated the near complete reduction of ATP in brine shrimps following killing by freezing, sodium cyanide, phenol and sodium azide. These results could be used to advantage, as the ATP bioassay could be used to differentiate between acute and chronic toxins. It is also possible to modify the test by using other target organisms, such as yeast cells or bacteria. The use of these organisms is mentioned in the literature as having been used in mycotoxin bioassay systems, these systems could be modified to produce a range of rapid ATP based bioassays suitable for different groups of mycotoxins.

Acknowledgements

This work was funded by The Chemical Safety of Food Division of the Ministry of Agriculture Fisheries and Food UK.

References

1. Watson DH, Lindsay DG. A critical review of biological methods for the detection of fungal toxins in foods and foodstuffs. J Sci Food Agric 1982;33:57-9.
2. Harwig J, Scott PM. Brine shrimp larvae as a screening system for fungal toxins. Appl Microbiol 1971;21:1011-16.

3. Cafarchia C, Celano GV, Tiecco G. Toxicity of mould cultures isolated from cheese. Industrie Alimentari 1994;33:744-46.
4. Logrieco A, Moretti A, Altomare C, Bottalico A, Carbonell Torres E. Occurrence and toxicity of *Fusarium subglutans* from Peruvian maize. Mycopathologica 1993;122:185-90.
5. Zietkiewicz L, Perkowski J, Chelkowski J. Mycotoxin production, pathogenicity and toxicity of *Fusarium* species isolated from potato tubers with dry rot injuries. Microbiologie Aliments Nutrition 1995;13:87-100.
6. Cheek S, Macdonald O. Statutory Controls to prevent the establishment of *Bemisia tabaci* in the United Kingdom. Pestic Sci 1994:42:135-42.
7. Hoke SH, Carley CM, Johnson ET, Broski FH. Use of solid phase extraction systems to improve the sensitivity of *Artemia* bioassays for trichothecene mycotoxins. J Assoc Off Anal Chem 1987;70:661-63.
8. Buckle AE, Sanders MF. An appraisal of bioassay methods for the detection of mycotoxins. Letters in Appl Microbiol 1990;10:155-60.

LUMINOUS MICROBIAL BIOSENSORS FOR THE SPECIFIC DETECTION OF HALOGENATED ORGANICS

Y Rozen, A Nejidat and S Belkin
The J. Blaustein Desert Research Institute, Ben Gurion University of the Negev
Sede Boker 84990, Israel

Introduction

This communication constitutes a preliminary report on the development of a novel type of bacterial biosensors, based upon the genetic fusion of *Vibrio fischeri lux* genes to promoters of bacterial dehalogenases.

Halogenated organics constitute approximately 60% of the organic compounds on the EPA priority pollutant list, and are probably the most prevalent class of environmental contaminants worldwide (1-3). They are released into the environment mostly as agricultural chemicals (herbicides, insecticides, fungicides) and due to industrial emissions (refrigerants, solvents, intermediates and degradation products); they are also generated by water chlorination, or, in some cases, by biological activities (1-3). While some halogenated organics are highly recalcitrant and are not amenable to biodegradation, many others can be fully or partially degraded by microorganisms. Almost universally, the first step in the process is cleavage of the halogen-carbon bond, an activity carried out by a group of enzymes known as dehalogenases (1-3). In most reported cases dehalogenases are inducible; while the nature of this inductive process is not always clear, in many cases the promoter sites have been identified and sequenced. In the present project, several such promoters were selected, amplified, and inserted into a multiple cloning site on a plasmid, upstream from *V. fischeri luxCDABE.* These plasmids were used to transform several bacterial strains, with the purpose of generating bacterial constructs which respond by increased bioluminescence to the presence of pre-determined target haloorganics, in a manner similar to that previously employed for the specific presence of heavy metals (5-7) or naphthalene (8), or to the non-specific induction by global stress factors (9, 10) or solvents (11). An *E. coli* strain capable of specific detection of 4-chlorobenzoic acid is presented here as an example.

Materials and Methods

Plasmid Construction: *Arthrobacter* SU (12) served as the source of the 4CBA dehalogenase promoter (the *fcbABC* operon). PCR primers were synthesized to allow the amplification of a 1,700 bp DNA fragment which probably included, along with the promoter, regulatory elements allowing substrate recognition. The PCR product, obtained with plasmid pAS5 as a template, was ligated into a plasmid containing the promoter-less *luxCDABE* operon of *Vibrio fischeri* (4).

The plasmid thus constructed (pASU) was used in an attempt to transform three bacterial species: (a) the original source of the promoter, *Arthrobacter* SU; (b) *Pseudomonas putida*; (c) *Escherichia coli.* At present, the *Arthrobacter* transformation is not yet successful, whereas luminescence by the *Pseudomonas* transformants was not satisfactorily inducible. While efforts continue in both directions, the description that follows is limited to the *E. coli* transformants.

<u>Growth and measurement of bioluminescence</u>: *E. coli* cells were grown in LB medium containing 50 μg/mL of kanamycin at 26°C in a rotary shaker. Following overnight growth, cells were diluted 100 fold in the same medium, lacking the antibiotics but containing MOPS (100 mM, pH 7). After a 4-5 h growth under the same conditions, 50 μL of culture were added to each well of an opaque white microtiter plate (Dynatech, Germany), already containing a two-fold dilution series of the compounds to be tested in 50 μL LB. Luminescence was followed using a temperature-controlled (26°C) microtiter plate luminometer (Lucy1, Anthos Labtec Instruments, Salzburg, Austria). Luminescence values are presented in the instrument's arbitrary relative light units (RLU). Maximal response ratios are the highest ratios determined during a specified period of luminescence in the sample-containing wells to that in wells containing untreated cells.
<u>Chemicals</u>: Stock solutions (100 g/L) of all aromatics (excluding phenol) were prepared in 1 M NaOH, and pH adjusted with acetic acid. Other potential inducers were added directly to the medium.

Results and Discussion

E. coli cells (strain RFM443,) containing plasmid pASU emitted light in response to the presence of 4-chlorobenzoic acid (Fig. 1). After a short lag period, luminescence lasted for several hours with light intensity reaching a maximum after approximately 1 hour (Fig. 1A). The response was proportional to 4CBA concentrations up to 1 g/L; higher concentrations were inhibitory (Fig. 1B).

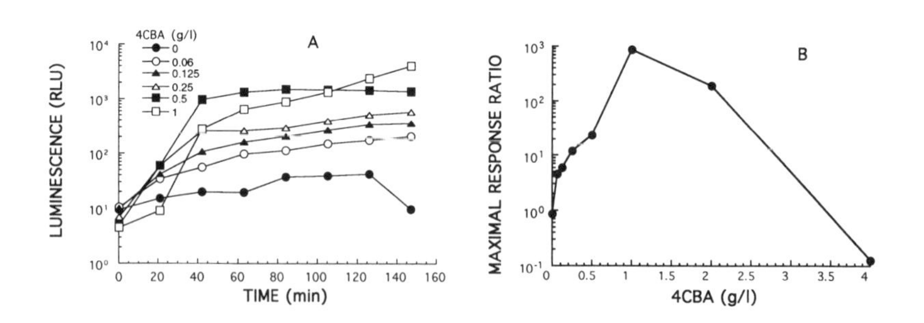

Figure 1. Induction by 4-chlorobenzoic acid of *E. coli* containing plasmid pASU. A - kinetics of light development; B - maximal response ratios obtained at each 4CBA concentration.untreated control.

An extensive screening of different organics has revealed that responses such as those demonstrated in figure 1 were limited to 4CBA and closely related compounds. Among the compounds which did not induce luminescence were chlorinated aliphatics (including dichloromethane, 1-chlorooctane, 1-chlorodecane, chloroacetate), non-halogenated solvents (methanol, ethanol, dimethylsulfoxide, phenol), and p-halogenated phenol derivatives (4-chloro-, 4-bromo- and 4-fluorophenol). Table 1 presents the maximal response ratios obtained for the positive compounds, all benzoic acid and its derivatives.

Substitution	Max. response Ratio	Min. detected concentration (g/L)*
none (benzoic acid)	66	1
2-chloro-	5	0.13
3-chloro-	28	0.13
4-chloro-	399	0.06
2-bromo-	7	0.25
3-bromo-	41	0.25
4-bromo-	120	0.13
4-fluoro-	145	0.25
4-iodo-	1	-

Table 1. Inductive properties of various benzoic acid derivatives.
*) minimal tested concentration causing at least a two fold increase in luminescence.

Clearly, the construct displayed a marked specificity towards 4CBA; with the chloride in a different position, or with a different halogen altogether, responses were much smaller. 4CBA was also the compound most sensitively detected; it should be noted, however, that apparent overall sensitivity was relatively low. This may be due to insolubility of the inducer in the assay medium, which may drastically reduce its actual effective concentrations. This point is presently under investigation. In addition, several options are considered for sensitivity enhancement, including genetic manipulations of membrane permeability. Another option for broadening the detection limit lies in the increased activity observed at lower pH values, as demonstrated in Figure 2. The improved performance, possibly due to an increase in the proton motive force or to facilitated diffusion of 4CBA, is clearly accompanied by a significant lowering of the detection threshold.

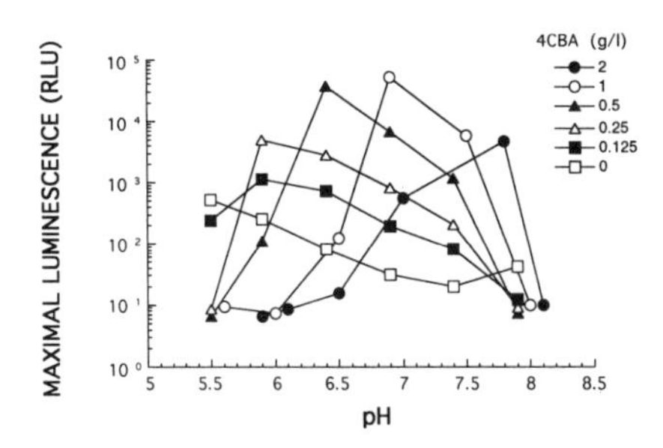

Figure 2. Effect of assay pH on *fcb::lux* expression.

In conclusion, we have presented a preliminary characterization of a plasmid which allows the rapid detection of a specific class of halogenated organics. With the correct choice of both dehalogenase promoters as well as host organisms, powerful biosensors of varying specificities can be constructed for real time detection of environmental pollutants.

Acknowledgments

Arthrobacter sp. strain SU, as well as the plasmid containing the isolated promoter fraction (pAS5), were a generous gift by K-H Gartemann. Research was supported in part by the Moria Foundation and by grant no. I-0442-168.09/95 of the German Israeli Foundation for Scientific Research and Development.

References

1. Belkin S. Biodegradation of haloalkanes. Biodegradation 1993;3: 299-313.
2. Fetzner S, Lingens F. Bacterial dehalogenases: biochemistry, genetics, and biotechnological applications. Microbiol Rev 1994; 58: 641-85.
3. Janssen DB, Pries F, van der Ploeg JR. Genetics and biochemistry of dehalogenating enzymes. Ann Rev Microbiol 1994;48:163-91.
4. Rogowsky PM, Close TJ, Chimera JA, Shaw JJ, Kado CI. Regulation of the *vir* genes of *Agrobacterium tumefaciens* plasmid pTiC58. J Bacteriol 1987;169:5101-12.
5. Selifonova O, Burlage R, Barkay T. Bioluminescent sensors for detection of bioavailable Hg (II) in the environment. Appl Environ Microbiol 1993;59:3083-90.
6. Corbisier P, Ji G, Nuyts G, Mergeay M, Silver S. *LuxAB* gene fusions with the arsenic and cadmium resistance operons of *Staphylococcus aureus* plasmid pI258. FEMS Microbiol Lett 1993;110:231-8.
7. Guzzo J, Guzzo A, DuBow MS. Characterization of the effects of aluminum on luciferase biosensors for the detection of ecotoxicity, Toxicol Lett 1992; 64/5:687-93.
8. King JMH, DiGrazia PM, Applegate B, Burlage R, Sanseverino J, Dunbar P, Larimer F, Sayler GS. Rapid, sensitive bioluminescent reporter technology for naphthalene exposure and biodegradation. Science 1990;249:778-81.
9. Van Dyk TK, Majarian WR, Konstantinov KB, Young RM, Dhurjati PS, LaRossa RA. Rapid and sensitive pollutant detection by induction of heat shock gene-bioluminescence gene fusions. Appl Environ Microbiol 1994;60:1414-20.
10. Belkin S, Smulski DR, Vollmer AC, Van Dyk TK, LaRossa RA. Oxidative stress detection with *Escherichia coli* bearing a *katG'::lux* fusion. Appl Environ Microbiol 1996; 62: 2252-6.
11. Selifonova OV, Eaton RW. Use of *ibp-lux* fusion to study regulation of the isopropylbenzene catabolism operon of *Pseudomonas putida* RE204 and to detect hydrophobic pollutants in the environment. Appl Environ Microbiol 1996;62:778-83.
12. Schmitz A, Gartemann KH, Fielder J, Grund E, Eichenlaub R. Cloning and sequence analysis of genes for dehalogenation of 4-chlorobenzoate from *Arthrobacter* sp. strain SU. Appl Environ Microbiol 1992;58:4068-71.

PRELIMINARY STUDY OF THE OPTIMUM CONDITIONS FOR A LACTATE SENSOR BASED ON BACTERIAL BIOLUMINESCENCE

DJ Min and JD Andrade
Dept. of Materials Science and Engineering, University of Utah, Salt Lake City, Utah 84112, USA

Introduction

The bacterial bioluminescence system can be applied as an enzyme sensor for biochemicals related to reduced nicotinamide adenine dinucleotide (NADH). A key problem is completeness of the conversion of reactants to products, because the efficiency of conversion affects the sensitivity. This problem is especially significant in multiple enzyme systems since each enzyme may have different optimal conditions. The selection of the optimal pH value, the effect of reactants and products on the enzyme activities, and the enzyme concentrations and their ratios are important in the optimization of enzyme sensors (1).

A lactate sensor was chosen as a model system because lactate is important in clinical analysis, food analysis, and sports medicine (2). Lactate monitoring using bacterial bioluminescence has many advantages, including simplicity and speed. The governing reactions are:

$$\text{lactate} + NAD^+ \xleftrightarrow{LDH} \text{pyruvate} + NADH + H^+ \quad (1)$$

$$NADH + FMN + H^+ \xrightarrow{OR} NAD^+ + FMNH_2 \quad (2)$$

$$FMNH_2 + RCHO + O_2 \xrightarrow{LF} FMN + RCOOH + H_2O + \text{light} \quad (3)$$

where NAD is ß-nicotinamide adenine dinucleotide, LDH is lactate dehydrogenase, FMN is flavin mononucleotide, OR is NADH:FMN oxidoreductase, $FMNH_2$ is reduced flavin mononucleotide, RCHO is decanal, and LF is bacterial luciferase. NADH formation is catalyzed by LDH. Light is emitted after the serial reactions by OR and LF. The light intensity is proportional to the rate of NADH formed, which is proportional to lactate concentration in the solution.

We will discuss optimal conditions for lactate analysis by bacterial bioluminescence and the interference reactants and products on the reactions.

Materials and Methods

Instrumentation: Bioluminescence was measured using a photon counting spectrofluorometer (I.S.S. Inc., Champaign, Ill., USA). The unit of light intensity is relative light units (RLU).

Reagents: All assays were performed in 500 mL solution of 0.1 mol/L phosphate buffer at room temperature. The final concentrations of NAD (Calbiochem, La Jolla, USA), lactic acid (Sigma, St. Louis, USA), FMN (Boehringer Mannheim, Indianapolis, USA), and decanal (Sigma, St. Louis, USA) in the assay are 0.25 mmol/L, 1 mmol/L, 10 µmol/L, and 0.001 %, respectively. In the interference experiments, the final concentrations of NAD, lactic acid, and pyruvate (Sigma, St. Louis, USA) were changed to the desired concentrations and the final concentration of NADH (Sigma, St. Louis, USA) was 0.1 mmol/L. The concentrations of stock solutions for LDH (Calbiochem, La Jolla, USA), LF (Sigma, St. Louis, USA), LF (Boehringer Mannheim, Indianapolis, USA), and OR (Boehringer Mannheim, Indianapolis, USA) were 2000 U/mL, 6 mg/mL, 7.5 mU/mL, and 5 U/mL, respectively. The amounts of the three enzymes were changed in each experiment. All reagents and enzymes were used without further purification.

Procedures: All assays were performed by adding reagent mixtures to the cuvettes containing LF, OR, and/or LDH in the different combinations. To test the interference in the reactions, the pH of reaction mixtures was titrated to 7.0 and 50 µL of Sigma LF stock solution was applied. Since Sigma LF already contained OR as a minor contaminant, it was not necessary to add OR. To obtain the optimal pH conditions, 50 µL each of Sigma LF and LDH stock solutions were added. For determining the optimal amounts of enzymes, the pH of reaction mixtures was adjusted to 7.6 and stock solutions of Boehringer LF, OR, and LDH were used.

Results and Discussion

Light intensity is dependent on the rates of the reactions in the bioluminescent system (3). The higher concentrations of reactants and/or higher activity of enzymes give higher light intensity. At a given concentrations of reactants, the light intensity depends solely on the enzyme activities. The effects of reactants and products in the serial lactate reactions were tested. Fig. 1 shows the effects of lactate, NAD, and pyruvate on the bioluminescent system. Although there was slight inhibition, it was not critical to the bioluminescence activity within the tested concentration range. Prahl et al. (4) found that a large excess of lactate, NAD, and pyruvate inhibit LDH activity, and an excess of NAD also inhibits LF activity. NAD concentration in their experiments was higher than our test range. The purity of reagents seemed to be important for these tests. The slight increase of light intensity at higher concentration of NAD and lactate was probably due to the impurity of NADH.

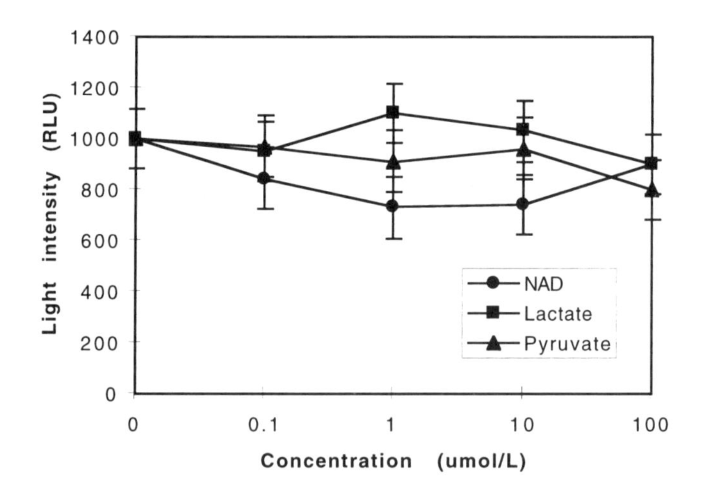

Figure 1. The effects of lactate, NAD, and pyruvate on the activity of the bioluminescent reactions. The reaction mixtures (pH 7.0) contain 0.1 mmol/L NADH, 10 µmol/L FMN, 0.001 % decanal, and 50 µL Sigma LF (6 mg/mL).

The optimum pH of bacterial bioluminescent reactions is known to be 7.0 and the reaction is inhibited at pH above 8.0 (5). However, the optimal pH of LDH reaction in the direction of NADH formation is 9.0-9.6 (6). The consumption of newly formed NADH in the LDH reaction by the other bioluminescent reactions allows sub-optimal pH conditions to be used, so the sensor can operate at pH below 8.0. An optimum pH of 7.6 was determined for the lactate sensor (Fig. 2). At pH below 7.6, NADH formation is likely inhibited; bioluminescent reactions may be inhibited at pH above 7.6. Prahl et al. (4) chose

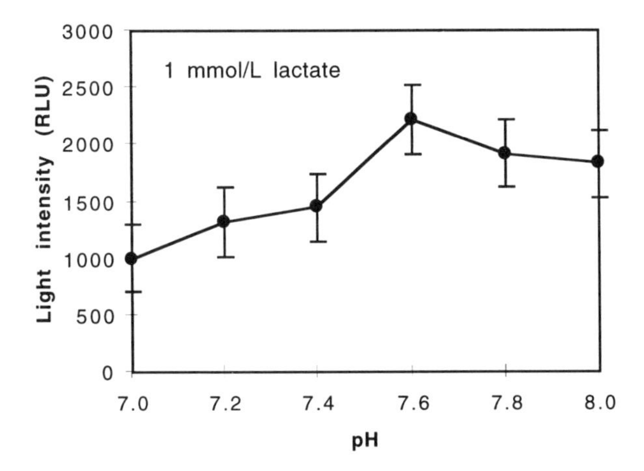

Figure 2. Light output as a function of pH in the lactate sensor. The reaction mixtures contain 1 mmol/L lactate, 0.25 mmol/L NAD, 10 μmol/L FMN, 0.001 % decanal, 50 μL (100 U) LDH, and 50 μL Sigma LF (6 mg/mL).

a pH of 7.4 for these reactions.

Low enzyme amounts result in a small dynamic range. Very high enzyme amounts are not economical for measuring low reactant concentrations. An optimum amount of enzyme for a given measurement range is important. Fig. 3 shows the effect of enzyme amount on light intensity. With 15 μL (75 mU) OR and 50 μL (100 U) LDH, saturation of light output occurred above 35 μL (0.26 mU) LF for 1 mmol/L lactate. The OR and LDH amounts used were experimentally determined from a set of optimizing experiments using 1

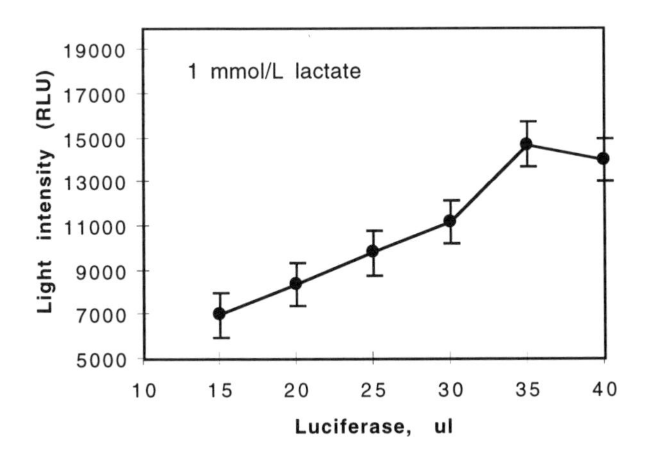

Figure 3. The effects of enzyme amounts on the light intensity in the bioluminescent lactate sensor. The reaction mixtures (pH 7.6) contain 1 mmol/L lactate, 0.25 mmol/L NAD, 10 μmol/L FMN, 0.001 % decanal, 50 μL (100 U) LDH, and 15 μL (75 mU) OR.

mmol/L lactate. With 75 mU OR and 0.26 mU LF, the light intensity did not change until 10 U LDH. Thus, the optimal ratio of LDH, OR, and LF was 10 U:75 mU:0.26 mU for 1 mmol/L lactate.

In summary, the inhibition effects of lactate, NAD, and pyruvate on the bioluminescent reactions were not critical. The optimum conditions for a lactate sensor based on bacterial bioluminescence (1 mmol/L lactate) were 10 U LDH, 75 mU OR, and 0.26 mU LF at pH 7.6.

This study is useful for the design of bioluminescent lactate sensors having different substrate ranges. The optimum enzyme amounts may depend on the lactate concentration to be measured.

Acknowledgements

We thank Prof. J. N. Herron for access to the photon counting spectrofluorometer. This work is partially supported by Protein Solutions, Inc., Salt Lake City, Utah via an STTR grant from the US National Science Foundation.

References

1. Blum LJ, Gautier SM, Coulet PR. Fiber-optic biosensors based on luminometric detection. In: Wagner G, Guilbault G, editors. Food biosensor analysis. New York: Marcel Dekker, Inc., 1994:101-21.
2. Palleschi G, Faridnia MH, Lubrano GJ, Guilbault GG. Determination of lactate in human saliva with an electrochemical enzyme probe. Anal Chim Acta 1991; 245:151-7.
3. Brolin S, Wettermark G, editors. Bioluminescence analysis. New York, VCH, 1992.
4. Prahl MS, Karp MT, Lovgren TN. Bioluminescent assay of lactate dehydrogenase and its isoenzyme-1 activity. J Applied Biochem 1984; 6:325-35.
5. Oda K, Yoshida S, Hirose S, Takeda T. Photon counting determination of ultratrace levels of nicotineamide adenine dinucleotide, reduced form (NADH) by use of immobilized luciferase. Chem Pharm Bull 1984; 32:185-92.
6. Tietz N, editor. Fundamentals of clinical chemistry. London, W. B. Saunders Co., 1987.

A METHOD FOR REAL-TIME DETECTION OF THE EFFECT OF DIFFERENT INHIBITORS ON REVERSE TRANSCRIPTASE ACTIVITY

S Karamohamed, M Ronaghi and P Nyrén
Dept. of Biochemistry and Biotechnology, The Royal Institute of Technology
S-100 44 Stockholm, Sweden

Introduction

Reverse transcriptase (RT) plays an essential role in the early steps of the replication cycle of retroviruses. The enzyme is responsible for the conversion of the viral genomic RNA into pro-viral DNA (1). RT has been the target of several anti-viral therapeutic agents used in treatment of different diseases (2, 3). 2'-Deoxyribonucleoside-5'-triphosphates (dNTP) modified at their 3'-hydroxyl position can act as terminators of DNA synthesis at an early stage of the viral replication cycle. 3'-Azido-2',3'-dideoxythymidine 5'-triphosphate (AZTTP), 2',3'-dideoxythymidine 5'-triphosphate (ddTTP), and other dideoxynucleoside interfere with the RT function (2, 3). Different assays have been developed, not only for the detection of retroviruses, but also for identifcation of new anti-retroviral substances (4-6). However, most of these methods are time-consuming and require electrophoretic steps. In the present study, we describe the use of a novel non-radioactive, non-electrophoretic and real-time approach for studies of the inhibitory effects of different nucleotides on RT activity.

Materials and Methods

Reagents: Bovine serum albumin; fraction V (BSA), adenosine 5'-phosphosulfate (APS), ATP sulfurylase (ATP: sulphate adenylyl transferase; EC 2.7.7.4), polyvinylpyrrolidone (PVP) (360,000), and dithiothreitol (DTT) (Sigma Chemical Co., St. Louis, MO, USA). Purified luciferase (EC 1.13.12.7), D-luciferin and L-luciferin (BioThema, Dalarö, Sweden). Avian myeloblastosis virus reverse transcriptase (AMV-RT), dNTP and 2',3'-dideoxnucleoside 5'-triphosphate (ddNTP) (Pharmacia Biotech, Uppsala, Sweden).

Oligonucleotides, DNA and RNA-templates: Poly(rA).p(dT)12-18 (A260 unit = 50 μg) (Pharmacia Biotech). E3PN (5'BGCTGGAATTCGTCAGACTGGCCGTCGTTT-TACAAC3'), E2PN (5'CGACGATCTGAGGTCATAGCTGTTTCCTGTGGAACTG-GCCGTCGTTTTACAACG3'), USPT (5'CGTTGTAAAACGACGGCCAGT3') and NUSP (5'GTAAAACGACGGCCAG3') were designed by us, synthesised, and purified by Pharmacia Biotech. The template (72 pmol E3PN) were hybridised to 80 pmol NUSP, in 20 mM Tris-HCl (pH 7.5), 8 mM $MgCl_2$ in a final volume of 32 μL. E3PN-NUSP was either used directly or immobilised (7) onto streptavidin-coated super paramagnetic beads, Dynabeads M280 (Dynal A. S., Oslo, Norway). The template (72 pmol E2PN) were hybridised to 80 pmol USPT as descibed above for E3PN.

Instrumentation: We used a LKB 1250 luminometer and a potentiometric recorder.

Real-time reverse transcriptase inhibition assay: The standard assay volume was 0.2 mL and contained the following components: 0.1 M Tris-acetate (pH 7.75), 2 mM EDTA, 10 mM magnesium acetate, 0.1% BSA, 1 mM DTT, 5 μM APS, 0.4 mg/mL PVP, 100 μg/mL D-luciferin, 4 μg/mL L-luciferin, 0.3 units/mL ATP sulfurylase and purified luciferase in an amount giving a response of 200 mV for 0.1 μM ATP. Either poly(rA).p(dT)12-18, E3PN-USP or E2PN-USPT was used as template. After addition of 10 nmol dTTP or 0.5 nmol dNTP, together with the desired inhibitor, the reaction was started by addition of AMV-RT.

ddGTP, ddCTP and ddATP as chain terminators: About 1 pmol template-primer (E2PN-USPT) was incubated with 13 U AMV-RT. The reaction was started by addition of the complementary base at a final concentration of 5 μM.

AZTTP as an inhibitor: When the effect of AZTTP on the AMV-RT was studied, 0.1 pmol poly(rA).p(dT)12-18 was pre-incubated with AZTTP in the presence of dTTP at a final concentration of 50 µM. The reaction was initiated by addition of 340 mU AMV-RT.

Solid-phase experiments: About 0.5 pmol of immobilised template-primer (E3PN-NUSP) was pre-incubated with AZTTP at final concentration of 3.5 µM. The reaction was initiated by addition of 5 U AMV-RT. After 10 min incubation at room temperature, the template-primer was washed. The AZTTP treated template was then used for a new primer extension experiment in the presence of 340 mU AMV-RT and 4 nmol dNTP.

Result and Discussion

The RT assay is based on the detection of the PPi formed in the RT-catalysed reaction by an enzymatic luminometric PPi detection assay (ELIDA) (8). In the ELIDA the PPi is converted to ATP by ATP sulfurylase and the ATP production is monitored by the firefly luciferase.

AZTTP as RT inhibitor: In Fig. 1, the rate of AMV-RT catalysed RNA-directed DNA-synthesis as a function of AZTTP concentration is shown. Seventy percent inhibition was obtained at a concentration of 1 µM AZTTP in the presence of 50 µM dTTP, and 95% at a concentration of 4 µM AZTTP.

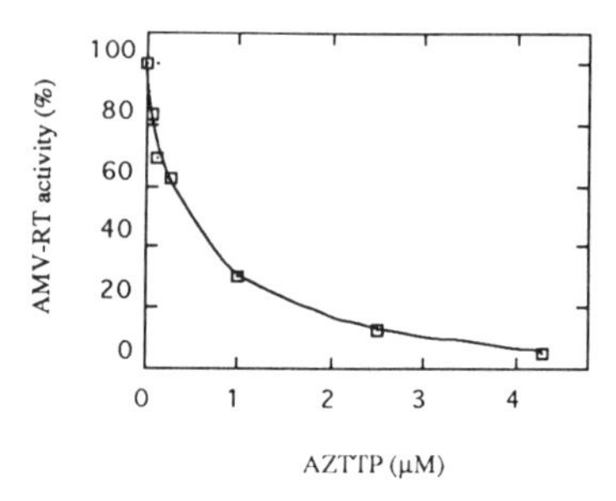

Fig. 1. Inhibition of AMV-RT activity by AZTTP. About 0.1 pmol poly(rA).p(dT) was incubated with 10 nmol dTTP and reactions were started by the addition of 340 mU AMV-RT. The PPi released were detected by the ELIDA.

ddGTP as chain terminating substance: In Fig. 2, a typical trace obtained from a ddGTP inhibition experiment is shown. After a calibration with PPi, the reaction was started by addition of dTTP. A signal corresponding to incorporation of one residue was detected. Then, dCTP was added and again a signal corresponding to incorporation of one residue was observed. The next base added was dATP. This time a signal corresponding to incorporation of five residues was noted (three A plus two C). Thereafter, ddGTP was added. This time the incorporation of one residue was noted. The following addition of dNTP did not result in any signal, which clearly shows that ddGTTP function as a chain terminator.

5' (19 bases)-GT
3' (19 bases)-CAAGTGTGTCCTTTGTCGATACTGGAGTCTAGCAGA 5'

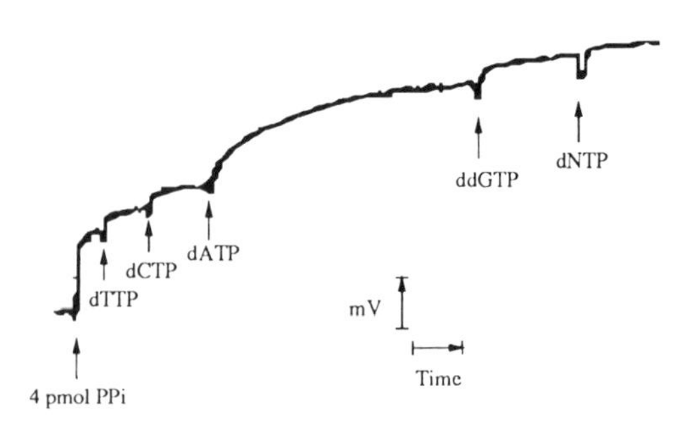

Fig. 2. Inhibition of AMV-RT activity by ddGTP. About 1 pmol of the template-primer (E2PN-USPT) was incubated with 13 U AMV-RT. The reaction was started by the addition of 4 nmol of the indicated nucleotides.

Effect of AZTTP as a chain terminator: AZTTP can inhibit AMV-RT either as a competitive inhibitor or as a chain terminator. To distinguish between these two mechanisms, the following experiment was performed (Fig. 3); biotinylated template-primer (E3PN-NUSP) was immobilised onto super paramagnetic beads.

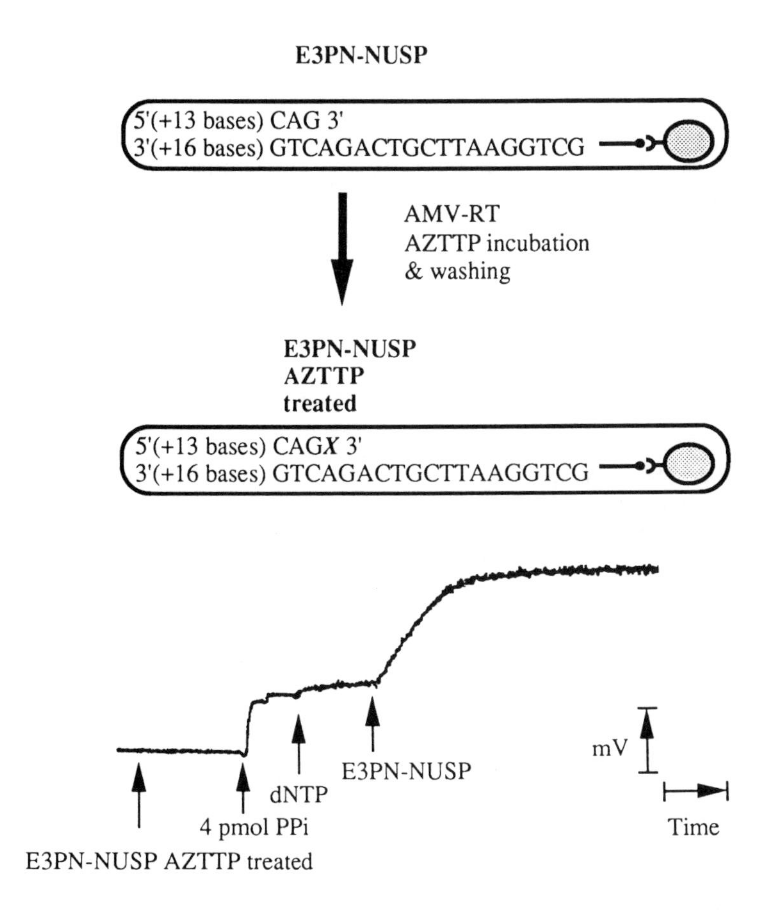

Fig. 3. Inhibition of AMV-RT activity by AZTTP. About 0.5 pmol immobilised E3PN-NUSP was pre-incubated for 10 min with 5 U AMV-RT and AZTTP at a final concentration of 3.5 µM. After washing the AZTTP treated template was incubated with 5 U AMV-RT in an new reaction mixture.The reaction was now started by addition of 4 nmol dNTP, followed by addition of 0.5 pmol fresh E3PN-NUSP. The PPi released were detected by the ELIDA. (***X*** = AZT monophosphate).

The template-primer was incubated with AZTTP in the presence of AMV-RT. After about 10 min incubation the beads were washed twice. To demonstrate the incorporation of AZTTP, the template-primer (E3PN-NUSP AZTTP treated) was incubated in another assay with dNTP in the presence of AMV-RT. No signal was observed, which shows that AZTTP has terminated all the available template-primers. As a control experiment an untreated template-primer was added. Now a signal corresponding to incorporation of 16 residues (full-extension of the primer) was detected. It can be noted that we were not able to detect base incorporation when AZTTP was added directly to the assay. The reason for the later observation might be the low incorporation rate for AZTTP.

Acknowledgements

The authors would like to thank Dr. Susan Cox and Dr. Sarah Palmer, the Swedish Institute for Infectious Disease Control, for the kind gift of AZTTP. This work was supported by grants from Magnus Bergvalls Stiftelse, Carl Tryggers Stiftelse för Vetenskaplig Forskning, Riksbankens Jubileumsfond, Procordia/Pharmacia AB Forsknings Stiftelse, and the Swedish Board for Technical Development.

References

1. Dvanov VA, Il'n YV. Reverse transcriptases and their biological role. J Mol Biol 1995;1:134-56.
2. Spence RA, Kait WM, Anderson KS, Johnson KA. Mechanism of inhibition of HIV-1 reverse transcriptase by nonnucleoside inhibitors. Science 1995;267: 988-93.
3. Arts EJ, Wainberg MA. Mechanisms of nucleoside analog antiviral activity and resistance during human immunodeficiency virus reverse transcription. Antimicro Agents Chemother 1996;527-40.
4. Pengsuparp T, Cai L, Constant H, Fong HHS, Lin L, Kinghorn AD et al. J Natural Products 1995; 58:1024-31.
5. Nakano T, Sano K, Odawara F, Saitoh Y, Otake T, Nakamura T et al. An improved non-radioisotopic reverse transcriptase assay and its evaluation. Japan J A Infect Diseases 1994;923-31.
6. Pyra H, Böni J, Schüpbach J. Ultrasensitive retrovirus detection by a reverse transcriptase assay based on product enhancement. Proc Natl Acad Sci 1994; 91: 1544-48.
7. Ståhl S, Hultman T, Olsson A, Moks T, Uhlén M. Solid phase DNA sequencing using the biotin-avidin system. Nucl Acids Res 1988;16:3025-3.
8. Nyrén P, Lundin A. Enzymatic method for continuous monitoring of inorganic pyrophosphate synthesis. Anal Biochem 1985;151:504-9.

DETERMINATION OF *RHIZOBIUM* NODULE OCCUPANCY BY LUMINESCENCE

Antonio J. Palomares, M. Enrique Vázquez and Angel Cebolla
Dpto. de Microbiología, Facultad de Farmacia, Universidad de Sevilla
41012 Sevilla, Spain

Introduction

Bioluminescence has been explored in reporter gene studies directly using genes that produce bioluminescent products, as the gene for firefly luciferase or indirectly via bioluminescent assays for the products of nonbioluminescent genes. The bioluminescent luciferase genes (*luc* and *lux*) have been the most popular and there are numerous publications detailing their aplication (1). A common technique used to monitor transcriptional activity is to fuse promoter sequences to reporter genes, such as *luc* genes. Such constructs allow for rapid, non destructive assays of transcriptional activation in bacteria which may be easily monitored by several procedures (2). The symbiotic interaction between bacteria of the genus *Rhizobium* and leguminous plants result in the formation of root nodules. These roots nodules represent highly organized structures in which nitrogen is converted to amonium by the bacteria allowing the plant to grow without an external supply of reduced nitrogen. The formation of a root nodule involves several stages determined by different sets of genes both in the micro and macrosymbiont. Concomitant with infection thread formation, mitosis is induced in the root cortex, leading to the formation of nodule primordia towars which infection threads grow. After the release of bacteria into the plant cells, they differentiate into N_2 fixing bacteroids while the primordium differentiates into a root nodule (3, 4). Quantification of nodule occupancy is required to determine the competitiveness of two or more strains. The ability of a strain to successfully produce and occupy nodules in the presence of one or more rival strains is defined as competitiveness. The ability to introduce a bioluminescent phenotype into bacteria that interact with plants provides a simple and highly sensitive system for monitoring the bacterial populations. In many conditions, light production from bioluminescent strains is linear with cell concentration. Therefore, is this also occurred with bacteria in nodule, the estimation of nodule occupanccy may be possible by measuring luciferase activity emited by symbiotic cells. This possibility has been positively tested in this work providing a new simple and rapid maner to measure competition for nodulation.

Materials and Methods

Bacterial strains and culture media. *Rhizobium meliloti* 1021 is a wild type, streptomycin and nalidixic acid resistants, Nod^+ and Fix^+ in alfalfa. *R. meliloti* CR201 is *R. meliloti* 1021 with a miniTn*5::luc* insertion in the chromosome (5). TY (yeast extract, 3g/L; tryptone, 5 g/L; $CaCl_2$, 0.84 g/L) medium was used.

Light emission measurements. For determination of the luciferase activity of intact cells in liquid cultures, 50µL of a cell suspension was placed in a tube, and 0.15 mL of 1mM luciferin-100 mM citric acid (pH5.0) was added. Alternatively, for measurement of the luciferase activity in cell extracts, 0.9 mL of the cell suspension was mixed with 0.1 mL of buffer to yield 0.1M potassium phosphate (pH8.0)-2 mM EDTA- 1mg of bovine serum albumin (BSA) per ml-5%glycerol (final concentration). The mixture was sonicated twice on ice for 30 s each time; 0.15 mL of 25 mM glycylglycine (pH 7.8)-10 mM $MgCl_2$-5 mM ATP-0.1 mM D-luciferin was then added to 50 µL of a sonicated sample. The specific enzymatic activity was reported as the peak height, in relative light units (RLU), relative to the cell mass estimated by measurements of the optical density at 600 nm of the culture. Light emission from bacterial colonies containing *luc* genes was detected either by plating cells onto a nitrocellulose filter or by blotting colonies onto filter paper. Filters were moistened with 500 to 700 μL of substrate solution (1 mM D-luciferin-0.1 M sodium citrate (pH5.0)). After difussion for a few minutes, light-emitting colonies were detected in the dark by dark adaptes eyes or photographically either by contact with Kodak OG-1 X-ray film or by reflex camera with Kodak Gold 400 ASA color film (2).

Plant assays. Alfalfa seeds (*Medicago sativa*) were surface sterilized and planted as described previously (6). Twelve seeds were planted in each assembled apparatus and 1 mL of *R. meliloti* culture, that had been grown to early stationary phase, was added to each. Plants were grown on nitrogen free plant medium. Nodules and bacteroids four weeks old were assayed for luciferase activity. As with the vegetative cells, a film assay of luciferase activity in the nodules was performed by placing the whole plant with the nodules over Kodak film. The film was exposed to the light produced by the luciferase reaction in the nodules for 20 minutes. Luciferase activity in bacteroids was measured after purification of the bacteroids by differential centrifugation. Nodules from three bottles of plants were picked and stored on ice. 2mL of cold 0.25 M mannitol /0.05 M Tris pH 7.8 was then added and the nodules were crushed with a glass rod. Bacteroids were pelleted by centrifugation and resuspended in 1mL of sonication buffer. Bacteroids luciferase activity was assayed in the same way as described for the bacterial assay (7).

Results and Discussion

The *R. meliloti* genetic tag has shown not to be affected either growth rate or the survival in sterile soil respect to the wild type strain. To test whether the insertion of the *luc* transposon had interrupted some gene involved in the symbiotic interaction, plant test with the marked strain were accomplished.

CR201:1021 (ratio)	Percentage of luminescent colonies isolated from nodules (%)
1 : 0	100
0.8 : 0.2	79.6
0.5 : 0.5	50.9
0.2 : 0.8	22.2
0 : 1	0

Table 1. Results of the plant test to assess the ability of the tagged *luc*-strain *Rm*CR201 to occupy alfalfa nodules versus the wild type *Rm*1021

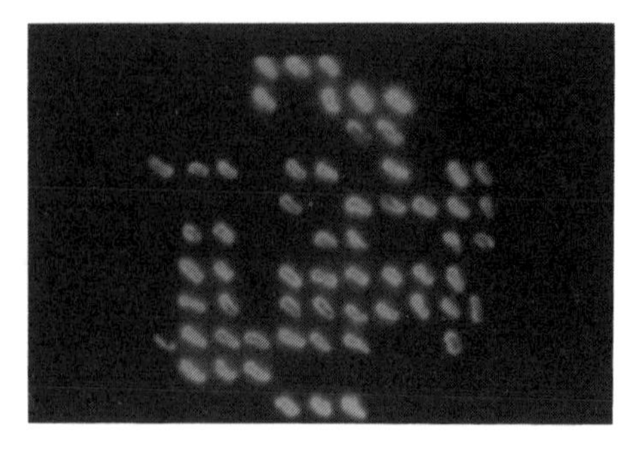

Figure 1. Proportion determination of nodule occupancy. In a mixture 1:1, from 108 colonies tested, 60 were luminescent.

Four weeks after inoculation, alfalfa plant inoculated with CR201 showed nodules similar to those originated by *R. meliloti* 1021. No differences were found in the size or aspect of plant inoculated by each strain. Therefore, the genes needed for symbiotic nitrogen fixation seemed to be intact in CR201. Due to a potential extra metabolic charge, the constitutive expression of the *luc* gene might cause a significant effect when competition with unmarked strain occurred. To test this possibility, alfalfa plants were mix-inoculated by suspension with different known proportions of CR201 and 1021(Table 1). The proportion of luminescent bacteria in each sample was measured by counting luminescent colonies. The percentage of luc^+ colonies coincided with the proportion of CR201 in the inocula. Therefore, no differences in the ability of the marked strain to develop effective nodules were observed. To achieve the experiment described above, bacterial cells were isolated from nodules and after letting to grow three days on solid medium to obtain visible colonies, these were tested for their ability to emit light (Figure 1). We have previously shown that concentration of free-living CR201 cells had a linear relationship with luciferase activity. In our experimental condition, the relationship between light emission and concentration of luciferase showed a linear relationship for at leats four magnitude orders (Figure 2).

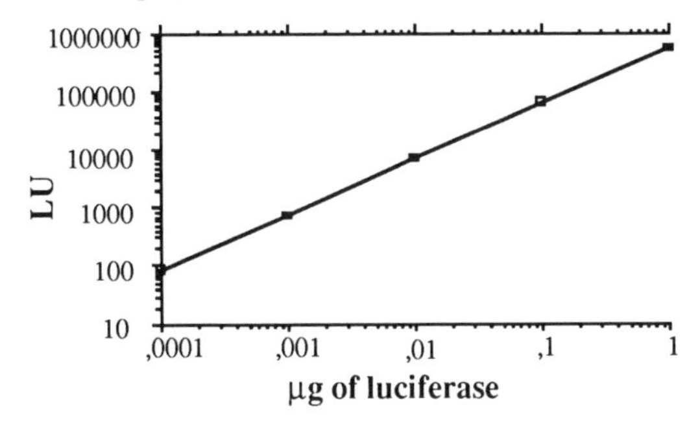

Figure 2. Response of the Luc assay to amount of luciferase. The purified enzyme and substrate were provide by Boehringer Mannheim.

Since the expression of luciferase in CR201 is constitutive, it is expected that the amount of reporter enzyme per cell does not differ much with the number of cells. Therefore, providing the same amount of bacterial mass, the luciferase activity from the bacteroid suspensions may be proportional to the percentage of *luc*-tagged strain in the nodules. We tested whether the luciferase activity of bacteroids suspensions could give an estimation of the relative occupation of nodules by luminescent bacterium. A linear relationship between the light emission and the proportion of luminescent viable-cells could be registered (Figure 3). Since there is no luciferase activity from suspensions of nodules without *luc*-strain, only the light units from the suspension with 100% of tagged bacteroids should be required to elaborate the linear function. Direct measurements of luminescence of cells isolated from nodules provides many advantages over other methods to estimate nodule occupancy.Since it did not require plating, it is less laborious and less time-consuming than any marker phenotypes that are only shown in colonies. A *nolC-lacZ* reporter has also been described for direct assessment of competition for nodulation (9). In that work, nodules were stained histochemically with X-gal counting how many nodules developed blue color. Our methods is simpler and has a better statistic worth because the samples are mixtures of bacteroids from as many nodules as wanted.

The existence of more that a strain per nodule is expected not to affect the estimation,

negative falses in the colorimetric assay.

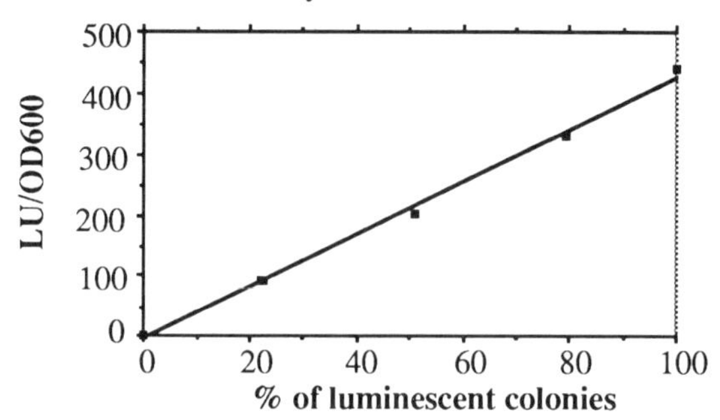

Figure 3. Luminescence from the *R. meliloti* bacteroid suspension prepared from nodules. Data are the means of two independent experiments in duplicate, except the data corresponding to the 100% of CR201 which was made twice more (r=0.998).

One disappointed result in our study was that the sensitivity of the assay was lower than our expectation. This observation might be due to higher stability of the reporter enzyme in free-living bacteria than in bacteroid, although differences in gene expression should not be rule out. In any case, under our experimental conditions less than 5% of CR201 in nodules may be assessed. Although one main aspect of the assay (to measure an enzymatic reaction to estimate relative mass of a particular bacteroid) could be accomplished by cells expressing other reporter genes (i.e., β-glucuronidase or β-galactosidase), neither of other systems provided the sensitivity nor simplycity of the bioluminescent assay.

Acknowledgements

This work was founded by grants BIO90-0521 and BIO93-0427.

References.

1. Campbell AK, Kricka LJ, Stanley PE, editors. Bioluminescence and Chemiluminescence: Fundamentals and Applied Aspects. Chichester, England: John Wiley & Sons, 1994.
2. Cebolla A, Vazquez ME, Palomares AJ. Expression vectors for the eukaryotic luciferases as bacterial markers with different colors of luminescence. Appl Environ Microbiol 1995; 61:660-8.
3. Hirsch AM. Developmental biology of legume nodulation. New Phytologist 1992;122:211-37.
4. Schultze M, Kondorosi E, Ratet P, Buire M, Kondorosi A. Cell and molecular biology of *Rhizobium*-plant interactions. International Review of Cytology 1994; 156:1-75.
5. Cebolla A, Ruiz F, Palomares AJ. Stable tagging of *Rhizobium meliloti* with the firefly luciferase gene for environmental monitoring. Appl Environ Microbiol 1993; 59:2511-9.
6. Palomares AJ, DeLuca M, Helinski DR. Firefly luciferase as a reporter enzyme for measuring gene expression in vegetative and symbiotic *Rhizobium meliloti* and other gram-negative bacteria. Gene 1989; 81:55-64.
7. Cebolla A, Ruiz F, Palomares AJ. Expression and quantification of firefly luciferase under control of *Rhizobium meliloti* symbiotic promoters. J Biolumin Chemilumin 1991; 6:177-84.
8. Krishnan HB, Pueppke SG. A *nolC-lacZ* fusion in *Rhizobium fredii* facilitates direct assessment of competition for nodulation of soybean. Can J Microbiol 1992; 38:515- 9.

EUKARYOTIC LUCIFERASE MARKERS AS MONITORING METHODS FOR PLASMID TRANSFER ANALYSIS IN SOIL ENVIRONMENT

M. Enrique Vázquez and Antonio J. Palomares
Dpto. de Microbiología , Facultad de Farmacia, Universidad de Sevilla, 41012 Sevilla, Spain

Introduction

The deliberate release of organisms may have unpredictable consequences for the environment, and extensive scientific research studying microbial interaction and specially the fate of introduced strains and their recombinant DNA is necessary. Most of the experiments aimed at assessing genetic interactions in soil have focused on the detection of conjugal transfer, which is regarded as the major process responsible for gene transfer. Concern has been expressed not only about the survival of the GMMs in the environment but also about the dissemination of their genetically engineered DNA sequences. Indigenous microorganisms acting as recipient strains for the genetically engineered DNA sequences acquire new information, and the acquisition of plasmids sometimes seems to cause physiological alterations.The aim of this study has been to establish a model for the plasmid transfer analysis. We have developed constructions with the lambda promoter, λP_R, and the luciferase genes *luc* and *lucOR*. This promoter is regulated by the thermosensible repressor *c*I857 to 28°C. When the temperature is increased (about 42°C), the repressor protein is inactived and the promoter, λP_R, is induced. We have carried out a fusion system with luciferase genes in our laboratory: *c*I857-λP_R::*luc* and *c*I857-λP_R::*lucOR*, which proved that was inducible by temperature in *E. coli*, with the peculiarity that the repressor was not functional in others gram negative bacteria, i.e., *Pseudomonas putida*, *Rhizobium meliloti*, *Agrobacterium tumefaciens, Alcaligenes faecalis*, etc., so, the expression was constitutive to 28°C and we could detect light either by *luc* or *lucOR* genes when the plasmids were mobilized to others backgrounds. Due to this fact, it is possible to use the colors of both luciferase genes (green and orange) to realize transfer experiments. We have studied the transfer frequency from RK2 derivative plasmids, incompatibility group incP (pACR4 and pACR18) and from R388 derivative plasmids, incompatibility group incW (pET2), either *in vitro* in the laboratory or microcosms of sterile and nonsterile soil.

Materials and Methods

Instrumentation. Photographics methods either by contact with Kodak OG-1 X-ray film or by reflex camera with Kodak Gold 400 ASA color film were used.

Bacterial strains, plasmid, media and antibiotic selection. The donor organisms used for different plasmids were; *E. coli* S17-1 for non-conjugative plasmids pACR4 and pACR18 (RK2 derivatives), *E. coli* CC118 to carry the helper plasmid pRK600 in triparentals matings and *E. coli* CMN for conjugative plasmid pET2 (R388 derivative). The recipient were *Pseudomonas putida* 2440 or indigenous bacteria. The donors, recipients and helpers were growth in standard LB medium. The antibiotics: tetracycline, chloramphenicol, kanamycin, nalidixic acid and cycloheximide, to avoid the fungus growth, were used to standard concentrations. The cells were washed and diluted with saline solution (NaCl 0.85%) to use in mating experiments.

Matings on filters in solid agar media. The donor organisms were grown in LB broth until late logaritmic phase at 28°C. Recipient organism was grown in LB for 16-18 h at 28°C. Donors and recipients (1:3) were combined in a mating patch on LB agar and incubated for 14-16 h at 28°C. The corresponding dilutions were plated in a selective

medium. The transconjugants were selected by light detection at 28°C for *luc* and *lucOR* gene expression once were transfered to *P. putida* 2440.
Soil used. Gathered fresh soil was kept at 4°C in closed plastic bags. Before using it, was sieved in a 1 mm poro size strainer. Soil used in steril microcosms, was sterilized for 1 h at 120°C during 3 consecutive days. All the experiments were carried out to moisture content at field capacity of 15% for dry soil. Cells were introduced in soil at 10^8-10^9 CFU/g dry soil.
Mating in sterile soil. Microcosms consisted of 50 g of sandy loam soils in sterile plastic bags. Culture of 1 mL of donor and recipient organisms were washed 3 times with NaCl 0.85% and were resuspended on 0.5 mL of the same solution. The receptor cells was added to the soil surface and was shaked vigorously to get a concentration of 10^9 CFU/g dry soil. After 30 min, donor cells were added in the same way. Samples of 1 g were taken and added to 9 mL of NaCl 0.85%, shaked in a vortex, diluted and plated in differents selective media. Errors in overstimations of transfer frequency due to plate mating were eliminate because we used nalidixic acid to select the transconjugants (1). All the possibles transconjugants found in selective medium were proved in a real transference, by test of light expression, through X-ray films or reflex camera with Kodak film color 400 ASA.
Mating in nonsterile soil. The competence effect of indigenous bacteria were evaluated by the repetition of assays described above in nonsterile soil. The microcosms were incubated to 28°C for 48 h to stimulate the metabolic activity of indigenous organisms. These incubated microcosms were inoculated with donors and recipients, as we described for an sterile soil microcosms. For transfer analysis to indigenous bacteria, we used the same procedure as in sterile soil microcosms, inoculating solely donor microorganism, with a cells concentration approximately of 10^9 CFU/g dry soil.
Transconjugant analysis. Plasmid transfer to recipient strains were proved by the comprobation antibiotic resistance, growing them in a selective medium or through the ligh emission. Light emission in colonies that contains *luc* and *lucOR* genes was detected by blotting colonies onto filter paper. Filters were moistened with 400 µL of substrate solution (1 mM D-luciferin in 0.1 M sodium citrate [pH 5]). After a difussion period of a few minutes, light-emitting colonies were detected photographically with Kodak OG-1 X-ray films (2) or by reflex camera with Kodak Gold 400 ASA color film, through 30 min to 1h exposition (3).

Table 1. Transfer Frequencies.

Donor	Helper	Receptor	Transfer frequency [a]		
			in vitro	steril soil[b]	nonsterile soil[b]
CC118 (pACR4)	pRK600	*P. putida* 2440	$5x10^{-2}$	—	—
CC118 (pACR18)	pRK600	*P. putida* 2440	$3,5x10^{-2}$	—	—
CMN (pET2)		*P. putida* 2440	$2x10^{-3}$	$1,2x10^{-4}$	$1,6x10^{-6}$
S.17-1 (pACR4)		*P. putida* 2440	$4,5x10^{-2}$	$5x10^{-5}$	$3,1x10^{-7}$
S.17-1 (pACR18)		*P. putida* 2440	$4,1x10^{-2}$	$2,5x10^{-5}$	$5x10^{-7}$
CMN (pET2)		Indigenous	—	—	$3,2x10^{-7}$
S.17-1 (pACR4)		Indigenous	—	—	$1x10^{-6}$
S.17-1 (pACR18)		Indigenous	—	—	$1x10^{-6}$

[a] Transfer Frequency (ratio: Transconjugants number/Receptors total number)
[b] Highest value obtained.

Results.

Gene transfer on plates. All conjugations were done to calculate the transfer frequency from different plasmids with *luc* and *lucOR* fusions (Table 1). Triparental matings were tested with pACR4 and pACR18 plasmids (Tra^-, Mob^+) using a mobilizing or helper plasmid, pRK600. Direct mobilization from RK2 plasmids were carried out using *E. coli* S17-1 as donor strain that bear *tra* functions from RP4 in its chromosome. Conjugation of pET2 plasmid was carried out directly without using helper or donor strain from *E. coli* S17-1, because it's a conjugative plasmid (Tra^+, Mob^+).

Gene transfer in sterile soil microcosm. All the samples were incubated to 28°C after the inoculation of parental cells. When the gene transfer assays were realized in sterile soil, plasmid transfer was quite lower than in experiments *in vitro* (Tabla 1). The transconjugants increased up 10^3 CFU/g dry soil in the plasmid groups belonging to IncP between 24 h and 48 h. Transfer frequency was $1/10^5$ recipient cells. In the case of pET2 plasmid, belonging to IncW group, the transconjugants were 10^4 CFU/g dry soil, while the transfer frequency was of $1/10^4$ recipient cells (Figure 1). The transconjugants identification was done as we described above in Transconjugant analysis.

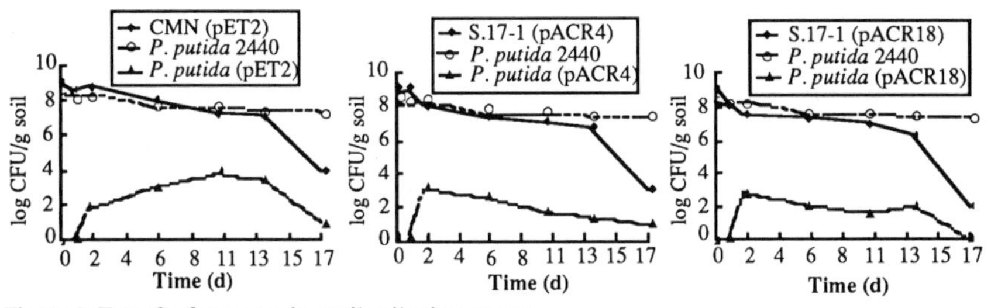

Figure 1. Transfer frequency in steril soil microcosms.

Gene transfer in nonsterile soil microcosms. These microcosms were inoculated and incubated with the same conditions than sterile soil microcosms. The transfer frequencies were in order to $1/10^7$ recipient cells in RK2 derivatives and $1/10^6$ recipient cells in R388 derivative (Table 1). The assays were repited adding nutrients to the soil. Under these conditions we could detect higher levels of transconjugants. The transconjugants were detected to 3 days from inoculation when we did not add nutrients to the soil (Figure 2).We could detect gene transfer to indigenous organisms with this methods based in *luc* and *lucOR* fusions. The experiments were done in the same way that previous assays. The nonsterile soil microcosms only were inoculated with donor organisms and we approximately detected transfer frequencies in order to $1/10^3$ recipient cells between the first and third day after inoculate the soils.

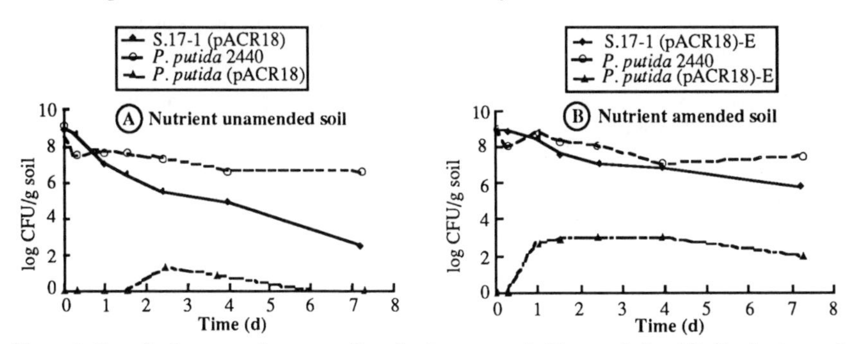

Figure 2. Transfer frequency in nonesterile soil microcosms. **A.** Unamended and **B.** Nutrient amended

Discussion.

We have constructed differents gene fusions with the regulated system from lambda promoter, λP_R, and thermosensible repressor *c*I857 to the *luc* and *lucOR* genes to use as marker genes. This system allowed us to identify the mobilization from different plasmids (RK2 and R388 derivatives), to recipient organisms. We comprobed a high transfer frequency in both plasmid groups in assays on solid agar media. Sterile soil does not simulate the normal soil environment but represents, however, a compromise between strict laboratory conditions and in situ experiments (4). We found in sterile soil microcosms a scarce survival in donor and recipient organisms. Lack of nutrients to be the main cause of the quick die-off of *E. coli* (5). The gene transfer in sterile soil was quite lower than on filters in plate (Table 1), perhaps due to physical separation between donor and recipient cells by the soil particles or because the parental cells had less growth activity in soil. Although useful, sterile soil studies do not evaluate true gene transfer rates since they do not take into account the influence of biotic stress. The influence of this stress caused that parental cell populations, above all the donor organisms, declined more quickly, approximately from third day. The poor survival in comparison with sterile soils is most probably caused by antagonism and competition with the indigenous microorganisms, by parasitism, and by predation by nematodes and soil protozoa (6). This poor survival of donors caused lowest detection levels of gene transfer in RK2 and R388 derivatives. The detected transconjugants number increased when the experiment was repeated, giving the introduced donor a competitive advantage by the addition of easily metabolizable nutrients, as LB broth. The dates showed that transference from pET2, pACR4 and pACR18 occur under ideal conditions of growth in nonesterile soils but the transfer frequency was very lower than transfer frequency in sterile soil and solid agar media. Most studies on conjugal plasmid transfer in soil have been based on the use of cointroduced donor and recipient cell populations (7). Transfer to indigenous soil bacteria has rarely been studied, and the few studies that have been described have been hindered by the lack of an adequate donor counterselection technique, thus increasing the detection limit. In our studies get to detect transference to indigenous microorganisms by the method developed in our laboratory based in bioluminiscence.

Acknowledgements

This work was founded by the Fundación Ramón Areces and grant BIO93-0427.

References.

1. Smit E, van Elsas JD. Determination of plasmid transfer frequency in soil: Consequences of bacterial mating on selective agar media. Curr Microbiol 1990;21:151-5.
2. Wood KV, DeLuca M. Photographic detection of luminiscence in *E. coli* containing the gene for firefly luciferase. Anal Biochem 1987;161:501-7.
3. Cebolla A, Vázquez E, Palomares AJ. Expression vectors for the use of eukaryotic luciferases as bacterial markers with different colors of luminescence. Appl Environ Microbiol 1995;61:660-8.
4. Trevors JT, Barkay T, Bourquin AW. Gene transfer among bacteria in soil and aquatic environments: a review. Can J Microbiol 1987;33:191-6.
5. Klein DA, Casida jr LE. *E. coli* die out from normal soil as related to nutrient availability and the indigenous microflora. Can J Microbiol 1967;13:1461-9.
6. Alexander M. Introduction to soil microbiology. New York: John Wiley & Sons, Inc. 1977.
7. Top E, Mergeay M, Springael D, Verstraete W. Gene escape model: Transfer of heavy-metal resistance genes from *Escherichia coli* to *Alcaligenes eutrophus* on agar plates and in soil samples. Appl Environ Microbiol 1990;56:2471-9.

ABOUT THE EFFECTS OF NUTRIENTS ON THE LUMINESCENT BACTERIA TEST

E Grabert[2], F Kössler[1]

[1]Federal Institute for Occupational Safety and Health, Dep. Occupational Medicine, Nöldnerstr. 40/42, D-10317 Berlin, Germany

[2]Dr. Bruno Lange GmbH Berlin, Willstätterstr. 11, D-40549 Düsseldorf, Germany

Introduction

The luminescent bacteria test was described in its current form with preserved luminescent bacteria as long ago as 1969 (1). When the test was standardized in Germany in the late 1980s the following observation was put forward as an objection to the suitability of the luminescent bacteria test as a method for detecting pollutants: If the nutrient medium (Yeast extract/peptone medium) in which the luminescent bacteria usually grow is used as a sample in the luminescent bacteria test, it is found that bioluminescence is inhibited and therefore toxicity is present. This cannot be the case, however, because the bacteria multiply rapidly in this medium. This observation was taken into account in the German standard DIN 38412 L34, L341 under "interferences" in the following sentence: "A pollutant independent reduction in bioluminescence is caused by non-degradable nutrients (e.G. urea, protein hydrolysate, yeast extract)" (2). In the international standard ISO CD 11348 the following sentence is included under "interferences": "An organic contamination of the sample by readily biodegradable nutrients (e.g. urea, peptone, yeast extract about ≥ 100 mg/L) may cause a pollutant-independent reduction in bioluminescence" (3). The fact that the authors are not aware of any scientific papers which substantiate the threshold values referred to in the DIN and ISO standards made them decide to test the bioluminescence-inhibiting properties of the above mentioned substances.

Materials and methods

The nutrients glycerol, glucose, lactose, yeast extract, malt extract, peptone and tryptone (protein hydrolysates) were tested in original concentrations of 150 and 8000 mg/L, together with the nutrient medium described in DIN 38412 L341 paragraph 6.1, and synthetic waste water as described in DIN 38 412 T24.

The luminescent bacteria test was carried out with liquid dried luminescent bacteria in accordance with DIN 38412 L341 paragraph 7.5.2. The LUMIStox measuring system of Dr. Lange GmbH Berlin, Germany, was used.

The Chemical Oxygen Demand (COD) was determined with the Dr. Lange cuvette Test.

In addition all samples were tested for long-term toxicity by means of the Dr. Lange LUMIS•24•tox test. This new test procedure determines the influence of samples on the bioluminescence and growth of luminescent bacteria (4).

Results and discussion

Because neither DIN 38412 L341 nor ISO CD 11348 defines the term "interference", the following model was used. A result of 0 to 10 % inhibition of bioluminescence in the luminescent bacteria test with nutrients as the sample is characterized as "no effect". If a nutrient causes more than 10 % inhibition of bioluminescence in a concentration at which the luminescent bacteria still multiply in the long-term test (4), this is characterized as interference. None of the tested nutrients was found to inhibit growth at concentrations tested (Tables 1 and 2). It can therefore be assumed that none of the tested nutrients has a toxic effect on the luminescent bacteria.

Nutrient	Test concentration. mg/L	% Inhibition of bioluminescence	Growth
Glycerol	75	0	+
Glucose	75	0	+
Lactose	75	0	+
Yeast extract	75	6	+
Malt extract	75	3	+
Peptone	75	5	+
Tryptone	75	4	+
Urea	75	2	+
Synthetic waste water	50 % (v/v)	16	+

Table 1: Influence of nutrients on bioluminescence in the luminescent bacteria test in accordance with DIN 38412 L341 and on growth (- = inhibition, + = growth) in the LUMIS•24•tox, Dr. Lange Berlin, Germany. The stock solutions were stored at 6°C for 24h at least.

The nutrients referred to in the DIN standard were first prepared in concentrations of 150 mg/L and subjected to the luminescent bacteria test to determine the degree of inhibition they cause. At the resulting test concentration of 75 mg/L the nutrients cause no interference in the luminescent bacteria test in accordance with DIN 38412 L341.

The individual nutrients where then tested to determine the lowest concentration at which interference occurs. A test concentration of 4000 mg/L was not exceeded (Table 2). The COD of the corresponding nutrient dilution was also determined. The fact that synthetic waste water caused 16 % inhibition of bioluminescence attracted attention. Beef extract was identified as the cause. Beef extract is a significant component of synthetic waste water that conforms to DIN 38412 T24 (110 mg/L). Two different beef extracts from different manufacturers were tested. In both cases a concentration of 100 mg/L beef extract was found to cause approximately 15 % inhibition. The LUMIS•24•tox test, Dr. Lange Berlin, Germany, failed to detect any inhibition of growth. It is not yet clear why beef extract should cause inhibition of bioluminescence and further research is necessary. The lowest test concentration of a nutrient to cause interference was 333 mg/L; this was malt extract. Because the test involves mixing equal volumes of luminescent bacteria suspension and sample, this corresponds to an original malt extract concentration of 666 mg/L. The COD at this concentration was 720 mg/L.

Nutrient	Test concentration mg/L	Test COD mg/L	% Inhibition of bioluminescence	Growth
Glycerol	4000	4130	8	+
Glucose	4000	4522	2	+
Lactose	4000	4443	3	+
Yeast extract	500	563	10	++
Malt extract	333	360	10	+
Peptone	1000	1307	9	++
Tryptone	1333	1787	9	++
Urea	4000	--	7	+
Nutrient medium	8,33 (%v/v)	859	10	++

Table 2: Test concentrations of nutrients at which 10 % or less inhibition of bioluminescence was determined in the luminescent bacteria test in accordance with DIN 38412 L341. The growth of the luminescent bacteria was determined with the Dr. Lange LUMIS•24•tox test (- = inhibition, + = growth, ++ = very considerable growth). The nutrient solutions were left to stand for 24 hours at 6 °C before being tested. COD = Chemical Oxygen Demand

Nutrient	Test concentration mg/L	% Inhibition of bioluminescence (fresh solution)	% Inhibition of bioluminescence (old solution)
Peptone	75	16	5
Tryptone	75	11	4

Table 3: Effect of ageing on amino acid hydrolysates. The inhibition properties of the amino acid hydrolysates were determined by carrying out the luminescent bacteria test in accordance with DIN 38412 L341 immediately after they were prepared (fresh solution) and after they had been left to stand for 24 hours in a refrigerator (approx. 6 °C) (old solution).

The study detected an ageing effect in protein hydrolysates. Freshly prepared peptone and tryptone solutions caused significantly higher levels of inhibition than solutions that had been left to stand in a refrigerator. The inhibition caused by the freshly prepared solutions decreased appreciably after only 24 hours (Table 3). The causes of the ageing effect are still unclear. It may be that there were substances in the industrial extracts which, in incompletely hydrated form, had an inhibitory effect on the bioluminescence.

This ageing effect may play a part in the different assessments of the effects of nutrients in the luminescent bacteria test.

An inhibitory effect of nutrients in relatively small concentrations on bioluminescence cannot be confirmed on the basis of this study. Pure nutrients such as glycerol, glucose, lactose and urea caused no notable inhibitions, even in high concentrations. Inhibitions at low concentrations was caused by industrial mixtures whose composition was not defined and not known in detail. It must however be emphasized that these inhibition effects first appear in conjunction with very high organic loads.

If these results are applied to the field of application of the luminescent bacteria test, it must be concluded that the test can be used in the waste water sector not only for the sewage treatment plant outflows but also for most inflows: In view of the fact that the outflows of 98 % of municipal sewage treatment plants in the Federal Republic of Germany have a COD of less than 120 mg/L, interference is unlikely. Some 2 % of the plants have more highly concentrated outflows, but scarcely of the order of magnitude of 700 mg/L. Even if the COD in sewage treatment plant outflows was only caused by nutrient mixtures there would be no interference; the concentrations are simply too low.

Sewage treatment plant inflows frequently have a COD of more than 700 mg/L. However, only part of the COD of the inflow is attributable to nutrients. In general a BOD (biological oxygen demand) fraction of 1/2 to 2/3 of the COD is assumed. The BOD represents the nutrient fraction. On this basis the original COD of an inflow would have to be at least 1,000 to 1,400 mg/L before there would be any incipient risk of nutrient interference in the luminescent bacteria test. This guideline calculation relates solely to a dilution factor G = 2 (undiluted sample in the test). If a dilution factor G = 64 is determined in the inflow, the original COD would have to be considerably higher than 10,000 mg/L before there would be any risk of nutrient-dependent interference in the luminescent bacteria test.

References

1. Kössler, F. Physiologische Studien zur Biolumineszenz, Untersuchungen über Wachstum, Atmung und Biolumineszenz von Leuchtbakterien unter dem Einfluß chemischer und physikalischer Faktoren. Habilitationsschrift, Humboldt Universität Berlin, Germany, 1969.
2. DIN 38412 German Standard methods for the examination of water, waste water and sludge: bioassays (Group L): determination of the inhibitory effect of waste water on the light emission of *Photobacterium phosphoreum*, luminescent bacteria waste water test (L34 and L341) Normenausschuß Wasserwesen im DIN Deutsches Institut für Normung e.V., Berlin, Germany, 1993.
3. ISO/CD 11348 Water quality - Determination of the inhibitory effect of water samples on the light emission of *Vibrio fischeri* (Luminescent bacteria test) Document ISO/TC 147/SC 5/WG 1 N146, 1994
4. Zieseniss K, Grabert E. A novel method for determining chronic toxicity with the LUMIStox luminescent bacteria test. In: Campbell AK, Kricka LJ and Stanley PE, editors. Bioluminescence and Chemiluminescence, Fundamentals and Applied Aspects. Chichester: Johne Wiley & Sons Ltd, 1994: 76-8

PART 8

Luminescence in Medicine and Disease, Clinical Chemistry and Microbiology

APPLICATION OF FLUORESCENCE, BIOLUMINESCENCE, AND CHEMILUMINESCENCE TECHNOLOGIES TO ANTIMYCOBACTERIAL DRUG SUSCEPTIBILITY TESTING

RC Cooksey
Division of AIDS, STD, and TB Laboratory Research,
National Center for Infectious Diseases, Centers for Disease Control and Prevention,
Atlanta, Georgia 30333 USA

Introduction

Effective management of infectious diseases depends upon expeditious isolation, identification, and antimicrobial susceptibility testing of the etiologic agent. These tasks become greater challenges for fastidious organisms and those with slow growth rates, such as some clinically important *Mycobacterium* species. Another key component of infection control which is also negatively impacted by long generation times is typing microbial strains to evaluate their relatedness. Standard in vitro antimicrobial susceptibility testing methods are generally performed by inoculating organisms from pure culture either to artificial media containing antimicrobial agents or to solid agar onto which papers disks impregnated with antimicrobics are placed after the surface is seeded. After a lawn of growth appears for the latter method, zones of inhibition surrounding the antimicrobic disks are measured and correlated with susceptible, intermediate, or resistant categories which have been determined to correlate with expected therapeutic outcomes (1). The former method is the standard for slowly growing mycobacteria and is referred to as the agar proportion method since the proportion of individual colony forming units (cfu's) that are resistant to an incorporated drug concentration may be compared to the total numbers of cfu's on drug-free control media (2). When this proportion exceeds 1% the mycobacterial isolate is considered resistant to the critical concentration. This quantitation of resistant mutants is nearly impossible to determine in liquid media interpreted only visually. The BACTEC system (Becton-Dickinson, Sparks, Maryland, USA), a liquid media susceptibility system which has surmounted this problem, is based upon the instrumental quantitation of $^{14}CO_2$ released from ^{14}C-palmitic acid during cellular metabolism (3). This process, or its inhibition in the presence of drugs, circumvents the need to wait for the appearance of visible growth, permits susceptibility results within 4 to 7 days, and offers automated inoculation and read sampling.

Fluorescence and bioluminescence technologies have been applied more recently to susceptibility testing or identification of specific drug-resistant strains of mycobacteria. The susceptibility methods, some of which are currently used in our Laboratory are based upon measurements of metabolic activities of cells which enables sensitivity of detection many orders of magnitude greater than standard turbidimetric methods. Identification of specific resistant strains is by fingerprinting techniques utilizing DNA hybridization or amplification techniques employing labeled probes or oligonucleotide primers.

Novel Methods

Bioluminescence: The action of naturally occurring firefly luciferase upon its substrate, luciferin, requires ATP as a cofactor which made this reaction originally useful in measurements of biomass. Quantitation of ATP using this enzymatic reaction also proved to be useful in susceptibility testing of mycobacteria (4). Reductions of ATP levels, and consequently visible light as measured by luminometry, were shown to correlate with inhibitory concentrations of antituberculosis drugs within 14 days. In efforts to develop even faster and more sensitive antimycobacterial drug testing methodologies, we cloned a commercially available firefly luciferase reporter gene (Clontech Laboratories, Palo Alto, California, USA) in a mycobacterial- *Escherichia coli* shuttle plasmid and Jacobs et al. cloned the same *lux* gene into a mycobacteriophage (5-7). The endogenous luciferase encoded by these vectors garnered susceptibility testing results for *Mycobacterium tuberculosis* complex and *Mycobacterium avium* complex organisms in as little as 3 days. The process required simply adding the luciferase substrate, d-luciferin, to drug-containing liquid cultures of strains previously infected with the reporter phage or harboring the luciferase plasmid. Decreases in bioluminescence associated with drug effects correlated with reductions in numbers of bacilli. The process was further facilitated by performance in microdilution plates read in a microplate luminometer and enabled additional characterizations of antimicrobial effects such as determinations of minimum inhibitory concentration (MIC) and minimum bactericidal concentration (MBC) (Figure 1). Our tentative definitions of an MIC and MBC are: 1) MIC is the lowest drug concentration that inhibits bioluminescence, as measured in relative light units (rlu's) by a microplate luminometer (Model ML300, Dynatech Laboratories, Chantilly, Virginia, USA), to levels that are ≤10% of drug-free control rlu's after incubation for 7 days; and 2) MBC is the lowest drug concentration that inhibits bioluminescence to levels that are ≤1% of time 0 (e.g., inoculum) rlu's.

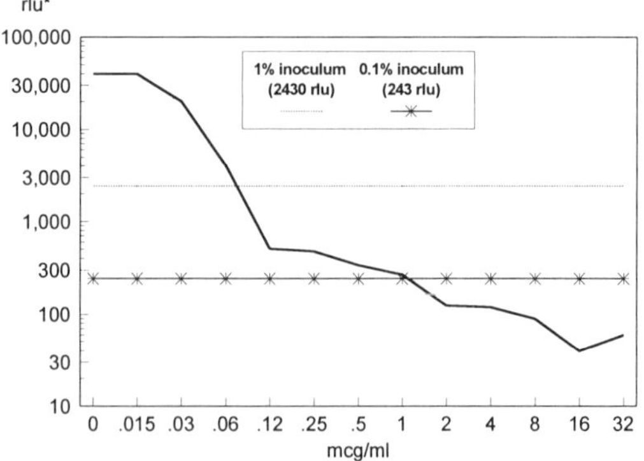

Figure 1. Determination of minimum bactericidal concentration (MBC) of rifampin against *Mycobacterium tuberculosis* strain $H_{37}Ra$ using luciferase assay. MBC (0.06 mcg/ml) was identified by reduction in bioluminescence to ≤1% of time 0 (inoculum) rlu's after 14-day incubation. * Relative light units X 10^{-4}

Fluorescence: Endogenous microbial esterases cleave compounds such as fluorescein diacetate (FDA) and derivatives such as sulfofluorescein diacetate to yield fluorescing compounds (e.g. fluorescein), and this enzymatic reaction has been exploited to evaluate the in vitro action of antimicrobial agents (8-10). Measurement of this activity using instrumentation such as microplate fluorometers or flow cytometers have enabled development of useful antimycobacterial drug testing methods. We are currently evaluating the usefulness of fluorescence based susceptibility testing of *M. tuberculosis* in microdilution plates. One obvious advantage of this procedure over our currently used luciferase microplate assay is that the construction of specific strains becomes unnecessary while not compromising speed, sensitivity, or specificity. Our continuing development of this assay, however, has necessitated our circumvention of the use standard mycobacterial broth media (7H9, GIBCO Laboratories, Madison, Wisconsin, USA) containing albumin-dextrose complex. This enrichment supports luxurious growth of mycobacteria but autofluoresces (emissions at ~485 nm) causing background interference with the specific fluorescein cleavage product. The media we developed, 5S broth, minimizes this non-specific fluorescence while supporting growth of mycobacteria, including slowly growing mycobacteria such as *M. tuberculosis*, for liquid culture-based susceptibility testing. Typically ~10^6 bacilli are inoculated into wells of microdilution plates containing broth and drug concentrations (180 µl), and after incubation for 3 days (for most standard antituberculosis drugs) fluorescein diacetate (FDA) in 5 µl ethanol is added (final well concentration is 20 µM) and, after an additional incubation of 30 min at 37°C, fluorescence is measured using a microplate fluorometer (Bio-Tek FL500, Bio-Tek Instruments, Winooski, Vermont, USA) set for "on the fly" reading, 30% sensitivity, 530-nm excitation, and 485-nm emission. Alternative instrument settings and other conditions such as increased amounts of FDA, the addition of methylene blue in various concentrations as a potential quenching agent of nonspecific fluorescence, increased time of incubation of organisms in drug broth prior to adding the substrate, and washing bacterial cells in phosphate buffered saline prior to adding FDA have been evaluated and found to offer no advantages for mycobacterial susceptibility testing. Standard antituberculosis drugs including streptomycin, rifampin, ciprofloxacin, isoniazid, and pyrazinamide (tested at pH 5.5) caused reductions in fluorescence to levels that were ≤10% of drug-free controls at MICs determined by conventional agar proportion testing and greater concentrations after 3 days (Figure 2). Seven days were required for determination of ethambutol MICs by this method. Inocula that consisted of mixtures of rifampin-resistant and rifampin-susceptible *M. tuberculosis* in 1:10 and 1:100 ratios were tested to evaluate the ability of the FDA microplate method to detect rifampin-resistant subpopulations. Reductions in fluorescence that were intermediate between fully resistant and fully susceptible populations were detectable after 7 days suggesting that this method may be a suitable alternative to the standard agar proportion method for at least semi-quantitation of resistant subpopulations of *M. tuberculosis.* The FDA microplate method is currently under development and is not performed for routine diagnosis of resistance in our Laboratory.

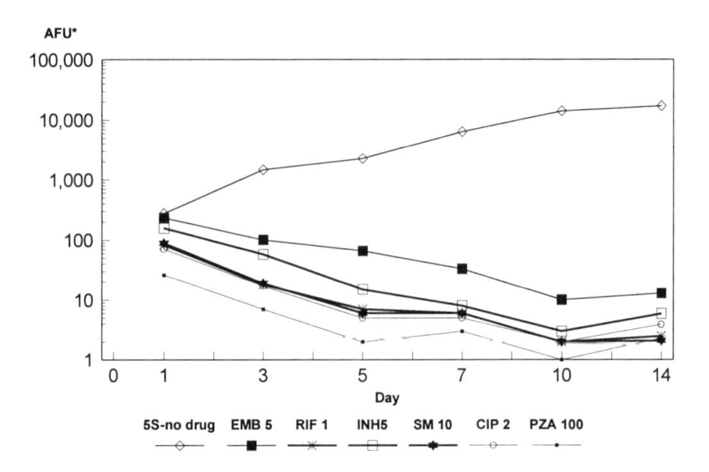

Figure 2. Susceptibility testing of *Mycobacterium tuberculosis* strain $H_{37}Rv$ in 5S medium using fluorescein diacetate (FDA) microplate method. EMB 5, 5 μg/ml ethambutol; RIF 1, 1 μg/ml rifampin; INH 5, 5 μg/ml isoniazid; SM 10, 10 μg/ml streptomycin; CIP 2, 2 μg/ml ciprofloxacin; PZA 100, 100 μg/ml pyrazinamide (pH5.5). * Arbitrary fluorescence units.

Another convenient fluorescence-based method useful for isolation and susceptibility testing of *M. tuberculosis* is the Mycobacteria Growth Indicator Tube (MGIT) system introduced commercially by the manufacturer of BACTEC (Becton Dickinson). An oxygen sensitive fluorescent compound is embedded in silicone in the bottom of broth culture tubes. Growth of organisms from processed specimens in media containing inhibitors of genera other than mycobacteria results in consumption of dissolved O_2 enabling fluorescence which is detectable by illuminating the tube with 365-nm ultraviolet light. Inhibition of mycobacterial growth in MGIT tubes containing active antimicrobial agents also suppresses fluorescence which enables susceptibility results to be determined in less than 10 days. The use of a fluorescence reader will help minimize subjective interpretation of intermediate fluorescence results. The MGIT system is most likely unsuitable for microdilution methods due to insufficient depth of media in microdilution plate wells.

Molecular methods: We are currently applying automated DNA sequencing with fluorescent oligonucleotide primers as a method for rapid and definitive detection of *M. tuberculosis* genes encoding resistance to streptomycin, rifampin, isoniazid, and pyrazinamide. Genomic regions associated with phenotypic resistance are amplified using polymerase chain reactions (PCR) and sequenced with *Taq* DyeDeoxy terminator cycle sequencing kits (Applied Biosystems Div. (ABI), Perkin-Elmer Corp., Foster City, California, USA) and run on an ABI Model 373 Automated Sequencer. We recently reported one specific mutation in the ribosomal 50S subunit-associated protein S12 to be responsible for clinically significant streptomycin resistance (in vitro MIC $\geq$ 10 μg/ml) (11). We have not, however, shown that automated sequencing is useful either for detection of resistant subpopulations dominated by wildtype sequences, or for resistance genes extracted directly from isolates in specimens (e.g., sputa).

The automated sequencer is also used in our Laboratory to identify specific multidrug resistant clones of *M. tuberculosis* using a PCR-based fingerprinting method (12). Routine laboratory epidemiology of tuberculosis is performed by analyses of restriction fragment length polymorphisms (RFLP) based upon genomic distribution of a *M.tuberculosis* complex-specific repetitive insertion element *IS*6110. Probes for *IS*6110 are prepared by PCR, labeled with a chemiluminescent tag (ECL kit, Amersham Corp., Arlington Heights, Illinois, USA), and hybridized to endonuclease restricted and denatured tuberculosis genomes bound to nylon membranes. Patterns of 1-20 bands appearing after autoradiography are stable, strain specific, and currently used as principal laboratory evidence by tuberculosis epidemiologists. For the fluorescent PCR method mixed linkers are ligated to genomic DNA digested with restriction endonuclease *Hha*I. The linker is an unphosphorylated duplex nucleic acid product with a guanine plus cytosine overhang at the 3' end compatible with *Hha*I fragments. One strand of the linker contains uracil enabling its enzymatic removal with uracyl *N*-glycosylase. Amplification using fluorescent primers specific for *IS*6110 and the linker is then performed and the products are electrophoresed in acrylamide gels along with fluorescent internal lane standards in the ABI automated sequencer loaded with software designed for analysis of gel band patterns in ranges of 37 to 400 nucleotide base pairs (GeneScan; ABI). The internal standard is used by the software to create a calibration curve of peak detection time to calculate sizes of PCR products in the gel. Although mixed-linker PCR for *IS*6110 is routinely performed in our Laboratory for epidemiological support, it is currently only of limited use for identification of specific drug-resistant clones of *M. tuberculosis*.

References

1. Woods GL, Washington JA. Antibacterial susceptibility tests: dilution and disk diffusion methods. In: Murray PR, Baron EJ, Pfaller MA, et al. Manual of clinical microbiology. Washington: ASM Press; 1995:p1327-35.
2. Kent PT, Kubica GP.Chapter 7, Antituberculosis chemotherapy and drug susceptibility testing. Public health mycobacteriology: a guide for the level III laboratory. Atlanta:U.S. Dept. of Health and Human Services, 1985:1985:159-84.
3. Middlebrook G, Reggiardo Z, Tigertt WD. Automatable radiometric detection of growth of *Mycobacterium tuberculosis* in selective media. Am Rev Respir Dis 1977;115:1066-9.
4. Beckers B, Lang HRM, Schimke D, Lammers A. Evaluation of a bioluminescence assay for rapid antimicrobial susceptibility testing of mycobacteria. Eur J Clin Microbiol Infect Dis 1985; 4:556-61.

5. Cooksey RC, Crawford JT, Jacobs WR Jr, Shinnick TM. A rapid method for screening antimicrobial agents for activities against a strain of *Mycobacterium tuberculosis* expressing firefly luciferase. Antimicrob Agents Chemother 1993; 37:1348-52.
6. Cooksey RC, Morlock GP, Beggs M, Crawford JT. A bioluminescence method for evaluating antimicrobial susceptibilities against *Mycobacterium avium.* Antimicrob Agents Chemother 1995;39:754-6.
7. Jacobs WR Jr, Barletta R, Udani R, Kalbut G, Sosne G, Keiser T et al. Rapid assessment of drug susceptibilities of *Mycobacterium tuberculosis* by means of luciferase reporter phages. Science 1993;S260:819-22.
8. Bercovier H, Resnick M, Kornitzer D, Levy L. Rapid method for testing drug-susceptibility of mycobacteria spp. and gram-positive bacteria using rhodamine 123 and fluorescein diacetate. J Microbiol Methods 1987;7:139-42.
9. Chand S, Lusunzi I, Veal DA, Williams LR. Rapid screening of the antimicrobial activity of extracts and natural products. J Antibiot 1994;47: 1295-1304.
10. Norden MA, Kurzynski TA, Bownds SE, Calister SM, Schell RF. Rapid susceptibility testing of *Mycobacterium tuberculosis* (H37Ra) by flow cytometry. J Clin Microbiol 1995;33:1231-7.
11. Cooksey RC, Morlock GP, McQueen A, Glickman, SE, Crawford JT. Characterization of streptomycin resistance mechanisms among *Mycobacterium tuberculosis* isolates from patients in New York City. Antimicrob Agents Chemother 1996;40:1186-8.
12. Butler WR, Haas WH, Crawford JT. Automated DNA fingerprinting analysis of *Mycobacterium tuberculosis* using fluorescent detection of PCR products. J Clin Microbiol 1996;34:1801-3.

NATIVE CHEMILUMINESCENCE OF NEUTROPHILS FROM SYNOVIAL FLUID OF PATIENTS WITH RHEUMATOID ARTHRITIS

J Arnhold, K Sonntag, J Schiller and K Arnold
Institute of Medical Physics and Biophysics, Medical Department,
University of Leipzig, Liebigstr. 27, 04103 Leipzig, Germany

Introduction

Although many details of aetiology and pathogenesis of rheumatoid arthritis remain unknown, it is well established that polymorphonuclear leukocytes (PMNs) accumulated in inflamed joint contribute to cartilage degradation. Stimulated cells release lysosomal enzymes, bactericidal proteins and generate reactive oxygen species. Previously we reported a marked native chemiluminescence (i.e. in the absence of any light amplifiers) in PMNs from synovial fluid of patients with rheumatoid arthritis upon stimulation with phorbol myristate acetate (PMA) (1,2). This emission has not been observed in PMNs from the peripheral blood of healthy volunteers under the same experimental conditions. Integral intensities of native chemiluminescence of patient cells correlate well with the activity of myeloperoxidase (MPO) in cell-free synovial fluid of the same patient (2). It was assumed that an oxidative stress in patient cells causes various oxidative reactions. Here, new data are given showing a relation between enhanced native chemiluminescence and lipid peroxidation. Moreover, myeloperoxidase activity in cell-free synovial fluids of patients with rheumatoid arthritis correlates well with the appearance of cartilage degradation products.

Materials and Methods

Instrumentation Native chemiluminescence of cell suspension was recorded using a luminometer AutoLumat LB 953 (Berthold, Wildbad, Germany).
NMR measurements were conducted on a Bruker AMX-NMR spectrometer operating at 300.13 MHz for ^{1}H.
Reagents 12-Phorbol-13-myristate acetate (PMA) was from Sigma (Deisenhofen, Germany). Pz-peptide was obtained from Bachem (Heidelberg, Germany). All other chemicals were purchased from Boehringer Ingelheim Bioproducts (Heidelberg, Germany). Hank's balanced salt solution (HBSS) was prepared every day and used without phenol red.
Cell preparation PMNs were isolated from freshly taken synovial fluid of knee joints of patients with rheumatoid arthritis and also from peripheral blood of healthy volunteers. The preparation included a Ficoll-Hypaque-density centrifugation, lysis of remaining red blood cells with distilled water and washing of cells with HBSS. Neutrophils were resuspended in HBSS at a concentration of 2 x 10^6 cells/mL and stored at 4 °C. They were used within 2h after preparation.
PMN chemiluminescence PMNs (10^5 cells/mL) were equilibrated with HBSS at 37 °C for five minutes. Then PMA (final concentration 1.6 x 10^{-8} mol/L) was added to stimulate cells. Chemiluminescence counts were determined over a period of 30 min.
Enzymatic activities MPO activity was determined in cell-free synovial fluid of patients with rheumatoid arthritis using guaiacol (3). Collagenase activity was also measured in these samples (4).

NMR measurements The intense water signal and broad resonances arising from proteins and polymeric carbohydrates in cell-free synovial fluids were suppressed by a combination of the Hahn spin-echo sequence (90°-τ-180°-τ-collect) and the application of continuous secondary irradiation at the water resonance frequency.

Results and Discussion

Fig. 1a shows five typical examples of the generation of a native chemiluminescence in PMA-stimulated neutrophil populations of synovial fluids from patients with rheumatoid arthritis and peripheral blood of healthy volunteers. This kind of chemiluminescence is usually found in neutrophil preparations from patients with rheumatoid arthritis, whereas neutrophils from peripheral blood of healthy volunteers only in seldom cases develop such a light emission (1).

It has been assumed that conditions of a long-lasting oxidative stress in inflamed joints favour an exhaustion of naturally occurring antioxidants in cell membranes and cause secondary oxidative reactions accompanied by a chemiluminescence (1). A further indication for this hypothesis was obtained measuring thiobarbituric acid (TBA) reactive products in PMA-stimulated neutrophils. TBA-reactive substances are typical products of a lipid peroxidation. Many reactive oxygen species generated by stimulated neutrophils are able to induce a lipid peroxidation (5,6).

Fig. 1b shows the accumulation of TBA-reactive products in the same neutrophil samples used in Fig. 1a. These products accumulate in neutrophils from synovial fluid in the same way as the native chemiluminescence raises. Three cell preparations with different patterns of native chemiluminescence have been selected in Fig. 1. Moreover, no light emission and no accumulation of TBA-reactive products were observed in all samples (n=6) of neutrophils from peripheral blood of healthy volunteers during 30 min after stimulation with PMA. Two examples of this group are given in Fig.1.

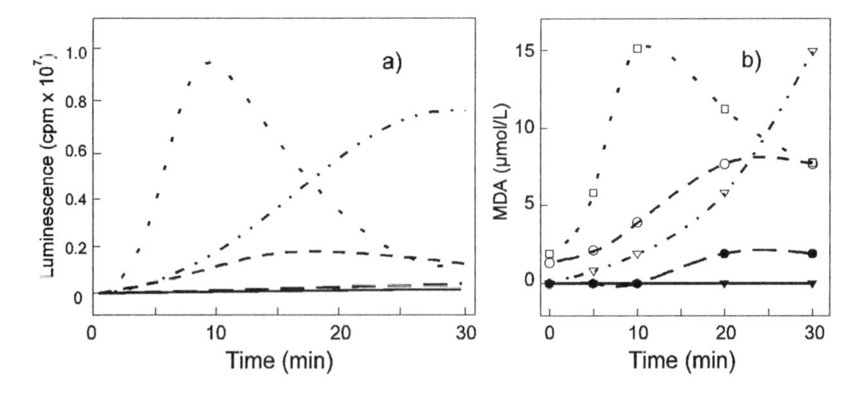

Fig. 1. Five selected examples of the generation of native chemiluminescence (panel a) and the accumulation of TBA-reactive products (panel b) in suspensions of neutrophils (10^5 cells in 1 mL Hank's medium) after stimulation with PMA ($1.6*10^{-8}$ mol/L, final concentration). Cells were obtained from synovial fluid of patients with rheumatoid arthritis (open symbols) and peripheral blood of healthy volunteers (closed symbols). Identical line styles are used for each sample in both panels.

These data indicate enhanced oxidative reactions in neutrophils from patient material. They are consistent with our previous result that a treatment of neutrophils with HOCl favors the developement of native chemiluminescence. This light emission correlates with the MPO activity in cell-free synovial fluids (2). This implies an important role of hypochlorous acid in alteration of cell properties during rheumatoid arthritis. HOCl itself causes a cartilage injury inducing in carbohydrates a breakdown of polymeric chains into smaller oligomeric units and a reaction with N-acetyl groups, whereby acetate is formed (7,8).

The degradation of cartilage polysaccharides was studied by means of NMR spectroscopy. Fig. 2a shows a typical ^{1}H-NMR spectrum of a cell-free synovial fluid from a patient with rheumatoid arthritis. The signal at 2.02 ppm contributes to methyl protons of relatively mobile N-acetyl groups of shorter oligomeric units of hyaluronic acid and smaller molecular mass polysaccharides like chondroitinsulfate and keratansulfate. In all synovial fluids there were also signals at 1.90 ppm for acetate.

A strong correlation was found between signal intensities and MPO acitivity in cell-free synovial fluids (Fig. 2b). The concentration of N-acetyl groups in NMR-detectable oligomers was related to the signal intensity of N-acetylglucosamine. The higher the MPO activity was, the more intense were the signals for acetate and N-acetyl groups. A similar correlation exists also between the intensity of ^{1}H-NMR signals in cell-free synovials fluids and the integral intensity of native chemiluminescence of PMA-stimulated neutrophils from synovial fluids of the same patients (data not shown).

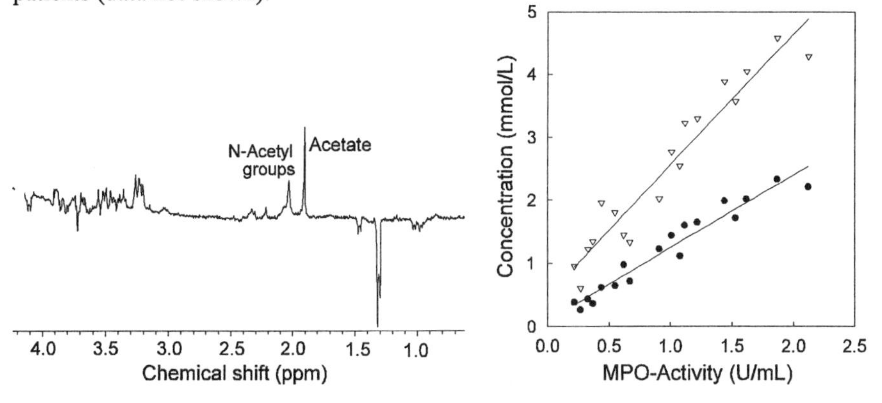

Fig. 2. ^{1}H-NMR spectrum of cell-free synovial fluid of a patient with rheumatoid arthritis (panel a) and concentrations of acetate (∇) and N-acetyl groups (•) as a function of myeloperoxidase activity in synovial fluids (panel b).

Proteolytic enzymes released from neutrophils and other cells in inflamed joints are assumed to contribute also to tissue injury in rheumatoid arthritis (9). Because a close correlation exists between native chemiluminescence of cells, myeloperoxidase activity and the accumulation of carbohydrate breakdown products, the question arises whether other enzymes are involved in these processes. Collagenase activities were determined in synovial fluids of patients with rheumatoid arthritis and compared with MPO activities (Fig. 3). At MPO activities lower than 0.6 U/ml there was a continuous increase in collagenase activity. However, synovial fluids with MPO activities higher than 0.6 U/ml were characterized by decreased values for collagenase activity. The higher the activity of myeloperoxidase was in this region of Fig. 3 the less active was collagenase.

Synovial fluids from patients with rheumatoid arthritis are characterized by enhanced MPO activity, the appearance of carbohydrate breakdown products and a marked native chemiluminescence of neutrophils upon stimulation with phorbol esters. Because hypochlorous acid, the product of the chlorinating activity of myeloperoxidase, causes the same properties in normal neutrophils and in carbohydrates, these results underline an important role of hypochlorous acid in the pathogenesis of rheumatoid arthritis.

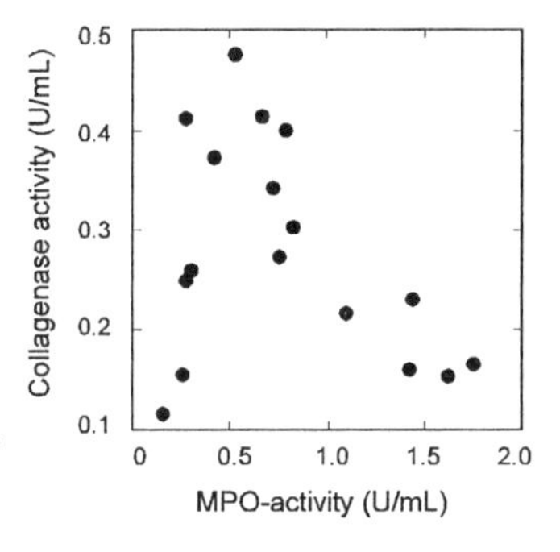

Fig. 3. Collagenase activities as a function of MPO activity in cell-free synovial fluid of patients with rheumatoid arthritis.

Acknowledgements

This work was supported by grants from the German Ministry of Research and Technology (01 ZZ 9103/9-R-6) and the Deutsche Forschungsgemeinschaft (INK 23/A-1).

References

1. Arnhold J, Sonntag K, Sauer H, Häntzschel H, Arnold K. Increased native chemiluminescence in granulocytes isolated from synovial fluid and peripheral blood of patients with rheumatoid arthritis. J Biolumin Chemilumin 1994; 9: 79-86.
2. Sonntag K, Arnhold J, Arnold K. Influence of myeloperoxidase on the generation of native chemiluminescence. In: Campbell AK, Kricka LJ, Stanley PE., editors. Bioluminescence and chemiluminescence. Fundamentals and applied aspects. Chicester, Wiley & sons, 1994: 187-90.
3. Klebanoff JF, Waltersdorff AM, Rosen H. Antimicrobial activity of myeloperoxidase. Meth Enzymol 1984; 105: 399-403.
4. Wünsch E, Heidrich HG. Zur quantitiven Bestimmung der Kollagenase. Z Physiol 1963; 173: 1281-6.
5. Stelmaszynska T, Kukovetz E, Egger G, Schaur RJ. Possible involvement of myeloperoxidase in lipid peroxidation. Int J Biochem 1992; 24: 121-8.
6. Panasenko OM, Arnhold J, Schiller J, Arnold K, Sergienko VI. Peroxidation of egg yolk phosphatidylcholine liposomes by hypochlorous acid. Biochim Biophys Acta 1994; 1215: 259-66.
7. Schiller J, Arnhold J, Gründer W, Arnold K. The action of hypochlorous acid on polymeric components of cartilage. Biol Chem Hoppe-Seyler 1994; 375: 167-72.
8. Schiller J, Arnhold J, Arnold K. NMR studies of the action of hypochlorous acid on native pig articular cartilage. Eur J Biochem 1995; 223: 672-6.
9. Weiss SJ, Peppin G, Ortiz X, Ragsdale C, Test ST. Oxidative autoactivation of latent collagenase by human neutrophils. Science 1985; 227: 747-9.

CHEMILUMINESCENCE IMAGING AS A BIOANALYTICAL TOOL

A Roda[1], P Pasini[1], M Musiani[2], C Robert[3], M Baraldini[4] and G Carrea[5]
[1]Dept of Pharmaceutical Sciences, [2]Inst of Microbiology, [4]Inst of Chemical Sciences, University of Bologna, Via Belmeloro 6, 40126 Bologna, Italy
[3]G.R.C.P.V.A. - Inst A. Bonniot, University Joseph Fourier, 38000 Grenoble, France
[5]Inst of Hormones Chemistry, C.N.R., Via Bianco 9, 20131 Milano, Italy

Introduction

Recently, new low-light imaging instrumentation based on the use of intensified Vidicon tube or CCD have become commercially available. These devices are characterized by a relatively low thermal noise, wide dynamic range, high sensitivity and spatial resolution as well as provide a quantitative relationship between photons detected and pixel addressed. Moreover, it has been previously proved that these instruments can be connected with optical microscopy reaching resolution of few micrometers and permitting to image a luminescent signal in single cells, tissue sections or organs (1-5).

The development of quantitative methodologies based on chemiluminescence-optical microscopy imaging are therefore required for ultrasensitive diagnostic purposes of several diseases. The image of bio-chemiluminescent molecules either directly or via immunological or genic probes allows to localize a particular antigen or to develop quantitative in situ hybridization techniques.

The main requirements to be superior to existing colorimetric methods are to reach a higher detectability and obtain quantitative information, thus allowing early diagnosis of a given disease and a therapeutic follow-up.

Methodological problems related to instrumental performance, data processing and the sample to be analyzed still limit a quantitative use of this principle. Critical points are the chemistry (kinetics, photochemistry) of the chemiluminescent system used and the sample to be imaged in terms of size, geometry, thickness, aspecific signal.

In order to evaluate the suitability of chemiluminescence-optical microscopy for quantitative imaging we used alkaline phosphatase (ALP) immobilized on calibrated nylon net (100±3 threads/cm, thread ∅: 20 μm) as a model system. The analytical performance of the instrument and its optimization were studied using the steady-state chemiluminescence of both ALP/dioxetanes and horseradish peroxidase (HRP)/enhanced chemiluminescence (ECL) systems. The chemiluminescent kinetics of nylon net coimmobilized coupled enzymes was also studied. Finally, the application of chemiluminescence-optical microscopy for immunological localization of cytokeratin in skin and in situ hybridization of viral genomes has been reported and discussed.

Materials and Methods

Instrumentation: The chemiluminescence signal was detected and analyzed using a low-light imaging device, Luminograph LB 980 (EG&G Berthold, Bad Wildbad, Germany) linked to a Model BH-2 light microscope (Olympus Optical, Tokyo, Japan) enclosed in a light-tight box. The luminograph consists of an intensified Saticon videocamera connected to a PC provided with software controlling the system and performing image processing and quantitative analysis. The chemiluminescent signal and the live image in transmitted light can be acquired, digitized and stored in the PC; processing and overlay functions allow the spatial distribution of the signal to be evaluated and quantified with a spatial resolution of 1 μm at x40 objective magnification (5).

Nylon net immobilized ALP: ALP was chemically immobilized on nylon net using a slightly modification of the method described by Hornby and Goldstein (6). Briefly, the nylon nets were first activated with triethyl oxonium tetrafluoroborate and, after treatment with 1,6-diaminohexane and glutaraldehyde, were allowed to react with the enzyme solution. Nets were stored in phosphate buffer containing 0.1% BSA at either 4°C or 25°C. The ALP activity and its spatial distribution were evaluated by placing a small portion (5x5 mm) of the net on a glass microscope slide and adding 20 μL of the chemiluminescent substrate.

Immunoenzymatic reaction: Immunohistochemical study was performed on acetone fixed skin cryosections (7 μm thick). Briefly, after inhibition of endogenous peroxidase and non specific immunoglobulin binding, the sections were incubated overnight at 4°C with a mouse monoclonal antibody against 56KD keratin (KL1, Immunotech, Marseille, France). After incubation with a second biotinylated antibody (Jackson, West Grove, USA) and then with a biotinylated peroxidase-streptavidin complex (Dakopatts, Glostrup, Denmark), the chemiluminescent substrate ECL (Amersham International, Amersham, UK) for peroxidase was used.

In situ hybridization: In situ hybridization was performed as previously described using digoxigenin (dig)-labelled DNA probes (7). Briefly, cytomegalovirus (CMV) infected human fibroblasts, Parvovirus B19 infected bone marrow cell smears and human papillomavirus (HPV) infected tissue cryosections were fixed in 4% paraformaldehyde, then overlaid with the hybridization mixture (50% deionized formamide, 10% dextran sulfate, 250 μg/mL of calf thymus DNA and 2 μg/mL of dig-labelled probe DNA in 2X SSC buffer). Samples and the hybridization mixture containing the dig-labelled probe were denaturated together by heating in an 85°C water bath for 5 min, and were then put to hybridize at 37°C for 3 h. After hybridization, samples were washed three times at stringent conditions. Hybridized probes were detected using polyclonal anti-digoxigenin antibody conjugated to ALP and chemiluminescent substrate for ALP.

Results and Discussion

The net immobilized ALP can be sharply detected and localized on the threads and the knots without appreciable aspecific signal in the solution sorrounding the net. Its activity can be calculated and expressed as ng/mm^2 using a calibration graph. The aspecific signal was relatively high but with low fluctuation, allowing to consider positive a signal corresponding to the background value plus five standard deviations (8). More complicated was when three enzymes were studied: the specific enzyme 3α-hydroxysteroid dehydrogenase (3α-HSD) coupled with the bacterial bioluminescent enzymes FMN-oxidoreductase and luciferase. Figure 1. The localization of 3α-HSD on nylon net was very poor with the specific enzyme alone immobilized and the two bioluminescent enzymes free in solution. It improved with two immobilized enzymes and became quite good with all the three enzymes co-immobilized. The image can be improved increasing the viscosity of the solution and the concentration of the indicator bioluminescent enzymes. Net system provides a good model to evaluate the role of the aspecific signal and to study and optimize the kinetics of coupled bioluminescent enzymatic reactions to detect enzymes or substrates on target specimens.

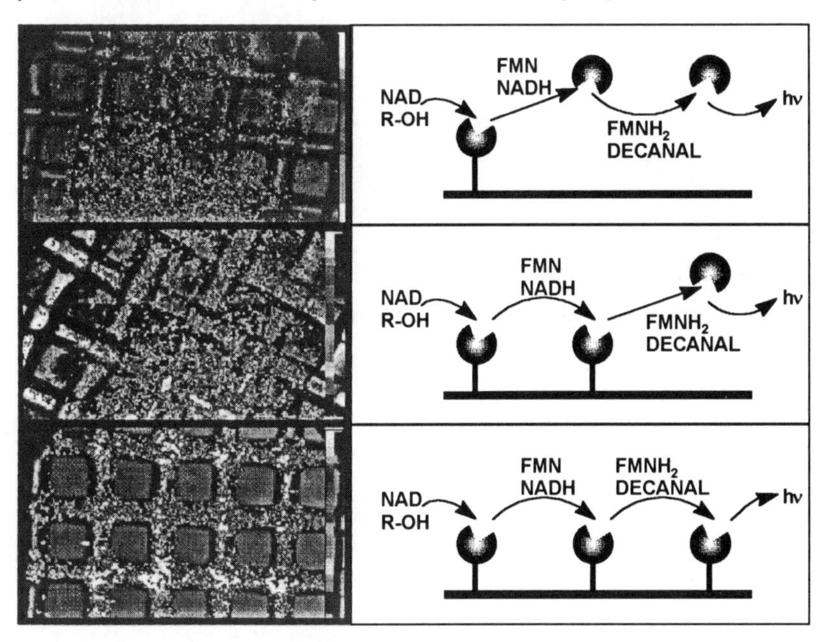

Figurel. 3α-HSD, FMN-oxidoreductase and luciferase coupled system. Effect of enzyme immobilization on luminescent image spatial resolution.

The chemiluminescence-optical microscopy system was used to detect and localize particular biomolecules of clinical interest via immunological or genic probes. The localization of cytokeratin in epithelial cells of skin tissue cryosections using a specific primary antibody and an amplified detection system is shown in Figure 2.

Another application was the detection of viral genomes by in situ hybridization reactions. Clinical samples from patients with different viral infections were analyzed. A strong and sharp signal was observed in the positive cells with a low background. The comparison with the conventional colorimetric detection showed a higher sensitivity and a good spatial resolution for the chemiluminescent substrates. Moreover, the photon emission can be measured and a relationship between the light output and the quantity of viral genomes was found (9), allowing a semiquantitative analysis.

These findings suggested that chemiluminescence in situ hybridization and immunoenzymatic assays could be a useful tool for the early diagnosis of viral infections and the therapeutic follow-up of patients. Furthermore, they could be applied for diagnostic purposes in various fields, such as genetic diseases and oncology.

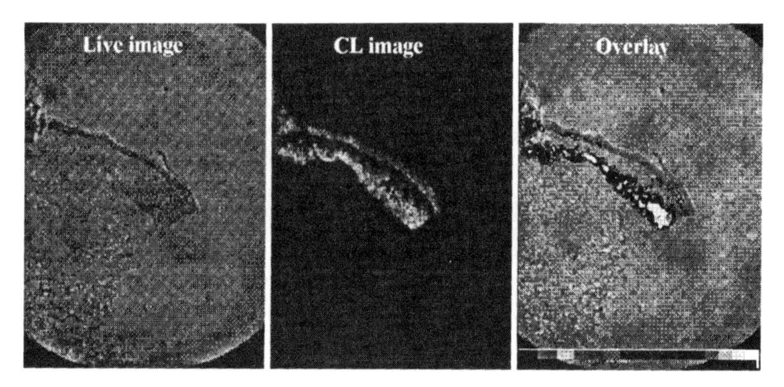

Figure 2. Immunoenzymatic localization of cytokeratin in skin cryosections.

In conclusion, low-light imaging-optical microscopy provides unprecedented capability for ultrasensitive in situ quantitative analysis of a given target molecule either directly or via immunological and genic biospecific reactions, with high spatial resolution allowing analysis at subcellular level.

References

1. Wick RA. Photon counting imaging: applications in biomedical research. BioTechniques 1989;7:262-8.
2. Hawkins E, Cumming R. Enhanced chemiluminescence for tissue antigen and cellular viral DNA detection. J Histochem Cytochem 1990;38:415-9.
3. Lorimier P, Lamarcq L, Labat-Moleur F, Guillermet C, Bethier R, Stoebner P. Enhanced chemiluminescence: a high sensitivity detection system for in situ hybridization and immunohistochemistry. J Histochem Cytochem 1993;41:1591-7.
4. Mueller-Klieser W, Walenta S. Geographical mapping of metabolites in biological tissue with quantitative bioluminescence and single photon imaging. Histochem J 1993;25:407-20.
5. Roda A, Pasini P, Musiani M, Girotti S, Baraldini M, Carrea G. Chemiluminescent low-light imaging of biospecific reactions on macro- and microsamples using a videocamera-based luminograph. Anal Chem 1996;68:1073-80.
6. Hornby WE, Goldstein L. Immobilization of enzymes on nylon. Methods Enzymol 1976;44:118-34.
7. Gentilomi G, Musiani M, Zerbini M, Gallinella G, Gibellini D, La Placa M. A hybrido-immunocytochemical assay for the in situ detection of cytomegalovirus DNA using digoxigenin-labeled probes. J Immunol Methods 1989;125:177-83.
8. Hooper CE, Ansorge RE, Rushbrooke JG. Low-light imaging technology in the life sciences. J Biolumin Chemilumin 1994;9:113-22.
9. Musiani M, Roda A, Zerbini M, Pasini P, Gentilomi G, Gallinella G et al. Chemiluminescent in situ hybridization for the detection of cytomegalovirus DNA. Am J Pathol 1996;148:1105-12.

LUMIGEN™ APS: NEW SUBSTRATES FOR THE CHEMILUMINESCENT DETECTION OF PHOSPHATASE ENZYMES

H Akhavan-Tafti, Z Arghavani, R DeSilva, RA Eickholt, RS Handley, BA Schoenfelner, S Siripurapu, K Sugioka and AP Schaap*
Lumigen, Inc., 24485 W. Ten Mile Rd., Southfield, MI 48034 USA

Introduction

Chemiluminescent substrates for detection and quantitation of alkaline phosphatase (AP) and AP conjugates have now found widespread commercial application in clinical analysis, forensic analysis, environmental analysis and in basic and applied research in the life sciences (1). Most detection methods utilize enzymatically triggerable dioxetanes as the chemiluminescent substrate, although other reagents based on enzymatic release of luminol (2), firefly luciferin (3), indole derivatives (4) or reducing agents (5) have been reported. Chemiluminescence emitted by triggerable dioxetanes takes the form of a gradual rise to a plateau while the concentration of a hydroxyaryl dioxetane builds to a steady state level. We have developed a new class of non-dioxetane chemiluminescent substrates for phosphatase enzymes, Lumigen™ APS which react rapidly with alkaline or acid phosphatase at ambient temperature to produce an almost instantaneous rise to a stable plateau. We have applied these new substrates in several assays to evaluate their speed, sensitivity and dynamic range for the detection of alkaline phosphatase.

Materials and Methods

Instrumentation: Immunoassays for TSH, HCG and estradiol were performed on an Immulite Automated Analyzer (Diagnostic Products Corp., Los Angeles, CA USA) using Immulite kits with the new detection reagent. The software was modified to allow a shorter substrate incubation and the substrate heater was disconnected. All data points are the average of triplicate tests except for blank readings which were the average of 5 tests of the sample diluent.
Reagents: Lumigen APS-1 and APS-2 will be commercially available from Lumigen. Purified calf intestine alkaline phosphatase (AP) was obtained commercially (Biozyme, San Diego, CA USA). Avidin-AP conjugate (Cappel, Durham, NC USA), human transferrin and bovine serum albumin (heat shocked) (Sigma, St. Louis, MO USA).
Procedures: Western blotting was performed according to the method of Laemmli (6).

Results and Discussion

We have discovered that a novel class of acridan compounds react with alkaline phosphatase in the presence of oxygen in alkaline buffer between pH 8 and 10 to generate easily detectable chemiluminescence. Reaction of either Lumigen APS-1 or APS-2 with alkaline phosphatase generates light originating from the excited state of N-methylacridone as shown by comparison of the chemiluminescence spectrum with the fluorescence of N-methylacridone. Light intensity reaches a plateau within seconds at room temperature, allowing easy detection and quantitation (Figure 1). Optimized formulations have been developed which achieve zeptomole sensitivity (10^{-21} moles) for detection of AP with a linear response over several orders of magnitude (Figure 2).

A likely mechanistic hypothesis which explains the observed chemiluminescent reaction involves enzymatic cleavage of the phosphate ester to produce an enolate which is converted to a dioxetanone by reaction with O_2 and expulsion of phenol or thiophenol. The precise order of steps and the identity of intermediates is under investigation. The observed organic reaction products are N-methylacridone and phenol or thiophenol as well as the corresponding ester or thioester as a by-product. Light is emitted from the singlet excited state of N-methylacridone. A surprising finding is that reaction of Lumigen APS-1 or APS-2 with a phosphatase enzyme at moderately basic pH produces far more intense light emission than is produced by autoxidation of the ester or thioester. In a control experiment, identical formulations were prepared containing either Lumigen APS-1 or the ester hydrolysis product phenyl N-methylacridan-9-carboxylate in alkaline buffer at the same concentration. Light intensity from the AP-triggered reaction of Lumigen APS-1 far exceeded the spontaneous light intensity of the ester. This finding suggests that formation of the ester or thioester in the enzymatic reaction represents a nonluminescent side reaction.

Scheme 1

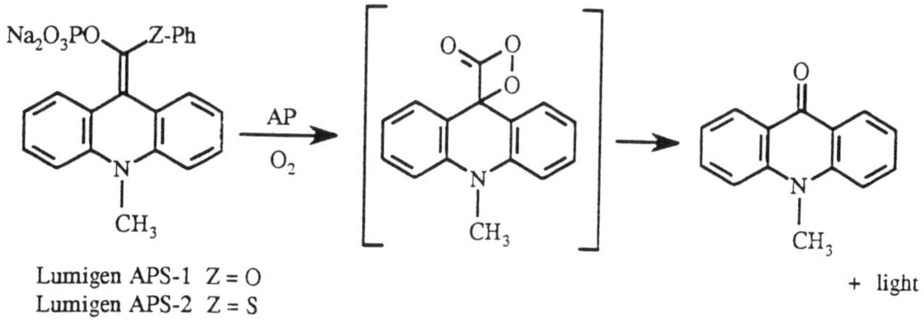

Na_2O_3PO Z-Ph; AP, O_2; CH_3 → + light

Lumigen APS-1 Z = O
Lumigen APS-2 Z = S

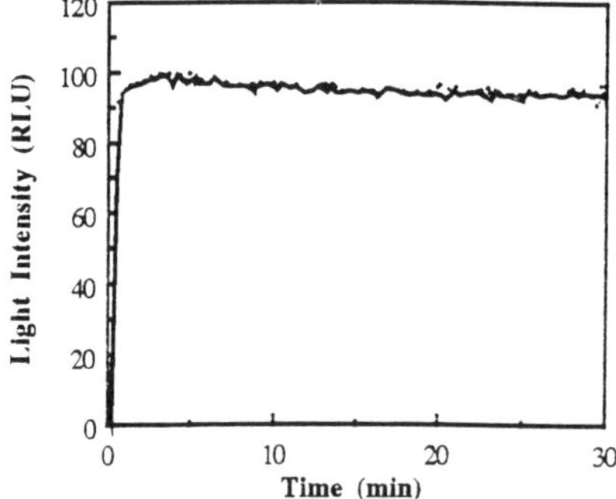

Figure 1. Typical chemiluminescence time profile of the reaction of AP with Lumigen APS-2 at room temperature.

One of the most noteworthy features of the new substrates is their time profile of light emission. Optimized formulations were prepared which rapidly reached a plateau of chemiluminescence intensity after addition of alkaline phosphatase (Figure 1). Plateau intensity was stable for at least 30 min. This kinetic profile was reproduced in reactions spanning several orders of magnitude of AP concentration. The shape of the chemiluminescence vs. time curve is consistent with a reaction scheme in which no long-lived intermediates are generated. Linearity of response was evaluated by measuring the plateau light intensity at 2.5 min from triplicate 100 µL aliquots of an optimized reagent containing Lumigen APS-2 in 0.1 mol/L tris buffer, pH 8.8 reacted with 10 µL aliquots of the enzyme solution containing from 8×10^{-22} to 8×10^{-16} moles of AP (Figure 2). The calculated detection limit (two σ over the background) was determined to be 3.5×10^{-21} moles of AP. Light emission characteristics, i.e. rise time, plateau intensity and duration, were relatively insensitive to temperature variation over the range 22-35 °C.

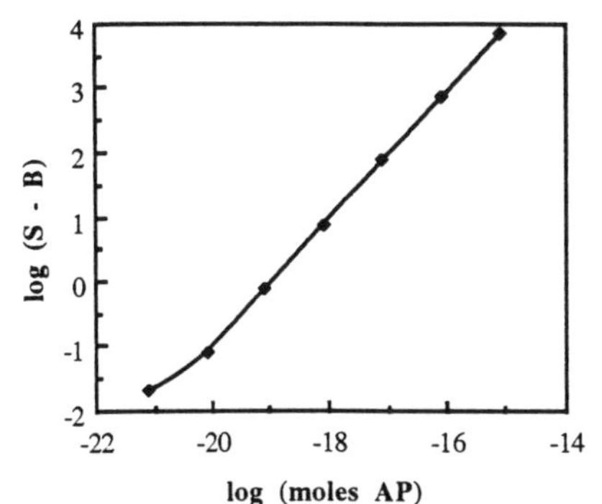

Figure 2. Detection of AP using a formulation containing Lumigen APS-2 at pH 8.8.

Immunoassays: A chemiluminescent sandwich immunoassay for thyroid stimulating hormone (TSH) was performed on an Immulite Automated Analyzer using an Immulite TSH Third Generation TSH Assay kit according to the manufacturer's protocol with modifications as described above and Lumigen APS-2 as the chemiluminescent substrate. A standard curve was constructed using dilutions of TSH calibrators supplied with the kit. Chemiluminescence measurements were made at 1.5 min after substrate introduction. The assay yielded a linear plot from 0.003 to 75 mIU/L. Independent tests on rapid reading luminometers indicated that the substrate could be read with as little as 10 s incubation time with no loss of sensitivity or accuracy.

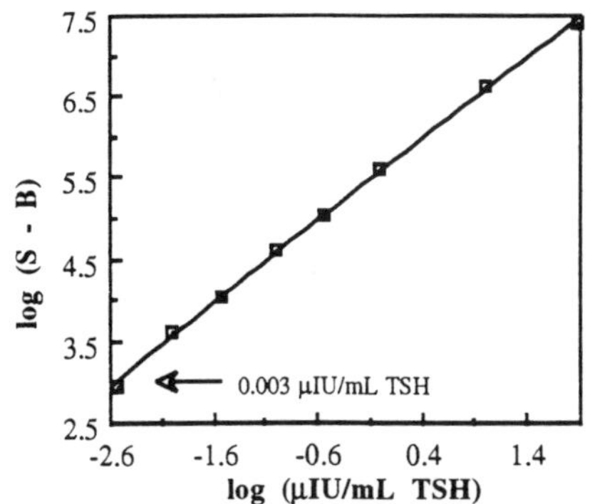

Figure 3. Chemiluminescent immunoassay of TSH using an Immulite Analyzer and Lumigen APS-2 detection reagent.

A sandwich immunoassay for human chorionic gonadotropin (hCG) was performed using an Immulite hCG assay kit according to the manufacturer's protocol substituting Lumigen APS-2 as the chemiluminescent substrate. Chemiluminescence measurements were made at 1.5 min after substrate introduction. The resulting assay yielded a linear response curve over the entire range tested, 1-6675 IU/L.

A competitive immunoassay for estradiol was performed using an Immulite estradiol assay kit according to the manufacturer's protocol substituting Lumigen APS-2 as the chemiluminescent substrate. Chemiluminescence measurements were made at 1.5 min after substrate introduction. The resulting assay yielded a linear response curve over the entire range of calibrators tested, 10-2000 ng/L.

The excellent sensitivity and broad dynamic range demonstrated in the enzyme assay and in immunoassays using Lumigen APS-2 prompted us to evaluate the detection of alkaline phosphatase conjugates on solid surfaces in Western and Southern blotting assays. We found that the rapid rise to maximum intensity and extended duration at constant intensity in solution carried over to membrane-based assay formats as well. Highly sensitive detection was possible on nylon, nitrocellulose and PVDF membranes. Lumigen APS reagents can also be used for nucleic acid hybridization assays and gene expression assays.

Western Blot: An optimized reagent containing Lumigen APS-2 was used in a Western blot assay of the protein human transferrin. Aliquots containing 5000, 1000, 180, 30 and 5 pg, respectively, of protein were electrophoresed and transferred to PVDF membranes. The membrane was blocked and then reacted sequentially with goat anti-human transferrin and rabbit anti-goat-AP conjugate. The membranes were soaked briefly in the detection reagent and exposed to X-ray film. The transferrin bands in all five lanes were detected after 10 min with a 2.5 min exposure. Light emission could be imaged for several hours. Similar results were obtained using nitrocellulose.

References

1. Beck S, Köster H. Applications of dioxetane chemiluminescent probes to molecular biology. Anal Chem 1990;62:2258-70.
2. Nakazono M, Nohta H, Sasamoto K, Ohkura Y. Chemiluminescent assays for ß-D-galactosidase and alkaline phosphatase using novel luminol derivatives as substrates. Anal Sci 1992;8:779-83.
3. Miska W, Geiger R. Luciferin derivatives in bioluminescence-enhanced enzyme immunoassays. J Biolumin Chemilumin 1989;4:119-28.
4. Albrecht S, Brandl H, Steinke M, Freidt T. Chemiluminescent enzyme immunoassay of prostate-specific antigen based on indoxyl phosphate substrate. Clin Chem 1994;40:1970-1.
5. Sasamoto II, Macda M, Tsuji A. Chemiluminescent assay of alkaline phosphatase using phenacyl phosphate. Anal Chim Acta 1995;306:61-6.
6. Laemmli UK. Cleavage of structural proteins during the assembly of the head of bacteriophage T4. Nature 1970;227:680-5.

EVALUATION OF THE BIOLUMINESCENCE-ENHANCED ZONA BINDING ASSAY

W Miska[1], T Monsees[1], DR Franken[2] and R Henkel[1]
[1] Centre for Dermatology and Andrology, Justus Liebig University Giessen, Gaffkystr. 14, 35385 Giessen, Germany
[2] Department of Obstetrics and Gynaecology, Tygerberg Hospital, Tygerberg 7505, South Africa

Introduction

The zona pellucida is the first of two physiological barriers for sperm entry into the ooplasm. In addition to the induction of the acrosome reaction, the zona pellucida prevents polyspermy and supports species specific recognition of spermatozoa. The zona facilitates sperm binding, which is mediated by O-linked carbohydrate side chains. Therefore, interaction of spermatozoa with the oocyte at the site of the zona pellucida plays a crucial role in fertilization and has been shown to be predictive for fertilization in vitro. For this purpose several zona binding assays have been developed. In a competitive zona binding assay described by (1) spermatozoa from the patient and the donor are marked with different fluorescent dyes and the ratio of the differently marked spermatozoa bound to the zona is calculated. The zona assay that has got practical importance in male factor infertility diagnosis and has been evaluated in an IVF program, is the hemizona assay (HZA) (2). In this assay, devitalized human oocytes are micro-bisected into two hemispheres and incubated with patient's and donor sperm, respectively. However, in both these zona binding assays relatively fresh zonae, which naturally are of changing quality, and an inverted microscope including micromanipulation equipment, is necessary. Therefore, the objective of this study was to develop an alternative method, which is simple, less time consuming, reproducible and sensitive. The latter was achieved by using the bioluminescent system of the firefly luciferase, which is the most effective bioluminescent system in nature (3). For the present, this new assay was established in the porcine system.

Material and Methods

D-Luciferin-O-ß-galactopyranoside (Lu-Gal) (Fig.: 1A) was synthesized from 2-cyano-6-hydroxybenzothiazole, 2,3,4,6-tetra-O-acetyl-α-D-galactopyranosyl bromide and D-cysteine. The product, 2-cyano-6-(2,3,4,6-tetra-O-acetyl-α-D-galacto-pyranosyloxy) benzothiazole, was cristallized from ethanol and deacylated at room temperature over night with dry methanol. Subsequent purification was performed by means of Biogel P2 and HPLC. Chemical characterization and investigation of the kinetic data of Lu-Gal were performed (4).

For isolation of porcine oocytes fresh ovaries were minced and filtered through nylon mesh nets of 200 µm and 80 µm, respectively. Subsequently, the residue from the 80 µm mesh net containing oocytes and some debris was washed with 100 mM PBS, pH 7.3, and separated by means of density gradient centrifugation in Percoll (40% - 20% - 10%) for 30 min at 2000 x g (5). The oocytes were taken from the

Percoll interphase (20% - 10%), washed in PBS once again to remove Percoll, and were stored until use in PBS/glycerol (1:1) at -20°C.

The zonae pellucidae were dissolved under visual control in 100 mM NaH_2PO_4, pH 2.0, at room temperature and degenerated ooplasm was removed by means of a micropipette. Subsequently, the zona solution was neutralized with 0.1 N NaOH. Afterwards, zona proteins were biotinylized with biotin-N-hydroxysuccinimide ester for 4 hours at room temperature and dialysed against 20 mM NaH_2PO_4, pH 7.5. Biotinylated zonae could be stored until use at -20°C.

Motile boar spermatozoa separated by means of glass wool filtration using SpermFertil® columns (Mello, Exeter, UK) or density gradient centrifugation in Percoll (80% - 70% - 55% - 40%). Afterwards, motile spermatozoa were washed in Tris-buffered medium (TBM; 5 mg BSA/mL, without Ca^{2+}) according to (6) and preincubated for 1 hour at 39°C. After final adjustment of sperm concentration to 10 - 30 x 10^6 spermatozoa/mL, 100 µL of this sperm suspension was incubated for 30 min at 39°C with a solution of 2.5 solubilized and biotinylized zonae (+/- addition of non-labelled, solubilized zona) plus streptavidin-ß-galactosidase. Non-bound enzyme was removed by centrifugation. The pellet was resuspended in 2.5 x 10^{-7} mol/L Lu-Gal in 50 mM K_2HPO_4, 50 mM KH_2PO_4, 1 mM $MgCl_2$, 100 mM ß-mercaptoethanol, pH 7.8, and incubated for 30 min. at 25°C. An aliquot of this suspension was mixed with the detection cocktail containing 41 mM HEPES, 5 mM $MgCl_2$, 2.6 mM ATP, 0.15 mM DTT, 0,05% BSA, luciferase 25 mg/mL and was transferred into a black microtiter plate. The emitted light was read using a MTP-Reader (Hamamatsu, Herrsching, Germany). The chemical reactions involved are illustrated in figure 2. Intra-species specificity of this novel bioluminescent assay was evaluated using the HZA.

In addition, inter-species specificity was determined. For this, human, bovine and hamster spermatozoa were separated by means of the swim-up technique and capacitated subsequently for 90 min. After washing the sperm concentration was adjusted to 10 - 30 x 10^6 sperm/mL. Subsequently, solubilized, labelled zona was added and the assay was run as described above.

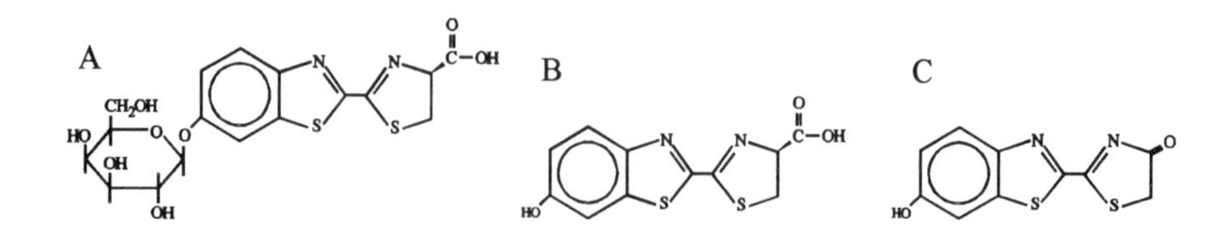

Figure 1: A: Bioluminogenic substrate D-luciferyl-O-ß-galactopyranoside (Lu-Gal)
B: D-Luciferin; Substrate for the enzyme luciferase.
C: Oxiluciferin; Product after reaction with luciferase.

Results

Synthesis of Lu-Gal (Fig.: 1A) resulted in purity better than 99,998% after HPLC compared to luciferin (Fig.: 1B). Analytical data and spectra were appropriate. In terms of the kinetic data, Lu-Gal is a highly sensitive and specific substrate for ß-galactosidase (Table 1). By using the described method for isolation of porcine zonae pellucidae, the analysis of the proteins by means of SDS-PAGE was according to results described in the literature. Using the given chemical conditions, non-labelled zona bound to boar spermatozoa competitively, resulting in high sensitivity and specificity. The addition of 10 non-labelled zonae could almost completely inhibit binding of labelled zonae (Fig. 3A). Comparing this bioluminescent system with the HZA, corresponding results were obtained. On the other hand, spermatozoa of other species showed no or only less binding to the porcine zona. A competitive displacement could not be observed, indicating the inter-species specificity of the assay (Fig. 3B).

A)

$$\text{Lu-Gal} \xrightarrow[\text{(ZP-bound)}]{\text{ß-galactosidase}} \text{D-Luciferin} + \text{galactose}$$

B)

$$\text{D-Luciferin} + \text{ATP} + O_2 \xrightarrow{\text{Luciferase}/MgCl_2} \text{Oxyluciferin} + \text{AMP} + \text{PP} + CO_2 + h \cdot \nu$$

Figure 2: Reaction scheme (A) of the cleavage of D-Luciferyl-O-ß-galactopyranoside (Lu-Gal) by means of ß-galactosidase to D-luciferin and galactose, and (B) the oxidation of D-luciferin to oxiluciferin (Fig. 1C) by means of luciferase. The emitted light was detected by means of a MTP-reader.

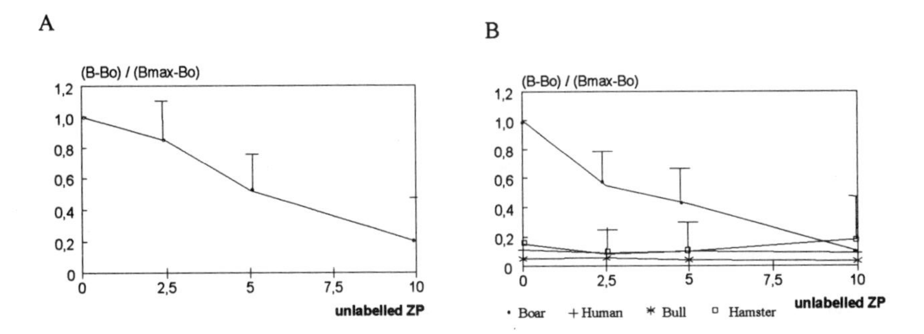

Figure 3: Intra- (A) and inter species specificity (B) of the bioluminescence-enhanced zona binding assay.

Table 1: Kinetic data of the hydrolysis of different D-luciferin derivates by carboxypeptidase N, ß-galactosidase and alkaline phosphatase.

Enzyme	Substrate	Km (mol/L)	k_{cat} (1/s)	k_{cat}/Km (L $mmol^{-1}xs^{-1}$)
Carboxypeptidase N	D-Luciferyl-N-α-arginine	$0{,}5 \times 10^{-4}$	11,8	236
ß-Galactosidase	D-Luciferin-O-ß-galactopyranoside	$2{,}9 \times 10^{-6}$	256	90.000
Alkaline phosphatase	D-Luciferin-O-phosphate	$4{,}3 \times 10^{-5}$	1.010	23.500

Conclusions

The presented data indicate high intra- and inter-species specificity of this novel, homogenous, bioluminescence enhanced zona binding assay. In contrast to the use of a conventional bioluminogenic substrate luciferin-O-phosphate (7), Lu-Gal showed an extreme lowe rate of spontaneous hydrolysis in aqueous solution as well as a very low back-ground luminescence. Compared to the conventional hemizona assay this assay is much less time-consuming. Moreover, due to the use of a homogeneous mixture of many zonae this assay shows a better reproducibility than the hemizona assay. Therefore, the transfer of this zona binding assay to the human system in andrological diagnosis seems to be possible on principle.

Acknowledgements

The work has been supported by the Federal Ministry of Education and Research, grant no. KY9105/6.

References

1. Liu DY, Lopata A, Johnston WIH, Baker HWG. A human sperm-zona pellucida binding test using oocytes that failed in vitro. Fertil Steril 1988;50:782-8.
2. Burkman LJ, Coddington CC, Franken DR, Kruger TF, Rosenwaks Z, Hodgen GD. The hemizona assay (HZA): development of a diagnostic test for the binding of human spermatozoa to the human hemizona pellucida to predict fertilization potential. Fertil Steril 1988;49:688-97.
3. McElroy WD, DeLuca M. Firefly luminescence. In: Burr JG (ed) Chemi- and Bioluminescence, New York, Marcel Dekker, Inc., 1985:387-400.
4. Geiger R, Schneider E, Wallenfels K, Miska W. A new ultrasensitive bioluminogenic enzyme substrate for ß-galactosidase. Biol Chem Hoppe-Seyler 1992;373:1187-91.
5. Hokke CH, Damm JBL, Penninkhof B, Aitken RJ, Kamerling JP, Vliegenthart FG. Structure of the O-linked carbohydrate chains of porcine zona pellucida glycoproteins. Eur J Biochem 1994;221:491-512.
6. Clarke RN, Johnson LA. Effect of liquid storage and cryopreservation of boar spermatozoa on acrosomal integrity and the penetration of zona-free hamster ova in vitro. Gamete Res 1987;16:193-204.
7. Geiger R, Miska W. Bioluminescence enhanced enzyme immunoasay. II. New ultrasensitive detection systems for enzyme immunoassays. J Clin Chem Clin Biochem 1987;25:31-8.

THE USE OF ADENYLATE KINASE FOR THE DETECTION AND IDENTIFICATION OF LOW NUMBERS OF MICROORGANISMS

MJ Murphy[1], *DJ Squirrell*[1], *MF Sanders*[2] *and R Blasco*[2]
[1]*PLSD Porton Down, Salisbury, SP4 0JQ, UK,*
[2]*Central Science Laboratory, Sand Hutton, York, YO4 1LV, UK*

Introduction

ATP bioluminescence has become an established technique for the rapid detection of microorganisms. The technique has a detection limit of 10^3 bacterial cells although in practice it is usually nearer 10^4. For many applications, such as in the personal care product and pharmaceutical industries, where lower numbers of organisms are of concern, ATP bioluminescence is insufficiently sensitive.

Adenylate kinase (E.C. 2.7.4.3) is an essential enzyme in virtually all cells. It catalyses the equilibrium reaction:

$$Mg^{++}.ATP + AMP \overset{AK}{\rightleftharpoons} Mg^{++}.ADP + ADP$$

By using ADP as the substrate to drive the reaction to the left, ATP can be generated. This ATP can then be measured in the usual way, using firefly luciferase (1). AK has a turnover number of approximately 40 000 per minute and represents about 0.1% of cell protein. In a typical bacterial cell there are around 1000 molecules of ATP for every 1 molecule of AK, allowing a 40-fold amplification of ATP per minute. We have previously shown (2) that by using AK as the cell marker rather than ATP, a 10-100 fold increase in sensitivity can be achieved if 5 minutes are allowed for the AK reaction. By increasing the reaction time, fewer than 10 organisms are detectable(3). This extends rapid microbiology via bioluminescence to applications where very low levels of contamination must be detected.

Many areas require a rapid, specific technique for particular contaminants rather than a generic "dirt" meter as provided by ATP bioluminescence or the more sensitive AK approach. At present selective media are often used to culture common contaminants, resulting in 18-36 hours before products can be confirmed as uncontaminated. We have developed two approaches, based on the AK method, to provide detection of specific organisms in less than 3 hours.

In the first approach, antibodies against target organisms are coupled to magnetic microbeads. When mixed with a contaminated product, the target cells are selectively removed from suspension. The beads with captured cells are then washed to remove unbound material, and a generic lysis step carried out to release AK. This is, in effect, an immunoassay where the cell itself provides the label.

The second approach involves selective lysis of the target population by bacteriophages. This method has been previously used with ATP bioluminescence for the detection of *Listeria monocytogenes* (4). The method could detect 2.5×10^5 cells and involved selective pre-enrichment of *Listeria* from food samples before addition of the bacteriophage. The total assay time was 15 hours, with the bioluminescent step taking 80 minutes. Using the AK approach, the need for pre-enrichment may be eliminated since lower numbers of organisms are detectable at the bioluminescent stage of the assay.

Materials and Methods

Instrumentation: Manual light output measurements were performed using a portable Multilite tube luminometer (Biotrace, Bridgend, UK). All bioluminescence procedures were carried out in a clean hood (Bassaire, Southampton, UK).

Reagents: The materials for determining AK were developed under contract and supplied by Celsis, Cambridge, UK. These were: 15mmol/L magnesium acetate; microbial AK assay reagent (a mixture of 2mmol/L ADP and a proprietary extractant); 2mmol/L ADP and Celsis Low Decay Rate (LDR) bioluminescence reagent. *Escherichia coli* (NCIMB 10243) and *Salmonella typhimurium* (NCIMB 13034) were obtained from the National Collection of Industrial and Marine bacteria, Aberdeen, UK. *E.coli* specific phage was obtained from the American Type Culture Collection, Rockville, Maryland, USA. Dynabeads (anti-Salmonella) were obtained from Dynal (UK) Ltd, Wirral, Merseyside, UK.

AK Determination: Samples for AK determination were prepared in phosphate buffered saline (PBS). 0.1mL magnesium acetate and 0.1mL of AK reagent were added to 0.1mL sample in a 3.5mL tube. After 5 minutes (also 30 minutes for the second magnetic bead assay) incubation at room temperature, 0.1mL LDR bioluminescence reagent was added and the light output measured, after a 5 second lag, using an integration time of 10 seconds.

Magnetic Bead Assay: A culture of *S.typhimurium* was grown overnight at 37°C in nutrient broth. The cells were harvested and washed with 3x1mL PBS. A standard AK assay was performed to determine approximate bacterial levels. 0.1mL bacterial dilution was added to 0.01mL Dynabeads (washed 3 times with PBS) in a 1.5mL eppendorf tube. This was incubated for 1 hour at 37°C on a mixing wheel. After incubation the beads were immobilised with a magnet and the supernatant removed. The beads were then washed 3 times with 0.5mL PBS. A standard AK assay was carried out on both the supernatant and the washed beads. Dilution plate counts of the initial samples were used to determine bacterial numbers.

Bacteriophage Mediated Cell Lysis Experiments: *E. coli* was grown overnight at 37°C with shaking in brain heart infusion (BHI) broth to provide a stock solution. The samples for analysis were made by an appropriate dilution of the stock in BHI broth to give a concentration of approximately 5×10^3 cells per mL. Samples were split into 2 equal 10mL fractions. 0.01mL of 10^{10} mL^{-1} phage was added to one, the other served as an uninfected control. These were incubated in a water bath at 37° C. At timed intervals 0.1mL volumes were removed and assayed, both as described above with extractant to determine total AK levels, and with ADP alone to determine the fraction of AK released by the lytic activity of the phage. ADP alone with the addition of neither extractant nor phage was used to measure extracellular AK which may have been released spontaneously. Cell numbers were determined from the uninfected sample using the pour plate method.

Results and Discussion

Capture and Non-Specific Lysis: The magnetic bead AK assay detected fewer than 1000 cells in the initial suspension, (5 minute amplification, total assay time 90 min). Results are shown in Figure 1. Since fewer than 100 free cells are detectable by a standard AK assay, there is potential for improvement. It was noticed that the background signals obtained from the bead fraction were higher than those found in assays on cells in suspension. This may be attributed to the blocking agent (BSA) added to the beads to prevent non-specific binding. BSA contains small levels of AK which contribute to background signals. Low AK containing blocking agents such as methylated BSA could be used to reduce this background. Alternatively lower concentrations of blocking agents could be used. Greater sensitivity may also be achieved by increasing the time allowed for the ATP generating reaction. Table 1 shows results from an experiment where a 30 min AK assay was carried out after the standard immunoassay.

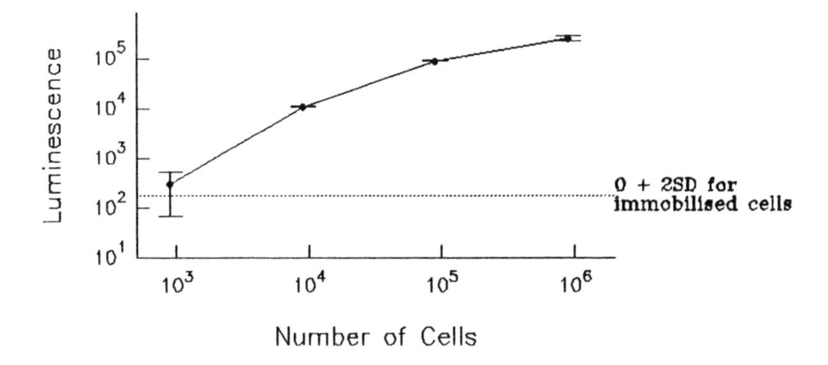

Figure 1, Magnetic Bead Assay for *Salmonella* using Adenylate Kinase with a 5 minute amplification time (n=3). Detection limit 600 cells .

Number of Cells Initially in Suspension	Supernatant Associated Luminescence	Bead Associated Luminescence	% Total Activity Associated with Beads
350 000	263 000	195 400	43%
35 000	36 000	36 400	50%
3 500	12 000	24 400	67%
350	5 000	17 900	78%

Table 1. Magnetic Bead Immunoassay for *Salmonella,* using a 30 minute Adenylate Kinase assay as the endpoint. Proportion of bacteria immobilised onto the beads.

By increasing the AK amplification time to 30 minutes, 350 cells could be detected with a signal to noise ratio of 3.6 to 1. The proportion of bacteria immobilised onto the beads was found to be dependent on the initial cell concentration. The beads were used at a constant 10^6 per tube. The manufacturers recommend that at least 4 beads are present for every target cell : these results suggest that a larger ratio is required. Further development of the method is needed to establish conditions to optimise the blocking of non-specific binding, incubation volumes, temperature and time for the immunoassay, and the incubation time for the AK assay itself. It is anticipated that the detection limits can be improved further and should approach that obtained from bacteria free in suspension ie. fewer than 100 cells.

Bacteriophage Mediated Cell Lysis: The results from triplicate assays using bacteriophage mediated cell lysis are shown in figure 2. This compares detergent and phage induced release of AK from infected and non-infected cells. The total amount of AK present, shown by detergent extraction, increased with time and remained relatively equal whether or not phage was present. It can be seen that AK began to be released from cells with added phage 25 to 30 minutes post infection. After 40 minutes, 30% of the total AK

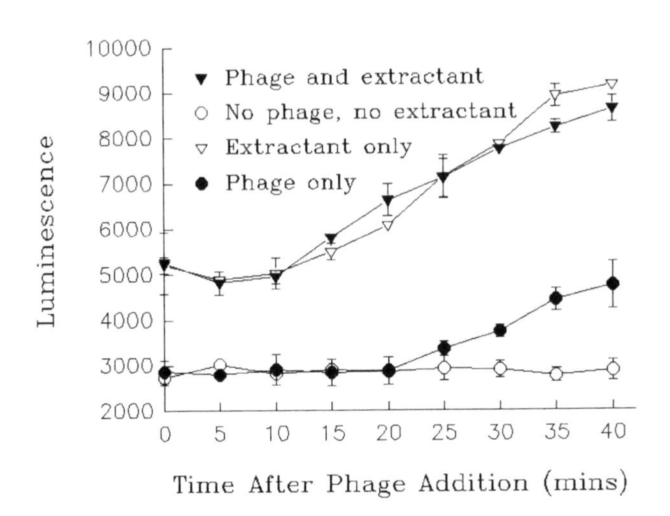

Figure 2, Comparison of the release of adenylate kinase from *E. coli* by phage and detergent extraction. Initial concentration 550 cells per 0.1mL sample grown at 37°C in Brain Heart Infusion broth. (n=3 ± 1SD)

content was released by the phage. This suggests that much of the cellular AK remained enclosed in some sort of membrane system, impermeable to AK or ADP but which could be disrupted by the detergent. 550 *E.coli* cells per 0.1mL sample were used and, at 35 mins post infection, gave a signal:noise ratio of 1.6:1 over the non-infected, non extracted control. A 5 min incubation AK assay was used as the end point giving a total assay time of less than 1 hour.

The results presented here illustrate that specific assays using AK as an end-poin4t determination are both quicker and more sensitive than conventional immunoassay techniques. A typical ELISA takes 2-3 hours to perform and can detect 10^4 to 10^6 cells per ml. By harnessing the cells intracellular component (AK) as the label, there is no need for a secondary, labelled antibody in the magnetic bead immunoassay. The use of bacteriophage allows assay to be tailored to specific applications since the phage can be chosen to be as specific or generic as required. Both techniques can detect fewer than 1 000 cells in 1 to 1.5 hours. They will only detect intact, viable organisms. There is some evidence that viable non-culturable cells can also be measured via their AK content.

References

1. Brolin SE, Borglund E, Agren A. Photokinetic microassay of adenylate kinase using the firefly luciferase reaction. J Biochemical and Biophysical Methods 1979;1:163-9.
2. Squirrell DJ, Murphy MJ. Adenylate kinase as a cell marker in bioluminescent assays. In: Campbell AK, Kricka LJ, Stanley PE, editors. Bioluminescence and Chemiluminescence; Fundamentals and Applied Aspects. Chichester: John Wiley and Sons,1995;486-9.
3. Squirrell DJ and Murphy MJ. Rapid detection of very low numbers of microorganisms using adenylate kinase as a cell marker. In: Stanley PE, Simpson WJ, Smither R. Editors. A Practical Guide to Industrial Uses of ATP-luminescence in Rapid Microbiology. Lingfield: Cara Technology (in press).
4. Sanders MF. A rapid bioluminescent technique for the detection and identification of *Listeria monocytogenes* in the presence of *Listeria innocua*. In: Campbell AK, Kricka LJ, Stanley PE. Editors. Bioluminescence and Chemiluminescence :Fundamentals and Applied Aspects. Chichester. John Wiley and Sons, 1995;454-7.

INTERACTION OF CARBOHYDRATES WITH HYPOCHLOROUS ACID AND HYDROXYL RADICALS

J Arnhold, J Schiller and K Arnold
Institute of Medical Physics and Biophysics, Medical Department, University of Leipzig, Liebigstr. 27, 04103 Leipzig, Germany

Introduction

Stimulated polymorphonuclear leukocytes generate a number of reactive oxygen species (ROS) including $O_2{\cdot}^-$, H_2O_2, HOCl, and ·OH. ROS play an important role in antimicrobicidal activity and also tissue injury in a number of pathologies. Especially hypochlorous acid and hydroxyl radical cause deleterious effects on surrounding tissue components. Hypochlorous acid is formed from hydrogen peroxide and chloride anions in a myeloperoxidase catalysed reaction. This enzyme is unique for neutrophils and monocytes. It is stored in azurophilic granules of these cells and released upon stimulation. Hydroxyl radical is derived from the Fenton reaction. Another source for hydroxyl radical is the disruption of peroxynitrite. Finally, hydroxyl radicals are also formed in the result of reaction of HOCl with superoxide anion radicals.

Reactions of ROS with polymeric carbohydrates are involved in the pathogenesis of inflammatory diseases (e.g. rheumatoid arthritis). Hyaluronic acid, chondroitinsulphate and keratansulphate are important constituents of articular cartilage. On the other hand, many cells are covered by polysaccharides, the glykocalix. Here, chemiluminescent methods are described for the determination of second order rate constants between different soluble carbohydrates and hypochlorous acid or hydroxyl radicals.

Materials and Methods

Instrumentation Luminescence measurements were performed using a luminometer AutoLumat LB 953 (Fa. Berthold, Wildbad, Germany).

Materials All mono- and polysaccharides of highest purity were obtained from Fluka (Switzerland). Hyaluronic acid from human umbilicial cords and chondroitinsulphate from bovine trachea were used. They were dialysed against 50 mmol/L isotonic phosphate buffer pH 7.4 to remove low-molecular components detectable by ^{1}H-NMR. Luminol was a product of Boehringer Mannheim (Germany). Phthalic hydrazide was from Aldrich (Germany). Ferrous chloride, sodium hypochlorite and hydrogen peroxide were purchased from Sigma (Deisenhofen, Germany).

Chemiluminescence measurements 950 µL of carbohydrate solutions and luminol or phthalic hydrazide in 0.14 mol/L NaCl, 0.05 mol/L phosphate pH 7.4 were coincubated at 37 °C. Then 50 µL of NaOCl (final concentration 2.5 $\times 10^{-6}$ mol/L) was added. Photons were counted over the next 10 s. To reveal effects of hydroxyl radicals on carbohydrate solutions first H_2O_2 (final concentration 2.5 $\times 10^{-6}$ mol/L) and then $FeCl_2$ (final concentration 2.5 $\times 10^{-6}$ mol/L) prepared freshly were injected. Photons were also counted for 10 s.

Results and Discussion

A chemiluminescence flash appears mixing luminol and sodium hypochlorite. Light intensity depends on the presence of other compounds competing with luminol for hypochlorous acid. This method has been used to compare reactivities of different proteins, amino acids and other molecules with hypochlorous acid (1). The inhibition of chemiluminescence intensities in the oxidation of luminol by hypochlorous acid by selected monomeric and polymeric carbohydrates is given in Fig. 1a. All curves show a sigmoidal shape of inhibition of chemiluminescence with increasing concentrations of carbohydrates. Only in the case of glucose light intensities were not changed.

Straight lines appear with an intercept at 1 on the ordinate if the quotient of I_O/I_C is plotted versus the carbohydrate concentration (Fig. 1b). I_O and I_C mean the luminescence intensities detected in the absence or presence of carbohydrates, respectively. Assuming the reaction of luminol as well as the reaction of carbohydrates with hypochlorous acid as first order reactions in respect to the concentration of hypochlorous acid the following equation can be derived:

$$\frac{I_o}{I_c} = 1 + \frac{k_b}{k_a[A]}[B].$$

k_a and k_b are the rate constants of luminol or carbohydrates, respectively, in its reaction with hypochlorous acid. [A] and [B] represent the concentrations of luminol or carbohydrates. Using a luminol concentration of 10^{-5} mol/L and $k_a = 5 \times 10^5$ $Lmol^{-1}s^{-1}$ (2) reaction constants k_b for different carbohydrates with HOCl can be calculated. These data are given in Tab. 1.

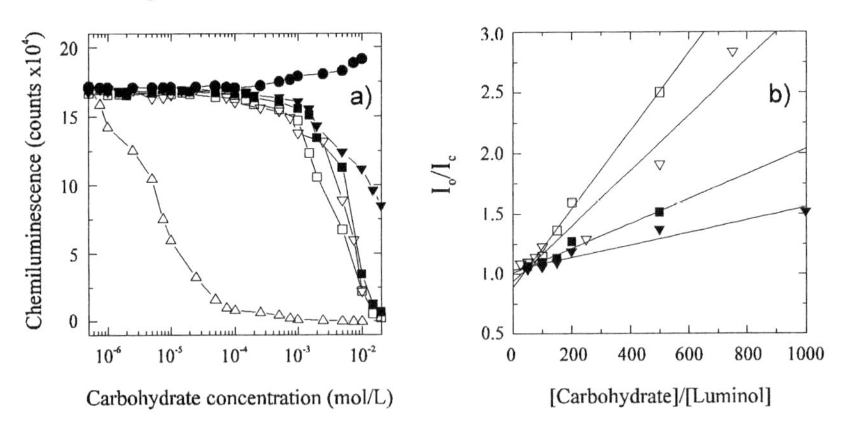

Fig. 1. Inhibition of chemiluminescence in oxidation of luminol by hypochlorous acid as a function of concentration of glucosamine (Δ), hyaluronic acid (□), chondroitinsulphate (■), saccharose (▼), N-acetylglucosamine (∇) and glucose (•). The concentration of dimeric and polymeric carbohydrates is related to the amount of monomers.

The highest rate constant was found for glucosamine from all substances tested. The amino group reacts easily with HOCl to form chloramine compounds. All substances containing N-acetyl groups show second order rate constants from 1.6 x 10^3 to 5 x 10^2 $Lmol^{-1}s^{-1}$. N-Acetyl groups react with HOCl to yield a transient chlorinated product from that acetate is splitted off (3). Additionally, a breakdown into smaller oligomeric units occurs in polymeric carbohydrates like hyaluronic acid and chondroitinsulphate.

Tab. 1. Second order rate constants for the reaction of carbohydrates with hypochlorous acid

Carbohydrate	Second order rate constant ($Lmol^{-1}s^{-1}$)
Glucose	no reaction
Glucuronic acid	no reaction
Glucosamine	9.75×10^5
N-Acetylglucosamine	1.15×10^3
N-Acetylgalactosamine	1.30×10^3
Saccharose	2.60×10^2
Hyaluronic acid	1.62×10^3
Chondroitinsulphate	5.15×10^2

To study the interaction of hydroxyl radicals with carbohydrates unsubsti-tuted phthalic hydrazide has been used instead of luminol. The oxidation of this compound by NaOCl or other oxidants is not accompanied by chemiluminescence. However, hydroxyl radicals cause a hydroxylation of the aromatic ring of phthalic hydrazide followed by a subsequent oxidation resulting in light emission (4). Ferrous chloride was added to a solution of phthalic hydrazide, carbohydrates and hydrogen peroxide to generate hydroxyl radicals. The chemiluminescence is inhibited by higher concentrations of carbohydrates (Fig. 2a) due to a competition of carbohydrates with phthalic hydrazide for hydroxyl radicals. As in oxidation of luminol by HOCl straight lines result with an intercept at 1 on the ordinate if the quotient of I_0/I_c is plotted versus the carbohydrate concentration (Fig. 2b). Using a concentration of phthalic hydrazide of 10^{-5} mol/L and a rate constant of 4.3×10^9 $Lmol^{-1}s^{-1}$ (5) for the reaction between phthalic hydrazide and hydroxyl radicals second order rate constants for the reaction of carbohydrates with hydroxyl radicals can be calculated. Results are given in Tab. 2.

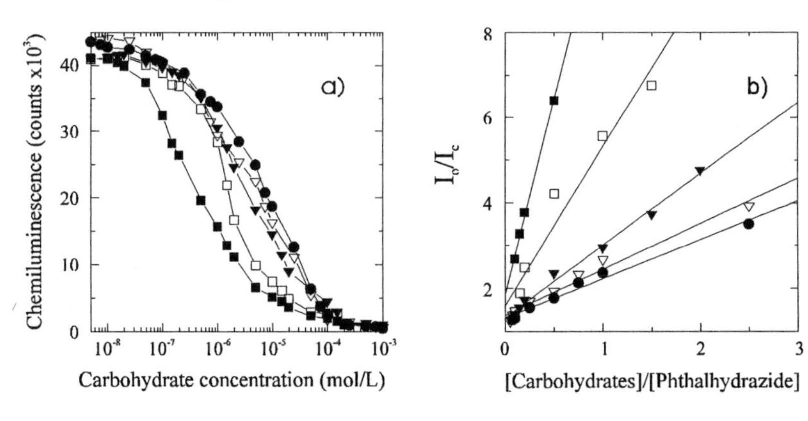

Fig. 2. Inhibition of chemiluminescence in oxidation of phthalic hydrazide by hydroxyl radicals generated in a Fenton reaction as a function of carbohydrate concentration. Symbols are the same as in Fig. 1.

All monosaccharides tested show an identical reactivity towards hydroxyl radicals with a rate constants of about 3.9-4.5 x 10^9 $Lmol^{-1}s^{-1}$. These values increase using saccharose and polymeric carbohydrates. It seems that the presence of glycosidic bonds and especially of sulphur containing functional groups enhances the sensitivity of carbohydrates to hydroxyl radicals.

Tab. 2. Second order rate constants for the reaction of carbohydrates with hydroxyl radicals

Carbohydrate	Second order rate constant ($Lmol^{-1}s^{-1}$)
Glucose	3.9×10^9
Glucuronic acid	4.0×10^9
N-Acetylglucosamine	4.6×10^9
N-Acetylgalactosamine	4.5×10^9
Saccharose	7.2×10^9
Hyaluronic acid	1.6×10^{10}
Chondroitinsulphate	3.9×10^{10}

The interaction of carbohydrates with hydroxyl radicals generated in a Fenton reaction was also studied using the luminol chemiluminescence (data not shown). However, an increase in light emission occurs at higher concentrations of carbohydrates coincubated. The chemilumines-cence is inhibited only at concentrations higher than 3-10 x 10^{-3} mol/L. The highest luminescence increase is found for chondroitinsulphate followed by hyaluronic acid. However, the complex chemistry of events in luminol oxidation makes an interpretation very difficult. Hydroxyl radicals cause two effects on luminol, a hydroxylation of the aromatic ring and the formation of semidiazaquinone radicals (6). Hydroxylated luminol acts as radical scavenger. The other pathway results in the formation of excited 3-aminophthalate. Carbohydrates and/or its radicals seem to interfere with luminol radicals influencing the chemiluminescence yield.

Although the chemistry of reactions leading to light emission in the oxidation of luminol or phthalic hydrazide is very complex it is possible to use these systems to study the interaction of carbohydrates with hypochlorous acid or hydroxyl radicals. These methods allow to detect differences in reactivities of various carbohydrates with reactive oxygen species. An understanding of these reactions is necessary to better know processes of tissue injury under pathological conditions.

Acknowledgements
This work was supported by grants (INK 23 A1-1, Graduiertenkolleg Molecular and cell biology of connective tissue) from the Deutsche Forschungsgemeinschaft.

References

1. Arnhold J, Mueller S, Arnold K, Sonntag K. Mechanisms of inhibition of chemiluminescence in the oxidation of luminol by sodium hypochlorite. J Biolumin Chemilumin 1993; 8: 307-13.
2. Baxendale JH. Pulse radiolysis study of the chemiluminescence from luminol (5-amino-2,3-dihydrophthalazine-1,4-dione). J Chem Soc Faraday Trans I 1973; 69: 1665-77.
3. Schiller J, Arnhold J, Gründer W, Arnold K. The action of hypochlorous acid on polymeric components of cartilage. Biol Chem Hoppe-Seyler 1994; 375: 167-72.
4. Reitberger T, Gierer J. Chemiluminescence as a means to study the role of hydroxyl radicals in oxidative processes. Holzforschung 1988; 42: 351-6.
5. Schiller J. unpublished results.
6. Merényi G, Lind J, Eriksen TE. Luminol chemiluminescence: chemistry, excitation, emitter. J Biolumin Chemilumin 1990; 5: 53-6.

INTERACTION OF NO, ENDOTHELIN AND OXALATE IN PATIENTS WITH SYSTEMIC INFLAMMATORY RESPONSE SYNDROME (S.I.R.S.)

S Albrecht, T Zimmermann[1], M Freidt, T Freidt[2], HD Saeger[1], W Distler
Dept of Gynecology and Obstetrics and [1]Dept of Surgery,
Technical University Dresden, Fetscherstrasse 74, D-01307 Dresden, Germany
[2]Medical Laboratory Bautzen, Töpferstrasse 17, D-02625 Bautzen, Germany

Introduction

The endothelium derived relaxing factor - NO - and the vasoconstrictor Endothelin play an important role in cases of acut inflammatory diseases, reperfusion phenomenon or atherogenesis. The authors developed high sensitive chemiluminometric methods for direct determination of plasma NO and Oxalate of septic patients. The synergism of Endothelin/NO/Oxalate was investigated to may be a potent novel regulation system to maintain the homeostasis of septic or polytraumatic patients.

Material and Methods

Determination of Endothelin

We used two commercial Immunoassays for the Endothelin measurement in human heparin plasma:

a) a competitive Radioimmunoassay with extraction (Fa. Laboserv, Giessen, Germany)
b) a competitive Enzymimmunoassay without extraction (Fa. Immundiagnostik, Bensheim, Germany).

The results of both methods showed a high correlation and were comparable (r = 0,97).

Determination of Nitric Oxide (NO)

We developed a high sensitive chemiluminometric method for direct determination of plasma NO by using the reaction with H_2O_2 and luminol [1]:

Reagent: Luminol was dissolved in HANK's solution up to an endconcentration of 10^{-4} mol/l. A concentrated H_2O_2 solution was added up to an endconcentration of 1 %.

Samples: We used 300 μl of human heparin plasma diluted with NaCl (0,9 %) in the ratio 1 + 9.

Measurement: After injection of 300 μl of the reagent we measured the integral light signal over a period of the first 10 s (Luminometer LB 9503, Fa. EG & G Berthold, Wildbad, Germany). The recorded measuring signal in RLU (rel. light units) correlates directly with the NO concentration in the sample. A relative standardization of the method can be achieved by bubbling of NO through the samples solutions.

Chemistry:

$$2\,NO + 3\,H_2O_2 \rightarrow \underset{\text{Peroxynitrite}}{2\,HOONO} + 2\,H_2O$$

$$2\,HOONO + \text{Luminol} \xrightarrow{\text{spontaneous}} \text{o-Aminophthalate} + \boxed{h\upsilon} + \text{products}$$

Determination of Oxalate

The determination of plasma oxalate was performed chemiluminometrically like previously described [2, 3]:

$$HOOC\text{-}COOH + H_2O_2 \xrightarrow{\text{Carbodiimide/Fluorescer}} 2\,CO_2 + 2\,H_2O + \boxed{h\upsilon}$$

Results and Discussion

In this study we invastigated 10 patients suffering from "systemic inflammatory response syndrome (S.I.R.S.) without lethal outcome over a period of 28 days. We found increased NO concentrations in all S.I.R.S. cases after 2 or 3 days.

The initial founded high Endothelin concentrations decreased by persistent NO production. The intracellular oxalate concentrations exceed that in the plasma by at least two or-orders of magnitude. In all cases of S.I.R.S. we found decreased intracellular (leucocytes) oxalate concentrations in comparison to healthy volunteers. The plasma oxalate concentration is on the other hand increased in patients with S.I.R.S.

The enzyme catalyzed oxalate metabolism known in the vegetable kingdom could indicate that an oxalate oxidase pathway may also be important in human metabolism:

$$HOOC\text{-}COOH + O_2 \xrightarrow{\text{Ox.-Oxidase}} 2\,CO_2 + H_2O_2$$

The reaction products have important biochemical functions, e.g. H_2O_2 forms a substantial component of the "respiratory burst" of cells capable of phagocytosis.

High oxalate levels correlated with low NO - concentrations in S.I.R.S.-patients - an indication of a possible protective role for oxalate to metabolize NO:

$$NO \xrightarrow{H_2O_2} \text{Peroxynitrite} \longrightarrow {}^1O_2\text{, products}$$

O_2-radical-dependend reactions causes the translocation of the transcription factor NF-kB into the cell nucleus - a main step of the cellular acute phase response to inflamma tory stimuli. Fig. 2 indicates the possible pathways to generate reactive radicals based on oxalate.

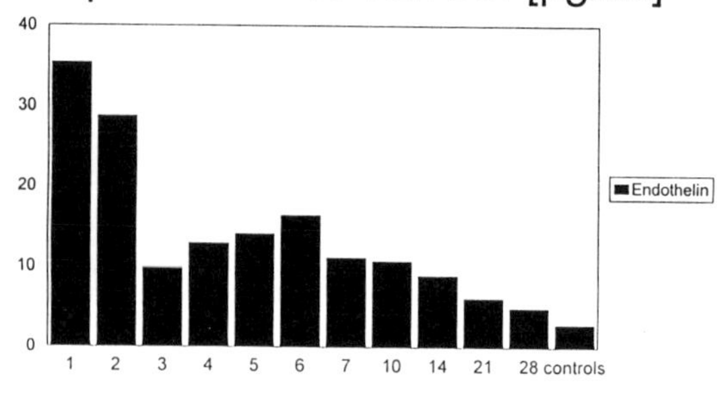

Fig.1: a) Plasma Endothelin concentrations in patients with S.I.R.S. (n = 10)

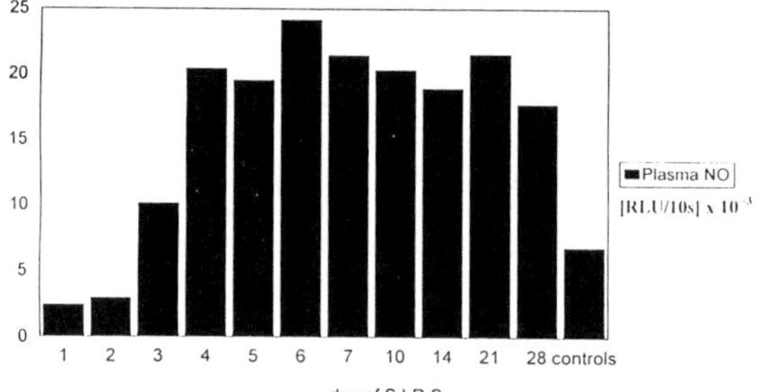

Fig.1: b) Relative NO-concentrations in patients with S.I.R.S. (n = 10)

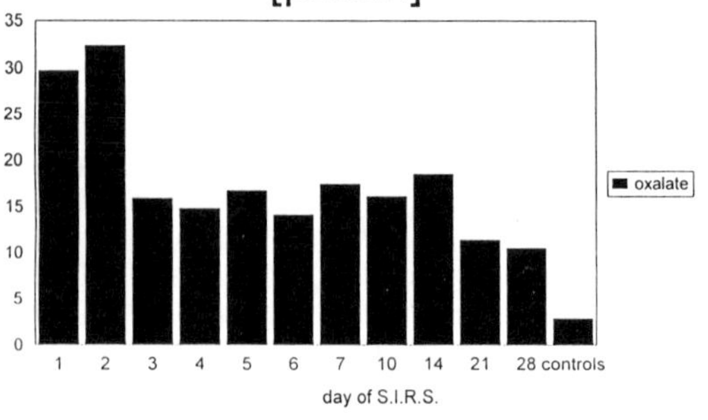

Fig.1: c) Plasma Oxalate concentrations in patients with S.I.R.S. (n = 10)

References

1. Kikuchi K, Nagano T, Hayakawa H, Hirata Y, Hirobe M. A novel chemiluminometric method for NO determinations. Anal Chem 1993;65:1794-1799.
2. Albrecht S, Hornak H, Freidt T, Böhm WD, Weis K, Reinschke A. Chemiluminometric determination of serum oxalate. Biolumin Chemolumin J 1993;8:21-24.
3. Albrecht S, Brandl H, Schönfels C. Human oxalate - only an endproduct of metabolism? Angew Chem Int Ed Engl 1994;33:1780-1781.

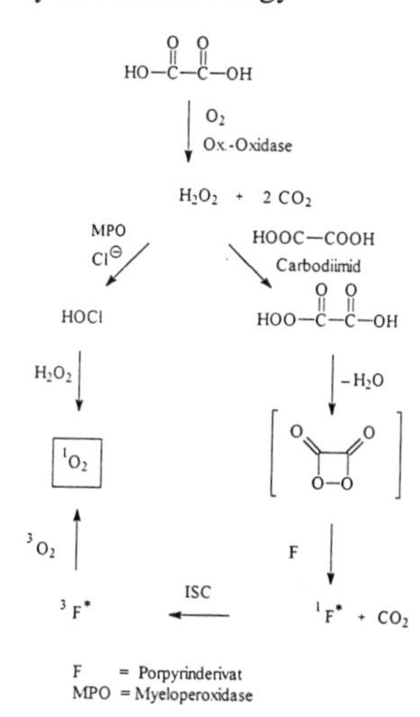

Fig.2: Oxalate pathways to generated reactive electronic exceeded states

NFκB - A NEW MEDIATOR OF SEPSIS

T Zimmermann[1], S Albrecht[2], A Bierhaus[3], JU Schilling[1], M Nagel[1], A Bunk[1], HD Saeger[1]
[1]Department of Visceral-, Thoracic and Vascular Surgery,
[2]Department of Gynaecology and Obstetrics,
[3]Department of Pathology , Technical University of Dresden, Germany

Introduction

The transcription factor NFκB is a mediator of the cellular acute response to pathogenic and inflammatory stimuli. NFκB activates a number of „immediate early genes" (Tissue factor, Endothelin, cytokines). That are described as involved in sepsis. Activation of NFκB in endotoxin-(LPS)treated cells could be shown in vitro (1,2). NFκB is localized in the cytoplasm in an inactive form, where it is associated with an inhibitor protein, IκB. This inhibitor retains the NFκB complex in the cytoplasm and inhibits DNA binding. Oxygene free radicals (beside other NFκB activators) cause in alteration in IκB, possibly by phosphorylation and/or activation of proteases. This allows NFκB to be released from the complex (3,4). NFκB then translocates to the nucleus, where it binds to the DNA recognition site and mediates gene transcription. To restore the NFκB system, IκB enters the nucleus, activately removes NFκB from its binding sites and transports NFκB back to the cytoplasm.(Fig.1)

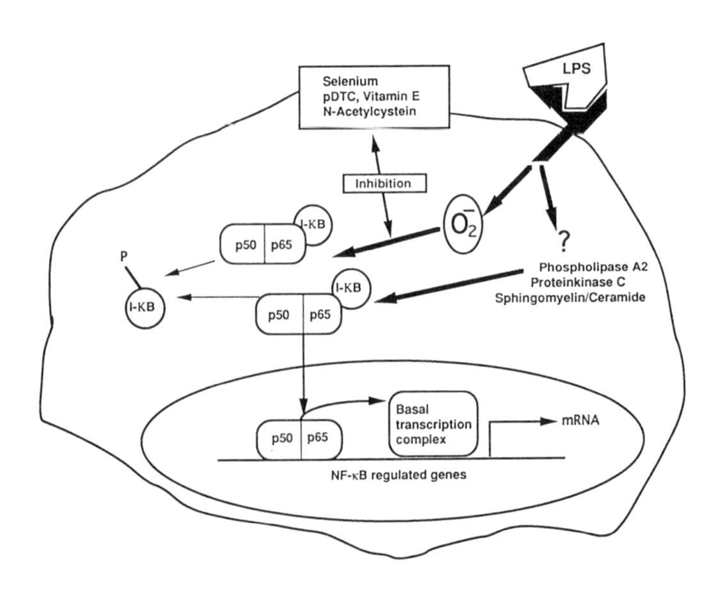

Figure 1 Mechanism of activation of NFκB in monocytes

Material and Methods

Laboratory Monitoring

1. Detection of NFκB in monocytes with „electrophoretic mobility shift assay" - EMSA-
2. Western-Blot-Technique for detection of NFκB in cytoplasmatic and nuclear extract of monocytes with „enhanced chemiluminescence"-ECL-system (Determination of NFκB translocation from the cytoplasm into the nucleus).

Patient samples: 40 patients suffering from „systemic inflammatory response syndrome"
(S.I.R.S.) and meet all inclusion and exclusion criteria

Inclusion criteria:	Exclusion criteria:
1) fever or hypothermia	1) pregnancy
2) tachycardia	2) massive blood loss
3) tachypnea	3) cardiogenic shock
4) hypotension	
5) evidence of organ dysfunction	

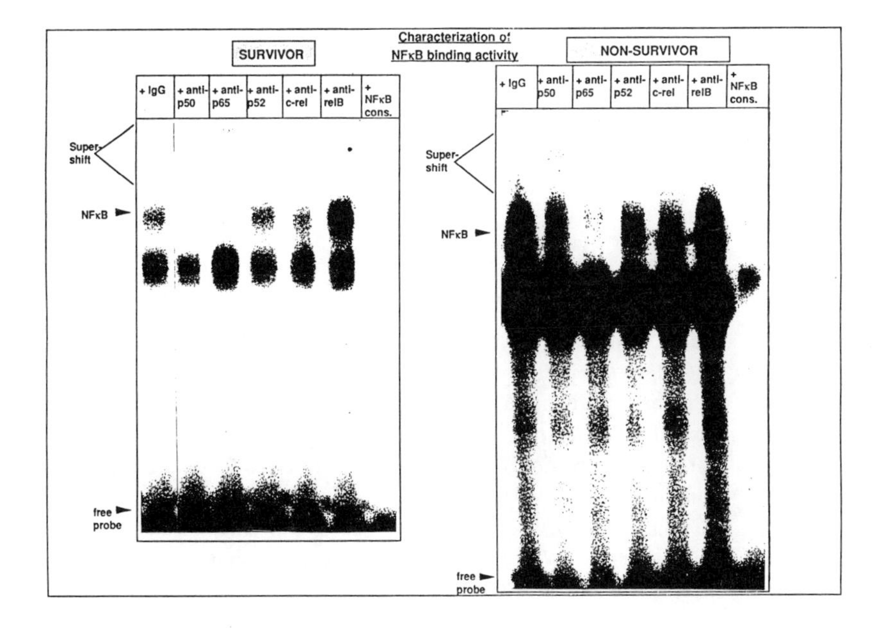

Figure 2 Characterization of NFκB binding activity in septic patients (survivor and non-survivor)

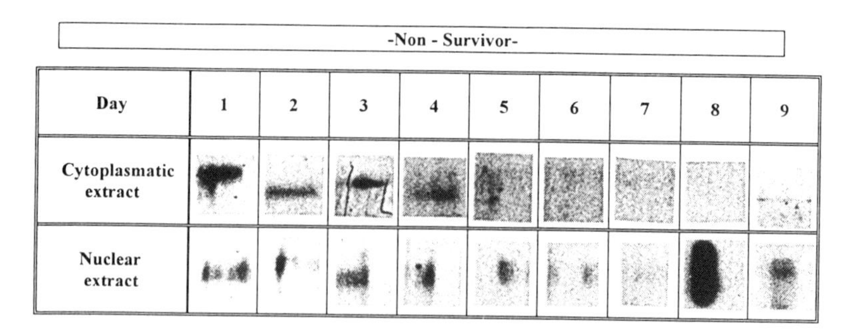

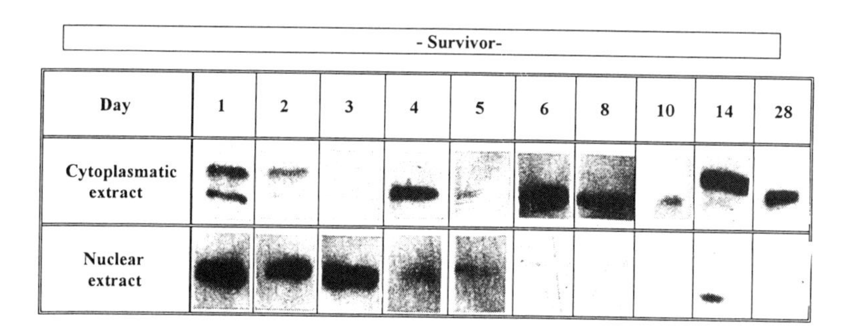

Figure 3 Detection of NFκB in monocytes by Western-Blot-Technique (with enhanced chemiluminescence technique)

Results

By EMSA we found a significant induction of an activation of NFκB in monocytes in septic patients. On day 1 we found an activation of NFκB in all patients (in comparisation with healthy volunteers). Successive antioxidative therapy caused a significant reduction of NFκB activation. (Fig.2) In contrast lethal outcome was often combined with increasing activation of NFκB. NFκB in active form is bound to the cytoplasmatic inhibitor IκB. Oxygene radical dependent phosphorylation and proteolytic degradation of IκB releases NFκB and causes translocation of NFκB from the cytoplasm into the nucleus. Western blots (enhanced CL) using cytoplasmatic and nuclear extract, collected from monocytes of septic patients, showed a sepsis induced translocation of NFκB into the nucleus. (Fig.3) This was suppressed by successive antioxidative therapy. In conclusion, sepsis causes an activation of NFκB, which could be proven in this study for the first time.

References

1. Oeth PA, Parry GCN, Kunsch C, Nantermet P, Rosen CA, Mackman N. Lipopolysaccharide induction of tissue factor gene expression in monocytic cells is mediated by binding of c-REL/p65 heterodimers to a kB-like site. Mol Cell Biol 1994; 14: 3772-81.
2. Zuckerman SH, Evans GF, Guthrie L. Transcriptional and post-transcriptional mechanisms involved in the differential expression of LPS-induced IL-1 and TNF mRNA. Immunology 1991; 73: 460-5.
3. Parry GCN, Mackman N. A set of inducible genes expressed by activated human monocytic and endothelial cells contain IkB-like sites that specifically bind c-REL-p65 heterodimers. J Biol Chem 1994; 269: 20823-25.
4. Nolan GP, Ghosh S, Liou HC, Tempst P, Baltimore D. DNA binding and I kappaB inhibition of the cloned p65 subunit of NFκB, a REL-related polypeptide. Cell 1991; 64: 961-69.

PHOLASIN® MONITORS NEUTROPHIL DEGRANULATION AND ACTIVATION

Jan Knight and Ann McCafferty
Knight Scientific Limited, 18 Western College Road, Plymouth, PL4 7AG, UK

Introduction

The use of chemiluminescence, especially when enhanced with luminol and lucigenin, to investigate phagocytosis and the respiratory burst has been well documented since Allen first recognised the phenomenon in phagocytosing polymorphonuclear leucocytes (PMN) (1). Luminol, which enters the cell, reacts mainly with myeloperoxidase (2) while lucigenin is believed to react more with extracellular superoxide. The sensitivity of lucigenin, however, is extremely low compared to Pholasin® (3,4) and reactions with luminol are not strictly comparable. Pholasin®, the photoprotein of *Pholas dactylus* (5), reacts with superoxide, hydroxyl and/or ferryl radical but not hydrogen peroxide. It also emits light with peroxidases acting as oxidases, as well as with hypochlorous and hypobromous acids and their derived chloramines and bromamines. Pholasin® does not cross the cell membrane, though it is likely for some molecules to become entrapped during the formation of the phagolysosome. The present report summarises some of the work currently underway with Pholasin®, aimed at identifying the species produced and events that occur during activation of the NADPH oxidase system and degranulation in the polymorphonuclear leucocyte (PMN).

Materials and Methods

Reagents: Pholasin®, Adjuvant-P™ (a novel chemiluminescent enhancer), f-met-leu-phe (fMLP) and phorbol myristate acetate (PMA) were supplied by Knight Scientific Ltd (KSL) under vacuum. The vials were reconstituted with Hank's balanced salt solution (HBSS) + 20mmol L^{-1} HEPES pH 7.4 (assay buffer) to obtain stock solutions of: Pholasin® 10 μg mL^{-1}, Adjuvant-P™, fMLP 10 μmol/L and PMA 16 μmol/L. Blood dilution medium (BDM) was assay buffer without calcium and magnesium. Cell removing liquid (CRL) was BDM with 0.01% gelatine. Superoxide dismutase (SOD) 100 U/mL, mannitol (MAN) 50mmol/L (Sigma, UK); desferrioxamine (DES) 50μg /mL (Desferal, Ciba) final concentrations.

Preparation of whole blood : To 2 mL BDM contained in stoppered polypropylene tubes, 20μL anticoagulated blood (lithium heparin 12 IU or potassium EDTA 1.2 mg/L) was added and the tubes inverted three times.

Preparation of leucocytes: Leucocytes were separated in 2 minutes from anticoagulated whole blood with a novel ABEL®-sep (Knight Scientific Ltd, Plymouth, UK) cell separation device (6) or density gradient medium (Neutrophil Isolation Medium (NIM), Los Alamos, Los Alamos, NM, US).

Whole blood assay (1mL total): 100μL diluted whole blood, 450μL assay buffer, 100μL Adjuvant-P™ into polypropylene luminometer cuvettes (Sarstedt, UK). Incubation (2 minutes at 37°C) and assay with mixing, in a Bio-Orbit 1251 luminometer (Bio-Orbit, Turku, Finland). Light emitted over 1 second was recorded every second; assay time 305s. At 5s, 250μL Pholasin® (2.5μg) was injected into the cuvette. At 65s 100μL of stimulant (10μmol/L fMLP, 1.6μmol/L PMA or a mixture) was injected. Plasma (100μL) replaced blood in the controls,

Cell assay (1mL total): As for whole blood assay but with 200μL cell suspension (usually 2.5 x 10^4 PMN), 350μL assay buffer, no Adjuvant-P™ and no mixing.

Cell assays with inhibitors: 100μL SOD, MAN or DES, were either injected at 305s and the assay run another 60s, or added to cuvettes before the start of the experiment.

Blood samples: Blood containing peroxidase deficient neutrophils (PDNs) were from anonymous donors; the PDNs were identified, from routine cell differential counting (Technicon H1 counter). Other samples were provided by known consenting donors.

Results and Discussion

Luminescent assays: Figure 1 is representative of 6 assays; Figure 2 of over 20. Replicate cuvettes have a coeficient of variation between 0.02 and 7.0. Figure 3 is from one patient undergoing dialysis with a low complement-activating membrane.

fMLP: The lag time, determined by Pholasin® for the onset of the respiratory burst in response to fMLP, is 2-3 seconds; a 5-10s lag has been reported by others (9) and this is consistent with the great sensitivity shown by Pholasin® for the detection of superoxide. After initiation, usually a single peak of luminescence is measured at about 20 seconds with the reaction generally tailing off after a further minute. However, a biphasic response to fMLP (not shown) is sometimes observed and work is currently underway to understand, both the underlying mechanism responsible for this second peak of luminescence as well as the conditions under which it comes about.

PMA: The response to stimulation with PMA (0.5-1μg) occurs within 20 seconds, is usually of greater magnitude than that to fMLP and usually reaches its peak within 4 minutes. The time to peak is dependent upon the concentration of PMA being fastest at 0.5-1μg/mL; maximum luminescence is the same between 0.01 and 1μg/mL (not shown); the signal is maintained for at least 15 minutes.

Combined stimulation: The fMLP and PMA pathways to activation and production of free radicals are usually independent. PMNs can respond to PMA after prior stimulation with fMLP (see figure 1) and cells can respond to both stimuli additively. There are, some individuals (the subject of another report) in which the response of their neutrophils to fMLP is the same as that observed when their cells are stimulated with a combination of fMLP and PMA.

Figure 1. Whole blood assays with fMLP, PMA or a mixture.

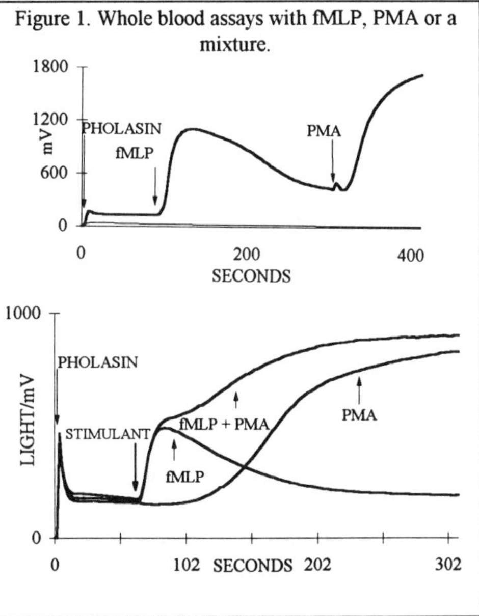

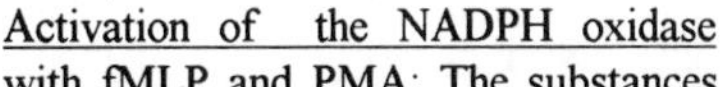

Activation of the NADPH oxidase with fMLP and PMA: The substances, fMLP and PMA, stimulate the initiation within the human neutrophil of a series of events leading to the activation of the NADPH oxidase system. This results in the respiratory burst and concomitant production of superoxide (7). It is initially the superoxide produced that is detected extracellularly by Pholasin® in assays with 1μL diluted whole blood or isolated leucocytes; later reactions involving degranulation processes and secondary production of other free radicals, possibly hydroxyl and /or ferryl radical, are monitored with Pholasin®. Activation with fMLP is through a receptor protein-tyrosine kinase and is calcium dependent (7). PMA, in contrast to fMLP, does not work through membrane receptors but migrates through the cell membrane to activate protein kinase C directly, an essentially step in the series of integrated signal transduction events that take place during activation (7).

Degranulation: PMA can activate the NADPH oxidase system of the secondary granule membrane. Pholasin® would initially detect extracellular superoxide from the activated NADPH oxidase as degranulation occurs and the secondary granules fuse with the plasma membrane. However, it is hypothesized that the luminescent signal of Pholasin® could occur by reactions between superoxide and Pholasin® as well as hydroxyl and/or ferryl radical, resulting from secondary reactions with and the iron of the secondary granule enzyme lactoferrin (8).

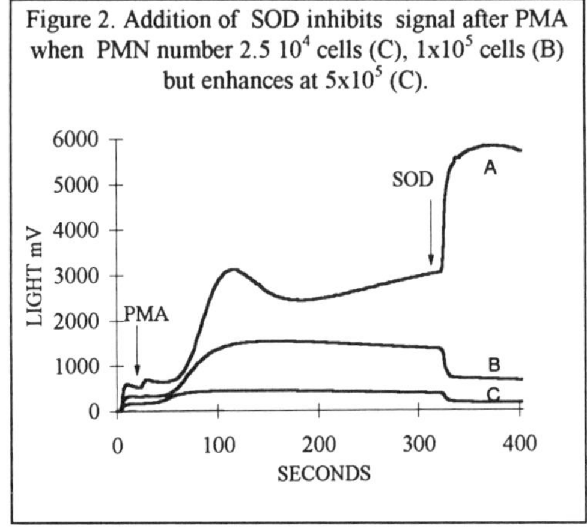

Figure 2. Addition of SOD inhibits signal after PMA when PMN number 2.5 10^4 cells (C), $1x10^5$ cells (B) but enhances at $5x10^5$ (C).

This hypothesis is supported by the observations (figure 2) that SOD inhibits the luminescent signal in response to PMA when PDN numbers are low (fewer than 10^5 cells/mL). However, the inhibition to SOD changes to enhancement with an increase in cell number, occurring usually at $5x10^5$ cells/mL. An explanation we suggest for the observations we have made on over 8 different cell preparations, is that at low concentrations of cells all the superoxide is either converted to hydrogen peroxide and subsequently used in hydroxyl and/or ferryl radical production or reacts directly with Pholasin®. At higher concentrations, when perhaps there is an accumulation of degranulated lactoferrin, the addition of SOD speeds up the dismutation of superoxide to hydrogen peroxide which then fuels the production of hydroxyl and/or ferryl radical via a superoxide driven Fenton reaction (9).

When DES was injected into the cuvette at peak luminescence of 8 different cell preparations (results not shown) there was an immediate response in which the signal was reduced to about prestimulation level in the cell range tested (2.5 x 10^4 cells/mL to 5 x 10^5 cells/mL). Experiments carried out in which DES was included in the cuvette before stimulation with PMA resulted in 82-96% inhibition (mean 88.5 $\pm$ 5 n=8); PDN and control cells responded in the same way. Addition of desferrioxamine, which complexes Fe^{3+}, prevents its reduction and subsequent participation in radical reactions (10). When SOD was treated in the same way there was a 43-73% overall inhibition (mean=54 $\pm$10 n=8). Mannitol, a hydroxyl radical scavenger, had little or no effect on the luminescent signal (inhibition mean 11.5% $\pm$14.9 n=8). It has been suggested that when hydroxyl radical scavengers fail to inhibit reactions thought to be fuelled by hydroxyl radicals that the species responsible in likely to be the ferryl radical (11). As hydrogen peroxide does not react with Pholasin® its role in enhancing the luminescence with Pholasin® can only come about as a result of secondary reactions.

That lactoferrin has a greater role than myeloperoxidase (MPO) (a primary granule enzyme) in the luminescence resulting after stimulation with PMA is supported by experiments with MPO deficient neutrophils (PDNs). In all PDNs tested (over 60) a luminescent response to PMA was observed. This is in contrast to luminol which mainly measures MPO activity (2). In 20 cell assays with 2.5 x 10^4 cells/mL PMN (PDNs compared to normal PMN) the rate and intensity of the luminescence was greater in 75% (mean 43% greater $\pm$16.9 n=15).

Degranulation: Blood stored in heparin often has higher concentrations of lactoferrin (12).We now consider we can identify when cells are prone to degranulation, as happens often during or after dialysis, by comparing the response to PMA in whole blood assays with Pholasin® using EDTA and heparin; the higher the PMA response in heparin compared to EDTA suggests a greater likelihood for degranulation to occur.

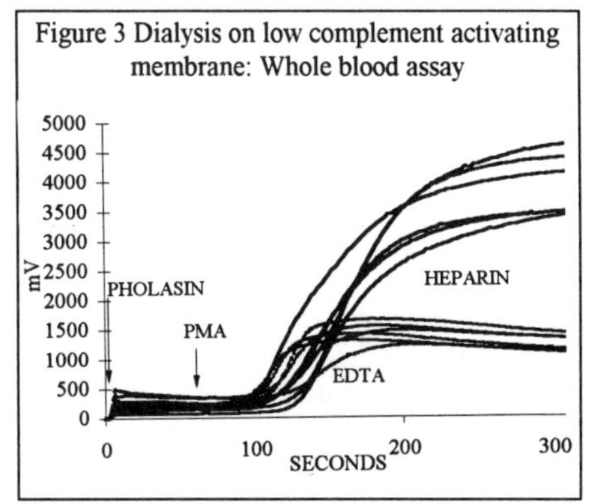

Figure 3 Dialysis on low complement activating membrane: Whole blood assay

Acknowledgements

We express thanks to Liz Noon, Taunton & Somerset NHS Trust, for selecting and supplying PDN bloods and matching controls. ABEL (analysis by emitted light) and Pholasin are registered trade marks of Knight Scientific Limited.

References

1. Allen RC, Stjernholm RL, Steele RH. Evidence for the generation of an electronic excitation state(s) in human polymorphonuclear leukocytes and its participation in bactericidal activity. Biochem Biophys Res Commun 1972;47:679-84.
2. McNally JA, Bell AL. Myeloperoxidase-based chemiluminescence of polymorphonuclear leukocytes and monocytes. J Biolumin Chemilumin 1996;11: 99-106.
3. Roberts PA, Knight J and Campbell AK Pholasin: a bioluminescent indicator for detecting activation of single neutrophils. Anal Biochem 1987;160:139-48.
4. Witko-Sarsat V, Nguyen AT, Knight J and Descamps-Latscha B Pholasin®: a new chemiluminescent probe for the detection of chloramines derived from human phagocytes. Free Rad Biol Med 1992;13:83-88.
5. Knight J. Studies on the biology and biochemistry of Pholas dactylus L. PhD thesis University of London, 1984.
6. Knight J, Knight R. Filtration arrangement. European Patent EPO 489602B1, US Patent US005266209A .
7. Segal, A.W. The NADPH oxidase of phagocytic cells is an electron pump that alkalinises the phagocytic vacuole. Protoplasm 1995;184:86-103.
8. Ambruso DR, Johnston RB. Lactoferrin enhances hydroxyl radical production by human neutrophils, neutrophil particulate fractions and an enzymatic generating system. J Clin Invest.1981; 67:352-59.
9. Borregaard N, Schwartz JH, Tauber AI . Proton secretion by stimulated neutrophils. Significance of hexose monophosphate shunt activity as source of electrons and protons for the respiratory burst. J Clin Invest. 1984; 83:757-76.
10. Aitken RJ, Harkiss D, Buckingham DW. Analysis of lipid peroxidation mechanisms in human spermatozoa. Mol Reprod Develop 1993;35:302-15.
11. Halliwell R, Gutteridege B. Free Radicals in Biology and Medicine 2nd edition. Oxford, Clarendon Press, 1989.
12. Német K, Solti V, Mód A, Pálóczi K, Segedi G, Hollán. Z Plasma lactoferrin levels in leukaemias. Clin lab Haemat 1987;9:137-45.

A CHEMILUMINESCENCE APPROACH TO STUDY THE REGULATION OF IRON METABOLISM BY OXIDATIVE STRESS

S Mueller[1] K Pantopoulos[2], MW Hentze[2] and W Stremmel[1]
[1]Dept. of Internal Medicine IV, University of Heidelberg, 69115 Heidelberg, Germany
[2]European Molecular Biology Laboratory, 69012 Heidelberg, Germany

The expression of key proteins involved in iron metabolism of vertebrate cells is controlled in the cytoplasm by specific mRNA-protein interactions. Two "iron regulatory proteins", IRP-1 and IRP-2, bind with high affinity and specificity to "iron responsive elements", IREs, hairpin structures in the untranslated regions of mRNAs encoding transferrin receptor, ferritin and the erythroid-specific form of aminolevulinate synthase [reviewed in (1)]. While both IRPs are activated by cellular iron deprivation, albeit by different mechanisms, IRP-1 was recently shown to specifically respond to oxidative stress in form of H_2O_2 (2, 3). Treatment of tissue culture cells with a bolus of 100 μmol/L H_2O_2 results in a complete induction of IRP-1 within 30-60 min. Several lines of evidence suggest that H_2O_2 mediates a signal transduction cascade which results in the rapid induction of IRP-1 by removal of a 4Fe-4S cluster. To study the underlying mechanism, we have applied a recently developed chemiluminescence assay for H_2O_2 (4). This non-enzymic assay is based on the chemiluminescence reaction of luminol and sodium hypochlorite. Luminol is oxidized by hypochlorite to a diazaquinone which specifically reacts with H_2O_2 forming an excited aminophthalate (5). The resulting chemiluminescence is very short ($<$ 2 s) and corresponds linearly to the H_2O_2 concentration down to 10^{-9} mol/L (6).

By employing this assay we found that exogenous H_2O_2 is rapidly decomposed when administered to cells in culture, most likely by intracellularly localized H_2O_2 degrading enzymes. The degradation rate of H_2O_2 negatively correlates with its potential to elicit IRP-1 induction. This finding raises the question of the threshold of extracellular H_2O_2 which suffices to activate IRP-1. Apparently, this requires a source of H_2O_2 at steady-state concentrations. By using the hypochlorite/luminol system we were able to generate steady-state levels of H_2O_2, by a titrated mixture of glucose, glucose oxidase and catalase (G/GOX/CAT). Preliminary experiments suggest that under these conditions IRP-1 activation can be observed by as low as $\sim$ 10 μmol/L H_2O_2. We suggest that this chemiluminescence approach may provide a useful tool to study other H_2O_2 controlled processes.

Materials and Methods

Instrumentation: Luminescence measurements were performed using a luminometer AutoLumat LB 953 (Fa. Berthold, Wildbad, Germany).

Reagents: Glucose, glucose oxidase, catalase, luminol, hypochlorite and H_2O_2 were from Sigma (Deisenhofen, Germany). Murine B6 fibroblasts were grown in DMEM supplemented with additives and 10% FCS. Treatments with H_2O_2 and G/GOX/CAT were performed in serum-free MEM.

Electrophoretic mobility shift assay (EMSA): EMSAs were performed as previously described, using a radiolabeled IRE probe (2).

Chemiluminescence determination of H_2O_2: For the determination of H_2O_2 a recently described, highly sensitive non-enzymic chemiluminescence technique was used (4, 7) with following modifications: Luminol was added to 500 μl of culture medium taken from the petri disc. NaOCl was finally added using an injector device located in light measuring position. The integral of luminescence peak was determined over 2 s. H_2O_2 concentration was directly determined from the peak integral according to a calibration curve. A **flow technique** was established to follow rapid kinetics and to adjust the H_2O_2 concentration in the G/GO/CAT system. Briefly, a solution of G/GO or CAT was aspirated via a peristaltic pump (4 ml/min). Luminol (10^{-4} mol/L) and hypochlorite (10^{-4} mol/L) were continuously added by a perfusor pump (6 ml/min) to the sample solution closed to the light detector thus monitoring the actual H_2O_2 concentration in real time. The photomultiplier was adjusted to the maximum intensity of the luminescence reaction. The luminescence integral was determined over 5 sec. After adding of H_2O_2 (10^{-5} mol/L) to the catalase solution the breakdown of H_2O_2 concentration could be followed. A magnetic stirrer was used to ensure a rapid mixing of the enzyme substrate solution.

Results and Discussion

Induction of IRP-1 correlates with the degradation rate of exogenous H_2O_2.

When B6 fibroblasts were treated with a bolus of 100 μmol/L H_2O_2 for 30 min and IRE-binding activity of cell extracts was analyzed on EMSA, we observed that the induction of IRP-1 surprisingly correlates with the volume of the culture medium (Fig. 1): It gradually decreases, as the volume is reduced from 10 to 5 and 2.5 ml (lanes 1-4). This effect is restored by applying higher H_2O_2 concentrations, 200 or 400 μmol/L respectively in a culture of 5 or 2.5 ml medium (lanes 5-6). In the bottom panel, cell extracts were treated with 2% 2-mercaptoethanol (2-ME), known to induce inactive IRP-1 in vitro (2), to confirm the equal loading of all lanes.

To investigate the dependence of IRP-activation on the total amount of the applied H_2O_2, rather than on the H_2O_2 concentration, we performed a chemiluminescence H_2O_2 assay. This assay reveals that the degradation rate of H_2O_2 clearly correlates with the total levels of H_2O_2 applied in the culture; the same concentration of H_2O_2 is much faster degraded in 2.5 ml medium than in 5, or 10 ml (Fig. 2). The H_2O_2-degrading activity is most likely intracellular, as 100 μmol/L H_2O_2 remains stable when incubated under the same conditions in the culture supernatant (data not shown). Thus, the dependence of IRP-1 induction on the total amount of the applied exogenous H_2O_2 is explained by the degradation of H_2O_2. This result raises the question on the minimal effective H_2O_2 concentration which suffices to induce IRP-1.

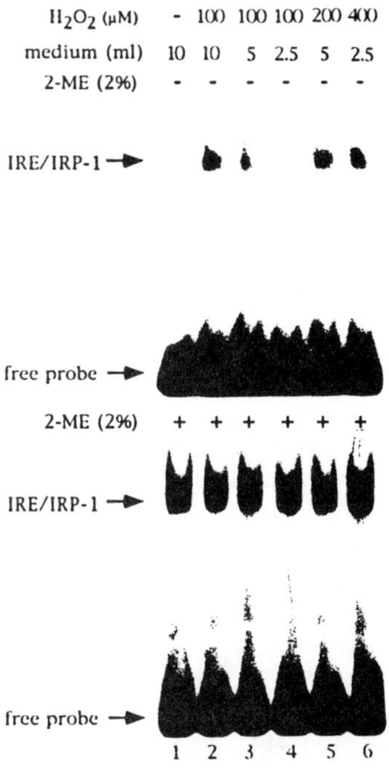

Fig. 1 Induction of IRP-1 by H_2O_2 depends on the volume of the culture medium

Generation of H_2O_2 at steady state levels

To address this question, we developed a system which generates H2O2 at steady-state levels by a mixture of glucose, glucose oxidase and catalase (G/GOX/CAT).

During the oxidation of glucose by glucose oxidase H_2O_2 is generated following a zero order kinetic with $dH_2O_2/dt = k_{gox}$ (k_{gox} = rate constant) if dioxygen and glucose are maintained at a constant concentrations (data not shown).

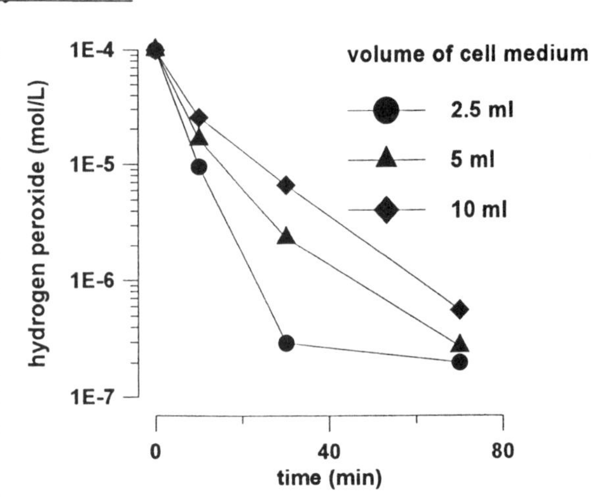

Fig. 2 H_2O_2 degradation by B6 fibroblasts depends on the volume of the culture medium

Accumulation of H_2O_2 can be controlled by adding appropriate amounts of catalase. Decomposition of H_2O_2 by catalase depends on H_2O_2 concentration and follows first order kinetics (8); H_2O_2 degradation rate is described by $dH_2O_2/dt = k_{CAT} \cdot [H_2O_2]$. Thus, steady-state levels of H_2O_2 are generated when $k_{GOX} = k_{CAT} \cdot [H_2O_2]$, and at a constant glucose and dioxygen concentration $[H_2O_2] = k_{GOX}/k_{CAT}$. Based on these considerations, we used saturating concentrations of glucose and calculated the amount of glucose oxidase and catalase required to generate steady-state levels of H_2O_2 in the micromolar range. In Fig. 3, a detection of H_2O_2 concentration at different time points by the hypochlorite/luminol assay is shown. After addition of catalase a first order decay of H_2O_2 is observed. The addition of glucose oxidase leads to a continuous increase of H_2O_2 concentration, until the steady state is reached. Under these conditions, steady-state levels of H_2O_2 in the micromolar range are maintained up to three hours.

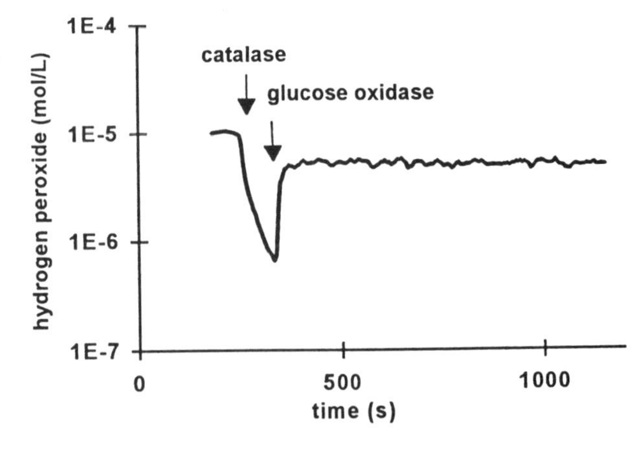

Fig. 3 Generation of steady-state levels of H_2O_2 by a G/GOX/CAT system

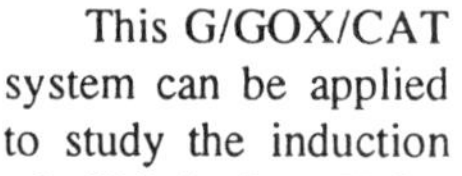

This G/GOX/CAT system can be applied to study the induction of IRP-1 by H_2O_2.

Control experiments show that treatment of B6 fibroblasts with G/GOX/CAT results in a clear induction of IRP-1. Preliminary data suggest that this induction observed upon treatment with as little as ~ 10 μmol/L H_2O_2, which is 5 times less to what previously estimated with bolus H_2O_2 (2, 3). The G/GOX/CAT system may provide valuable insights in the study of other H_2O_2-regulated pathways. Reactive oxygen intermediates in different forms have been implicated in specific regulation of various biological targets. In addition to H_2O_2, oxygen concentration and tension also appear to be crucial for the regulation of stress-response genes. The expression of the heme-binding protein hemopexin, which responds to oxidative stress, appears to be transcriptionally regulated by the oxygen tension and concentration rather by ROIs, including H_2O_2 (9). By increasing the amount of catalase in the G/GOX/CAT system, it is possible to adjust and maintain very low H_2O_2 concentrations. We suggest that this approach may help to distinguish between an H_2O_2- or a dioxygen-mediated regulation.

The study of oxidative stress and H_2O_2-mediated responses on signal transduction and gene regulation emerges as an interesting and expanding field. The correct choice of an H_2O_2 source may help to clarify relevant questions.

Acknowledgement

This work was supported by grants from the Deutsche Forschungsgemeinschaft STR 216/41.

References

1. Hentze MW, Kühn LC. Molecular control of vertebrate iron metabolism: mRNA-based regulatory circuits operated by iron, nitric oxide and oxidative stress. Proc Natl Acad Sci USA 1996; 93: 8175-82.
2. Pantopoulos K, Hentze MW. Rapid responses to oxidative stress mediated by iron regulatory protein. EMBO J 1995; 14: 2917-24.
3. Pantopoulos K, Weiss G, Hentze MW. Nitric oxide and oxidative stress (H_2O_2) control mammalian iron metabolism by different pathways. Mol Cell Biol 1996; 16: 3781-88.
4. Mueller S, Arnhold J. Fast and sensitive chemiluminescence determination of H2O2 concentration in stimulated neutrophils. J Biolum Chemilum 1995; 10: 229-37.
5. Merényi G, Lind J, Eriksen TE. Luminol chemiluminescence: Chemistry, Excitation, Emitter. J Biolum Chemilum 1990; 5: 53-6.
6. Arnhold J, Mueller S, Arnold K, Sonntag K. Mechanisms of inhibition of chemiluminescence in the oxidation of luminol by sodium hypochlorite. J Biolum Chemilum 1993; 6: 307-13.
7. Mueller S, Arnhold J. Detection of hydrogen peroxide in stimulated neutrophils using luminol and hypochlorite. In: Campbell AK, Kricka LJ, Stanley PE, editors. Bioluminescence and chemiluminescence: Fundamentals and Applied Aspects. Chichester, John Wiley & Sons, 1994: 242-5.
8. Aebi H. Catalase in vitro. Methods Enzymol 1984; 105: 121-6.
9. Kietzmann T, Immenschuh S, Katz N, Jungermann K, Muller EU. Modulation of hemopexin gene expression by physiological oxygen tensions in primary rat hepatocyte cultures. Biochem Biophys Res Commun 1995; 213: 397-403.

BIOLUMINESCENT ATP-METRY FOR INVESTIGATION OF THE PROCESS OF LYMPHOCYTE ACTIVATION

EG Bulanova[1], NA Romanova[2], LYu Brovko[2], VM Budayan[1], AA Yarilin[1] and NN Ugarova[2]
[1]Moscow Immunology Institute, Moscow 115478, Russia
[2] Dept of Chem., Lomonosov Moscow State University, Moscow 119899, Russia

Introduction

Bioluminescent (BL) methods of intracellular metabolites' assay become more and more popular for *in vitro* and *in vivo* evaluation of functional activity of the cells and response of the cell to the external factors. The advantages of the bioluminescent methods in comparison with the other ones are the following: 1) high sensitivity, that provides possibility to analyze very small samples up to single cells; 2) high specificity, that allows to analyze crude biological samples without isolating the particular metabolite and even to work with whole cells *in vivo*; 3) possibility to perform analysis and to monitor the changes in metabolite concentration in real time scale, beginning from minutes or even seconds to hours and days (1,2). Adenosine-5'-triphosphate (ATP) is the universal energy source of the living cell. The bioluminescent method of ATP assay is based on the reaction of luciferin oxidation catalyzed by firefly luciferase. This method is widely used in clinical and microbiological analysis to estimate the number of living cells in the sample, because in the similar conditions the ATP content in the cell remains almost constant (3). The fact of proportional dependence of intracellular ATP content on the number of cells in the sample was used by us in the bioluminescent assay for human lymphocyte blast transformation (4). The proposed method possesses all necessary criteria of immunological practice: high sensitivity and precision. In addition it is less labor-consuming and avoids using hazardous radioactive materials.

On the other hand the action of some external factors of physical or chemical nature on the cell results in the significant change of intracellular ATP content. It was shown in (5), that illumination of the *E.coli* cells with low-power He-Ne laser induces the rapid increase of intracellular ATP content. The same effect was observed in the cultivated fibroblasts after irradiation with low-power radioactive sources (6). The lymphocyte intracellular ATP level increases after their activation by phytohemagglutinin (PHA) before the stage of proliferation (7). The mechanism and extend of lymphocytes' activation depends on the nature of the activator and the receptors involved in the process as well as on the immune status of the whole organism. To investigate this process in more details bioluminescent monitoring of the intracellular ATP concentration in lymphocytes during the activation was chosen. Lymphocyte activators of different types were used: Platelet Derived Growth Factor (PDGF), Interleukin-2 (IL-2), human Growth Hormone (hGH), and phytohemagglutinin (PHA) to compare ATP time-courses and thus, the mechanisms of action of different immunomodulators.

Materials and Methods

Instrumentation. Luminometer CLIMBI, model LB-3P (Moscow, Russia) was used for bioluminescence monitoring.

Reagents. Bioluminescent ATP-Reagent "Immolum" (3) (Chem Dept, Lomonosov Moscow State University, Russia) was used for ATP assay. All other reagents were of analytical grade.

Methods. Samples of venous blood were obtained from healthy volunteers. The procedures of lymphocyte isolation and following activation were as described in (4). The intracellular ATP content in lymphocytes was measured in the control samples and in the samples containing PDGF (5 ng/mL), or IL-2 (10 ng/mL), or hGH (10 ng/mL, or PHA (15 μg/mL) in 1, 5, 15, 60, 120, 180 min after the addition of activator.

Results and Discussion

The time-courses of intracellular ATP content in lymphocytes of 15 healthy donors were measured after action of activators of different types. The similar dependencies were obtained with one type of activator for all samples. The difference between donors was no more than 15-25%. The typical patterns for activated and control samples are presented in Fig. 1.

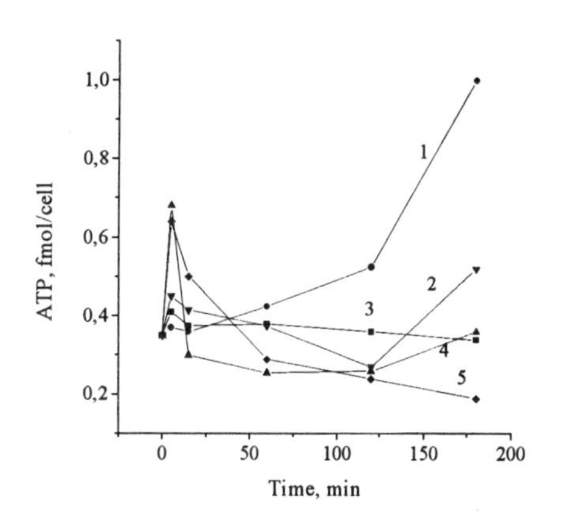

Fig.1. Time-course of intracellular ATP content in lymphocytes in control sample (3) and after addition of PHA (1), hGH (2), IL-2 (4) and PDGF (5)

Each point represents the mean value of intracellular ATP content for 6 independent measurements of the cells' sample. The standard deviation for each sample did not exceed 10-15%. For control sample (curve 3, Fig. 1) there were no significant changes in ATP content during the first 3 hours of incubation. For the samples containing activators two different patterns of ATP time-courses were observed. For the first group of activators (IL-2 and PDGF, curves 4 and 5, Fig.1, respectively), during the first five minutes of lymphocytes' activation significant (2-3 times) increase of intracellular ATP was observed, followed by slow ATP decrease up to initial values until the beginning of the proliferation step in 3 hrs. For the second group (PHA and hGH, curves 1,2, Fig.1, respectively) slight decrease of intracellular ATP was observed during first hour of the activation followed by 1.5-2.0 times growth of intracellular ATP content by 3 hr of activation prior to the proliferation. The observed difference in lymphocyte behavior during the activation by mitogens, antigens and other cells' activators provides us with the new approach for investigation of the metabolic processes involved in the immunological reaction of the organism. The proposed method of intracellular ATP monitoring may be used as a tool for express assay of the activity of immunomodulators *in vitro* for patient with immune deficit.

Acknowledgments.
The research was supported by Russian Foundation of Fundamental Science

References

1. Ugarova NN, Brovko LYu, Kutuzova GD. Bioluminescence and bioluminescent analysis: recent development in the field. Biochemistry (Moscow, translated from Russian) 1993; 58: 976-92.
2. Campbell AK. Chemiluminescence, principles and application in biology and medicine. Chichester: Horwood/VCH, 1988.
3. Brovko LYu, Trdatyan IYu, Ugarova NN. Optimization of bioluminescent assay of microbial biomass. Appl Biochem Microbiol (transl. from Russian) 1991; 27: 134-41.
4. Bulanova EG, Budagyan VM, Romanova NA, Brovko LYu, Ugarova NN. Bioluminescent assay for human lymphocyte blast transformation. Immunology Letters 1995; 46: 153-5.
5. Brovko LYu, Romanova NA, Ugarova NN. Bioluminescent assay of bacterial AMP, ADP, and ATP with the use of a coimmobilized three-enzyme reagent (adenylate kinase, pyruvate kinase, and firefly luciferase). Anal Biochem 1994; 220: 410-4.
6. Parkhomenko IM, Perishvily GV, Turovetzki VB, Kudryashov YB, Rubin AB, Brovko LY. Influence of low doses of ionized radiation on intracellular pH, ATP content and synthetic activity of cultivated cells of Chinese hamstèr. Radiobiologia (in Russian) 1993; 33: 104-9.
7. Bulanova EG, Brovko LYu, Rozenkov VY, Romanova NA, Butakov AA, Ugarova NN. Investigation of the intracellular adenosin triphosphate in mononuclear cells of human peripheral blood with the help of bioluminescent method. Immunologia (in Russian) 1994; 3: 55-7

BIOLUMINESCENCE *IN VITRO* CHEMOSENSITIVITY AND c-erbB-2 AMPLIFICATION IN HUMAN OVARIAN CARCINOMAS

[1]P Pinzani, [1]R Sestini, [1]G Vona, [2]E Mini , [2]T Mazzei, [3]G Amunni, [1]M Pazzagli
[1]Dept of Clinical Pathophysiology, Clinical Biochemistry Unit, [2]Dept of Preclinical and Clinical Pharmacology, Chemotherapy Unit, University of Florence; Clinic of Obstetrics and Gynecology , Florence, Italy

Introduction

Oncogenesis and tumor progression from normal cells to highly malignant cells refractory to therapy are associated with multiple molecular changes that often include the unregulated expression of growth factor receptors or of other elements such as oncogenes involved in the cellular signaling pathways (1). Some of these changes have been associated with poor prognosis in several human tumors and therefore they are likely to affect response to therapeutic agents. The receptors most frequently implicated in human cancers have been members of the epidermal growth factor or c-erbB receptor family. The c-erbB-2 oncogene is a potentially useful prognostic marker whose relevance for diagnosis and therapy in human breast and ovarian cancer has been recently focused (2). Its potential role in modulation of chemotherapeutic drug sensitivity has also been suggested from retrospective analysis of results of several clinical studies (3,4). If confirmed, these findings could have important implications in patient management and treatment decisions. While some *in vitro* experimental models providing evidence that oncogenes may influence drug sensitivity in cell lines have been already reported (5-9), evidences demostrating an association between oncogenes and drug sensitivity/resistance in human tumor cell primary cultures are still lacking. We evaluated c-erbB-2 amplification in 12 primary ovarian carcinomas which were previously studied in regard to in vitro sensitivity to some of the drugs most commonly used in the treatment of ovarian cancer.

Materials and Methods

Chemosensitivity Test. The *in vitro* tumor chemosensitivity assay (TCA) was performed as previously reported (10). Briefly, tumor specimens were enzymatically dissociated and the cell suspension was assessed for viability by Trypan blue exclusion, counted and then adjusted to $1\text{-}3x10^5$ cells/mL in Complete Assay Medium (CAM). The Tumor Chemosensitivity Assay (TCA) was prepared in 96-well, round-bottom polypropylene microplates (Costar 3790). Each microplate was used to test 4 single agents or drug combinations at 6 serial dilutions corresponding to 200-6.25% of a standard test drug concentration (TDC) in triplicate. TDC values were determined by pharmacokinetic and clinical information (11). Standard TDC values were 15.8 µg/mL for carboplatin (CBDCA), 0.5 µg/mL for epirubicin (EPI), 6.8 µg/mL for taxol (TAX) and 6.25 µg/mL for mafosfamide (MAF, a stable analogue of cyclophosphamyde generating 4-hydroxycyclophosphamyde i.e. the active metabolite of cyclophosphamyde). Cultures were incubated at 37°C for 6 days. ATP was then extracted and measured in a Berthold LB 952 luminometer using 0.05 mL culture extract injected with 0.05 mL luciferin-luciferase counting reagent. A 10-sec count integration time with 4-sec delay was used for light emission measurement.

Competitive PCR. DNA samples were obtained by standard phenol-chloroform procedure. c-erbB-2 amplification was measured by a competitive PCR method as described by Sestini et al. (12) using ß-globin as the single copy reference gene.

Subjects. The tumor chemosensitivity test and competitive PCR assay method for the measurement of c-erbB-2 amplification were applied to study 12 patients (mean age: 56.8 yr; range: 28-75 yr) affected by primary ovarian carcinoma afferent to the Clinic of Obstetrics and Gynecology of Florence.

Results and Discussion

Three out of twelve DNA samples (25%) undergone to competitive PCR demonstrated c-erbB-2 oncogene degree of amplification higher than 3.

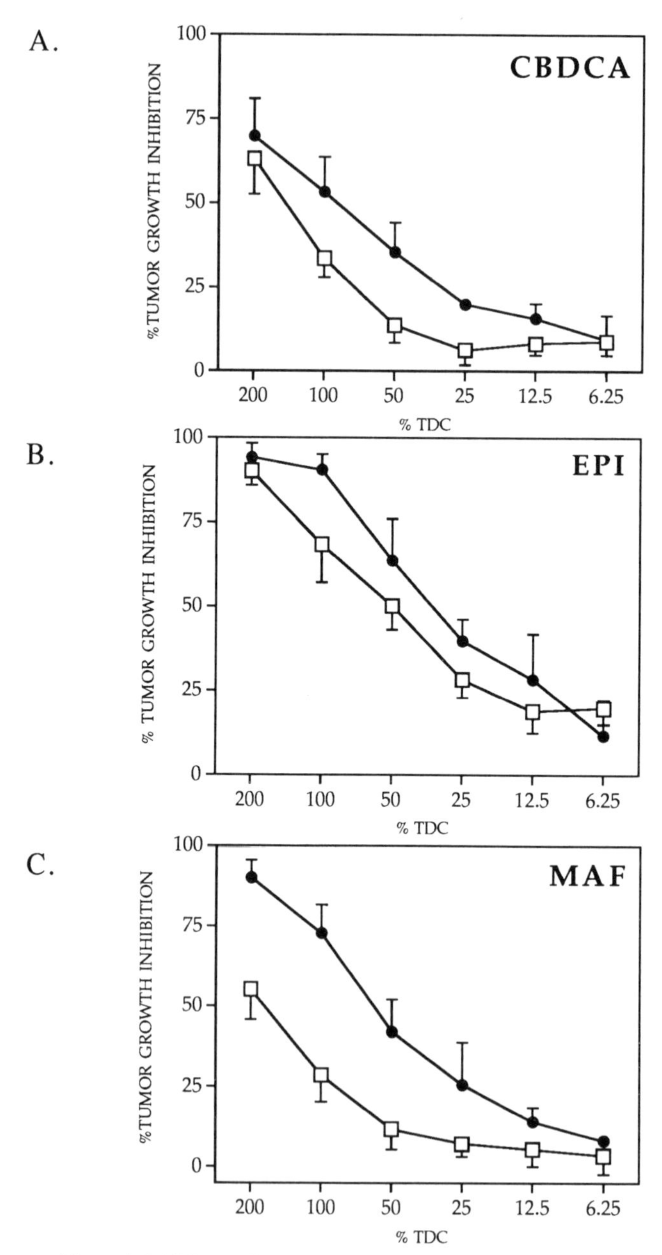

Figure 1. Inhibitory effect of carboplatin (CBDCA, panel A), epirubicin (EPI, panel B) and mafosfamide (MAF, panel C) on the growth of cells obtained from tumor explants of patients affected by primary ovarian carcinoma with (●, n=3) or without (❐, n=9) oncogene c-erbB-2 amplification as assessed by the bioluminescence TCA assay.

We studied the in vitro chemosensitivity levels of tumor cell suspensions from the same 12 patients to CBDCA, EPI, MAF and TAX both as single agent and in combination (CBDCA+EPI, CBDCA+MAF, EPI+MAF and CBDCA+EPI+MAF). Results of TCA are expressed as mean±SD of the two groups individuated on the basis of c-erbB-2 amplification (chemosensitivity mean curve of patients which show c-erbB-2 amplification versus that of patients with no oncogene amplification). Higher chemosensitivity to CBDCA, EPI and MAF was found in subjects with c-erbB-2 amplification (Figure 1). This result of higher chemosensitivity in subjects with c-erbB-2 amplification was confirmed when testing the same drugs in combination. In Figure 2 chemosensitivity dose-response curves for the triple combination of CBDCA, EPI, MAF (the most commonly used therapeutic regimen in ovarian carcinoma treatment) are reported for both groups under study. Similar results were obtained using the other combination CBDCA+EPI, CBDCA+MAF, EPI+MAF (data not shown). Results obtained for taxol indicate, on the contrary, a relative higher chemosensitivity in subjects with no amplification of c-erbB-2 oncogene (data not shown).

Interpretation of these results is somehow contradictory since oncogene activation has been so far associated with enhanced chemoresistance in certain cell culture models (5,6). On the other hand, in clinical studies it has also been reported that patients whose tumor overexpresses c-erbB-2 may derive the greatest benefit from higher doses of chemotheraphy (3). It is however conceivable that c-erbB-2 oncogene amplification may reflect an increase in the tumor proliferative activity of ovarian tumor explants. Actively proliferating cells could be in fact more sensitive to chemotherapeutic agents, in particular with drugs that react with DNA causing its structural and functional alteration. In conclusion, these preliminary results, obtained in a limited number of ovarian carcinomas, seem to confirm an important, even if contradictory, correlation between oncogene alteration and tumor chemosensitivity.

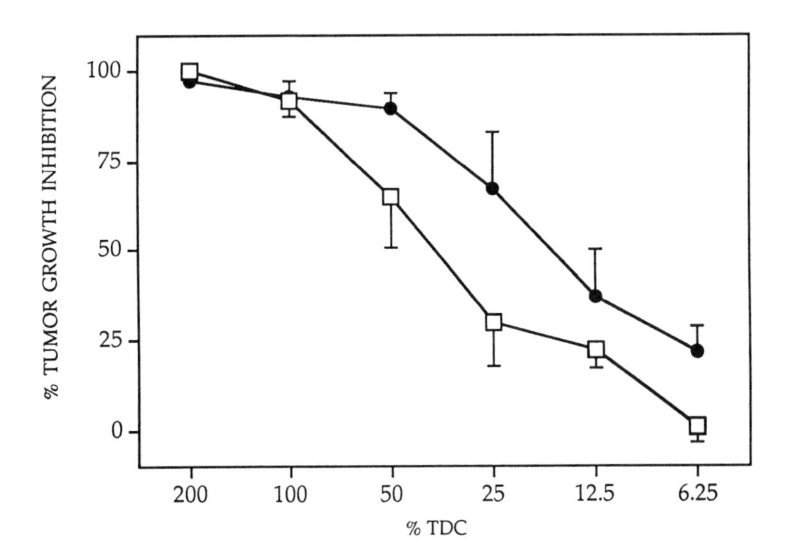

Figure 2. Inhibitory effect of the triplex association CBDCA+EPI+MAF on the growth of cells obtained from tumor explants of patients affected by primary ovarian carcinoma with (●, n=3) or without (□, n=9) oncogene c-erbB-2 amplification as assessed by the bioluminescence TCA assay.

References

1. Aaronson SA. Growth Factors and Cancer. Science 1991; 254:1146-52.
2. Harris AL, Nicholson S, Sainsbury R, Wright C, Farndon J. Epidermal growth factor receptor and other oncogenes as prognostic markers. Monogr Natl Cancer Inst 1992; 11:181-7.
3. Muss HB, Thor AD, Berry DA, Kute T, Liu ET, Koerner F, et al. c-erbB-2 expression and response to adjuvant theraphy in women with node-positive early breast cancer. N Engl J Med1994; 330:1260-66.
4. International (Ludwig) Breast Cancer Study Group. Prognostic importance of c-erbB-2 expression in breast cancer. J Clin Oncol 1992; 10:1049-56.
5. Niimi S, Nakagawa K, Yokota J, et al. Resistance to anticancer drugs in NIH3T3 cells transfected with c-myc and/or c-H-ras genes. Br J Cancer 1991; 63:237-41.
6. Sklar MD. Increased resistance to cis-diammine dichloroplatinum (II) in NIH 3T3 cells transformed by ras oncogenes. Cancer Res 1988; 48:793-7.
7. Toffoli G, Viel A, Tumiotto L, et al. Sensitivity pattern of normal and Ha-ras transformed NIH3T3 fibroblasts to antineoplastic drugs. Tumori 1989; 75:423-8.
8. Tsai CM, Levitzki A, Wu LH, Chang KT, Cheng CC, Gazit A, et al. Enhancement of chemosensitivity by Tyrphostin AG825 in High-p185^{neu} expressing non-small cell lung cancer cells. Cancer Research 1996; 56:1068-74.
9. Pietras RJ, Fendly BM, Chazin VR, Pegram MD, Howell SB, Slamon DJ. Antibody to HER-2/ *neu* receptor blocks DNA repair after cisplatin in human breast and ovarian cancer cells. Oncogene 1994; 9:1829-38.
10. Andreotti PE, Cree IA, Kurbacher CM, Hartmann DM, Linder D, et al. Chemosensitivity testing of human tumors using a microplate adenosine triphosphate luminescence assay: Clinical correlation for cisplatin resistance of ovarian carcinoma. Cancer Res 1995; 55:5276-82.
11. Alberts DS, Chen HSG Tabular summary of pharmacokinetic parameters relevant to *in vitro* drug assay. In: Salmon S editor. Cloning of Human Tumor Stem Cells. New York: Liss, 1980: 351.
12. Sestini R, Orlando C, Zentilin L, Gelmini S, Pinzani P, Bianchi S, et al. Measuring c-erbB-2 Oncogene Amplification in Fresh and Paraffin-Embedded Tumors by Competitive Polymerase Chain Reaction. Clin Chem 1994; 40:630-6.

AGE-RELATED SERUM ANTIOXIDANT CAPACITY IN HEALTHY SUBJECTS EVALUATED BY AN ENHANCED CHEMILUMINESCENCE ASSAY

P Pinzani [1], E Petruzzi [2], R Sestini[1], C Orlando[1], M Pazzagli[1]
Dept. of Clinical Pathophysiology, [1]Clinical Biochemistry ; [2]Institute of Gerontology - University of Florence, Florence, Italy.

Introduction

Many biological compounds act as antioxidants and they include vitamins such as vitamin C and vitamin E, enzymes (gluthatione peroxidase, catalase, superoxide dismutase), proteins (transferrin, ceruloplasmin) and other molecules such as uric acid. The modes of action of many of the natural antioxidants that are found in biological fluids and tissues have been extensively studied (1). The antioxidants act cooperatively in vivo so as to provide greater protection to the organism against radical damage than it could be provided by any single antioxidant acting alone. Several methods have been proposed to study the mechanisms of antioxidant protection against free radical-induced injury. Our attention has been focused on the measurement of the Total Antioxidant Capacity (TAC) in body fluids based on enhanced chemiluminescence (2) that seems to provide a simple and reproducible method of assessing free radical scavenging antioxidants in aqueous solution. Since cellular damages by oxy-radicals are considered as important factors involved in the phenomenon of biological aging (3), we have investigated serum antioxidant capacity in a group of healthy subjects in the age range 18 - 105 year.

Materials and Methods

Reagents. Enhanced chemiluminescence immunoassay signal reagent and HRP conjugate (mouse IgG HRP linked whole antibody from sheep) were purchased by Amersham International. The standard Trolox (6-hydroxy-2,5,7,8-tetramethylchroman-2-carboxylic acid) was obtained from Aldrich Chemical Co., UK. Proteins were removed from serum samples by ultrafiltration using Microcon-10 microconcentrators (Amicon, MA, USA).

Assay principle and procedure. This assay has been developed based on a chemiluminescent reaction catalized by horse-radish peroxidase conjugate which determines oxidation of the chemiluminescent substrate luminol in the presence of hydrogen peroxide. The use of p-iodophenol as enhancer of the chemiluminescent signal determines a more intense, prolonged and stable light emission. Radical scavenging antioxidants can interfere with the costant production of free radical intermediates and hence stop light emission. The signal will be restored after antioxidants consumption and the time period of suppression will be directely related to the amount of antioxidants.

The assay procedure used in this study is a modification of that already reported by Whitehead et al. (2). Briefly, 5 µL of a 1:2000 dilution of HRP conjugate were added to 50 µL of signal reagent (tablets A and B dissolved in the signal buffer) and 445 µL of water. After the light emission became stable, the addition of 20 µL of Trolox standard solutions (from 5 to 40 µmol/L) or of 1:50 prediluted samples will suppress light emission. The kinetics of the reaction is recordered by a chart recorder connected to the luminometer. The time in seconds from sample addition untill a 10% of recovery of the initial light output is measured and compared to the time value for the Trolox calibrant to express the total antioxidant capacity as µmol/L of Trolox equivalents. This procedure implies the determination of TAC both on whole serum (WTAC) and on its deproteinated fraction (DPTAC), obtained by ultrafiltration.

Instrumentation. A single-tube luminometer Berthold BIOLUMAT LB 9500 was used in the study. Kinetics of light emission were recorded by a LKB 2010-RECORDER 2-Channel.

Subjects. Subjects of both gender were recruited by advertisement. Young and elderly healthy subjects were selected by the criteria of JUNIEUR and SENIEUR protocol (4). We have measured WTAC and DPTAC in 30 normal control subjects (15 men and 15 women, age range 18-91 yr.) and in 12 healthy centenarians (3 men and 9 women, age range 100-105 yr.). All subjects gave oral and written informed consent. Blood was drawn after an overnight fast. Fresh blood samples, after centrifugation, were processed for the total antioxidant capacity. Additional biochemical tests such as glucose, uric acid, bilirubin, alcaline phosphatase, etc. (see Table 1) were also determined.

Results and Discussion

WTAC and DPTAC values vs. age are reported in Figure 1A e B. Evaluation of the results by linear regression analysis showed a significant negative relationship between WTAC and DPTAC and subject age ($r=.582$, $p<0.01$ for WTAC; $r=.444$, $p<0.01$ for DPTAC; $n=42$). On the other hand if we exclude WTAC and DPTAC values of centenarians, the relationship between antioxidant capacity values and subject's age is not significant (data not shown).

The significant reduction of antioxidant capacity in both fractions (whole and deproteinated) may indicate the involvement of filterable as well as proteic factors in the decrement of antioxidant activity in centenarians. The methodological approach we have used is not able to indicate which are the biochemical changes responsible for the observed reduction in the antioxidant capacity in centenarians. We have also measured some standard biochemical tests which may affect the antioxidant capacity in serum (see Table 1). We can observe that total proteins were in the normal range, whereas total serum iron was lower than that in the control adult population; additional changes include bilirubin which shows a significant decrease. In particular, we observe that serum urate levels which account for 60-70% of TAC (5, 6), were unchanged in our centenarian subjects.

The significant reduction in both WTAC and DPTAC related to the centenarians seems to indicate the presence of an imbalance between pro- and antioxidants. It remains to be clarified if the reduced antioxidant capacity in centenarians can play some physiopathological role in the aging process and eventually to verify the role that a supplementary antioxidant therapy may have on the life span even if these data confirm the observation of Nohl H who recently reported: *Although it cannot be stated conclusively that oxidative stress is ultimately required to initiate the transition of a biological system in equilibrium to disequilibrium, its existence as an accompanying event of ageing has been established* (3).

Acknowledgements

This work is supported by the University of Florence

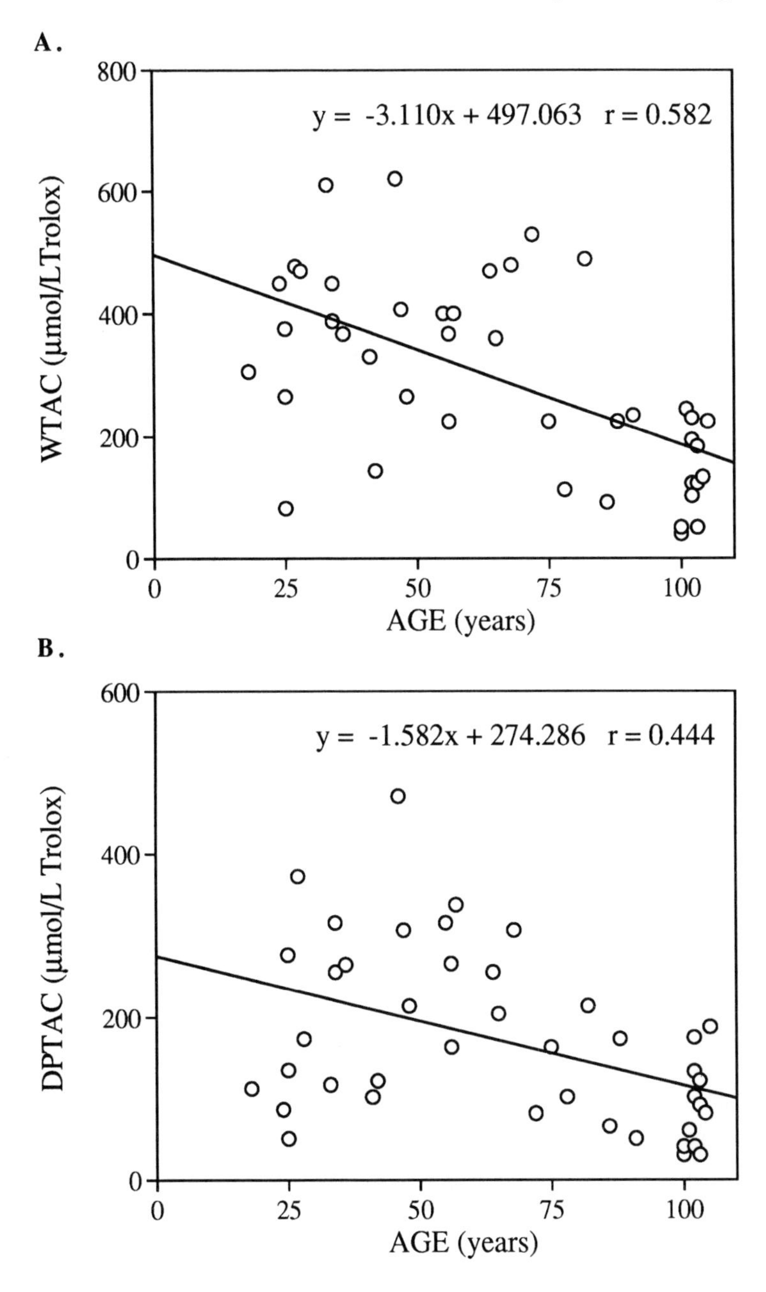

Figure 1. Relationship between WSAC (A) and DPSAC (B) and subject age.

References

1. Gutteridge JMC. Lipid peroxidation and Antioxidants as Biomarkers of Tissue Damage. Clin Chem 1995;41:1819-28.
2. Whitehead TP, Thorpe GHG, Maxwell SRJ. Enhanced chemiluminescent assay for antioxidant capacity in biological fluids. Anal Chim Acta 1992; 266:265-77.
3. Nohl H. Involvement of free radicals in ageing: a consequence or cause of senescence. Brit Med Bull 1993; 49(3):653-67.
4. Ligthart GJ, Corberand JX, Geertzen HGM, Meinders E, Knook DL, Hijmans W. Necessity of the assessment of health status in human immunogerontological studies: evaluation of the senieur protocol. Mech Ageing Dev 1990; 28: 47-55.
5. Wayner DDM, Burton GW, Ingold KU, Barclay LRC, Locke SJ. The relative contributions of vitamin E, urate, ascorbate and proteins to the total peroxyl radical-trapping antioxidant activity of human blood plasma. Biochim Biophys Acta 1987; 924:408-19.
6. Maxwell SRJ. PhD Thesis, Birmingham University, 1996.

TABLE 1. Some biochemical parameters measured in normal adults and healthy centenarians' serum samples.

TEST	NORMAL ADULTS Mean±SD (n=23)	CENTENARIANS Mean±SD (n=12)	LABORATORY REFERENCE RANGE
AST (GOT) (IU/L)	28.92±4.44	22.66±8.31	5 - 40
ALT (GPT) (IU/L)	6.77±2.45	9.55±4.06	5 - 40
UREA (g/L)	0.47±0.16	0.43±0.13	0.1 -0.5
GLUCOSE (g/L)	0.88±0.096	0.92±0.24	0.65 -1.1
CREATININE (mg/dL)	1.28±0.36	1.03±0.25	0.6 -1.5
CHOLESTEROL (mg/dL)	199.69±55.34	185.46±42.18	160 - 220
TRIGLICERIDES (mg/dL)	105.0±13.33	128.77±52.34	50 - 170
TOTAL PROTEINS (g/dL)	7.89±0.90	7.1±0.73	6.0 - 8.0
URATE (mg/dL)	4.98±2.07	5.5±2.1	3.5 -6.5
TOTAL BILIRUBIN (mg/dL)	0.89±0.40	0.57±0.26*	0.2 - 1.0
DIRECT BILIRUBIN (mg/dL)	0.15±0.084	0.096±0.113	0.0 - 0.3
TOTAL SERUM IRON (mg/dL)	81.08±44.92	46.2±22.7*	60 - 160

*p<0.05

ULTRAWEAK LUMINESCENCE
OF DOMESTIC ANIMALS' SPERMATOZOA

M.Godlewski[1], T.Kwiecinska[1], D.Wierzuchowska[1], Z.Rajfur[1], A.Laszczka[2], B.Szczesniak-Fabianczyk[2] ,J.Slawinski[3]

[1]Institute of Physics & Computer Sciences, Pedagogical University, 30-084 Krakow, Podchorazych 2, Poland

[2]National Research Institute of Animal Production, Department of Physiology of Reproduction, 32-083, Balice-Krakow, Poland

[3]Institute of Chemistry and Technical Electrochemistry, Poznan University of Technology, 60-095 Poznan, Poland

Introduction

It is well known that deep freezing is the best method of long lasting semen preservation. This method is well developed in the case of bull spermatozoa. However there are some restrictions of applying this method to spermatozoa of other domestic animals like ram and boar.

Our previous results (1,2,3,4,5) allow to assume that measurement of ultraweak luminescence (UL) could serve as a new tool for further detailed investigation of processes which take place during deep freezing preservation of semen. As the first step further investigation of UL from bull spermatozoa was undertaken.

Materials and Methods

Fresh bull semen was collected at Artificial Insemination Station in Zabierzow near Krakow and was evaluated for volume, concentration and motility. Part of the semen was undertaken standard cycle of conservation process by deep freezing in liquid nitrogen. Fresh/thawed spermatozoa were separated from seminal plasma/diluent by centrifugation (400g, 10min.). Cells were resuspended in 0.9% NaCl and centrifugation procedure was repeated. Final concentration of cells in 0.9% NaCl was $50\text{-}60*10^6$cells/mL.

Chemicals of analytical grade were obtained from POCH (Gliwice, Poland) and Sigma Chemical Company. Bidistilled water was produced using all-glass apparatus.

Measurements were performed by means of computerised equipment working in single photon counting mode, employing cooled FEU38 photomultiplier sensitive in the region 350-800nm. A glass cuvette containing 4mL of cell suspension was placed in light-tight compartment in which temperature was stabilised at 44 ± 0.1^oC. During the measurements suspension was stirred by bubbling of oxygen (5L/h).

Cut-off filters placed between the sample and photomultiplier were used to determine emission spectra which are presented as corrected for filters' transmission and photomultiplier sensitivity. During the measurements chemicals were injected to the sample through a light-tight injection valve.

Results and Discussion

Time course curves of the UL intensity from fresh cells and cells after a deep freezing cycle were measured. Before measurements spermatozoa were incubated 3h at 44^oC. Two curves, as an example, are presented in fig.1.

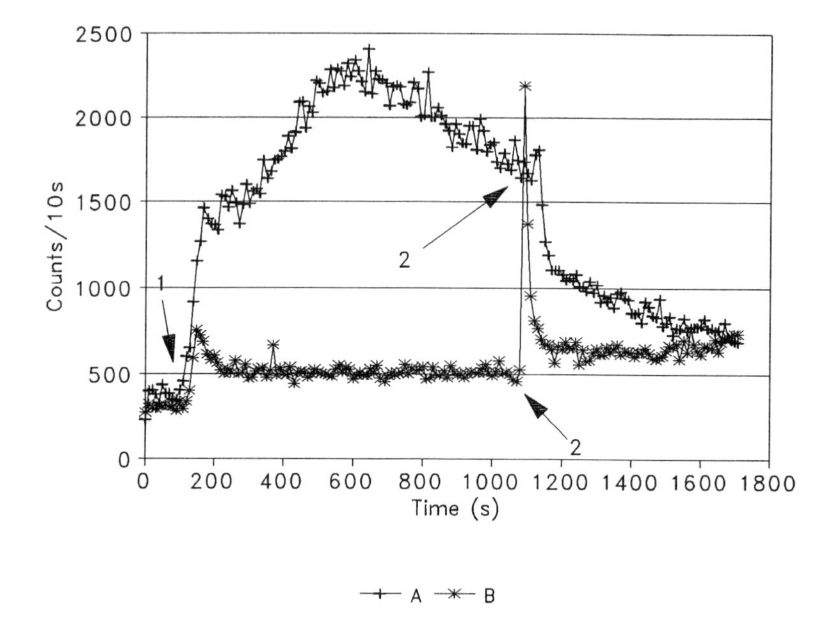

Fig.1. The time course of ultraweak luminescence from fresh (A) and after a deep freezing cycle (B) bull spermatozoa. Arrows indicate injection of Fe-ascorbate system (1) and $FeSO_4$ (2).

The process of cells' membrane lipid peroxidation was intensified by injection of the redox-cycling Fe^{+2}-ascorbate system (final concentration: Fe-0.045mmol/L, asc.-0.22mmol/L). After 1000s $FeSO_4$ solution (Fe final concentration 0.96mmol/L) was injected. The time course of UL from fresh cells is different then from the cells after freezing cycle. Their shapes are repeatable.

Emission spectra of fresh cells and cells after deep freezing cycle are presented on Fig.2. For the first time it has been possible to measure spectra of the intrinsic UL from spermatozoa without luminescent probes. Emission spectra give information about the energetics of chemiexcitation and photon emission. There are no statistically significant differences between UL spectra from fresh cells and cells after deep freezing cycle.

Results of these measurements, together with our previous results: spermatozoa motility, vitality and UL intensity dependence on time of cells' incubation at $44^{o}C$ (1,2,3), confirm that the main source of the observed UL are radical reactions which are associated with lipid peroxidation of cell membranes and that these reactions play very important role in the degradation of semen quality during long time preservation.

These measurements will be completed for different phases of deep freezing process and also for ram and boar spermatozoa (already partially performed).

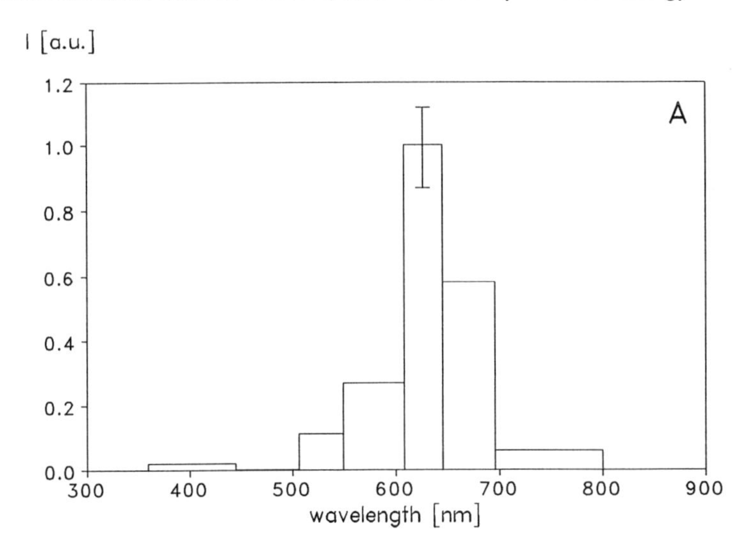

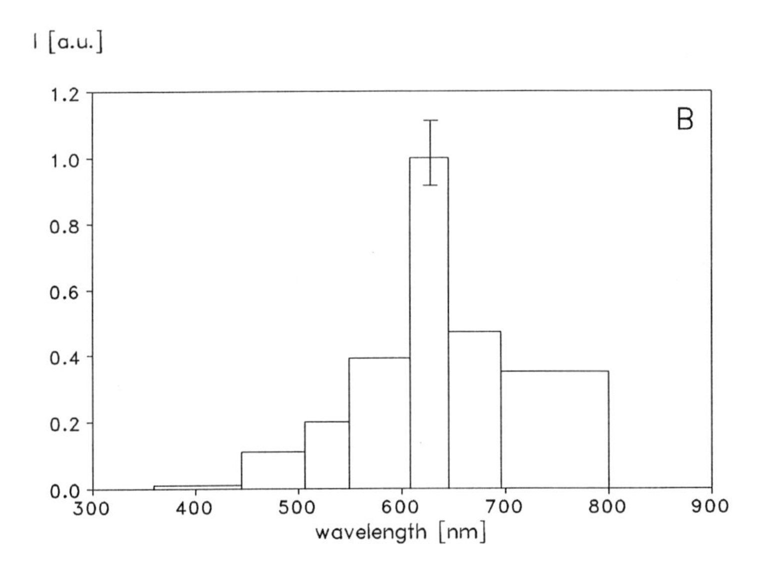

Fig.2. Emission spectra for fresh (A) and after deep freezing cycle (B) spermatozoa. UL induced by injection of Fe-ascorbate system. Error bars indicate maximal experimental error value.

Acknowledgements
This work was partially supported by the State Committee for Scientific Research (KBN) in Warsaw, Grant No 5 PO6D 047 09 and Grant No BS-27/M/94 from the Cracow Pedagogical University.

References

1. Laszczka A, Godlewski M, Kwiecinska T, Rajfur Z, Sitko D, Szczesniak-Fabianczyk B, at al. Ultraweak luminescence of spermatozoa, Curr Top Biophys 1995;19;20-31.
2. Godlewski M, Kwiecinska T, Laszczka M, Rajfur Z, Slawinski J, Szczesniak-Fabianczyk B, at al. Diagnostic value of ultraweak chemiluminescence of cells subjected to the oxidative stress, Curr Top in Biophys (in press).
3. Laszczka A, Ezzahir A, Godlewski M, Kwiecinska T, Rajfur Z, Sitko D, et al. Iron-induced ultraweak chemiluminescence and vitality of bull spermatozoa, In: Szalay AA, Kricka LJ, Stanley PE, editors. Bioluminescence and Chemiluminescence: Status Report. Chichester: John Wiley & Sons, 1993; 523-7.
4. Ezzahir A, Godlewski M, Kwiecinska T, Sitko D, Slawinski J, Szczesniak-Fabianczyk B, et al. Iron induced chemiluminescence of bull spermatozoa, Appl Biol Commun 1992; 2/3; 139-44.
5. Ezzahir A, Kwiecinska T, Godlewski M, Sitko D, Rajfur Z, Slawinski J, et al. The influence of white light on photo-induced luminescence from bull spermatozoa, Appl Biol Commun 1992; 2/3; 133-7.

ULTRAWEAK PHOTON EMISSION OF CELLS SUBJECTED TO ENVIRONMENTAL STRESS

M.Godlewski[1], T.Kwiecinska[1], D.Wierzuchowska[1], Z.Rajfur[1], A.Laszczka[2], B.Szczesniak-Fabianczyk[2], J.Slawinski[3], M.Guminska[4], T.Kedryna[4],

[1]Institute of Physics & Computer Sciences, Pedagogical University, 30-084 Kraków, Podchorążych 2, Poland

[2]National Research Institute of Animal Production, Department of Physiology of Reproduction, 32-083, Balice-Krakow, Poland

[3]Institute of Chemistry and Technical Electrochemistry, Poznan University of Technology, 60-095 Poznan, Poland

[4]Institute of Biochemistry, Collegium Medicum, Jagiellonian University, Krakow, Kopernika 8, Poland

Introduction

Environmental influences which perturb homeostasis of biological objects can be treated as environmental stress. Values of many biological and physiological indices change under environmental influences. Changes of these parameters provide partial, specific information about the influence of environmental stress factors on biological objects.

As ultraweak photon emission (UPE) is functionally linked to biochemical/metabolic processes in cells it should give holistic information about biological effect of stress factors. Here we report research on UPE of ram spermatozoa undergoing oxidative stress.

Materials and methods

Fresh ram ejaculates, collected at National Research Institute of Animal Production in Balice-Krakow were initially diluted with modified Goetze solution and then twice centrifuged (400g, 10min) and resuspended in modified Goetze diluent. Final concentration of cells was 50-60*10^6cells/mL. Suspension was incubated at 40°C. In regular periods of time UPE intensity, ATP level, vitality (V), motility (M) and oxygen consumption were measured. UPE measurements were performed by means of single photon counting method. During UPE measurements cell suspension was constantly stirred and oxygenated. The lipid-peroxidation linked UPE was initiated by addition of the redox-cycling ascorbate-Fe^{2+} system (final concentration: 0.22mmol/L ascorbate, 0.045mmol/L Fe^{2+}). Chemicals of analytical grade were obtained from POCH (Gliwice, Poland) and Sigma Chemical Company. Bidistilled water was produced using all-glass apparatus. The sperm motility was assessed for forward and backward movement with an experimental error of 10%. The vitality of cells were estimated by supravital staining with eosin and nigrosin according to Blom (1). Oxygen consumption was monitored using a Clark electrode (YSI 5300 Biological Oxygen Monitor). ATP level in spermatozoa was assayed by the enzymatic/spectrophotometric (2) or luminometric (3) methods.

Results and Discussion

UPE intensity increase with the elongation of incubation process. Three kinetic curves: before, after 150min. and 390min. of incubation are presented in Fig.1. Motility and vitality of spermatozoa substantially decrease with incubation time (Fig.2A). The dependence of ATP level, oxygen consumption and total number of counts (which is the sum of counts over one thousand seconds after Fe-ascorbate addition) on incubation time is illustrated in Fig.2B. The rate of oxygen consumption decreases while the level of ATP and total number of counts increase with incubation time.

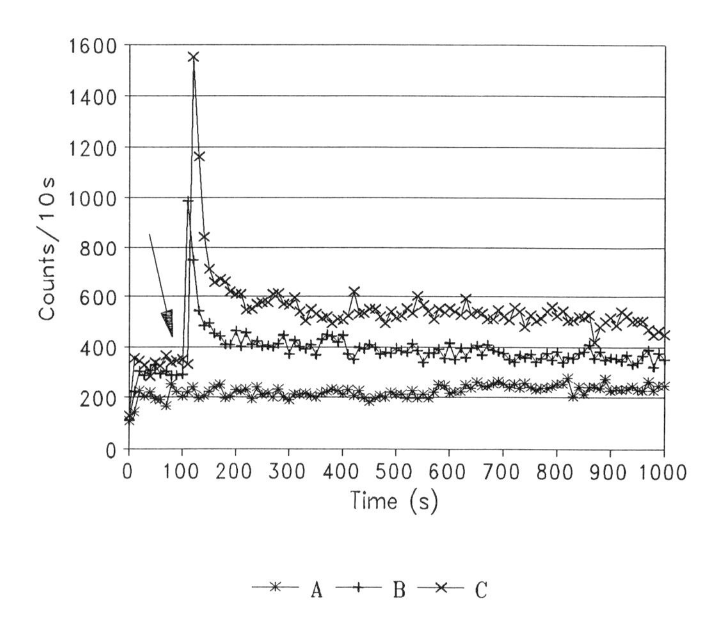

Fig.1. Photon emission kinetic curves for ram spermatozoa: (A)-before incubation, (B)-after 150min. and (C)-after 390min. of incubation. The arrow indicate the moment of Fe-ascorbate injection.

It can be presumed that the increase of ATP level could be caused by the decrease of the energy consumption as a result of motility reduction. Thus ATP molecules produced in this conditions are accumulated in the cells. The primary cause of the observed discrepancy between the increase in ATP level and intensity of UL and decrease in the M and V values is most likely due to the perturbation of the cell membrane integrity. The injury of the ATP-ase (Na,K) and the flagellar motility system may particularly account for the increasing level of ATP and decrease in its consumption.

Changes of the physiological state of cells due to the incubation process, the length of which can be treated as a stress factor, are reflected in the observed physiological/biological indices. These changes are correlated with the photon emission intensity. This fact is not specific for ram spermatozoa only. Our previous results (4,5,6,7) obtained for bull spermatozoa and yeast cells confirm that environmental stress factors (e.g. osmotic stress, illumination with strong light, formaldehyde presence) influence photon emission parameters.

Results of the parallel investigation of the biological and biochemical parameters and UPE indicate vital association of UPE parameters with physiological state of the cell. Therefore UPE can be expected to serve as a biophysical tool for studying environmental stress influence on biological objects.

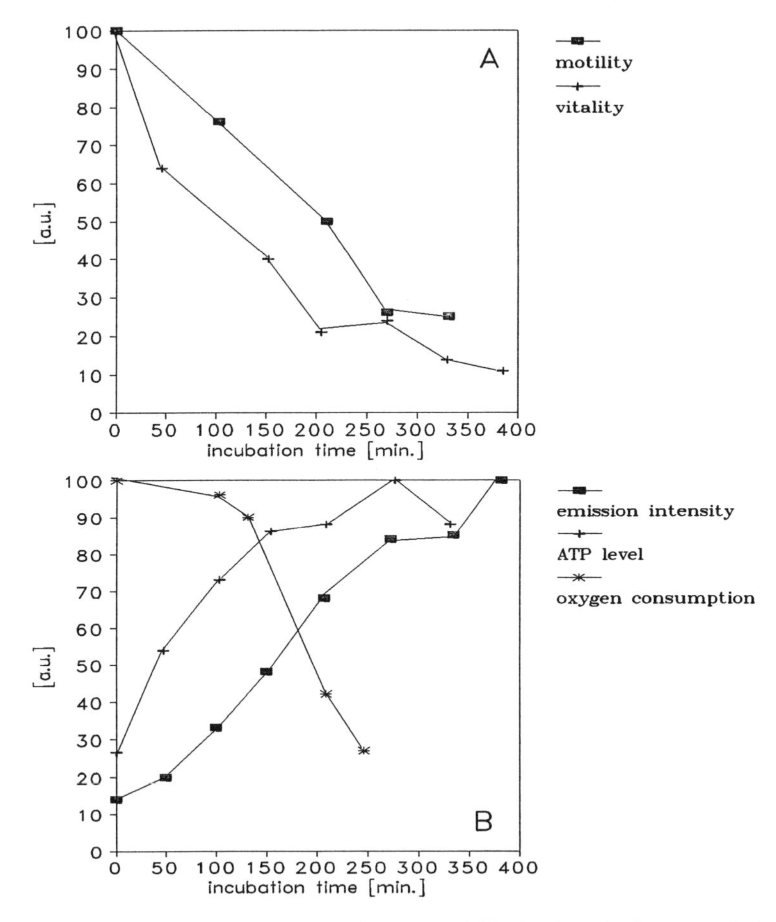

Fig.2. Motility, vitality, oxygen consumption rate, ATP level and photon emission intensity (integrated over 1000s after addition of ascorbate-Fe system) dependence on duration of incubation at 40°C for ram spermatozoa.

Acknowledgements
This work was partially supported by the State Committee for Scientific Research (KBN) in Warsaw, Grant No 5 PO6D 047 09 and Grant No BS-27/M/94 from the Cracow Pedagogical University.

References

1. Blom E, A rapid staining method using eosin-negrosin to distinguish between live and dead spermatozoa, Nord. Vet. Med. 1950; 2; 58-61.
2. Adam H, Adenosine-5'-triphosphate. Determination with phosphoglicerate kinase, in: Bergmeyer HU, editor. Methods of enzymatic analysis, Academic Press, N.J.; 1965; 539-43.
3. Spielmann H, Jacob-Muller U, Schulz P, Simple assay of 0.1-1.0 pmol of ATP, ADP and AMP in single somatic cells using purified luciferin luciferase, Anal Bioch 1981; 113; 172-8.
4. Slawinski J, Ezzahir A, Godlewski M, Kwiecinska T, Rajfur Z, Sitko D, et al. Stress-induced photon emission from perturbed organisms, Experientia 1992; 48; 1041-57.
5. Laszczka A, Ezzahir A, Godlewski M, Kwiecinska T, Rajfur Z, Sitko D, et al. Iron-induced ultraweak chemiluminescence and vitality of bull spermatozoa, In: Szalay AA, Kricka LJ, Stanley PE, editors. Bioluminescence and Chemiluminescence: Status Report. Chichester: John Wiley & Sons, 1993; 523-7.
6. Ezzahir A, Godlewski M, Kwiecinska T, Sitko D, Slawinski J, Szczesniak-Fabianczyk B, et al. Iron induced chemiluminescence of bull spermatozoa, Appl Biol Commun 1992; 2/3; 139-44.
7. Ezzahir A, Kwiecinska T, Godlewski M, Sitko D, Rajfur Z, Slawinski J, et al. The influence of white light on photo-induced luminescence from bull spermatozoa, Appl Biol Commun 1992; 2/3; 133-7.

EVALUATION OF BILE ACIDS ANTIOXIDANT ACTIVITY USING ENHANCED CHEMILUMINESCENT ASSAY

A Roda[1], P Pasini[1], C Russo[1], M Baraldini[2], G Feroci[2], LJ Kricka[3] and AM Gioacchini[1]
[1]Dept of Pharmaceutical Sciences, [2]Inst of Chemical Sciences, University of Bologna, Via Belmeloro 6, 40126 Bologna, Italy
[3]Dept of Pathology and Laboratory Medicine, University of Pennsylvania, Philadelphia, PA 19104, USA

Introduction

Free radicals are highly reactive oxidants able to damage important biological molecules such as nucleic acids, proteins and lipids with consequent impairment of cellular functions. The involvement of oxygen free radicals in the development of several pathological states including cancer, rheumatoid arthritis, atherosclerosis and post-ischemic reoxygenation injury of organs has been reported (1).

Different endogenous antioxidant systems that counteract oxidative stress or reduce the effects of free radical damage have been discovered and studied. They include intracellular scavengers like superoxide dismutase and catalase and extracellular chain breaking or preventative antioxidants (2-4).

Bile acids (BA) include various molecules, synthesized from cholesterol, which are known to be important components in lipid digestion and absorption of the products of digestion (5). Recently, potential antioxidant properties of BA, both as monomers and aggregates to form micelles, have been hypothesized (6), but systematic studies assessing the activity of individual natural occurring BA as oxygen free radical scavengers have not been published.

Several bio-chemiluminescence assays for antioxidants have been described (7-9). The enhanced chemiluminescent assay based on horseradish peroxidase and a luminol-oxidant-enhancer reagent provides a simple and sensitive new tool for antioxidant activity studies (10). In this system, the addition of antioxidants to a glowing chemiluminescent reaction temporarily interrupts light output. Light emission is then restored after an interval of time that is linearly related to the molar quantity of antioxidant added.

The aim of our work was to study the antioxidant capacity of bile acids with different number, position and orientation of the hydroxy groups, using the enhanced chemiluminescence assay. Unconjugated, glycine and taurine amidated BA were analyzed at a final concentration ranging from 1 to 28 mM, thus exploring their effect as monomers and when aggregated to form micelles.

Materials and Methods

<u>Instrumentation</u>: The chemiluminescent emission was detected and analyzed using a low-light imaging device, Luminograph LB 980 (EG&G Berthold, Bad Wildbad, Germany). It consists of an intensified Saticon videocamera connected to a PC provided with software controlling the system and performing image processing and quantitative analysis.

Reagents: Cholic acid (CA), deoxycholic acid (DCA), chenodeoxycholic acid (CDCA), ursodeoxycholic acid (UDCA), glycocholic acid (GCA), glycodeoxycholic acid (GDCA), glycochenodeoxycholic acid (GCDCA), glycoursodeoxycholic acid (GUDCA), taurocholic acid (TCA), taurodeoxycholic acid (TDCA), taurochenodeoxycholic acid (TCDCA), tauroursodeoxycholic acid (TUDCA), ursocholic acid (UCA), hyocholic acid (HCA), hyodeoxycholic acid (HDCA) were purchased from Sigma Chemical (St. Louis, MO, USA), as horseradish peroxidase (type VI-A, 1100 IU/mg). The enhanced chemiluminescent luminol reagent (ECL) was purchased from Amersham International (Amersham, UK).

Procedure: The luminescent assay was performed in black polystyrene microtiter wells (Dynatech Laboratories, Chantilly, VA, USA). The chemiluminescent cocktail was prepared by adding 100 µL of a 1/10.000 v/v dilution of a stock horseradish peroxidase solution (1 mg/mL in 0.1 mol/L Tris-HCl buffer, pH 8.6) to 5 mL of ECL reagent; 100 µL of the chemiluminescent mixture were dispensed in duplicate into the wells containing 50 µL of an aqueous solution of the BA at different concentrations. The microtiter plate was put in a light-tight box, the luminous emission from all wells was immediately measured as photons/s/pixel and the kinetics of the light output monitored for 1 h. Distilled water was used as negative control and vitamin C and the water soluble tocopherol analogue Trolox were used as positive controls. The results were expressed as cumulative light output over 30 and 60 min and the % light inhibition compared to the total light output of the control was calculated.

Results and Discussion

All the BA studied inhibited the light emission from the peroxidase-ECL reaction. Typical behaviour of some BA as a function of concentration is shown in Figure 1. The light inhibition occurred immediately after the addition of these compounds and, after a specific period of time, the light emission resumed reaching levels similar to those for the control or in some cases (eg CDCA or CA) higher levels.

The extent of the light quenching and the kinetics of the resumption of light differed among the BA studied. BA with a low critical micellar concentration (CMC) were the most potent. For BA with a similar CMC, BA with a high aggregation number of the micelles and with a low surface tension at the CMC were the most effective antioxidants. Among unconjugated BA, DCA and CDCA were the most potent, followed by UDCA, CA and UCA. The respective glycine conjugated BA showed a similar or slightly higher activity, while the taurine conjugates were significantly more potent than the corresponding free BA; this is consistent with lower CMC values of the conjugated compared to the free BA (5).

The effect was concentration-dependent, with the percent inhibition increasing with concentration for a given BA. Usually, at low concentration below the CMC, BA are ineffective or enhance the light output compared to the control values, while a significant reduction of the steady-state light output is achieved approaching the CMC of each BA.

The structure of the BA and BA micelles suggested that the light inhibition may derive from a physical interaction of BA micelles with oxygen reactive species, generated in the peroxidase-ECL reaction, via hydrophobic interaction in the micelle inner domain, thus reducing the reactivity of the oxygen species and facilitating the $^{\bullet}O_2^-$ dismutation. This postulated interaction fitted with the observed light inhibition and restoration, and with the dose-dependent effect of the BA. Moreover, gas chromatograpgy-mass spectrometry studies demonstrated that BA were unmodified after the interaction with oxygen radicals.

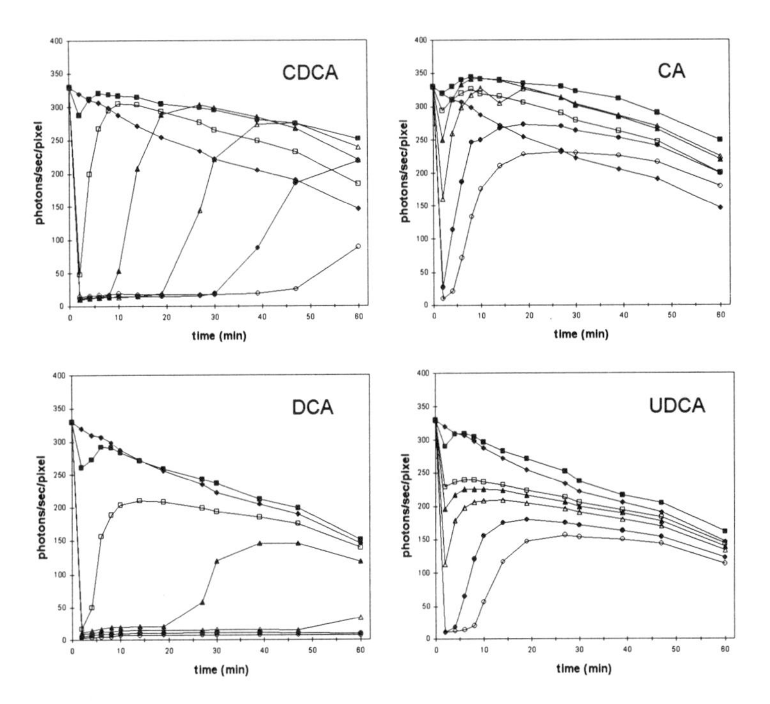

Figure 1. Kinetic profiles of the light emission from the peroxidase-ECL reaction in the presence of unconjugated bile acids as a function of concentration. ◆ Control, ■ 1.1, □ 5.7, ▲ 11.4, Δ 17.4, ● 22.8, O 28.6 mM.

Oxy-radicals, particularly $^{\bullet}O_2^-$, were trapped by the BA micellar domain and the light was inhibited since the radical-mediated formation of the unstable intermediate emitting light did not occur. When all the binding sites were saturated, the light emission resumed due to the presence of $^{\bullet}O_2^-$ free in solution that was available to react with luminol radicals. As the BA analytical concentration was raised the number of micelles and binding sites increased, the quenching was higher and the time to restore the light emission was longer. This was due to an increased number of micelles in solution, as the analytical concentration increased while the monomeric BA concentration remained practically constant.

In addition, the increasing effect with high aggregation number of the micelles further supported the concept of insertion of oxygen into micelles as a mechanism of light quenching and antioxidant activity. This interaction was further demonstrated by repeating the addition of a BA once the light was restored after the initial inhibition: the quenching occurred again followed by restoration of light emission.

Our data demonstrate that BA are effective antioxidants at concentrations approaching their CMC. Further studies are required to explain the finding that at lower concentrations many BA, particularly those with a 7α hydroxyl (eg CA and CDCA), are promoters of oxygen free radical production.

As regards the biological implications of these effects, the micelle interaction with reactive oxygen species could be a physiological mechanism of defence against BA toxicity in the intestinal content due to their detergent properties. Alternatively, alteration in bile acid organ distribution, concentration and composition, which is known to lead to cell membrane damage (11,12), could be associated with oxygen free radical generation.

References

1. Cross CE, Halliwell B, Borish ET, Pryor WA, Ames BN, Saul RL et al. Oxygen radicals and human disease. Ann Intern Med 1987;107:526-45.
2. Fridovich I. Superoxide dismutases. Adv Enzymol 1974;41:35-97.
3. Deisseroth A, Dounce AL. Catalase: physical and chemical properties, mechanism of catalysis and physiological role. Physiol Rev 1970;50:319-75.
4. Halliwell B, Gutteridge JMC. The antioxidants of human extracellular fluids. Arch Biochem Biophys 1990;280:1-8.
5. Hofmann AF, Roda A. Physicochemical properties of bile acids and their relationship to biological properties: an overview of the problem. J Lipid Res 1984;25:1477-89.
6. DeLange RJ, Glazer AN. Bile acids: antioxidants or enhancers of peroxidation depending on lipid concentration. Arch Biochem Biophys 1990;276:19-25.
7. Pascual C, Romay C. Effect of antioxidants on chemiluminescence produced by reactive oxygen species. J Biolumin Chemilumin 1992;7:123-32.
8. Metsä-Ketelä T. Luminescent assay for total peroxyl radical-trapping capability of plasma. In: Stanley PE, Kricka LJ, editors. Bioluminescence & Chemiluminescence: current status. Chichester: John Wiley & Sons, 1991:389-92.
9. Mashiko S, Iwanaga S, Hatate H, Suzuki N, Seto R, Hara Y et al. Antioxidative activity of bioactive compounds: measurement by *Cypridina* chemiluminescence method. In: Szalay AA, Kricka LJ, Stanley PE, editors. Bioluminescence & Chemiluminescence: status report. Chichester: John Wiley & Sons, 1993:247-51.
10. Whitehead TP, Thorpe GHG, Maxwell SRJ. Enhanced chemiluminescent assay for antioxidant capacity in biological fluids. Anal Chim Acta 1992;266:265-77.
11. Sokol RJ, Devereaux M, Khandwala R, O'Brien K. Evidence for involvement of oxygen free radicals in bile acid toxicity to isolated rat hepatocytes. Hepatology 1992;17:869-81.
12. Craven PA, Pfanstiel J, Saito R, DeRubertis FR. Actions of sulfasalazine and 5-amino salycilic acid as reactive oxygen scavengers in the suppression of bile acid-induced increases in colonic epithelial cell loss and proliferative activity. Gastroenterology 1987;92:1998-2008.

PART 9

Green Fluorescent Protein, Aequorin and Obelin

GFP AS A MARKER FOR A NUCLEAR PORE COMPLEX PROTEIN

G Imreh, H Söderqvist, M Kihlmark and E Hallberg
Dept of Biochemistry, Stockholm University, Stockholm S-106 91, Sweden

Introduction

The eukaryotic chromatin is separated from the cytoplasm by the nuclear envelope (Fig. 1) consisting of an outer and an inner nuclear membrane. For a review see (1). At numerous circumscribed points the outer and inner membranes fuse to form the nuclear pores. The fusion points delineating the pores are referred to as the pore membrane and harbor the NPCs (Nuclear Pore Complexes), the exclusive sites of nucleo/cytoplasmic exchange of proteins and RNA.

Integral proteins of the pore membrane, such as, POM121 (**po**re **m**embrane protein of 121 kDa) are thought to be essential for pore formationand assembly of the NPC. POM121 was recently identified and its cDNA cloned and sequenced (2). POM121 has a short luminally exposed N-terminal tail (Fig. 1), a single transmembrane segment and a large C-terminal portion adjoining the NPC. The C-terminal also containins a characteristic NPC protein signature, suggesting that it might interact directly with other NPC proteins (3).

Integral membrane proteins are first synthesized on the rough endoplasmic reticulum and then targeted to their specific cellular membranes by lateral diffusion and/or vesicular transport. We have investigated how POM121 is targeted to the nuclear pores by overexpressing deletion mutants of rat POM121 in COS (monkey) cells and selectively detecting the overexpressed protein by indirect immunofluorescence using species specific POM121 antibodies (4). In the present paper we demonstrate the usefulness of the GFP (Green Fluorescent Protein) from *Aequorea victoria* as a marker for intracellular sorting of POM121.

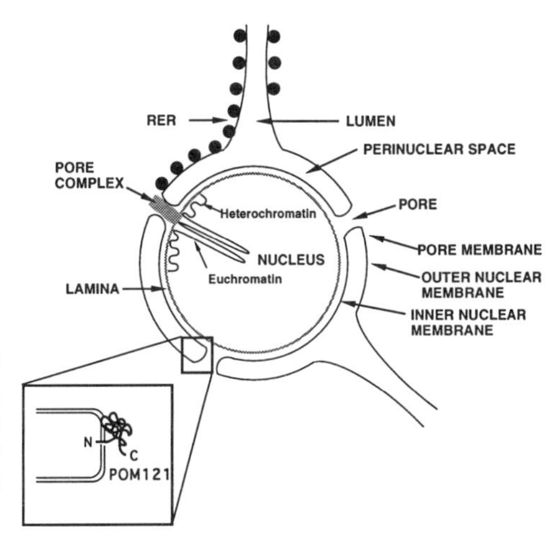

Fig. 1. Schematic representation of a eukaryotic cell nucleus. The outer nuclear membrane is continuous with the rough endoplasmic reticulum (RER) and connected with the inner nuclear membrane via the pore membrane. The pore complexes are embedded in the nuclear pores. The nuclear pore complex protein POM121 have been hypothetically inserted in the pore membrane (inset).

Materials and Methods

Instrumentation: The GFP fluorescence was examined using a Zeiss Axiophot microscope equipped with a cooled CCD camera (C4880 digital, Hamamatsu Photonics, Japan).

Reagents: Plasmid purification kit were purchased from Quiagen Inc., Chatsworth, CA, USA. Cell culture media and lipofectamine were obtained from Gibco Laboratories, Grand Island, NY, USA. Hoechst 33258 and *p*-phenylene diamine were obtained from Sigma chemical co, St. Louis, MO, USA.

Procedures: In order to fuse the genes encoding rat POM121 and GFP, restriction sites were generated by PCR using modified primers, for details see (6). The clone c11 (2), cDNA encoding full length POM121 inserted in pBluescript II sk+, was used as a template for amplification of a modified portion of the POM121 gene in order to replace c11. The modified plasmid was cleaved using an *XbaI* site and a novel *NheI* site in order to incorporate a portion of the pGFP 10.1 plasmid (positions 726-1718) (5) containing the entire coding region of GFP and novel *NheI* and *XbaI* restriction sites to match those of the modified c11 plasmid. The resulting DNA encodes a fusion protein where the initiating methionine of GFP is replaced with an alanine which is fused to the C-terminal lysine-1199 of POM121. Unique *NcoI* and *BstB1* restriction sites in the GFP gene were used to exchange amino acids 54 -206 of the wild type GFP with the corresponding region of the mutated forms. The fusion genes were cleaved by *SalI* and *XbaI* and incorporated in the eukaryotic expression vector pcDNA1Neo (Invitrogen, San Diego, CA, USA), which was cut by *XhoI* (complementary to *SalI*) and *XbaI*.

Culture and transfection of COS-1 cells (African green monkey kidney cells) were performed as described in (4,6). 8 μg plasmid DNA in 10 mL culture medium per 100 mm Petridish were used. After 5 h transfection using the cationic lipid lipofectamine in serum free medium the cell cultures were allowed to incubate in normal culture medium for an additional 12-72 h at 25°C or 37°C, respectively. Coverslips containing transfected COS cells were fixed, permeabilized and mounted in 1 mg/mL *p*-phenylenediamine in 90 % glycerol, pH 8.0

Nuclei from transfected COS cells were prepared as described (10), incubated with 0.5 μg/mL Hoechst 33258 and mixed with an equal volume of 1 mg/mL *p*-phenylenediamine in 90 % glycerol, pH 8.0 for mounting under a 13 mm coverslip, which was sealed using nail polish.

The expression of POM121-GFP was monitored by GFP's natural fluorescence using an epifluorescence microscope or by Western blotting using anti-POM121 antibodies (2).

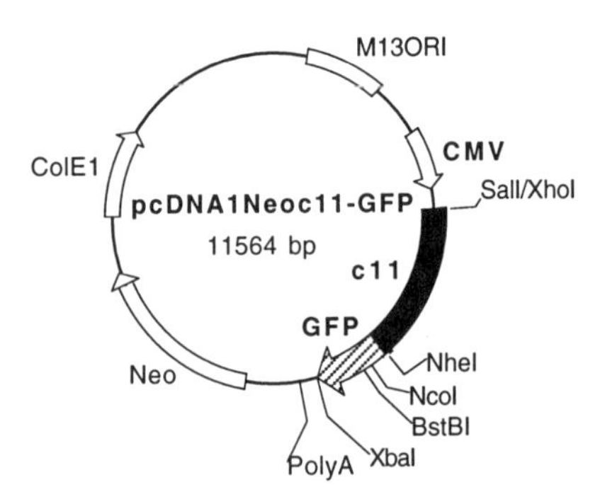

Fig. 2. Plasmid pcDNA1Neo-c11-GFP. The *c11::gfp* gene fusion, encoding GFP fused to the C-terminus of POM121 (*c11*), is inserted in the MCS using the XhoI and XbaI sites. The plasmid carries an SV40 origin of replication, a gene for neomycin resistance and a CMV promoter. The GFP chromophore is contained between the NcoI and BstBI sites (amino acids 54 -206 of GFP), which may thus be used for exchange with mutant chromophores.

Results and Discussion

We used PCR to generate restriction sites for fusion of the 5' end of cDNA encoding GFP (5) to the 3' end of cDNA encoding POM121 (Fig. 2) in order to create a chimerical POM121-GFP fusion gene. A DNA fragment (encoding amino acids 54 -206 of GFP) of the wild type GFP (5) was exchanged with the corresponding nucleotides of different GFP mutants (7-9) using the NcoI and BstBI restriction sites in the GFP coding region. The fusion genes were incorporated into the eukaryotic expression vector pcDNA1Neo, which was used to transfect monolayer COS cells. The cells were incubated at 25°C or 37°C for periods between 12-72 hours after transfection (see materials and methods).

Although expression levels, as judged by Western blotting (not shown), were similar the GFP mutants gave rise to a more intense fluorescence as compared to the wild type at both temperatures (Table 1). However, only the double and triple mutants gave rise to detectable signals in low expressing cells at 37°C. In our system the GFP[F64L,S65T] double mutant gave rise to the most intense signal, clearly detectable in living or fixed cells after 12 h and remaining for at least 72 h, and were hence used throughout this study.

Mutations	excitation maximum	emission maximum	Refernce	25°C	37°C
Wild type	396 (475) nm	508 nm	5	+	-
I167T	475 nm	510 nm	7	++	-
S65T	488 nm	511 nm	8	++	-
S54T,V68L,S72A	(395) 488 nm	510 nm	9	+++	+++
S65G,S72A	(395) 488 nm	510 nm	9	+++	+++
F64L,S65T	(395) 488 nm	510 nm	9	++++	++++

Table 1. An overview of the GFP variants fused to POM121 and overexpressed in COS cells incubated at 25°C and 37°C, respectively. The two columns to the right denotes the relative intensities of the GFP fluorescence estimated by fluorescence microscopy using a BP 485/20 nm excitation filter and a LP 520 nm emission filter.

In the low expressing cells the GFP fluorescence appeared in a finely punctate pattern over the surface of the nucleus and as an intense rim in equatorial sections of the nucleus. The distribution of the fluorescence obtained is consistent with a localization of the POM121-GFP fusion proteins at the nuclear pores (6). Parallel immunostaining suggests that the overexpressed POM121-GFP molecules can be detected in amounts similar to those of endogenous POM121, using the instrumentation described in materials and methods. Isolated nuclei from COS cells overexpressing POM121-GFP display an intense rim fluorescence over the nuclear envelope (Fig.3) and may thus be an excellent starting material for biochemical analysis of overexpressed proteins.

In conclusion, POM121 is the first example where GFP have been used as a marker for an integral nuclear pore protein. The results from this study suggest that the GFP[F64L,S65T] mutant (9) is an ideal marker for studies of intracellular trafficking of nuclear pore proteins at physiological temperatures. Since GFP fluoresces without fixation or addition of cofactors, a stably transfected cell line expressing POM121-GFP would serve as an excellent model for studies of nuclear envelope dynamics in living cells. Such experiments are presently under way.

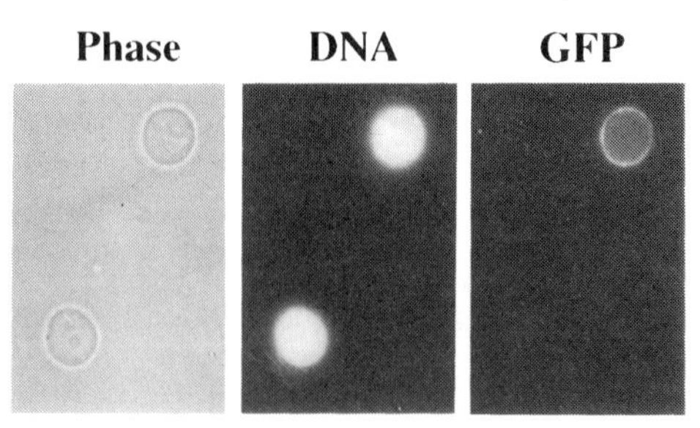

Fig. 3. Purified nuclei from transfected COS cells. The three panels show the same field of nuclei purified from COS-1 cells overexpressing a chimerical POM121-GFP[F64L,S65T] fusion protein and viewed by phase contrast optics (left panel), DNA staining by Hoechst 33258 viewed in the UV channel (middle panel), or GFP fluorescence viewed in the FITC channel (right panel). A nucleus from a transfected COS cell (upper) display a continuos rim fluorescence from the numerous overlapping nuclear pores in this equatorial section of the nuclear envelope, whereas a nucleus from a non expressing cell (lower) does not. From (10) with permission from Academic Press.

Acknowledgements

We wish to thank Dr. Chalfie and Dr. Falkow for kindly donating the wild type and mutant GFP genes. This investigation was supported by the Swedish Natural Science Council, Åke Wibergs Stiftelse, Stiftelsen Lars Hiertas Minne.

References

1. Panté N, Aebi U. Toward the molecular details of the nuclear pore complex. J Struct Biol 1994;113:179-89.
2. Hallberg E, Wozniak RW, Blobel G. An integral membrane protein of the pore membrane domain of the nuclear envelope contains a nucloporin-like region. J Cell Biol 1993;122:513-21.
3. Söderqvist H, Hallberg E. The large C-terminal region of the integral pore membrane protein, POM121, is facing the nuclear pore complex. Eur J Cell Biol 1994;64:186-91.
4. Söderqvist H, Jiang W-Q, Ringertz N, Hallberg E. Formation of nuclear bodies in cells overexpressing the nuclear pore protein POM121. Exp Cell Res 1996;225:75-84.
5. Chalfie M, Tu Y, Euskirchen G, Ward WW, Prasher DC. Green fluorescent protein as a marker for gene expression. Science 1994;263:802-5.
6. Söderqvist H, Imreh G, Kihlmark M, Thölix T, Ringertz N, Hallberg E. Intracellular distribution of the nuclear pore protein POM121 fused to GFP (Green Fluorescent Protein) from *Aequorea victoria*. Exp. Cell Res. (submitted).
7. Heim R, Prasher DC, Tsien R. Wavelength mutations and posttranslational autooxidation of green fluorescent protein. Proc Natl Acad Sci 1994;91:12501-4.
8. Heim R, Cubitt AB, Tsien RY. Improved green fluorescence. Nature 1995;373:662-3.
9. Cormack BP, Valdivia RH, Falkow S. FACS-optimized mutants of the green fluorescent protein (GFP). Gene 1996;173:33-38.
10. Kihlmark M, Hallberg E. Isolation of nuclei and nuclear envelopes. In: Celis J, editor. Cell Biology. A Laboratory Handbook. San Diego, CA, USA: Academic Press, Inc. 1997 (submitted).

LOCALIZATION OF CELLS ACTIVE IN GENE EXPRESSION AND GENE TRANSFER IN INTACT BACTERIAL BIOFILMS

C Sternberg, BB Christensen, S Møller, JB Andersen and S Molin
Department of Microbiology, Technical University of Denmark, DK-2800 Lyngby Denmark

Introduction

Microbial communities consisting of many different species are commonly found as biofilms attached to solid or semi-solid surfaces in natural settings as well as in many applied and man-made contexts. In fact, bacterial life in general is normally organized in surface bound consortia rather than as free-flowing planctonic cells in liquid. The understanding of this bacterial life-form requires detailed investigations of the processes and the composition of such communities. For the basic understanding it is essential to characterize the specific metabolic and proliferative activities in different positions of the community in order to understand the ways that the overall performance is coordinated; analogously, it is highly significant in specific applications of microbial communities (e.g. in bioreactors) to map the specific activities of the relevant community members and monitor the spread of new properties in order to interfere and optimize the community performance.

Two types of activity are in particular interesting: 1) Activation of genes related to specific environmental signals (substrates, stress, oxygen, pH etc.), and 2) gene transfer introducing new metabolic elements of significance for the overall processing of the available substrates. In simple test tube settings there are many molecular methods available for analysis of specific gene expression and gene transfer. We report here on two new approaches allowing sensitive and localized monitoring of gene expression and gene transfer *in situ* in multi-species communities based on microscopic inspection of fluorescently tagged bacteria.

Materials and Methods

Strains and media: Biofilm communities are cultivated as a mixture of seven bacterial species isolated from a toluene degrading bioreactor (1). Three of the seven species are capable of utilizing toluene as sole carbon source, *Pseudomonas putida* R1 (SMO116), *Acinetobacter sp.* SMO112, and an untyped gram-negative strain, D8 (SMO125). The remaining four species are unable to use toluene as sole carbon source. *P.putida* KT2442 (a gift from Dr. Victor de Lorenzo), modified by the insertion of the *lacI*^{Q1} gene into the chromosome was used as donor in plasmid transfer experiments. Cells were cultivated in four channel flow-chambers (2) and maintained on FAB media (1 mmol/L MgCl, 0.1 mmol/L $CaCl_2$, 0.01 mmol/L Fe-EDTA (Sigma E6760, Sigma St. Louis, MO, USA), 0.15 mmol/L $(NH_4)SO_4$, 0.33 mmol/L Na_2HPO_4, 0.2 mmol/L KH_2PO_4, 0.5 mmol/L NaCl). As carbon source was used either 2 mmol/L Na_3-citrate or 0.5 mmol/L benzylalcohol. Luria-broth (LB) was used for plating experiments.

Plasmids: The gene encoding green fluorescent protein (Gfp) was obtained as an enhanced version (E*gfp*-mut3) from pGFPmut3 (R. Valdivia pers. comm.). The Egfp-mut3 protein emits up to 20 times more fluorescence per mole than Gfp (3). E*gfp*-mut3 was inserted into the transposon of pVDL20 (4), downstream of the Pm promoter from the TOL-meta pathway genes. The resulting plasmid, pJBA26 carries a Tn*5*-mini-transposon capable of transposing into a recipient strain while not co-transposing the transposase gene. Further, the transposon carries the regulator gene *xylS*, required for activation of Pm. For gene transfer experiments, E*gfp*-mut3 was inserted downstream of the P*lac*$_{A1/04/03}$ promoter (5) in a Tn*5*-mini-transposon. This monitor cassette was inserted into the TOL plasmid by transposition (TOL-Plac-E*gfp*).

Instrumentation: Flow chambers were constructed of Plexiglas. The growth chambers were 4 mm wide, 1 mm deep and 40 mm long. The chambers were covered with glass coverslips which were glued to the Plexiglas with a silicone adhesive (3M Super Silicone Sealant, 3M St.Paul, MN) to allow microscopic examination (2). Media was flowed through the chambers at a constant flow rate of 0,2 mm/s. All microscopic inspections were performed on a Zeiss Axioplan epi-fluorescence microscope (Carl Zeiss Jena GmbH, Jena, Germany) fitted with a Photometrics KAF1400 CCD (Charge coupled device) camera (Photometrics, Tucson, AZ) or on a Leica TCS4D confocal microscope (Leica Lasertechnik GmbH, Heidelberg, Germany). In the Zeiss microscope we used the standard FITC (Zeiss #9) filter set for visualization of Gfp, while in the Leica microscope we utilized the blue laser line (488 nm) for excitation and used a LP515 emission filter.

Procedures: Embedding: Flow chamber communities were fixed in 2% paraformaldehyde, followed by a 3 min wash in PBS. Whole chambers were embedded in 20% acrylamide (200:1 acrylamide:bis-acrylamide) by gently filling the chambers with embedding solution using a peristaltic pump. Probing of ribosomal RNA with oligonucleotide probes was performed on acrylamide stabs as previously described (6)

Results and Discussion

The bacterial community under study consists of seven strains and forms a highly complex 3D structure after several days of growth in the flow chamber. Using ribosomal probes, a tendency to form micro colonies consisting entirely or almost entirely of single species was observed (S.Møller, manuscript in preparation). One of the seven bacteria in the community, *P.putida* R1, was selected for further examinations. The strain is able to degrade toluene derivatives, such as benzylalcohol and benzoate. However, preliminary results indicate that this strain does not harbor the TOL plasmid. We introduced the monitor transposon from pJBA26 harboring a fusion between the TOL-Pm promoter and the gene encoding Gfp into the chromosome of R1 (R1-Pm-E*gfp*). The activator of this monitor system is the presence of benzoate, or derivatives of benzoate, such as 3-methyl-benzoate (3MB), which results in Gfp+ (green fluorescent) cells. A microbial community was established replacing the original *P. putida* R1 strain with the tagged strain. FAB minimal growth medium supplied with citrate was used.

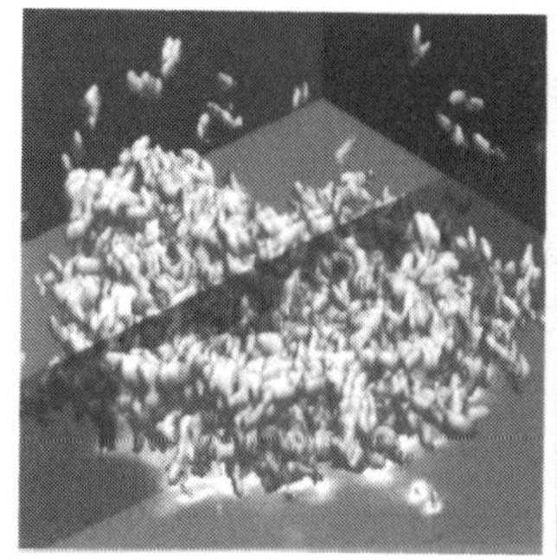

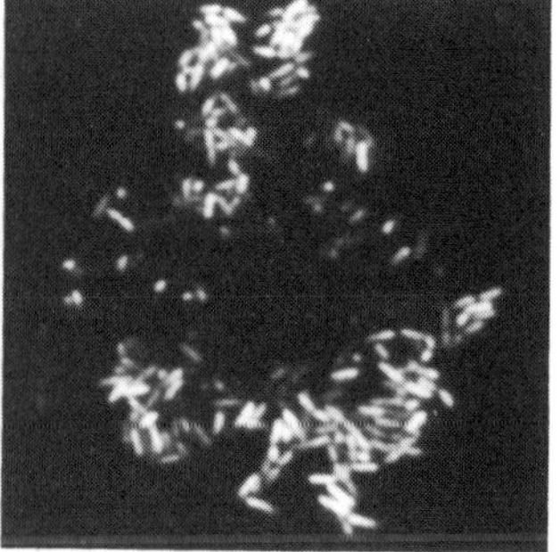

Fig. 1. A. Three dimensional reconstruction of a stack of confocal images of cells expressing Gfp from the TOL-Pm promoter. The cone is viewed from the substratum. B. A single confocal image. Note the cells with less intense fluorescence near the cavity of the cone.

No observable effect of the strain replacement in overall community composition could be detected after several days of growth when comparing the community composed of the R1-Pm-E*gfp* and the remaining 6 strains with a parallel system inoculated by the 7 wt strains. No fluorescence was detected in either system. After three days of growth, the substrate was shifted to one containing 5 mmol/L 3MB and allowed one additional day of growth. In the community inhabited by R1-Pm-E*gfp*, well-defined micro-colonies emitting green fluorescence could be detected (Fig. 1). Of particular interest was dome shaped structures consisting of green fluorescent cells (i.e. R1-Pm-E*gfp*) forming a cavity where cells were swimming freely. We examined several such domes by the confocal microscope (Fig. 1), and it was clear that towards the void of the domes cells were less green than towards the outside. The cause of this difference in fluorescence is unclear, but several explanations are possible. Oxygen limitation within the domes, resulting in lesser extent of Gfp maturation may be one explanation, or substrate limitation could prevail inside the domes. This remains to be investigated further.

For gene transfer experiments, we have constructed a special donor strain, derived from *P.putida* KT2442. The donor strain has a *lacI* gene inserted into the chromosome. Further, we introduced the conjugative monitor plasmid, TOL-Plac-E*gfp* in this strain. When grown on any substrate, this strain, 2442/TOL-Plac-E*gfp*, is

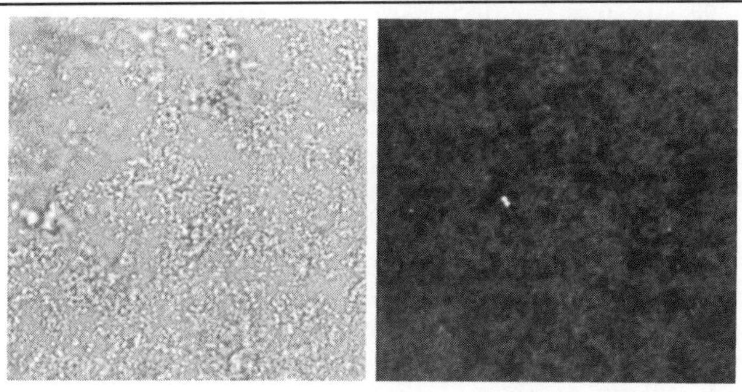

Fig. 2. A. A micrograph of a mature biofilm using conventional transmitted light. B. Same view as in A, using epi-fluorescence. One bacterium lights up, indicating the reception of the gfp-tagged TOL plasmid.

non-fluorescent. When plated on 1 mmol/L IPTG (isopropyl-β-D-thiogalactopyranoside) LB-plates, cells become brightly fluorescent. As an additional control, the tagged TOL plasmid was conjugated into a control strain which did not host the *lacI* gene resulting in Gfp positive cells. To investigate the *in situ* gene transfer in the seven strain biofilm, we inoculated the 2442/TOL-Plac-E*gfp* together with the seven wt strains in a flow cell. The evolution of green fluorescence was followed by microscopic inspection. Also, the number of ex-conjugants, recipients and donors were followed by plate counts from effluent media. One day after inoculation only very few ex-conjugants could be detected (Fig. 2). After 5 days, the biofilm was harvested by fixation and embedded in acrylamide. The acrylamide blocks were probed by two ribosomal probes targeted at *Pseudomonas spp.* and *Acinetobacter spp.* (Fig. 3) and with an eubacterial probe (data not shown). The probing revealed that ex-conjugants were primarily *Pseudomonas* (i. e. R1), while other species did not receive the plasmid in any detectable extent. Further it was found that the overall conjugation frequency was much lower than what might have been expected when the numbers of donors and recipients are taken into consideration. Less than 1% of the available recipients had received the TOL-Plac-E*gfp* plasmid. This finding is different from our previous experiments on conjugation on semi-solid surfaces (7), where we observed an efficiency of up to 25% conjugation relative to the number of recipients after 24 hours of conjugation.

We have in this communication presented molecular approaches towards specific monitoring of gene expression and gene transfer, as these important processes take place in complex microbial communities. From this presentation of the methods and tools it is our hope that their application to a number of different scenarios involving bacterial growth and performance in composite communities is obvious. Wherever it is considered important to monitor and understand the basis for microbial activity in natural or man-made settings methods like those described above should prove their relevance.

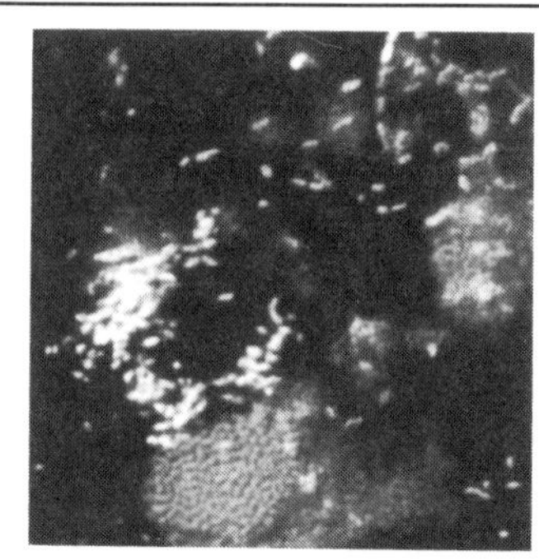

Fig. 3. A biofilm containing Gfp+ ex-conjugants (bright rods). *Pseudomonas* cells (rods), and *Acinetobacter* (gray spheres) are visualized by ribosomal probing.

Acknowledgments

The authors wish to thank Victor de Lorenzo and Raphael Valdivia for the gift of strains and plasmids.

References

1. Møller S, Pedersen AR, Poulsen LK, Arvin E, Molin S. Activity and three-dimensional distribution of toulene degrading *Pseudomonas putida* in a multispecies biofilm assessed by quantitative in situ hybridization and scanning confocal microscopy. Appl Environ Microbiol, Submitted.
2. Woolfaardt GM, Lawrence JR, Robarts RD, Caldwell SJ, Caldwell DE. Multicellular organization in a degradative biofilm community. Appl Environ Microbiol 1994;60:434-6.
3. Cormack BP, Valdivia RH, Falkow S. FACS-optimized mutants of the green fluorescent protein (GFP). Gene 1996;173:33-8.
4. de Lorenzo V, Cases I, Herrero M, Timmis KN. Early and late responses of TOL promoters to pathway inducers: Identification of postexponential promoters in *Pseudomonas putida* with a *lacZ-tet* bicistronic reporters. J Bacteriol 1993;175:6902-7.
5. Lanzer H, Bujard H. Promoters largely determine the efficiency of repressor action. Proc Natl Acad Sci USA 1988;8973-7.
6. Poulsen LK, Ballard G, Stahl DA. Use of rRNA fluorescence in situ hybridization for measuring the activity of single cells in young and established biofilms. Appl Environ Microbiol. 1993;59:1354-60.
7. Christensen B, Sternberg C, Molin S. Bacterial plasmid conjugation on semi-solid surfaces monitored with the green fluorescent protein (GFP) from *Aequorea victoria* as a marker. Gene 1996;173:59-65.

THE THREE-DIMENSIONAL STRUCTURE OF GREEN FLUORESCENT PROTEIN

Fan Yang[1], *Larry G. Moss*[2], *and George N. Phillips, Jr.*[1]
[1]*Department of Biochemistry and Cell Biology and the W.M. Keck Center for Computational Biology, Rice University, Houston, TX 77005-1892, USA*
[2]*Division of Endocrinology, Department of Medicine, Tufts University School of Medicine and the New England Medical Center, Boston, MA 02111, USA*

Introduction

Green fluorescent protein, GFP, is a spontaneously fluorescent protein isolated from coelenterates, such as the Pacific jellyfish, *Aequorea victoria* (1). Its role is to transduce, by energy transfer, the blue chemiluminescence of another protein, aequorin, into green fluorescent light (2). The molecular cloning of GFP cDNA (3) and the demonstration by Chalfie that GFP can be expressed as a functional transgene (4) have opened exciting new avenues of investigation in cell, developmental and molecular biology. Fluorescent GFP has been expressed in a variety of organisms (5). GFP can also function as a protein tag, as it tolerates N- and C-terminal fusion to a broad variety of proteins many of which have been shown to retain native function. The enormous flexibility as a noninvasive marker in living cells allows for numerous other applications such as a cell lineage tracer, reporter of gene expression and as a potential measure of protein-protein interactions (6).

Green fluorescent protein is comprised of 238 amino acids. Its wild-type absorbance/ excitation peak is at 395 nm with a minor peak at 475 nm with extinction coefficients of roughly 30,000 and 7,000 L/mol-cm, respectively (7). The emission peak is at 508 nm. Interestingly, excitation at 395 nm leads to decrease over time of the 395 nm excitation peak and a reciprocal increase in the 475 nm excitation band (5). This presumed photoisomerization effect is especially evident with irradiation of GFP by UV light. Analysis of a hexapeptide derived by proteolysis of purified GFP led to the prediction that the fluorophore originates from an internal Ser-Tyr-Gly sequence which is post-translationally modified to a 4-(*p*-hydroxybenzylidene)-imidazolidin-5-one structure (8). Studies of recombinant GFP expression in *E. coli* led to a proposed sequential mechanism initiated by a rapid cyclization between Ser^{65} and Gly^{67} to form a imidazolin-5-one intermediate followed by a much slower (hours) rate-limiting oxygenation of the Tyr^{66} side chain by O_2 (9). Combinatorial mutagenesis suggests that the Gly^{67} is required for formation of the fluorophore (10). While no known co-factors or enzymatic components are required for this apparently auto-catalytic process, it is rather thermosensitive with the yield of fluorescently active to total GFP protein decreasing at temperatures greater than 30° C (11). However, once produced GFP is quite thermostable.

Physical and chemical studies of purified GFP have identified several important characteristics. It is very resistant to denaturation requiring treatment with 6 mol/L guanidine hydrochloride at 90° C or pH of <4.0 or >12.0. Partial to near total renaturation occurs within minutes following reversal of denaturing conditions by dialysis or neutralization (12). Over a nondenaturing range of pH, increasing pH leads to a reduction in fluorescence by 395 nm excitation and an increased sensitivity to 475 nm excitation (13).

The availability of *E. coli* clones expressing GFP has led to extensive mutational analysis of GFP function. Truncation of more than 7 amino acids from the C-terminus or more than the N-terminal Met lead to total loss of fluorescence(14). Screens of random and directed point mutations for changes in fluorescent behavior have uncovered a number of informative amino acid substitutions. Mutation of Tyr^{66} in the fluorophore to His results in a shift of the excitation maximum to the UV (383 nm) with emission now in the blue at 448 nm (9). A $Tyr^{66}Trp$ mutant is blue-shifted albeit to a lesser degree. Both changes are associated with a severe weakening of fluorescence intensity compared to wild type GFP. Mutation of Ser^{65} to Thr, Ala, Cys or Leu causes a loss of the 395 nm excitation peak with a major increase in blue excitation (10, 15). When combined with Ser^{65} mutants, mutations at other sites near the fluorophore such as $Val^{68}Leu$ and $Ser^{72}Ala$ can further enhance the intensity of green fluorescence produced by excitation at 488 nm (10, 16). However, amino acid substitutions significantly outside this region also affect the protein's spectral character. For example, $Ser^{202}Phe$ and $Thr^{203}Ile$ both cause the loss of excitation in the 475 nm region with preservation of 395 nm excitation (4, 9, 17). Ile^{167} Thr results in a reversed ratio of 395 to 475 nm sensitivity (5), while $Glu^{222}Gly$ is associated with the elimination of only the 395 nm excitation (17). Another change, $Val^{163}Arg$, increases the temperature tolerance for functional GFP expression (7).

Because GFP *in crystallum* exhibits a nearly identical fluorescence spectrum and lifetime to that for GFP in aqueous solution (18) and fluorescence is not an inherent property of the isolated fluorophore, the elucidation of its three-dimensional structure will help provide an explanation for the generation of fluorescence in the mature protein, as well as the mechanism of autocatalytic fluorophore formation. Furthermore, the development of fluorescent proteins with additional emission and excitation characteristics would dramatically expand their biological applications. Here we describe the structure derived from a crystal of wild-type, recombinant *A. victoria* green fluorescent protein.

Materials and Methods

Green fluorescent protein was purified from *E. coli* containing plasmid TU#58, bearing the wild-type *A. victoria* green fluorescent protein (4) as described elsewhere (19). Gel filtration columns run at 10 mmol/L phosphate showed predominately a 2-fold higher molecular weight species. The protein was crystallized in sitting drop vapor diffusion wells at room temperature using 58% 2-methyl-2,4-pentanediol, 50 mmol/L morpholino ethane sulfonic acid, 0.1% sodium azide at pH 6.8. The protein concentration was typically 20-30 mg/ml. Crystals grew as square bipyramids up to 0.5 mm on a side. The space group was determined to be $P4_12_12$ with a=b= 89.23 Å and c= 119.78 Å at room temperature.

Multi-wavelength anomalous dispersion (MAD) data (19) were taken at the Brookhaven synchrotron. The resulting MAD-phased map was solvent flattened and two-fold averaged. The polypeptide chain was traced for one of the barrels beginning from the seleniomethionines and extending in each direction, helped by the recognition of the modified tyrosine in the middle of the barrel as Tyr^{66}, the nucleus of the fluorophore. Refinement (using X-PLOR (20)) with the native data collected at room temperature gives an R-factor at 1.9 Å of 0.21 with an R-free of 0.26, with good geometry (rms bond and angle deviations from ideality of 0.013 Å and 1.8°, respectively) and tight restraint of the non-crystallographic symmetry. Coordinates have been deposited at the Brookhaven Protein Data Bank.

Results

The MAD electron density maps of GFP were very clear, revealing a dimer of two quite regular β-barrels with 11 strands on the outside of cylinders (Figure 1 and 2).

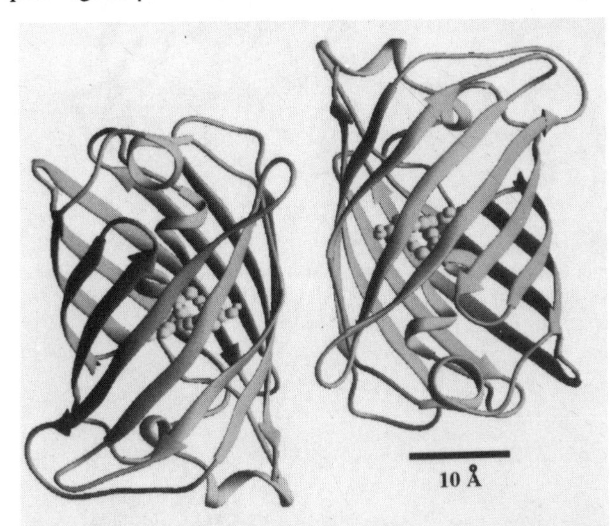

Figure 1.The overall shape of the protein and its association into dimers. Eleven strands of β-sheet form the walls of a cylinder. Short segments of α-helices cap the top and bottom of the 'can' and also provide a scaffold for the fluorophore which is near geometric center of the can. This folding motif,with β-sheet outside and helix inside, represents a new class of proteins. Two monomers are associated into a dimer in the crystal and in solution at low ionic strengths. This view is along the two-fold axis.

These cylinders have a diameter of about 30 Å and a length of about 40 Å. Inspection of the density within the cylinders revealed modified tyrosine side chains as a part of an irregular α-helical segment. Small sections of α-helix also form caps on the ends of the cylinders. This motif, with a single α-helix inside a very uniform cylinder of β-sheet structure, represents a new protein class, which we call the β-can.

The fluorophore is highly protected, located on the central helix within a couple of Ångstroms of the geometric center of the cylinder. The pocket containing the fluorophore has a surprising number of charged residues in the immediate environment (Figure 3). The environment around the fluorophore includes both apolar and polar amino acid side chains and immobilized water molecules. Phe^{64} and Phe^{46} are near the fluorophore and separate the single tryptophan, Trp^{63} from direct contact with fluorophore (closest distance of 13 Å). The crystallographic contacts are all rather tenuous, consisting of a few amino acids side chains for each. The non-crystallographic symmetry is maintained by extensive contacts and thus is likely to be the source of the dimerization seen in solution studies. The dimer contacts are fairly

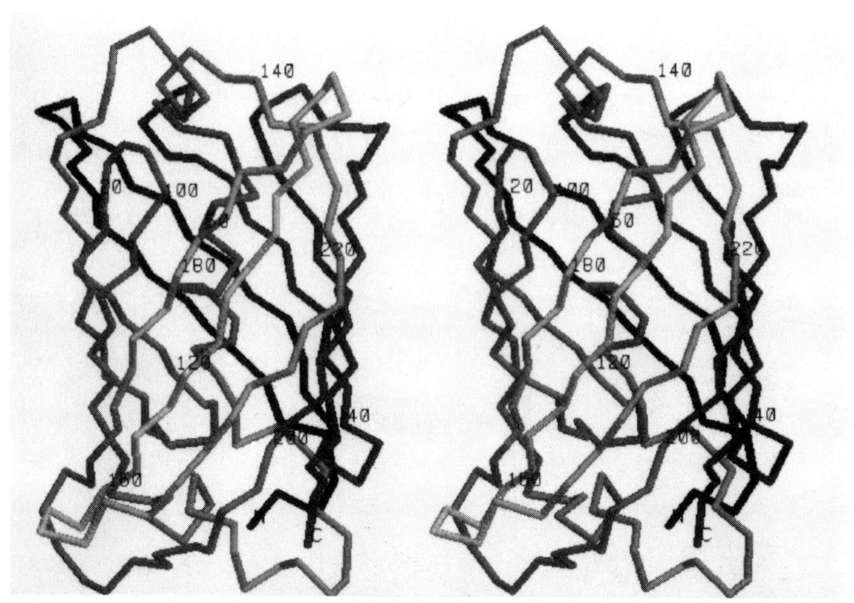

Figure 2. Stereo view of a monomer, with shades that vary slowly as a function of the distance along the polypeptide chain. The termini and Cα atoms of every 20th amino acid are marked just to the upper right of each atom. Figure porduced with RasMol.

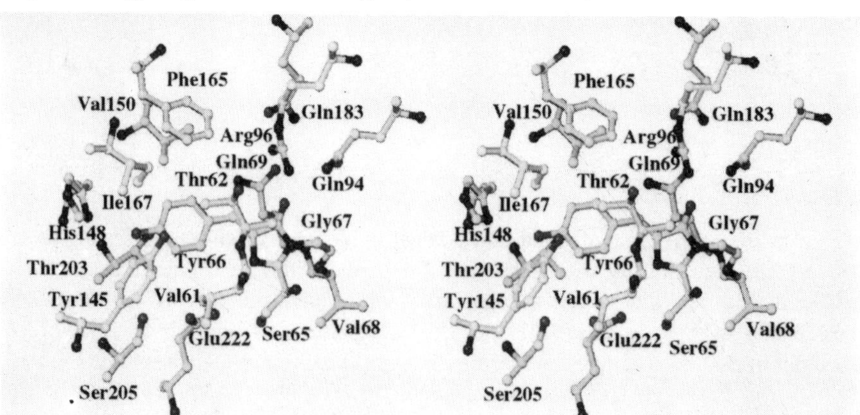

Figure 3. Stereo view of the fluorophore and its en-vironment. His^{148}, Gln^{94}, Arg^{96}, and Glu^{222} can be seen on opposite ends of the fluorophore and probably stabilize resonant forms. Water molecules, charged, polar, and non-polar side chains all contact the fluorophore in various ways.

tight and consist of a core of hydrophobic side chains from each of the two monomers and a wealth of hydrophilic contacts.

Discussion

The remarkable cylindrical fold of the protein, which we have named the β-can, seems ideally suited for the function of the protein. The strands of β-sheet are tightly fitted to each other like staves in a barrel, and form a regular pattern of hydrogen bonds. Together with the short α-helices and loops on the ends, the barrel structure forms a single compact domain and does not have obvious clefts for easy access of diffusable ligands to the fluorophore. This fold, taken with the observation that the fluorophore is near the geometric center of the molecule explains the observed protection of the fluorophore from collisional quenching by oxygen ($K_{bm} < 0.004$ L/mol-s) (21) and hence reduction of the quantum yield. Perhaps more seriously, photochemical damage by the formation of singlet oxygen through intersystem crossing is reduced by the structure. The tightly constructed β-barrel would appear to serve this role well, as well as to provide overall stability and resistance to unfolding by heat and denaturants.

The location of certain amino acid side chains in the vicinity of the fluorophore also begins to explain the fluorescence and the behavior of certain mutants of the protein. At least two resonant forms of the fluorophore can be drawn, one with a partial negative charge on the benzyl oxygen of Tyr^{66}, and one with the charge on the carbonyl oxygen of the imidizolidone ring. Interestingly, basic residues appear to form hydrogen bonds with each of these oxygen atoms, His^{148} with Tyr^{66} and Gln^{94} and Arg^{96} with the imidizolidone (Figure 4).

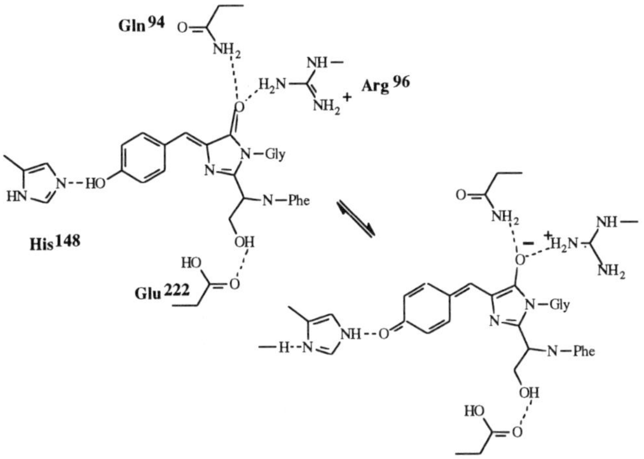

Figure 4. Schematic diagram of the resonant forms of the fluorophore with nearby basic amino acids, His^{148}, Gln^{94}, and Arg^{96} and the acid, Glu^{222}. The bases appear to stabilize anionic oxygen atoms at opposite ends of the fluorophore and the acid forms a hydrogen bond withthe hydroxyl of Ser^{65}.

These bases presumably act to stabilize and possibly further delocalize the charge on the fluorophore. Recent kinetic studies have suggested that proton transfers may play an important role in the excitation/fluorescence process, (22) and Youvan and Michel-Beyerle, personal communication). Based on the structure, the carbonyl oxygen of the imidazolidone, arising originally from the backbone of Tyr^{66} , is a likely candidate for the acceptor, with Arg^{96} the donor. Another possibility is the Tyr^{66} distal oxygen-His^{148} donor/acceptor pair. Disruption of either this interaction by mutations at position 66 result in only the high energy absorption peak and a blue-shifted emission band, whereas disruption of the Ser^{65} hydroxyl-Glu^{222} interactions result in a red-shifted absorption maximum and an unchanged emission spectrum.

Most of the other polar residues in the pocket form an extensive hydrogen-bonding network on the side of Tyr^{66} that requires abstraction of protons in the oxidation process. It is tempting to speculate that these residues help abstract the protons. As for the mutants, atoms in the side chains of Thr^{203}, Glu^{222}, and Ile^{167} are in van der Waals contact with Tyr^{66}, so their mutation would have direct steric effects on the fluorophore and would also change its electrostatic environment if the charge were changed, as suggested previously (17). It seems likely that other mutations of the residues identified to be near the fluorophore would also have effects on the absorption and/or emission spectra.

Mutations in regions of the sequence adjacent to the fluorophore, i.e. in the range of positions 65-67, have been systematically explored (10), some having significant wavelength shifts and most suffer a loss of fluorescence intensity. For example, mutation of the central Tyr to Phe or His shifts the excitation bands but there is an overall loss of intensity. Secondary mutations to compensate for the deleterious intensity effects may also now be possible. The Ser^{65}Thr mutant is particularly interesting because of its reported increase in fluorescence intensity (9, 15). The mechanism for increased fluorescence may be reduced collisional quenching, as the additional methyl group may make for better packing in the interior of the protein. On the other hand, the effect has been suggested to be through improved conversion of the tyrosine to dehydrotyrosine. However, the fact that we see significantly altered structure relative to standard protein conformations in the wild-type argues against a dramatic increase in cyclization and/or oxidation. This effect is most likely produced by increased expression and/or folding of the protein. The report of improvements in GFP by DNA shuffling (23), comprising mutations Phe^{99}Ser, Met^{153}Thr and Val^{163}Ala, as numbered in the TU#58 system, are difficult to explain simply based on the structure. Positions 153 and 163 are on the surface of the protein and may exert their effects through improved solubility and/or reduced aggregation. The Phe-Ser mutation at first glance would appear to destabilize the core of the protein and we have no idea how it would improve the system, except perhaps by allowing more complete formation of the fluorophore through different folding kinetics.

The mechanism of activation of the fluorophore from ordinary protein structure is consistent with a non-enzymatic cyclization mechanism like that of Asn-Gly deamidation (24) followed by oxidation of the tyrosine to dehydrotyrosine, as previously suggested. The role of molecular oxygen in this mechanism and in GFP fluorescence is paradoxical, however. Molecular oxygen is proposed to be needed for oxidation of tyrosine to form an extended aromatic system, but oxygen must also be excluded from regular interactions with the fluorophore or else collisional quenching of the fluorescence or damaging photochemistry will occur. The low bimolecular quenching rate suggests that the protein's design sacrifices efficient fluorophore formation for stability and higher quantum yields once fully formed.

The N- and C-termini truncation studies and the fluorescent fusion products are now understandable, given the structure of the protein. Since the C-terminus loops back outside the cylinder and the last seven or so amino acids are disordered it shouldn't be critical to have them present and further addition would seem to be easily tolerated. These residues do not form a stave of the barrel. The role of the N-terminus is a little less clear, as the first strand in the barrel does not begin until amino acid 10 or 11 Thus barrel formation does not require the N-terminal region. The N-terminal segment, is however, an integral part of the 'cap' on one end of the protein, and may be essential in folding events or in protecting the fluorophore. Again, extensions at the N-terminus would not disrupt the motif structure of the protein. The pH dependence of the excitation bands at 395 nm and 475 nm (13) is almost certainly due to His^{148}, whose Nδ atom is 3.3 Å from the Tyr^{66} hydroxyl oxygen atom of the fluorophore, although NMR pKa measurements or mutagenesis studies would be needed for confirmation.

The dimer we see as the asymmetric unit in the crystal is likely to be the same one formed in solution, since the ionic strength of the crystallization buffer is low, and we see dimers at low (<100 mmol/L) ionic strengths in solution. Thus, it is not surprising to us to see the large number of hydrophilic dimer contacts. The smaller hydrophobic patch could conceivably be involved in physiological interactions with aequorin, as there would be a natural advantage to close proximity for efficient energy transfer. Control of the dimerization will be important for fluorescence resonance energy transfer (FRET) studies of protein-protein interactions using GFP, as one would not want to induce association and hence resonance energy transfer between the differently colored GFP proteins by mechanisms other that of the target protein interactions. Mutants may also be developed for reduction of aggregation during expression and hence fewer problems with inclusion bodies.

Thus the three-dimensional structure of GFP has provided a physico-chemical basis of many observed features of the protein, including its stability, protection of it fluorophore, behavior of mutants, and dimerization properties. The structure will also allow directed mutation studies to complement random and combinatorial approaches.

Acknowledgments

We thank the following for financial support: the Robert A. Welch Foundation, the W. M. Keck Center for Computational Biology, and the NIH (AR40252 to GNP, and GRASP Center DK34928 and DK34447 to LGM).

References

1. Morin JG, Hastings JW. Energy transfer in a bioluminescent system. J. Cell Physiol. 1971;77:313-8.
2. Ward WW, in Photochemical and Photobiological Reviews, K. Smith, Editor. 1979, Plenum: NY. p. 1-57.
3. Prasher DC, Eckenrode VK, Ward WW, Prendergast FG, Cormier MJ. Primary structure of the *A. victoria* green-fluorescent protein. Gene 1992;111:229-33.
4. Chalfie M, Tu Y, Euskirchen G, Ward WW, Prasher DC. Green fluorescent protein as a marker for gene expression. Science 1994;263:802-5.
5. Cubitt AB, Heim R, Adams SR, Boyd AE, Gross LA, Tsien RY. Understanding, improving and using green fluorescent proteins. TIBS 1995;20:448-55.
6. Mitra RD, Silva CM, Youvan DC. Fluorescence resonance energy transfer between blue-emitting and red-shifted excitation derivatives of the green fluorescent protein. Gene 1996;173:13-7.

7. Kahana J, Silver PA, in Current Protocols in Molecular Biology, F. Ausabel, *et al.*, Editors. 1996, Green and Wiley: NY. p. 9.7.22-9.7-28.
8. Cody CW, Prasher DC, Westler WM, Prendergast FG, and Ward WW. Chemical structure of the hexapeptide chromophore of the Aequorea green-fluorescent protein. Biochemistry 1993;32:1212-8.
9. Heim R, Prasher DC, Tsien RY. Wavelength mutations and posttranslational autoxidation of green fluorescent protein. Proceedings of the National Academy of Sciences of the United States of America 1994;91:12501-4.
10. Delagrave S, Hawtin RE, Silva CM, Yang MM, Youvan DC. Red-shifted excitation mutants of the green fluorescent protein. Biotechnology 1995;13:151-4.
11. Lim CR, Kimata K, Oka M, Nomaguchi K, Kohno K. Thermosensitivity of a green fluorescent protein utilized to reveal novel nuclear-like compartments. J Biochem (Tokyo) 1995;118:13-17.
12. Ward WW, Bokman SH. Reversible denaturation of Aequorea green-fluorescent protein: physical separation and characterization of the renatured protein. Biochemistry 1982;21:4535-40.
13. Ward WW, Prentice H, Roth A, Cody C, Reeves S. Spectral perturbations of the *Aequorea* green fluorescent protein. Photochem. Photobiol. 1982;35:803-808.
14. Dopf J, Horiagan TM. Deletion mapping of the *Aequorea victoria* green fluorescent protein. Gene 1996;173:39-44.
15. Heim R, Cubitt AB, Tsien RY. Improved green fluorescence. Nature 1995;373:663-664.
16. Cormack BP, Valdivia RH, Falkow S. FACS-optimized mutants of the green fluorescent protein (GFP). Gene 1996;173:33-38.
17. Ehrig T, O'Kane DJ, Prendergast FG. Green-fluorescent protein mutants with altered fluorescence excitation spectra. FEBS Lett. 1995;367:163-6.
18. Perozzo MA, Ward KB, Thompson RB, Ward WW. X-ray diffraction and time-resolved fluorescence analyses of Aequorea green fluorescent protein crystals. J. Biol. Chem. 1988;263:7713-6.
19. Yang F, Moss LG, Phillips GN, Jr. The molecular structure of green fluorescent protein. Nature Biotechnology 1996;in press. 14: 1246-1251
20. Brunger AT, X-PLOR Version 3.1: A system for X-ray crystallography and NMR. 1992, New Haven: Yale University Press.
21. Rao BDN, Kemple MD, Prendergast FG. Proton nuclear magnetic resonance and fluorescence spectroscopic studies of segmental mobility in aequorin and a gren fluorescent protein from aequorea forskalea. Biophys. J. 1980;32:630-2.
22. Chattoraj M, King BA, Bublitz GU, Boxer SG. Ultra-fast excited state dynamics in green fluorescnet protein: Multiple states and proton transfer. Proc. Natl. Acad. Sci. USA 1996;93:8362-7.
23. Crameri A, Whitehorn EA, Tate E, Stemmer WPC. Improved green fluorescent protein by molecular evolution using DNA shuffling. Nature Biotech. 1996;14:315-9.
24. Wright HT. Nonenzymatic deamidation of asparaginyl and glutaminyl residues in proteins. Crit Rev Biochem Mol Biol 1991;26:1-52.

MONITORING BIOFILM-INDUCED PERSISTENCE OF *MYCOBACTERIUM* IN DRINKING WATER SYSTEMS USING GFP FLUORESCENCE

AA Arrage[1,2] *and DC White*[2,3]
[1]*Microbial Insights, Inc., Knoxville, TN, 37922, USA*
[2]*Center for Environmental Microbiology, Univ. of Tennessee, Knoxville, TN, 37932, USA*
[3]*Oak Ridge National Laboratory, Oak Ridge, TN, 37831, USA*

Introduction

Biofilms are ubiquitous in drinking water distribution systems. They are relatively impervious to mitigation treatments targeted for suspended cells, and have been implicated in blooms of coliforms and in harboring pathogens in otherwise properly maintained systems (1-6). To investigate the impact of biofilms on pathogen persistence in potable water, the attachment and retention of a *gfp*-transformed *Mycobacterium smegmatis* (MS) strain was monitored in laminar flowcells exposed to different concentrations of chlorination. On-line GFP fluorescence was used to measure non-destructively MS biomass levels in monoculture and mixed-culture biofilms.

Materials and Methods

Bacterial Strains and Growth Media: The *Mycobacterium smegmatis* strain was kindly provided by Dr. Vojo Deretic, University of Texas Health Sciences Center (7). The triculture mixed-species biofilm was composed of three bacterial species isolated from corroded copper drinking-water pipes. The bacteria were identified by fatty acid profiles as *Acidovorax* sp., *Bacillus* sp., and *Pseudomonas* sp. (data not shown).

M. smegmatis was maintained on enriched Middlebrook 7H9 media supplemented with 25 ug/mL kanamycin. The three drinking water isolates were maintained on tryptic soy agar. All media reagents were purchased from Difco Laboratories (Detroit, MI, USA).

Test System: A 1:1000 dilution of tryptic soy broth (TSB) in distilled deionized water was pumped through laminar flowcells in a once-through design. Each flowcell contained five stainless steel coupons inserted flush with the bottom of the chamber. Three hollow screws with a quartz disk attached at the end were positioned above each coupon and provided viewports to measure biofilm fluorescence (8). Fifty mL of a 48 h MS culture (~8.7 x 10^7 cells ml^{-1}) were injected into each flowcell and flow was stopped for 2 h to allow for cell attachment. The triculture was maintained in a chemostat using diluted TSB. A flow line from the chemostat was used to introduce the triculture cells into the flowcells for 4 h. Subsequently, the inoculation line was clamped off and sterile media flow was resumed at a rate of 10 mL min^{-1}. All experiments were performed in triplicate.

Biofilm fluorescence was measured using a fluorometer equipped with a fiber-optic attachment (Spex Industries Inc., Edison, NJ, USA). Tryptophan (ex. 295 nm; em. 340 nm) and GFP (ex. 395 nm; em. 509 nm) fluorescence readings were used to determine the total biofilm and MS biomass respectively. Background fluorescence levels were subtracted from all subsequent readings (8,9).

Cell Enumerations: Viable MS cells were determined by plating suspensions of biofilm material onto Middlebrook 7H9 agar plates and measuring colony forming units. The MS colonies were morphologically distinct from the other bacterial components of the biofilms.

Total cell counts were determined by staining biofilm material with acridine orange as described by Arrage et al. (8) and examined under epifluorescent illumination.

Disinfection: MS biofilms and mixed-species biofilms were exposed to 0, 1, and 5 ppm total chlorine. Media amended with chlorine was pumped through the flowcells beginning 1 h before MS inoculation and continuing through the remainder of the experiment. For experiments involving mixed-species biofilms, the triculture cells were inoculated into the flowcells and allowed to form biofilms for 96 h prior to chlorine and MS addition.

Results and Discussion

The addition of MS into sterile flowcells resulted in the detection of GFP fluorescence at 3 h post-inoculation (1 h after the resumption of media flow), which decreased to background levels after 12 h (Fig. 1). This result was presumably due to the wash-out of unattached cells and occurred at all chlorine treatments by 24 h. Upon termination of the experiment, microscopic examination of the substratum revealed cell densities of approximately 400 cells cm^{-2} which did not vary significantly with chlorine concentration.

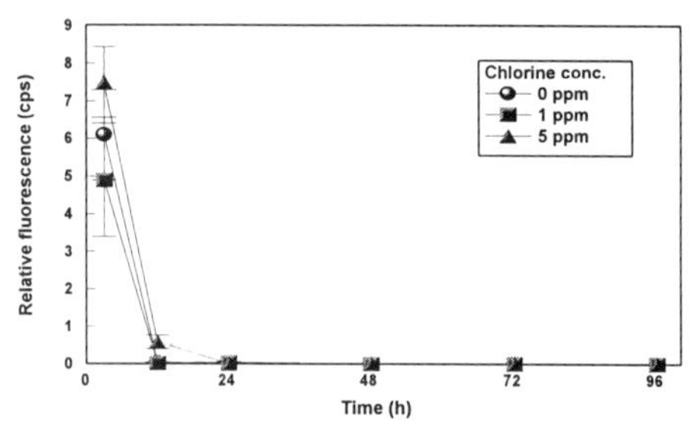

Fig. 1. On-line GFP fluorscence from MS cells inoculated into sterile flowcells.

When MS was inoculated into flowcells containing a 96 h triculture biofilm, GFP fluorescence was an order of magnitude greater than that recorded from the sterile environment (Fig. 2). The release of MS cells was more gradual in the presence of an established biofilm with GFP fluorescence detected up to 96 h post-MS inoculation. At this timepoint, MS fluorescence at the 0 and 1 ppm chlorine levels was 3-fold greater than that measured at 5 ppm chlorine. The on-line data was

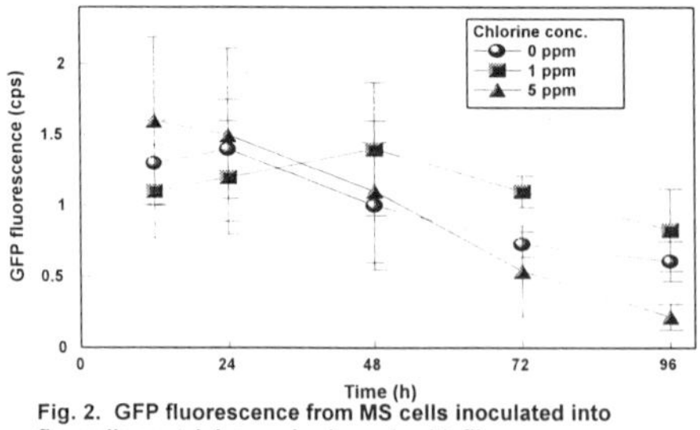

Fig. 2. GFP fluorescence from MS cells inoculated into flowcells containing a mixed-species biofilm.

supported by epifluorescent microscopic counts of GFP-expressing MS cells recovered from biofilms at the end of the experiment (Table 1). Although there was an inverse relationship between chlorine

Chlorine conc.	Microscopic cell counts (10^5 cells cm^{-2})	Viable cell counts (10^5 cells cm^{-2})	Viability Index (viable/total)
0 ppm	43 ± 8.5	7.8 ± 1.2	0.18
1 ppm	9.1 ± 6.3	1.3 ± 0.53	0.14
5 ppm	2.8 ± 1.9	0.55 ± 0.11	0.20

Table 1. Numbers of *M. smegmatis* cells recovered from mixed-species biofilms after 96 h exposure to chlorine.

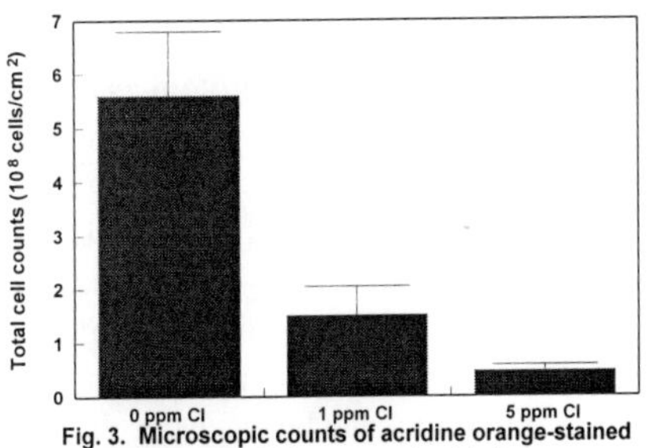

Fig. 3. Microscopic counts of acridine orange-stained biofilm cells after 96 h exposure to chlorine.

and the total number of MS cells, the proportion of those cells that were viable was independent of the chlorine concentration (Table 1).

The results of this study suggest that MS attachment and retention was dependent on the amount of biofilm biomass present on the substratum, rather than on the chlorine concentration. There was a marked decrease in biofilm development when triculture cells were exposed to chlorine (Fig. 3). This may have resulted in a less adherent surface for planktonic MS cells. It has been shown that initial colonizing bacterial species can modify surfaces and enhance the subsequent attachment of succeeding microorganisms (10).

This study has demonstrated the utility of using a fluorescent tag to selectively identify a specific cell population within a multi-species biofilm. It has been widely reported that biofilms can mitigate the effectiveness of biocides and antibiotics on target organisms (1,4). Using an organism which has been transformed to express the tag allows for the ability to track its response to perturbations in its environment (i.e. chlorine) in real time which is not always possible when using lethal fluorescent or colorimetric stains.

Acknowledgments

We thank R. Kirkegaard and A. Cardwell for excellent technical assistance. This study was funded by the National Water Research Institute project #HRA 699-510-94.

References

1. Cargill KL, Pyle BH, Sauer RL, McFeters GA. Effects of culture conditions and biofilm formation on the iodine susceptibility of *Legionella pneumophila*. Can J Microbiol 1992;38:423-8.

2. LeChevallier MW, Evans TM, Seidler RJ. Effect of turbidity on chlorination efficiency and bacterial persistence in drinking water. Appl Environ Microbiol 1981;42:159-67.
3. LeChevallier MW, Cawthon CD, Lee RG. Factors promoting survival of bacteria in chlorinated water supplies. Appl Environ Microbiol 1988;54:649-54.
4. LeChevallier MW, Cawthon CD, Lee RG. Inactivation of biofilm bacteria. Appl Environ Microbiol 1988;54:2492-9.
5. Rogers J, Dowsett AB, Dennis PJ, Lee JV, Keevil CW. Influence of plumbing materials on biofilm formation and growth of *Legionella pneumophila* in potable water systems. Appl Environ Microbiol 1994;60:1842-51.
6. Van Der Wende E., Characklis WG, Smith DB. Biofilms and bacterial drinking water quality. Water Res 1989;23:1313-22.
7. Dhandayuthapani S, Via LE, Deretic V. Green fluorescent protein as a marker for gene expression and cell biology of mycobacterial interactions with macrophages. Mol Microbiol 1995;17:901-12.
8. Arrage AA, Vasishtha N, Sundberg D, Bausch G, Vincent HL, White DC. On-line monitoring of antifouling and fouling-release surfaces using bioluminescence and fluorescence measurements during laminar flow. J Ind Microbiol 1995;15:277-82.
9. Angell P, Arrage AA, Mittelman MW, White DC. On-line non-destructive biomass determination of bacterial biofilms by fluorometry. J Microbiol Methods 1993;18:317-27.
10. James GA, Beaudette L, Costerton JW. Interspecies bacterial interactions in biofilms. J Ind Microbiol 1995;15:257-62.

MUTANTS OF GREEN FLUORESCENT PROTEIN (GFP) WITH ENHANCED FLUORESCENCE CHARACTERISTICS

Brendan P. Cormack, Raphael H Valdivia and Stanley Falkow
Dept. of Microbiology and Immunology, Stanford University Medical School, Stanford, CA, 94305-5402, USA

Introduction

The green fluorescent protein (GFP) of the jellyfish *Aequorea victoria* absorbs light with an excitation maximum of 395 nm, and fluoresces with an emission maxima of 510 nm (1, 2). Since fluorescence does not require any cofactors, GFP is extremely useful as a marker for gene expression and as a tag in studying protein localization in a variety of organisms (3). The presence of GFP can be monitored using standard fluorescein isothiocyanate (FITC) excitation-emission filter sets by virtue of a minor absorption peak at 470 nm (see fig.1). The resulting fluorescence, however, is less intense than that resulting from optimal excitation. Mutations in GFP which shift the excitation maxima from 395 nm to around 490 nm have been reported and these proteins do fluoresce more intensely when excited at 488 nm (3). Here we describe isolation of GFP mutants that show much enhanced fluorescence when excited at 488 nm. We constructed a library of mutant GFP molecules using an oligo-directed, codon-based mutagenesis method (4). The mutagenesis strategy permitted the simultaneous mutagenesis of a 20 aa region surrounding the chromophore. We then used fluorescence activated cell sorting (FACS) to screen this library for GFP mutants with enhanced fluorescence when excited at 488 nm. We isolated three distinct GFP mutants. These mutants have markedly shifted excitation maxima and, in addition, fold more efficiently than wt GFP in *Escherichia coli*. These mutants show enhanced fluorescence in a wide variety of prokaryotic and eukaryotic organisms, and should be broadly applicable.

Materials and Methods

Instrumentation: FACS analysis was carried out using a FACStar^{plus} (Becton Dickinson, Cockeysville, MD, USA). Spectra were measured with a SPEX fluorolog fluorimeter.

Procedures: In mutagenizing GFP, we targeted a region corresponding to aa 55-74 using oligo-directed mutagenesis. The oligonucleotide sequence and the synthesis method we used is described in detail elsewhere (4). In essence, at the step in oligo synthesis corresponding to a codon, the oligo-synthesis column is dismantled, and the silica matrix divided and placed into two new columns. One of these columns is subjected to three rounds of synthesis with an equimolar mix of the four nucleotides. The other column receives three rounds of wt nucleotides. The matrix in the two columns is combined and the process repeated for the desired number of codons. Thus, if a particular codon is mutagenized at all, all three nucleotides for that codon are randomized, giving an equal probability of any codon being substituted for the wt codon. Taking advantage of standard methods, this oligonucleotide was used to make a library of GFP mutants carried on a high copy vector, and expressed from an IPTG-inducible *tac* promoter (cloning details in ref.4).

Lysates of the GFP expressing strains were made using a French press; fractionation of soluble GFP from GFP in inclusion bodies was done by centrifugation at 17000xG.

Results and Discussion

Mutagenesis of GFP: The chromophore of GFP is made by postranslational modification of three amino acids, 65-67. In making our library, we mutagenized the

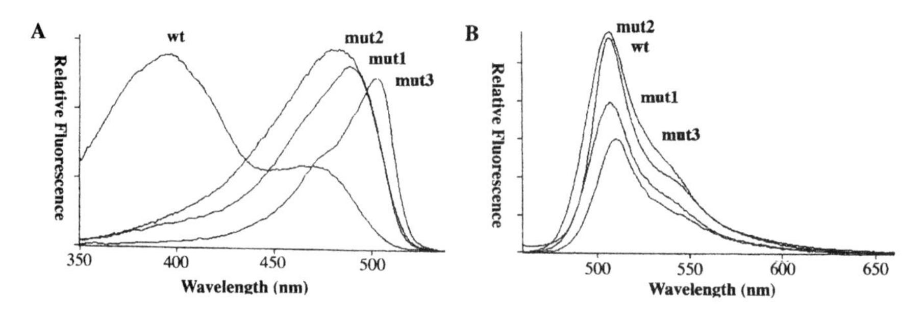

Figure 1. Excitation and emission spectra for wt and mutant GFPs. Spectra were measured using 1.0 nm band widths, and corrected using standard correction files supplied by the manufacturer. (A) Excitation spectra, with emission recorded at 540nm. (B) Emission spectra with excitation wavelength of 450 nm.

20 amino acids (55-74) around the chromophore using oligo-directed mutagenesis. Because we used a codon-directed strategy, as described elsewhere (4), the mutant library is compact and yet highly representative. The library contained $6x10^6$ independent clones. Theoretically, with this number of clones, the library contains all possible single and double amino acid changes for the 20 amino acid region mutagenized; in addition approximately 2% of all possible triple amino acid changes are represented.

Isolation of GFP mutants: The *gfp* genes were cloned behind an IPTG-inducible *tac* promoter on a puc-based expression vector. To recover the most fluorescent GFP mutants in the library, we induced the pool of mutants for 2.5 hours with IPTG. When we analyzed the population for fluorescence using FACS, we found that a sub population of the cells were highly fluorescent. We recovered these, amplified them by growth in L-broth, and repeated the FACS sort, recovering the most intensely fluorescing 0.5% of this already enriched population. From this pool, we analyzed 50 strains in detail. After induction, individual bacterial strains fluoresced between 10- and 110-fold more intensely than a control strain expressing wt GFP. We chose the 12 most fluorescent strains, which were approximately 100 fold more fluorescent than strains expressing wt GFP (table 1). Sequence analysis of the corresponding GFPs revealed that in this group of 12, three distinct mutants were represented. GFPmut1 has a double substitution: F64L,S65T. GFPmut2 has a triple substitution: S65A, V68L, S72A. GFPmut3, represented by ten of the twelve mutants, has the same double aa substitution: S65G S72A. Thus, in all three mutants Ser^{65} is mutated and always in the context of at least one other mutation. After induction with IPTG, bacterial strains expressing the three mutant GFPs show detectable fluorescence within 8 minutes, and reach half-maximal fluorescence within 25 minutes.

Characterization of mutant GFPs: The difference in fluorescence intensity between strains expressing the wt and mutant GFPs could in principle be due to any of a number of factors: increased protein expression, more efficient protein folding, increased A_{488nm}, or faster chromophore formation. It is unlikely that there would be any significant improvement in quantum efficiency for the chromophore, since for wt GFP, it is already between 0.7 and 0.8 (2). After induction with IPTG, the amount of protein produced by the wt and mutant strains is the same. There is, however, a significant effect of the mutations on protein folding. Consistent with what has been previously observed (5), we observed that >90% of the wt GFP is found in inclusion bodies as non-fluorescent insoluble protein. By contrast, when expressed under

	Excitation max. (nm)	Emission max. (nm)	Fluorescence E. coli	Fluorescence soluble protein	Solubility Index
wt	395	508	1	1	.08
mut1	488	507	45	35	.65
mut2	481	507	90	19	.80
mut3	501	511	80	21	.80

Table 1. Characteristics of the mutant GFPs. The fluorescence in *E. coli* was measured for single cells from logarithmic cultures using FACS. Measurements of fluorescence per unit of soluble protein were made on equal amounts of soluble GFP, as determined by Coomassie straining of SDS-PA gels. The Solubility Index is the ratio of soluble GFP to total GFP for a given number of cells.

identical conditions, 70% of GFPmut1 and 85% of GFPmut2 or GFPmut3 is soluble (table 1). This makes a significant contribution to the increased fluorescence of the bacteria expressing the mutant GFPs. Indeed, while bacteria expressing the mutant proteins are between 50 and 100 fold more fluorescent than strains expressing wt GFP, when bacterial lysates equalized for soluble GFP are excited at 488 nM, the mutant GFPs are between 20 and 35 times as fluorescent as wt GFP.

We used fluorescence spectroscopy to analyze the excitation and emission spectra of the mutant and wt GFPs. We found that the emission spectras were very similar (with the maximas being 508 nm for wt and 507 or 511 for the mutants). By contrast, the excitation spectras were significantly altered, with the maximas being shifted from 395 nm for wt GFP to between 480 nm and 501 nm for the mutants (table 1). This shift in absorption is responsible for most of the increased fluorescence. It seems clear that the relative effects of the mutations on absorption and protein solubility are different for the three mutants. Specifically, when equal amounts of soluble protein is measured, mutants 2 and 3 are about 2 fold less fluorescent than mutant 1; however, bacteria expressing mutants 2 and 3 are about 2 fold more fluorescent than those expressing mutant 1. This suggests that of the three mutant GFPs, the $\underline{A}_{488nm}$ for mutant 1 is in fact the largest; however in bacteria GFPmut1 folds less efficiently than GFPmut2 or mut3, resulting in an relative decrease in fluorescence for the GFPmut1 expressing strains. It also suggests that, at least for bacterial expression, a combination of mutations at Ser^{72} or Val^{68} with the mutations F64L,S65T present in GFPmut1, might result in even higher overall fluorescence. This remains to be tested.

These mutant GFPs are useful in a broad range of organisms. We have expressed them in a number of bacterial species. In the gram-negative bacteria *Yersinia pseudotuberculosis, Salmonella typhimurium,* and *Legionella pneumophila,* mutant 2 gives approximately the same increase in fluorescence over wt GFP as in E. coli. In the gram positive*Mycobacterium marinum* or *Mycobacterium smegmatis,* GFPmut2 is approximately 10-20 fold better than wt GFP (R.V. and L. Ramakrishnan, unpublished data). The three mutants are approximately 30-45 fold more fluorescent than wt in *Saccharomyces cerevisiae* (6). In mammalian cells, a codon optimized version of GFPmut1, EGFP, is as much as 350 fold more fluorescent than wt GFP (S. Kain, pers. comm.). GFPmut2 has been successfully expressed in the malaria parasite, *Plasmodium falciparum* (K. Haldar, pers. comm.). Protein fusions to GFPmut2 in *E. coli* (7) as well as mammalian cells (K. Haldar pers. comm.) have worked significantly better than the same fusions to either wt GFP or the S65T mutant GFP.

Other published mutations in the C-terminus of the protein (8) affect the solubility of GFP. It has been proposed that these C-terminal mutations alter an exposed hydrophobic face of the protein that interacts *in situ* with aequorin. Since hydrophilic residues are not substituted for hydrophobic residues in any of our mutants, it does not seem likely that our mutations will map to this proposed hydrophobic surface. It will be interesting to see in the X-ray crystal structure where F^{64}, Val^{68} and Ser^{72} map with respect to the chromophore. The proximity of our mutations to the chromophore raises the interesting possibility that the effect on solubility of our mutations is the indirect result of changes in chromophore structure or synthesis. Perhaps steps in the chromophore synthesis are necessary for the folding of the protein as a whole.

We (R.H.V. and S.F.) are using these GFP mutants and FACS to separate bacterial cells on the basis of differential GFP expression levels, allowing identification of novel promoters that respond to complex environmental cues in bacterial-host interactions. Moreover, the fast kinetics of chromophore assembly in these mutant GFPs allows accurate assessment of the kinetics of transcriptional induction for these same promoters. The novel characteristics of the mutant GFPs that make them useful in microbial systems should also make them broadly useful in other systems.

Acknowledgments
We thank Mark Troll for fluorimetric analysis. This work was supported by a grant from the NIH (AI 36396) and by an unrestricted gift from Lederle-Praxis Biologicals. B.P.C. is supported by a Helen Hay Whitney postdoctoral fellowship.

References

1. Morise JG, Shimomura O, Johnson FH,Winant J, Intermolecular energy transfer in the bioluminescent system of *Aequorea*. Biochemistry 1974; 13:2656-62.
2. Ward WW, Cody CW, Hart RC, Cormier MJ, Spectrophotometric identity of the energy-transfer chromophores in *Renilla* and *Aequorea* green fluorescent proteins. Photochemistry and Photobiology 1980; 31:611-15.
3. Cubitt AB, Heim R, Adams SR, Boyd AE, Gross LA, Tsien RY, Understanding, improving and using green fluorescent proteins. Trends Biochem Sci 1995; 20:448-55.
4. Cormack BP, Valdivia RH, Falkow S, FACS optimized mutants of the green fluorescent protein (GFP). Gene 1996; 173:33-38.
5. Heim R, Prasher DC, Tsien RY, Wavelength mutations and posttranslational autoxidation of green fluorescent protein. Proc. Natl. Acad. Sci. USA 1994; 91: 12501-4.
6. Cormack BP, Bertram G, Egerton M, Gow NAR, Falkow S, Brown AJP, Yeast Enhanced Green Fluorescent Protein (yEGFP): a marker for gene expression in *Candida albicans*. Microbiology 1996; in press
7. Ma, X, Ehrhardt, DW, Margolin W, Co-localization of cell division proteins FtsZ and FtsA to cytoskeletal structures in living *Escerichia coli* cells using green fluorescent protein. Proc. Natl. Acad. Sci. USA 1996; in press.
8. Crameri A, Whitehorn EA, Tate, E Stemmer WPC, Improved green fluorescent protein by molecular evolution using DNA shuffling. Nature Biotechnology 1996; 14: 315-319.

OPTIMIZATION OF GFP AS A MARKER FOR DETECTION OF BACTERIA IN ENVIRONMENTAL SAMPLES

A Unge, R Tombolini, A Möller and JK Jansson
Dept of Biochemistry, Stockholm University, Stockholm S-10691 Sweden

Introduction

There are several environmental applications of genetically modified microorganisms (GMMs) in nature. These applications include bioremediation of toxic waste, plant protection, plant fertilizer production, and others. It is important to have efficient methods to monitor released GMMs for the purpose of risk assessment as well as to judge their product performance.

Considerable effort has recently been spent on the development of molecular markers as tags to distinguish engineered bacteria from the natural microbial population (1,2). The ideal marker gene should have the following characteristics:

1. It should be specific for the tagged bacteria, in order to distinguish the tagged organisms from the natural microbial population. Since there are thousands of distinct microbial genotypes in a single gram of soil, greater than 90% of which have never been isolated or identified, specificity is not a simple task to achieve.
2. It should be detectable using sensitive assays. Since one cell is in theory sufficient to build up a population in nature, the detection method should be capable of detecting low cell numbers (i.e. <10 cells gram^{-1} soil).
3. It should be quantifiable in order to relate back to the number of cells (biomass) present in a sample.
4. It should enable tagged cells to be detected *in situ*, without the need for substrate addition, or sample extraction.

The bioluminescent markers, *lux* (encoding bacterial luciferase) and *luc* (encoding firefly luciferase) are particularly useful due to their sensitivity, specificity and ease of detection (3,4). One problem with the bioluminescent marker genes has been the dependence of the luciferase light reaction on addition of a specific substrate. Additionally, the light output per cell varies depending on the metabolic status of the cells, since the luciferase enzyme reactions are energy dependent (3).

Recently, we have developed the green fluorescent protein (GFP) as an alternative tag for environmental microorganisms (5). Unlike luciferase enzymes, the detection of GFP does not require any exogenous substrate, cofactors (6) or chemical energy, but only ultraviolet or blue light, making it potentially more applicable as a marker than the luciferase system.

In addition, GFP seems to be ideally suited for *in situ* detection of specific bacteria in environmental samples. There are numerous methods developed for sensitive detection of fluorescent cells that are easily adaptable to detection of *gfp*-tagged cells. We have adapted flow cytometry and epifluorescence microscopy for monitoring GFP fluorescence in tagged bacteria. For example, a *gfp*-tagged pseudomonad was detected in liquid cultures by flow cytometry, in soil suspensions by CCD-enhanced microscopy and on root surfaces by laser confocal microscopy (5).

Another advantage of GFP as a marker is its stability during starvation conditions, which is a common state for bacteria in natural environments. A carbon-starved culture of *gfp*-tagged *Pseudomonas fluorescens* had a stable GFP fluorescence output, even over extended incubation periods, demonstrating the applicability of *gfp* as a tag for monitoring of bacteria in environmental samples (5).

Finally, the fluorescence intensity of GFP is so strong that even bacteria with a single copy of *gfp* can be detected as individual fluorescing cells by epifluorescence microscopy (5). In some cases a brief enhancement with a CCD camera was used to enhance the fluorescence image. To date, the *gfp*-marker is unsurpassed in sensitivity for *in situ* detection of single cells by any other known marker system.

Here we describe the extension of these studies to include detection of gram positive bacteria and the construction of a double *gfp* cassette for obtaining even greater GFP fluorescence in tagged cells.

Materials and Methods

Bacterial strains, growth conditions: *Arthrobacter* sp. A-6 was previously isolated from soil based on its ability to utilize 4-chlorophenol as its sole source of carbon (JK Jansson, unpublished data). *Arthrobacter* sp. A-6 was routinely grown on minimal salts medium containing 4-chlorophenol at a concentration of 150 μg mL^{-1}. Kanamycin was added at a concentration of 35 μg mL^{-1} for growth of the mutant. Cultures were incubated at 25 °C while shaking (200-300 rpm).

P. fluorescens A506 wt and *gfp*-tagged cells were grown at 30 °C in LB with rifampicin (100 μg mL^{-1}), or rifampicin plus kanamycin (200 μg mL^{-1}), respectively.

Cloning strategy for pUTgfp2: pUTgfp2 was constructed by isolating a *Not*I fragment containing a *gfp* cassette from pUC18Notgfp (5) consisting of mutant P11 *gfp* (7) under control of the *psb*A promoter and a T7gene10 ribosome binding site. The cassette was inserted in two adjacent copies into *Not*I digested pUTminiTn*5* (8), resulting in pUTgfp2.

Electroporation: Electroporation of *P. fluorescens* cells was carried out as previously described (5). *Arthrobacter* sp. A-6 cells were electroporated using similar conditions to that described for Pseudomonads, except for the following differences. After electroporation, the cells were incubated in LB at 28 °C before plating on selective minimal medium with 4-chlorophenol (150 μg mL^{-1}) and kanamycin (35 μg mL^{-1}).

Direct detection of fluorescent colonies: Fluorescent colonies were detected in a dark room upon exposure to a "black light blue" lamp (Philips, cat. no. 73411; Eindhoven, The Netherlands) as previously described (5).

Epifluorescent microscopic detection of cells: Slides were prepared as previously described (5). Fluorescence in single cells was detected using an Axiophot Epifluorescence Microscope (Zeiss, Oberkochen, Germany) with an excitation BP 480/20, Chromatic beam splitter FT510 and emission filter LP520 (Zeiss filter set, cat. 4873-16). Digital CCD-enhanced images were obtained using a cooled C4880 CCD (Hamamatsu Photonics, K.K., Hamamatsu City, Japan) connected to the Zeiss Axiophot Epifluorescence Microscope with a C-mount.

Flow cytometry: Samples for analysis by flow cytometry were washed in 0.22 μm filtered 1.5X PBS, diluted and resuspended in the same buffer. The bacterial cells were injected into the flow cytometer within a few minutes after suspension. The flow cytometer used was a FACScalibur (Becton Dickinson, Oxford, UK) equipped with an 15-mW, air-cooled argon-ion laser as the excitation light source (488 nm).

Fluorescence in the range of 515-545 nm was detected via a fluorescence detector with a photomultiplier tube voltage of 600 V and logarithmic gain. Forward scatter (FS) was collected by a diode with an amplification factor of 10 and processed in log gain. Side scatter (SS) was detected in log gain by a photomultiplier tube set at 450 V. 10,000 events were collected for each measurement.

Results and Discussion

We constructed *gfp*-cassettes containing one or two copies of *gfp* with the strong *psbA* constitutive promoter, known to be expressed in a variety of bacterial types, plus a ribosome binding site from T7gene10 (Fig. 1). Plasmids pUTgfp (5) and pUTgfp2 are transposon delivery systems used for stable chromosomal integration of one or two copies of *gfp*, based on the mini-Tn*5* delivery vector, pUT (8).

These vectors were introduced into gram negative *Pseudomonas fluorescens* A506 and gram positive *Arthrobacter* sp. A-6 by electroporation and green fluorescent colonies were obtained. Initially, there was a long lag period before green fluorescent colonies were observed. However, once green colonies were picked out and restreaked, they remained consistently green.

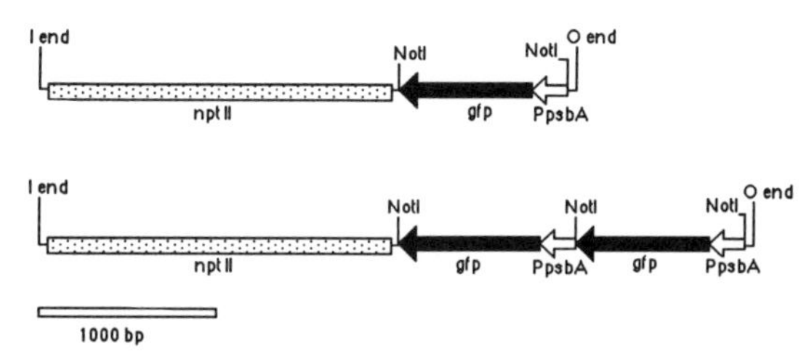

Figure 1 . *gfp* cassettes for chromosomal integration. Upper, pUTgfp (6); lower, pUTgfp2

Both single and double *gfp* (Fig. 1) were used to chromosomally tag *Pseudomonas fluorescens* A506. Cells tagged with a single chromosomal integration of *gfp* were faintly visible by epifluorescent microscopy and had a fluorescence intensity that was above background levels by flow cytometry (Fig. 2). However, the fluorescence intensity, as visualized by microscopy, was much greater in cells having two copies of *gfp* in their chromosome. The enhanced fluorescence intensity of cells containing two *gfp* copies was also confirmed by flow cytometry (Fig. 2). We routinely found that the fluorescence intenstity of *P. fluorescence* A506 cells tagged with double *gfp* was much higher than with a single *gfp* insertion, with mean relative fluorescence intensity values of 1.05 and 1.8, respectively. We previously reported a highly fluorescent mutant with a single chromosomal integration of *gfp* (5), although these bright mutants are rare, and probably result due to positioning of *gfp* on the microbial genome, i.e. location near enhancer sequences (5).

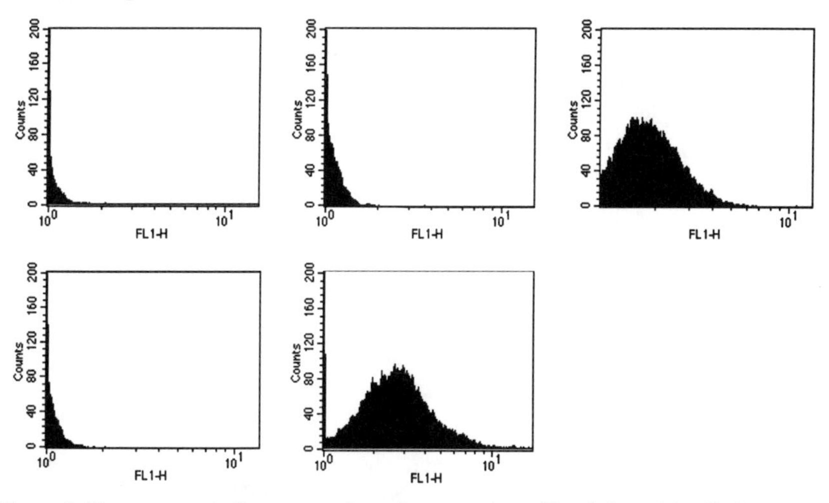

Figure 2. Flow cytometric fluorescence intensity comparisons. Top, left to right: *P. fluorescens* A-506 (wt), *P. fluorescens* A506::gfp *and P. fluorescens* A506::gfp2; Bottom, left to right: *Arthrobacter* sp. A-6 (wt) *and Arthrobacter* sp. A-6::gfp2. The data consists of histograms of distribution of fluorescence per cell of approximately the same number of events (number of cells gated).

Based on the results with the pseudomonads, we used pUTgfp2 as a vector to tag the gram positive microorganism, *Arthrobacter* sp. A-6 with two *gfp* copies (Fig. 2). *Arthrobacter* sp. A-6 was previously isolated based on its ability to catabolize high concentrations of the pollutant compound, 4-chlorophenol (Jansson, unpublished). The ultimate goal is to use GFP fluorescence as a method to monitor the survival of the *Arthrobacter* strain during bioremediation of 4-chlorophenol contaminated sites.

Fluorescence intensity was higher for *Arthrobacter* cells tagged using pUTgfp2 when compared to the *P. fluorescence* cells tagged with the same construct, with a mean relative intensity value of 2.05, compared to 1.8 for the pseudomonads (Fig. 2). The tagged cells were easily observed by epifluorescence microscopy in soil samples and with CCD enhancement much of the background fluorescence from soil particles was eliminated (Fig. 3).

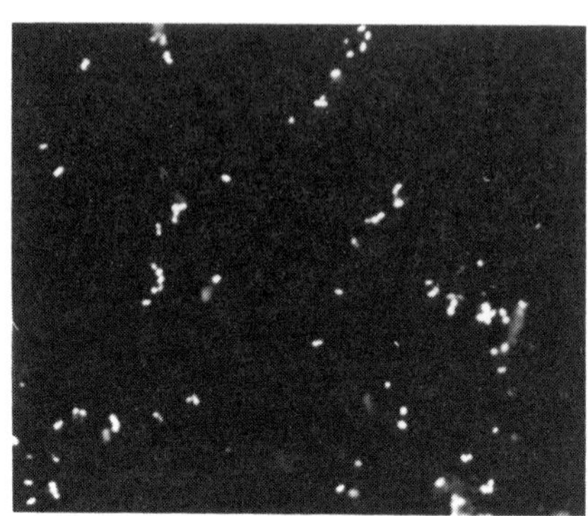

Figure 3. CCD enhanced image of *Arthrobacter* sp. A-6::gfp2 cells in soil

Although *gfp*-tagged *Arthrobacter* cells were brightly fluorescent when observed either as single cells or as colonies, they were impaired in their ability to grow on 4-chlorophenol. By comparison, a *luc*-tagged *Arthrobacter* strain grew and degraded 4-chlorophenol similarly to the wild type (Möller and Jansson, unpublished). Therefore, either the *gfp* gene integration, or the GFP protein product somehow interferes deleteriously with the ability of *Arthrobacter* sp. A-6 to biodegrade chlorophenol.

In conclusion, the development of GFP as marker for monitoring of bacteria in nature is at the developmental stage. We previously reported that a single chromosomal integration of *gfp* was sufficient for microscopic and flow cytometric detection of single cells (5). Here we demonstrate the enhancement of GFP fluorescence by chromosomal integration of two *gfp* copies. In addition, we demonstrate the use of double *gfp* to tag the gram positive microorganism, *Arthrobacter* sp. A-6. The results of this study indicate that GFP is promising as an *in situ* marker of tagged bacteria in environmental samples.

Acknowledgements
This work was funded by the Carl Tryggers Foundation, the Swedish Environmental Protection Agency and the Swedish Research Council for Engineering Sciences.

References

1. Jansson JK. Tracking genetically engineered microorganisms in nature. Curr Opin Biotech 1995;6: 275-83.
2. Prosser JI. Molecular marker systems for the detection of genetically modified microorganisms in the environment. Microbiol 1994;140: 5-17.
3. Möller A, Gustafsson K, Jansson JK. Specific monitoring by PCR amplification and bioluminescence of firefly luciferase gene-tagged bacteria added to environmental samples. FEMS Microbiol Ecol 1994;15:193-206.
4. Möller A, Norrby AM, Gustafsson K, Jansson JK. Luminometry and PCR-based monitoring of genetically modified cyanobacteria in Baltic Sea microcosms. FEMS Letters 1995;129:43-50.
5. Tombolini R, Unge A, Davey ME, de Bruijn FJ, Jansson JK. Flow cytometric and microscopic analysis of GFP-tagged *Pseudomonas fluorescens* bacteria. FEMS Microbiol Ecol 1996;in press.
6. Chalfie M, Yuan T, Euskirchen G, Ward WW, Prasher DC. Green fluorescent protein as a marker for gene expression. Science 1994;263:802-05.
7. Heim R, Prasher DC and Tsien R. Wavelength mutations and posttranslational autoxidation of green fluorescent protein. Proc Natl Acad Sci 1994;91:12501-04.
8. Herrero M, de Lorenzo V, Timmis K. Transposon vectors containing non-antibiotic resistance selection markers for cloning and stable chromosomal insertion of foreign genes in gram-negative bacteria. J Bacteriol 1990;172:6557-67.

AEQUORIA GREEN FLUORESCENT PROTEIN: STRUCTURAL ELUCIDATION OF THE CHROMOPHORE

H Niwa[1], *T Matsuno*[1], *S Kojima*[1], *M Kubota*[1], *T Hirano*[1], *M Ohashi*[1], *S Inouye*[2], *Y Ohmiya*[3] *and FI Tsuji*[4]
[1]*Dept. of Appl. Physics and Chem., The University of Electro-Commun., Chofu, Tokyo 182, Japan,*
[2]*Yokahama Res. Center, Chisso Corp., Yokahama, Kanagawa 236, Japan,*
[3]*Dept. of Chem., Fac. of Educ., Shizuoka Univ., Shizuoka 422, Japan,*
[4]*Marine Biol. Res. Div., 0202, Scripps Institution of Oceanography, UCSD, La Jolla, CA 92093, USA*

Introduction

On mechanical or electrical stimulation, the jellyfish *Aequorea victoria* emits a greenish light (λ_{max} 508 nm) from the margins of its umbrella. The bioluminescence system of the jellyfish consists of two proteins, a calcium binding photoprotein named aequorin and a green fluorescent protein (GFP). Aequorin is made up of apoaequorin (apoprotein), molecular oxygen, and a coelenterate-type of luciferin, called coelenterazine. On binding calcium ions, the aequorin reaction, *in vitro*, yields blue light with a maximum at 469 nm, a blue fluorescent protein (BFP), and carbon dioxide. BFP is a non-covalent complex of apoaequorin and coelenteramide, the oxidation product of coelenterazine. In the luminous tissue of the jellyfish, aequorin is considered to be closely associated with GFP. On binding calcium ions, a radiationless energy transfer occurs from the singlet excited state of BFP, generated from aequorin, to the grand state GFP. The singlet excited state of GFP then emits the green light. GFP is made up of 238 amino acid residues in a single polypeptide chain and has a fluorescent chromophore which emits green light on irradiation with long ultraviolet light (1). On denaturation or enzymatic digestion, GFP loses its characteristic green fluorescence. Under some conditions, the denaturation of GFP is reversible and recovery of fluorescence takes place with renaturation of the protein. The chromophore of GFP is believed to be formed through the post-translational modification of the polypeptide chain. Recent studies have indicated that oxygen is required for the formation of the chromophore (2, 3). Fig. **1** shows the post-translational cyclization followed by oxidation, of the tripeptide Ser^{65}–Tyr^{66}–Gly^{67}, leading to the formation of an imidazolone ring system responsible for the fluorescence of GFP (4). However, previous studies have reported that the model compounds possessing the proposed imidazolone ring system have no fluorescent properties (5, 6). The precise chromophore structure and the mechanism for the energy transfer are still unclear. In order to establish the chemical structure of the GFP chromophore, we have reinvestigated the spectral properties of chromopeptides of recombinant GFP derived by enzymatic digestion and a chemically synthesized model compounds **1** and **2** possessing the proposed chromophore structure (5). Detailed mass analyses of the digested chromopeptides were caried out.

Materials and Methods

The model compound **1** and **2** were synthesized according to the reported procedure (7, 8, 9). Recombinant GFP with a histidine-tag (His-GFP) was produced in *Escherichia coli* and purified by nickel-chelate affinity chromatography as described before (3, 10). The denatured His-GFP was digested with lysyl endopeptidase and then with proteinase K. The digest was purified by reversed-phase HPLC. The fractions containing the chromophore peptides were lyophilized and stored at -20 °C

Editors' Note:
Please also see:- Cody *et al*, Biochemistry 1993; 32:1212-1218 for additional information.

until used. Ultraviolet-visible absorption spectra were measured with a Hitachi model 320 spectrophotometer. Fluorescence emission spectra were recorded on a Hitachi model F4010 spectrofluorimeter. An all quartz tube (5 mm diam.) immersed in liquid N_2 in an all quartz Dewar was used for low temperature (77 K) fluorescence spectra measurement. ESI mass spectra were measured on a JEOL HX110/HX110 tandem mass spectrometer, equipped with an electrospray ion source. MALDI mass spectra were measured on a Brücker REFLEX TOF mass spectrometer. A linear TOF mode was performed for the measurement of molecular weight and for the PSD fragment ion mass analysis Nitrogen pulse laser (337 nm) was used for ionization. α-Cyano-4-hydroxycinnamic acid was used for the matrix.

Results and Discussion

Mass spectrometry The two-step, enzymatic digestion of His-GFP gave a small chromopeptide. The linear TOF mode MALDI mass spectrum of the chromopeptide gave a $[M+H]^+$ ion peak at m/z 781.3, indicating this chromopeptide was closely related to the heptapeptide, Thr^{63}–Phe^{64}–Ser^{65}–Tyr^{66}–Gly^{67}–Val^{68}–Gln^{69} having an average mass of 801.9 Da. The mass difference of 20.6 Da was attributed to the loss of one water molecule (18 Da) and two hydrogens (2 Da) during the chromophore formation as shown in Fig. **1**. The detailed PSD fragment ion analysis of the chromopeptide revealed the amino acid sequence of the peptide and showed the chromophore to be formed from the tripeptide Ser^{65}–Tyr^{66}–Gly^{67} of the polypeptide chain.

UV-vis spectrometry His-GFP has two absorption maxima at 398 nm and 476 nm in 0.1 M ammonium bicarbonate at pH 8. In 0.1 M HCl, His-GFP exhibits a single absorption with a maximum at 382 nm, while a single maxima at 447 nm was observed in 0.1 M NaOH. The isosbestic point was 405 nm during the pH change, being similar to that of the native GFP. The same spectral behavior was observed for the lysyl endopeptidase-digested His-GFP fragment, indicating that the chemical structure of the chromophore moiety in His-GFP was preserved even after denaturation followed by the enzymatic digestion. The clear overlapping of the UV-vis absorption spectra of the model compound **1** in 2-propanol (Fig. **2-B**) with those of the lysyl endopeptidase fragment (Fig. **2-A**) indicated that the structure of the GFP chromophore consist of a 4-(4-hydroxyphenylmethylidene)imidazol-5-one ring system.

Fluorescence spectrometry The fluorescence property of GFP vanished completely on denaturation or enzymatic digestion. The model compound **1** and **2** exhibited essentially no fluorescence in any fluid solvent examined. However, when ethanol solutions of **1** and **2** were frozen with liquid N_2 (77 K), **1** and **2** in ethanol glass became strongly fluorescent. For example, the fluorescent emission maxima of **1** in ethanol and in ethanol containing 1 M HCl (1% v/v) were 435 nm and 437 nm, respectively. In ethanol glass containing 1 M NaOH (1% v/v), the fluorescence of **1** was a bluish-green with a maximum at ~490nm (Fig. **3**, **b**). Similarly, the lysyl endopeptidase fragment in ethanol containing 1 M NaOH (1% v/v) exhibited strong fluorescence at 77 K with a maximum at ~475 nm, coinciding almost exactly with that of **1** (Fig.**3**, **a**). These findings support that (i) the GFP chromophore contains the same 4-(4-hydroxyphenylmethylidene)imidazol-5-one ring system as **1** and **2**, (ii) the light emitter of the GFP fluorescence is the singlet excited state of the phenolate anion of the chromophore, (iii) the fluorescence from the chromophore is highly dependent on the environment around the chromophore, and the inhibition of the cis and trans isomerization of the exo-double bond of the chromophore moiety accounts for the efficient light emission. Thus the characteristic greenish fluorescence of GFP at room temperature may be due to a restricted molecular motion of the chromophore in the tightly packed peptide environment.

Fig.1. A plausible process for the formation of the GFP chromophore.

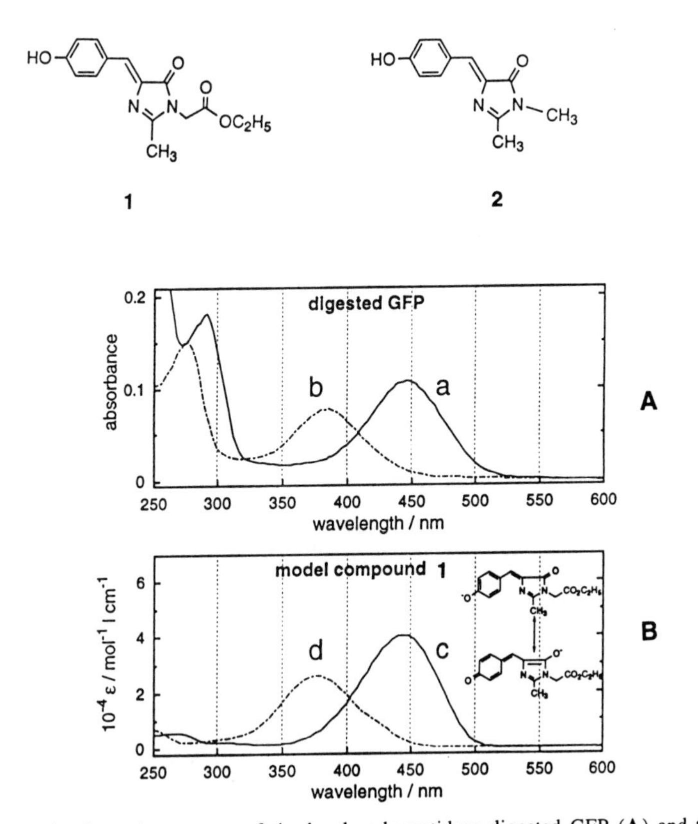

Fig. 2. UV-vis absorption spectra of the lysyl endopeptidase digested GFP (**A**) and the model compoumd **1** (**B**): **a**, the digested GFP in 0.1 M NaOH; **b**, the digested GFP in 0.1 M HCl; **c**, **1** in 2-propanol containing 1.0 M NaOH (2.5% v/v); **d**, **1** in 2-propanol containing 1.0 M HCl (2.5% v/v).

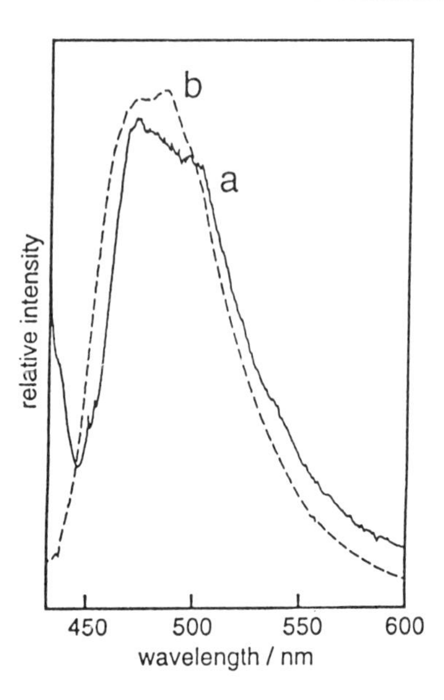

Fig. 3. Fluorescence spectra of the lysyl endopeptidase digested GFP (**a**) and the model compound **1** (**b**) in the ethanol glass containing 1.0 M NaOH (1%v/v) at 77 K.

References

1. Prasher DC, Eckenrode VK, Ward WW, Prendergast FG, Cormier MJ. Primary structure of the *Aequorea victoria* green-fluorescent protein. Gene 1992; 111: 229–233.
2. Inouye S, Tsuji FI. Evidence for redox forms of the *Aequorea* green fluorescent protein. FEBS Letters 1994; 351: 211–4.
3. Heim R. Prasher DC, Tsien RY. Wavelength mutation and posttranslational autoxidation of green fluorescent protein. Proc Natl Acad Sci USA 1994; 91:12501–4.
4. Cody CW, Prasher DC, Westler WM, Prendergast FG, Ward WW. Chemical structure of the hexapeptide chromophore of *Aequorea* green-fluorescent protein. Biochemstry 1993; 32: 1212–8.
5. Shimomura O. Structure of the chromophore of *Aequorea* green fluorescent protein. FEBS Letters 1979; 104: 220–2.
6. McCapra F, Razavi Z, Neary AP. The fluorescence of the chromophore of the green fluorescent protein of *Aequorea* and *Renilla*. J Chem Soc Chem Commun 1988; 790–1.
7. Devasia GM. A new method for the synthesis of unsaturated 2,4-disubstituted 2-imidazolin-5-ones. Tetrahedron Letters 1976; 571–2.
8. Devasia GM, Shafi PM. A covenient synthesis of unsaturated 2,4-disubstituted 2-imidazolin-5-ones. Indian J Chem Sect B 1981; 20B: 657–660.
9. Dalla Croce P, La Rosa C. Synthesis, steric configuration, and reactivity of 4-arylidene-2-methyl-1,4-dihydroimidazol-5-ones. J Chem Research (S) 185; 360–1.
10. Inouye S, Tsuji FI. *Aequorea* green fluorescent protein: expression of the gene and fluorescence characteristics of the recombinant protein. FEBS Letters 1994; 341: 277–280.

THE THREE-DIMENSIONAL STRUCTURE OF GREEN FLUORESCENT PROTEIN RESEMBLES A LANTERN

[1,2]C.-K. Wu, [1]Z.-J. Liu, [1]J. P. Rose, [3]S. Inouye,
[3]F. Tsuji, [4]R. Y. Tsien, [5]S. J. Remington, [1]B.-C. Wang,
[1]*Department of Biochemistry and Molecular Biology, University of Georgia, Athens, GA 30602,* [2]*Department of Crystallography, University of Pittsburgh, Pittsburgh, PA 15260,* [3]*Marine Biology Research Division 0202, Scripps Institution of Oceanography, University of California at San Diego, La Jolla, CA 92093,* [4]*Department of Pharmacology, University of California, San Diego, La Jolla, CA 92093,* [5]*Institute of Molecular Biology, University of Oregon, Eugene, OR 97403.*

Introduction

The greenish luminescence (λ_{max} = 508 nm) from the jellyfish, *Aequorea victoria*, is the result of a chemical reaction initiated by the binding of Ca^{2+} to aequorin, a Ca^{2+}-binding protein, followed by a physical interaction between the reacted aequorin and its closely associated green fluorescent protein (GFP) (1). In the *in vitro* reaction, aequorin emits a bluish light at λ_{max} = 470 nm, whereas in the *in vivo* reaction, the animal emits a greenish light at λ_{max} = 508 nm, due to a resonance energy transfer from the excited state of aequorin to GFP. This characteristic greenish light can also be reproduced *in vitro* in the absence of aequorin by excitation with other lights, including the UV or microscope lights. As a result of the discovery of the cloning and the heterologous expression of its cDNA the GFP has attracted widespread interest since the striking fluorescence produced by GFP can be used as a marker for gene expression, cell lineage and as a tag for fusion proteins.

Wild type GFP has two absorption maxima at about 395 nm and 475 nm. A recombinant GFP (238 a.a.) with an N-terminal 37 residue His-tag has fluorescence absorption maxima at 395 nm and 478 nm, whereas the emission maximum was at 507 nm. Thus the recombinant His-GFP and the wild type GFP have similar fluorescence characteristics (2). We have recently crystallized and determined the three-dimensional structure of the His-GFP. The His-GFP crystals belong to the hexagonal space group $P6_122$ with cell constants of a = b = 77.0 Å, c = 330.4 Å and γ = 120° and there are two GFP molecules per asymmetric unit. These crystals belong to a crystallographic space group that is different from those previously reported for GFPs (3,4) which were the orthorhombic space group $P2_12_12_1$ (3) and the tetragonal space group $P4_12_12$ (4). We report here the structure of the dimeric GFP molecules in the hexagonal crystals and a comparison with the monomeric GFP molecules in the orthorhombic crystals (3).

Materials and methods

Sample preparation: The recombinant protein used in this study was prepared according to the reported procedures (2). The protein sample (275 residues) consists of the wild-type GFP (238 residues), a His-tag (34 residues) containing 6 histidines and a 3 residue linker at the fusion junction. The His-tag was not removed from the protein prior to crystallization since the removal of the His-tag would produce non-homogenous proteins (5).

Crystallization: The His-GFP was crystallized at room temperature from Hepes buffer pH 7.5 containing 2.0 M ammonium sulfate, 0.05 M calcium acetate and 8% PEG 400. The crystals diffract to 2.5 Å resolution. A native data set was collected on an R-Axis IV area detector using CuKα radiation and processed with program HKL (6). The overall R_{sym} for the data set, 90% complete to 2.7 Å, is 10.8%.

Structure determination: The structure was solved by molecular replacement with program AMoRe (7), using the S65T GFP structure (3) as the search model and refined with program X-PLOR (8). The current R-value is 22.8% (Rfree = 37.6) for 15,537 independent reflections (> 2 σF) from 10 Å to 2.7 Å resolution. The rms (root mean squares) deviation from ideal bond length is 0.02 Å. The structure comparison was carried out with program X-PLOR (8).

Result and discussion

The overall structure: The GFP molecules in the present crystal form are dimers (Figure 1). This is different from that observed in the orthorhombic crystal form (3) in which GFP has been crystallized as monomers. The subunits (Figures 1, 2) in the present dimeric molecule have the same distinctive structural feature, the novel 11-stranded β barrel with a central helix, as observed in the orthorhombic (3) and the tetragonal (4) crystal forms.

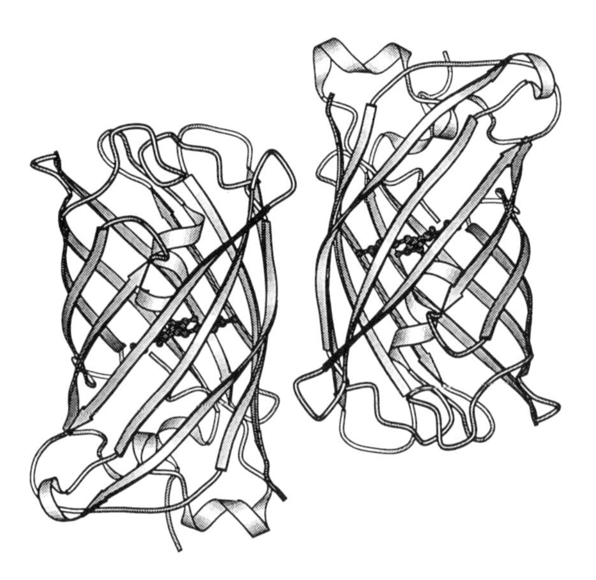

Figure 1. A ribbon drawing of a GFP dimer as viewed along a local two-fold symmetry axis. Extensive hydrophilic interactions are found at the monomer-monomer interface. Produced by Molscript (9).

GFP resembles a lantern: GFP looks very much like a lantern - a protective shield (β-strands) on the outside and a light-generating element (chromophore) at the center. We propose to call this structural fold a "β-lantern". Such a name has the advantage of making clear the relationship between structure and function in GFP. Since the chromophore requires no direct interaction with other molecules in order to absorb and emit light, there is no need for it to be exposed for easy access. Its location in the center of a protective shell (β-strands) makes a perfect structural sense in terms of its functional role.

Monomeric and dimeric GFP: A comparison of the atomic models for the current dimeric GFP and the monomeric GFP observed in the orthorhombic crystals (3) gave an rms deviation of 1.26 Å and 1.01 Å respectively between all atoms and between the main-chain atoms of the two structures. These deviations are slightly larger than those of 0.99 Å and 0.59 Å found between the above corresponding sets of atoms within a pair of monomers in dimers of the current structure. To further illustrate the difference, a comparison between the chromophore structures in the monomeric (light) and in the dimeric GFP (dark) was made (Figure 3). A locational shift of the chromophore with respect to the interface strands is observed. The significance of this shift is being analyzed.

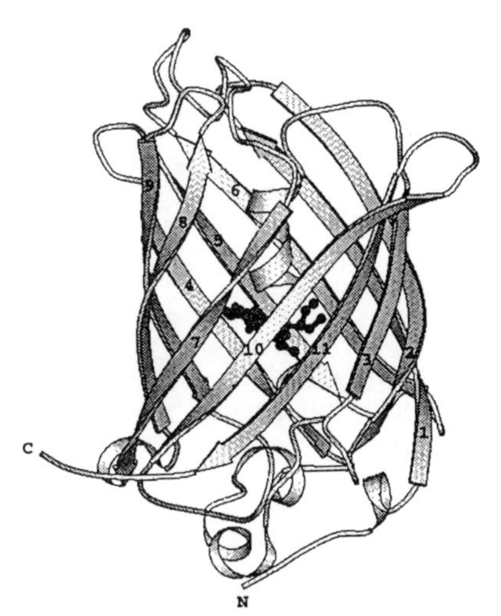

Figure 2. A GFP monomer. Most of the monomer-monomer contact residues are located on β-strands 7, 8 and 10. Produced by Molscript (9)

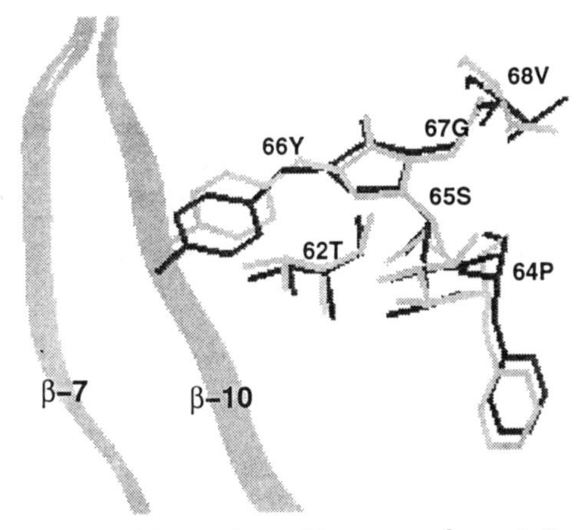

Figure 3. Relative locations of chromophore with respect to β-strands 7 and 10 in GFP monomers (3) (light) and dimers (this work) (dark). Produced by O (10).

Interactions at the dimer interface: The dimers are formed through extensive hydrophilic interactions between β-strands 7, 8 and 10 (Figure 2) of one monomer to the corresponding strands (7, 8 and 10) of the other monomers of the dimers. A total of 18 possible intermolecular hydrogen bonds is observed at the dimeric interface (Table 1). Numerous hydrophilic interactions were also observed in the GFP dimers of the tetragonal crystals (4) and they may be similar to those observed here. Four of these 18 hydrogen bonding interactions are replaced by protein-water interactions in the orthorhombic crystals (3) where GFP existed as monomeric form.

Table 1: Possible hydrogen bonding interactions (< 3.4 Å) at the dimer interface

Dimers as observed in the hexagonal crystals							Corresponding envir. in orthorhombic crystals (3)			
Monomer A		...	Dist. (Å)	...	Monomer B		...	Dist. (Å)	...	water
OE1	Glu142'	...	3.4	...	OD1	Asn149*				
	,,	...	3.2	...	ND2	,,	...	3.2	...	HOH
N	Tyr145'	...	2.9	...	OG	Ser147*	...	3.2	...	HOH
	,,	...	3.4	...	OE1	Gln204*				
O	,,	...	2.7	...	OG	Ser147*	...	3.3	...	HOH
OD1	Asn146'	...	3.0	...	N	,,				
N	Ser147'	...	2.8	...	OD1	Asn146*				
OG	,,	...	3.3	...	N	Tyr145				
	,,	...	3.1	...	O	,,				
OD1	Asn149'	...	3.4	...	OE1	Glu142*				
NH1	Arg168'	...	3.4	...	OD1	Asn170*				
OD1	Asn170'	...	2.9	...	NH1	Arg168*				
OE1	Gln204'	...	3.0	...	N	Leu207				
NE2	,,	...	3.1	...	O	,,				
	,,	...	3.2	...	O	Tyr143				
N	Leu207'	...	2.9	...	OE1	Gln204*				
O	,,	...	3.1	...	NE2	,,				
OD1	Asp210'	...	3.3	...	OH	Tyr39*	...	2.4	...	HOH

* Hydrophilic interaction observed in the tetragonal crystals (4).

The His-tag: The His-tag is disordered since no electron density can be seen beyond the '-1' residue. The current crystal structure however, has observable electron density from residues -1 to 232 which is more than residues 2 to 228 observed in the orthorhombic crystals (3) and residues 10 to 228 observed in the tetragonal crystals (4). The His-tag has evidently reduced the flexibility of the N-terminal residues in GFP.

References:

1. Morin J, Hastings J. Energy transfer in a bioluminescent system. J Cell Physi. 1971;77:313-318.
2. Inouye S, Tsuji FI. Aequorea green fluorescent protein: Expression of the gene and fluorescence characteristics of the recombinant protein. FEBS Letters 1994;341:277-280.
3. Ormö M, Cubitt A, Kallio K, Gross L, Tsien RY, Remington SJ. Crystal structure of the Aequorea victoria green fluorescent protein. Science. 1996;273:1392-5
4. Yang F, Moss LG, Phillips GN Jr. The Molecular structure of green fluorescent protein. Nature Biotech. 1996;14:1246-1251.
5. Inouye S, Tsuji FI. Evidence for redox forms of the Aequorea green fluorescent protein. FEBS Letters 1994;351: 211-4.
6. Otwinowski, Z., Minor, W. The HKL Program Suite, in preparation.
7. Navaza J. AMoRe: An automated package for molecular replacement. Acta Cryst. 1994; A50:157-163
8. Brünger AT. X-PLOR, version 3.1. 1992. Yale Univ. Press, New Haven, CT, USA.
9. Kraulis PJ. MOLSCRIP: a program to produce both detailed and schematic plots of protein structures. J. Appl Cryst. 1991;24:946-950.
10. Jones TA, Zou JY, Cowan SW, Kjeldgaard M. Improved methods for building protein models in electron density maps and location of errors in these maps, Acta Cryst. 1991;A47:110-119.

SPECTRAL ANALYSIS AND PROPOSED MODEL FOR GFP DIMERIZATION

MW Cutler and WW Ward
Dept. of Biochemistry and Microbiology, Rutgers University,
Cook College, New Brunswick, NJ 08903, USA

Introduction

Bioluminescence in *Aequorea victoria* involves aequorin that emits light upon calcium binding and green-fluorescent protein (GFP) that absorbs energy from aequorin and reemits light at a longer wavelength (1,2). The GFP chromophore (derived from the primary sequence by cyclization of an internal peptide, serine-dehydrotyrosine-glycine) (3) gives rise to absorption maxima at 395 nm and 470 nm. Fluorescence polarization measurements and measurements of oxygen quenching and oxygen quenching anisotropy initially indicated that the chromophore is rigidly held withing a conformationally inflexible domain (4,5). However, later work showed that the GFP chromophore is conformationally flexible and readily accessible to a variety of external perturbants, resulting in substantial spectral changes (6). The spectral changes observed with *Aequorea* GFP (with variation in pH, temperature, ionic strength and protein concentration) go through sharp isosbestic points, suggesting a simple two-state interconversion between two spectral forms of the chromophore--one absorbing maximally at 395 nm and the other at 470 nm (6). Increasing protein concentration and increasing ionic strength decrease the 470 nm peak by as much as 75% with no denaturation or loss of fluorescence. Increasing pH, temperature and organic solvent concentration can increase the 470 nm peak up to 360% with the final result being denaturation or loss of fluorescence (6-9).

Materials and Methods

Native GFP was purified to homogeneity from crude extracts of *Aequorea victoria* as previously described (10,11). Concentrated samples were prepared by ammonium sulfate precipitation or ultrafiltration. Absorption spectra were performed at room temperature on a Spectronic 2000 spectrophotometer. Fluorescence excitation spectra (uncorrected) were performed at room temperature on an Aminco-Bowman spectrofluorometer using 1.0 cm fluorometric cuvettes. Circular dichroism measurements were obtained on an Aviv 60DS CD spectrometer.

Results and Discussion

Aequorea GFP varies in average molecular weight with increasing protein concentration and ionic strength--the result of hydrophobically driven dimerization (11). Dimerization is accompanied by perturbations in the absorption spectrum (four-fold suppression of the 470 nm shoulder) providing an easy and non-invasive assay of conditions that promote self-association (Fig. 1a,b). A high ratio of absorbance (A_{395}/A_{470}=7.5) indicates dimer while a low ratio (A_{395}/A_{470}=1.8) indicates monomer. Three stages in the formation of GFP dimers exist. Stage 1 (low absorbance ratio, infinite dilution)

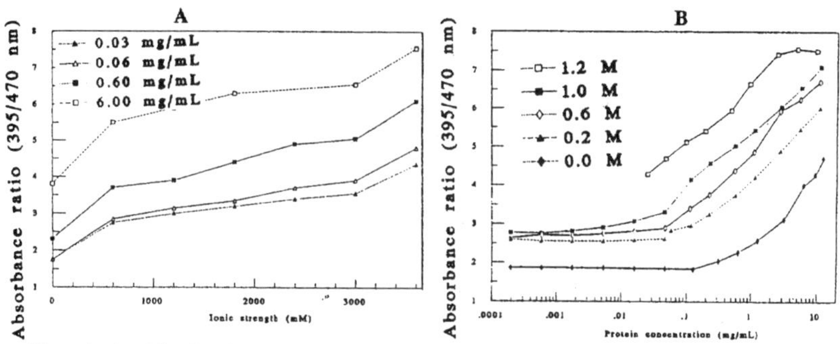

Figure 1. **A.** The absorbance ratio (395/470 nm) of *Aequorea* GFP at differing ionic strengths for four concentrations. Samples of GFP at 0.03, 0.06, 0.6 and 6.0 mg/mL were prepared and the concentration of ammonium sulfate in the sample was varied to achieve the desired ionic strengths. **B.** The absorbance ratio of *Aequorea* GFP at different protein concentrations. Pure samples of GFP were prepared at room temperature at different concentrations in a buffer of 10 mM Tris-EDTA. The concentration of ammonium sulfate in the buffer was varied to achieve concentrations of 0, 0.2, 0.6, 1.0 and 1.2 M. Absorption spectra were recorded at room temperature on a Bausch & Lomb Spec 2000 for GFP samples in the concentration range of 0.05 to 75 mg/mL. Fluorescence excitation spectra were recorded on an Aminco-Bowman spectrofluorometer in a 1 cm path length for GFP concentrations in the range 0.0002 to 0.05 mg/mL.

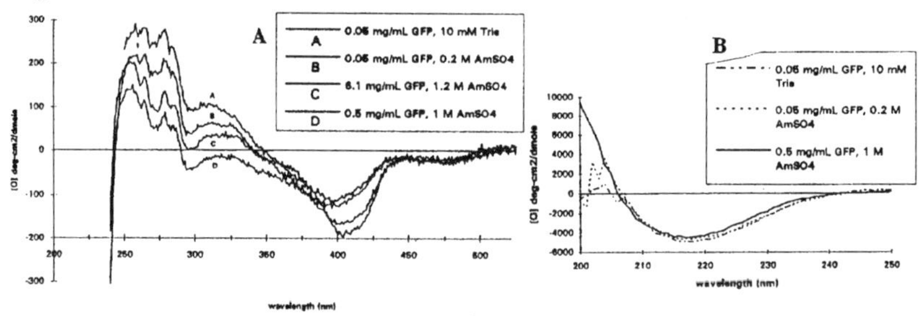

Figure 2. **A.** Near UV and visible circular dichroism spectra of *Aequorea* GFP under various conditions. **B.** Far UV circular dichroism spectra of *Aequorea* GFP under various conditions. Pure samples of GFP at differing protein (0.05, 0.5, 6.1 mg/mL) and salt concentrations were scanned on an Aviv 60DS CD spectrophotometer at room temperature. Spectra represent the average of three runs.

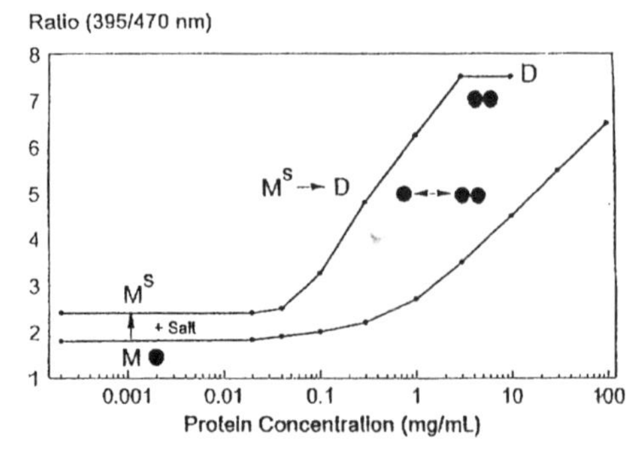

Figure 3. Schematic diagram of the change in absorbance ratio (395/470 nm) as a function of protein and salt concentration outlining the proposed model for GFP dimerization. M = monomer, M^s = monomer in ammonium sulfate, D = dimer in ammonium sulfate.

represents 100% monomer. An increase in ionic strength at this stage produces a change in the absorbance ratio that can be attributed to perturbations in the hydration sphere around the GFP monomer that are communicated indirectly to the buried chromophore. Changes in the CD spectrum in the visible region (Fig. 2a) are consistent with salt-induced changes in the chromophore environment. But, absence of significant changes in the far UV spectrum (Fig. 2b) rule out salt-induced changes in the protein secondary structure. Stage 2 is defined by a rapid increase in the absorbance ratio with increasing protein concentration. Circular dichroism spectra (Fig. 2a) show large changes in the visible and near UV regions during this phase, but the absence of any significant changes in the far UV indicates that the secondary structure remains unperturbed in going from monomer to dimer (Fig. 2b). Stage 3 (100% dimer) is characterized by a high absorbance ratio unaffected by any further increase in protein concentration. CD data at stage 3 indicate further spectral changes in the near UV and visible range, but, at such high protein concentration, far UV spectra could not be obtained. Concomitant with dimer formation, changes appear in the intensity and/or position of as many as eight CD transitions in the near UV and visible spectrum, the most obvious being the main CD transition in the 395 nm (chromophore) region.

A schematic model for the three stages of GFP dimer formation is depicted in Figure 3. The area indicated by M represents the absorbance ratio corresponding to 100% monomer. Addition of salt to monomeric GFP (M^s) results in a small increase in the absorbance ratio reflecting an interaction of salt ions with the protein surface or with the protein hydration sphere. Since the magniture of the absorbance change correlates with the Hofmeister and chaotropic ion series, it is more likely that this effect is derived from changes in the water structure surrounding the protein. In the second stage, both M and M^s go through a large increase in absorbance ratio that represents a shift in the equilibrium from monomer to dimer. Addition of an antichaotropic salt ($M^s \rightarrow D$) shifts the curve to lower protein concentrations suggesting hydrophobically driven dimerization. This stage is accompanied by large CD spectral changes in the visible and near UV regions but not in the far UV. Spectral perturbations of the chromophore must be communicated from the dimer interface to the interior of the protein where the chromophore is buried. The final area in the model, designated D, represents the absorbance ratio of 100% dimer.

At GFP concentrations near 25 mg/ml (calculated for *Aequorea* photocytes) a salt concentration isotonic with sea water would be expected to drive the equilibrium close to 100% dimer (11). By greatly suppressing the 470 nm absorption band, dimerization would significantly reduce the efficiency of *in vivo* trivial energy transfer from aequorin (emission max. $\simeq$ 470 nm) to GFP, to the extent that over 99% of the blue light emitted from aequorin would go untrapped by GFP within the photogenic mass. We conclude from this work that GFP and aequorin must form a complex (heterotetramer) that can transfer energy by a radiationless mechanism. We hypothesize that aequorin and GFP independently form homodimers by hydrophobic interactions. This is followed by the formation of a heterotetramer, stabilized by electrostatic interactions, that acts as the *in vivo* light emitting complex. Such a complex has, in fact, been isolated by the chromatographic method of Hummel and Dreyer (11).

References

1. Shimomura O, Johnson FH, Saiga Y. Extraction, purification and properties of aequorin, a bioluminescent protein from the luminous hydromedusa, *Aequorea*. J Cell Comp Physiol 1962;59:228-39.
2. Morin JG, Hastings JW. Energy transfer in a bioluminescent system. J Cell Physiol 1971;77:313-8.
3. Cody CW, Prasher DC, Westler WM, Prendergast FG, Ward WW. Chemical structure of the hexapeptide chromophore of the *Aequorea* green-fluorescent protein. Biochemistry 1993;32:1212-18.
4. Rao BDN, Kemple MD, Prendergast FG. Proton nuclear magnetic resonance and fluorescence spectroscopic studies of segmental mobility in aequorin and green-fluorescent protein from *Aequorea forskalea*. Biophys J 1980;32:630-2.
5. Wampler JE, Hori K, Lee JW, Cormier MJ. Structured bioluminescence. Two emitters during both the in vitro and in vivo bioluminescence of the sea pansy, *Renilla*. Biochemistry 1971;10:2903-9.
6. Ward WW, Prentice HJ, Roth AF, Cody CW, Reeves SC. Spectral perturbations of the *Aequorea* green-fluorescent protein. Photochem Photobiol 1982;35:803-8.
7. Robart FD, Ward WW. Solvent perturbations of *Aequorea* green-fluorescent protein. Photochem Photobiol 1990;51:92S.
8. Bokman SH, Ward WW. Renaturation of *Aequorea* green-fluorescent protein. Biochem Biophys Res Commun 1981;101:1372-80.
9. Ward WW, Bokman SH. Reversible denaturation of *Aequorea* green-fluorescent protein: physical separation and characterization of the renatured protein. Biochemistry 1982;21:35-40.
10. Roth AF. Purification and protease susceptibility of the green-fluorescent protein of *Aequorea aequorea* with a note on *Halistaura*. M.S. thesis. Rutgers Univ., New Brunswick, NJ 1985.
11. Cutler MW. Characterization and energy transfer mechanism of the green-fluorescent protein from *Aequorea victoria*. Ph.D. Thesis. Rutgers Univ., New Brunswick, NJ 1995.

A COMPUTATIONAL ANALYSIS OF THE PREORGANIZATION AND THE ACTIVATION OF THE CHROMOPHORE FORMING HEXAPEPTIDE FRAGMENT IN GREEN FLUORESCENT PROTEIN

Marc Zimmer, Bruce Branchini, John O. Lusins*

Chemistry Department, Connecticut College, New London, CT 06320, USA

Introduction

The green bioluminescence of the jellyfish *Aequorea victoria* and sea pansy *Renilla reniformis* is due to a chromophore found in the non-catalytic green fluorescent protein (GFP). The gene for GFP has been cloned, and it has been shown that recombinant GFP is very similar or identical to the native protein. This has attracted wide interest due to the possibility for continuous *in situ* monitoring of gene expression, protein movement and cell development. The advantages of GFP as a reporter protein are its natural fluorescence, stability, small size (238 aa's) and its heterologous expression, while the disadvantages are that the onset of fluorescence after expression is slow and that it has two excitation peaks. These disadvantages have been partially overcome by the creation of a series of mutants.

The chromophore is formed from the hexapeptide 64Phe-Ser-Tyr-Gly-Val-Gln69, and after being made by internal cyclization of the tripeptide Ser-Tyr-Gly, it remains covalently attached to GFP. Although diffraction quality crystals have been obtained, no crystal structure has been solved yet.

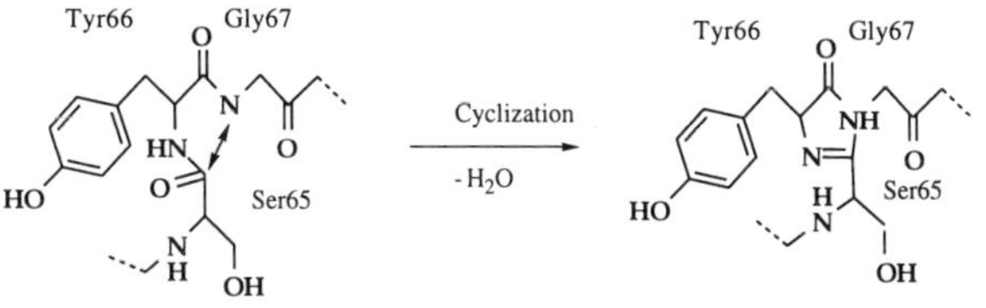

Figure 1: First step in the formation of the GFP chromophore.

One of the most interesting features of GFP is the formation of its chromophore. At present the mechanism of its formation is unknown. However, Heim and coworkers (1) have proposed the biosynthetic mechanism shown in Figure 1. We suggest that the posttransciptional chromophore formation occurs due to the presence of low energy conformations which have very short intramolecular distances between the carbonyl carbon of Ser-65 and the amide nitrogen of Gly-67. We also show that Arg-73 can hydrogen bond to the carbonyl oxygen of Ser-65 activating the corresponding carbonyl carbon for attack by the nucleophilic lone pair of the Gly-67 amide nitrogen.

Materials and Methods

The AMBER* force field as implemented in MacroModel v5.5(2) was used for all molecular modeling. It uses a 6,12 Lennard Jones hydrogen bonding treatment and an improved protein backbone parameter set(3).

Dihedral Monte Carlo (MC) searches were undertaken in which all rotateable bonds were varied. During the search procedure minimization continued until convergence was reached or until 1,500 iterations had been performed. Minimization occurred "in vacuo" and a derivative convergence criterion of 0.05kJ/mol was used. Structures within 50kJ/mol of the lowest energy minimum were kept and a usage directed method was used to select structures for subsequent MC steps. All conformations within 50 kJ/mol of the lowest energy conformation were combined, and subjected to a further 10,000 iterations in a continuum of solvent using the GB/SA model(4).

Results and Discussion

Non-enzymatic posttranscriptional backbone modifications are very rare(5), and therefore the mechanism of the GFP chromophore formation is of major importance. The mechanism shown in Figure 1 does not explain why the same tripeptide found in other proteins does not cyclize. We have addressed these questions using computational methods to model the chromophore forming region of GFP.

Preorganization of the chromophore forming hexapeptide. Proteins are extremely flexible and the number of conformations available to a peptide with N amino acid residues is estimated to be 10^N. It is therefore impossible to *de novo* calculate the structure of proteins even if they are as small as GFP. The chromophore forming hexapeptide contains 23 flexible dihedral angles, this is close to the upper limit of current conformational searching methods. Although it is not possible to find all the low energy minima of the hexapeptide fragment, or to unequivocally find the global minimum, it is possible to determine the low energy conformational families the hexapeptide fragment can adopt in the protein. The conformational motifs presented in this paper, which account for the non-enzymatic posttranscriptional chromophore formation, are based on such a low energy conformational family.

This family of low energy conformations was found having a very short distance (<3.1Å) between the carbonyl carbon of Ser-65 and the nitrogen of Gly-67.

The significance of this short distance is shown in Figure 1. In order for the cyclization leading to chromophore formation to occur, the distance between the carbonyl carbon of Ser-65 and the nitrogen of Gly-67 must be short. This family of low energy conformations with the geometry required for cyclization makes up about 15% of all the conformations, and although it is similar to an inverse γ turn it is still a unique turn, which we have called the "tight turn". The same low energy, tight turn conformational family is also found for the hexapeptide fragment with the S65T mutation.

The preceding results and discussion show that GFP can adopt a conformation in which the chromophore forming residues are preorganized in a way that facilitates the cyclization depicted in Figure 1.

Does the "tight turn" conformation occur in other proteins? No other peptides with the hexapeptide sequence (SFYGVQ) were found in the Protein Database (PDB). Therefore our conformational analysis was extended to the chromophore forming tripeptide, Ser-Tyr-Gly, to see whether it forms the "tight turn". None of its low energy conformations are folded in the manner required for cyclization. However, the tetrapeptide, Phe-Ser-Tyr-Gly has a low energy "tight turn" conformation. A PDB database search confirmed that in some cases the Phe-Ser-Tyr-Gly tetrapeptide is found in the calculated "tight turn" conformation.

Why isn't the chromphore formed in other Proteins? The fact that some other proteins, which contain the Phe-Ser-Tyr-Gly sequence, have the same "tight turn" as found in GFP, poses the question, why these peptides do not form the imidazolone moiety.

The presence of arginine significantly affects the dissociation pathways of peptides and hinders the mass spectral analysis of proteins(6). It has recently been proposed(7) that hydrogen bonding between the proton from the arginine side chain guanidinium group and the carbonyl oxygen on the adjacent amino acid residue is responsible for activating the carbonyl carbon, leading to the subsequent formation of certain rearrangement ions, which are observed in their mass-spectra.

GFP contains six arginines that can potentially hydrogen bond with the Ser-65 carbonyl, and activate it for attack by the Gly-67 amine lone pair, in a manner analogous to that proposed above. A partial conformational analysis of the decapeptide, 64FSYGVQCFSR73, showed that Arg-73 can loop back and hydrogen-bond to the Ser-65 carbonyl, as shown in Figure 2. A STO-3-21G* single point ab-initio calculation confirmed that the Ser65 carbonyl is activated by hydrogen bonding to Arg73 in this conformation. Due to the great number of conformations available to the decapeptide it was not possible to complete a full conformational search. Therefore, although the conformation shown in Figures 2 is a low energy structure (within 10 kJ/mol of the lowest energy conformation we found), it is quite possible that a whole section of conformational space has been missed and that the Ser65 is activated by one of the other Arg's present in GFP. To test our computational based hypothesis on the arginine assisted formation of GFP, we are currently preparing arginine GFP mutants.

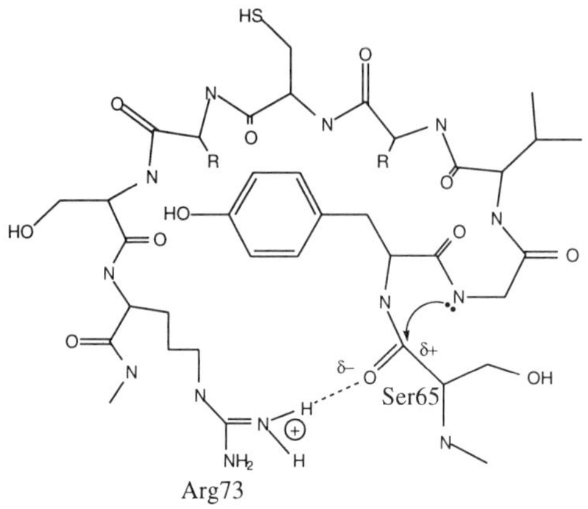

Figure 2: Proposed hydrogen bonding between Arg73 and Ser65 that activates the Ser65 carbonyl for attack from Gly67.

References

1. Heim R, Prasher DC, Tsien RY. Wavelength mutations and posttranslational autoxidation of green fluorescent protein. Proc Natl Acad Sci USA 1994, 91, 12501-4.
2. Mohamadi F, Richards NGF, Guida WC, Liskamp R, Lipton M, Caulfield C, Chang G, Hendrickson T, Still WC. MacroModel-An integrated software system for modeling organic and bioorganic molecules using molecular mechanics. J Comp Chem 1990, 11, 440-67.
3. McDonald Q, Still WC. AMBER* torsional parameters for the peptide backbone. Tetrahedron Lett. 1992, 33, 7743-6.
4. Still, WC, Tempczyk A, Hawley RC, Hendrickson T. Semianalytical treatment of solvation for molecular mechanics and dynamics. J Am Chem Soc 1990, 112, 6127-9.
5. Bayer A, Freund S, Nicholson G, Jung G. Posttranslational backbone modifications in the ribosomal biosysnthesis of the glycine-rich antibiotic microcin B17. Angew Chem Int Ed Engl 1993, 32, 1336-9.
6. Tang X-J, Thibault P, Boyd RK. Fragmentation reactions of multiply-protonated pepitides and implications for sequencing by tandem mass spectrometry with low-energy collision-induced dissociation. Anal Chem 1993, 65, 2824-34.
7. Vachet RW, Asam MR, Glish GL. Secondary interactions affecting the dissociation patterns of arginine-containing peptide ions. J Am Chem Soc 1996, 118, 6252-6.

MONITORING OF GENE EXPRESSION WITH GREEN FLUORESCENT PROTEIN

CR Albano, L Randers-Eichhorn, Q Chang, WE Bentley and G Rao
University of Maryland, Baltimore, MD 21250, USA

Introduction

For the first time, there is a marker that one can use to visualize gene expression in real time. Since its cloning and commerical availability, the use of GFP as a reporter gene has become very prevalent along with the creation of a number of altered spectra mutants (1-4). Our approach toward the application of GFP is to exploit the fluorescent characteristics of GFP as a tool for on-line measurements. By correlating the fluorescence intensity with GFP quantity, fusion proteins containing GFP as a reporter could be quantitated simply by spectrofluorimetry, thereby alleviating the need for the time consuming steps of sample preparation, gel electrophoresis, gel staining and quantitation.

Reporter assays are a common molecular tool used to evaluate gene expression of a particular gene by fusing it to another 'reporter gene'. The 'reporter gene' is one that can be quantitated and since the two genes are fused, the expression of the reporter gene should accurately reflect the expression of the gene of interest. The attractiveness and versatility unique to GFP as a reporter is that there is no need for the sacrifice or fixation of the organisms or the addition of chemical substrates for visualization. This revolutionary trait makes GFP as a reporter gene unique in allowing for in vivo fluorescence visualization.

Additionally, since it has been successfully expressed in a variety of hosts, the mechanism for fluorescence appears to be self-contained. It is the primary structure of the protein that gives GFP its fluorescent properties (4). It has the advantages of not requiring a co-factor and generally does not appear to place a deleterious metabolic burden on the host. Our data suggest that GFP fluorescence measurements can be quantitatively related to the levels of recombinant protein produced.

Two constructs of GFP were analyzed, pGFP under the lac promoter and pBAD-GFP under the arabinose promoter. Through a time course experiment presented here, we show how the fluorescence intensity measurements relate to protein concentration in both the pGFP and pBAD-GFP constructs.

Materials and Methods

pGFP Construct: The cDNA construct cloned by Prasher named TU#60 (5) is available for commerical distribution under the name of pGFP (Clontech, Palo Alto, CA). GFP is under the lac promoter within a pUC19 backbone and inducible with isopropyl-β-D-thiogalactoside (IPTG).

pBAD-GFP Construct: Through the technique of DNA shuffling (7), Crameri et al. constructed a GFP mutant from pGFP that was more fluorescent than the Clontech construct (1). The GFP mutant, named pBAD-GFP (Affymax, Palo Alto, CA) was placed under the arabinose promoter and inducible with arabinose.

Bacterial Transformation: *E. coli* strain JM105 was transformed separately with pGFP (Clontech) and pBAD-GFP (Affymax). Transformants of each construct were selected by growth on Luria Bertani (LB) plates with 100 ug/mL ampicillin and either 1 mM IPTG or 0.2% arabinose to induce fluorescence when excited by a hand held UV light (model UVGL-25, UVP Inc., San Gabriel, CA).

Time Course: A 3% inoculum of overnight saturated cultures for both transformants were used to start fresh cultures grown in 1 L shake flasks containing 200 mL LB with 100 ug/mL ampicillin at 30°C with shaking (300 rpm) until the optical density reached 0.4. At this point, pGFP and pBAD-GFP cultures were induced with 1 mM

IPTG or 0.2% arabinose respectively. Negative control cultures for both pGFP and pBAD-GFP were substituted with an equal volume of distilled water. Five mL aliquots were taken from the four cultures at various time points over the next 24 hours. The aliquots were used for optical density, fluorescence intensity measurements and protein gel analysis. Optical density was measured at 600 nm in a Milton Roy Spectronic 401 spectrophotometer. Fluorescence intensity measurements were analyzed in a Perkin-Elmer MPF-66 Spectrofluorimeter (Oak Brook, IL) at an excitation wavelength of 395 nm and emission at 509 nm.

Sample Preparation: Sample preparation of aliquots for protein gel analysis consisted of centrifugation for 2 min 4°C at 12,000 g, resuspension in 1 mL 100 mM $NaPO_4$, pH 8.0 and sonication for 10 sec on ice with a Fisher sonic dismembrator model 300 (Fisher Scientific, Pittsburgh, PA). After sonication, the lysate was centrifuged for 20 min 4°C at 12,000 g and stored at -70°C until tested. Total protein measurements were determined by the Bradford method using the Micro Protein Determination kit (Sigma Chemical Co., St. Louis, MO).

Electrophoresis: Samples were standardized to 0.1 ug/uL and denatured at 95°C for 5 min in Laemmli buffer (Sigma). A 4/15% stacking SDS-polyacrylamide gel was loaded with a total protein of 1 ug/well for the pGFP samples and 0.5 ug/well for the pBAD-GFP samples. Electrophoresis using a Bio-Rad Mini-Protean II apparatus was carried out at 150 V for 2 h and proteins were transferred using a Mini-Trans Blot cell (Bio-Rad) for 2 h at 100 V to nitrocellulose (Schleidher & Schuell, Keene, NH) in cold Towbin transfer buffer (25 mM Tris, 193 mM glycine, 20% methanol).

Western Blot: Following transfer, protein was visualized by probing with a 1:2000 dilution of polyclonal anti-rGFP (Clontech) and a 1:30000 dilution of goat anti-rabbit IgG conjugated to alkaline phosphatase and developed with an alkaline phosphatase conjugate substrate kit according to product directions (Bio-Rad, Hercules, CA). Samples were calibrated to known concentrations of pure GFP (Clontech) on the Western blot. The pure GFP was diluted to a concentration of 0.05 ug/uL and sonicated for 10 sec. GFP concentrations of 0.5, 0.25 and 0.125 ug were loaded as a calibration curve for each Western blot.

Image Analysis: The image was acquired using a CCD camera, light box, (OPELCO, Washington, DC) and NIH Image software. Analysis was performed on a Macintosh IIsi using the public domain NIH Image program (written by Wayne Rasband at the U.S. National Institutes of Health and available from the Internet by anonymous ftp from zippy.nimh.nih.gov or on floppy disk from NTIS, 5285 Port Royal R., Springfield, VA 22161, part number PB93-504868).

Results and Discussion

A quantitative comparison of fluorescence intensity for the pGFP and pBAD-GFP transformants is shown in Figure 1. Samples were taken over a 24 hour time course and analyzed on a Perkin-Elmer spectrofluorimeter at an excitation wavelength of 395 nm and emission wavelength of 509 nm. A maximum difference of 24 fold was seen between the two constructs at the 3 hour timepoint. Also note that the pGFP construct reached maximum fluorescence much slower than the pBAD-GFP construct. This reflects the difference in cyclization time for each construct.

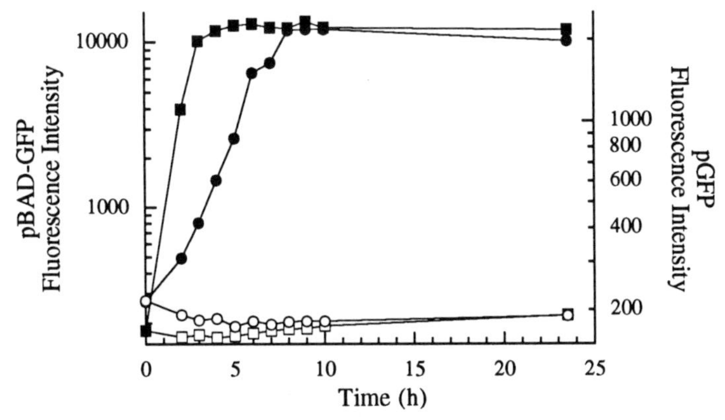

Figure 1: Microbial fluorescence intensity comparison of pGFP (closed circles) and pBAD-GFP (closed squares) transformants. Fluorescence intensity of uninduced cultures also shown, pGFP (open circles) and pBAD-GFP (open squares).

Figures 2 and 3 show the time course plot of fluorescence intensity and GFP concentration determined by SDS-PAGE and Western blot. Both cultures have a delay from the presence of GFP detected by Western blot to the appearance of fluorescence. This is demonstrated in Figure 2 with the fluorescence intensity curve represented by a dotted line shifted to the left 3.5 hours. This shows that the profile of the GFP concentration and fluorescence intensity curves are superimposible. The pBAD-GFP construct is reported to have a cyclization time of 95 min (1); correspondingly this is reflected in Figure 3. The shifted fluorescence intensity curve is also represented here by a dotted line, shifted to the left by 95 minutes.

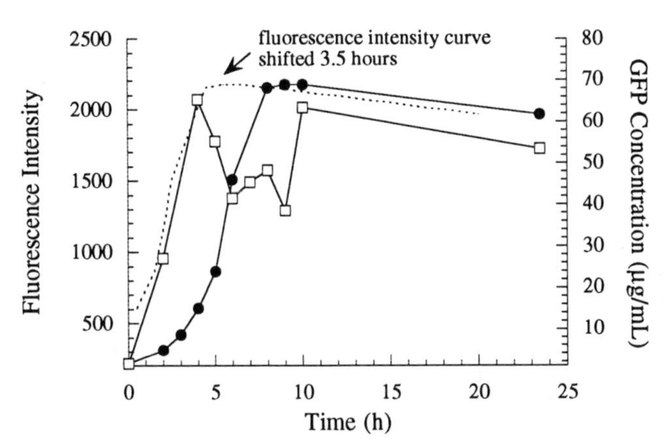

Figure 2: Time course comparison of GFP concentration (open squares) and microbial fluorescence intensity (closed circles) for pGFP. Dotted line represents the fluorescence intensity curve shifted to the left 3.5 hours to correlate with GFP concentration.

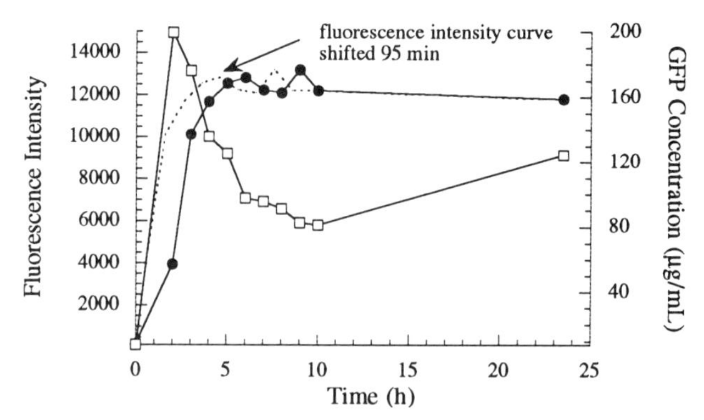

Figure 3: Time course comparison of GFP concentration (open squares) and microbial fluorescence intensity (closed circles) for pBAD-GFP. Dotted line represent the fluorescence intensity curve shifted to the left 95 min to correlate with GFP concentration.

These experiments provide a basis for using the fluorescent properties of GFP as a monitor of gene expression on-line. It is conceivable that GFP could be used as a reporter gene to measure recombinant protein concentration simply by the increase in fluorescence. Fluorescence intensity could be monitored over time to reflect when the recombinant protein was at the desired concentration or used to evaluate different protocols for optimization. These experiments show that not only is the fluorescence of GFP quantifiable but it is an accurate reflection of the GFP concentration. This revolutionary application of a reporter assay bypasses the time consuming steps of sampling, sample preparation, gel electrophoresis and Western blot.

Acknowledgments
We would like to thank Tracey R. Pulliam for providing us with the pGFP transformant and *E.coli* JM105 stock. This work was supported by NSF grant (BCS 9157852) along with matching funds from Genentech, Inc.

References

1. Crameri A, Whitehorn EA, Tate E, Stemmer WPC. Improved green fluorescent protein by molecular evolution using DNA shuffling. Nature Biotech 1996;14:315-9.
2. Delagrave S, Hawtin RE, Silva CM, Yang MM, Youvan DC. Red-shifted excitation mutants of the green fluorescent protein. Bio/Technology 1995;13:151-4.
3. Ehrig T, O'Kane DJ, Prendergast FG. Green fluorescent protein mutants with altered fluorescence excitation spectra. FEBS Letters 1995;367:163-6.
4. Heim R, Prasher DC, Tsien RY. Wavelength mutations and post-translational autoxidation of green fluorescent protein. Proc Natl Acad Sci USA 1994;91:12501-4.
5. Prasher DD, Eckenrode VK, Ward WW, Prendergast FG, Cormier MJ. Primary structure of the Aequorea victoria green-fluorescent protein. Gene 1992;111:229-33.
6. Shimomura O, Johnson FH, Saiga Y. Extraction, purification and properties of Aequorin, a bioluminescent protein from the luminous hydromedusan, Aequorea. J Cell Comp Physiol 1962;59:223-39.
7. Stemmer WPC. DNA shuffling by random fragmentation and reassembly: In vitro recombination for molecular evolution. Proc Natl Acad Sci USA 1994;91:10747-51.

IMAGING OF LUCIFERASE FUSION GENE EXPRESSION IN TRANSFORMED CELLS AND EMBRYOS

G Wang[1], S Cseh[1] and Y Wang[2] and AA Szalay[1]
Center for Molecular Biology and Gene Therapy, [1]Dept. of Microbiology and Molecular Genetics & [2]Dept. of Biochemistry, School of Medicine, Loma Linda University, Loma Linda, California 92350, USA

Introduction

The capability of bioluminescent organisms to glow in the dark has fascinated humans for centuries. The light emission from fire flies, the jelly fish *Aequorea*, and bacterial colonies were captured by photography which resulted in dramatic pictures such as glowing trees illuminated by swarms of fireflies. With the advancement of gene cloning and gene sequencing, a number of genes and cDNAs which encode a variety of luciferases and proteins capable of light emission were isolated, cloned and provided as useful tools to cell biologists. In the early 1980s, the structural genes *luxA* and *B* encoding the α and β subunits of bacterial luciferase from *Vibrio harveyi* were identified (1). The structural gene cassette *luxA&B* was linked to the *nif* promoter (nitrogenase promoter) and introduced into the *Rhizobium* genome by site specific recombination (2). Cross sections made from root nodules allowed the microsope aided low light imaging of the nitrogen fixing bacteria in the infected host plant cell based on light emission.

In 1990 White and colleagues (3) reported the first success of imaging firefly luciferase expression in single mammalian cells. Recently, the introduction of GFP cDNA gene into a variety of cell types made possible the simple, elegant identification of individual cells based on emission of green fluorescent light (4). However, the strength of promoter activation and levels of gene expression based on fluorescent light emission is difficult to quantify.

Here we describe experiments which show the number of GFP molecules required in individual mammalian cells and early stage embryos which can be detected by microscope-aided camera imaging systems. Further, we describe experiments using *Renilla* luciferase - GFP fusion gene cassettes for visualization of individual transformed cells and embryos based on GFP activity. Once the desired transformants are identified, the luciferase-GFP fusion protein will allow us in subsequent experiments to quantify promoter strength based on GFP linked *Renilla* luciferase activities.

Materials and Methods

Cell culture conditions: Mouse fibroblast L-M (TK^-) cells [ATCC ccl-1.3, derived from a subline of a 5-bromo-2-deoxyuridine (BudR) resistant strain of the L-M mouse fibroblast cell line], were cultured in an F-12 Nutrient Mixture (HAM, Gibco), supplemented with 10% fetal bovine serum (Defined, HyClone) and an antibiotic-antimycotic mixture (Sigma), at 37°C in an atmosphere of 5% CO_2. The cells were grown to 70% confluency, and fresh medium was added prior to microinjection.

Embryo harvesting, culture and implantation: Experimental procedures were carried out essentially as described by Brigid Hogan, et. al. in the laboratory manual "Manipulating the Mouse Embryo" (5).

Preparation of microinjection materials: When GFP was used, the purified protein (2mg/mL, provided by D.J. O'Kane) was serially diluted with filtered injection buffer (10 mM Tris pH7.4, 0.2 mM EDTA, 100 mM NaCl), and kept at 4°C. DNA for microinjection were prepared as follows: Plasmid DNA was first purified to remove endotoxic compounds (EndoFree Plasmid Maxi kit, Qiagen), then linearized with a restriction endonuclease. The linearized plasmid DNA was further diluted to 1 ng/μL with injection buffer, filtered and used directly for microinjection.
Conditions for microinjection of mammalian cells and mouse embryos: Microinjection of mammalian cells and mouse embryos were performed using an automatic micromanipulator (Model 5171, Eppendorf) and a transjector (Model 5246, Eppendorf) system. Injection needles used were either purchased from Eppendorf Co. (Femtotips & Femtotips II) or self-made (Borosilicate glass with filament, I.D. 0.78 mm, Sutter Instrument Co.) using micropipette puller (Sutter Instrument Co., MP-97). The injection volume ranged from 2 to 10 picoliter upon each injection.
Assay of GFP in mammalian cells and embryos in vivo under fluorescent microscope: GFP activity was analyzed and detected in injected cells and embryos using an FITC filter system combined with a Xenon light as a UV excitation source. The resulting fluorescence image was directly photographed or sent through a video camera to an image analysis system.
Image analysis system: The image analysis system for detection of GFP in living organisms consists of an inverted fluorescent microscope (Zeiss, Axiovert 100TV), a CCD video camera (Hamamatsu, C2400), a monitor (Sony), an image intensifier (Hamamatsu, M4314) and a computer connected with a Digital Color Printer (Sony, UP-D8800), and a Polaroid Digital Palette. Both video images and fluorescence images were collected and analyzed using MetaMorph software, and finally montaged using Photoshop software programs.
Generation of transformed ES cells: The embryonic stem cells (ES) used in this study were from Dr. J. Mann. The STO feeder cells for supporting the growth of ES cells were from L. Robertson. Hygromycin B resistant feeder cell line was generated by transfecting STO with plasmid pCEP4 which contains the hygromycin B resistance gene. ES cells were cultured by standard procedures (5). After transformation, ES cells were cultured for two days in medium without selection, then placed into hygromycin B (200 μg/mL) containing selection medium.

Results and Discussion

We have used microinjection techniques to deliver known amounts of purified GFP protein into mouse fibroblast and L293 cells. After injecting 2 picoliters of an aqueous solution of GFP into the cytoplasm, we were able to visualize individual cells under the camera system. Figure 1a, shows cells injected with approximately 40-50,000 GFP molecules which allows instant visualization. Figure 1b, shows injection of purified GFP protein into the cytoplasm of one cell stage mouse embryos 12 hrs after fertilization. Increased amounts of GFP lead to early cell death and inhibits cell division and further embryo development. The GFP activity was maintained for up to 96 hrs. During prolonged embryo culture, the injected GFP is maintained after the first division and appears with equal intensity in 2-cell stage embryos, 36 hrs after injection. We find that increased levels of GFP will arrest embryo development at the 2-cell stage. These findings indicate that the GFP may be used for identification of individual cells but should be used with caution for temporary measurement of cellular processes based on GFP protein detection.

a.

b.

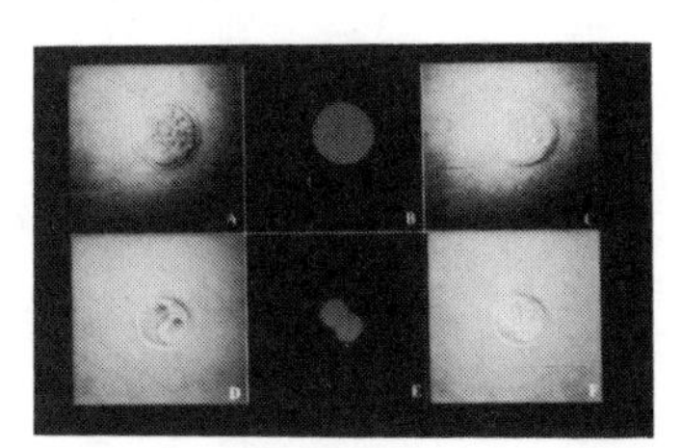

Figure 1: Injection of mouse fibroblast cells and mouse embryos with purified GFP and fluorescence image analysis a) LM-TK- cells, b) mouse embryos

In further experiments we injected linearized DNA into cultured mouse fibroblast cells L-M(TK-) as well as into one-cell stage embryos. We used several plasmids which contained the GFP gene and the *Renilla* luciferase gene as separate transcriptional units and in the form of a *Renilla* luciferase - GFP fusion gene cassette. For expression of these genes, different promoters were selected. Due to space limitations, only one experiment will be presented. By light microscopy cells (nuclei) injected with cytomegalovirus promoter (CMV) RG gene fusions (1 femtogram DNA per nucleus), showed green fluorescent light emission 24 hrs after injection . This light emission represents transient gene expression, (Figure 2a). Plasmid DNA constructs (3 femtograms) were injected into the male pronuclei of one cell stage mouse embryos. Figure 2b shows the expression of GFP protein based on green fluorescence 36 hours after DNA injection. By this time embryo development had proceeded to the 4-cell stage. To our surprise, the CMV promoter RG fusion did not give light emission in embryos, while the human β- actin promoter resulted in strong promoter activation. Visualization based on light emission can be monitored throughout advanced developmental stages up to blastocyst stage. Two-cell stage embryos containing the β-actin RG fusion gene cassettes were implanted into mice recipients, 20 newborn animals were obtained and the presence of the RG fusion DNA construct in their genomic DNA is presently under analysis.

In contrast to embryos, mouse ES cells electroporated with DNA (CMV promoter linked to the RG fusion gene cassette) showed strong fluorescent light emission 2 days after culture. Intense light emission from the ES cells allows the identification of transformed ES colonies and permits their transfer to culture conditions for obtaining permanent transformants. Figure 3 shows 2 individual ES colonies cultured on feeder layers in the presence of hygromycin. Once the stable transformed ES cells are identified, the GFP based light emission may aid investigators in the injection of transformed ES cells into recipient blastocysts followed by implantation to recipient mice. We expect that GFP activity will serve as a useful tool for identification of transformed tissues in chimeric animals.

a. 

b. 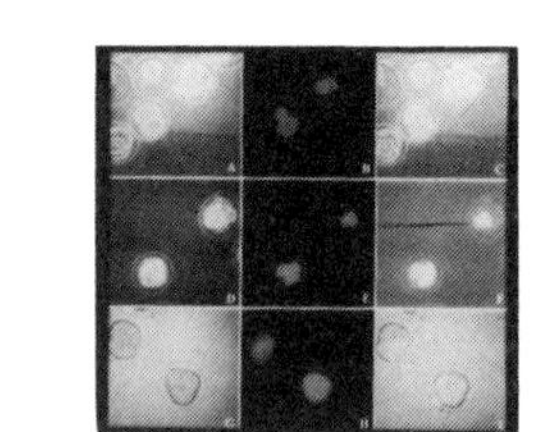

Figure 2: Expression of RG fusion genes in microinjected cells and mouse embryos. a) mouse fibroblast cells, b) mouse embryos at different stages of development

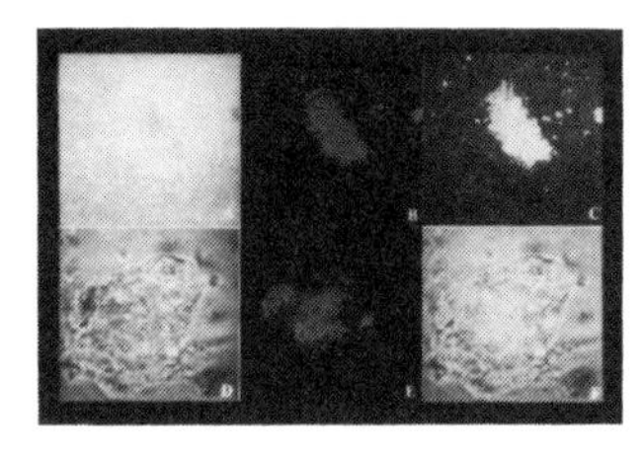

Figure 3: Visualization of transformed ES cells in culture. A&D two individual transformed ES colonies, B&E their green fluorescence under UV excitation, C&F overlay image

Acknowledgments

The authors would like to thank Dr. W.H.R. Langridge, Center for Molecular Biology and Gene Therapy, Loma Linda University for critical review of the manuscript and expert advice during the microscope image analysis experiments and Dr. D.J. O'Kane for GFP and advice in image visualization.

References

1. Engebrecht J, Simon M, Silverman M. Measurement gene expression with light. Science 227:1345-7.
2. Legocki RP, Legocki M, Baldwin TO, Szalay AA. (1986) Bioluminescence in soybean root nodules: Demonstration of a general approach to assay gene expression in vivo using bacterial luciferase. Proc Natl Acad Sci USA, 83:9080-4.
3. White MRH, Craig FF, Watmore D, McCapra F, Simmonds AC. Applications of the direct imaging of firefly luciferase expression in single intact mammalian cells using charge-coupled device cameras. In: Stanley PE, Kricka LJ, editors. Bioluminescence and Chemiluminescence: Current Status. Chichester: John Wiley & Sons, 1991:357-60.
4. Chalfie M, Tu Y, Euskirchen G, Ward WW, Prasher DC. Green fluorescent protein as a marker for gene expression. Science 1994; 263:802-5.

THE *RENILLA* LUCIFERASE-MODIFIED GFP FUSION PROTEIN IS FUNCTIONAL IN TRANSFORMED CELLS

Y Wang[1,3], G Wang[1], DJ O'Kane[2] and AA Szalay[1]
Center for Molecular Biology and Gene Therapy,
[1]Dept. of Microbiology and Molecular Genetics
[3]Dept. of Biochemistry, School of Medicine, Loma Linda University, Loma Linda CA 92350, USA
[2]Dept. of Pathology and Molecular Medicine, Mayo Clinic, Rochester, MN 55905, USA

Introduction:

The cDNA from *Renilla* luciferase (ruc) has been isolated and sequenced (1). By providing appropriate promoters, the cDNA gene cassettes were expressed in bacteria, transformed plant cells, and mammalian cells (2,3). The documented high efficiency of the *Renilla* luciferase is a useful and novel trait for a marker enzyme for gene expression studies. The usefulness of the green fluorescent protein from the jellyfish *Aequorea victoria* was documented as a reporter in prokaryotes and animal cell systems (4). The UV light stimulated GFP fluorescence does not require cofactors and the gene product alone is sufficient to allow detection of living cells under the light microscope. Bioluminescence in *Renilla* reniformis is produced by the reaction between coelenterazine and oxygen, a reaction which is catalyzed by *Renilla* luciferase. The reaction yields blue light with an emission wavelength maximum of 480 nm using the purified enzyme. In *Renilla* reniformis cells, this reaction is shifted toward the green with a maximum of 509 nm. This wavelength transition is due to an energy transfer to a green fluorescent protein. In this paper, we describe the engineering of a novel protein with dual functions combining characteristics for *Renilla* luciferase and GFP. This task is accomplished by construction of a fusion gene between the cDNA of *Renilla* and the cDNA of the "humanized" *Aequorea* GFP (5).

Materials and Methods

Vectors and Cells: The vectors used for cloning and expression of the gene constructs in *E. coli* and mammalian systems were pGEM-5zf(+) (Promega) and pCEP4 respectively. The cDNA of *Renilla* luciferase and "humanized" GFP were in plasmids designed pCEP4-RUC provided by Dr. Cormier and pTR- β actin-gfp_h provided by Dr. Muzyczka respectively. The former cDNA is under transcriptional control of the CMV promoter and later the β-actin promoter. *E. coli* strains used include DLT101 and DH5α. Mammalian cell line LM-TK^- is used as a recipient for gene expression.

Primers: The following five primers were designed for cloning of RG and GR gene constructs: double underlines indicate the restriction sites, the start codons are in bold form, DNA sequences in bold italics are used to remove stop codons from both ruc and gfph genes, Shine-Dalgarno sequences are underlined. RUC5: 5' CTGCAG (PstI) AGGAGGAATTCAGCTTAAAG**ATG** 3'; RUC3: 5' GCGGCCGC (NotI) ***TTG*** TTCATTTTTGAGAAC 3'; GFP5: 5' GGGGTACC (KpnI) CC**ATG**AGCAAGGGCGAG GAACT 3'; GFP3: 5'GGGGTACC (KpnI) C***CTT***GTACAGCTCGTCCATGCCA3'; GFP5a 5' CCCGGG (SmaI) AGGAGGTACCCC**ATG**AGCAAG 3'.

Construction of *Renilla* GFP fusion (RG cassette) and GFP *Renilla* fusion (GR cassette): The RG and GR fusion cassettes were constructed by removing the stop codon, adding restriction sites and SD sequences to the 5' of the cDNAs using PCR methods. The pGEM-T system (Promega) was used for cloning PCR products. The primers were designed that the downstream cDNA is in frame with the upstream cDNA. The intergenic sequence of RG is GCGGCCGCGCCACC (15 bp), and GR is GGGTACCAGATCGAATTCAGCTTAAAG (27 bp). The RG and GR cassettes were under the transcriptional control of the T7 promoter in pGEM-5zf(+) vector and CMV promoter in the pCEP4 vector, which were used for expression in *E. coli* and mammalian cells, respectively.
Visualization of GFP in vivo: *E. coli* strain DH5α was transformed with the plasmids pGEM-5zf(+)-RG and pGEM-5zf(+) GR. Positive colonies were identified and cultured in 37°, LB medium with 100 μg/mL of ampicillin selection. After 12 hours incubation, one drop of *E. coli* culture was put on a slide and visualized by fluorescent microscopy at 1000 x magnification. LM-TK$^-$ cells were transfected with plasmids pCEP4-RG and pCEP4-GR using calcium phosphate methods. The culture dishes were monitored using an inverted fluorescent microscope 12 hours after the transfection.
Luciferase Assay: Before and after IPTG induction, an aliquot of transformed *E. coli* was used for luciferase assay in a Turner TD 20e luminometer. The results were recorded as relative light units. The mammalian cells were harvested 36 hrs after transfection and measured for luciferase activity.
Spectrofluorimetry: An SPEX fluorolog spectrofluorimeter operated in the ratio mode was used to detect the corrected emission spectra. Fluorescence emission was measured by UV excitation at 390 nm. Bioluminescence emission was recorded with the excitation beam blocked following the addition of 0.1 μg of coelenterazine in acidified methanol. Five spectra were averaged for each sample over the wavelength range from 400 to 600 nm.
Protein isolation and immuno detection: *E. coli* 1ml (OD_{600}=1.0) was harvested and 400μl of cell suspension buffer (0.1M NaCl, 0.01 M Tris-HCl pH 7.6, 0.001M EDTA, 100μg/mL PMSF) and 100μl of loading buffer (50mM Tris-HCl pH6.8, 2% SDS, 10% glycerol, 5% 2-mercaptoethanol) were added. The samples were boiled for 4 min and loaded on a 7.5% - 20% gradient SDS-polyacrylamide gel. Polyclonal anti-*Renilla* luciferase was used as the primary antibody, goat peroxidase conjugated anti-IgG (anti-rabbit) was employed as the secondary antibody.

Results and Discussion

The construction of two fusion genes is described in detail in Figure 1. Fusion gene 1 contains the *Renilla* cDNA linked at its 3' end to a 15 nucleotide linker sequence with 5 amino acid coding capacity followed by the 5' end of the intact GFP cDNA (RG cassette). In fusion gene 2, the cDNA of GFP is linked to a bridging peptide of 9 amino acids in length followed by 5' end of the *Renilla* cDNA (GR cassette). Both gene cassettes 1 and 2 were placed into a prokaryotic expression vector pGEM - 5zf (+) (Figure 1a) and into the pCEP4 eukaryotic expression vector (Figure 1b) and transformed into *E. coli* and into different mammalian cell lines and microinjected into mouse embryos. Only cells containing the RG fusion gene cassette gave strong fluorescence while the GR gene cassette showed minimal response to UV light excitation under the light microscope. Figure 2a shows individual *E. coli* cells transformed with the RG construct exhibiting strong green fluorescence under oil immersion. Figure 2b shows the strong green fluorescence emitted by mammalian cells transformed with the RG construct but no fluorescence with the GR construct (data not shown). In contrast, luciferase measurements (Figure 2c, d) show that cells transformed with the GR construct have significant luciferase activity, which is reduced by 3-fold in the GR construct containing cells. Analysis of the fusion proteins in SDS polyacrylamide gels followed by immunodetection of luciferase and GFP confirms the presence of a 62 KDa band indicative of the full length fusion protein in the RG lane (data not shown).

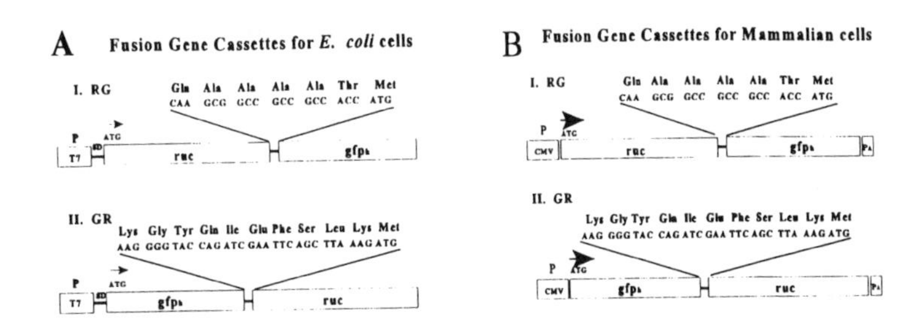

Figure 1: Construction of *Renilla* luciferase and "humanized" gfp fusion gene cassettes for gene expression in *E. coli* (A) and in mammalian cells (B)

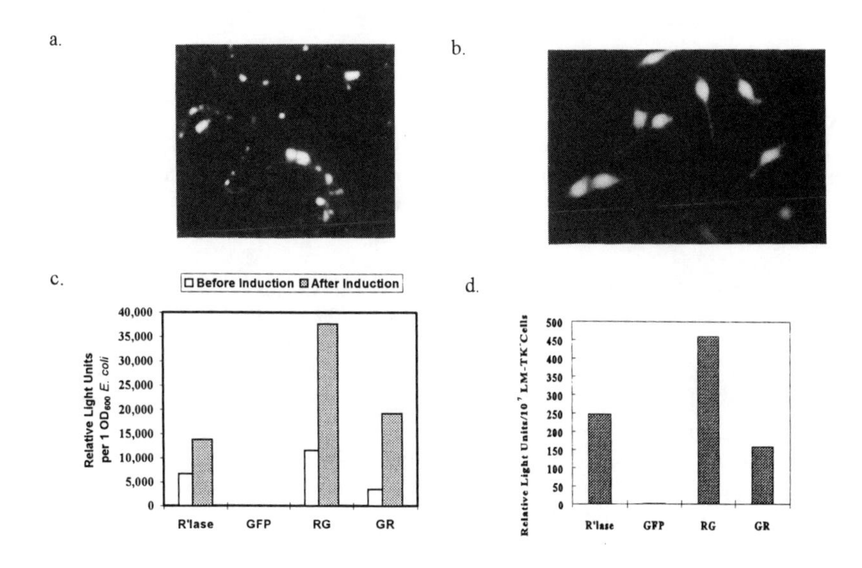

Figure 2: Analysis of GFP activity in transformed cells by fluorescence microscopy and fluorescence image system. a: In *E. coli* cells b: In LM-TK- mouse fibroblast cells c: Luciferase activity with and without promoter induction in RG and GR transformed in *E. coli* cells. d: In RG and GR transformed LM-TK- mouse fibroblast cells

However, we did not detect a band with the expected molecular weight of the GR fusion protein, but instead we find a band in the position of a 34 KDa *Renilla* luciferase (data not shown). Therefore, the lack of GFP activity in GR transformed cells may be due to incorrect folding and degradation, or due to the requirement for a pre-GFP terminus which is not available in the GR fusion protein. Data obtained from spectrofluorimetric measurements indicate that there is energy transfer (from 480 nm to 510 nm) between *Renilla* luciferase and GFP in the RG fusion containing bacterial as well as mammalian cells (Figure 3). Cells containing the GR fusion show only one emission peak at 480 nm which indicate *Renilla* luciferase activity. In addition, RG fusion containing cells show a 510 nm emission peak upon excitation at 390 nm. This activity is not found in GR construct transformed cells. Based on this data, we conclude that the 62 KDa, RG polypeptide exhibits both *Renilla* luciferase and GFP activity in living cells. This bifunctional polypeptide may become a useful tool for identification of transformed cells at the single cell level based on fluorescence. Simultaneously, it may allow quantifying of promoter activation in transformed tissues and transgenic organisms by measurement of luciferase activity.

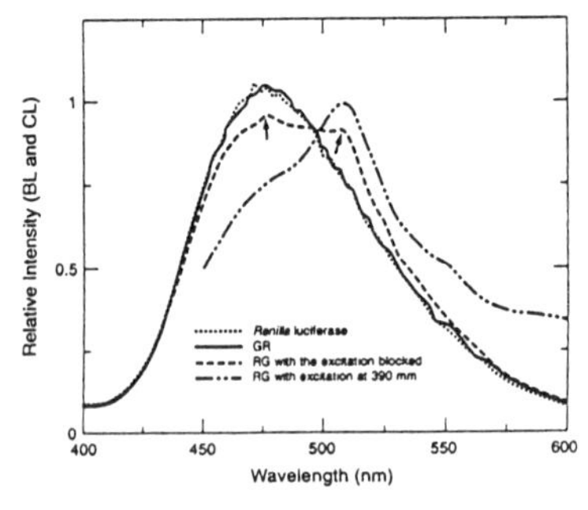

Figure 3: Light emission from transformed *E. coli* cells

Acknowledgments

The authors would like to thank Dr. W.H.R. Langridge, Center for Molecular Biology and Gene Therapy, Loma Linda University for critical review of the manuscript and expert advice during the microscope image analysis experiments.

References

1. Lorenz WW, McCann RO, Longiaru M, Cormier MJ. Isolation and expression of a cDNA encoding *Renilla reniformis*. Proc Natl Acad Sci USA 1991; 88:4438-42.
2. Mayerhofer R, Langridge WHR, Cormier MJ, Szalay AA. Expression of recombinant *Renilla* luciferase in transgenic plants results in high levels of light emission. The Plant Journal 1995; 7:1031-8.
3. Lorenz WW, Cormier MJ, O'Kane DJ, Hua D, Escher A, Szalay AA. Expression of the *Renilla reniformis* luciferase gene in mammalian cells. J Biolumin Chemilumin 1995; 11:31-7.
4. Chalfie M, Tu Y, Euskirchen G, Ward WW, Prasher DC. Green fluorescent proteins a marker for gene expression. Science 1994;263:802-5.
5. Zolotukhin S, Potter M, and Hauswirth WW, Guy J, Muzyczka N. A "Humanized" green fluorescent protein cDNA adapted for high-level expression in mammalian cells. J Virology 1996;70:4646-54.

PREPARATION OF STABLE COVALENT CONJUGATES OF RECOMBINANT AEQUORIN WITH PROTEINS AND NUCLEIC ACIDS

NL Stults, HN Rivera, J Burke-Payne and DF Smith
SeaLite Sciences, Inc. 3000 Northwoods Parkway, Suite 200, Norcross, GA 30071, USA

Introduction

Aequorin, the 22,000 kDa calcium dependent photoprotein derived from the circumoral ring of the jellyfish *Aequorea victoria*, catalyzes the oxidation of coelenterate luciferin to oxyluciferin resulting in a 2 sec flash of blue light (λ_{max} 469 nm). Aequorin produces a direct, high quantum yield signal which can be detected at the subattomol (10^{-19} mol) level in tube and microplate luminometers. While this bioluminescent protein has been used for many years to monitor intracellular levels of calcium, the potential utility of using aequorin as a nonradioisotopic reporter molecule has only been recently realized due the availability of AquaLite®, the recombinant protein (1) and to the development of a conjugation method which yields highly active and stable AquaLite conjugates with a variety of binding reagents (US Patent Nos. 5,422,266 and 5,486,455). Here we describe the sulfhydryl:maleimide coupling strategy used for the preparation of stable AquaLite conjugates of proteins and nucleic acids with high photoprotein activity and their performance in immunoassays and quantitative DNA detection, respectively. This approach has also been used successfully to prepare an antibody conjugate of recombinant obelin (2), a calcium dependent photoprotein from the hydroid *Obelia longissima* which is closely related to aequorin and also utilizes coelenterate luciferin.

Materials and Methods

Instrumentation: Luminescence measurements were made using a Berthold AutoClinilumat Model LB 952 T/16 tube luminometer (Wallac, Gaithersburg, MD) or a Dynatech ML3000 Microplate luminometer (DYNEX, Chantilly, VA).
Reagents: All reagents were of the highest quality available. Synthetic coelenterazine was obtained from Molecular Probes (Eugene, Oregon).
Conjugation: AquaLite is routinely sulfhydryl activated by treatment with Traut's reagent (2-iminothiolane) for 30 min at 25°C in Hepes buffered saline containing 2 mM EDTA (HBS/E), pH 8.0. After the reaction is terminated by addition of excess amine containing compound (Tris or lysine), the reaction mixture is desalted and the sulfhydryl content is determined using Ellman's reagent (dithionitrobenzene). Binding reagents such as antibodies or 5' amine-labeled oligonucleotides are routinely activated with a heterobifunctional crosslinker such as SMCC (succinimidyl 4-(N-maleimidomethyl)cyclohexane-1-carboxylate) in HBS/E, pH 8.0 or similar buffer for 30-60 min at 25°C. After quenching the reaction, the maleimido activated binding partner is desalted and the maleimide content determined by back titration with mercaptoethylamine. Conjugation is accomplished by mixing the activated components at a 2.5 to 10 fold molar excess of activated AquaLite to activated binding partner for 60 min at 25°C. The reaction is quenched by addition of excess sulfhydryl containing compound (DTT or cysteine), and the conjugate is then purified using an appropriate combination of gel filtration (Sephacryl S200HR or HPLC TosoHaas G3000SW) and ion exchange chromatography (DEAE Sepharose or HPLC DEAE 5PW). DNA conjugates are prepared essentially as described above using maleimido activated oligonucleotides modified at the 5' end with an amino group. Purification of AquaLite DNA conjugates is accomplished by ion exchange chromatography on Q-Sepharose.
Bioluminescent TSH Assay: Calibrators (200 µL) and anti-TSH antibody AquaLite conjugate (100 µL at 3-4 million RLU) were incubated together in polystryrene tubes coated with an anti-intact TSH monoclonal antibody for 2 h at 25°C. After washing,

the tubes were read for 2 sec in the Berthold luminometer upon injection of 300 μL of calcium trigger buffer.

Hybridization immunoassay for γinterferon (γIFN) cDNA: Linearized plasmid containing mouse cytokine γIFN cDNA insert was used as a source of γIFN cDNA. PCR amplification of dilutions of the linearized plasmid was carried out using 0.02 U/μL Taq polymerase (Life Technologies, Gaithersburg, MD) and 0.5 μM each of antisense and biotinylated-sense primer. Denaturation, annealing, and elongation were at 96°C, 54°C, and 74°C for 30 sec, 30 sec, and 20 sec, respectively using an Idaho Technology Rapidcycler hot air thermal cycler (Idaho Falls, ID) for 20 cycles. PCR product (399 bp) was alkali denatured, neutralized, and added to hybridization buffer containing 1 million RLU AquaLite labeled γIFN probe (18 mer). The hybridization was carried out at 45°C for 90 min in streptavidin coated microplate wells (MicroCoat, Penzburg, FRG). After washing, the wells were read for 1 sec in a Dynatech ML3000 microplate luminometer upon injection of 200 μL calcium trigger buffer.

Results and Discussion

Ruling out the use of a variety of coupling methods, we attempted to prepare AquaLite conjugates using heterobifunctional sulfhydryl:maleimide chemistry. This approach was fully expected to be problematic given that each of the three cysteine groups in aequorin likely plays a role in calcium dependent light production (3). Therefore, chemical modification of such critical sulfhydryl groups with maleimide groups would be expected to inactivate AquaLite. Derivatization of AquaLite with a heterobifunctional reagent such as SMCC which contains the primary amine reactive N-hydroxysuccinimide and the thiol reactive maleimide resulted in prompt inactivation of photoprotein activity (data not shown). This observation was consistent with the report that an aequorin fab fragment conjugate prepared using apoaequorin activated with SMCC retained only 10 % of the activity of underivatized aequorin (4).

Reversing the coupling strategy by reacting the sulfhydryl groups of AquaLite with a maleimido activated binding partner such as an antibody was also predicted to yield a conjugate of compromised photoprotein activity. We speculated that addition of exogenous sulfhydryl groups to AquaLite with Traut's reagent would help preserve photoprotein activity of resulting conjugates by reducing the likelihood of the coupling reaction occurring through the critical endogenous sulfhydryl groups. Table 1 illustrates the differences between antibody conjugates made with unmodified AquaLite and sulfhydryl activated AquaLite. In this case the antibody used was against human thyroid stimulating hormone (TSH) and the resulting conjugates were evaluated in a simultaneous sandwich immunoassay for TSH using serum calibrators of known concentration. These data demonstrate that the conjugate prepared with unmodified AquaLite exhibited inferior assay performance and had a specific activity 10 % that of the conjugate prepared with sulfhydryl activated AquaLite.

The addition, on the average, of one sulfhydryl group to AquaLite was found to preserve photoprotein activity of conjugates prepared with maleimido activated binding partners. The protective effect of additional sulfhydryl groups may be a result of reducing the probability that a critical sulfhydryl group reacts with the maleimido activated binding partner. Furthermore, the site at which the sulfhydryl group is added may be readily accessible to participate in the coupling reaction. This conjugation strategy has been shown to have universal applicability to antibodies, fab fragments, streptavidin, haptens, and oligonucleotides. The use of a variety of sulfhydryl and maleimide activating reagents in addition to SMCC and Traut's reagent has also been successfully demonstrated (data not shown).

Table 1. Comparison of Anti-TSH Antibody AquaLite Conjugates

Conjugate	Specific Activity[1] (RLU/OD)	RLU Bound at mIU/L TSH[2] 0	0.5	50
SH-AquaLite	4.2×10^{11}	232 (6.7)[3]	7163 (16.5)	333368 (2.1)
AquaLite[4]	3.1×10^{10}	266 (4.5)	1603 (7.2)	106068 (12.7)

[1]The specific activities were compared based on RLU per optical density (OD) unit at 280 nm.
[2]mIU/L is milli International Units per L.
[3]%CVs (coefficient of variation) of duplicate determinations are given in parenthesis.
[4]6.5 fold more of this conjugate was required for the TSH assays to have similar levels of input RLUs.

The utility and ease of use of conjugates of AquaLite with antibodies and streptavidin has been previously described (5,6,7). The performance characteristics of the tube based bioluminescent clinical TSH immunoassay (AquaLite® TSH) can best illustrate the sensitivity attained with such conjugates. This simultaneous immunometric assay has an analytical sensitivity (limit of detection) in the subnormal range at 0.001 mIU/L which is approximately equivalent to 0.9 attomol (10^{-18} mol). The functional sensitivity of the assay (defined as the concentration at which the interassay precision is ≤ 20 %) is 0.017 mIU/L which supports a third generation sensitivity claim (interassay CV = 20 % at 0.01-0.02 mIU/L) (7,8). A typical TSH calibration curve is shown in Figure 1.

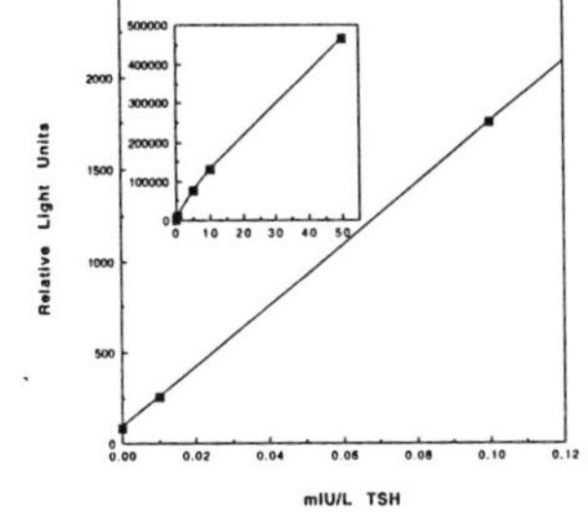

Figure 1. Typical calibration curve for AquaLite TSH assay.

Conjugates of AquaLite and antibodies directed against hapten labeled DNA probes such as digoxigenin, fluorescein, and dinitrophenol (DNP) are proving useful for quantitative hybridization immunoassays using streptavidin coated microplates to capture biotinylated target DNA. An AquaLite conjugate with sheep anti-digoxigenin fab fragment (Boehringer Mannheim, Indianapolis, IN) has been used for the quantitative detection of hepatitis B virus PCR product (9) and cytokine RT-PCR (reverse transcriptase) products (10). More recently, we have applied the conjugation strategy to the preparation of direct conjugates of AquaLite with 5'-amine labeled oligonucleotide probes. Because of their inherent sensitivity, both the indirect and direct conjugates can be used to detect PCR products during the linear phase of the amplification reaction. Figure 2 illustrates the use of an AquaLite labeled probe to detect γinterferon (γINF) PCR product (399 bp) after 20 cycles of amplification. In addition, AquaLite labeled probes can be used at hybridization temperatures of up to 50°C for 1-2 h without any significant loss in bound RLU signal.

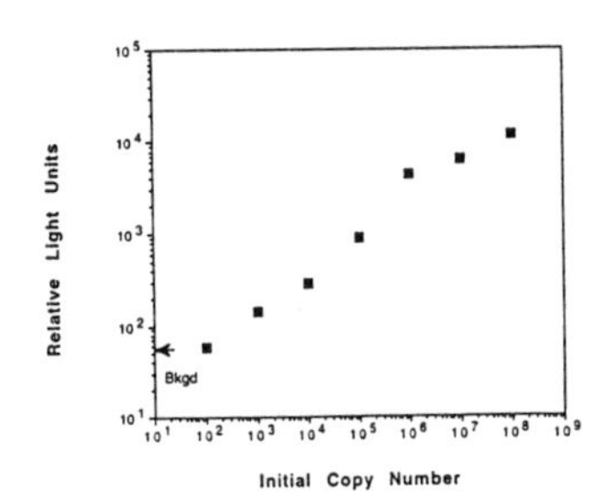

Figure 2. Quantitative detection of mouse γinterferon PCR product using AquaLite labeled probe.

The data in Table 2 show that, without any optimization, a highly functional anti-TSH antibody conjugate was prepared with charged obelin using the same sulfhydryl:maleimide chemistry used for the preparation of AquaLite conjugates. While the RLU signals are somewhat less than observed with a corresponding AquaLite conjugate, the assay is linear and has broad dynamic range. Since AquaLite and obelin share the same mechanism of light production and a high degree of homology, we expect this conjugation strategy to be successful for all calcium dependent photoproteins.

Table 2. Performance of Obelin Antibody Conjugate in TSH Assay

TSH (mIU/L)[1]	RLU Bound
0.0	126
0.1	437 (1.0)[2]
0.5	1653 (1.1)
5.3	16781 (7.6)
10	32549 (0.1)
50	213431 (1.3)

[1]mIU/L is milli International Units per L.
[2]%CVs are given in parenthesis.

Acknowledgements
We thank Dr. Eugene Vysotski for the gift of recombinant obelin, and Carla Pretto and Beverly Ahlburg for technical assistance. The polymerase chain reaction (PCR) process is covered by US patents issued to Hoffman-LaRoche.

References

1. Cormier MJ, Prasher DC, Longiaru M, McCann RO. The enzymology and molecular biology of the Ca^{2+}-activated photoprotein aequorin. Photochem Photobiol 1989;49:509-12.
2. Illarionov BA, Markova SV, Bondar, VS, Illarionova VA, Vysotski ES. Cloning, expression and nucleotide sequence investigation of apoobelin cDNA from the hydroid *Obelia longissima*. In: Szalay A, Kricka L, Stanley P, editors. Bioluminescence and Chemiluminescence: Status Report Chichester, UK, Wiley and Sons 1993;183-5.
3. Kurose K, Inouye S, Sakaki Y, Tsuji FI. Bioluminescence of the Ca^{2+}-binding photoprotein aequorin after cysteine modification. Proc Natl Acad Sci 1989;86:80-4.
4. Erikaku T, Zenno S, Inouye, S. Bioluminescent immunoassay using a monomeric fab'-photoprotein aequorin conjugate. Biochem Biophys Res Comm 1991;174:1331-6.
5. Smith DF, Stults N, Mercer WD. Bioluminescent immunoassays using streptavidin and biotin conjugates of recombinant aequorin. Am Biotech Lab April 1995.
6. Smith DF, Stults NL. AquaLite, a bioluminescent label for immunoassay and nucleic acid detection: quantitative analyses at the attomol level. In: Cohn GE, Soper SA, Chen CHW, editors. Ultrasensitive Biochemical Diagnostics. Proc SPIE 1996;2680:156-66.
7. Rigl CT, Rivera HN, Patel MT, Ball RT, Stults NL, Smith DF. Bioluminescence immunoassays for human endocrine hormones based on AquaLite®, a calcium activated photoprotein. Clin Chem 1995;41:1363-4.
8. Sgoutas DM, Tuten TE, Verras AA, Love A, Barton EG. AquaLite® bioluminescence assay of thyrotropin in serum evaluated. Clin Chem 1995;41:1637-43.
9. Stults NL, Rivera HN, Ball RT, Smith DF. Bioluminescent Hybridization Immunoassay for digoxigenin labeled PCR products based on AquaLite®, a calcium activated photoprotein. J NIH Res 1995;7:74.
10. Siddiqi A, Jennings VM, Kidd MR, Actor JK, Hunter RL. Evaluation of electrochemiluminescence-based assays for quantitating specific DNA. J Clin Lab Anal 1996;In press.

CHARACTERIZATION OF RECOMBINANT OBELIN AS AN INTRACELLULAR Ca^{++} INDICATOR

VA Illarionova[1], BA Illarionov[1], VS Bondar[1], ES Vysotski[1] and JR Blinks[2]
[1]Institute of Biophysics, Russian Academy of Sciences, Siberian Branch, Krasnoyarsk 660036, Russia
[2]Friday Harbor Laboratories, University of Washington, Friday Harbor, WA 98250, USA

Introduction

To date, six Ca^{++}-regulated photoproteins (1) have been isolated and studied from organisms in the phyla *Cnidaria* and *Ctenophora* - aequorin, halistaurin, obelin, phialidin, mnemiopsin, and berovin. Of these, only aequorin has been used widely as a biological calcium indicator so far, primarily because of the scarcity or instability of the other photoproteins (2). Although aequorin has been used successfully to measure cytosolic free calcium concentration in a great variety of animal and plant cells, it does have a number of disadvantages that have limited its utility. Among the most important of these are that aequorin is too slow to follow the most rapid intracellular calcium transients without distortion, and that physiological concentrations of magnesium antagonize the effects of calcium and slow the kinetics even further. Other photoproteins do not necessarily share these defects in equal measure, and with application of recombinant technology, it should be possible to overcome previous limitations of supply. During the past ten years, cloning and sequence analysis have been carried out on cDNAs for four photoproteins: aequorin (3,4), clytin (phialidin) (5), mitrocomin (halistaurin) (6), and obelin (7). In principle, any of these photoproteins could now be used as biological calcium indicators. However, to select among them one needs information about some of their key properties, particularly calcium sensitivity, the speed of response to sudden changes of [Ca^{++}], and the influence of Mg^{++} on both of these characteristics. Here we report the results of studies of these properties carried out on recombinant obelin produced with a cDNA isolated from the hydroid *Obelia longissima.*

Materials and Methods

Materials. Coelenterazine was obtained from Molecular Probes Inc. Dithiothreitol (DTT) and Triton X-100 were purchased from Sigma Chemical Co. BioGel P6 and Chelex 100 resins were obtained from BioRad. EGTA and PIPES were purchased from Fluka.

Growth of *E.coli* cells and expression of recombinant obelin. The two-plasmid system was used to express the gene encoding apoobelin essentially as described (8). *E.coli* C600/pGP1-2, containing the plasmid pOL110, was grown as described (9).

Purification and activation of recombinant obelin. *E.coli* cells were disrupted by sonication at 0°C in 20 mmol/L Tris-HCl buffer pH 7.0. The suspension was centrifuged and the supernatant discarded. The pellet was sequentially washed with solutions: (a) 150 mmol/L NaCl, 20 mmol/L Tris-HCl pH 7.0 (×2); (b) 1% Triton X-100, 20 mmol/L Tris-HCl pH 7.0 (×3); (c) 5 mmol/L $CaCl_2$, 20 mmol/L Tris-HCl pH

7.0 (×1). To extract the apoobelin the final pellet was suspended in 6 mol/L urea, 5 mmol/L $CaCl_2$, 10 mmol/L DTT, 20 mmol/L Tris-HCl pH 7.0 and stored at 4°C overnight, then centrifuged. The pellet was discarded, and the supernatant was used for apoobelin purification. High purity apoobelin (according to SDS-PAGE) was obtained by chromatography on a Mono P HR 5/20 column (Pharmacia, Sweden). The apoobelin was converted to active obelin by incubation with 6 μmol/L coelenterazine at 4°C overnight in the presence of 25 mmol/L EDTA and 10 mmol/L DTT. EDTA was removed from the obelin solution by running the sample through a 0.9 x 30 cm bed of BioGel P6 gel filtration medium (Pharmacia K9/30 plastic column) equilibrated and eluted with 150 mmol/L KCl, 5 mmol/L PIPES, pH 7.0, which had previously passed (×3) through freshly washed beds of Chelex-100 chelating resin. The fractions containing photoprotein were identified by their ability to emit light on the addition of calcium. The early fractions containing photoprotein were used for the determination of Ca^{++} concentration-effect curves and for kinetic measurements.

Ca^{++} concentration-effect curve and rapid-mixing kinetic measurements. Ca-EGTA buffers were used for Ca^{++} concentrations below 10^{-5} mol/L; simple dilutions of $CaCl_2$ (in Chelex-treated solution of 150 mmol/L KCl, 5 mmol/L PIPES pH 7.0) for higher Ca^{++} concentrations. When measurements were to be made in solutions containing Mg^{++}, the obelin was pre-equilibrated with the same $[Mg^{++}]$. For the Ca^{++} concentration-effect curves, peak light intensity was measured after 10 μL of photoprotein solution was forcefully injected into 1 mL of the test solution, as previously described (2). The kinetics of light responses after sudden exposure to higher Ca^{++} concentration were examined with a rapid-mixing stopped-flow apparatus similar to that described by Gibson and Milnes (10). Dead time was approximately 2 ms at the air pressure used. EDTA-free photoprotein samples were mixed with solutions containing 10 mmol/L $CaCl_2$, so that the mixed solution had a $[Ca^{++}]$ of 5 mmol/L, which is saturating. All measurements were made at 20°C. Light was conducted through fiberoptic probes to an EMI 9635B photomultiplier, the output of which was recorded with a digital oscilloscope. The oscilloscope was triggered at the moment when flow was stopped (time zero on the horizontal axis), but it was set to recall and display the photomultiplier signal for a brief period before the triggering signal. Thus the display includes the rise of luminescence as mixing begins, and a brief period of steady luminescence reflecting constant-velocity flow (see Figure 1a).

Results and Discussion

Figs. 1a&b show the time course of the light response from recombinant obelin after the photoprotein was suddenly exposed to a saturating Ca^{++} concentration. The rise of light (Fig.1a) after flow is stopped can be accurately described by a single exponential with a rate constant of 379 s^{-1}. This is almost 4 times the rate for aequorin under the same conditions. The decay of luminescence after the peak (the consumption of obelin) is subject to more complex laws: at least three phases can be distinguished on a semi-logarithmic plot (Fig.1b). Even in a concentration as high as 10 mmol/L, Mg^{++} had almost no effect on the rise of light emission (rate constant 331 s^{-1}) (Fig.1a). But Mg^{++} was not without effect on recombinant obelin. 10 mmol/L Mg^{++} reduced the peak light emission by more than half (Fig.1b). This decrease did not reflect a reduction in the quantum yield: the reduction in peak light emission was accompanied by a prolongation of the declining phase of the flash; the total light yield was not significantly altered.

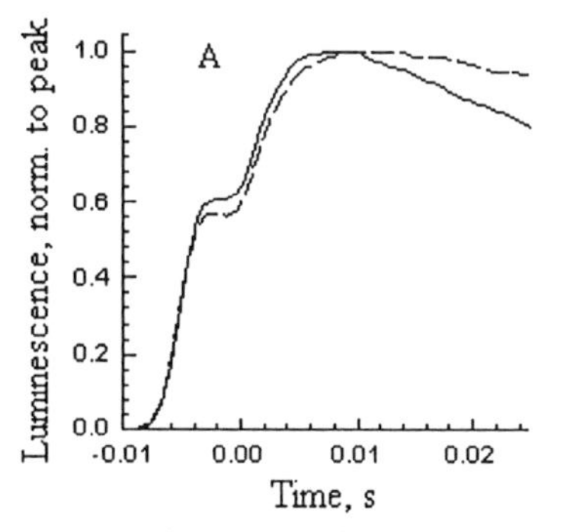

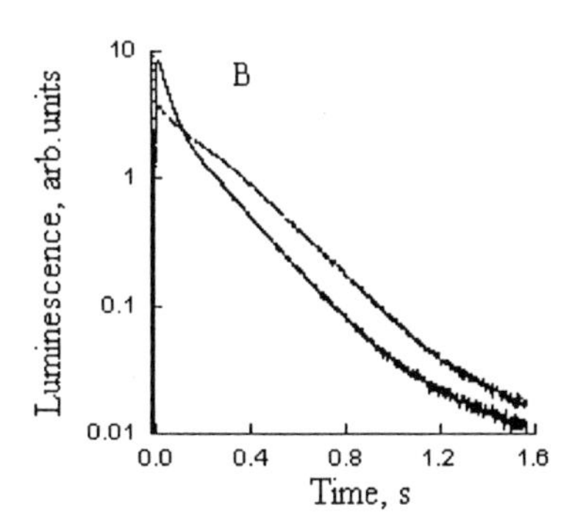

Figure 1. Time course of light emission after rapid mixing of recombinant obelin and calcium. The solid and dashed lines represent the light response of recombinant obelin without and with 10 mmol/L Mg^{++} respectively. Note in Fig.1a that flow is stopped at time zero; the initial rise of luminescence is registered during the period in which mixing begins; the plateau at about 60% of peak light intensity reflects constant flow velocity; the last phase of the rise is the portion of interest.

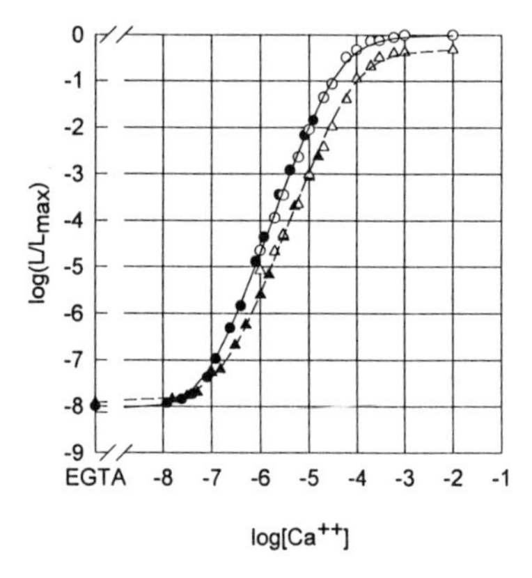

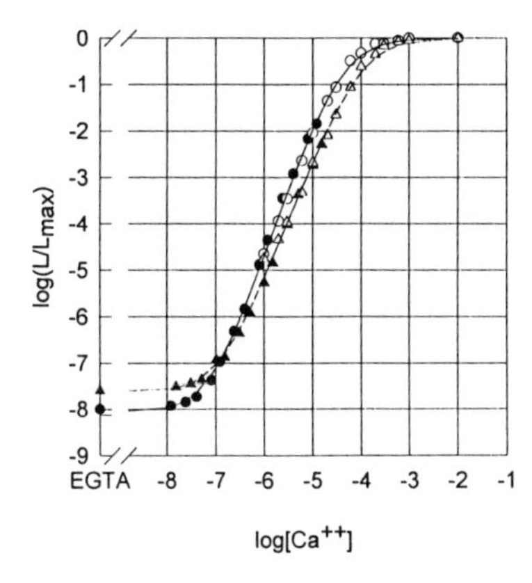

Figure 2. Ca^{++} concentration-effect curve for recombinant obelin. The two panels show the same results plotted in two ways. In the left panel (Fig.2a), both sets of measurements (with and without Mg^{++}) are plotted in relation to Lmax recorded in the absence of Mg^{++}. In the right panel (Fig.2b), each curve is plotted in relation to the peak luminescence (L_{max}) determined in that curve. Since Mg^{++} substantially reduces L_{max} for obelin, the lower portion of the curve is artificially elevated if the reduced L_{max} is used in the calculation of L/L_{max}.
(o) - without Mg^{++}; (Δ) - 10 mmol/L Mg^{++}. Filled symbols: Ca-EGTA buffers (2 mmol/L); open symbols: dilutions of $CaCl_2$. The lines were drawn with the two-state model of Allen et al. (11).

The Ca^{++} concentration-effect curve is sigmoid on a log-log plot, with maximum slope about 2.5, and spans 8 decades of light intensity. Note the very low level of Ca^{++}-independent luminescence for recombinant obelin. The two-state model (11) provided a good fit to the data with $K_R=1.13\times10^7$ M^{-1} and $K_{TR}=362$ in the absence of Mg^{++}. One desirable characteristic of a photoprotein with respect to its potential use as an intracellular Ca^{++} indicator is a lack of sensitivity to Mg^{++}. Figure 2 demonstrates the effect of 10 mmol/L Mg^{++} on the Ca^{++} concentration-effect curve of recombinant obelin. This relatively high concentration of Mg^{++} has no effect on the Ca^{++}-independent luminescence, and very little effect on the sensitivity of the photoprotein to Ca^{++}, but reduces L_{max} by more than half.

Acknowledgments. This work was supported by grant 96-04-48489 from the Fundamental Research Foundation of the Russian Academy of Sciences and NIH grants HL12186 and TW00412. The authors are grateful to Michelle Woodbury for expert technical assistance.

References

1. Blinks JR, Wier WG, Hess P, Prendergast FG. Measurement of Ca^{2+} concentration in living cells. Prog Biophys Molec Biol 1982; 40:1-114.
2. Blinks JR. Use of calcium-regulated photoproteins as intracellular Ca^{2+} indicator. Methods Enzymol 1989;172:164-203.
3. Prasher D, McCann RO, Cormier M.J. Cloning and expression of the cDNA coding for aequorin, a bioluminescent calcium-binding protein. Biochem Biophys Res Comm 1985;126:1259-68.
4. Inouye S, Noguchi M, Sakaki Y, Takagi Y, Miyata T., et al. Cloning and sequence analysis of cDNA for the luminescent protein aequorin. Proc Natl Acad Sci USA 1985;82:3154-8.
5. Inouye S, Tsuji FI. Cloning and sequence analysis of cDNA for the Ca^{2+}-activated photoprotein, clytin. FEBS Lett 1993;315:343-6.
6. Fagan TF, Ohmiya Y, Blinks JR, Inouye S, Tsuji FI. Cloning, expression and sequence analysis of cDNA for the Ca^{2+}-binding photoprotein, mitrocomin. FEBS Lett 1993;333:301-5.
7. Illarionov BA, Bondar' VS, Illarionova VA, Vysotski ES. Sequence of the cDNA encoding the Ca^{2+}-activated photoprotein obelin from the hydroid polyp *Obelia longissima.* Gene 1995;153:273-4.
8. Tabor S. Current Protocols in Molecular biology. New York, Green Publishing and Wiley Interscience, 1995:16.2.1-16.2.5.
9. Frank LA, Illarionova VA, Vysotski ES. Use of proZZ-obelin fusion protein in bioluminescent immunoassay. Biochem Biophys Res Comm 1996;219:475-9.
10. Gibson QH, Milnes L. Apparatus for rapid and sensitive spectrophotometry. Biochem J 1964;91:161-71.
11. Allen DG, Blinks JR, Prendergast FG. Aequorin luminescence: relation of light emission to calcium concentration - a calcium-independent component. Science 1977;195:996-8.

REMOVAL OF ESSENTIAL LIGAND IN N-TERMINAL CALCIUM-BINDING DOMAIN OF OBELIN DOES NOT INACTIVATE THE PHOTOPROTEIN OR REDUCE ITS CALCIUM SENSITIVITY, BUT DRAMATICALLY ALTERS THE KINETICS OF THE LUMINESCENT REACTION

VA Illarionova[1], BA Illarionov[1], VS Bondar[1], ES Vysotski[1] and JR Blinks[2]
[1]Photobiology Lab, Institute of Biophysics, Krasnoyarsk 660036, Russia
[2]Friday Harbor Labs, Univ. of Washington, Friday Harbor WA 98250, USA

Introduction

The Ca-regulated photoproteins obtained from various luminous coelenterates emit blue light at a rate that varies with [Ca^{2+}], without involvement of any other cofactors (1). Because the reaction is sensitive to calcium and the photoproteins are nontoxic in most biological systems, Ca^{2+}-regulated photoproteins such as aequorin and obelin have been widely used as intracellular Ca^{2+} indicators (2). Although the organic chemistry of the luminescent reaction of Ca^{2+}-regulated photoproteins has been largely established, it is still not clear how the binding of calcium regulates that reaction. One essential question has to do with the stoichiometry of calcium binding. The log-log plot of light intensity as a function of calcium concentration has a maximum slope of about 2.5, implying that more than two calcium ions are involved in the regulation of light emission (2). That the number is three was strongly suggested by the findings first that the amino acid sequence of native aequorin contains three stretches corresponding to Ca^{2+}-binding sites of the EF hand type (3), and later that the primary protein structures deduced from the nucleotide sequences of cDNA coding for several other photoproteins also contain three EF-hand sites (4). However, it has recently been suggested that only two Ca^{2+}-binding sites are involved in the regulation of aequorin luminescence (5-6). A powerful approach to determining the importance of the three putative Ca^{2+}-binding sites is to modify them in specific ways by site-directed mutagenesis, and to study the effects of these changes on the properties of the photoprotein. Here we report the effects of a modification that would be expected to greatly reduce the calcium affinity of the binding site closest to the N-terminus of recombinant obelin.

Materials and Methods

Instrumentation: An FPLC system from Pharmacia (Sweden) was used for protein purification. Rapid-mixing stop-flow measurements were carried out with an apparatus similar in principle to that described by Gibson and Milnes (7). The photometer used for the determination of Ca^{2+} concentration-effect curves is illustrated in ref. 11 (Fig. 7).

Materials: Coelenterazine was obtained from Molecular Probes, Inc. Restriction enzymes and T4 DNA ligase were purchased from Fermentas (Vilnius, Lithuania). The oligonucleotide 5'-ACAATTCCAGCGAGG-3' that was designed for mutagenesis of apoobelin cDNA was purchased from Vektor (Novosibirsk, Russia).

Procedures: Site-directed mutagenesis of apoobelin cDNA was performed according to Kunkel (8). DNAs for wild type and mutant apoobelin were expressed in *E.coli* by means of the T7 RNA polymerase/promoter system (9). Cultivation of recombinant *E.coli* and isolation of apoobelin were carried out essentially as described previously (10). The apoobelin was purified in two stages by FPLC on a Mono P HR 5/20 column (Pharmacia); the two chromatography runs differed in that 6 mol/L urea was present in all solutions during the first, while the urea was omitted in the second. In other respects the two runs were the same; all solutions contained 5 mmol/L $CaCl_2$, 5 mmol/L dithiothreitol, and 20 mmol/L Tris-HCl (pH 7.0), and the protein was eluted from the column with a sodium acetate gradient ascending from 0 to 0.5 mol/L. Almost all of the protein in the product ran as single band on SDS-PAGE. The procedures for charging the apoprotein with coelenterazine(10), preparation and handling of chelator-free photoprotein (11), and determination of Ca^{2+} concentration-effect curves (11) were as described earlier.

Results and Discussion

The majority of EF-hand-type Ca^{2+}-binding sites contribute seven oxygen ligands to coordinate the metal ion. Among those ligands the invariant glutamate in the "-Z" position of the EF loop coordinates the Ca^{2+} with a bidentate side chain carboxylic group (12). Hence the substitution of glycine for the glutamate ought to lead to a considerable change in the affinity of the site for calcium. This substitution was generated in the N-terminal Ca^{2+}-binding domain of apoobelin as depicted in Fig. 1. To avoid the possibility that aspartate-40 might mimic the function of glutamate-41, Asp40 was replaced with Ala by means of the same oligonucleotide.

```
30(+X)                                      40  41(-Z)
-Asp Ile Asn Gly Asn Gly Lys Ile Thr Leu Asp Glu-
                                          ↓   ↓
                                         Ala Gly
```

Figure.1. The amino acid substitutions generated in the N-terminal Ca^{2+}-binding site of apoobelin. The numbers indicate the position of corresponding amino acid in apoobelin amino acid sequence.

When purified and charged with coelenterazine under standardized conditions, the resultant variant (mutant) of obelin had a total light yield approximately 40% of that of wild type obelin. For a given total light yield, the calcium-independent luminescence was increased 13-fold in the mutant, while peak light in saturating $[Ca^{2+}]$ (Lmax) was reduced by a factor of 20. The kinetics of light emission after rapid mixing with saturating $[Ca^{2+}]$ (5 mmol/L) were strikingly altered by the mutation (see Fig. 2). The risetime was essentially unchanged (rate constants of 379 s^{-1} for wild type obelin and 362 s^{-1} for the mutant), but the decline became strikingly biphasic, with a rapid initial decline (halftime ~50ms) to about 30% of the peak, followed by a much slower one (halftime 5.6 s). The fact that photoprotein utilization takes place in two very distinct steps in the mutant suggests that a conformational transition required for luminescence takes place very much more slowly in the mutant than in the wild-type photoprotein. This may reflect a great reduction in the binding of calcium to the N-terminal binding site. In other words, this site may be contributing to the overall luminescent reaction as a normal site would in the presence of a very much lower calcium ion concentration.

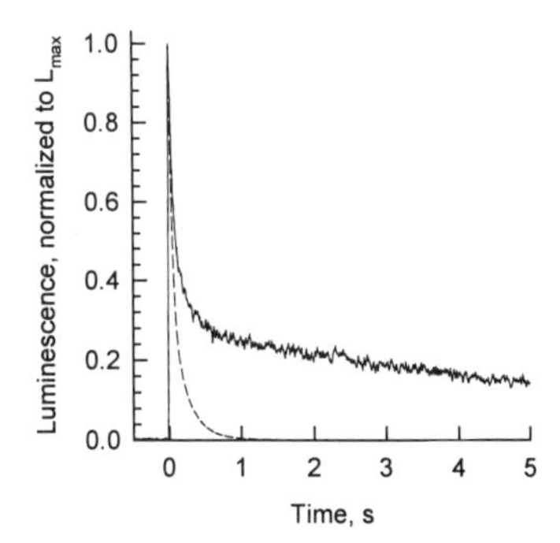

Figure 2. Time course of light emission after rapid mixing of calcium with either recombinant wild type obelin (dashed line) or mutant obelin (solid line).

Figure 3 shows the Ca^{2+} concentration-effect curves for the two photoproteins. Both curves are sigmoid on a log-log plot and have three phases: a Ca^{2+}-independent phase, a logarithmic phase, and a saturating phase. The curves span about eight orders of magnitude of light intensity for wild-type obelin, and six orders for the mutant. In the logarithmic phase the Ca^{2+} concentration-effect curve has a maximum slope of 2.5 for the wild-type and 1.74 for the mutant obelin. The difference in the slopes of the curves may indicate that three calcium ions initiate the luminescent reaction of the wild-type obelin and two Ca^{2+} initiate the luminescence of the mutant obelin. However, it should be noted that in spite of a reduction of the steepness of the Ca^{2+} concentration-effect curve for the mutant obelin, and a great reduction in peak luminescence, the range of calcium concentrations over which luminescence increased was approximately the same in the mutant as in the wild-type obelin (Fig. 3). 10 mmol/L Mg^{2+} had little effect on Ca^{2+} sensitivity or risetime in either protein, but substantially decreased Lmax and slowed decay in both (data not shown).

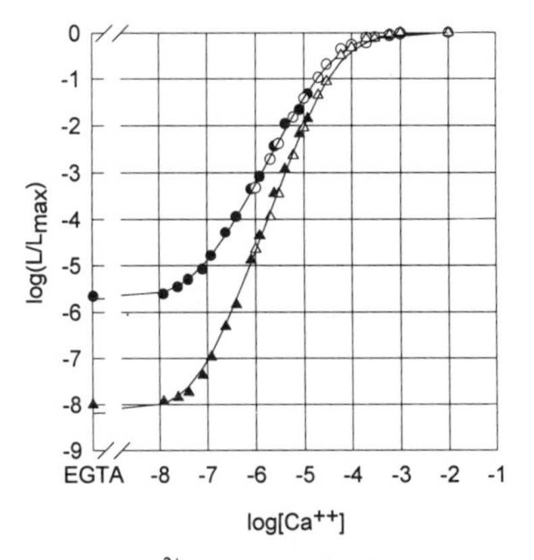

Figure 3. Ca^{2+} concentration-effect curve for recombinant wild type obelin (Δ) and mutant obelin (o). Filled symbols: Ca-EGTA buffers (2 mmol/L), open symbols: dilutions of $CaCl_2$.

Our results are consistent with the expectation that removal of the "-Z" ligand from the N-terminal EF-loop of obelin would greatly reduce the affinity of this loop for Ca^{2+}. They also make it quite clear that the N-terminal binding site plays an important role in the regulation of obelin luminescence.

Acknowledgements
This work was supported by grants HL12186 and TW00412 from the National Institutes of Health, and grant 96-04-48489 from the Fundamental Research Foundation of the Russian Academy of Sciences. The authors are grateful to Michelle Woodbury for expert technical assistance.

References

1. Shimomura O, Johnson FH, Saiga Y. Extraction, purification and properties of aequorin, a bioluminescent protein from the luminous hydromedusan, Aequorea. J Cell Comp Physiol 1962; 59: 223-40.
2. Blinks JR, Prendergast FG, Allen DG. Photoproteins as biological calcium indicators. Pharmacol Rev 1976; 28: 1-93.
3. Charbonneau H, Walsh KA, McCann RO, Prendergast FG, Cormier MJ, Vanaman TC. Amino acid sequence of the calcium-dependent photoprotein aequorin. Biochem 1985; 24: 6762-71.
4. Tsuji FI, Ohmiya Y, Fagan TF, Toh H, Inouye S. Molecular evolution of the Ca^{2+}-binding photoproteins of the hydrozoa. Photochem Photobiol 1995; 62: 657-61.
5. Shimomura O. Luminescence of aequorin is triggered by the binding of two calcium ions. Biochem Biophys Res Commun 1995; 211: 359-63.
6. Shimomura O, Inouye S. Titration of recombinant aequorin with calcium chloride. Biochem Biophys Res Commun 1996; 221: 77-81.
7. Gibson QH, Milnes L. Apparatus for rapid and sensitive spectrophotometry. Biochem J 1964; 91: 161-71.
8. Kunkel TA. Rapid and efficient site-specific mutagenesis without phenotypic selection. Proc Natl Acad Sci USA 1985; 82:488-92.
9. Tabor S, Richardson CC. A bacteriophage T7 RNA polymerase/promoter system for controlled exclusive expression of specific genes. Proc Natl Acad Sci USA 1985; 82: 1074-8.
10. Bondar VS, Sergeev AG, Illarionov BA, Vervoort J, Hagen WR. Cadmium-induced luminescence of recombinant photoprotein obelin. Biochim Biophys Acta 1995; 1231: 29-32.
11. Blinks JR Use of calcium-regulated photoproteins as intracellular Ca^{2+} indicators. Methods Enzymol 1989; 172: 164-203.
12. Strynadka NCJ, James MNG Crystal structures of the helix-loop-helix calcium binding proteins. Annu Rev Biochem 1989; 58: 951-98.

OBELIN AS A CARRIER PROTEIN AND REPORTER ENZYME FOR IN VITRO SYNTHESIZED SMALL BIOACTIVE POLYPEPTIDES: NEW APPROACH TO OBTAIN SARCOTOXIN

BA Illarionov[1], SV Matveev[2], VS Bondar[1], VA Skosyrev[2] and YB Alakhov[2]
[1]Photobiology Lab, Institute of Biophysics, Krasnoyarsk 660036, Russia
[2]Pushchino Branch of Institute Bioorganic Chemistry, Pushchino 142292, Russia

Introduction

The use of gene fusion expression systems has become an increasingly popular method of producing different proteins in different systems. The maltose-binding protein (1), glutathione-S-transferase (2), thioredoxin (3), and synthetic “ZZ” domain based on the “B” IgG-binding domain of protein A (4) - all these fusion systems have proven successful in producing biologically active proteins. However, no one of them provides a method for rapid monitoring of protein accumulation.

Sarcotoxin is a small 39mer polypeptide produced by a larvae of flesh-fly *Sarcophaga peregrina* to prevent an uncontrolled propagation of bacteria. The folded polypeptide monomer is comprised from two amphiphilic α-helixes (5). Several associated monomers supposed to be assembled in bacterial membrane facing with their hydrophobic sides inside the membrane and with hydrophilic ones inside the complex, thus forming a hydrophilic hole. The attempts to obtain recombinant sarcotoxin from bacteria or eukaryotic cultured cells were unsuccessful due to its properties. Moreover, the attempt to synthesize sarcotoxin in cell-free system was also unsuccessful probably due to aggregation of translation product and inhibition of translation system (Matveev S, Alakhov Y, unpublished observations).

It was shown earlier that Ca^{2+}-activated photoprotein obelin can be used as a reporter of effectivity of *in vitro* translation (6). Obelin retained its luminous activity when bearing genetically engineered polypeptide redundancy at its N-terminus (7). The purpose of this work was to improve the translation effectivity of sarcotoxin and to develop a method to test its accumulation in cell-free translation system as part of a fusion protein with obelin.

Materials and Methods

Instrumentation: The luminescence was measured with luminometer BLM 8801 (SKTB Nauka, Russia).

Materials: Coelenterazine was obtained from Molecular Probes, Inc. The SP6 RNA polymerase and Pvu II restriction enzyme were generous gifts from Dr. N.I. Matvienko. All other restriction enzymes and T4 DNA ligase were obtained from Fermentas (Vilnius, Lithuania) or from Sibenzyme (Novosibirsk, Russia). [^{14}C]Leu (specific activity 319 mCi/mmol) was from Amersham.

Procedures: Gene for fusion protein sarcotoxin/apoobelin was prepared as follows. The C-terminus of synthetic gene for sarcotoxin from pDS10 was modified by means of PCR so that stop codon of sarcotoxin gene was replaced by ATG one and ClaI restriction site was generated just downstream from it. After cloning in pGem-T (Promega, USA), the expected nucleotide sequence of PCR product was confirmed

(8). The gene for apoobelin was inserted downstream from the modified gene for sarcotoxin using ClaI and SphI restriction sites. The resulting plasmid was designated pSO13. The *in vitro* synthesis of mRNA on either pNOV (6) or pSO13 was performed as described earlier (9). The wheat germ extract (BBL, Latvia) was used for *in vitro* translation (6). Sodium dodecyl sulfate-polyacrylamide gel electrophoresis (SDS-PAGE) (10) was used to separate and analyse the *in vitro* synthesized polypeptides. In 60 minutes after start of translation, five microliters from translation mixture containing [^{14}C]Leu-labeled proteins were placed in a well of 12.5% gel slab. After electrophoresis was finished the gel was stained, dried and exposed to X-ray film. The regions of the gel that corresponded to the radioactive protein bands as well as remained PAAG from each lane were cut out, soaked in an absolute ethanol for overnight, and radioactivity was counted in standard toluene-POP-POPOP mixture. The calculation of amount of *de novo* synthesized protein, the reactivation of apoobelin as well as fusion protein, and luminescence measurements were performed as described earlier (6). The binding of the fusion protein to bacteria was tested using *Escherichia coli* K37 or *Bacillus subtilis* cells taken at a logarithmic phase of growth (OD_{590}=0.5). Cells were centrifuged, washed with 0.9% NaCl and resuspended in 0.9% NaCl to a concentration of 10^7 cells/ml. 30 μl of translation mixture containing reactivated obelin or fusion protein was added to 0.5 ml of each sample of cell suspension or to a 0.5 ml of 0.9% NaCl (controle), shaken and incubated 10 minutes at room temperature. Then cells were span down at 3500 rpm, room temperature. Neutral EDTA solution was added to supernatant to final concentration of 5 mmol/L. Pellet was washed once with 0.5 ml of 0.9% NaCl, 0.3 mmol/L EDTA, and centrifuged again. EDTA solution was added to supernatant to 5 mmol/L and both supernatant samples were combined. Pellet was resuspended in100 μl of reaction buffer (100 mmol/L Tris-HCl, pH 8.8, 10 mmol/L EDTA). Supernatant and pellet fractions were used immediately for luminescence measurement.

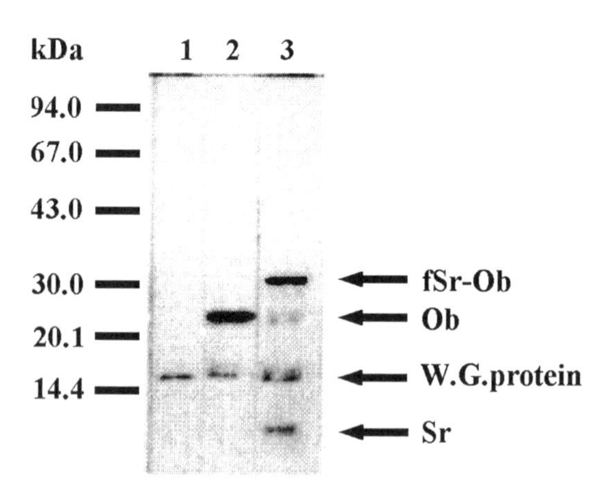

Figure 1. Radioautograph of PAAG after electrophoresis of translation products. No mRNA added to translation mixture (lane 1), apoobelin mRNA (lane 2), fusion protein mRNA (lane 3).

Results and Discussion

The mRNAs for apoobelin (synthesized on pNOV) or fusion protein (pSO13) were used for protein synthesis in wheat germ cell-free translation system. The products were analysed by SDS-PAGE (Fig. 1). It is obvious that translation system contained some endogenous mRNA that gives the protein band (W.G. protein) of 16 kDa in each lane. Apoobelin mRNA directed the synthesis of protein (Ob) with molecular mass of 23 kDa that is in good agreement with calculated apoobelin molecular mass (11). There are two additional protein bands in the lane 3. The fact that the upper band (fSr-Ob) of 30.5 kDa is somewhat higher than calculated molecular mass of sarcotoxin/apoobelin fusion protein (27 kDa) may be explained by abundance of charged amino acids in sarcotoxin. This band represents about 60% of total synthesized protein. The second band of 23 kDa represents apoobelin. The third band (Sr) is of approximately 6 kDa that is in good agreement with calculated molecular mass of sarcotoxin (4.6 kDa). Because there are no open reading frames located

Table. Synthesized proteins and their luminescence.

Band	Count in band, cpm/µl	Concentration of protein, corrected on endogenous Leu, pmol/µl of translation mix.	Luminescence of proteins synthesized and reactivated in translation mixture		Specific luminescent activity of protein, normalized to obelin specific activity
			LU/µl	LU/pmole	
Fig. 1, lane 1					
W.G. protein	82.7±8.2			0.0	0.0
Total in line	120.5±6.0		0.0		
Fig. 1, lane 2					
Obelin	313.5±8.4	720±19		1.71±0.01	1.00
W.G. protein	80.7±8.2				
Total in line	430±4.3		1230±12		
Fig. 1, lane 3					
f Sr-Ob	143±4.2	287±9		0.59±0.02	0.35
Obelin	20.1±6.1	46±12		1.14±0.01	1.00
W.G. protein	73.0±8.4				
Sr	21.5±6.1	150±40			
Total in line	292±6.1		222.0±8		

downstream from SP6 promoter that would encode the polypeptide of similar molecular mass, and apoobelin mRNA does not direct the synthesis of this protein (see Fig. 1, lane 2), it is more than likely that this band represents sarcotoxin. The amount of synthesized protein from each translation mixture as well as luminescence were determined. The results are shown in the Table. Obviously the fusion protein is synthesized in cell-free translation mixture effectively, however not as effectively as apoobelin alone. Its luminescence activity is comparable to that of obelin. The reactivated fusion protein was tested for binding to *E.coli* or *B.subtilis* cells (Fig. 2).

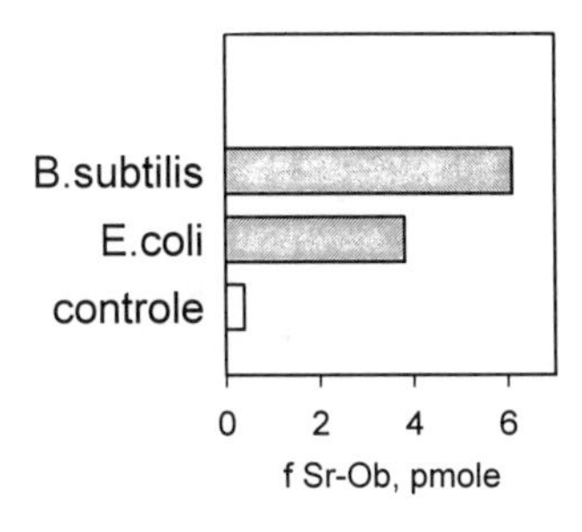

Figure 2. The amount of reactivated sarcotoxin/obelin fusion protein bound to 5×10^6 bacterial cells.

The binding of fusion protein to *B.subtilis* is as twice as higher than to *E.coli*.These results illustrated that in contrast to previous attempts to synthesize sarcotoxin in cell-free system, the described approach turned out to be successful, giving substantial amount of sarcotoxin as part of the fusion protein. The resulting fusion protein retained the capabilities to emit light and to bind to bacterial cells. The former property is very useful for monitoring of fusion protein accumulation in cell-free system.

Acknowledgments

The authors thank Viktoria Illarionova for expert technical assistance in cloning of PCR product and Dr. Eugene Vysotski for helpful discussion of subject of investigation. This work was supported in part by the grant from Ministery of Science of Russia.

References

1. Guan C, Li P, Riggs PD, Inouye H. Vectors that facilitate the expression and purification of foreign peptides in *Escherichia coli* by fusion to maltose-binding protein. Gene 1987; 67: 21-30.
2. Smith DB, Johnson KS. Single-step purification of polypeptides expressed in *Escherichia coli* as fusions with glutathione S-transferase. Gene 1993; 4: 220-9.
3. LaVallie ER, DiBlasio EA, Kovacic S, Grant KL, Schendel PF, McCoy JM. A thioredoxin gene fusion system that circumvents inclusion body formation in the *E.coli* cytoplasm. Bio/Technology 1993; 11: 187-93.
4. Nilsson B, Moks T, Jansson B, Abrahamsen L, Emblad A, Holmgren E, et al. A synthetic IgG-binding domain based on staphylococcal protein A. Prot Engineering 1987; 1: 107-13.
5. Iwai H, Nakajima Y, Natori S, Arata Y, Shimada I. Solution conformation of an antibacterial peptide, sarcotoxin I A, as determined by ^{1}H-NMR. Eur J Biochem 1993; 217: 639-44.
6. Matveev SV, Illarionov BA, Vysotski ES, Bondar VS, Markova SV, Alakhov YB. Obelin mRNA - a new tool for studies of translation in cell-free systems. Anal Biochem 1995; 231: 34-9.
7. Frank LA, Illarionova VA, Vysotski ES. Use of proZZ-obelin fusion protein in bioluminescent immunoassay. Biochem Biophys Res Commun 1996; 219: 475-9.
8. Maxam AM, Gilbert W. Sequencing end-labeled DNA with base-specific chemical cleavages. Methods Enzymol 1980; 100: 499-560.
9. Gurevich VV, Pokrovskaya ID, Obukhova TA, Zozulya SA. Preparative *in vitro* mRNA synthesis using SP6 and T7 RNA polymerases. Anal Biochem 1991; 195: 207-13.
10. Laemmly UK. Cleavage of structural proteins during the assembly of the head of the bacteriophage T4. Nature (London) 1970; 227: 680-5.
11. Illarionov BA, Bondar VS, Illarionova VA, Vysotski ES. Sequence of the cDNA encoding the Ca^{2+}-activated photoprotein obelin from hydroid polyp *Obelia longissima*. Gene 1995; 153: 273-4.

BIOLUMINESCENT IMMUNOASSAY OF ALPHAFETOPROTEIN WITH Ca^{2+}-ACTIVATED PHOTOPROTEIN OBELIN

LA Frank and ES Vysotski
Institute of Biophysics, Russian Academy of Sciences, Siberian Branch, Krasnoyarsk 660036, Russia

Introduction

Obelin is a Ca^{2+}-activated bioluminescent photoprotein that has been isolated from marine polyp *Obelia longissima* (1). The luminescent reaction is initiated by Ca^{2+}-binding to protein. The protein is built up from a single polypeptide chain of a relatively small size (M_r = 22.2 kDa) and presents a complex of apoobelin, coelenterazine and oxygen. Like in case of a well studied photoprotein aequorin, one may suggest the luminescent reaction to be a result of an intramolecular reaction of coelenterazine oxidation, yielding coelenteramide, CO_2 and blue light (2). The cloning and sequence analysis of cDNA for apoobelin have revealed obelin to consist of 195 amino acid residues and to have three EF-hand structures characteristic of Ca^{2+}-binding sites (3). The gene of apoobelin has been overexpressed in *E.coli* (4). The purified recombinant apoobelin converted to obelin with a high yield upon incubation with synthetic coelenterazine (5).

It has been demonstrated that the assay of photoproteins, and obelin particularly, is highly sensitive, non-hazardous, the reaction being rapid and simple (6). It makes the photoproteins attractive for immunoassay systems application as the reporter proteins - such enzymes as alkaline phosphatase and horseradish peroxidase. In this report we describe an application of the recombinant obelin for immunoassay using avidin-biotin system. For this purposes obelin was derivatized with biotin by the reaction with N-hydroxysucinimidil-ε-aminocapronoil-biotin. These derivatives, in combination with avidin, can be used for detection of biotinylated targets immobilized on microtiter wells. We applied this technique to detect alphafetoproteins (AFP) which level monitoring is widely used for early cancer diagnostics and pre-natal diagnostics of some genetic diseases. The obtained results were compared to those obtained for radioimmunoassay and were found to be similar.

Materials and Methods

Instrumentation: The luminescence was measured with Luminoscan v1.30 microtiter wells luminometer (Labsystems, Finland).

Reagents: Synthetic coelenterazine was provided by Molecular Probes, Inc. (Oregon, USA). All other reagents were of highest quality commercial grade.

Procedures: Recombinant obelin was isolated and reactivated according to (5). Obelin was biotinylated using a 5-fold molar excess of N-hydroxysuccinimidil-ε-aminocapronoil-biotin (Bio~Su) for 30 min at room temperature in 0.1 mol/L $NaHCO_3$, 5 mmol/L EDTA, pH 9.0. The reaction was stopped by addition of 0.1 mol/L Gly and the mixture was dialyzed against 0.05 mol/L phosphate-buffered saline pH 7.0 (PBS), containing 5 mmol/L EDTA. AFP-antibodies (AFP-AB) were biotinylated in a similar manner using 100-fold molar excess of Bio~Su. The number of biotin fragments, included into protein molecules was determined according to (7).

Detection of biotinylated obelin: Aliquots (100 μL) containing different amounts of biotinylated recombinant obelin (Obe~Bio) were placed in the wells of a microtiter plate. Light production was monitored for 2 s upon injection of 100 μL 100 mmol/L Tris pH 8.8, containing 300 mmol/L $CaCl_2$.

Bioluminescent AFP detection on microtiter wells. Aliquots of the AFP-AB in PBS (5 μg per mL, 100 μL) were placed into opaque microtiter wells (Dinatech, USA) and stored at 37°C for 1 h. After washing (PBS, 0.1% Tween 20) a free surface was blocked with BSA (0.5% solution, 4°C, overnight). After washing, aliquots of AFP solution (control or calibrated human sera) were placed into the wells and incubated at 37°C, for 1 h. Subsequent incubations with the biotinylated AFP-AB (2.5 μg per mL), avidin (2.5 μg per mL), Obe~Bio (2.5 μg per mL) were carried out the same way. The last washing solution contained 5 mmol/L EDTA. A bound obelin was detected just after injection of 0.3 mol/L $CaCl_2$ in the Luminoscan v 1.30 luminometer.

Results and Discussion

The recombinant obelin amino groups can be modified with commercially available N-hydroxysuccinimide ester derivatives of biotin. The loss of specific activity was not more than 30% at the described conditions. The analysis revealed 2-2.2 mol biotin fragments per every obelin molecule, and 6-8 Bio-fragments per every antibody molecule. Biotinylation of apoobelin prior to regeneration with coelenterazine leads to the practically full loss of activity at any tested conditions.

The measurement of light production of Obe~Bio in the Luminoscan v1.30 exhibited a wide range of linearity, as shown in Fig. 1.

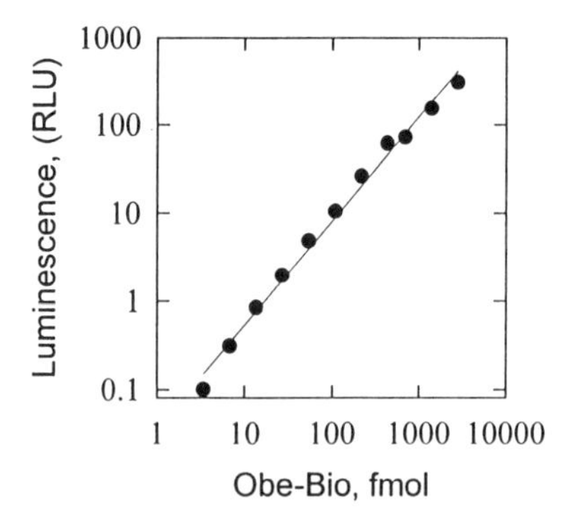

Figure 1. Standard curve of biotinylated obelin light production measured by Luminoscan v1.30 microplate luminometer.

Biotinylated obelin derivatives are stable at frozen-melting procedure, the solution doesn't loose activity when stored at 4°C for 2 weeks.

A microtiter-based immunoassay was carried out according to the following scheme:

$$\text{AFP-AB} + \text{AFP} + \text{AFP-AB}\sim\text{Bio} + \text{Avi} + \text{Obe}\sim\text{Bio} \xrightarrow{Ca^{2+}} \text{light}$$

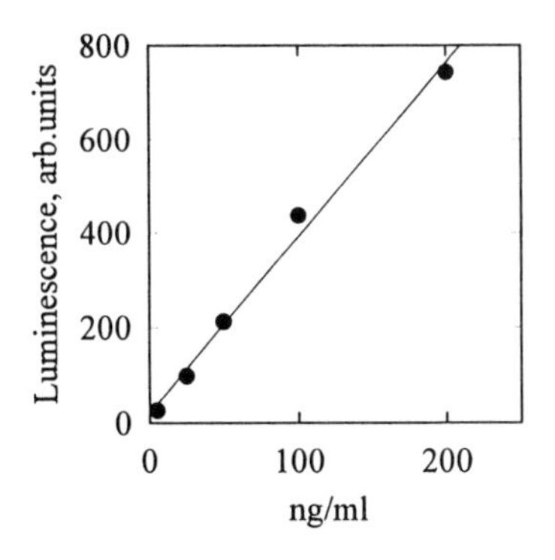

Figure 2. Bioluminescent immunoassay of AFP

Figure 3. Radioimmunoassay of AFP

Fig. 2 shows the dependence of obelin luminescence on AFP concentration. Each point represents the mean of 10 independent determinations. By comparison, Fig. 3 demonstrates the results of radioimmunoassay of AFP.

The main characteristics of both types of analysis are given in the Table 1. They seem to be closely related.

Table 1

Immunoassay of Alphafetoproteins

	Tag	
	Isotope (^{125}I)	Bioluminescent
Tag bounding in AFP absence (%)	1.5	0.5
Sensitivity (ng/ml)	0.8-2.2	2-2.5
"Finding" (%)*	90-110	98-105

*The kind of test to determine a preliminary set concentration in a probe

Biotinylated obelin may be successfully used as a sensitive tag for immunoassay of any other molecules through avidin-biotin system. The sensitivity it can be measured with, coupled with linearity of light emission over many log units of concentration, makes obelin a very promising tool for analysis in a quite simple and rapid manner, with a photon-counting device used.

Acknowledgments
The authors thank Dr. A.I. Petunin and A.V. Navdaev for providing the AFP radioimmunoassay data. This work was supported by the grant from Federal Research

Program 08.05 “Newest Methods of Bioengineering”, subprogram “Engineering in Enzymology”.

References

1. Bondar’ VS, Trofimov KP, Vysotski ES. Physico-chemical properties of the hydroid polyp *Obelia longissima* photoprotein. Biokhimiya 1992;57:1481-90 (in Russian).
2. Shimomura O, Jonson FH. Introduction to the bioluminescence of medusa, with special reference to the photoprotein aequorin. Methods Enzymol 1978;57:271-91.
3. Illarionov BA, Bondar' VS, Illarionova VA, Vysotski ES. Sequence of the cDNA encoding the Ca^{2+}-activated photoprotein obelin from the hydroid polyp *Obelia longissima.* Gene 1995;153:273-4.
4. Illarionov BA, Markova SV, Bondar’ VS, Vysotski ES, Gitelson JI. Cloning and expression of calcium-activated photoprotein obelin cDNA from the hydroid polyp *Obelia longissima.* Doklady Academii Nauk 1992;326:911-3 (in Russian).
5. Vysotski ES, Trofimov KP, Bondar’ VS, Frank LA, Markova SV, Illarionov BA. Mn^{2+}-activated luminescence of the photoprotein obelin. Arch Biochem Biophys 1995;316:92-9.
6. Stults NL, Stocks NF, Rivera H, Gray J, McCann RO, O’Kane D, Cummings RD et al. Use of recombinant biotinylated aequorin in microtiter and membrane-based assays: purification of recombinant apoaequorin from *Escherichia coli.* Biochemistry 1992;31:1433-42.
7. Green NM. Spectrophotometric determination of avidin and biotin. Methods Enzymol 1970;18:418-24.

PART 10

Luminescent Reporter Genes in Cell Biology and Analytical Applications

SINGLE-CELL BIOLUMINESCENCE AND GFP IN BIOFILM RESEARCH

RJ Palmer Jr, [1]C Phiefer, [2]R Burlage, [1]G Sayler, and [1,2]DC White
[1]Ctr Env Biotech/Univ. Tenn, 10515 Research Dr/Suite 300, Knoxville, TN, 37932, USA
[2]Env Sci Div, Oak Ridge National Lab, Oak Ridge, TN, 37831, USA

Introduction

In recent years, it has become apparent that bioluminescence is one of several metabolic processes in bacteria that are controlled by population dynamics, and that the population-size trigger (quorum sensing) is widespread throughout the bacterial kingdom (1). Thus, quorum sensing is a "global" regulation mechanism demonstrated by many different genera and species, and bioluminescence is the quorum-sensing-regulated metabolic process about which we have the most information. In concert, rapid advances in understanding gene regulation have arisen from the development of bioreporter systems in which the synthesis of a foreign biomolecule, often an enzyme, is linked to the transcriptional activity of a particular "gene-of-interest". The enzyme is produced when the gene is activated, and the enzyme activity can be monitored in various ways that are less laborious than direct probing for the product of the gene-of-interest. Luciferase activity (bioluminescence) is frequently used as a reporting system in eukaryotic and in prokaryotic cells, although the eukaryotic reporter systems normally rely on luciferase of eukaryotic origin. Certain fluorescent proteins can also be used a bioreporters when their sequences, inserted into the operon of the gene-of-interest, are transcribed and translated along with the natural gene product. Bioluminescence and fluorescent proteins are therefore powerful tools for detection of gene expression in living cells in real time (*e.g.*, 2).

Attachment of bacteria to substrata (formation of biofilms) is a trigger that, like quorum sensing, can induce changes in physiology and thus in gene regulation (3,4); the additional trigger of population size could become important in the developmental biology of biofilms as single cells develop into microcolonies. We are therefore interested in the development and application of reporters to bacteria in biofilms - from single cells to multiple layers of cells tens-of-microns in thickness. The important criteria in selection of these systems are spatial resolution (in x, y, and z) and temporal resolution (rapid response to induction and rapid response to cessation of gene activity). For reporter systems based on fluorescence, (e.g., GFP-linked reporters), spatial resolution criteria have been largely fulfilled through the development of confocal microscopy and of digital deconvolution microscopy. However, most fluorescence based systems have poor temporal response, at least when one considers downshifts in gene activity. Luciferase-based reporters have good temporal response but, to our knowledge, have not yet been applied in a three-dimensional manner.

There have been several reports that include the terminology "single bacterial cells" in connection with luciferase as a reporter. With few exceptions (*e.g.*, 5), these papers deal with detection of light from colonies grown on plates (inferring that the colony arose from a single cell), with non-quantitative detection (unprocessed CCD data), or with detection of light not conclusively demonstrated to be colocalized with a single bacterial cell. The present contribution demonstrates the types of applications in which we have interest and shows how true single-cell bioluminescence and GFP bioreporters can be used in bacterial biofilm research.

Materials and Methods

<u>Bacterial Strains:</u> *Vibrio fischeri* (ATCC 7744) and *Vibrio harveyi* (ATCC 14126) were maintained, respectively, on Seawater Complete and Marine agar media. Frozen

stock cultures are maintained as reserves should dark mutants arise. *Escherichia coli* DH5α was maintained on LB agar in the presence of ampicillin.

Establishment of Bacterial Biofilms in Flowcells: For all microscopy work, bacterial cells were allowed to adhere and to grow attached in glass flowcells. The flowcells (6) are essentially perfusion chambers constructed of microscopy coverslips; the volume of each chamber is 200 μL and the thickness (from top to bottom) is 1 mm. Normally, an inoculum is injected and allowed to adhere for 10-30 minutes prior to initiation of flow through the chamber. Medium flowrates are low (< 5 mL/hour) except when the chambers are cleared of unattached organisms.

Instrumentation: Photon-counting was performed using a Hamamatsu VIM-CCD camera mounted on a Zeiss Axioplan microscope. A 100 x Plan-Apo oil-immersion lens (NA 1.4) delivered light directly to the camera. Hamamatsu Argus 50 controller hardware (including center-of-gravity board) and software ran on a P90 IBM-compatible with 32 MB RAM and a 1.2 GB removable optical drive. Standard transmitted-light images were captured using the Argus software and a Dage MTI 70 camera.

Results

Photon-counting Demonstrates Variations in Bioluminescence Within and Between Strains: Figure 1A shows a transmitted-light image, and Figure 1B a photon-counting image ("slice" image), of *Vibrio harveyi* cells soon after (20 minutes) attachment inside a flowcell. It is clear that not all cells are producing light and, among those that are, a large variation in light output exists. The arrows mark identical cells in each image; the marked cells are those producing high amounts of light.

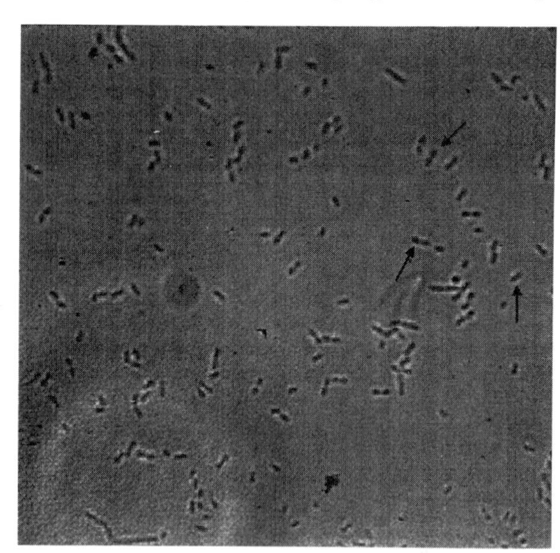

Figure 1A. Transmitted light image of *V. harveyi* cells attached in a glass flowcell.

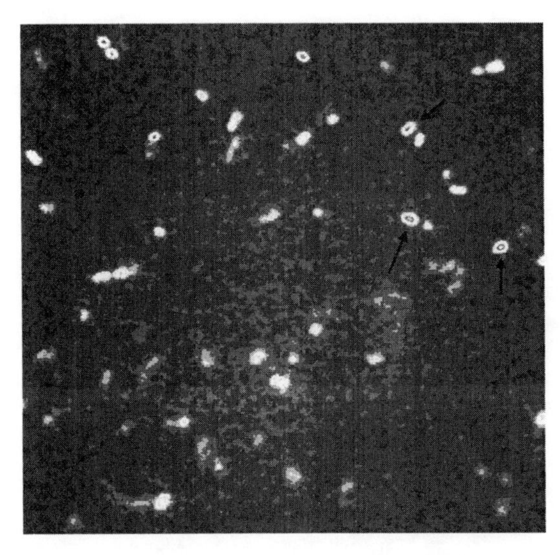

Figure 1B. Photon-counting ("slice") image of field shown in 1A. Arrows indicate identical cells. These cells are also the brightest cells in the photon-counting image (appear red in the original pseudocolored image created by the program).

Figure 2 shows that microcolonies and single cells of *V. fischeri* respond to the presence of autoinducer when attached in the flowcell. Absolute light levels from the brightest *V. fischeri* and *V. harveyi* cells were similar. However, a much lower proportion (roughly 20%) of *V. harveyi* cells emit high levels of light than in *V. fischeri* (roughly 60%; data not shown).

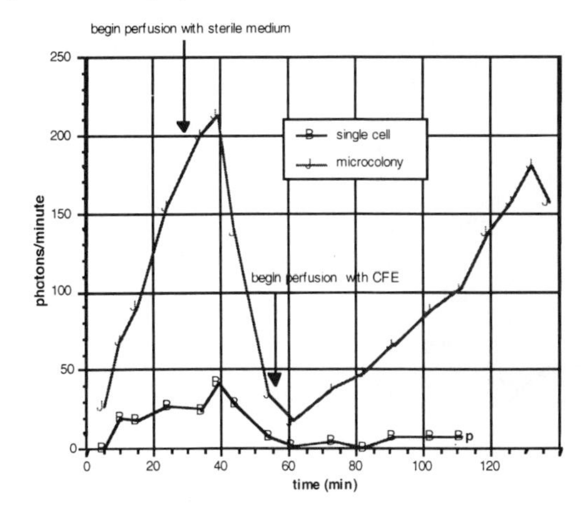

Figure 2. Timecourse of light emission from a microcolony and a single cell of *V. fischeri* grown in a flowcell. Composition of the perfusion medium was changed at times indicated by arrows. CFE = Cell Free Extract (filtered medium containing autoinducer).

Expression of *lux* and of GFP may be incompatible in the same cell: We have inserted a plasmid containing the lux cassette (under control of an Hg-detoxification promoter; light emitted in the presence of Hg) and a plasmid containing the GFP sequence (under control of the lacZ promotor; GFP formed in the presence of IPTG) into E. coli (7). Figure 3A shows a transmitted-light image of cells in the presence of both inducers. Figures 3B and 3C show, respectively, GFP-containing cells and light-emitting cells. Expression of one of the two reporters was seen in several cells. However, no cell expressed both reporters.

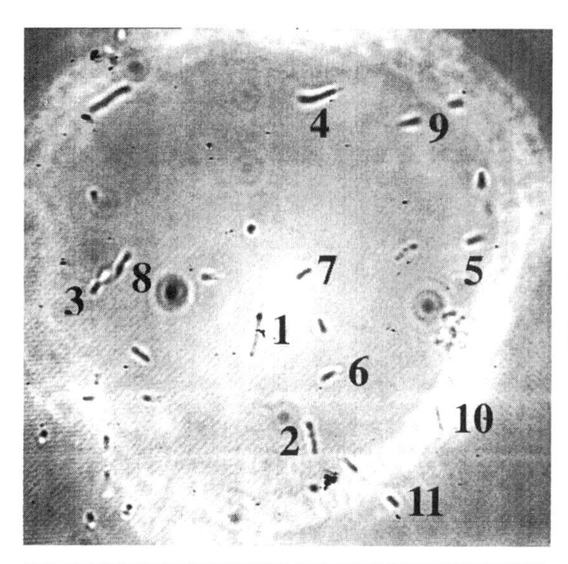

Figure 3A. Transmitted-light image of E. coli cells containing a GFP-bearing plasmid and a lux-bearing plasmid. Both bioreporters were induced.

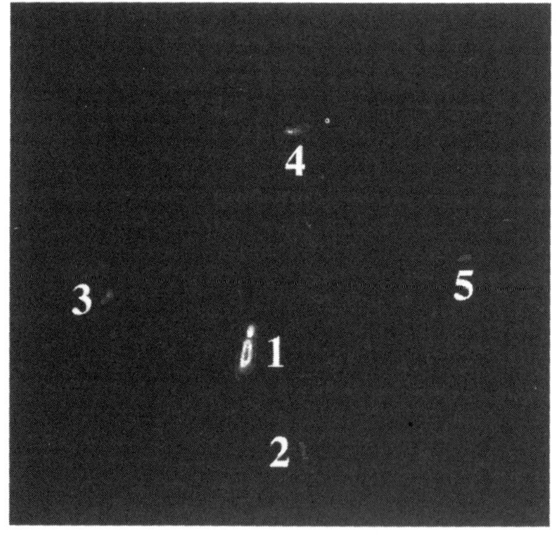

Figure 3B. Epifluorescence micrograph of field shown in 3A. Numbers indicate identical cells in both images. GFP fluorescence was detected using the photon-counting camera in slice mode.

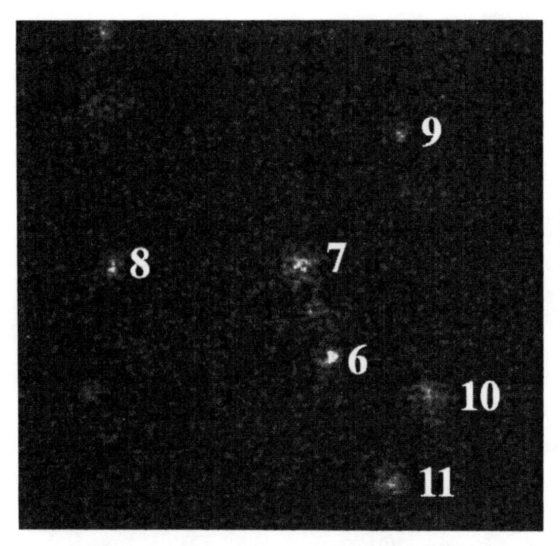

Figure 3C. Photon-counting image (slice mode) of field shown in 3A. Numbers indicate identical cells in both fields. Note that no GFP-containing cells (Fig. 3B) appear in the photon-counting image or vice versa.

Discussion

Using flowcells and a combination of microscopy techniques, we can unequivocally identify single bacterial cells that express bioluminescent and fluorescent bioreporters, and we can quantitate the light produced by these cells. At the present time, our methodology is limited to detection in two dimensions (x and y). We are currently extending these techniques by incorporating the dimension of depth (z) and time (t) to create four-dimensional detection systems useful for study of gene expression in bacterial biofilms.

We have shown that, for attached cells, bioluminescence output within a bacterial strain can vary greatly from cell to cell. Similar non-quantitative data for *Photobacterium phosphoreum* has been presented (8), although those bacterial cells were grown in liquid culture then transferred to an agar-coated slide plate for imaging; we have seen that such transfer severely inhibits bioluminescence in *V. harveyi*. We suggest that a difference in bioluminescence from cell to cell also occurs in batch culture and that the wide range of output from cell to cell in *V. harveyi* is a simple explanation for the empirical observation that batch cultures of *V. harveyi* are much dimmer that those of *V. fischeri*. Furthermore, these data indicate that careful interpretation is required for bioluminescence data normalized to factors such as optical density or cell number.

To date, we have been unable to detect GFP and luciferase activity in single cells that theoretically should produce both; such double labeling would be desirable from the standpoint of having a positional marker as well as a gene-expression indicator in cells in a biofilm. The two processes (synthesis and post-translational processing of GFP; synthesis, post-translational processing, and supplemental requirements for light production by luciferase) may be incompatible in a single cell. Both processes require significant amounts of cellular energy to be directed toward synthesis of proteins not required for normal cellular metabolism. The concentration of GFP required for epifluorescent detection using standard cameras is quite high and may in some way preclude normal cellular metabolism. Use of extremely sensitive photon-counting methods will allow fluorescent protein bioreporters to be detected at concentrations lower than those currently thought to be necessary.

Acknowledgements
This work was supported though grants from the Office of Naval Research.

References

1. Fuqua WC, Winans SC, Greenberg EP. Quorum sensing in bacteria: the LuxR-LuxI family of cell density-responsive tanscriptional regulators. J Bacteriol 1994; 176:269-75.
2. Kay SA, editor. Optical imaging of gene expression and signaling in living cells. Abstracts of a meeting held at Cold Spring Harbor Laboratory, March, 1996.
3. Davies, DG, Geesey GG. Regulation of the alginate biosynthetic gene algC in *Pseudomonas aeruginosa* during biofilm development in continuous culture. App Environ Microbiol 1995; 61:860-7.
4. Valeur A, Tunlid A, Odham G. Differences in lipid composition between free-living and initially adhered cells of a Gram-negative bacterium. Arch Microbiol 1988; 149:521-6.
5. Elhai J, Wolk CP. Developmental regulation and spatial pattern of expression of the structural genes for nitrogenase in the cyanobacterium *Anabaena*. EMBO J 1990; 9:3379-88.
6. Palmer RJJr, Caldwell DE A flowcell for the study of plaque removal and regrowth. J Microbiol Meth 1995; 24:171-82.
7. Khang Y, Burlage RS, Palmer R. A combined fluorescent/bioluminescent bioreporter for the detection of mercury (II) in environmental samples. In: Kay SA, editor. Optical imaging of gene expression and signaling in living cells. Cold Spring Harbor Laboratory, 1996; 17.
8. Hill PJ, Eberl L, Molin S, Stewart GSAB. Imaging of bioluminescence in single bacteria. In: Campbell AK, Kricka LJ, Stanley PE, editors. Bioluminescence and Chemiluminescence - Fundamentals and Applied Aspects. New York: John Wiley & Sons, 1994: 629-32.

COMBINED LUMINESCENT ASSAYS FOR MULTIPLE ENZYMES

I Bronstein[1], CS Martin, CEM Olesen, and JC Voyta
Tropix, Inc., 47 Wiggins Avenue, Bedford, MA 01730 USA

Introduction

The measurement of multiple enzyme activities in a single combined assay offers increased accuracy, precision and throughput compared to performing individual assays. Reporter gene assays (1) are widely used in both biomedical and pharmaceutical research, in the study of gene regulation and identification of factors that affect gene expression, including screening of combinatorial chemical and natural product libraries. For accurate quantitation of gene expression from a reporter construct, a second reporter enzyme is frequently used to monitor transfection efficiency. Pharmaceutical screening strategies may require multiple reporter assays to distinguish the effect of a potential therapeutic on the target transcription factor from a non-specific effect on gene expression. Furthermore, multiple drug targets can be screened in cells bearing multiple reporter enzyme constructs.

Chemiluminescent reporter gene assays utilizing 1,2-dioxetane substrates offer highly sensitive enzyme detection with a wide dynamic range. 1,2-Dioxetane substrates have been incorporated into reporter gene assays for the reporter enzymes β-galactosidase (β-gal), β-glucuronidase (GUS), and placental alkaline phosphatase (PLAP) (2-5). These reporter assays provide extremely sensitive reporter enzyme detection with wide dynamic range, similar to that achieved with luminescent assays for the luciferase reporter enzyme (6) β-Gal has been used in conjunction with luciferase (7) to normalize transfection efficiency, and the similar detection sensitivities enable a single extract dilution to be assayed for both. β-Gal has also been used together with placental alkaline phosphatase (8,9). Combined luminescent assays for β-gal/luciferase, GUS/luciferase and PLAP/luciferase have been developed. The Dual-Light™ system (5,10) incorporates luciferin and Galacton-Plus™ substrates, for the firefly luciferase and β-gal reporter enzymes. The enzyme activities are quantitated sequentially in a single tube in the same sample of extract from cells cotransfected with both reporter plasmids.

Dual detection of two protein antigens on a single membrane blot can be performed using β-gal and alkaline phosphatase-conjugated antibodies, followed by sequential incubation with the 1,2-dioxetane substrates Galacton-*Star*™ (11) and CSPD®. With the Galacton-Star substrate for β-gal, the enzymatic reaction and subsequent light-producing decomposition reaction proceed at the same pH, enabling its use for membrane-based detection of β-galactosidase. Thus, two protein antigens can be probed essentially simultaneously, without the need for membrane stripping.

Other potential applications include detection of multiple analytes in immunoassay systems, detection of multiple probes on nucleic acid blots and triple enzyme assays. 1,2-Dioxetane chemiluminescent substrates provide highly sensitive detection of enzymes and enzyme labels and have been incorporated into several dual detection assay formats, together or with luciferase, that provide sensitive, accurate and efficient enzyme and analyte detection.

---

[1] Author of correspondence

Materials and Methods

Reagents and Instrumentation: Purified *E. coli* β-galactosidase (G-5635, EC 3.2.1.23), purified mollusk β-glucuronidase (G-7896, EC 3.2.1.31) and purified human placental alkaline phosphatase (P-3895, EC 3.1.3.1) were obtained from Sigma (St Louis, MO). Purified firefly (*P. pyralis*) luciferase (EC 1.13.12.7) was obtained from Analytical Luminescence Laboratory (Lansing, MI). Opaque white microplates (Microlite 2™) were from Dynatech Laboratories, Inc. (Chantilly, VA). The following reagents are products of Tropix, Inc. (Bedford, MA): Galacton-Plus™, Lysis Buffer, Buffer A, Buffer B and Accelerator-II (components of the Dual-Light™ system), CSPD®, Glucuron™, Galacton-*Star*™, Sapphire-II™, Tropifluor™ PVDF, I-Block™, goat anti-mouse IgG+IgM-alkaline phosphatase conjugate, Nitro-Block-II™. Light emission was measured in a Dynatech ML2250 Microtiter® plate luminometer.

β-Galactosidase/Luciferase Assay (Dual-Light™): Purified enzymes were diluted in Lysis Buffer (0.1 mol/L potassium phosphatase, pH 7.8, 2 mL/L Triton X-100) containing 0.5 mmol/L dithiothreitol and 1 g/L bovine serum albumin (BSA) Fraction V (A-3059; Sigma). A 10-μL aliquot of diluted enzyme was added to each microplate well. Subsequently, 25 μL of Buffer A was added to each well, and the plate was placed in the luminometer. The relative light units (RLUs) obtained with luciferase were measured for 5 s, beginning 2 s after the injection of 100 μL of Buffer B containing luciferin and Galacton-Plus substrates. The RLUs for β-gal were determined 30 min after addition of Buffer B and initiated by injection of 100 μL of Accelerator-II. Light emission was measured for 5 s after a 2 s delay.

β-Glucuronidase/Luciferase Assay: Purified enzymes were diluted in Lysis Buffer (0.1 mol/L potassium phosphatase, pH 7.8, 2 mL/L Triton X-100) containing 1 g/L BSA. A 10-μL aliquot of diluted enzyme was added to each microplate well. Subsequently, 25 μL of Buffer A was added to each well, and the plate was placed in the luminometer. The RLUs obtained with luciferase were measured for 5 s, beginning 2 s after the injection of 100 μL of Buffer B containing luciferin and Glucuron substrates. The RLUs for β-glucuronidase were determined 60 min after the addition of Buffer B and initiated by injection of 100 μL of Accelerator-II. Light emission was measured for 5 s after a 2 s delay.

Placental Alkaline Phosphatase/Luciferase Assay: Luciferase was diluted in Lysis Buffer (0.1 mol/L potassium phosphatase, pH 7.8, 2 mL/L Triton X-100) containing 1 g/L BSA. PLAP was diluted in 50 mmol/L Tris-HCl pH 8.0, 150 mmol/L NaCl. A 10-μL aliquot of diluted enzyme was added to each microplate well. Subsequently, 25 μL of Buffer A was added to each well, and the plate was placed in the luminometer. The relative light units obtained with luciferase were measured for 5 s, beginning 2 s after the injection of 100 μL of Buffer B (containing only luciferin). The RLUs for PLAP were determined by injection of 100 μL of (2 mol/L diethanolamine, 0.8 mmol/L CSPD, 200 mL/L Sapphire-II, 30 mmol/L EDTA, 30 mmol/L L-homoarginine) 30 min after injection of Buffer B and then incubating for 30 min followed by measurement in "Glow" mode.

Immunoblot Detection: SDS-PAGE minigels were prepared with dilutions of human brain extract (Clontech, Palo Alto, CA) and human serum and electrotransferred to PVDF membrane. All antibodies are diluted in Blocking Buffer (0.2% I-Block, 0.1% Tween-20 in PBS). Blots were blocked in Blocking Buffer for 1 h and then incubated simultaneously with monoclonal anti-β-actin (diluted 1:5000; Sigma) and polyclonal anti-human transferrin (diluted 1:5000; Boehringer Mannheim, Indianapolis, IN) for 1 h. Blots were washed 2 x 5 min with Blocking Buffer and then incubated simultaneously with goat anti-mouse IgG+IgM-alkaline phosphatase conjugate (diluted 1:10,000) and goat anti-rabbit

IgG-β-gal conjugate (diluted 1:2500; Southern Biotechnology Associates, Birmingham, AL) for 1 h. Blots were washed as described above, and then 2 x 2 min in 0.1 mol/L sodium phosphate, pH 8.0. Blots were incubated for 5 min in 0.1 mmol/L Galacton-*Star* with 5% Nitro-Block-II in sodium phosphate buffer and immediately imaged. After 45 min, blots were then washed 2 x 1 min in 0.1 mol/L diethanolamine, pH 10.0, 1 mmol/L $MgCl_2$, incubated for 5 min in 0.25 mmol/L CSPD in diethanolamine buffer and imaged again. Blots were imaged on XAR-5 X-ray film (Kodak, Rochester, NY).

Results and Discussion

Dual enzyme assays have been developed for the quantitation of two enzyme activities from a single sample in a single reaction vessel. A similar assay format is used for dual quantitation of β-gal/luciferase, GUS/luciferase and PLAP/luciferase: sample containing enzyme is added to a well or tube, followed by the addition of Buffer A containing buffer salts and components necessary for the enhanced luciferase reaction. Injection of Buffer B containing luciferin and Galacton-Plus or Glucuron initiates an immediate luminescence signal from the luciferin/luciferase reaction, which decays with a half-life of approximately one minute. Production of light signal from either Galacton-Plus or Glucuron during the luciferase signal measurement is negligible due to the low pH and absence of polymeric enhancer. Following measurement of the luciferase signal, the reaction mixture is further incubated for 30-60 min. Accelerator-II, containing Sapphire-II at alkaline pH is then injected, and the light signal intensity from the Galacton-Plus/β-gal or Glucuron/GUS reaction is measured. At this point, the residual luciferase signal is extremely low due to its fast decay and effective quenching by the accelerator formulation.

The Dual-Light assay is utilized for the quantitation of purified luciferase and β-gal. Detection of 10^{-15} to 10^{-8} g of luciferase and β-gal is achieved with a dynamic range of seven orders of magnitude of enzyme concentration (Figure 1). Detection

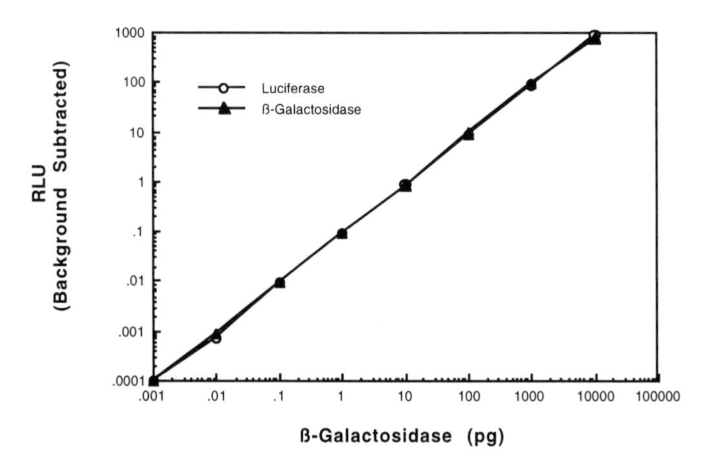

Figure 1. Dual luminescent detection of luciferase and β-galactosidase.

of 2 fg (0.032 amol) of luciferase and 8 fg (0.059 amol) of β-galactosidase was obtained with a signal-to-noise ratio of 2. The level of signal contributed by one enzyme during measurement of the other was determined by assaying individual enzyme dilution series with the dual assay protocol (see (10)). In the absence of luciferase, the presence of up to 1 ng of β-gal does not produce signal above the luciferase reaction background. In the absence of β-gal, the presence of less than 1 ng of luciferase does not contribute any residual signal above the β-gal reaction background. At luciferase concentrations of 1 ng or above, residual luciferase signal is measurable above the β-gal background; however, this level of signal is not interfering except when β-gal concentration is extremely low. Dynamic range over five orders of magnitude is maintained. The light signal from the decomposition of Galacton-Plus persists as a steady glow that decays with a half-life of approximately 180 min. The Dual-Light system was used to assay for luciferase and β-gal from transiently transfected N20.1 cell lysates and the results compared to individual assays for each enzyme (10). Identical results were obtained compared to the individual assays, demonstrating that the dual detection system achieves equivalent sensitivity to separate assays.

Quantitation of purified luciferase and GUS is performed in a similar dual assay format. Detection of 10^{-15} to 10^{-8} g of luciferase was achieved, with a dynamic range of seven orders of magnitude (Figure 2). Detection of 10^{-13} to 10^{-8} g of GUS is achieved with a dynamic range of five orders of magnitude. The signal interference between the two enzymes is similar to that obtained with the β-gal/luciferase combination. GUS activity does not interfere with the luciferase

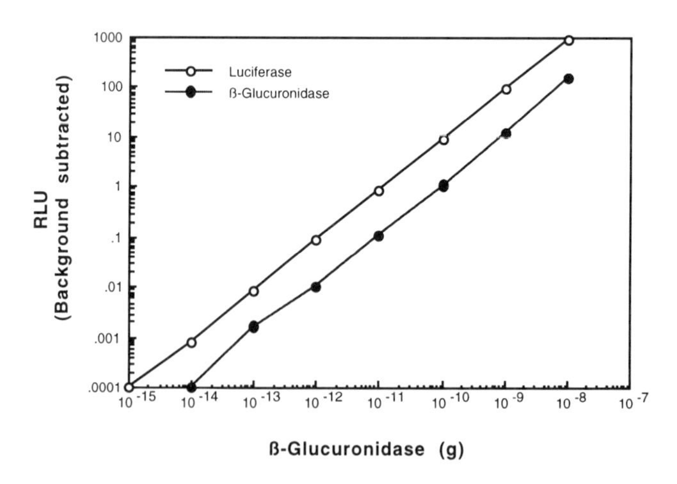

Figure 2. Dual luminescent detection of luciferase and β-glucuronidase.

signal at any concentration, while greater than 1 ng of luciferase exhibits some interference when very low levels of GUS are detected (data not shown). Luciferase and GUS are widely used reporter enzymes for gene expression studies in plants and have been used together for normalization of transformation efficiency (12).

For the PLAP/luciferase assay, CSPD substrate is added later, with a high pH buffer, rather than with Buffer B. Alkaline phosphatase is active at the higher pH, and "glow" light emission kinetics results. Quantitation of luciferase and PLAP in this dual assay format enables detection of 10^{-15} to 10^{-8} g of luciferase and 10^{-13} to 10^{-7} g of PLAP (Figure 3).

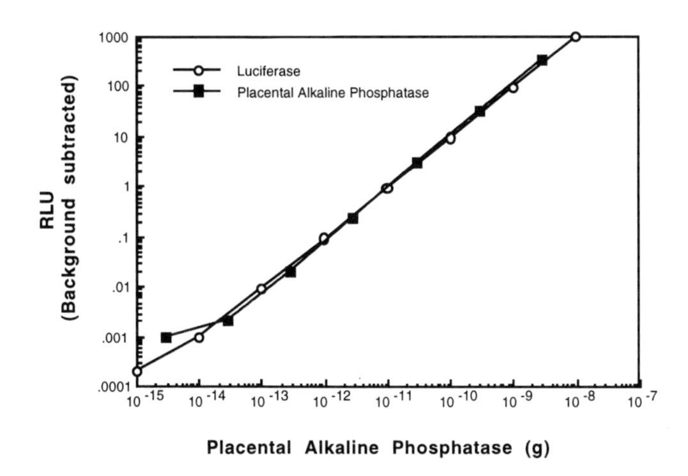

Figure 3. Dual luminescent detection of luciferase and placental alkaline phosphatase.

Dual enzyme-label immunoblot detection was performed using both β-gal and alkaline phosphatase-labeled secondary antibody conjugates, followed by sequential incubation in Galacton-*Star* and CSPD 1,2-dioxetane substrates (Figure 4). Blots were incubated simultaneously with two primary antibodies (rabbit anti-transferrin and mouse anti-β-actin) and simultaneously with both secondary antibody-enzyme conjugates. Blots are developed and imaged first with Galacton-*Star*. Following addition of CSPD substrate at pH 10, the light emission from Galacton-*Star* persists at the higher pH. This dual enzyme-label immunoblot detection protocol enables sequential imaging of two protein species with simultaneous immunodetection, eliminating the need for blot erasure. The success of immunoblot stripping protocols is highly variable, depending on membrane type, the specific protein epitope and detection method.

Dual reporter enzyme assay systems will provide wide-spread benefit for both research and pharmaceutical screening applications, enabling highly sensitive detection, and increased efficiency and accuracy. 1,2-Dioxetane substrates have proved to be extremely versatile reagents for both solution-based and membrane-based chemiluminescent detection protocols and have enabled the development of these various dual enzyme detection assay formats.

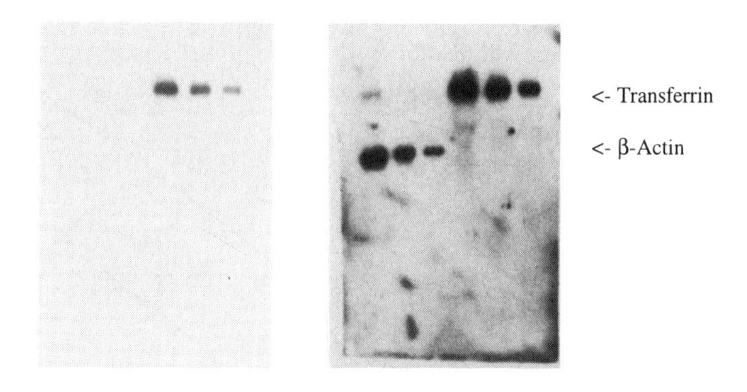

+ Galacton-Star/Nitro-Block-II + CSPD

Figure 4. Dual chemiluminescent immunodetection with alkaline phosphatase and β-galactosidase-conjugates. Lanes 1-3: Human brain extract (7.5, 2.5, 0.8 μg/lane). Lanes 4-6: Human serum (1:3 dilutions from 0.3 μL/lane). The blot was developed sequentially with Galacton-*Star* and CSPD and initially imaged for 10 min, 30 min following Galacton-*Star* incubation. The blot was subsequently imaged for 10 s, 20 min following CSPD incubation (1 h after Galacton-*Star* incubation).

References

1. Alam J, Cook JL. Reporter genes: application to the study of mammalian gene transcription. Anal Biochem 1990;188:245-54.
2. Jain VK, Magrath IT. A chemiluminescent assay for quantitation of β-galactosidase in the femtogram range: application to quantitation of β-galactosidase in lacZ-transfected cells. Anal Biochem 1991;199:119-24.
3. Bronstein I, Fortin JJ, Voyta JC, Juo R-R, Edwards B, Olesen CEM, et al. Chemiluminescent reporter gene assays: sensitive detection of the GUS and SEAP gene products. BioTechniques 1994;17:172-8.
4. Bronstein I, Fortin J, Voyta JC, Olesen CEM, Kricka LJ. Chemiluminescent reporter gene assays for β-galactosidase, β-glucuronidase and secreted alkaline phosphatase. In Campbell AK, Kricka LJ, Stanley PE, editors. Bioluminescence and Chemiluminescence: Fundamental and Applied Aspects. Chichester, UK: John Wiley, 1994;20-3.
5. Bronstein I, Martin CS, Fortin JJ, Olesen CEM, Voyta JC. Chemiluminescence: sensitive detection technology for reporter gene assays. Clin Chem 1996;42:1542-6.
6. Brasier AR, Tate JE, Habener JF. Optimized use of the firefly luciferase assay as a reporter gene in mammalian cell lines. BioTechniques 1989;7:1116-22.
7. Fulton R, Van Ness B. Luminescent reporter gene assays for luciferase and β-galactosidase using a liquid scintillation counter. BioTechniques 1993;14:762-3.
8. O'Connor KL, Culp LA. Quantitation of two histochemical markers in the same extract using chemiluminescent substrates. BioTechniques 1994;17:502-9.
9. O'Connor KL, Culp LA. Topological and quantitative analyses of early events in tumor formation using histochemically-tagged transformed 3T3 cells. Oncol Rep 1994;1:869-76.

10. Martin CS, Wight PA, Dobretsova A, Bronstein I. Dual luminescence-based reporter gene assay for luciferase and β-galactosidase. BioTechniques 1996;21:520-4.
11. Martin CS, Olesen CEM, Liu B, Voyta JC, Shumway JL, Juo R-R, Bronstein I. Continuous sensitive detection of β-galactosidase with a novel chemiluminescent 1,2-dioxetane. In Hastings JW, Kricka LJ, Stanley PE, editors. Bioluminescence and Chemiluminescence: Molecular Reporting with Photons. Chichester, UK: John Wiley, 1997; this volume.
12. Leckie F, Devoto A, De Lorenzo G. Normalization of GUS by luciferase activity from the same extract reduces transformation variability. BioTechniques 1994;17:52-7.

APPLICATIONS OF LUMINOUS OXIDATIVE STRESS BIOSENSORS: UNDERSTANDING DISINFECTANTS MODE OF ACTION

S Belkin[1] and S Dukan[2]
[1]The J. Blaustein Desert Research Institute, Ben Gurion University of the Negev, Sede Boker 84990, Israel
[2]Institut Jacques Monod, CNRS-Université Paris 7, 2 place Jussieu, 75251 Paris Cedex 05, France

Introduction

Uncombined chlorine, in the form of unionized hypochlorous acid (HOCl), is a very strong oxidant and a potent bactericidal agent, and chlorination is the most widely used method for disinfection of water and wastewater. Nevertheless, the mechanism by which HOCl exerts its lethal effects on microorganisms has never been fully elucidated experimentally. While many reports document specific and general damages, in most cases it is difficult to differentiate between primary and secondary effects of exposure to free chlorine.

A closely related subject which has also been poorly investigated, is the physiological aspects of bacterial resistance to free chlorine. Most reports on bacterial antioxidative metabolism concentrate on two major oxidants: superoxides and peroxides. Two *E. coli* global regulatory circuits which appear to be dedicated to the fight against these oxygen species are *soxRS* and *oxyR*, respectively. Other global regulators involved are the *rpoH* , *rpoS* , *soxQ* , *fur*, *arcA*, *fnr* and possibly other circuits (1,2).

In order to begin to decipher the mode of action of free chlorine, we have attempted to identify genes induced by sub-lethal HOCl exposure (3). Several genes, belonging to different regulatory systems known to be involved in the defenses against oxidative stress, were selected; their induction was monitored using a set of plasmids on which their promoters were fused to the bioluminescence operon of *Vibrio fischeri*. The parental plasmid used, pUCD615 (4), contained *luxCDABE* genes downstream from a multiple cloning site, into which promoters of the selected genes were inserted. Members of this set of plasmids have been previously described (4-7), and the advantages of their uses as general and specific reporters for bacterial stress were documented.

Materials and Methods

Bacterial strains: All strains were *E. coli* K-12 derivatives. Construction of the different plasmids was reported elsewhere.

Growth and hypochlorous acid challenge conditions: *E. coli* cells were grown in LB medium containing 50 or 100 μg/mL of kanamycin monosulfate or ampicilin, respectively, at 26°C in a rotary shaker. Following overnight growth, cells were diluted to 10^7/mL in the same medium, lacking the antibiotics, and allowed to grow for a few generations under the same conditions. At an early exponential phase the cells were washed with 0.05 M phosphate buffer (pH 7.0), and resuspended in the same buffer at a cell density of 10^8/mL. Samples (1 mL) were distributed into glass test tubes and a freshly prepared hypochlorous acid stock solution was added to the desired final concentration. After 20 minutes of incubation at 26°C in the dark with gentle shaking, free chlorine was quenched by the addition of sterile sodium thiosulfate to 0.5 mM. Culturable bacteria were assayed by plating on LB plates after serial dilutions in phosphate buffer. Preliminary experiments have indicated that under these conditions, HOCl concentrations lower then 1 mg/L resulted in no loss of viability and a small (<50%) decrease in culturability.

Measurement of bioluminescence: luminescence was monitored at 26°C using a microtiter-plate luminometer (Lucy1, Anthos Labtec Instruments, Salzburg, Austria), following addition of 75 μL of cells (untreated or treated as described above) to 25 μL of a four-fold concentrated LB solution containing different additions, where required. All experiments were conducted in duplicate, in opaque white microtiter plates (Dynatech, Germany). Results are presented either as the averaged relative light units of the instrument (RLU), or as the response ratios (luminescence of the treated cells over that of untreated controls).

Results and Discussion

Five different *E. coli* global defense circuits were tested for their short-term (up to 150 min) response to sub-lethal concentrations of HOCl (Table 1). Surprisingly, the *oxyR* regulon did not appear to be induced, in spite of the similarities between hypochlorous acid and hydrogen peroxide (8). Similarly, no induction was observed by the promoter of the gene encoding for the universal stress protein (*uspA*) or by that of *recA*., the SOS protease (9).

Circuit	Gene tested	Response
oxyR	*katG*	-
"Universal stress"	*uspA*	-
SOS	*recA*	-
soxRS	*micF*	+
rpoH (heat shock)	*grpE, dnaK, lon*	+

Table 1. Circuits induced by HOCl exposure

Heat shock induction: Following chlorine exposure of an *E. coli* strain that contained a plasmid in which the promoter of the heat shock gene *grpE* is fused to the structural part of the *Vibrio fischeri lux* operon (6), a very clear induction took place (Fig. 1a). The response was dose-dependent, with a maximum at 0.2 mg/L. Activity peaked after 40 to 60 min, and then declined. The same pattern was displayed following the exposure to HOCl of *E. coli* cells in which *lux* genes were fused to two other heat shock promoters, *lon* and *dnaK*. Figure 1b presents the results for *grpE* and *dnaK* in a pair of isogenic hosts, one of them defective in *rpoH*, the gene coding for the heat shock regulator σ^{32} (10). In addition to the identity in the response pattern of the two promoters, the data in Figure 1b clearly displays the dependency of the reaction to free chlorine on *rpoH*. The responses of the *lon'::lux* fusion were similar (not shown).

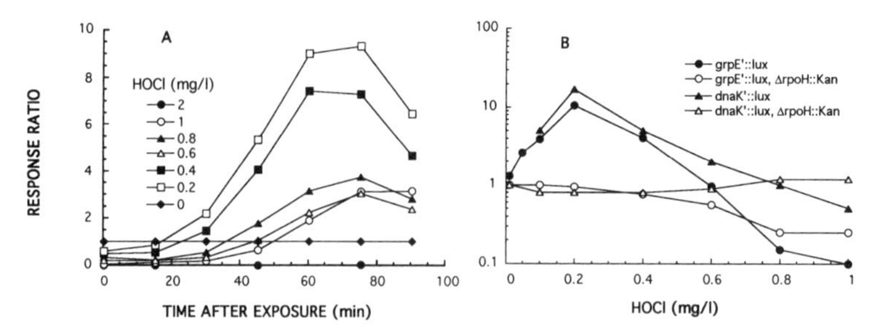

Figure 1. Luminescent response to HOCl exposure. A - kinetics of light development by an *E. coli* strain bearing a *grpE'::lux* fusion. B - effect of *rpoH* mutations on *grpE'::lux* and *dnaK'::lux* expression.

<u>SoxR activation:</u> Another gene clearly induced by HOCl exposure is *micF*. The rate of light production was maximal (ca. 20 fold) at 1 mg/L of hypochlorous acid and about 120 min after its addition. The response was *soxR*-dependent: no induction was observed in a *soxR* mutant (Fig. 2). The same *soxR* mutation did not affect the activation of other *lux* fusions, such as *katG'::lux* induced with H_2O_2 (not shown).

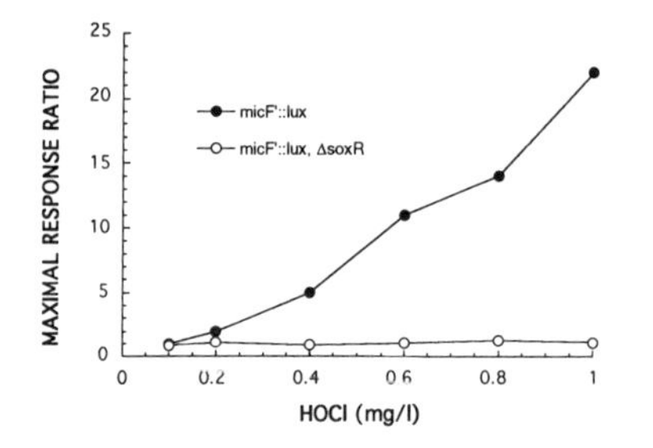

Figure 2. *micF'::lux* is induced by HOCl in a *soxR*-dependent manner.

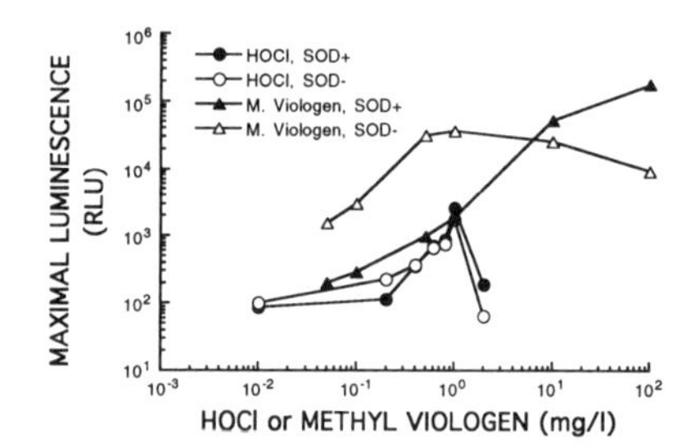

Figure 3. *micF'::lux* induction by m. viologen, but not by HOCl, is strongly affected by SOD mutations.

<u>The HOCl effect is not superoxide-mediated</u>: The effects of superoxide dismutase mutations on *micF'::lux* induction were tested, in order to indirectly assess the involvement of superoxide oxygen in hypochlorous acid stress. From Figure 3 it is clear that the double mutation in the genes for two forms of this enzyme, *sodA* and *sodB*, had a minor effect on HOCl induction of *micF*. This is in contrast to the very dramatic effect these mutations had on the induction of the same promoter by methyl viologen, a redox cycling agent which generates superoxide radicals. These results suggest that *micF* induction by HOCl is not superoxide mediated, and point at the possibility of either a direct activation of *soxR* by HOCl or its derivative, or indirect activation mediated by an affected cellular component.

In summary, the results presented above demonstrate for the first time free chlorine activation of bacterial *rpoH* (heat shock) and *soxRS* circuits, and shed some new light on the little-understood effects of this oxidant on bacterial metabolism; much more work is needed in order to understand its actual interactions with these regulons or any other circuit it may activate. Clearly, plasmids containing *lux* fusions to promoters of interest are very powerful tools for such studies.

References

1. Demple, B. Regulation of bacterial oxidative stress genes. Ann Rev Genet 1991;25:315-37.
2. Compan I, Touati D. Interaction of six global transcription regulators in expression of manganese superoxide dismutase in *Escherichia coli* K-12. J Bacteriol 1993;175:1687-96.
3. Dukan S, Dadon S, Smulski DR, Belkin S. Hypochlorous acid activates the heat shock and *soxRS* systems of *Escherichia coli*. Appl Environ Microbiol; in press.
4. Rogowsky PM, Close TJ, Chimera JA, Shaw JJ, Kado CI. Regulation of the *vir* genes of *Agrobacterium tumefaciens* plasmid pTiC58. J Bacteriol 1987;169:5101-12.
5. Belkin S, Smulski DR, Vollmer AC, Van Dyk TK, LaRossa RA. Oxidative stress detection with *Escherichia coli* bearing a *katG'::lux* fusion. Appl Environ Microbiol 1996; 62: 2252-6.
6. Van Dyk TK, Majarian WR, Konstantinov KB, Young RM, Dhurjati PS, LaRossa RA. Rapid and sensitive pollutant detection by induction of heat shock gene-bioluminescence gene fusions. Appl Environ Microbiol 1994;60:1414-20.
7. Van Dyk TK, Smulski DR, Reed TR, Belkin S, Vollmer AC, LaRossa RA. Responses to toxicants of an *Escherichia coli* strain carrying a *uspA'::lux* genetic fusion and an *E. coli* strain carrying a *grpE'::lux* fusion are similar. Appl Environ Microbiol 1995;61: 4124-7.
8. Dukan S, Touati D. Hypochlorous acid stress in *E. coli*: genes involved in protection, DNA damage, common features and differences with resistance to hydrogen peroxide. J Bacteriol; in press.
9. Walker GC. The SOS response of *Escherichia coli*. In: Neidhardt FC, Curtis R, Ingraham JL, Lin ECC, Low KB, Magasanik B, Reznikoff WS, Riley M, Schaechter M, Umbarger HE, editors. *Escherichia coli* and *Salmonella typhimurium*: Cellular and Molecular Biology, 2nd edition. Washington, DC: ASM, 1996:1400-16.
10. Neidhardt FC, VanBogelen RH. Heat shock response. In: Neidhardt FC, Ingraham JL, Low KB, Magasanik B, Schaechter M, Umbarger HE, editors. *Escherichia coli* and *Salmonella typhimurium*: Cellular and Molecular Biology. Washington, DC: ASM, 1987:1334-45.

CHEMILUMINESCENCE DETERMINATION OF CATALASE AT PHYSIOLOGICAL H_2O_2 CONCENTRATIONS

S Mueller, HD Riedel and W Stremmel
Department of Internal Medicine IV, University of Heidelberg, 69115 Heidelberg, Germany

Catalase, which decomposes H_2O_2 to water and O_2, is a widely distributed enzyme. Although investigated since many decades its physiological function and regulation are still poorly understood (1). Despite the increasing interest in studying H_2O_2 generation and removal in cells, an accurate method to determine catalase activity at physiological H_2O_2 concentrations has not been described so far.

Several methods have been developed to determine catalase activity (2). The UV-spectrophotometric assay is the most commonly used but it is associated with some serious disadvantages. First, studies at concentrations below 10^{-4} mol/L H_2O_2, i.e. at physiological and non-toxic levels, are impossible due to the relatively low H_2O_2 sensitivity of the assay. Second, the unphysiological high substrate concentrations in the millimolar range result in rapid catalase inactivation (usually after 30 s) (2). Finally, molecular oxygen may be liberated in gaseous form leading to disturbance of absorbance (2).

Here, we describe a novel chemiluminescence assay for the determination of catalase activity at submicromolar H_2O_2 concentrations. Luminol is oxidized by hypochlorous acid to diazaquinone which is further converted by H_2O_2 to an excited aminophthalate via an alpha-hydroxy-hydroperoxide (3, 4). Recent experiments have shown that the short luminescence signal of this reaction linearly depends on H_2O_2 down to nanomolar concentrations (5-7). Using the luminol-hypochlorite system we now show the possibility to follow the catalase-mediated exponential decay of H_2O_2 at submicromolar H_2O_2 concentration. This experimental approach allows the rapid and simple determination of catalase activity in crude cell homogenates and cell suspensions without enzyme inactivation, loss of cell viability and gaseous oxygen liberation. We also show that this technique can be applied to glutathione peroxidase (GPO). Finally, an experimental model is demonstrated to determine either catalase or GPO activity at physiological H_2O_2 concentrations in fresh hemolysates. Our data indicate that catalase contributes the major part in the detoxification of H_2O_2 in the human erythrocyte.

Materials and Methods

Instrumentation: Luminescence measurements were performed using a luminometer AutoLumat LB 953 (Fa. Berthold, Wildbad, Germany).

Reagents: Luminol, catalase, GPO, glutathione (GSH), hydrogen peroxide, sodium azide, aminotriazole, NADPH, iodoacetamide, sodium hypochlorite and sodium azide were from Sigma (Deisenhofen, Germany)

Preparation of human erythrocytes and hemolysates: Fresh human red cells were prepared as previously described (8). Preparation of hemolysates was performed by hypotonic lysis diluting the red cell stock solution with tridistilled water 1:10.

Reference determination of catalase activity: The activity of catalase (k) was determined at pH 7.4 following the breakdown of H_2O_2 at 240 nm using the spectrophotometric method described by Aebi (2)

Chemiluminescence determination of catalase activity: For most experiments a **flow technique** was established to follow rapid kinetics. Catalase solutions, diluted liver homogenates or cell suspensions were aspirated via a peristaltic pump (4 ml/min). Luminol (10^{-4} mol/L) and hypochlorite (10^{-4} mol/L) were continuously added by a perfusor pump (6 ml/min) to the sample solution closed to the light detector thus monitoring the actual H_2O_2 concentration in real time. The photomultiplier was adjusted to the maximum intensity of the luminescence reaction. The luminescence integral was determined over 5 sec. After adding of H_2O_2 (10^{-5} mol/L) to the catalase solution the breakdown of H_2O_2 concentration could be followed. For low catalase activity an **injection technique** is recommanded as descibed in (7). Catalase activity was calculated as the rate constant k of the exponential decay of H_2O_2 by linear regression analysis (2). In experiments with hemolysates the catalase activity k in s^{-1} was related to (g) of hemoglobin per liter (L).

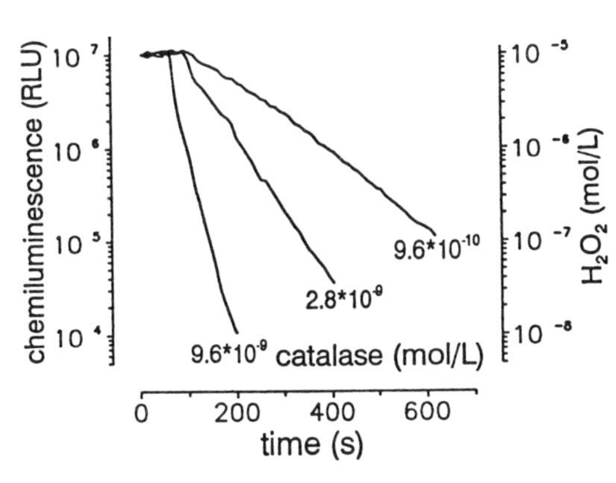

Fig. 1 First order decay of H_2O_2 by different concentrations of catalase

Results and Discussion

Determination of catalase

Injection of NaOCl to luminol and H_2O_2 results in a luminescence peak linearly dependent on the actual H_2O_2 concentration (7). H_2O_2 removal by catalase can be determined by starting the luminescence reaction after various incubation times of catalase. Upon addition of NaOCl a fast luminescence peak of less than 2 s is observed. Whereas the peak integrals clearly depend on the time of catalase incubation the curve profiles remain unchanged (not shown). The quantitative relationship between luminescence intensity and incubation time of catalase is demonstrated in Fig. 1 using luminol and NaOCl by a flow technique. The semilogarithmic plot shows an exponential decay (r^2=0.992) down to 10^{-8} mol/L H_2O_2. This kinetic is typical for catalase since the enzyme is not saturable up to molar H_2O_2 concentrations (2). Consequently, catalase activity is described by the rate constant (k) of first order decay of H_2O_2: $k = \ln (S_1/S_2)/dt$, where dt is the measured time interval,

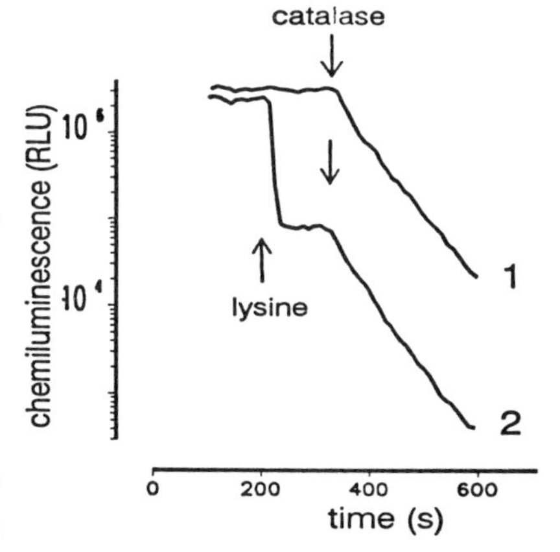

Fig. 2 Influence of 10^{-4} mol/L lysine on the determination of catalase activity

S_1 and S_2 are H_2O_2 concentrations at time t_1 and t_2, respectively (2). Since in this equation the ratio rather than absolute values of H_2O_2 concentrations is of importance, k can be calculated directly from the luminescence intensities. Depending on the catalase dilution and the number of data collected k is determined by linear regression analysis with a standard deviation of 1 - 3%. Since HOCL is known to react efficiently with many targets such as sulfhydryl and amino groups of proteins and other compounds a competition of these substrates with luminol for HOCl may occur (9). Thus, the effect of compounds containing functional groups was investigated.Fig. 2 shows an example where lysine is added to a H_2O_2 solution. Addition of lysine apparently decreases luminescence intensity whereas the exponential decay of the catalase mediated H_2O_2 decomposition is not changed in comparison to the control. Alanine, GSH, human serum albumin, and cysteine were found to act in similar fashion (data not shown). Hence, determination of the rate constant k by luminol and hypochlorite is not affected by functional groups which scavenge hypochlorite and, thus, the method is well suited for tissue homogenates and cell suspensions.

H_2O_2 decomposition in erythrocytes: importance of catalase and GPO activity at physiological H_2O_2 concentrations

An example of H_2O_2 decomposition in fresh hemolysate is demonstrated in Fig. 3A. In agreement with data obtained with purified catalase, degradation of H_2O_2 follows an exponential kinetics. The activity is completely blocked by the catalase inhibitor sodium azide (Fig. 3B) as shown by a renewed application of H_2O_2 (3C). Similar results are observed with intact erythrocytes. No loss of enzyme activity or cell viability is observed after repetitive injections of 10^{-5} mol/L H_2O_2 within 30 minutes (data not shown). The very low H_2O_2 concentrations further lead to the advantage that all oxygen generated during the catalytic reaction is completely dissolved in the medium. Thus, troublesome gaseous liberation of oxygen and its consequences upon absorption are avoided. Recent experiments have further shown that the system NaOCL/luminol also allows the determination of H_2O_2 decomposition by GPO. Sensitivity and measuring range are mainly limited by the concentration of the substrate GSH as discussed in Fig. 2. H_2O_2 removal can be observed down to 10^{-6} mol/L H_2O_2 at 2 mmol/L GSH representing the intracellular GSH concentration within the human erythrocyte (10). In contrast to catalase, purified GPO shows a saturation kinetic for H_2O_2. Fig. 3 also shows the activity of GPO in fresh hemolysate. After inhibition of catalase by sodium azide (Fig. 3B) 2 mmol/L GSH is added and the flow system is re-calibrated (Fig. 3D). If 10^{-4} mol/L H_2O_2 is added (Fig. 3E) it is continuously decomposed. To better demonstrate the

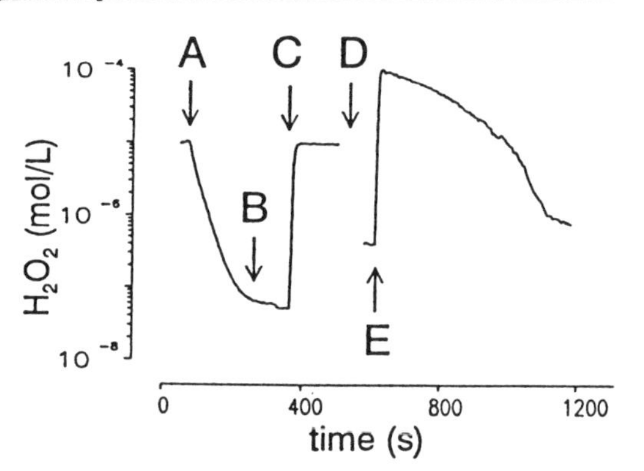

Fig. 3 Determination of catalase and GPO activity in fresh hemolysate at physiological H_2O_2 concentrations (see text)

time course for Fig. 3E a 10 times higher concentrated hemolysate than in Fig. 3A-C was used. In contrast to catalase activity the H_2O_2 decay corresponds to linear kinetics (r^2 = 0.99) down to 10^{-5} mol/L H_2O_2. Below this concentration the kinetics becomes exponentially (non saturated for H_2O_2). The Michaelis-Menton constant of GPO for H_2O_2 is found to be $4*10^{-6}$ mol/L H_2O_2. The procedure described in Fig. 3A-E allows the determination of catalase and GPO activity in one experiment at physiological H_2O_2 concentrations. Due to the different kinetics of both enzymes catalase was thought only to act at very high H_2O_2 concentrations. Our experiments show that catalase is the predominant enzyme within the erythrocyte. Even presuming a strict zero order kinetic for H_2O_2, GPO would remove only more than 50% below $3.7*10^{-7}$ mol/L H_2O_2. However, since the kinetics becomes nonlinear below 10^{-5} mol/L H_2O_2 GPO reaches only 21% of the rate of catalase. These results support recent indirect measurements by Gaetani et al who found that human GPO decomposes only 17% of the rate at which human catalase simultaneously removed H_2O_2 (10).

We believe that this first chemiluminescence assay for catalase activity is a useful tool to investigate the role of catalase under physiological conditions in various biological systems. It is furthermore assumed that the assay can be extended to other enzymes involving H_2O_2 either as a substrate or product.

References

1. Clerch LB. A 3' untranslated region of catalase mRNA composed of a stem-loop and dinucleotide repeat elements binds a 69-kDa redox-sensitive protein. Arch Biochem Biophys 1995; 317: 267-74.
2. Aebi H. Catalase in vitro. Methods Enzymol 1984; 105: 121-6.
3. Eriksen TE, Lind J, Merenyi G. Oxidation of luminol by chlorine dioxide: formation of 5-aminophthalazine-1,4-dione. J Chem Soc Faraday Trans 1981; 77: 2125-35.
4. Merényi G, Lind J, Eriksen TE. Luminol chemiluminescence: Chemistry, Excitation, Emitter. J Biolum Chemilum 1990; 5: 53-6.
5. Arnhold J, Mueller S, Arnold K, Sonntag K. Mechanisms of inhibition of chemiluminescence in the oxidation of luminol by sodium hypochlorite. J Biolum Chemilum 1993; 6: 307-13.
6. Mueller S, Arnhold J. Detection of hydrogen peroxide in stimulated neutrophils using luminol and hypochlorite. In: Campbell AK, Kricka LJ, Stanley PE, editors. Bioluminescence and chemiluminescence: Fundamentals and Applied Aspects. Chichester, John Wiley & Sons, 1994: 242-5.
7. Mueller S, Arnhold J. Fast and sensitive chemiluminescence determination of H_2O_2 concentration in stimulated neutrophils. J Biolum Chemilum 1995; 10: 229-37.
8. Giulivi C, Hochstein P, Davies KJ. Hydrogen peroxide production by red blood cells. Free Radic Biol Med 1994; 16: 123-9.
9. Arnhold J, Hammerschmidt S, Wagner M, Mueller S, Arnold K, Grimm E. On the action of hypochlorite on human serum albumine. Biomed Biochim Acta 1990; 49: 991-7.
10. Gaetani GF, Ferraris AM, Rolfo M, Mangerini R, Arena S, Kirkman HN. Predominant role of catalase in the disposal of hydrogen peroxide within human erythrocytes. Blood 1996; 87: 1595-9.

REAL-TIME SEQUENCE-BASED DNA ANALYSES USING BIOLUMINESCENCE

P Nyrén, S Karamohamed and M Ronaghi
Dept of Biochemistry and Biotechnology, The Royal Institute of Technology, S-100 44 Stockholm, Sweden

Introduction

The firefly-luciferase system is widely used for detection of ATP. The firefly assay of ATP may also be used for the detection of metabolites and enzymes participating in ATP-converting reactions. For instance, if the firefly-luciferase system is coupled to ATP sulfurylase, inorganic pyrophosphate (PPi) can be detected (1). The enzymatic inorganic pyrophosphate detection assay (ELIDA) can be used as an alternative to ATP measurements for cell enumeration (2), and also for detection of cell lysis and cell-lysing activity (3). There are also many other possible applications for the ELIDA, such as for analysis of DNA polymerase activity (4). The greatest advantage of using PPi detection for analyses of enzymes acting on nucleic acids is the possibility to follow their activity in real-time. With a real-time method the initial events during nucleic acid modifying reactions can be studied in detail. The availability of simple and real-time methods for analysis of nucleic acids is of great importance in many different fields such as genetic diseases, cancer diagnosis, and analysis of infectious diseases. In Table 1, are some of the possible applications for the ELIDA presented. In this paper, the use of the ELIDA for real-time DNA analyses is exemplified with four different applications: i) a method for detection of exonuclease activity, ii) a method for detection of idling activity, iii) a method for detection of RNA-polymerase activity, and iv) a method for real-time DNA sequencing.

Table 1 Possible applications for the ELIDA in the field of nucleic acids analyses

Exonuclease assay
Idling assay
RNA polymerase assay
DNA polymerase assay
- Detection of known single-base changes
- Real-time DNA sequencing approach

DNA ligase assay
Reverse transcriptase assay
Terminal deoxynucleotidyl transferase assay

Materials and Methods

Reagents: D-luciferin, L-luciferin, and purified luciferase (BioThema, Dalarö, Sweden). Exonuclease-deficient (exo$^-$) Klenow DNA polymerase (Amersham, UK). T4 DNA polymerase, T7 RNA polymerase, NTPs, and dNTPs (Pharmacia Biotech, Uppsala, Sweden), ATP sulfurylase, and adenosine 5'-phosphosulfate (APS) (Sigma Chemical Co., St. Louis, MO, USA). Streptavidin-coated super paramagnetic beads, Dynabeads M280 (Dynal AS, Norway).

Oligonucleotides: The oligonucleotides used in this study were designed by us, and synthesised and purified by Pharmacia Biotech, Uppsala, Sweden. The 43-base-long oligonucleotide T7RN (5'GCTGGAATTCGTCAGACTGGCCGTCGTTTTACAA-CGTCTCCCTATAGTGAGTCGTATTAGGTACC3'), was used, together with E3PN (see below), for detection of RNA polymerase activity. The 5' biotinylated 35-base-long oligonucleotide E3PN (5'GCTGGAATTCGTCAGACTGGCCGTCGTT-TTACAAC3'), was hybridised to either NUSPT: 5'GTAAAACGACGGCCAGT3', or NUSPA: 5'GTAAAACGACGGCCAGA3'.

TTACAAC3'), was hybridised to either NUSPT: 5'GTAAAACGACGGCCAGT3', or NUSPA: 5'GTAAAACGACGGCCAGA3'.
Instrumentation: We used a LKB 1250 luminometer and a potentiometric recorder.
Detection of exonuclease and idling activity: The template (10 pmol E3PN) was hybridised to 10 pmol mismatch primer (NUSPA) in 20 mM Tris-HCl (pH 7.5) and 8 mM $MgCl_2$ in a final volume of 10 µL. One pmol of the DNA-fragment, and 400 pmol of dNTPs were added to the assay solution. The reaction was started by adding 1.5 units of either T4 DNA polymerase, or exonuclease deficient (exo-) Klenow DNA polymerase. The PPi released, due to exonuclease activity followed by nucleotide incorporation, was detected by the ELIDA as described in Ref. 1 with the exception that 0.4 mg/ml polyvinylpyrrolidone (360,000) was added to the assay buffer. The reaction was carried out at room temperature.
Detection of RNA-polymerase activity: The template was constructed by primer extension, catalysed by Klenow polymerase, of biotinylated E3PN hybridised to T7MR (containing the T7 promotor) and immobilised onto streptavidin-coated super paramagnetic beads. The obtained 65-base-long immobilised double-stranded oligonucleotide was used as template for real-time detection of RNA polymerase activity. About five pmol of the immobilised DNA-fragment and 1 nmol NTPs were added to the assay solution. The RNA polymerase reaction was started by adding 28 units RNA polymerase. The PPi released due to nucleotide incorporation was detected by the ELIDA as described above. The reaction was carried out at room temperature.
Real-time DNA sequencing: The oligonucleotide E3PN hybridised to the sequencing primer NUSPT was used as template for real-time DNA sequencing. The DNA-fragments were incubated with (exo-) Klenow DNA polymerase. The sequencing procedure was carried out by stepwise elongation of the primer strand upon sequential addition of the different dNTPs. The PPi released due to nucleotide incorporation was detected by the ELIDA as described above. About two pmol of the DNA-fragment, and 3 pmol (exo-) Klenow DNA polymerase were added to the assay solution. The sequencing reaction was started by adding 40 pmol of one of the nucleotides. The reaction was carried out at room temperature.

Result and Discussion

In the following, four of the applications presented in Table 1, are described. All methods relies on the detection of the PPi released in the different nucleic acid modifying reactions. The PPi formed in the reactions are converted to ATP by ATP sulfurylase and the ATP production is continuously monitored by the firefly luciferase. The reactions occurring in the ELIDA are:

$$\text{PPi} + \text{APS} \xrightarrow{\text{ATP-sulfurylase}} \text{ATP} + SO_4^{2-} \quad [1]$$

$$\text{ATP} + \text{luciferin} + O_2 \xrightarrow{\text{luciferase}} \text{AMP} + \text{PPi} + \text{oxyluciferin} + CO_2 + h\nu \quad [2]$$

Exonuclease assay For detection of exonuclease activity the extension rates on a DNA-template hybridised to a mismatch (at the 3'-termini) primer was analysed by the ELIDA. In Fig. 1, is the primer extension rate after the addition of (exo-) Klenow and T4 DNA polymerase, respectively, shown. No activity was observed in the presence of (exo-) Klenow, whereas strong activity was observed after the subsequent addition of T4 DNA polymerase. The mismatch (A:A) is very difficult to extend for both polymerases, however, as the exonuclease activity of the T4 DNA polymerase cleaves the mismatch termini, the primer can be easiely extended.

Idling activity Idling activity is a cycle of nucleotide addition followed by excission. The activity is observed for some polymerases when they reach the end of a template. The polymerase then add an extra base (3'-overhang) to the second strand followed by exonuclease cleavage of this base. This activity continuous until all nucleotides are used up. In Fig. 1, the idling activity is observed as the continuous increase in PPi production after all primers were fully extended.

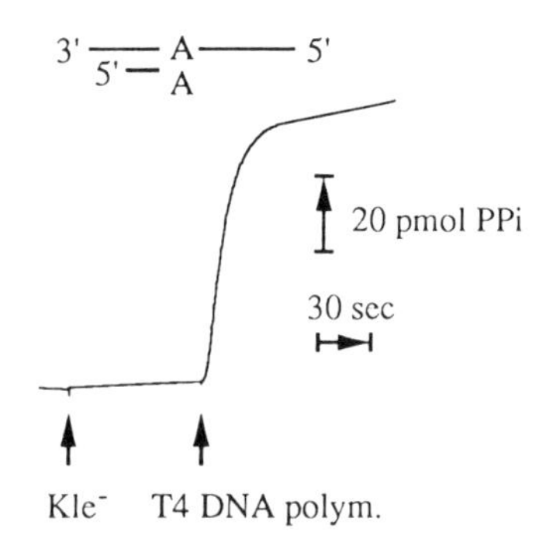

Figure 1. Real-time detection of exonuclease and idling activity. One pmol of the of the template/primer (E3PN/NUSPA) was incubated with 400 pmol of deoxynucleotides. The reactions were started by addition of either 1.5 units T4 DNA polymerase or (exo⁻) Klenow as indicated and the PPi released were detected by the ELIDA.

RNA polymerase assay Fig. 2 shows a typical trace from the ELIDA during T7 RNA polymerase catalysed RNA synthesis on a 65-base-long immobilised double-stranded template. The activity continues as long as there are nucleotides present. This assay can be used for: i) analyses of the efficiency of different promotors, ii) transcription-regulation studies, and perhaps also for iii) DNA-sequencing on double-stranded DNA.

DNA sequencing in real-time In the method for real-time DNA sequencing a primer hybridised to a single-stranded DNA template was stepwise elongated upon sequential addition of different nucleotides.

Real-time signals in the ELIDA, proportional to the amount of nucleotide incorporated, were observed when complementary bases were incorporated (Fig. 3). The sequencing procedure was started by addition of dCTP. A signal corresponding to incorporation of one residue was observed (2 pmol of template were used and a signal corresponding to 2 pmol PPi was detected). The next base added was dTTP, again a signal corresponding to incorporation of one residue was detected. Thereafter, dGTP was added. Also this time the incorporation of one residue was noted. The assay system was then calibrated by the addition of 2 pmol PPi. At this point three different nucleotides are present in the assay, so the subsequent addi-

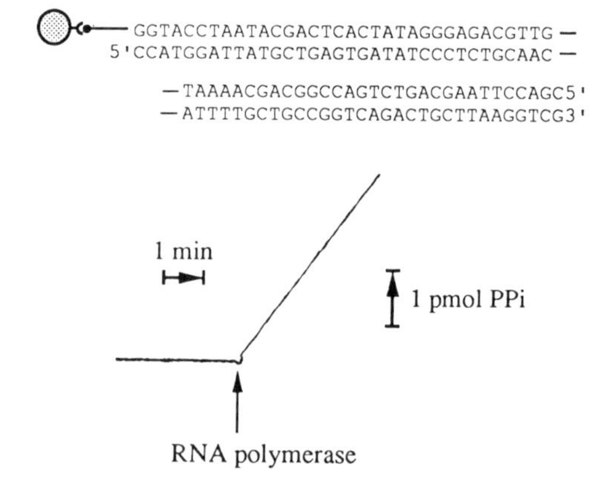

Figure 2. Real-time detection of RNA polymerase activity. About five pmol of the immobilised template (immobilised E3PN/T7MR extended by Klenow DNA polymerase) was incubated with 1 nmol of nucleotides. The reaction was started by addition of 28 units T7 RNA polymerase as indicated and the PPi released were detected by the ELIDA.

tion of dATP gave a very strong signal, corresponding to full extension of the primer (12 bases). No signal was observed when non-complementary bases were added (not shown).

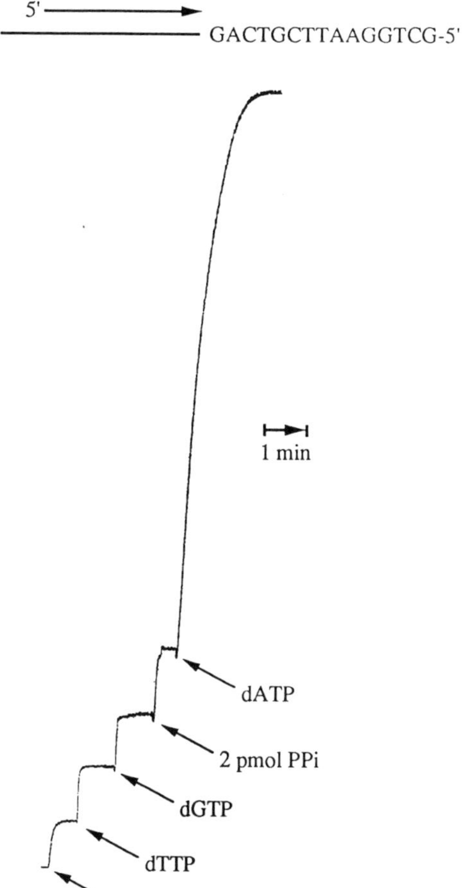

Figure 3. Real-time DNA sequencing performed on a 15-base-long single-stranded template. About two pmol of the template/primer (E3PN/NUSPT) were incubated with 3 pmol (exo$^-$) Klenow. The reaction was started by the addition of 40 pmol of the indicated deoxynucleotide and the PPi released was detected by the ELIDA.

References

1. Nyrén P, Lundin A. Enzymatic method for continuous monitoring of inorganic pyrophosphate synthesis. Anal Biochem 1985;151: 504-9.
2. Nyrén P, Edwin V. Inorganic pyrophosphatase-based detection systems I. Detection and quantification of cell lysis and cell-lysing activity. Anal Biochem 1994;220:39-45.
3. Nyrén P, Edwin V. Inorganic pyrophosphatase-based detection systems II. Detection and enumeration of cells. Anal Biochem 1994;220:46-52.
4. Nyrén P, Lundin A. Enzymatic method for continuous monitoring of DNA polymerase activity. Anal Biochem 1987;167: 235-8.

PART 11

Applications of Chemiluminescence

EFFECTS OF POLYMERS ON THE BORONIC ACID ENHANCED HORSERADISH PEROXIDASE-LUMINOL-HYDROGEN PEROXIDE REACTION

X Ji and LJ Kricka
Department of Pathology and Laboratory Medicine, University of Pennsylvania Medical Center, Philadelphia, PA 19104, USA

Introduction

Soluble hydroxy-polymers stabilize light emission from boronic acid enhanced horseradish peroxidase (HRP) catalyzed chemiluminescent oxidation of luminol (1). In this study we describe results of screening studies of different hydroxy-polymers and the effects of polymer concentration and molecular weight.

Materials and Methods

Luminol (Aldrich, Milwaukee, WI) was purified as the sodium salt by recrystallization from sodium hydroxide (2). 4-Biphenylboronic acid, hydrogen peroxide (30 % w/v), and hydroxypropyl cellulose were purchased from Aldrich. 4-Iodophenylboronic acid was synthesized by Cookson Chemicals Ltd (Southampton, UK).

<u>Screening of Polymers</u> Polymer solutions (0.1% w/v) were prepared in Tris buffer (0.1 M, pH 8.6). 2-Hydroxyethy methacrylate (Polysciences Inc., Warrington, PA), poly(tetramethylene ether glycol (Polysciences) (PTEG, MW 650, 2900), and hydroxypropyl cellulose (Aldrich) (HPC, MW 100000, 370000, 1000000) were solubilized as described previously (3). Poly(ethylene glycol) (PEG, MW 200, 18500), poly(propylene glycol) (PPG, MW 400, 2000, 4000) and dextran (MW 15000-20000) were from Polyscience. Luminol (1.25 mM) - peroxide (2.7 mM) solutions were prepared in the different polymer buffers and in Tris buffer (control). Luminol-peroxide (100 ml), 4-biphenylboronic acid (10 ml, 1 mM) or 4-iodophenylboronic acid (10 ml, 10 mM), and HRP Type VI-A (5 fmoles, Sigma, St Louis, MO) were mixed together in a microwell and light emission measured for up to 40 min using an ML-3000 microtiter plate luminometer (Dynatech Laboratories, Chantilly, VA).

<u>Effect of Polymer Concentration</u> Poly(propylene glycol) in Tris buffer (0-0.1 % w/v) was used to prepare the luminol (1.25 mM) - peroxide (2.7 mM) reagent. Luminol-peroxide (100 ml), 4-biphenylboronic acid (10 ml), and HRP (5 fmoles) were mixed together and the light emission measured for 40 min.

Results and Discussion

<u>Screening of polymers</u> The hydroxy polymers stabilized light emission in the boronic acid-enhanced chemiluminescent oxidation of luminol and Table 1 shows the effect of different polymers on the 4-biphenylboronic acid enhanced luminol oxidation catalyzed by HRP Type VI-A. The most significant improvement in light emission were obtained >15 min after initiation of the reaction. HPC, PEG, and PTEG were superior to dextran and PPG in terms of increase in light emission signal (>45-fold increase at 30 min) but these polymers also increased the background light emission >2-fold.

<u>Effect of polymer molecular weight</u> Figure 1 shows the effect of different molecular weight preparations of HPC and PPG on the 4-iodophenylboronic acid and 4-biphenylboronic acid enhanced luminol oxidation catalyzed by HRP Type VI-A, respectively. Increasing the molecular weight of the polymer had no major effect on the stabilization of the light emission for any of the combinations of polymer or enhancer tested.

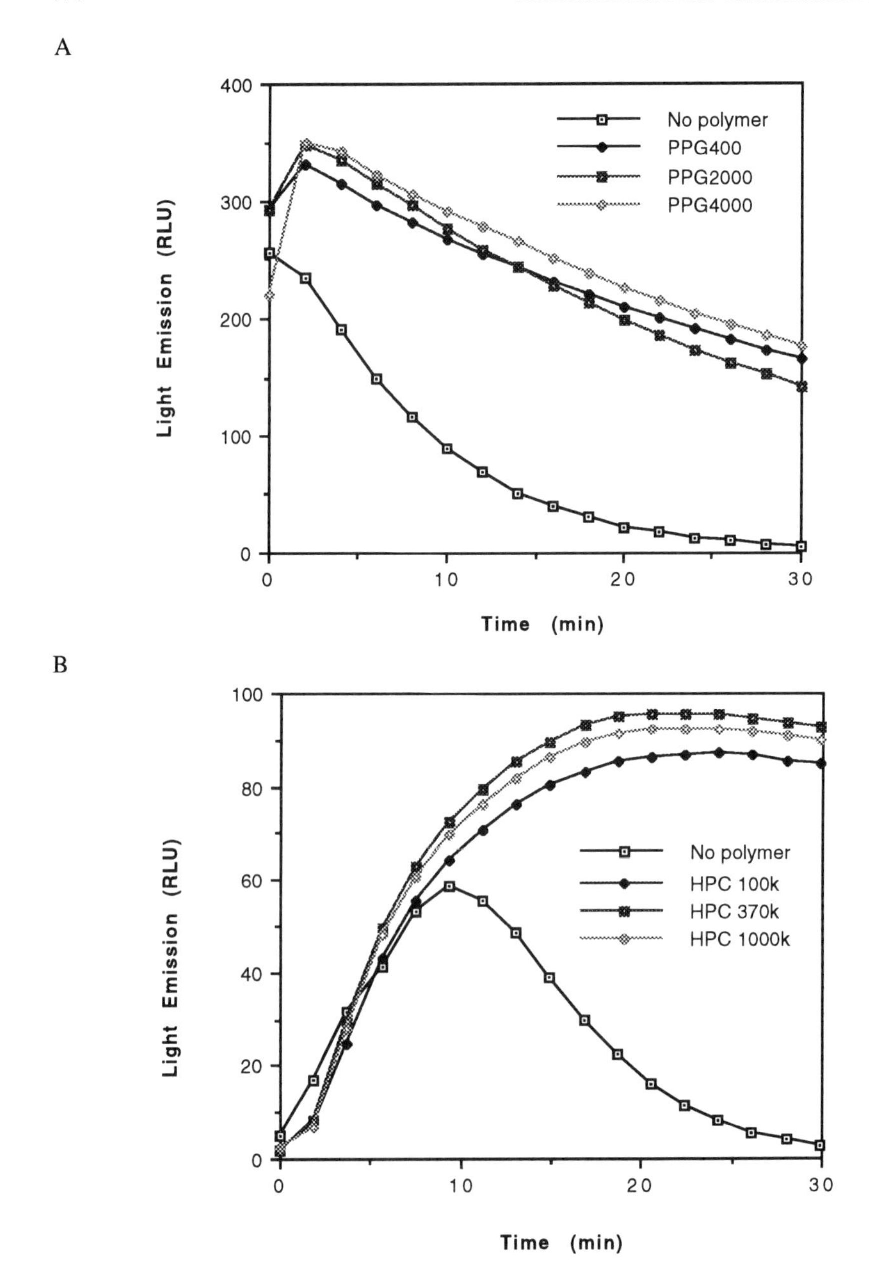

Figure 1. A, Effect of poly(propylene glycol) molecular weight on the 4-biphenylboronic acid enhanced luminol oxidation catalyzed by HRP Type VI-A. B, Effect of hydroxypropyl cellulose molecular weight on the 4-iodophenylboronic acid enhanced luminol oxidation catalyzed by HRP Type VI-A.

Table 1. Effect of hydroxy polymers on 4-biphenylboronic acid enhanced horseradish peroxidase-luminol-hydrogen peroxide reaction

Polymer (MW)	ratio	Light emission 2 min	15 min	30 min
Hydroxypropyl cellulose (HPC, 100,000)	B_p/B_c	1.13	2.93	4.37
	S_p/S_c	1.24	6.45	47.61
Poly(ethylene glycol) (PEG, 18,500)	B_p/B_c	0.72	2.24	2.79
	S_p/S_c	0.47	6.45	47.5
Poly(tetramethylene ether glycol) (PTEG, 2,900)	B_p/B_c	0.96	2.52	3.37
	S_p/S_c	1.16	6.51	48.55
Dextran (15,000-20,000)	B_p/B_c	0.83	1.45	1.68
	S_p/S_c	1.01	2.47	8.43
Poly(propylene glycol) (PPG, 400)	B_p/B_c	1.13	1.67	1.64
	S_p/S_c	0.75	4.32	11.09

B_p, test background; B_c, control background; S_p, test signal; S_c, control signal

Effect of polymer concentration The polymer-mediated stabilization effect was concentration dependent and is illustrated by the poly(propylene glycol)-mediated stabilization of light emission in a 4-biphenylboronic acid enhanced HRP (Type VI-A) - catalyzed luminol oxidation (Figure 2). Maximum light emission was achieved at a polymer concentration of approximately 0.01% w/v and further increases in polymer concentration did not increase the light emission.

The mechanism of the stabilization of light emission from the boronic acid enhanced HRP catalyzed oxidation of luminol is not known. It has many characteristics in common with the phenol enhanced reaction (eg. pH and concentration dependence, isoenzyme specificity) but rate constant data for reactions between boronic acid enhancers and HRP-I and HRP-II have not explained observed differences in the potency of phenols and boronic acids (4). We tested a diverse range of polymers with varying numbers of hydroxy substituents (eg, dextran, 3 hydroxy groups per repeat unit vs PEG, 2 hydroxy groups per molecule) and obtained stabilization in all cases. Potent stabilization and signal enhancement were observed for relatively low molecular weight polymers (eg, PPG, 400) and high molecular weight polymers (eg, HPC, 100,000). Boronic acids can form complexes with hydroxy compounds and this may be a component of the mechanism. The decay in light emission observed in ECL reactions, especially at high peroxidase concentrations, is thought to be due to inactivation of HRP by reactive species formed in the oxidation reaction. In this study the stabilized signal is interpreted as a decrease in the deactivation of HRP and complex formation between the boronic acid and the hydroxy polymer may be involved.

Signal stability is an important and desirable feature of chemiluminescent assays. This work indicates that polymers can advantageously stabilize chemiluminescent signals in HRP catalyzed reactions, and this may have analytical utility in blotting assays and in immunoassays. Others have noted an effect of polymers such as Tween 20 (polyoxyethylenesorbitan monolaurate) on HRP catalyzed chemiluminescent reactions but no mechanistic details have been published (5).

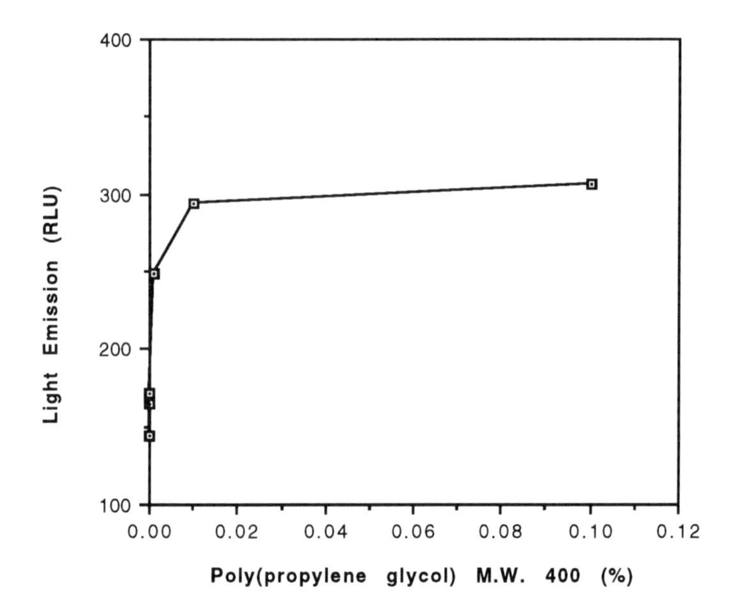

Figure 2. Concentration dependence of poly(propylene glycol)-mediated 4-biphenylboronic acid enhanced reaction (t = 5 min)

Acknowledgments

The financial support of the British Technology Group (BTG) is gratefully acknowledged.

References

1. Ji X, Kricka LJ. Effect of polymers on enhanced chemiluminescent assays for peroxidase, peroxidase labels and antioxidants. J Biolumin Chemilumin in press.
2. Ham G, Belcher R, Kricka LJ, Carter TJN. Stability of trace iodine solutions. Anal Lett 1979;12:535-41.
3. Ulfelder KJ, Schwartz HE, Hall JM, Sunzeri FJ. Restriction fragment length polymorphism analysis of ERBB2 oncogene by capillary electrophoresis. Anal Biochem 1992;200:260-7.
4. Sun W, Ji X, Kricka LJ, Dunford HB. Rate constants for reactions of horseradish peroxidase compounds I and II with 4-substituted arylboronic acids. Can J Chem 1994;72:2159-62.
5. Akhavan-Tafti, HM. Enzyme-catalyzed chemiluminescence from hydroxyaryl cyclic diacylhydrazide compounds. PCT Patent Appl. WO 94/10337. 1994.

STRUCTURE - ACTIVITY RELATIONSHIPS OF BORONIC ACIDS

X Ji and LJ Kricka
Department of Pathology and Laboratory Medicine, University of Pennsylvania Medical Center, Philadelphia, PA 19104, USA

Introduction

The chemiluminescent horseradish peroxidase catalyzed oxidation of luminol is enhanced by firefly luciferin (1, 2), 4-substituted phenols (3, 4), substituted naphthols (4), aromatic amines (5), phenylboronic acids (6, 7), and a series of aromatic molecules (eg, 4-hydroxyacetanilide) (8). Aryl boronic acids represent one of the latest classes of compound that are effective as enhancers. In an effort to understand the relationship between enhancement activity and structure, we have undertaken an extensive screening study of a range mono-, di, and polysubstituted boronic acid derivatives.

Materials and Methods

Luminol (Aldrich, Milwaukee, WI) was purified as the sodium salt by recrystallization from sodium hydroxide as described previously (9). Light emission was measured using either a Berthold LB9500C (Berthold EG & G, Nashua, NH) luminometer, or an Amerlite analyzer (Amerlite Diagnostics PLC, Amersham, UK) using opaque white microwell strips (Dynatech Laboratories, Chantilly, VA). Horseradish peroxidase (HRP type VI-A) and hydrogen peroxide (30 % w/w) solution were purchased from Sigma (St Louis, MO).

Screening Procedure Derivatives were screened for their ability to act as enhancers by determining their influence on light emission from the HRP catalyzed oxidation of luminol at pH 8.6. The luminol-hydrogen peroxide reagent was prepared as follows: sodium luminol (12.5 mg) was dissolved in 50 mL of Tris buffer (0.1 mol/L, pH 8.6), and 15.5 μL of hydrogen peroxide (30% w/w) was mixed with 0.5 mL of Tris buffer (0.1 mol/L, pH 8.6). These two solutions were combined and diluted (1:10 dilution). Stock solutions of different boronates were prepared in dimethyl sulfoxide (DMSO) and dilutions prepared in Tris buffer (0.1 mol/L, pH 8.6). In order to test the effect of the boronate on the background light emission, either 10 μL of boronate, or as a control, 10 μL of Tris buffer (0.1 mol/L, pH 8.6) was placed in a cuvette. The chemiluminescent reaction was initiated by addition of 100 μL of luminol-hydrogen peroxide reagent. Light emission was monitored for 20 min. In order to test of the effect of boronate on the signal in the presence of HRP, luminol-hydrogen peroxide reagent (100 μL) was mixed with either 10 μL of boronate, or as a control, 10 μL of Tris buffer (0.1 mol/L, pH 8.6). The chemiluminescent reaction was initiated by addition of 10 μL of HRP. Light emission was monitored for 20 min.

Results

The qualitative results of the screening study are summarized in Table 1. The majority of the boron compounds tested reduced the blank light emission from a luminol-peroxide reagent. However, the magnitude of this reduction was insufficient to provide a two-fold increase in the S/B compared to the control reaction (arbitrary measure of significant enhancement).

Aryl boronic acids The parent compound, phenylboronic acid had no significant effect on blank or signal, in contrast to the various substituted phenylboronic acids that acted as enhancers. The majority of the 4-substituted phenylboronic acids were effective enhancers of light emission (except 4-methoxy and 4-phenoxy), but in contrast none of the 2- or 3- substituted derivatives tested enhanced light emission.

Table 1. Screening of boron compounds for enhancer activity in the horseradish peroxidase catalyzed oxidation of luminol

Compound	Blank decreased	Signal increased	S/B* increased	Source
ARYL BORONIC ACIDS				
Mono-substituted				
phenylboronic acid	Y	N	N	b
2-biphenylboronic acid	Y	N	N	b
2-chlorophenylboronic acid	Y	N	N	e
2-(methylthiomethyl)phenylboronic acid	Y	N	N	e
2-tolueneboronic acid	Y	N	N	b
3-aminophenylboronic acid	Y	N	Y	a
3-chloroacetylaminophenylboronic acid	Y	N	N	e
3-chlorophenylboronic acid	Y	N	Y	e
3-(2-methylbutoylamino)phenylboronic acid	Y	N	N	e
3-nitrophenylboronic acid	Y	N	N	b
4-bromophenylboronic acid	Y	Y	Y	c
4-(4'-bromodiphenyl)di-n-butoxyborane	Y	Y	Y	e
4-biphenylylboronic acid	Y	Y	Y	b
4,4'-bis(phenylboronic acid)	Y	Y	Y	e
4-(3-boron-4-hydroxyphenylazo)benzoic acid	Y	N	Y	b
4-chlorophenylboronic acid	Y	Y	Y	e
N-(4-chlorophenyl)-4-aminophenyl-boronic acid	Y	Y	Y	e
4-iodophenylboronic acid	Y	Y	Y	d
4-methoxybenzeneboronic acid	Y	N	N	c
4-methylphenylboronic acid	Y	Y	Y	e
4-(phenoxy)benzeneboronic acid	Y	N	N	b
1,4-phenyldiboronic acid	Y	N	N	e
trans-4-(3-propenoic acid)phenylboronic acid	Y	Y	Y	d
4-(trimethylsilyl)benzeneboronic acid	Y	Y	Y	b
Disubstituted				
2,3-dichlorophenylboronic acid	Y	N	N	e
2,4-dichlorobenzeneboronic acid	Y	N	N	b
2,5-dichlorophenylboronic acid	Y	N	N	e
2-hydroxy-5-(3-(trifluoromethyl)-phenzylazo)benzenboronic acid	N	Y	N	b
3,4-dichlorophenylboronic acid	Y	N	N	e
3,5-dichlorophenylboronic acid	Y	Y	Y	c
4-chloro-3-nitrophenylboronic acid	Y	N	N	e
4-carboxy-3-nitrophenylboronic acid	Y	Y	Y	e
5-bromo-2-methoxybenzeneboronic acid	Y	N	Y	b
Polysubstituted				
2,4,6-trichlorophenylboronic acid	Y	N	N	e
3-amino-2,4,6-trichlorophenylboronic acid	Y	Y	Y	e

Fused polycyclic boronic acids				
1-naphthaleneboronic acid	Y	N	N	b
1-thianthreneboronic acid	Y	N	N	b

Table 1 continued

2-benzimidazolylphenylboronic acid	Y	N	N	e
4-dibenzofuranboronic acid	Y	N	N	b
6-hydroxy-2-naphthaleneboronic acid	Y	N	Y	b
BORANES AND MISCELLANEOUS BORON COMPOUNDS				
di-(3,4,6-trichlorophenoxy)-3,4,6-trichlorophenylborane	Y	N	N	e
di-(3,5-dichlorophenoxy)-3,5-dichlorophenylborane	Y	N	N	e
4-chlorophenyl-di-(4-chlorophenoxy)borane	Y	Y	Y	e
di-(1-naphthoxy)-1-naphthylborane	Y	N	N	e
2-bromomethylphenyl-di-(2-bromomethylphenoxy)borane	Y	N	N	e
1-butaneboronic acid	N	N	N	a
2-phenyl-1,3,2-dioxaborinane	Y	N	N	b
boroglycine	Y	N	N	b
tetraphenylboron	Y	N	N	b
pentaerythritol borate	Y	N	N	b
diphenylisobutoxyborane	Y	N	N	e
di-3-nitrophenylboronic acid, calcium salt	Y	N	N	e
4-bromophenyl-di-n-butoxyborane	Y	Y	Y	e
methyl-(2-tolylboronic acid)sulfoxide	Y	N	N	e

* A two-fold increase in S/B compared to the control in the absence the test compound was considered significant; Y=yes, N=no. a, Sigma (St Louis, MO); b, Aldrich (Milwaukee, WI); c, Lancaster Synthesis Inc (Windham, NH); d, Cookson Chemicals Ltd (Southampton, UK); e, U.S. Borax Research Corporation (Anaheim, CA).

Only one of the 2,5-and one of the 3,4- disubstituted derivatives and none of the 2,3- or 2,4-disubstituted aryl boronic acids had enhancer properties. One of the two polysubstituted derivatives available for testing (2,3,4,6 substitution pattern) had enhancer properties. Few enhancers were discovered among the other boron compounds (boranes, borinanes, and borates) that we tested.

Discussion

Substitution was a critical factor for aryl boronic acid enhancers. The electronic properties of substitutents was also a determinant of enhancer activity in some of the series of boronates tested. For example, for the 4-substituted boronates, the Hammett constant (σ^{+}_{p}) correlated with enhancer properties. A σ^{+}_{p} value for a 4-substituted aryl boronic acid $\geq$ -0.31 conferred enhancer properties (σ^{+}_{p} Br = +0.15, I = +0.14, Ph = -0.18, Me = -0.31). 4-Substituted boronates with high negative σ^{+}_{p} (eg, 4-OMe σ^{+}_{p} = -0.78) did not act as enhancers. Similar analysis for a series of 3-substituted boronates failed to reveal analogous correlations. 3-Amino (σ^{+}_{p} = -0.16), 3-chloro (σ^{+}_{p} =

to luminol and pKa of the boronic acid, may help to define the molecular features that lead to enhancer activity.

References

1. Whitehead TP, Kricka LJ, Thorpe GHG. Enhanced luminescent or luminometric assay. European Patent Publication 116454. 1987.
2. Whitehead TP, Thorpe GHG, Carter TJN, Groucutt C, Kricka LJ. Enhanced luminescence procedure for sensitive determination of peroxidase labelled conjugates in immunoassay. Nature (London) 1983;305:158.
3. Thorpe GHG, Kricka LJ, Moseley SB, Whitehead TP. Phenols as enhancers of the chemiluminescent horseradish peroxidase-luminol-hydrogen peroxide reaction: application in luminescence monitored enzyme immunoassays. Clin Chem 1985;31:1335.
4. Thorpe GHG, Kricka LJ. Enhanced chemiluminescent reactions catalyzed by horseradish peroxidase. Methods Enzymol 1986;133:331-54.
5. Kricka LJ, O'Toole AM, Thorpe GHG, Whitehead TP. Enhanced luminescent or luminometric assay. U.S. Patent Publication 4729950. 1988.
6. Kricka LJ, Ji X. 4-Phenylylboronic acid: a new type of enhancer for the horseradish peroxidase catalyzed chemiluminescent oxidation of luminol. J Biolumin Chemilumin 1995;10:49-54.
7. Kricka LJ. Chemiluminescent enhancers. PCT Patent Application WO 93/16195. 1993.
8. Thomas RK. Chemiluminescent composition containing cationic surfactants or polymers and 4'-hydroxyacetanilide, test kits and their use in analytical methods. U.S Patent 5,279,940. 1994.
9. Ham G, Belcher R, Kricka LJ, Carter TJN. Stability of trace iodine solutions. Anal. Lett 1979;12:535-41.

STABILIZED PHENYL ACRIDINIUM ESTERS FOR CHEMILUMINESCENT IMMUNOASSAY

M Kawaguchi[1] and N Suzuki[2]
[1]R/D Center, International Reagents Co., Kobe 651-22, Japan
[2]Dept of Food Science, National University of Fisheries, Shimonoseki 759-65, Japan

Introduction

Chemiluminescent acridinium-9-carboxylic acid derivatives, first introduced by McCapra and Woodhead et al. (1), have been increasing importance particularly for immunodiagnostics, because of not only their high sensitivity but also short assay time and cost effectiveness. Stabilized aryl ester (2,3) or N-sulfonylcarboxamide (4,5) provides long reagent shelf compatible with full automated immunoassay system. With respect to the structure of acridinium ester, McCapra described two significant points for the leaving groups (3,6); (1) Introduction of the electron withdrawing group onto the phenolic ring to have high luminescence efficiency; (2) Sterically hindering groups introduced into neighboring the carbonyl carbon protects the ester from the hydrolysis in an aqueous solution. He also referred to the possibility of the effect of peri-methyl group onto the acridinium ring, but did not show the data in detail. We now want to describe the most stable acridinium compound versatile in chemiluminescent immunoassay system.

Materials and Methods

Instrumentation: Chemiluminescence (CL) was measured by photon counting luminometer (Nichion, Lumicounter 2500-S; Photomultiplier: Hamamatsu Photonics, H3460-54, l_{max} (sens) 390 nm) equipped with a microprocessor controlled pipetter and interfaced to an NEC personal computer to evaluate light emission kinetics (Nichion, LC2500N.EXE Software). ^{1}H-NMR, IR, Mass spectra were recorded with 60-MHz (Hitachi, R-1200) and 400-MHz NMR spectrometers (Bruker, AM-400), FT-IR spectrometer (JASCO, FTIR-5300) and FAB mass spectrometer (JEOL, SX-I02A), respectively.

Synthesis of Acridinium Salts by Esterification and Methylation: The substituted or unsubstituted acridine-9-carboxylic acids were esterified with the corresponding phenolic compounds in anhydrous pyridine in the presence of three equivalents of 4-toluenesulfonyl chloride. The acridine-9-carboxylates purified by chromatography were methylated with methyl fluorosulfonate in anhydrous dichloromethane in the presence of 2,6-di-tert-butylpyridine to give the chemiluminescent acridinium salts (compds 1-17). N-Hydroxysuccinimide esters of acridinium salts for labeling ligands (antigens or antibodies) were synthesized by the procedure reported previously (compds 18-22) (1,7).

Conjugates and Solid Phase Preparation for an Immunoassay: Anti human carcinoembryonic antigen (CEA) mouse monoclonal antibody (IgG), cloned in our laboratory, were coupled with five kinds of N-hydroxysuccinimide-activated acridinium esters (compds 18-22) by partially modifying the method of Weeks et al. (7). Anti CEA goat IgG was immobilized physically inside the polystyrene tube (NUNC-Immunotube, Maxisorp Startube, 40 x 10.5 mm) by placing 0.3 mL of 0.03 mg/ml IgG solution at 4 ^{o}C over night, followed by washing and blocking by 0.1% bovine serum albumin in 0.1 mol/L sodium phosphate buffer solution at pH 7.0.

Results and Discussion

The general structure of the acridinium esters prepared in this work is shown in Figure 1. Initial attempts of methylating the acridines using methyl fluorosulfonate gave the desired N-methylated products in fluctuating yields (60-80%), which was caused by partial protonation of the nitrogen atom on the acridine ring probably due to water, a contaminant in the reaction mixture. Attempts of the methylation with ten equivalents of methyl fluorosulfonate in the presence of 2,6-di-tert-butylpyridine allowed to eliminate undesirable protonation and the reaction was completed within 8 h at room temperature to give the methylated products in excellent yields (> 90%). 2,6-Di-tert-butylpyridine fluorosulfonate was easily removed by recrystallization from methanol and ether.

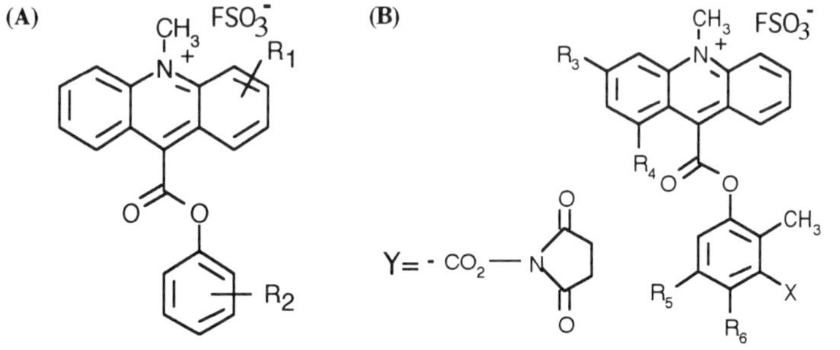

Figure 1. Acridinium Esters 1-17 (A) and N-Hydroxy-succinimide Esters 18-20 (B). 18: $R_3=R_4=CH_3$; $R_5=X=H$; $R_6=Y$, 19: $R_3=R_5=X=H$; $R_4=CH_3$; $R_6=Y$, and 20: $R_3=R_4=CH_3$; $R_6=H$; $R_5=Y$; $X=NO_2$.

Compounds 1-17 were dissolved in DMF and serially diluted 10 times to a final concentration of 5 nmol/L in 0.1 mol/L sodium phosphate buffer solution with 0.1% bovine serum albumin at pH 7.0. A 0.02 mL aliquot of each diluted sample was added to 0.04 mL of 0.4 mol/L HNO_3 with 1.2% hydrogen peroxide. After 30 s, 0.15 mL of 0.25 mol/L NaOH was injected and the light output was integrated for 5 s. The maximum light emissions were observed within 0.4 s for all compounds tested. The results of the CL efficiency and heat stability in an aqueous solution were summarized in Table 1.

Within experimental error, CL output on the luminometer did not differ among the compounds except compound 4 whose leaving group has high pka (5). In the case of the unsubstituted acridinium having dimethyl groups on the ortho-position of phenolic ring gives higher heat stability (compds 5,6) and mono-methyl substitution improves stability a little compared with the unsubstituted phenol (2,3). Halides gave less stability than methyl group in spite of their bulkiness. On the other hand, the introduction of a methyl group onto the peri-position of the acridinium ring causes significant stabilization of the ester which has no substituent on the ortho-position of the phenolic ring (compds 9-11). And the combination of the peri-methyl substitution with ortho-methyl or trifluoromethyl group on the phenolic ring are more effective (compds 12-15), while that with the electron withdrawing group causes reduced stability (compds 14,15). Based on these observation, three kinds of N-hydroxy-succinimide-activated acridinium esters were prepared (Fig. 1B). For the comparisons, the known fluorosulfonates (compds 21 & 22)(1,2) were also synthesized.

Four molecules of the acridinium ester are estimated to conjugate with one anti CEA IgG molecule by adjusting the reaction ratio of each compound in a buffer solution (pH 7.0) and then being observed by UV-visible absorption spectra in diluted HCl. Each conjugate was diluted to the similar concentration giving 20,000 counts per 0.02 mL with 0.1 mol/L sodium phosphate buffer solution at pH 7.0 containing 0.1% bovine serum albumin to evaluate the deterioration.

Figure 2 shows that the heat stability of the conjugate labeled with the brand-new three compounds are superior to that labeled with the known ones. The immunoassay was performed according to the following assay protocol. The 0.02 mL of CEA standard solutions or human patient sera and 0.3 mL of 10 mmol/L phosphate buffered saline at pH 7.0 (PBS) with 0.1% bovine serum albumin were placed into the antibody immobilized polystyrene tube and vortexed at 37 °C for 10 min. After washing by 0.1% Tween 20 in PBS, the conjugate solution (0.3 mL) was diluted to the similar concentration, which gave light emission of 80,000 counts/0.02 mL with PBS containing 0.1% bovine serum albumin, placed and vortexed again for 10 min at 37 °C. After washing by the same manner, each tube was placed in the cell-chamber of the luminometer. It was found that this method provides good precision (not lower than 5% of C.V.), linearity (up to 250 ng/mL), high sensitivity (0.5 ng/mL of detection limit) and correlates well with commercially available CEA assay kits based on ELISA.

In order to investigate further into superiority, the patient serum (CEA; 101 ng/mL) was assayed without calibration. The ready-to-use conjugate solution stored at 25 °C gave consistent CL output over 22 days as shown in Fig. 3. These characteristics will be so advantageous for the full automated chemiluminescent immunoassay system.

Acknowledgments
The authors would like to thank Dr. S. Takemura of Intern. Reagents Co. for helpful discussions on chemical synthesis, Dr. T. Shingu for NMR spectra, and Mr. K. Ishibashi for immunoassay.

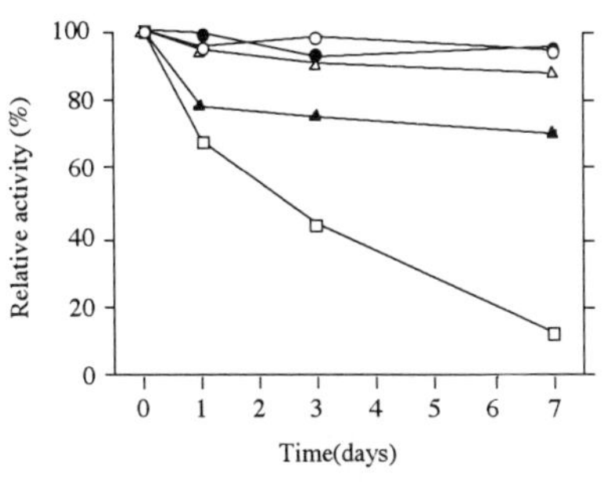

Figure 2. Stability of the conjugate stored at 37℃. The conjugates were prepared with the acridinium esters,compound 18;(○),compound 19;(●),compound 20;(△),compound 21;(▲),compound 22;(□)

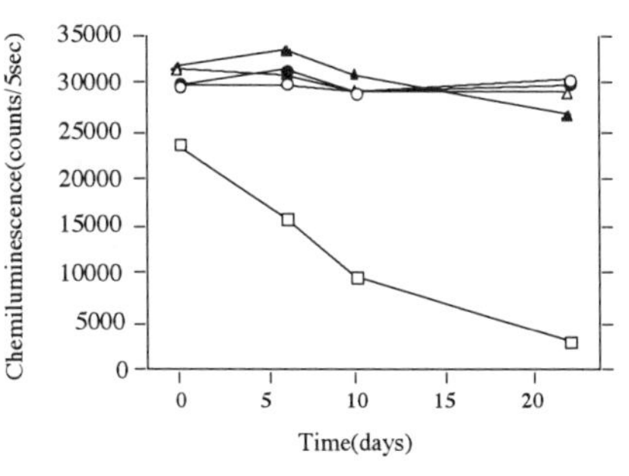

Figure 3. Stability of the conjugates stored at 25℃.The chmiluminescenct immunoassay of CEA was performed by using the conjugates,to be prepared with the acridinium esters,compound 18(○),compound 19(●),compound 20(△), compound 21(▲),compound 22(□)

Table 1. The CL efficiency of the acridinium salts and the heat stability in an aqueous solution.

Compound	R1	R2	Initial CL	Stored time (days) 1	4	7
			counts/mole X10[15]	remained activity (%)		
1	H	H	6493	62	16	8
2	H	2-Methyl-5-nitro	7101	79	30	15
3	H	2-Methyl-4-benzyloxycarbonyl	7009	71	25	13
4	H	2,6-Dimethyl	494	102	93	92
5	H	2,6-Dimethyl-4-benzyloxycarbonyl	6994	106	83	71
6	H	2,6-Dimethyl-4-nitro	7123	110	98	84
7	H	2.6-Dibromo-4-mehtyl	7524	78	41	24
8	H	2,4,6-Triiodo	6762	86	50	28
9	1-Methyl	H	6748	94	63	54
10	1,3-Dimethyl	H	6634	89	65	56
11	1,3-Dimethyl	4-Acetyl	7755	92	60	45
12	1-Methyl	2-Methyl-5-nitro	8705	99	96	97
13	1,3-Dimethyl	2-Methyl-5-nitro	8634	98	102	99
14	1,3-Dimethyl	2-Nitro	5307	49	5	2
15	1-Methyl	2-Bromo	6959	89	61	43
16	1-Methyl	2-Trifluoromethyl	8330	95	100	103
17	1,3-Dimethyl	2-Trifluoromethyl	7952	94	96	101

References

1. Weeks I, Beheshti I, McCapra F, Campbell AK, Woodhead S. Acridinium esters of high-specific-activity labels in immunoassay. Clin Chem 1983; 29: 1474-9.
2. Law S-J, Miller T, Piran U, Klukas C, Chang S, Unger J. Novel poly-substituted aryl acridinium esters and their use in immunoassay. J Biolum Chemilum 1989; 4: 88-98.
3. McCapra F. Hydrolytically stable chemiluminescent labels and their conjugates, and assays therefrom by adduct formation. US Patent 5338847; Chem Abstr 1994; 122: 55905s; Preparation of hydrolytically stable acridiniumcarboxylates as chemiluminescent labels and assays therefrom. US Patent 5284951; Chem Abstr 1994; 121: 157542t.
4. Kinkel T, Luebbers H, Schmidt E, Molz P, Skrzipczyk HJ. Synthesis and properties of new luminescent acridinium-9-carboxylic acid derivatives and their application in luminescence immunoassays (LIA). J Biolum Chemilum 1989; 4: 136-9.
5. Mattingly PG. Chemiluminescent 10-methylacridinium-9-(N-sulfonylcarboxamide) salts. Synthesis and kinetics of light emission. J Biolum Chemilum 1991; 6: 107-14.
6. McCapra F, Watmore D, Sumun F, Patel A, Beheshti I, Ramakrishnan K et al. Luminescent labels for immunoassay - From concept to practice. J Biolum Chemilum 1989; 4: 51-8.
7. Weeks I, Sturgess M, Brown RC, Woodhead JS. Immunoassays using acridinium esters. Methods Enzymol 1986; 133: 366-87.

LOW MOLECULAR WEIGHT THIOL ANTIOXIDANT ACTIVITY IN ORAL FLUIDS

ILC Chapple[1], GHG Thorpe[2], GI Mason[1] and JB Matthews[1]
[1]Dept of Dentistry, The University of Birmingham, Edgbaston, Birmingham UK,
[2]The Wolfson Applied Technology Laboratory, The University of Birmingham UK

INTRODUCTION

The human inflammatory periodontal diseases are amongst the most common of chronic diseases to affect adults. In the United Kingdom, 69% of adults have early signs of disease and only 5% are completely free from clinical signs of inflammation (1). The periodontal complex comprises alveolar bone, periodontal ligament, root cementum and the overlying gingival (gum) tissues (figure 1). Destructive periodontal disease is still a major cause of tooth loss and potential cause of morbidity in medically compromised groups of patients, where a focus of infection and subsequent bacteraemia are a major risk. The polymorphonuclear leukocyte (PMNL) has, primarily, a protective role in the periodontium. However, evidence is emerging from several studies (2,3) that in early onset forms of periodontitis PMNL's are functionally activated and exhibit increased free radical production. There appears to be a delicate balance between inflammatory and immune cell hypofunction, where unchecked pathogens cause direct tissue damage, and hyperfunction, where host defence cell products elaborated in an effort to eliminate pathogens, inadvertently cause substantial collateral host tissue damage (4).

The mouth possesses an epithelial tissue barrier, that separates the internal systems from the external environment. The barrier (figure 1) is called the junctional epithelium (JE) and is permeable to external (bacterial) material passing into the adjacent connective tissue (CT) and blood stream and to products of internal defence systems (fluid, leukocytes, antibodies and cytokines) passing outwards. To assist this vulnerable barrier in protecting the underlying host tissues from damage by the products of bacterial plaque, a fluid is produced from beneath the gingival margin called gingival fluid (GF). The JE is similar in its location and function to the alveolar epithelium in the lungs, where alveolar lining fluid, rather than GF, bathes and protects the lung epithelium. Local production of the antioxidant reduced glutathione (GSH) has been reported in alveolar lining fluid at high concentrations of 400μmol/L (5), with levels being raised in smokers and deficient in patients suffering from pulmonary fibrosis (6) and acute respiratory distress syndrome (7).

Our group has developed an enhanced chemiluminescent assay for determining total antioxidant defence in biological fluids (8). In this paper we report its use (in medicine & disease) with saliva and sub-microlitre volumes of GF (collected non-invasively on filter paper strips) (9), in the investigation of local (saliva) and peripheral (serum) antioxidant (AO) capacities of patients with and without periodontitis. Studies are also reported, which investigate the nature of an antioxidant response seen consistently in GF and saliva, but not detectable in serum from the same patients.

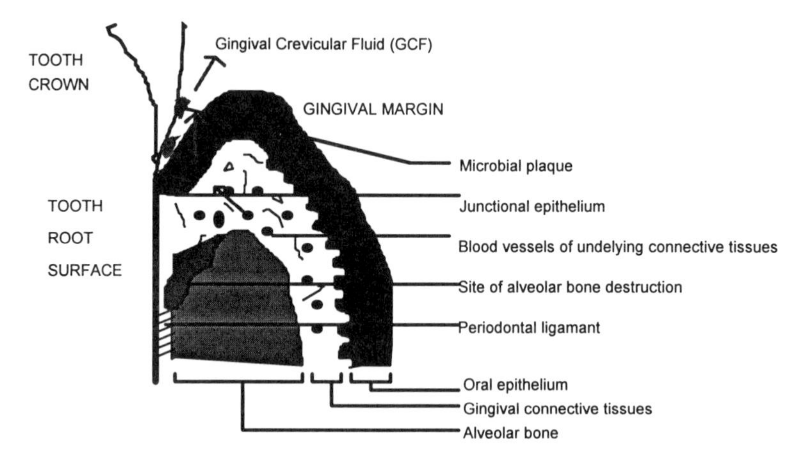

Figure 1. Longitudinal section through a tooth and the adjacent periodontal tissues demonstrating the junctional epithelial (JE) barrier.

Materials and Methods

Patient Volunteers: For the saliva study, the control group comprised 16 patients who were periodontally healthy and 19 patients who demonstrated previous periodontal disease, as determined by clinical and radiological examination. For GF studies, 70 periodontal sites were sampled from 7 patients with disease and 81 sites from 12 periodontally healthy patients.

Fluid Samples: The GF collection was performed for 30 seconds and strips were eluted into 100μL of a PBS buffer by standing at room temperature for 30 minutes. At this stage, collection papers were removed. Saliva samples were collected by asking patients to rinse their mouths with sterile water for 1 minute, followed by a further short rinse and a period of rest for 3 minutes. Stimulated saliva was then collected over a 5 minute period by rolling a sterile marble around the mouth and expectoration. Salivary and GF volumes were measured, and all samples were stored under liquid nitrogen until assay. Serum was obtained from venous blood. Saliva samples were diluted 1:5 and serum samples 1:10 with assay buffer prior to analysis.

Cell Preparations: PMNL's ($4x10^6$ cells/mL) were obtained from the peripheral blood of 5 volunteers (10), incubated with cytochalasin-B and stimulated both aerobically and anaerobically using F-MET-LEU-PHE (FMLP = 2μL 10^{-7} mol/L per 2mL PMNL prep). Supernatant and cytosolic solutions were assayed for a specific AO response. The PMNL intracellular reducing agents, taurine (30μL of $1x10^{-3}$ mol/L), hypotaurine (12μL of $2x10^{-5}$ mol/L), β-NADPH (1μL of $33x10^{-11}$ mol/L $\pm$ $2x10^{-5}$ mol/L oxidised glutathione [GSSG]) were also assayed after addition to 5μL of 1:10 serum. Finally, cysteamine (2μL of 1mmol/L) and cysteine (1μL of 1mmol/L), were assayed in the same manner, with and without pre-incubation with 2μL of 10mmol/L N-ethylmalemide, a thiol inhibitor. Periodontal bacteria (*Fusobacterium nucleatum 369, Veilonella parvula 10790, Streptococcus sanguis I,* 10^9cells/mL) with variable reducing capacities were assayed in triplicate, by adding 20-100μL of suspension to the assay.

Assay Procedure: The assay has been reported previously (8), and is based upon the HRP-catalysed oxidation of luminol by H_2O_2 or perborate, which can be inhibited by solutions containing chain-breaking antioxidants. Light emission from the reaction can be enhanced by the addition of ρ-iodophenol, which prolongs and intensifies light output. The reaction is dependent upon the continuous production of radicals derived from luminol, enhancer and oxygen and the signal can therefore be temporarily suppressed by radical scavenging antioxidant species. The AO capacity of the solution under test, is calculated from standard curves run with a water soluble vitamin-E calibrant (Trolox) and using a fixed recovery point, such as the time taken to reach 10% or 20% of the initial peak signal (T10% or T20%). The AO capacity of the test sample is thus expressed in μmol/L Trolox equivalents. The reaction was performed at room temperature using a BioOrbit 1250 luminometer (Labtech International, Sussex, UK), interfaced with a PC running software which allowed simultaneous recording / display of up to 5000 data points over a period up to 1 hour.

Results and Discussion

Results for the saliva study (table 1) demonstrated no differences between serum AO capacities for the non-periodontitis (NP) and periodontitis (P) groups (range = 340-700 μmol/L; $p > 0.2$), between rates of salivary flow (range = 0.5-6.5 mL/min; $p > 0.4$); or AO production per minute of stimulated saliva flow (range = 20-500 μmol/min; $p > 0.8$). However, when salivary AO capacity was expressed as a concentration, levels for the periodontitis patients (175 ± 53μmol/L) were lower than the non-periodontitis group (254 ± 110μmol/L;$p<0.01$). A kinetically distinct two-step AO response, not seen in serum was found in saliva and more prominently in GF (figure 2a&b). This was reproduced from the cytosol of anaerobically stimulated PMNL's and by β-NADPH, cysteamine and L-cysteine when added to serum.

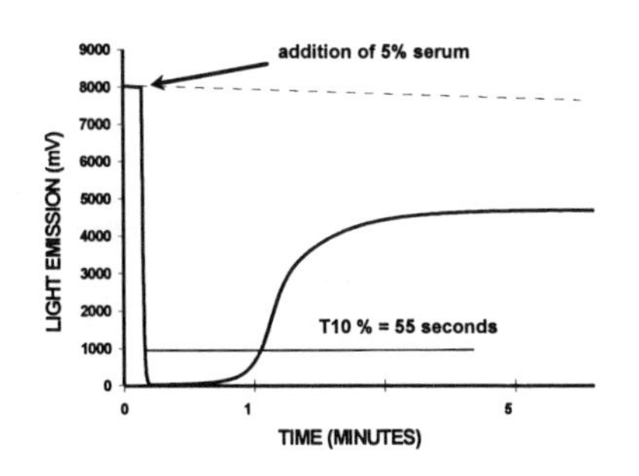

Figure 2a - Serum Response

(---- is signal decay if serum sample not added)

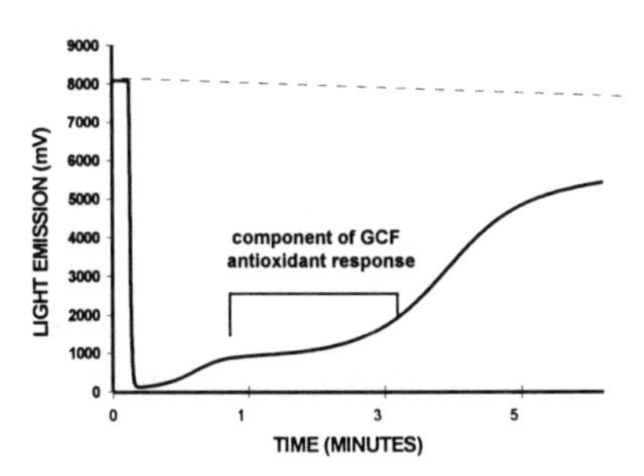

Figure 2b - GF Response

(---- is signal decay if GCF sample not added)

Oxidation of β-NADPH by addition of GSSG removed the two-step response, but this removal could not be repeated with GF samples. N-ethylmalemide removed the response from cysteamine and L-cysteine samples and also from GF samples. No response was elucidated from any of the bacterial samples. Filtration experiments

(Microcon 10: Amicon Inc., Beverly, MA, USA), demonstrated that the thiol responsible for the salivary and GF AO response was below 10Kd in size.

Patient	Salivary antioxidant (AO) capacity		Serum AO capacity
	μmol/min Trolox equiv. (mean n=3)	μmol/L Trolox equiv. (mean n=3)	μmol/L Trolox equivalents (mean n=3)
Non-periodontitis (n=16)	152 ± 129	254 ± 110	495 ± 95
Periodontitis group (n=19)	112 ± 101	175 ± 53	480 ± 77

Table 1. Serum and salivary capacities in patients with and without periodontal disease.

Alterations in local defence systems against free radicals, may reflect an underlying susceptibility to periodontal disease or reflect increased levels of local radical production by phagocytes during episodes of progressive disease. The unusual local antioxidant response reported in this study, but not detected with serum, may play an important role in the maintenance of periodontal, or indeed, general oral health. Further analysis of GF and tissue antioxidants is currently in progress, in an effort to determine the cellular source and identity of the low molecular weight thiol detected.

References

1. Todd JE, Lader D. 1988 adult dental health survey of the United Kingdom. Office of Population Censuses and Surveys, Social Division, 1988:131-44.
2. Asman B, Engstrom PE, Olsson T, Bergstrom K. Increased luminol enhanced chemiluminescence from peripheral granulocytes in juvenile periodontitis. Scand J Dent Res 1984;92:218-23.
3. Shapira L, Borinski R, Sela MN, Soskolne A. Superoxide formation and chemiluminescence of peripheral polymorphonuclear leukocytes in rapidly progressive periodontitis patients. J Clin Perio 1991;18:44-8.
4. Chapple ILC. The role of free radicals and antioxidants in the pathogenesis of inflammatory diseases: emphasis on the periodontal diseases. J Clin Mol Path 1996, in press.
5. Bernard GR. N-acetylcysteine in experimental and clinical acute lung injury. Amer J Med 1991;S3C:54S-59S.
6. Cantin A, North SL, Hubbard RC, Crystal RG. Normal alveolar epithelial lining fluid contains high levels of glutathione. J Appl Physiol 1987;63:152-7.
7. Pacht ER, Timerman AP, Lykens MG, Merola AJ. Deficiency of alveolar fluid glutathione in patients with sepsis and the adult respiratory distress syndrome. Chest 1991;100:1397-403.
8. Whitehead TP, Thorpe GHG, Maxwell SRJ. Enhanced chemiluminescent assay for antioxidant capacity in biological fluids. Anal Chim Acta 1992;266:265-77.
9. Chapple ILC, Cross IA, Glenwright HD, Matthews JB. Calibration and reliability of the Periotron 6000 for individual gingival crevicular fluid samples. J Perio Res 1995;30:73-9.
10. Kalmar JR, Arnold RR, Warbington ML, Gerdner MK. Superior leukocyte separation with a discontinuous one-step Ficoll-Hypaque gradient for the isolation of human neutrophils. J Immunol Meth 1988;110:275-81.

NEW CHEMILUMINESCENT ASSAY OF β-D-GALACTOSIDASE BASED ON LIGHT EMISSION OF INDOLE DERIVATIVE

H Arakawa, M Maeda and A Tsuji
School of Pharmaceutical Sciences, Showa University, Shinagawa-ku, Tokyo 142, Japan

Introduction

β-D-galactosidase (β-gal) is used as label enzyme in enzyme immunoassay (EIA) and DNA-probe assay and is usually detected by fluorimetric and colurimetric method. Recently, high sensitive bio- and chemi-luminescent method have developed for detecting the β-gal label used in EIA and DNA probe (1-4). In the present paper, we have developed a novel chemiluminescent assay of β-gal based on chemiluminescence of indol. 5-Bromo-4-chloro-3-indolyl-β-D-galactopyranoside (X-gal) was used as substrate for β-gal and also as light emitter. X-gal is hydrolyzed by β-gal to liberate free indoxyl followed by oxidation to indigo dye and simultaneously produces hydrogen peroxide (H_2O_2). H_2O_2 is reacts with the residual X-gal in the presence of horseradish peroxidase (HRP), to emit light.

The measurable range of β-gal obtained by the proposed method was 6 x 10^{-14} mol/L to 6 x 10^{-11} mol/L, the detection limit was 3 a mol / assay. The coefficient of variation (CV, n=5) was examined at each point of the standard curve. The mean CV percentage was 5.5 %. This assay system was applied to enzyme immunoassay of thyroxine using β-gal as label enzyme.

Materials and Methods

The chemiluminescence intensity was measured by an Aloka Luminescence Reader (Tokyo, Japan). β-gal (E.C 3.2.1.23) and HRP (E.C 1.11.1.7.)were purchased from Boehringer-Mannheim-Yamanouchi, Co.(Tokyo, Japan). X-gal were purchased from Wako Pure Chemical Industries, Ltd. (Tokyo, Japan). Thyroxine(T_4), anti T_4 antibody coated solid phase and β-gal-thyroxine conjugate were donated by Dainippon Pharmaceutical Co., LTD. (Osaka, Japan). Other chemicals were of analytical reagent grade.

Procedures : 1. CL assay of β-gal A 50 μL of sample solution in 0.05 mol/L phosphate buffer (PB pH7.0) containing 0.1% bovine serum albumin (BSA) was added to a microtest tube, followed by addition of 100 μL of 1 mmol/L X-gal solution in 0.05 mol/L PB (pH8.5) containing 0.1 %BSA and 0.5 mmol/L $MgCl_2$. The reaction mixture was incubated at 37℃ for 30 min and then 50 μL of 4.5 μmol/L HRP in 0.05 mol/L phosphate buffer (PB pH7.0) was added. The CL intensity was measured by an Aloka luminescence reader for the 6s interval from 15 to 21 s after addition of the CL reagent solution.

2. CL EIA of thyroxine The standard procedure (MARKIT Thyroxine) was slightly improved in order to apply the proposed chemiluminescence method to the determination of β-gal activity at the final step in the assay. The T_4 standard solution, T_4-β-D-galactosidase conjugate solution and insolubilized anti-T_4 antibody suspension were added to a test tube and incubated for 30 min at 37℃. After incubation, 2 ml saline solution was added to each tube, which was centrifuged at 3000 rpm for 10 min.

The supernates were decanted and the precipitate was suspended by adding 250 μL of 0.01 mol/L PB (pH8.5) containing 0.5 mmol/L $MgCl_2$. A 50 μL of the suspension was transferred to another tube and the enzyme activity (bound fraction) was assayed by the chemiluminescence method described above.

Results and Discussion

β-gal has been widely used as the label enzyme in EIA and DNA probe assay. A sensitive assay of β-gal is very important for developing more sensitive these assays. A sensitive chemi- and bioluminescence method for β-gal have been reported (1,2,4) . We also developed a sensitive chemiluminescent assay of β-gal using X-gal as a substrate (3,5). The principle is that X-gal produce hydrogen peroxide by hydrolysis of β-gal, which was detected by adding isoluminol and micro peroxidase. Philbrook et al (6) reported that indol itself emit light by oxidation of $K_2S_2O_8$ in alkaline solution. As the preliminary test, it was studied whether various indole derivatives (X-gal, 5-bromo-4-chloro-3-indolyl phosphate (BCIP) and indolyl phosphate), which are used as substrate for alkaline phosphatase and β-gal, emits chemiluminescence in the presence of hydrogen peroxide and HRP. Among these indole derivatives, X-gal emitted intense chemiluminescence. Therefore, it is a possible to develop chemiluminescent assay of β-gal in the presence of HRP and the excess X-gal, that is, isoluminol used in the previous our papers (3,5) is not required in this assay. The a possible mechanism for this assay is shown in Fig.1.

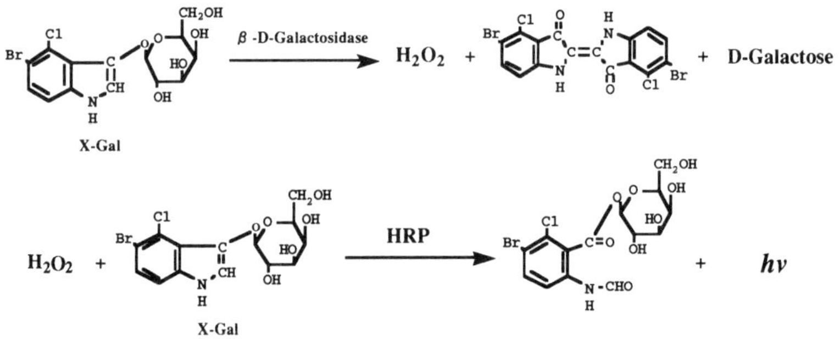

Figure 1. A possible mechanism for the proposed assay

CL assay of β-gal

The effects of various factors (concentrations of the reagents used, pH, reaction time) on CL assay of β-gal were examined to establish the optimal assay conditions and the optimum was as described in methods.

Standard curve of β-gal was obtained according to the procedure. As shown in Fig.2, a working curve was obtained in the range of 6 x 10^{-14} mol/L to 6 x10^{-11} mol/L. The CV percentages (n=5) were 5.5%. The detection limit (blank+2SD) was 3x10^{-18} mol/assay. The sensitivity obtained by the propsed method is one order lower than the value obtained by the method of the previous paper (3, 5). However, the sensitivity is corresponding to the chemiluminescent assay for β-gal using AMPGD (2).

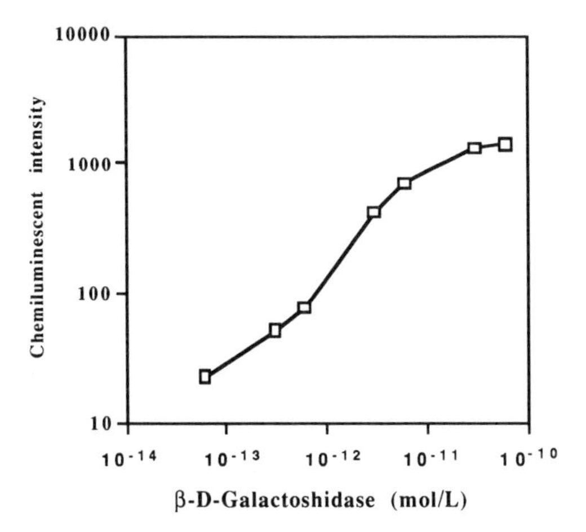

Figure 2. Standard curve of β-D-galactoshidase

Application of CL assay of β-gal in enzyme immunoassay

The new chemiluminescent assay for β-gal was applied to detect the β-gal conjugate in EIA. Immunoassay was carried out according to the method described in procedure. The measurable range for T_4 is 32 to 320 nmol/L. This sensitivity sufficed to determine T_4 in serum clinical sample. Figure 3 illustrates the standard curve of T_4 by this CL EIA.

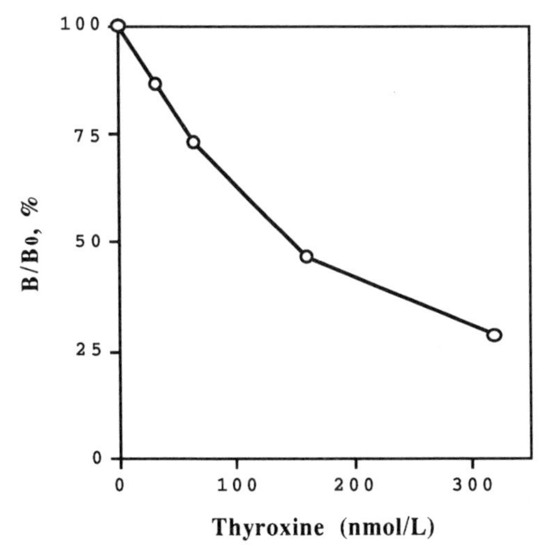

Figure 3. Standard curve of T4 by CL-EIA

In conclusion, CL assay of β-gal based on indole chemiluminescence have developed. X-gal was used as substrate for enzyme and light emitter. This assay systems of β-gal are sensitive, simple , and inexpensive. The CL assay of β-gal has been applied to the EIA for thyroxine. Further studies on the application of this system are in progress in our laboratory.

References

1. Tanaka K, Ishikawa E. A highly sensitive bioluminescent assay of β-D-galactoshidase from Esherichia coli using 2-nitro-phenyl-β-D-galactopyranoside as a substrate. Anal Lett, 1986; 19:433-44.
2. Bronstein I, Edwards B, Voyta JC. 1,2-Dioxetanes: novel chemiluminescent enzyme substrates. Applications to immunoassay. J Biolumin Chemilumin 1989;4:99-111.
3. Arakawa H, Maeda M, Tsuji A. Chemiluminescent assay of various enzymes using indoxyl derivatives as substrate and its applications to enzyme immunoassay and DNA probe assay. Anal Biochem 1991;199:238-42.
4. Maeda M, Shimizu S, Tsuji A. Chemiluminescence assay of β-D-galactoshidase and its application to competitive immunoassy for 17α-hydroxyprogesterone and thyroxine. Anal Chim Acta, 1992; 266:213-7.
5. Arakawa H, Ikegami T, Maeda M, Tsuji A. Chemiluminescent enzyme immunoassay for Alpha-fetoprotein using β-D-galactoshidase as label and 5-bromo-4-chloro-3-indolyl-β-D-galactopyranoside as substrate. J Biolumin Chemilumin 1993;8:135-9.
6. Philbrook GE, Maxwell MA, Taylor RE, Totter JR. The chemiluminescence of certain indoles. Photochem Photobiol 1965;4:869-876.

CHEMILUMINESCENCE TRIGGERED BY ESTERASES: A METHOD FOR THE DETERMINATION OF LIPASE ACTIVITY

LH Catalani[1], NG Malta[1] and A Campa[2]
[1]Instituto de Química and [2]Faculdade de Ciências Farmacêuticas
Universidade de São Paulo, CP 05599-970, São Paulo, 05599-970, Brazil

Introduction

The determination of lipase and other esterase activities in biological fluids and tissues is of great importance to medicine and fundamental research. In clinical chemistry, determinations of serum pancreatic lipase is a powerful tool in the diagnosis of acute pancreatitis (1,2). The available methodology for lipase activity has several problems such as long incubation times, unstable and non reproducible substrates, need for very sensititve devices and lack of discrimination between active and inactive enzyme (3). Here we describe a sensitive and specific method to determine lipase activity based on the the hydrolysis of 2-methyl-1-propenyl laurate (MPL) and 2-methyl-1-propenyl acetate (MPA) coupled with the horseradish peroxidase/H_2O_2/O_2 oxidative system.

The horseradish peroxidase catalyzed oxidation of isobutanal leads to formation of electronically excited triplet acetone which produces chemiluminescence in a normally aerated medium. The true substrate of this oxidation, which runs at expense of hydroperoxides like H_2O_2, has been shown to be the enol form of the aldehyde (4).

We have recently shown that liver esterase can trigger chemiluminescence in the hydrolysis of 2-methyl-1-propenyl benzoate (MPB) in the presence of the HRP/H_2O_2/O_2 system (5).

ESTERASE
H_2O
+
HRP/H_2O_2/O_2
+ HCOOH

The luminescence was shown to be acetone phosphorescence which is protected against oxygen deactivation and is proportional to the esterase activity. The addition of ~10 µmol/L of sodium 9,10-dibromoanthracene-2-sulfonate (DBAS) caused a 30-fold enhancement of the luminescence. Furthermore, we found that the peroxidase used can be substituted by a commercial HRP-IgG conjugate, and the light intensity can be regulated by the amount of conjugate used (5). This important finding extends the application of this system to immunoassays.

Our goal with the use of MPL and MPA is to confer specificity to the hydrolase activity, in order to achieve more accurate measurements of lipase in the serum and tissues.

Materials and Methods

Instrumentation: Light emission was measured on an in-house constructed photon-counter using a EG&G Model 1121A discriminator (EG&G PARC, Princeton, USA) coupled to a EMI-9658RM photomultiplier tube (Thorn-EMI, Middlesex, England) in a Fact 50MK3 housing-cooler (Thorn-EMI, Middlesex, England).

Reagents: The 2-methyl-1-propenyl laurate synthesis was adapted from literature (6). Its identity was confirmed by conventional ^{1}H- ^{13}C-NMR and mass spectrometry. Porcine pancreas lipase Type II, horseradish peroxidase Type VI, bile salts and lecithin were from Sigma.

Procedure: unless otherwise stated the standard reaction mixture contained 50 mmol/L phosphate buffer pH 7.4, 3.0 μmol/L HRP, 0.1 mmol/L H_2O_2, 2 mmol/L bile salt, 4 mmol/L MPL (added with 2 minutes of vortex mixing) and 5-100 U/mL of lipase. The assays were performed at 33°C or 37°C and were started by addition of lipase.

Results and Discussion

Analogous to the hydrolysis of MPB, the enol form of isobutanal is also released by action of wheat lipase and pancreatic lipase on MPA. In the presence of the $HRP/H_2O_2/O_2$ system, acetone phosphorescence is emitted with an intensity proportional to the amount of lipase (6). Furthermore, we observed that purified hydrolases can be substituted by serum. While correlation of the amount of serum with integrated light intensity was detected, the previous administration in vivo of intravenous heparin resulted in an increase of total light, providing evidence that lipoprotein lipase also recognizes and can hydrolyze MPA.

The lauryl derivative (dodecyl) proved to be much less soluble than MPA. Nevertheless, the light intensity observed, even with a milky solution, was much more intense. Likewise, the controls without HRP, lipase or substrate indicated the emission is dependent on all the reactants. Figure 1 shows the emission profile of different amounts of pancreatic lipase.

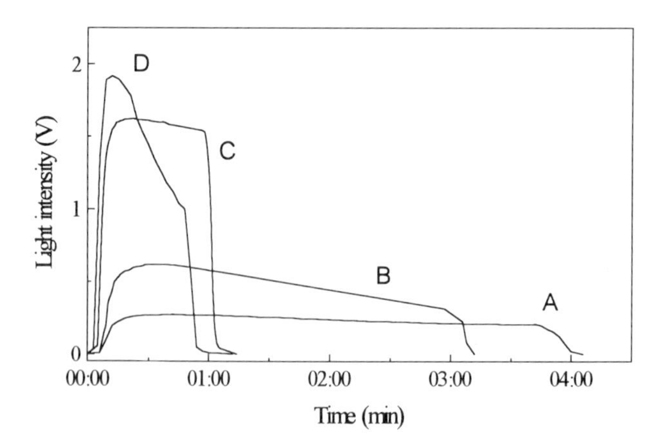

Figure 1: Emission profile of MPL/pancreatic lipase + $HRP/H_2O_2/O_2$ system using standard conditions at 37°C. [lipase] (in U/mL) = (A) 2.3; (B) 3.9; (C) 9.3; (D) 14.

The lengthening of the emission profile with the decrease of lipase concentration indicates that the hydrolysis of MPL is the rate limiting step. Bile salts were used as surfactants to increase MPL solubility. Although it is widely accepted that bile salts inhibit lipase activity, under the low concentration used we did not observe inhibition (7). The best results, however, were found using lecithin as co-surfactant. When 31 mg/L of lecithin was added, the MPL solution became clear and the maximum intensity increased 1.5 times. Sonication also proved to be an efficient method to dissolve MPL. After 20 minutes sonication of substrate stock solution, the maximun intensity increased ten fold.

Figure 2 shows the linear dependence of the maximum intensity with lipase concentration, demonstrating that this method can be used to determine lipase activity.

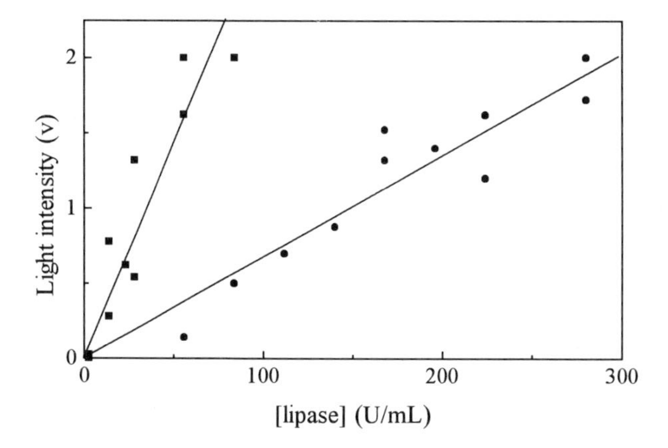

Figure 2: Maximun light intensity dependence with lipase concentration using standard conditions at (●) 33°C and (■) 37°C.

The assay range for lipase concentration at 37°C extends from 10 U/mL linearly to 0.5 U/mL. The lipase Type II used is a crude extract and cannot be used to establish a definitive working range for this method and the system is stillfar from optimized. For instance, the addition of sensitizers like DBAS or chlorophill-*a* can also enhance the emission, as in the case of MPB (5) and MPA (6). Furthermore bile salt and lecithin concentrations and temperature could also be optimized.

The light emission profile observed for MPL substrate offer some advantages in relation to that observed with MPA. As shown in Figure 1, the maximum emission is reached a few seconds after lipase addition and can be easily recorded. Furthermore the reaction is faster than the conventional reactions used to determine lipase activity.

The 2-methyl-1-propenyl ester linked to a long carbon chain allows a substrate organization more similar to the natural substrate of lipases, *i.e.*, triglyceride emulsion. An emulsion formed by substrate and high bile salt concentration in the presence and absence of colipase could discriminate between pancreatic lipase from lipoprotein lipase, carboxyl esterase and other non specific serum hydrolase activities.

Acknowledgments
We thank FAPESP, CNPq, CAPES and FINEP for financial support. We also thank Dr. Thérèse Wilson for critically reading this manuscript.

References

1. Moss DW, Henderson AR, Kachmar JF. Enzymes. In: Tietz NW, editor. Textbook of clinical chemistry. Philadelphia: WB Sauders, 1986:735-42.
2. Stein EA. Lipids, lipoproteins and apolipoproteins. In: Tietz NW, editor. Textbook of clinical chemistry. Philadelphia: WB Sauders, 1986:860-1.
3. McNeely MDD. Lipase. In: Kaplan LA, Pesce AJ, editors. Clinical chemistry. Theory, analysis and correlations. St. Louis: C.V.Mosby, 1989:930-3.
4. Dunford HB, Baader WJ, Bohne C, Cilento G. Peroxidase-catalyzed formation of triplet acetone and chemiluminescence from isobutyraldehyde and molecular oxygen. Biochem Biophys Res Commun 1984;122:28-32.
5. Yavo B, Campa A, Catalani LH. Esterase coupled with the H_2O_2/horseradish peroxidase system triggers chemiluminescence from 2-methyl-1-propenybenzoate: a potential analytical tool for esterase analysis. Anal Biochem 1996;234:215-20.
6. Campa A, Andrade AC, Catalani LH. Chemiluminescence triggered by hydrolase activity in a horseradish peroxidase/H_2O_2-coupled assay. Photochem Photobiol 1996;63:742-5.
7. Morgan RGH, Hoffman NE. The interaction of lipase, lipase cofactor and bile salts in triglyceride hydrolysis. Biochim Biophys Acta 1971;248:143-8.

CHEMILUMINESCENT HALOALKOXY-SUBSTITUTED DIOXETANES: PROPERTIES AND APPLICATIONS

H Akhavan-Tafti, Z Arghavani, RA Eickholt, RS Handley, MP Perkins, KS Rashid and AP Schaap*
Lumigen, Inc., 24485 W. Ten Mile Rd., Southfield, MI 48034 USA

Introduction

The first chemiluminescent dioxetane substrate for alkaline phosphatase, 4-methoxy-4-(3-phosphoryloxyphenyl)spiro[1,2-dioxetane-3,2'-tricyclo[3.3.1.$1^{3,7}$] decane], disodium salt, (1) has found widespread use in medical diagnostics and life science research. Various modifications to the structure of this dioxetane have been reported in attempts to improve its properties for use in chemiluminescent assays. Substitution of one or more of the adamantyl ring hydrogens by halogen, alkoxy or carboxyl groups has been reported (2). Substitution of one or more of the phenyl ring hydrogens by a halogen or alkoxy group also has a significant effect on decay kinetics and chemiluminescence quantum yield presumably by affecting the electron transfer properties of the aromatic fragment (3,4). The effect of substitution of other groups in place of the methoxy group on the dioxetane ring has, however, been little studied.

Materials and Methods

Instrumentation: Luminescence measurements were performed using either a LabSystems Luminoskan microplate reader with white opaque 96-well strips or a Turner Designs TD-20e tube luminometer.

Reagents: Dioxetane 11 is commercially available from Lumigen. Dioxetanes 2-5 and 12 were prepared using a reaction scheme featuring a titanium-mediated coupling of a haloalkyl ester and adamantanone as previously described (5). Dioxetanes 6-10 were prepared by a new process in which the haloalkoxy group is substituted for an alkythio or arylthio group in a sulfur-substituted dioxetane. The latter reaction is described in a companion presentation. Dioxetane 13 was prepared using the methods described in reference 5 starting with 3-amino-4-chlorobenzoic acid.

Results and Discussion

Our hypothesis that the presence of electron-withdrawing groups in an alkoxy substituent may influence the electron-transfer process associated with the chemiluminescent decomposition of the dioxetane has led to the synthesis of novel 1,2-dioxetane substrates for alkaline phosphatase which have short half-lives for the grow-in kinetics of light emission. The dioxetanes were prepared by dye-sensitized photooxygenation of a vinyl ether precursor or by a new substitution reaction using a sulfur-substituted dioxetane precursor. The haloalkoxy-substituted dioxetanes are represented by the formula

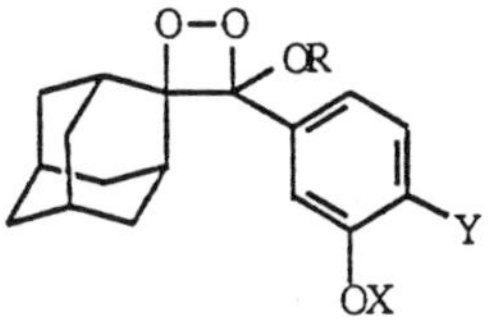

O–O OR Y OX

where OR is a fluorine or chlorine-substituted haloalkoxy group of 1-4 carbon atoms, X is a protecting group which is removed to form an unstable aryloxide-substituted intermediate and Y is H or Cl as shown in Table 1. The haloalkoxy group contains 1-7 halogens but is unsubstituted at C-1.

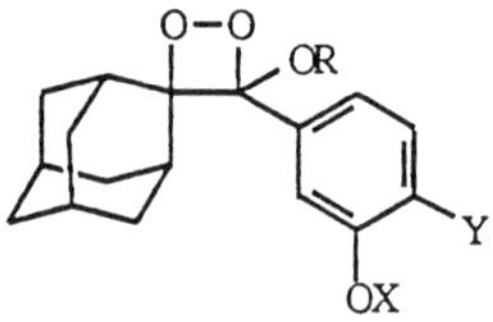

Dioxetane	R	X	Y
1	CH_3	H	H
2	CH_2CF_3	H	H
3	CH_2CHF_2	H	H
4	CH_2CH_2F	H	H
5	CH_2CH_2Cl	H	H
6	CH_2CHCl_2	H	H
7	CH_2CCl_3	H	H
8	$CH(CF_3)_2$	H	H
9	$CH_2CF_2CF_3$	H	H
10	$CH_2CF_2CF_2CF_3$	H	H
11	CH_3	PO_3Na_2	H
12	CH_2CF_3	PO_3Na_2	H
13	CH_2CF_3	PO_3Na_2	Cl

Table 1. Dioxetane Compounds Prepared

Comparison of Rates of Base-Induced Decomposition of Hydroxy Dioxetanes: The first order decay of chemiluminescence of dioxetanes 1-10 in 0.2 M 2-methyl-2-amino-1-propanol buffer, pH 9.6 containing 1.0 mg/mL of the enhancer 1-(tri-n-octylphosphoniummethyl)-4-(tri-n-butylphosphoniummethyl)benzene dichloride at 37 °C was measured in a luminometer. Dioxetanes with at least one fluorine or two chlorine atoms exhibited significantly faster decay kinetics and higher maximum chemiluminescence intensities than the structurally related methoxy analog 1. Only the monochloro-dioxetane 5 showed a slower half-life than dioxetane 1.

Dioxetane	$t_{1/2}$ (min) 37 °C
1	15.9
2	2.0
3	2.8
4	6.1
5	17.1
6	4.1
7	3.7
8	0.16
9	3.4
10	4.9

Table 2. Kinetics of Light Emission from Hydroxy Dioxetanes

Several effects are apparent from these results. Increasing halogenation at C-2 of the haloalkoxy group causes a progressive increase in the rate of decomposition, both in the fluorine series 4, 3 and 2 and in the chlorine series 5, 6 and 7. Compound 8 with a branched alkoxy group effectively containing 6 fluorine atoms at C-2 decayed 100 times faster than 1. Halogen substitution at positions farther out on the alkyl chain than C-2 appeared to slow the rate of decomposition as seen from a comparison of 3, 9 and 10, each of which have 2 fluorine atoms at C-2.

The half-life of decay of chemiluminescence ($t_{1/2}$) of dioxetanes 1 and 2 in 0.2 M 2-methyl-2-amino-1-propanol buffer, pH 9.6 containing 0.88 mM Mg^{+2} and 1.0 mg/mL of 1-(tri-n-octylphosphoniummethyl)-4-(tri-n-butylphosphonium-methyl)benzene dichloride correlate with the times required to reach the maximum light intensity (I_{max}) in the alkaline phosphatase-triggered decomposition of the corresponding phosphate dioxetanes 11 and 12 in the same formulation. The times to reach 95% of maximum light intensity from alkaline phosphatase-triggering of dioxetanes 11 and 12 were 32 and 6 min, respectively (Figure 1). The half-life of decay of luminescence of a hydroxy dioxetane is, therefore, useful for predicting the grow-in kinetics of light emission for phosphatase triggering of the corresponding phosphate dioxetane. In particular, hydroxy dioxetanes which show a faster $t_{1/2}$ than dioxetane 1 indicate that the corresponding phosphate dioxetanes are expected to reach I_{max} more quickly than dioxetane 11.

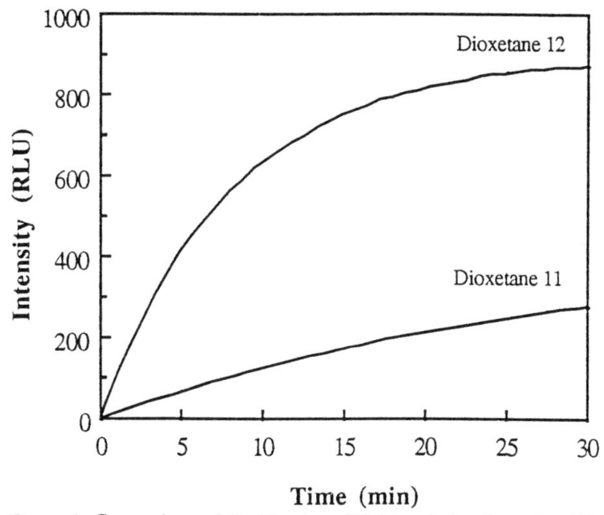

Figure 1. Comparison of the kinetics of light emission from the AP-triggered decomposition of dioxetanes 11 and 12 at 37 °C.

Additionally, a formulation containing dioxetane 12 achieved a higher plateau light intensity compared to dioxetane 11 (Figure 1). The phosphate ester of the trifluoroethoxy-substituted dioxetane provides a sensitive reagent for the chemiluminescent detection of alkaline phosphatase. A solution assay resulted in a detection limit of 1.25 zeptomol of AP (blank + 2σ). Dioxetanes substituted with the haloalkoxy group are useful for the detection of enzyme-labeled analytes in immunoassays, DNA probe assays and blotting applications.

With this knowledge in hand, we sought next to test whether the known chemiluminescence enhancing effect of chlorine substitution on the benzene ring could be combined with the accelerated reaction kinetics from haloalkoxy substitution to produce a dioxetane with unique properties. This seemed possible since the former effect presumably operates through the π system while the effect of the haloalkoxy group should be of a different nature since the halogen substituents are separated from the π system of the incipient carbonyl cleavage product by 3 σ bonds. To this end, we prepared phosphate dioxetane 13 and compared its AP-triggered chemiluminescence plateau intensity and rise time relative to dioxetane 12. Solutions of either dioxetane 12 or 13 in 0.2 M 2-methyl-2-amino-1-propanol buffer, pH 9.6 containing 0.88 mM Mg^{+2} and 1.0 mg/mL of the enhancer were reacted with 8 x 10^{-17} moles of AP at 37 °C. The relative chemiluminescence profiles in Figure 2 demonstrate the combined operation of the two effects in dioxetane 13. Unexpectedly and in contrast to the results of previous studies in which chlorine substitution on the benzene ring slowed chemiluminescence kinetics (3,4), the rise time to Imax is actually shorter for 13 than for 12.

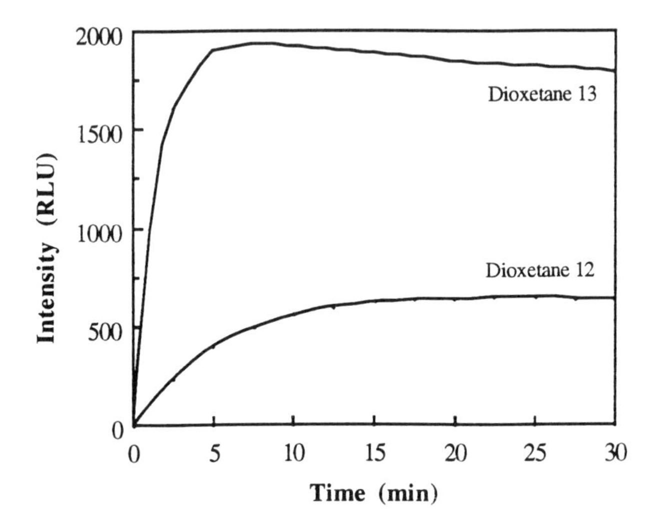

Figure 2. Comparison of the kinetics of light emission from the AP-triggered decomposition of dioxetanes 12 and 13 at 37 °C.

References

1. Schaap AP. Chemical and enzymatic triggering of 1,2-dioxetanes. Photochem Photobiol 1988;47S:50S.
2. Bronstein I, Juo R, Voyta J, Edwards B. In: Stanley PE, Kricka LJ, editors. Novel chemiluminescent adamantyl 1,2-dioxetane enzyme substrates. Bioluminescence and Chemiluminescence Current Status. Chichester: Wiley, 1991:73-82.
3. De Ladonchamps A. I. Arylsulfatase-induced chemiluminescence from 4-methoxy-4-(3-sulfatephenyl)spiro(1,2-dioxetane-3,2'-adamantane) II. Base-induced chemiluminescence from 4-(2-chloro-5-hydroxyphenyl)-4-methoxyspiro(1,2-dioxetane-3,2'-adamantane). Masters Thesis, Wayne State University, Detroit, 1990.
4. Edwards B, Sparks A, Voyta J, Bronstein I. In: Campbell AK, Kricka LJ, Stanley PE, editors. New chemiluminescent dioxetane enzyme substrates. Bioluminescence and Chemiluminescence Fundamentals and Applied Aspects. Chichester: Wiley, 1994:56-9.
5. Akhavan-Tafti H, Arghavani Z, Eickholt R, Rashid K. Novel 1,2-dioxetane compounds with haloalkoxy groups, methods of preparation and use. PCT Application WO 96/15122 1996.

CHEMILUMINESCENT DETECTION OF OXIDASE ENZYMES BY PEROXIDASE-MEDIATED OXIDATION OF ACRIDAN COMPOUNDS

H Akhavan-Tafti, R DeSilva, Z Arghavani, RA Eickholt, RS Handley, BA Schoenfelner and AP Schaap*
Lumigen, Inc., 24485 W. Ten Mile Rd., Southfield, MI 48034 USA

Introduction

We have previously shown (1) that acridancarboxylic acid derivatives are efficiently oxidized by a peroxidase enzyme and a peroxide to produce chemiluminescence. This reaction serves a sensitive method for chemiluminescent detection of peroxidase activity and for peroxidase conjugates in enzyme-linked assays. Enzyme-catalyzed acridan chemiluminescence can also function as a sensitive reporter for the presence of peroxide. Since oxidase enzymes and their substrates undergo a reaction with molecular oxygen to produce hydrogen peroxide, we have investigated the use of the acridan/peroxidase chemiluminescent system to detect hydrogen peroxide generated by various oxidase enzymes. Numerous oxidase enzymes or their substrates are of clinical and biological importance. In addition, glucose oxidase has been employed as a label in immunoassays. The coupled enzymatic reaction presented here represents a new chemiluminescent method for the detection of oxidase enzymes and their substrates.

Materials and Methods

Reagents: Horseradish peroxidase, type VI (HRP) and glucose oxidase, type VII.S (GOD) were from Biozyme (San Diego, CA USA). Choline oxidase and alcohol oxidase were obtained from Sigma (St. Louis, MO USA). The acridan compound 2,3,6-trifluorophenyl 3-methoxy-10-methylacridan-9-carboxylate was prepared as described previously (2). ß-D-Glucose and choline chloride were from Sigma.
Instrumentation: Luminescence measurements were performed on a LabSystems Luminoskan microplate reader using white opaque 96-well strips.
Procedures: In general, solutions of the analyte (10-100 µL) were manually pipetted into the plate. The reagent containing the acridan and HRP was then automatically injected using the instument's dispenser and reaction timing commenced. All assay results are the average of triplicate determinations at each concentration of analyte.

Results and Discussion

Coupled enzyme assays have been developed utilizing either a one-step or a two-step reaction. In the one-step reaction, all reagents are mixed together and the two enzymatic reactions proceed concurrently. In the two-step reaction, the oxidase enzyme and substrate undergo a preliminary incubation and are then reacted with the peroxidase, acridan substrate and enhancer. The amount of enhancer, nonionic surfactant and acridan compound have been fixed in all assays at levels previously shown to provide excellent sensitivity for detection of HRP. Reaction pH and reaction time were set after examining the time dependence of light emission over the range of analyte concentrations studied. The amount of oxidase enzyme or substrate, respectively, was set, after preliminary evaluations, in order to provide the widest dynamic range for determining the other reactant.

Conceptually, the assay can be depicted as shown in the scheme below in which either the substrate or the oxidase enzyme is the analyte. Additionally, the oxidase enzyme may be a conjugate used as a detectable label for a high affinity binding agent. The acridan (Lumigen PS-1) is enzymatically oxidized to produce an intermediate acridinium ester which reacts with peroxide to generate light.

Oxidase Enzyme + Substrate $\longrightarrow$ H_2O_2

$\xrightarrow[\text{Enhancer}]{\text{HRP, } H_2O_2}$ $\xrightarrow{H_2O_2}$ $h\nu$

Lumigen PS-1

Glucose Oxidase The following assay conditions for a one-step assay for GOD were used. Ten µL portions of GOD serial dilutions containing between 2.12×10^{-13} mol and 2.12×10^{-19} mol were added to the wells. The detection reagent consisted of 0.01 mol/L tris buffer, pH 8.0, 50 µmol/L Lumigen PS-1, 100 µmol/L 4-phenylphenol, 0.11 mol/L glucose, 0.025% (v/v) Tween 20 and 1.4×10^{-11} mol/L HRP. A 100 µL portion of the detection reagent was added and the reaction solution incubated at ambient temperature. The time profile of light intensity varied significantly over the range of GOD concentrations tested. A five minute incubation period was chosen empirically for light intensity measurement and provided the widest dynamic range of response. A typical dose-response curve, expressed as the log of the signal corrected for the blank (S-B) vs. the log of the amount of enzyme is shown in Figure 1.

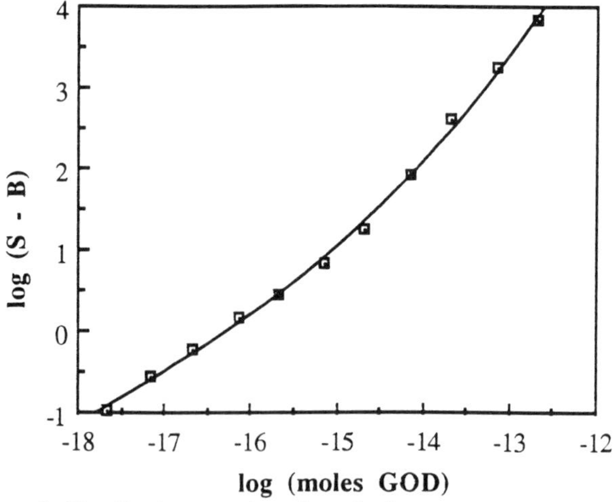

Figure 1. Chemiluminescence intensity at 5 min vs. amount of glucose oxidase using a reagent containing glucose, HRP and Lumigen PS-1

The detection limit using this limited number of replicates is between 2 and 7 amol/assay when defined as the signal equal to the blank + 2 σ. This level of sensitivity compares well with the value of 30 amol/assay reported in another chemiluminescent GOD assay in which the H_2O_2 produced was detected with peroxyoxalate luminescence (3). Diluting the concentration of HRP in the detection reagent by ten-fold introduced a marked curvature into the dose-response curve. However, when light intensity was measured at 15 min, the result was nearly the same as in Figure 1. Conducting the reaction at pH 7.0 produced a triphasic dose-response curve in which light intensity was nearly constant when the amount of GOD was in the range of 2 x 10^{-15} mol and 2 x 10^{-16} mol.

Glucose A two step assay was developed in which glucose standards in 0.01 mol/L phosphate buffer, pH 6.0 were incubated for 15 min with 5.3 x 10^{-6} mol/L GOD at ambient temperature. A 100 µL aliquot was combined with 100 µL of detection reagent and light intensity integrated for 75 s. The detection reagent consisted of 0.01 mol/L tris buffer, pH 8.0, 50 µmol/L Lumigen PS-1, 100 µmol/L 4-phenylphenol, 0.025% (v/v) Tween 20 and 1.4 x 10^{-11} mol/L HRP. As shown by the results in Figure 2, light intensity displayed a monotonic increase with the amount of glucose over the range 6-0.003 mmol/L (108-0.054 mg/dL).

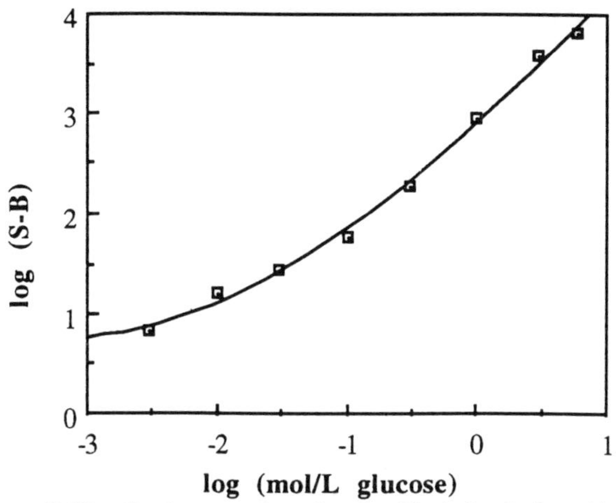

Figure 2. Chemiluminescence intensity after 15 min incubation with 5.3 µmol/L GOD followed by reaction with a reagent containing HRP and Lumigen PS-1 vs. amount of glucose.

Choline Oxidase A one step assay for choline oxidase was developed using the following assay conditions. Ten µL portions of choline oxidase (ChO) serial dilutions in water containing between 4.8 x 10^{-12} mol and 4.8 x 10^{-16} mol were added to the wells. The detection reagent consisted of 0.01 mol/L tris buffer, pH 8.0, 50 µmol/L Lumigen PS-1, 100 µmol/L 4-phenylphenol, 14 mmol/L choline chloride, 0.025% (v/v) Tween 20 and 1.4 x 10^{-11} mol/L HRP. A 100 µL portion of the detection reagent was added and the reaction solution incubated at ambient temperature for five min. Light intensity displayed a monotonic increase with the amount of ChO over the range 4.8 x 10^{-12} mol to 4.8 x 10^{-15} mol /assay. The use of choline iodide as the substrate resulted in the total quenching of light emission.

Chemiluminescent detection of choline and acetylcholine with a three enzyme system system employing acetylcholinesterase, choline oxidase, HRP and either luminol or a dimethylaminonaphthalene hydrazide (4) has previously been reported. Comparison with the present work is difficult since our results were based on using a limiting amount of ChO rather than the substrate.

Ethanol A two step assay was developed in which ethanol standards in the range 0.001 - 0.3 % (v/v) in 0.01 mol/L phosphate buffer, pH 7.5 were incubated for 15 min with 1.0 mg/L alcohol oxidase at ambient temperature. A 100 µL aliquot was combined with 100 µL of detection reagent and light intensity measured after 1.25 and 5 min. The detection reagent consisted of 0.01 mol/L tris buffer, pH 8.0, 50 µmol/L Lumigen PS-1, 100 µmol/L 4-phenylphenol, 0.025% (v/v) Tween 20 and 1.4 x 10^{-11} mol/L HRP. Under these conditions, light intensity was linear with the amount of ethanol over the range 0.001-0.1 % as shown by the results in Figure 3.

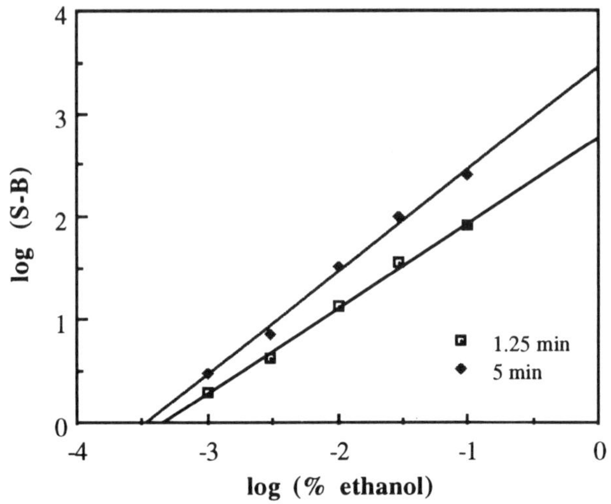

Figure 3. Chemiluminescence intensity after 15 min incubation with 1 mg/L alcohol oxidase followed by reaction with a reagent containing HRP and Lumigen PS-1 vs. amount of ethanol.

The methods described here show that a variety of oxidase enzymes or their substrates including analytes of clinical importance can be quantified with good sensitivity and dynamic range. Cholesterol has also been measured using these procedures. The method is applicable to the analysis of a variety of analytes in various media such as human serum and food products.

References

1. Akhavan-Tafti H, DeSilva R, Arghavani Z, Eickholt RA, Handley RS, Schaap AP. Lumigen PS: new chemiluminescent substrates for the detection of horseradish peroxidase. In: Campbell AK, Kricka LJ, Stanley PE, editors. Bioluminescence and Chemiluminescence Fundamentals and Applied Aspects. Chichester: Wiley, 1994:199-202.
2. Akhavan-Tafti H, DeSilva R, Arghavani Z, Schoenfelner BA. Novel aryl N-alkylacridancarboxylate derivatives useful for chemiluminescent detection. PCT Application WO95/23971 1995.
3. Tsuji A, Maeda M, Arakawa H. Chemiluminescent enzyme immunoassay a review. Anal Sci 1989;5:497-506.
4. Ternaux JP, Chamoin MC. Enhanced chemiluminescent assays for acetylcholine. J Biolumin Chemilumin 1994;9:65-72.

OXYGEN FREE RADICAL DETECTION IN ISOLATED AND PERFUSED RAT LIVER BY CHEMILUMINESCENCE IMAGING

P Pasini[1], A Gasbarrini[2], S De Notariis[3], B Nardo[4], M Bernardi[3] and A Roda[1]
[1]Dept of Pharmaceutical Sciences, [3]Patologia Medica I, [4]Clinica Chirurgica II, University of Bologna, Via Belmeloro 6, 40126 Bologna, Italy
[2]Patologia Medica, Catholic University of Rome, Largo Gemelli 1, 00168Rome, Italy

Introduction

Oxygen free radicals (OFR) are known to be involved in many pathophysiological states such as inflammation, ischemia-reperfusion injury of organs, atherosclerosis, rheumatoid arthritis, cancer, aging (1).

Chemiluminescence is a tool for oxygen free radical formation studies because a weak spontaneous photon emission is associated with OFR-related oxidative processes. This phenomenon has been shown in whole organs, isolated cells, tissue homogenates and subcellular organelles (2). Chemical light enhancers are employed to increase luminescent output intensity by interaction with OFR; in particular, the acridinium salt lucigenin has been shown to react specifically with superoxide radical $\bullet O_2^-$ to form N-methylacridone in the excited state which reverts to its ground state with the emission of energy as light (3).

Isolated and perfused organs represent a widely used model allowing a simplified, though reliable, approach to the study of various biologic functions including the effect of post-ischemic reoxygenation (4). Several studies utilising chemiluminescence to monitor OFR formation in perfused tissue have been performed (5-7).

In these works chemiluminescent emission was detected using photomultiplier tube-based photon counting systems, allowing just to measure the light signal. Improvements in photon detection technology provided new ultrasensitive photon imaging devices, based on CCD or Vidicon videocameras, able not only to quantify but also localize the light signal thus permitting its spatial distribution on a target surface to be evaluated (8-10).

The aim of our study was to develop a chemiluminescent method for the *real time* quantitative detection and localization of oxygen free radicals in isolated and perfused rat livers exposed to ischemia-reperfusion. We used the chemiluminescent reaction of lucigenin with $\bullet O_2^-$ and a videocamera-based detection system for image quantitative analysis.

Materials and Methods

Instrumentation: The chemiluminescent emission was detected and analyzed using a low-light imaging device, Luminograph LB 980 (EG&G Berthold, Bad Wildbad, Germany). It consists of an intensified Saticon videocamera connected to a PC provided with software controlling the system and performing image processing and quantitative analysis. The chemiluminescent signal and the live image in transmitted light can be acquired, digitized and stored in the PC; processing and overlay functions allow the spatial distribution of the signal to be evaluated with a spatial resolution of 240 μm in the standard lenses configuration, while a measure function quantifies the light emission. The samples were placed in a light-tight box supplied with a temperature control system.

A Model 312 Minipuls 3 peristaltic pump (Gilson, Villiers le Bel, France) adjusted to a flow of 18 mL/min was used to drive the perfusion medium.

Reagents: bis-N-methylacridinium nitrate (lucigenin) and superoxide dismutase were purchased from Sigma Chemical Co. (St. Louis, MO, USA).

Procedure: Livers were isolated from anesthetized Wistar male fed rats weighting 200-250 g. Portal vein was cannulated with a 18-gauge catheter, the organ was washed with Ringer lactate solution and exposed to 1 h ischemia at 37°C. It was then placed in the luminograph chamber thermostated at 37°C and perfused for 1 h through the portal catheter with oxygenated Krebs Henseleit buffer, containing 10 μM lucigenin; the perfusion solution was maintained at 37°C. The chemiluminescent signal from the organ surface was continuously monitored and live images were recorded at the beginning and at the end of the experiment.

In some experiments, 30 IU/mL superoxide dismutase, a specific scavenger of $\bullet O_2^-$, was added to perfusion medium after 20 min reperfusion.

Chemiluminescent signals during ischemia and during reperfusion of non ischemic livers were acquired as negative controls.

Results and Discussion

A chemiluminescent signal was observed a few minutes after the reperfusion of ischemic livers. The superimposition of the chemiluminescent and the live images, acquired successively during reperfusion, allowed to localize the luminescent emission and showed that it started around the hepatic vasculare pedicle, then diffused progressively to the whole organ surface. Figure 1.

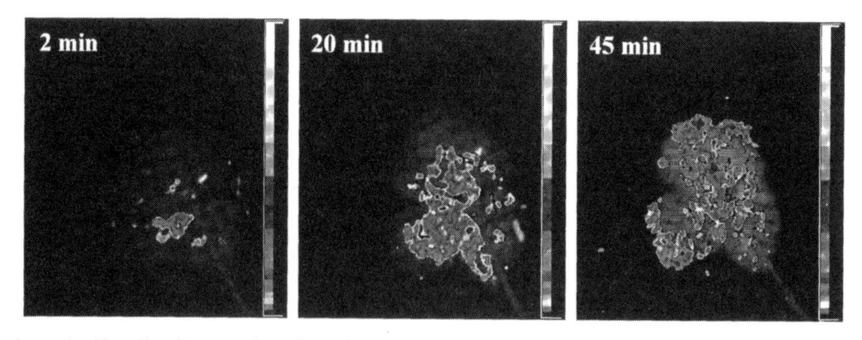

Figure 1. Chemiluminescent imaging of oxygen free radical formation in isolated and perfused rat liver. Spatial distribution of the luminescent signal in ischemic liver at different times after reperfusion.

The light output kinetics was evaluated by measuring the photon flux/s/organ surface at consecutive 5 min intervals: the signal increased reaching the maximum value after 15-20 min, then it slowly decayed disappearing after 50-60 min. The peak of luminescence could correspond to a burst of superoxide radical production and the following chemiluminescence decrease suggested that the rate of superoxide production declined.

The addition of 30 IU/mL superoxide dismutase, a specific scavenger of $\bullet O_2^-$, to the perfusion medium caused a rapid decrease of the light output, showing that the observed signal was really dependent on superoxide radical formation.

Luminescence emission was not recorded during ischemia accordingly to a lack of O_2, which is required for oxygen radical formation. Also during reperfusion of non ischemic livers light production was not detected, probably because the postulated conversion during hypoxia of xanthine dehydrogenase to its oxygen radical-producing form, xanthine oxidase, did not occurr (11).

The light emission may correspond to superoxide radical formation on the liver surface, but previous studies postulated that it reflects metabolism of the whole organ and correlate with more global parameters such as lipid peroxidation (12).

In conclusion, we showed the possibility to assess *in real time* the rate and spatial distribution of oxygen free radical formation on the surface of intact organs. This system opens new perspectives in the study of the physiopathogenesis of oxidative stress and provides a suitable model for the screening of substances with antioxidant activity. In addition, it could represent a useful tool to test the effect of storage conditions and media in order to preserve organs for transplantation.

References

1. Cross CE, Halliwell B, Borish ET, Pryor WA, Ames BN, Saul RL et al. Oxygen radicals and human disease. Ann Intern Med 1987;107:526-45.
2. Campbell AK. Chemiluminescence: principles and applications in biology and medicine. Chichester: Ellis Horwood Ltd, 1988.
3. Allen RC. Biochemiexcitation: chemiluminescence and the study of biological oxygenation reactions. In: Adam W, Cilento G, editors. Chemical and biological generation of excited states. New York: Academic Press, 1982:309-44.
4. Van Ness K, Sorrentino D, Berk PD. Isolated liver perfusion: a method to study the whole organ. In: Muraca M, editor. Methods in biliary research. Boca Raton, FL: CRC Press, 1995:245-54.
5. Boveris A, Cadenas E, Reiter R, Filipkowski M, Nakase Y, Chance B. Organ chemiluminescence: noninvasive assay for oxidative radical reactions. Proc Natl Acad Sci USA 1980;77:347-51.
6. Kumar C, Okuda M, Chance B. Luminol enhanced chemiluminescence of the perfused rat heart during ischemia and reperfusion. FEBS Lett 1990;272:121-4.
7. Nunes FA, Kumar C, Chance B, Brass CA. Chemiluminescent measurement of increased free radical formation after ischemia/reperfusion. Mechanisms of free radical formation in the liver. Dig Dis Sci 1995;40:1045-53.
8. Wick RA. Photon counting imaging: applications in biomedical research. BioTechniques 1989;7:262-8.

9. Mueller-Klieser W, Walenta S. Geographical mapping of metabolites in biological tissue with quantitative bioluminescence and single photon imaging. Histochem J 1993;25:407-20.
10. Roda A, Pasini P, Musiani M, Girotti S, Baraldini M, Carrea G et al. Chemiluminescent low-light imaging of biospecific reactions on macro- and microsamples using a videocamera-based luminograph. Anal Chem 1996;68:1073-80.
11. Amaya Y, Yamazaki KI, Sato M, Noda K, Nishino T. Proteolytic conversion of xanthine dehydrogenase from the NAD-dependent type to the O_2-dependent type. J Biol Chem 1990;265:14170-5.
12. Cadenas E, Boveris A, Chance B. Low level chemiluminescence of biological systems. In: Pryor WA, editor. Free radicals in biology. New York: Academic Press, 1984:211-42.

CHEMILUMINESCENCE IN SITU HYBRIDIZATION FOR THE DETECTION OF VIRAL GENOMES

P Pasini[1], M Musiani[2], A Roda[1], M Zerbini[2], G Gentilomi[2], M Baraldini[3], G Gallinella[2] and S Venturoli[2]
[1]Dept of Pharmaceutical Sciences, [2]Inst of Microbiology, [3]Inst of Chemical Sciences, University of Bologna, Via Belmeloro 6, 40126 Bologna, Italy

Introduction

Early diagnosis of viral infection is highly recommended, especially for prompt treatment with antiviral agents in order to curb progression of the disease. Sensitive, specific, reliable and rapid methods are required to check the presence of virus in biological samples, particularly for viruses not growing in cell coltures which cannot be detected by isolation procedures.

Recently, a significant improvement has been achieved by the development of hybridization techniques, such as blot hybridization and in situ hybridization (ISH), and polymerase chain reaction (PCR) for the detection of nucleic acids (1). Blot hybridization and PCR allow to detect viral nucleic acids in body fluids or extracted from tissue and cellular digests, while ISH is performed on tissue sections or cellular smears with the preservation of cell and tissue morphology, providing information about the virus localization (2).

ISH assays have been developed using either radioactive or non-radioactive probes (3,4). Non-radioactive probes revealed by colorimetric immunoenzymatic systems are largely used, despite lower sensitivity compared to radioisotopic ones. Recently, chemiluminescent substrates have been proposed as a more sensitive alternative to colorimetric ones for many assays in which small amounts of analyte have to be revealed (5-7). Furthermore, the continuing improvements in ultrasensitive photon-imaging devices permitted quantitation and localization of the chemiluminescent emissions on a target surface with high spatial resolution (8-11).

The aim of the present study was the development of chemiluminescence in situ hybridization assays for the detection and localization of viral DNAs. In particular, we set up methods for Cytomegalovirus (CMV), which grows slowly in cell coltures, and Parvovirus B19 and Human Papillomavirus (HPV) not efficiently growing in vitro. We tried to combine the spatial resolution and localization of the signal of in situ hybridization, the sensitivity of chemiluminescent substrates and the specificity of digoxigenin (dig)-labelled probes constructed in our laboratory.

Briefly, in our system a probe corresponding to a specific sequence of the target nucleic acid was constructed using dig-labelled dUTP nucleotides. The labelled probe was hybridized with the target nucleic acid and the hybrid revealed with anti-dig antibody conjugated to alkaline phosphatase (ALP) and a chemiluminescent substrate for ALP.

Materials and Methods

Instrumentation: The chemiluminescent emission was detected and analyzed using a low-light imaging device, Luminograph LB 980 (EG&G Berthold, Bad Wildbad, Germany) linked to a Model BH-2 light microscope (Olympus Optical, Tokyo, Japan) enclosed in a light-tight box. The luminograph consists of an intensified Saticon videocamera connected to a PC provided with software controlling the system and performing image processing and quantitative analysis. The chemiluminescent signal and the live image in transmitted light can be acquired, digitized and stored in the PC; processing and overlay functions allow the spatial distribution of the signal to be evaluated with a spatial resolution of 1 μm at x40 objective magnification, while a measure function quantifies the light emission.

Samples: CMV was detected in infected coltured fibroblast monolayers and paraffin embedded and frozen tissue sections from AIDS patients with associated CMV infection, previously tested by colorimetric ISH. Parvovirus B19 was detected in bone marrow cell smears from patients with aplastic crisis or hypoplastic anemia, previously tested by colorimetric ISH, dot blot hybridization, nested PCR. HPV was detected in paraffin embedded skin and mucosa specimens from patients with different HPV associated pathologies, previously tested by PCR.

In situ hybridization reaction: In situ hybridization was performed as previously decribed using dig-labelled probes (12). Briefly, cell monolayers, cell smears and frozen tissue sections were fixed in 4% paraformaldehyde, then treated with pronase, and dehydrated by ethanol washes; paraffin embedded tissue sections were dewaxed by incubation in xylene and washed in absolute ethanol, then treated with pronase and dehydrated by ethanol washes. Dehydrated samples were overlaid with 10 μL of the hybridization mixture (50% deionized formamide, 10% dextran sulfate, 250 μg/mL of calf thymus DNA and 2 μg/mL of dig-labelled probe DNA in 2X SSC buffer). Samples and the hybridization mixture containing the dig-labelled probe were denatured together by heating in an 85°C water bath for 5 min, and were then put to hybridize at 37°C for 3 h. After hybridization, samples were washed three times at stringent conditions and then incubated with polyclonal anti-dig Fab fragments conjugated to ALP.

Chemiluminescence detection: Cell and tissue samples were overlaid with 20 μL of Lumi-Phos® Plus (Lumigen, Inc, MI, USA), containing the chemiluminescent substrate for alkaline phosphatase adamantil-1,2-dioxetane phenyl phosphate. After 20 min incubation in the dark, the substrate solution was removed and the chemiluminescent emission recorded.

Results and Discussion

The light emission from CMV-infected cultured fibroblasts fixed at various times after infection was measured, thus following the virus replication cycle. Increasing values of emitted light, expressed as photon flux/s/infected cells, were observed at 48, 60, 72, 96 h after infection respectively, showing that this chemiluminescent technique provides semi-quantitative data. This is relevant from a therapeutic point of view since it should allow to analyze the viral DNA content in clinical specimens and monitor the efficacy of antiviral therapy.

Control experiments to assess the specificity of the method were performed: mock infected cells were treated with labelled probe; CMV-infected cells were treated with unlabelled probe; CMV-infected cells were treated with control labelled probe (plasmid pACYC 184 DNA). The previously described detection system was used and no signal was observed in the three experiments, showing that the chemiluminescence ISH reaction detects CMV nucleic acid sequences specifically.

The preliminary results obtained with in vitro infected cultured cells prompted us to test our system on clinical samples in order to assess its suitability for diagnostic purposes. Cellular smears and both frozen and paraffin embedded tissue sections were examined.

Lung and adrenal gland tissue sections from AIDS patients with confirmed CMV infection were analyzed and we could detect a strong signal with a sharp localization in some positive parenchimal cells clearly standing against the negatively reactive background.

Bone marrow cell smears from patients with aplastic crisis or hypoplastic anemia were used to detect Parvovirus B19 genome. Figure 1. Multiple samples from the same patient were processed and the viral infection revealed by chemiluminescent or colorimetric detection, using a colorimetric substrate for ALP, to compare sensitivity. Chemiluminescent detection proved more sensitive than colorimetric one, allowing to find more positive cells/counted cells. Control experiments proved that the detected signal was specific.

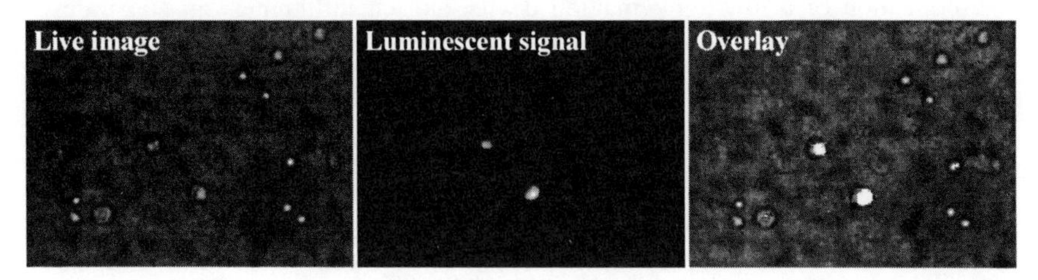

Figure 1. Chemiluminescence in situ hybridization for the detection of Parvovirus B19 genome in bone marrow cells.

HPV infection was studied on samples from patients affected by different dermatologic pathologies. In particular, a skin sample from a patient with cutaneous Bowen's disease gave further evidence of the semi-quantitative data provided by chemiluminescence ISH. Increasing values of emitted light corresponding to increasing numbers of viral genome copies were observed from the basal to the superficial layers of the epithelium. This is consistent with the finding that the virus entered the organism localizes in undifferentiated basal epithelial cells and replicates within them following their differentiation process and migration towards epithelium surface.

In conclusion we can say that the chemiluminescence in situ hybridization assays developed for the detection of viral DNAs proved sensitive, semi-quantitative, able to localize viral genomes specifically and suitable for different types of clinical samples. They could represent a useful tool for a sensitive and specific diagnosis of viral diseases; moreover, the analysis can be performed also on archival samples allowing retrospective studies. In addition, the method can be easily adapted to detect any genic sequence inside the cells provided the specific labelled probe and can be applied in different fields such as genetic diseases and oncology to study gene expression in cells and evaluate tumoral gene activity.

References

1. Chou S. Newer methods for diagnosis of cytomegalovirus infection. Rev Infect Dis 1990;12:S727-36.
2. Wilcox JH. Fundamental principles of *in situ* hybridization. J Histochem Cytochem 1993;41:1725-33.
3. Morris RG, Arends MJ, Bishop PE, Sizer K, Duvall E, Bird CC. Sensitivity of digoxigenin and biotin labelled probes for detection of human papillomavirus by in situ hybridization. J Clin Pathol 1990;43:800-6.
4. Nascimento JP, Hallam NF, Field AM, Clewley JP, Brown KE, Cohen BJ. Detection of B19 Parvovirus in human fetal tissue by in situ hybridization. J Med Virol 1991;33:77-82.
5. Holtze HJ, Sagner G, Kessler C, Schmitz G. Sensitive chemiluminescent detection of digoxigenin labeled nucleic acid: a fast and simple protocol and its applications. BioTechniques 1992;12:104-13.
6. Girotti S, Musiani M, Pasini P, Ferri E, Gallinella G, Zerbini ML et al. Application of a low-light imaging device and chemiluminescent substrates for quantitative detection of viral DNA in hybridization reactions. Clin Chem 1995;41:1693-7.
7. Martin CS, Butler L, Bronstein I. Quantitation of PCR products with chemiluminescence. BioTechniques 1995;18:908-12.
8. Wick RA. Photon counting imaging: applications in biomedical research. BioTechniques 1989;7:262-8.
9. Hawkins E, Cumming R. Enhanced chemiluminescence for tissue antigen and cellular viral DNA detection. J Histochem Cytochem 1990;38:415-9.
10. Lorimier P, Lamarcq L, Labat-Moleur F, Guillermet C, Bethier R, Stoebner P. Enhanced chemiluminescence: a high sensitivity detection system for in situ hybridization and immunohistochemistry. J Histochem Cytochem 1993;41:1591-7.
11. Roda A, Pasini P, Musiani M, Girotti S, Baraldini M, Carrea G et al. Chemiluminescent low-light imaging of biospecific reactions on macro- and microsamples using a videocamera-based luminograph. Anal Chem 1996;68:1073-80.
12. Gentilomi G, Musiani M, Zerbini M, Gallinella G, Gibellini D, La Placa M. A hybrido-immunocytochemical assay for the in situ detection of cytomegalovirus DNA using digoxigenin-labeled probes. J Immunol Methods 1989;125:177-83.

DEVELOPMENT OF A CHEMILUMINESCENCE ASSAY FOR THE DETECTION OF T-CELL RECEPTOR GENE REARRANGEMENTS

Kathleen Tenner, Margaret Karst,
Steven Thibodeau and Dennis O'Kane
Department of Laboratory Medicine and Pathology
Mayo Foundation, Rochester, MN 55905, USA

Introduction

Detection of T-cell receptor (TcR) gene rearrangements by Southern blotting is invaluable for differentiating neoplasms from benign proliferative processes (1,2). Isotopic labeling with ^{32}P, which is reliable and sensitive, is frequently used in Southern blotting. Routine clinical application, however, is limited by rapid radioactive decay of the probes. This results in extra personnel time required for frequent probe labelling and quality control procedures. Many non-isotopic methods have been developed for nucleic acid detection (3). The use of chemiluminescent substrates, combined with hapten-labeled probes and secondary enzyme or avidin complexes, has proven to be as sensitive as their isotopic counterparts (4). This sensitivity is required for detecting rearrangements of TcR, a single copy gene. We have evaluated several commercial products and describe the system with the best signal-to-noise ratio when used in an existing TcR gene rearrangement assay.

Materials and Methods

<u>Southern Blotting</u>: Human genomic DNA was extracted by salt precipitation using the Puregene kit (Gentra Systems & Corp., Minneapolis MN) according to manufacturer's directions. Sample DNA (2.5 μg) was digested overnight with *Eco*R1 and separated by electrophoresis on a 0.8% agarose gel at 55 V. After electrophoresis, the DNA gel was denatured (0.4 mol/L NaOH, 0.6 mol/L NaCl), neutralized (0.5 mol/L Tris-HCl, 1.5 mol/L NaCl, pH 7.5), and transferred overnight to a Magnagraph positively charged nylon membrane (MSI, Westboro MA). The membrane was baked at 70°C for 2 h and stored in the dark at room temperature until use.
<u>Probe Labeling</u>: A 4.4 Kb $J\beta_2$ probe insert was isolated from plasmid by *Eco*R1 digestion and separated by electrophoresis on a 0.6 % low melting point (LMP) agarose gel. The probe was heat-denatured and labeled by the following methods: i. random priming with digoxigenin or biotin using High-Prime mix (Boehringer-Mannheim, Indianapolis IN); ii. direct-labeling with HRP (Amersham, Arlington Heights IL); and iii. non-enzymatic labeling with psoralen by the Rad-Free system (Schleicher & Schuell, Keene NH) or with biotin by the Fast-Tag system (Vector Laboratories, Inc., Burlingame CA) according to manufacturers' directions.
<u>Hybridization and Stringency Washes</u>: The membrane-bound DNA was pre-hybridized for 2 h at 45°C in 20 mL of pre-warmed hybridization buffer (50% formamide, 5 X SSC, 1 X Denhardt's, 0.6% SDS, 10% dextran sulfate, 0.2 mg/mL yeast t-RNA). Fifty nanograms (2.5 ng/mL) of denatured, labeled $J\beta_2$ probe was added to the pre-hybridization buffer and incubated overnight in a roller bottle at 45°C. The membrane was washed twice in 2 X SSC, 0.1% SDS, once in 0.2 X SSC, 0.1% SDS, and once in 0.1 X SSC, 0.1% SDS at 60°C, 30 min each.
<u>Chemiluminescence Signal Generation and Detection</u>: After high stringency washes the membrane was equilibrated for 5 min in 1 X TBS and blocked for 1.5 h in 50 mL of 1 X TBS, 0.5% Kodak blocker (Kodak, Rochester NY), 0.1% SDS at 37°C. The membrane was incubated for 0.5 h with anti-digoxigenin or streptavidin coupled with either HRP or ALP and washed (1 X TBS, 0.5% SDS) four times, 10 min each. Signal was generated with Lumi-Phos Plus (Lumigen, Southfield MI), CSPD (Tropix, Bedford MA), or PS3 (Lumigen) substrates and detected by a 5 min exposure with a CCD camera cooled to -35°C (10).

Results

Dot blot dilution series were used to determine the efficiency of biotin, digoxigenin, HRP and non-enzymatic labeling of Jβ_2 probe which had been separated by electrophoresis in LMP agarose. Dilution series of labeled control DNA were compared to ten-fold dilution series of labeled J$\beta_2$probe. In our hands, the digoxigenin and biotin High-Prime labeling procedures were the most efficient and were investigated further for sensitivity of detection using three different signal systems. ALP conjugate was detected with the dioxetane-based substrates Lumi-Phos Plus or CSPD. HRP conjugate was detected with PS3, an acridan-based substrate. Lumi-Phos Plus and PS3 resulted in the most intense signals (Figure 1), which were tested further using digoxigenin- and biotin-labeled probes for Southern blot detection.

When used for visualizing TcR gene rearrangements by Southern blotting, the biotin-HRP system used with PS3 substrate produced a more intense signal than the other probe-conjugate substrate combinations (Figure 2). Even an overnight incubation with Lumi-Phos Plus (Figure 2b) did not produce an intensity comparable to a 5 min incubation with PS3 (Figure 2a). The biotin HRP-PS3 system enabled resolution and identification of bands indicating a polyclonal benign disease process, as well as a clonal gene rearrangement from bands of a normal patient (Figure 2a). This high resolution was striking because the amount of DNA used in our procedure is approximately 5 times less than what is typically reported in Southern blotting procedures (Table 1). The sensitivity of chemiluminescence detection was assessed by analyzing decreasing amounts of normal control DNA by Southern blot. The biotin-HRP system was able to detect the germline restriction fragment in 0.31 μg of total genomic DNA (Figure 3). Comparison of the optimized HRP-PS3 chemilumigraph (5 min exposure) with a ^{32}P autoradiograph (overnight exposure) on identical Southern blots demonstrated that the PS3 system is as sensitive as isotopic detection (Figure 2d).

In addition to the high sensitivity obtained with chemiluminescence detection, turn-around time and laboratory testing efficiency is improved. A comparison of times required for isotopic and chemiluminescence detection methods (Table 2) illustrate these time savings. Batch labeling and a long shelf-life of non-isotopic probes results in decreased personnel time spent on the tedious task of probe labeling. High-Prime labeling also saves personnel time because hands-on time required for this method is approximately 5 min as compared to 2 h, each week required for ^{32}P-labeling of probes.

Discussion

In the TcR gene rearrangement detection procedure we developed, chemiluminescence detection proved to be as sensitive as radioactive detection. As well as a high level of sensitivity, the use of chemiluminescent methods offers several advantages over radioactive methods. Time spent by personnel on probe labeling is decreased because of batch labeling and long-term storage capacity of hapten-labeled probes. Signal generation is completed within 5 minutes verses up to 3 days with radioactive detection. Furthermore, biohazards and disposal fees that coincide with isotope use are eliminated. Chemiluminescence detection methods are a convenient, cost effective, and safe alternative to isotopic detection in the Southern blot analysis of TcR gene rearrangements.

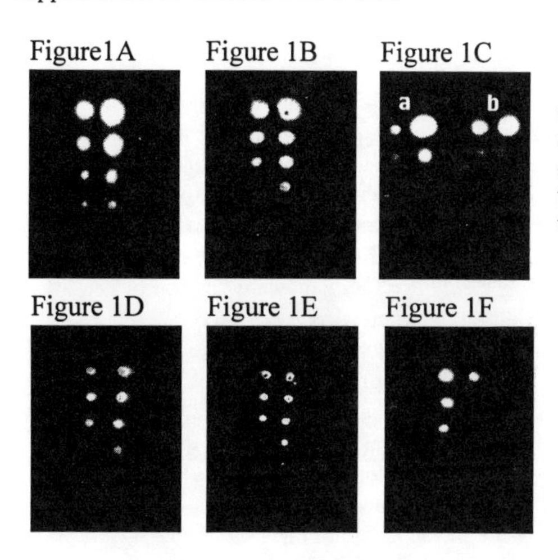

FIGURE 1A-F. Dot blots of dilution series of labeled control DNA and ten-fold dilution series of labeled Jβ_2 detected with chemiluminescent substrates for estimating the efficiency of labeling. Lane I, from the top: 1 ng, 50 pg, 10 pg, 2 pg, and 0.5 pg of labeled control DNA. Lane II, from the top: 1:10, 1:100, 1:1000, 1:10,000, and 1:100,000 dilution of labeled Jβ_2. Jβ_2 probe biotin-labeled using High-Prime mix and detected with PS3 (1A) and Lumi-Phos Plus (1B). Dot blot identical to 1A, but Jβ_2 probe: biotin-labeled by High-Prime mix (1C-a) and Fast Tag (1C-b); digoxigenin-labeled with High-Prime mix and detected by PS3 (1D) and Lumi-Phos Plus (1E); direct-labeled with HRP and detected with PS3 (1F).

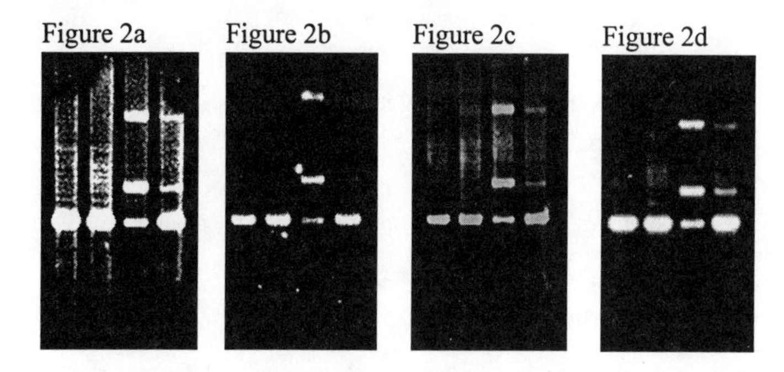

FIGURE 2a-d. Chemilumigrams of Southern blots of normal patient DNA (lane 1), polyclonal benign disease process (lane 2), 50% TcR clonal gene rearrangement (lane 3), and 10% TcR clonal gene rearrangement (lane4) probed with biotin-labeled Jβ_2 probe and detected with PS3 (a) and Lumi-Phos Plus (b). Chemilumigram of Southern blot identical to 2a, but probed with digoxigenin-labeled Jβ_2 and detected with PS3 (c). 2d. Autoradiogram of lot identical to 2a, but probed with ^{32}P-labeled Jβ_2 probe.

Table 1

Amount of DNA Used in Southern Blots

Reference#	Micrograms/lane	Detection Limit (μg)
6	10.0	NA
5	10.0-15.0	0.20
4	10.0	0.25
8	5.0	0.50
Current Study	2.5	0.31

Table 1. Reported values of the amount of DNA used per lane in Southern blotting.

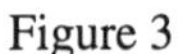

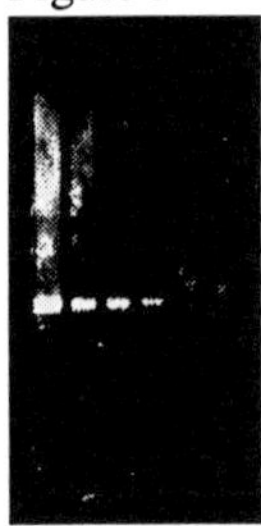

FIGURE 3. Sensitivity limit of chemiluminescence detection of *Eco*R1-digested germline control DNA with biotin-labeled $J\beta_2$ probe and PS3 detection. Lanes 1-6: 2.5 μg, 1.25 μg, 0.62 μg, 0.31 μg, 0.15 μg and 0.07 μg *Eco*R1-digested germline control DNA.

Table 2

Comparison of Time Required for Southern Blot Deetection Methods

	PS3	^{32}P
Probe Labeling	5 min - Batch	2 h - Weekly
Hybridization	17 h	17 h
Stringency Wash	2 h	2 h
Signal Generation	2.5 h	-----
Exposure	5 min	1-3 days

Table 2. A comparison of the time required for each procedure in isotopic and chemiluminescence Southern blot detection methods.

References

1. Cossman J, Uppenkamp M. T-cell gene rearrangements and the diagnosis of T-cell neoplasms. Clinics Lab Med 1988; 8: 45-64
2. Cossman J, Uppenkamp M, Sundeen J, Coupland R, Raffeld M. Molecular genetics and the diagnosis of lymphoma. Arch Pathol Lab Med. 1988; 112: 117-27
3. Holtke HJ, Sagner G, Kessler C, Schmitz G. Sensitive chemiluminescent detection of digoxigenin-labeled nucleic acids: a fast and simple protocol and its applications. Biotechniques 1992; 12: 104-13
4. Hodges KA, Kosciol CM, Rezuke WN, Abernathy EC, Pastuszk WT, Tsongalis GJ. Chemiluminescent detection of gene rearrangement in hematologic malignancy. Ann Clin Lab Sci 1996; 26: 114-18
5. Edwards AM, Hunter SV, Hankin RC. Gene rearrangement analysis by nonorganic extraction and chemiluminescent detection. Lab Med 1993; 24: 629-33
6. Engler-Blum G, Meier M, Frank J, Müller GA. Reduction of background problems in nonradioactive northern and southern blot analyses enables higher sensitivity than ^{32}P-based hybridizations. Anal Biochem 1993; 210: 235-44
7. Akhavan-Tafti H, Schaap PA, Arghavani Z, DeSilva R, Eickholt RA, Handley RS, Schoenfelner BA, Sugioka K, Sugioka Y. CCD camera imaging for the chemiluminescent detection of enzymes using new ultrasensitive reagents. J Biolumin Chemilumin 1994; 9: 155-64
8. Hopfenbeck JA, Holden JA, Wittwer CT, Kjeldsberg CR. Digoxigenin-labeled probes amplified from genomic DNA detect T-cell gene rearrangements. Am J Clin Pathol 1992; 97: 639-44
9. Kricka LJ. Nonisotopic DNA Probe Techniques. San Diego: Academic Press; 1992
10. Martin CS, Bronstein I. Imaging of chemiluminescent signals with cooled CCD camera systems. J Biolumin Chemilumin 1994; 9: 145-53

CHEMILUMINESCENCE SCREENING ASSAY FOR LEUKOCYTES IN URINE

BJ Hallaway, ME Copeman, BS Stevens, TS Larson, DM Wilson, DJ O'Kane
Dept. Laboratory Medicine and Pathology, Mayo Foundation for Medical Education and Research, Rochester, Minnesota 55905, USA

Introduction

High levels of white blood cells (i.e. leukocytes) in urine may be an indication of urinary tract infection (UTI) or cystitis. Clinical methods used to detect leukocytes in urine include visual microscopic analysis, leukocyte esterase detection with dipsticks, and automated imaging (1-3). These methods are time-consuming, expensive, prone to error, and cannot screen out normal samples efficiently. A rapid, inexpensive screening procedure that identifies patient urines containing abnormally high levels of leukocytes and distinguishes these from normal urines is highly desirable. Leukocytes, including monocytes, granulocytes and polymorphonuclear forms, express myeloperoxidase (MPO). This enzyme has been assayed with luminol derivatives (4,5). However, luminol derivatives react with other heme proteins found in urine, such as hemoglobin (Hb) and myoglobin (Mb) (6,7). A new acridan substrate for peroxidase activities has been reported (8) which may offer increased analytic sensitivity and specificity for detection of leukocytes in urine.

The purposes of this study was to determine first if MPO could be detected with an acridan substrate; and second if MPO could be used as a surrogate marker for urinary white blood cells without interference from other heme proteins present, such as Hb.

Materials and Methods

Reagents: PS-1 acridan substrate was obtained from Lumigen Inc, Southfield, MI. Methimazole, ofloxacin, ascorbic acid, ammonium chloride, EDTA, Hb, bilirubin, and lactoperoxidase (LPO) were purchased from Sigma Chemicals (St. Louis, MO). Glycerol was purchased from Aldrich Chemical Company (Milwaukee, WI). Sodium chloride was purchased from Fisher Scientific (Fair Lawn, NJ). MPO and Mb were purchased from Calbiochem (La Jolla, CA). Thyroid peroxidase (TPO) was purchased from O.E.M. Concepts, Inc (Toms River, NJ). Eosinophil peroxidase (EPO) was obtained from Dr. G.J. Gleich (Mayo Foundation, Rochester, MN).
Procedures: A normal urine matrix for the peroxidases and leukocyte controls was prepared by obtaining 5 L of pooled urine. The urine was centrifuged at 4000 rpm for 10 min to remove the urine sediment. The supernatant was poured through a 0.2 μm filter unit attached to vacuum to retain any remaining sediment. The filtered urine was frozen in aliquots at -80°C. A stored aliquot was thawed each day and used as the matrix for the MPO and leukocyte controls.

Leukocyte control samples were prepared from a buffy coat obtained from 40 mL of EDTA treated whole blood from a normal donor. The blood was centrifuged for 10 min at 3700 rpm to separate plasma and cells. Plasma and erythrocytes were discarded. The buffy coat was washed and centrifuged for 10 min with 0.9% saline to remove residual plasma which was discarded. Contaminating erythrocytes were lysed by treatment for 10 min with 156 mmol/L NH_4Cl containing 10 mmol/L $NaHCO_3$ and 0.12 mmol/L EDTA. The buffy coat was sedimented at 4000 rpm for 10 min and the lysing solution discarded. This step was repeated three times. The buffy coat pellet was reconstituted in 10 mmol/L Tris HCL buffer containing 10 mmol/L EDTA buffer and 40% glycerol. The buffy coat leukocytes were frozen in aliquots at -80°C. The cells were thawed once, diluted in filtered urine and used as a positive control each day.

The PS-1 substrate was prepared by mixing a 40:1 dilution of solution A to B. Once the substrate was mixed, the bottle was covered with aluminum foil to protect it from light and used within 4 h.

MPO Assay Method: Urine samples were diluted 1:4 (v/v) into distilled water to minimize inhibitory effects observed with hyperosmolal urines. Two µL of diluted urine was added to 12 x 75 mm polystyrene test tube containing 50 µL of PS-1 substrate and incubated at 37°C for 5 min. MPO activity was determined by CL emission using a Turner TD-20e luminometer maintained at 37°C. Leukocyte counting was performed manually by using a hemacytometer chamber (Kova Glasstic slide, Hycor, Irvine, CA) on each urine sample to determine the number of cells/mL.

Results and Discussion

Luminol derivatives react with a wide variety of heme proteins. This lack of specificity limits its application in urinalysis where Hb and Mb may be present with peroxidases. The reactivity of PS-1 with heme proteins was assessed to determine if this substrate had more restricted specificity. Several peroxidase activities were detected with PS-1 substrate including EPO, MPO, LPO, and TPO. Addition of 50 µmol/L methimazole, an inhibitor of EPO (9), reduces activities of EPO, TPO and LPO by >90%, while MPO was inhibited by only 50% (Table 1). The addition of methimazole to the PS-1 assay allows MPO activity to be detected selectively. Hb and Mb at concentrations of <10 µg/mL and <2 µg/mL respectively did not interfere with the assay.

Linear standard curves were obtained with the MPO standard and the buffy coat leukocytes diluted in filtered urine (Figure 1). Detergent present in the PS-1 substrate lyses the buffy coat leukocytes to liberate MPO without additional treatment.

CL emission from PS-1 substrate results in a glow emission that lasts for several minutes. MPO activity in urine samples were compared at room temperature and 37°C to find the most stable reaction conditions in order to minimize the effects of timing. At room temperature, light emission reaches a maximum at 15 min but does not attain constant emission. The kinetics of CL emission from MPO at 37°C was followed over time using several urine samples diluted 4-fold in water (Figure 2). Constant CL emission, less than 5% change in RLU, was found to occur between 3 and 20 min. Assays were performed subsequently at 37°C and CL emission determined at 5 min.

Hyperosmolal urine samples (>600 mOsm) inhibit MPO activity (10). Consequently, the effects of urine osmolality on the acridan CL assay was investigated. Higher osmolality urine samples did not give as high a CL signal as lower osmolality samples Filtered urine samples with osmolalities ranging from 260 to 1053 mOsm were spiked with the same amount of MPO. Serial dilutions (1:2 to 1:16) of each spiked urine sample were made with water (Figure 3) and CL emission of the diluted samples was determined. The quenching effect of hyperosmolal urines was minimized at a 4-fold dilution.

Recovery studies were performed on 10 urine samples. Known amounts of MPO were added to five normal urine samples. Five urine samples, which contained leukocytes, had additional MPO added. Percent recoveries ranged from 91.2 to 135% in all samples.

Urine samples may contain many epithelial cells, casts, crystals, and numerous artifacts but none of these appear to interfere with the CL emission from MPO. Several possible interfering substances were added to urine and tested for inhibitory effects. Uric acid, ascorbic acid, bilirubin, and trolox did not affect assay performance. Hb at concentrations of >10 µg/mL and Mb at concentrations >2 µg/mL were detected by the PS-1 substrate as the CL emission was above the normal urine cutoff. Samples grossly populated with bacteria caused slight inhibition of the MPO activity.

A patient study was performed using samples from 144 males and 237 females to compare manual microscopic leukocyte counts (Kova chambers) with MPO CL emission with the PS-1 substrate (Figure 4). Arbitrary set points to define abnormal urines were established from

Concentration (µg/mL)	50µmol/L Methimazole Without	With	% Inhibition
Myeloperoxidase	305	130	58%
Eosinophil Peroxidase	764	74	91%
Thyroid Peroxidase	1.67	0.127	93%
Lactoperoxidase	22	0.92	96%
Hemoglobin	0.914	0.84	9%
Myoglobin	3.18	1.95	39%

Table 1. Peroxidase activities in the presence of PS-1 with and without Methimazole.

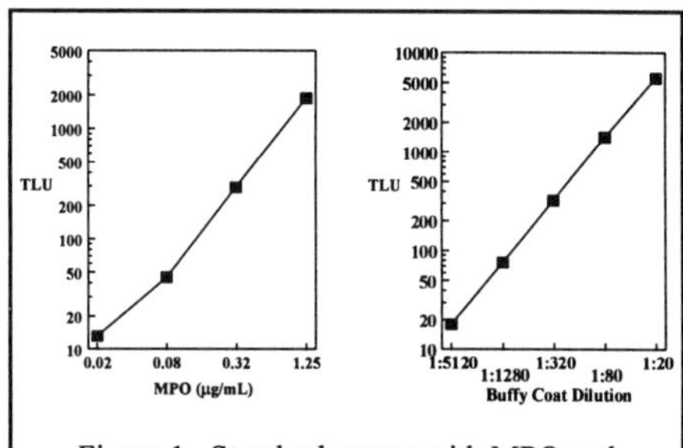

Figure 1. Standard curves with MPO and leukocytes from buffy coat.

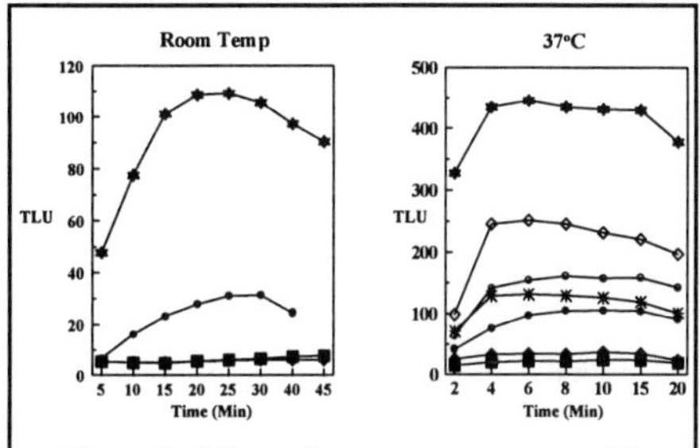

Figure 2. Effect of assay temperature on CL emission.

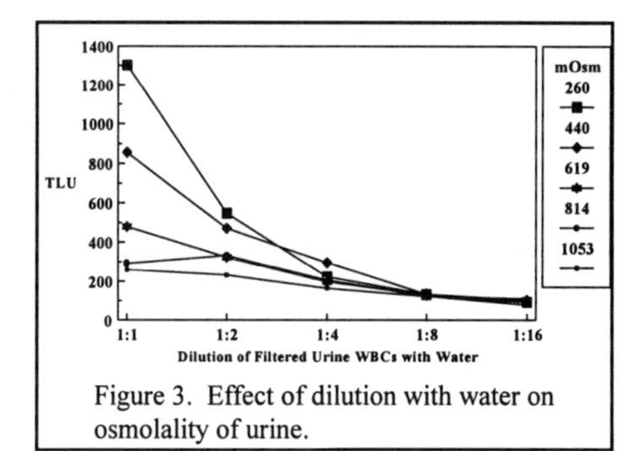

Figure 3. Effect of dilution with water on osmolality of urine.

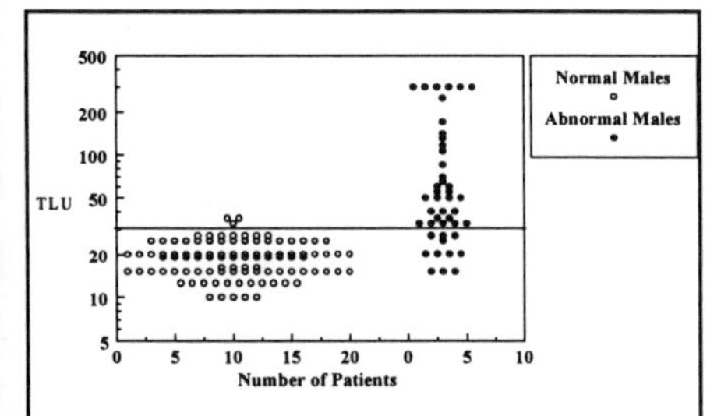

Figure 4a. Normal and abnormal urines as defined by the CL emission of 30 TLU and <10,000 WBC/mL for males.

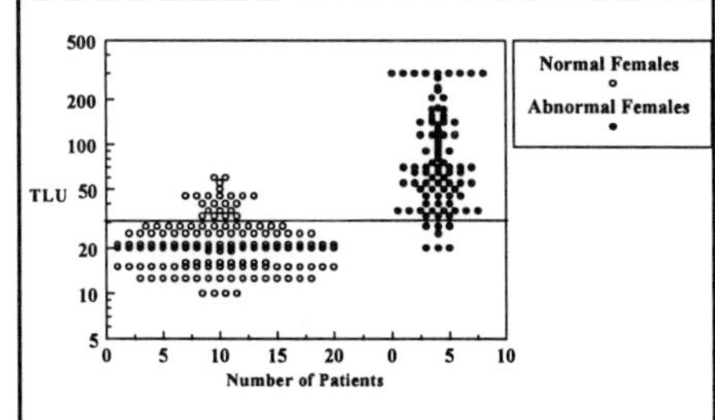

Figure 4b. Normal and Abnormal urines as defined by the CL emission of 30 TLU and <16,000 WBC/mL for females.

literature values as >10,000 WBC/mL for males and >16,000 WBC/mL for females. Using a CL emission cutoff at 30 TLU clinical sensitivity for males was 86% and 88% for females. Apparent false negative rates were 25% (males) and 7.4% (females). None of the patients with false negative results had UTIs as ascertained by inspection of the patients' medical records.

Existing assays for detecting leukocytes in urine involve manual methods or semi-automated imaging which are slow. Preliminary results indicate that this screening test would reduce the number of normal patient samples entering the laboratory work-flow by approximately 80%. This screening assay would decrease costs and improve turn-around time for emergency situations awaiting urinalysis results with greater clinical sensitivity than provided by currently utilized dipsticks.

References

1. Holland DJ, Bliss KJ, Allen CD, Gilbert GL. A comparison of chemical dipsticks read visually or by photometry in the routine screening of urine specimens in the clinical microbiology laboratory. Pathology 1995;27:91-6.
2. Bartlett RC, Zern DA, Ratkiewicz I, Tetreault JZ. Reagent strip screening for sediment abnormalities identified by automated microscopy in urine from patients suspected to have urinary tract disease. Arch Path Lab Med 1994;118:1096-101.
3. McGinley M, Wong LL, McBride JH, Rodgerson DO. Comparison of various methods for the enumeration of blood cells in urine. J Clin Lab Anal 1992;6:359-61.
4. Briheim G, Stendahl O, Dahlgren C. Intra- and extracellular events in luminol-dependent chemiluminescence of polymorphonuclear leukocytes. Infect Immun 1984;45:1-5.
5. Roschger P, Graninger W, Klima H. Simultaneous detection of native and luminol-dependent luminescence of stimulated human polymorphonuclear leukocytes. Biochem Biophys Res Commun 1984;123:1047-53.
6. Olsson T, Bergstrom K, Thore A. A sensitive method for determination of serum hemoglobin based on iso-luminol chemiluminescence. Clin Chim Acta 1982;122:125-33.
7. Olsson T, Bergstrom K, Thore A. Chemiluminescent immunosorbent assay of serum myoglobin based on the luminol reaction. Clin Chim Acta 1984;138:31-40.
8. Akhavan-Tafti H, Sugioka K, Arghavani Z, DeSilva R, Handley RS, Sugioka Y, et al. Lumigen (TM) PS: Chemiluminescent detection of horseradish peroxidase by enzymatic generation of acridinium esters. Clin Chem 1995;41:1368-69.
9. Taurog A, Dorris ML. Myeloperoxidase-catalyzed iodination and coupling. Arch Biochem Biophys 1992;296:239-46.
10. Matsumoto T, Takahashi K, Kubo S, Haraoka M, Tanaka M, Mizunoe Y, et al. Preservation of chemiluminescence of human leukocytes subjected to hyperosmolality by ofloxacin and DR-3355. Inter J Exp Clin Chemother 1991;4:125-31.

CHEMILUMINESCENCE (CL) ASSAY OF SERUM ALKALINE PHOSPHATASE

BJ Hallaway, MF Burritt, DJ O'Kane
Dept. Laboratory Medicine and Pathology, Mayo Foundation for Medical Education and Research, Rochester, Minnesota 55905, USA

Introduction

Clinical assay of total serum alkaline phosphatase (sALP) is usually performed colorimetrically using p-nitrophenyl phosphate in alkaline conditions. Due to recent changes of HCFCA regulations, fewer colorimetric ALP assays are performed on random-access instruments. This change allows the possibility of using a chemiluminescence (CL) method for routine, clinical sALP assay. More-sensitive bioluminescence and chemiluminescence assays for ALP have been described recently (1-4). The most popular assays utilize substituted dioxetane-phosphate substrates. However, several problems exist with these assays which must be overcome before it would be feasible to implement a CL sALP assay. First, loss of assay linearity is observed as the serum matrix is diluted to compensate for the excess sensitivity of these assays. Second, the time to achieve a stable glow emission with dioxetane-phosphates is longer than with colorimetric procedures and this increases turn-around time. Third, the incorporation of polymeric enhancers to increase the CL signal causes precipitation of the serum matrix. These problems might be avoided if a fluorescent protein matrix could be developed and used with optimized conditions to perform a rapid CL sALP assay. The goal of the study was to determine if fluorescein-labeled serum could be used as a combined sample matrix diluent and fluorescent enhancer to replace the polymeric enhancers used in ALP assays.

Materials and Methods

Reagents: The following dioxetane-phosphate substrates were used: CSPD (Tropix, Bedford, MA); AMPPD (BioRad, Richmond, CA); and Lumi-Phos Plus (Lumigen, Southfield, MI). Fluorescein isothiocyanate, sodium bicarbonate and carbonate, magnesium chloride, hemoglobin, bilirubin and sodium azide were purchased from Sigma Chemicals (St. Louis, MO).

Procedures: A matrix for serum samples was prepared by heat-inactivating a unit of serum (200 mL) for several hours at 55°C to denature alkaline phosphatase. Aliquots of the heat-inactivated serum were saved at -20°C for linearity studies. Fluorescein isothiocyanate on celite was added to the remaining heat-inactivated serum at a concentration of 10 g/L and mixed together on a rotator for 4 h at RT. The FITC-labeled serum was centrifuged for 10 min to remove the celite and the supernatant was dialyzed (Spectra/Por membrane MWCO 50000) against 10 mmol/L bicarbonate buffer, pH 8 overnight (for 3 changes) to remove the free FITC. The FITC-labeled serum was heat-inactivated a second time for 1 h at 55°C in the presence of 2mmol/L EDTA to remove ALP-bound zinc and dialyzed in bicarbonate buffer for two days with at least 6 changes to remove the excess EDTA. The dialyzed FITC-labeled serum was stored in aliquots at -20°C.

Phosphatase substrates were prepared in 0.1 mol/L Na_2CO_3 buffer containing 1 mmol/L $MgCl_2$ and 0.05% NaN_3 at pH 10. CSPD and AMPPD (25 mmol/L) as received from the manufacturers, were added at a concentration of 2 µL/mL and the FITC-labeled serum was added at a concentration of 50 µL/mL to the carbonate buffer. These substrates were prepared immediately prior to use in the assay. Lumi-Phos Plus (100 µL) was used as supplied by the manufacturer.

sALP Assay Method: Dioxetane-phosphate/FITC/Na_2CO_3 buffer (100 µL) was added to a 12 x 75 mm polystyrene test tube and incubated at 37°C, or other temperatures as indicated, for 2 min. Serum or serum dilution (2.5 µL) was added to the buffered substrate and incubated at the temperature indicated for 10 min. sALP activity was detected by CL emission using a Berthold

953 Luminometer maintained at 37°C, where appropriate.

Results and Discussion

CL emission from 1,2-dioxetane-phosphates results in a glow-emission that can persist for several minutes to several hours. The immediate goal for determination of sALP was to define conditions of assay where constant CL emission (<5% change in RLU) would be attained rapidly and maintained for a long time. This required optimizing conditions of buffer pH, temperature, and substrate. Serum (2.5 µL) was added to 100 µL of five different substrates/enhancer combinations and the CL emission was followed over time (Figure 1). Lumi-Phos Plus produced extremely high CL values but never attained constant CL emission even after several hours incubation. AMPPD yielded lower CL emission than Lumi-Phos Plus but constant CL emission was never attained. Constant CL emission was attained rapidly with CSPD and was stable from 10 to 30 min. The addition of FITC-labeled serum to the CSPD substrate increased the CL emission by 3-fold while providing a stable matrix for serum sample dilutions. CSPD with FITC-labeled serum was used in subsequent assays.

Assay temperature had a profound effect on the CL emission. Serum (2.5 µL) was added to CSPD in carbonate buffer containing FITC-labeled serum (100 µL) at pH 10 and incubated at 23°C, 30°C, 37°C and 42°C (Figure 2). Constant CL emission was attained in 30 min at 23°C, 20 min at 30°C, and 8 min at 37°C. CL emission peaked immediately at 42°C but rapidly declined. The optimum assay temperature for sALP assay was 37°C which was utilized in subsequent procedures.

The effect of pH on the sALP assay was tested by diluting CSPD in carbonate buffer containing FITC-labeled serum at pH 9.0, 9.5, 9.7, 10.0 and 10.5 (Figure 3). CL emission was higher at pH 9.0, 9.5 and 9.7 but remained constant for only 5 min. This was in contrast to the results obtained at pH 10.0 where CL emission remained constant for up to 25 min. At pH 10.5 the CL emission was lower and only stable for 15 min. The optimum assay pH was 10.0 because of the lengthy duration of constant CL emission.

The CL sALP assay was tested for the effects of substances known to interfere with the colorimetric sALP assay. Known concentrations of hemoglobin, bilirubin and triglycerides were added to serum samples and tested for interference. Bilirubin and triglycerides did not interfere with sALP detection by CL. Hemoglobin at >5 mg/mL concentration was found to decrease the CL emission from sALP. Consequently, noticeably hemolyzed serum samples were diluted in heat-inactivated serum to minimize the inhibitory effect of hemoglobin and then re-run in the assay.

The linearity time course of sALP was tested by serially-diluting a serum sample from 1:2 to 1:128 in heat-inactived serum and following the CL emission over time. Serum dilutions (2.5 µL) were added to CSPD in carbonate buffer containing FITC-labeled serum (100 µL) at pH 10.0 at 37°C. The serial-dilutions in serum maintained constant CL emission for the same length of time and produced a linear dose-response.

Serum samples with sALP levels ranging from 93 to 1337 U/L were used to assess the intra- and inter-assay precision of the assay. Intra-assay precision was determined by running 10 replicates of 8 different serum samples within the same assay (Table 1). The coefficient of variation (C.V.) ranged from 3.7 to 7.3%. Inter-assay precision was determined on duplicate measurements of 5 different sera run over 5 consecutive days using thawed aliquots of sera and substrate each day. The C.V. ranged from 2.9 to 6.5%.

A recovery study was run by mixing 3 different sets of 2 sera samples together in concentrations of 2:1, 1:1, and 1:2 (Table 2) then tested in the assay. The observed values of CL emission were compared with the expected values and calculated for percent of recovery. Recoveries ranged from 97.1 to 105%.

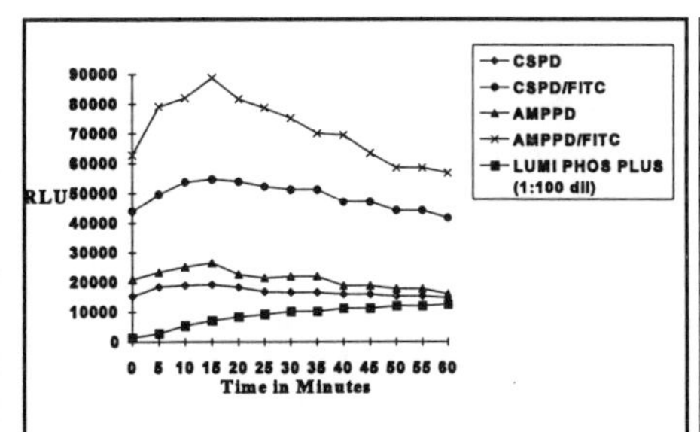

Figure 1. Kinetics of CL emission with five different substrate/enhancer combinations.

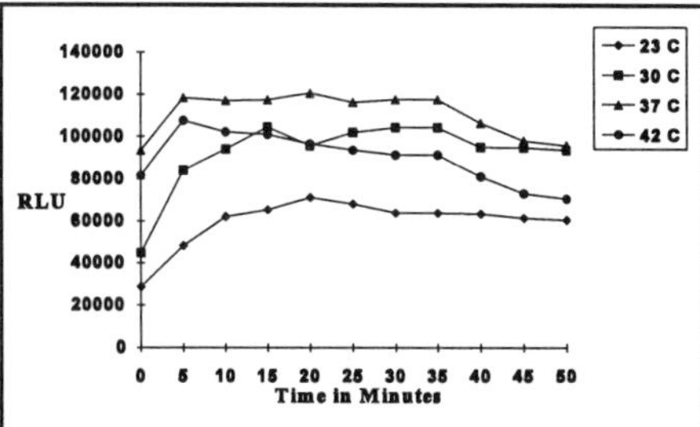

Figure 2. Effect of assay temperature on CL emission of sALP.

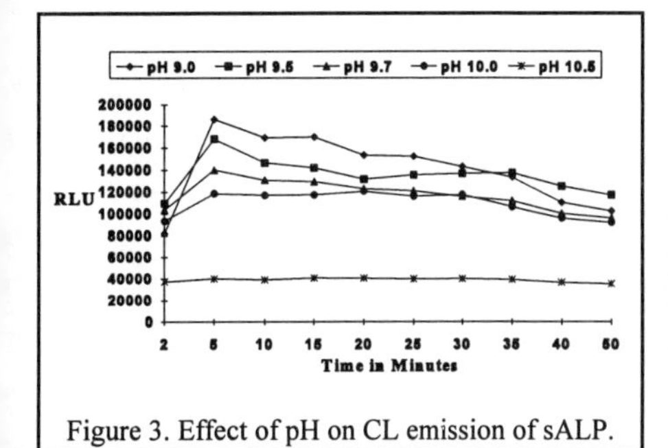

Figure 3. Effect of pH on CL emission of sALP.

Mean Serum ALP (U/L)	Intra-Assay %CV	Inter-Assay %CV
93	4.3	2.9
120	6.4	6.5
198	3.5	ND
280	3.7	ND
420	4.0	5.9
623	7.3	3.7
993	5.6	5.0
1337	4.2	ND

Table 1. Intra- and Inter-assay precision.

	Observed U/L	Expected U/L	%O/E
Undiluted High Patient	352		
2:1	162	163	99.4
1:1	203.9	210	97.1
1:2	269.8	257	105
Undiluted Low Patient	69		
Undiluted High Patient	521		
2:1	227.2	225	101
1:1	297.3	300	99.1
1:2	366.6	373	98.3
Undiluted Low Patient	78		
Undiluted High Patient	1208		
2:1	551	535	103
1:1	692.4	703	98.5
1:2	879.7	871	101
Undiluted Low Patient	199		

Table 2. Recovery study of sALP.

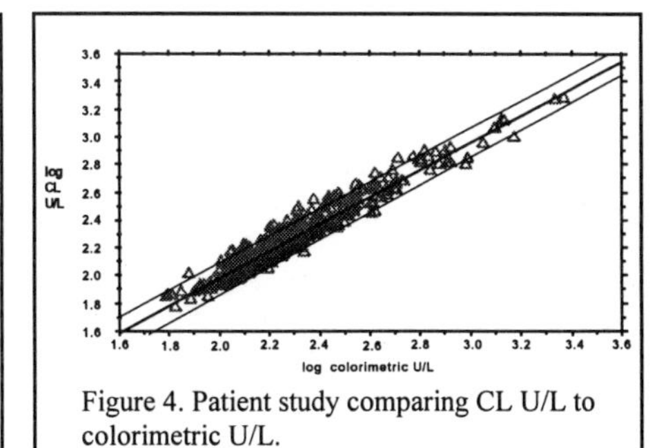

Figure 4. Patient study comparing CL U/L to colorimetric U/L.

A patient comparison study was performed on 498 serum samples (Figure 4). Results from the CL sALP assays were obtained approximately 4 h after the clinical colorimetric assays were performed. The patient samples were run in duplicate and the values of the CL RLU compared well with the clinical method. Twenty five values were expected to fall outside the 95% confidence intervals: twenty seven values were observed. The calculated conversion factor was 342.1 RLU = 1 IU/L under the conditions of assay. The calculated regression formula was IU/L (CL) = 0.978 x IU/L (colorimetric) + 0.096 with $r^2 = 0.931$.

The CL assay for sALP is relatively fast, easy to perform and compares well with the colorimetric method. The C.V.s for the CL assay are low with good recoveries. The addition of FITC-labeled serum improves the assay linearity by minimizing the matrix effects and serves as a fluorescence enhancer as well. This assay could be converted easily to random access instrumentation or a microtiter plate format to accommodate large numbers of samples.

References

1. Girotti S, Ferri E, Ghini S, Budini R, Patrono D, Incorvara L, et al. Chemiluminescent determination of alkaline phosphatase activity in serum. Anal Lett 1994;27:323-5.
2. Bronstein I, Kricka LJ. Clinical applications of luminescent assays for enzymes and enzyme labels. J Clin Lab Anal 1989;3:316-22.
3. Vant Erve Y, Voyta JC, Edwards B, Kricka LJ, Bronstein I. Influence of reaction conditions of the chemiluminescent dephosphorylation of AMPPD. In: Szalay AA, Kricka LJ, Stanley PE, editors. Bioluminescence and Chemiluminescence. John Wiley & Sons, 1993;306-11.
4. Girotti S, Ferri E, Ghini S, Roda A, Navarro J, Ortega E. Chemiluminescent analysis of ALP and lactopod in milk: data obtained with various reagents. In: Campbell AK, Kricka LJ, Stanley PE, editors. Bioluminescence and Chemiluminescence. John Wiley & Sons, 1994;44-8.

CONTINUOUS SENSITIVE DETECTION OF β-GALACTOSIDASE WITH A NOVEL CHEMILUMINESCENT 1,2-DIOXETANE

CS Martin, CEM Olesen, B Liu, JC Voyta, JL Shumway, RR Juo and I Bronstein[1]
Tropix, Inc., 47 Wiggins Avenue, Bedford, MA 01730 USA

Introduction

A new chemiluminescent 1,2-dioxetane substrate, Galacton-*Star*™, has been developed that now enables detection of β-galactosidase (β-gal) or β-gal-conjugated molecules in both solution-based and membrane blotting applications. 1,2-Dioxetane substrates, including Galacton™ and Galacton-Plus™, have been widely used in solution assays for highly sensitive quantitation of β-gal reporter enzyme activity (1-3). In contrast to Galacton and Galacton-Plus, Galacton-*Star* can be employed in an assay format in which the enzymatic deglycosylation and light-producing reaction proceed at the same pH. A luminescent reaction with continuous light signal emission is initiated upon addition of substrate to enzyme with concurrent enzymatic production and subsequent decomposition of the unstable light-generating anion. Glow kinetics, or constant light emission, is important for signal detection on membranes and imaging with X-ray film. Other 1,2-dioxetanes, such as CSPD® and CDP-*Star*™ substrates for alkaline phosphatase, have been widely used for biomolecule detection on membranes. The development of Galacton-*Star* now enables the use of β-gal enzyme labels in membrane-based applications. In addition, Galacton-*Star* simplifies solution-based assays for β-gal, including reporter gene assays or immunoassays, performed in a single-step reaction format. The resulting glow kinetics eliminate the need for instruments with injection capabilities.

Materials and Methods

Solution Assays: Dilutions of purified *E. coli* β-galactosidase (prepared in 0.1mol/L potassium phosphate pH 7.8, 2 mL/L Triton X-100, 1 g/L bovine serum albumin (BSA)) were assayed with 0.2 mmol/L Galacton-*Star*, 50 g/L Sapphire-II™ enhancer in 0.1 mol/L sodium phosphate, pH 7.0, 1 mmol/L $MgCl_2$. Light emission was measured as a 5 s integral in a Berthold LB952T (Wallac, Inc., Gaithersburg, MD).
Immunoblot Detection: Dilutions of human brain extract (Clontech, Palo Alto, CA) were separated by SDS-PAGE, followed by electrotransfer to nitrocellulose (BA-S 85, 0.45 μm; Schleicher & Schuell, Keene, NH), PVDF (Tropifluor™; Tropix, Bedford, MA) and neutral nylon (Biodyne A, 0.2 μm; Pall BioSupport, Glen Cove, NY) membranes. Blots were blocked in Blocking Buffer (0.2% I-Block™, 0.1% Tween-20 in PBS) for 1 h, incubated with a monoclonal anti-β-actin (diluted 1:5000 in Blocking Buffer; Sigma, St. Louis, MO) for 1 h, washed, incubated with β-gal-conjugated goat anti-mouse IgG+IgM (American Qualex, La Mirada, CA) and washed. Washes were 2 x 5 min with Wash Buffer (1 mL/L Tween-20 in PBS) for nitrocellulose and nylon or Blocking Buffer for PVDF. Blots were then washed 2 x 5 min with 0.1 mol/L sodium phosphate and incubated for 5 min in 0.1 mmol/L Galacton-*Star* with 50 g/L Nitro-Block-II™ enhancer (for nitrocellulose and PVDF) in 0.1 mol/L sodium phosphate at the pH optimum. Blots were imaged on XAR-5 X-ray film (Kodak, Rochester, NY).

Results and Discussion

The structure of Galacton-*Star* substrate is shown in Figure 1. Detection of purified β-galactosidase in a solution assay was performed with Galacton-*Star* substrate (Figure 2). With this substrate, 10 fg of purified enzyme is detected at a signal/noise ratio of 2,

[1] Author of correspondence

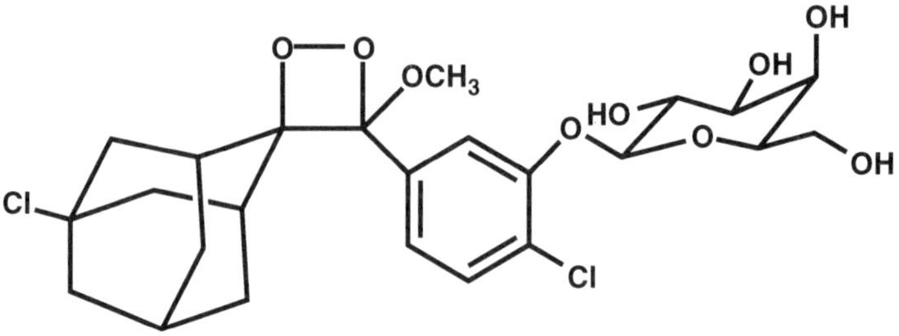

Figure 1. Galacton-*Star*.

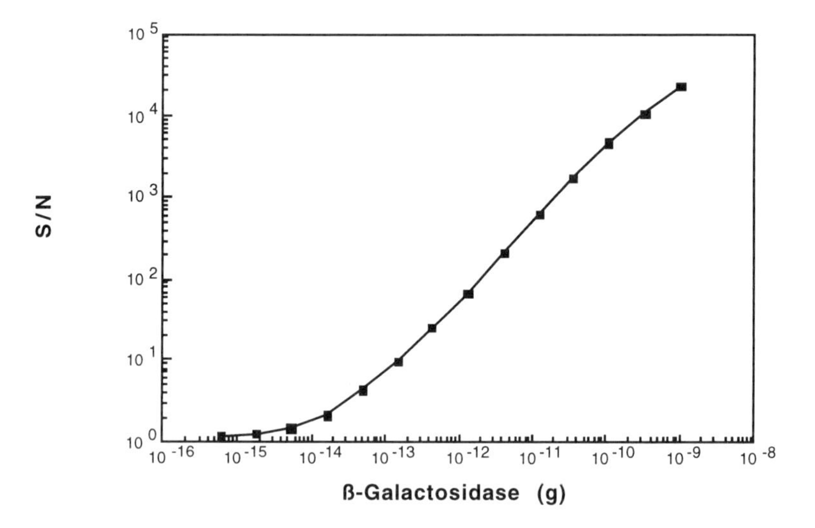

Figure 2. Detection of purified β-galactosidase with Galacton-*Star*. Light emission was measured 20 min after substrate addition.

with a dynamic range of five to six orders of magnitude of enzyme concentration. The kinetics of light emission obtained with Galacton-*Star* in this assay are shown in Figure 3. Maximum light emission is reached 10-20 min after substrate addition and remains nearly constant for over 60 min.

Galacton-*Star* has been utilized in immunodetection with β-galactosidase-conjugated secondary antibodies on nitrocellulose, PVDF and neutral and positively-charged nylon membranes. Figure 4 demonstrates immunodetection with Galacton-*Star* on nitrocellulose, PVDF and neutral nylon membranes. The pH optimum for detection on nitrocellulose is 8.0, 7.5-8.0 on PVDF, and 7.5 on neutral and positively-charged nylon. The use of Nitro-Block-II is necessary on nitrocellulose and is recommended for PVDF. The use of an enhancer is not required with nylon membranes. On membranes, light emission with Galacton-*Star* reaches maximum intensity within 1-2 h and continues for up to 24 h. Dual enzyme-label immunoblot detection can also be performed using β-galactosidase and alkaline phosphatase-labeled secondary antibody conjugates, followed by sequential incubation in Galacton-*Star* and CSPD 1,2-dioxetane substrates (4).

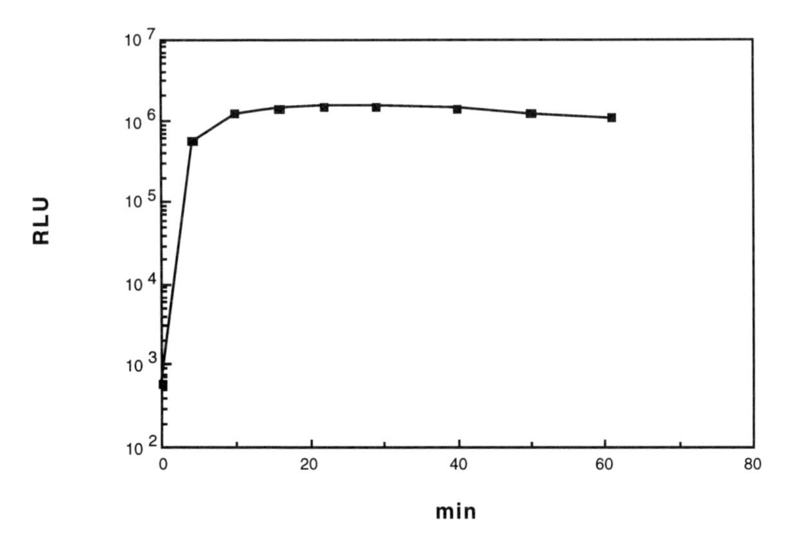

Figure 3. Kinetics of light emission with Galacton-*Star*.

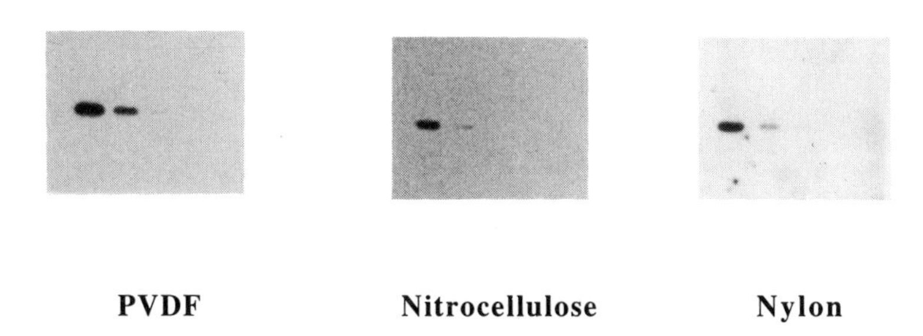

Figure 4. Immunodetection of β-actin with β-galactosidase-conjugated secondary antibody and Galacton-*Star* on PVDF, nitrocellulose and neutral nylon membranes. X-Ray film exposures were for 30 min, 5 min following substrate incubation (PVDF and nitrocellulose) or for 18 h (nylon).

The development of the Galacton-*Star* substrate for β-galactosidase has, for the first time, enabled chemiluminescent detection of β-galactosidase-labeled conjugates in membrane blotting applications. Galacton-*Star* can be used for immunodetection on nitrocellulose, PVDF and nylon membranes, in addition to the detection of β-galactosidase or β-galactosidase-labeled conjugates in solution-based assays, including reporter gene assays and ELISA formats. Chemiluminescence quantification of β-galactosidase has been utilized in several important research applications, including reporter analysis in mammalian cultured cell (5-7) and tissue extracts (8,9), yeast extracts (10), coliform detection (11) and quantitative yeast two hybrid system assays (12). The glow kinetics achieved with Galacton-*Star* eliminates the need for instrumentation with injection capabilities and therefore simplifies assay formats, particularly for high-throughput reporter screening methods. The new chemiluminescent Galacton-*Star* substrate is a versatile reagent for highly sensitive, fast and simple detection of β-galactosidase in both solution and membrane-based assay formats.

References

1. Bronstein I, Fortin J, Voyta JC, Olesen CEM, Kricka LJ. Chemiluminescent reporter gene assays for β-galactosidase, β-glucuronidase and secreted alkaline phosphatase. In: Campbell AK, Kricka LJ, Stanley PE, editors. Bioluminescence and Chemiluminescence: Fundamental and Applied Aspects. Chichester, UK: John Wiley, 1994;20-3.
2. Jain VK, Magrath IT. A chemiluminescent assay for quantitation of β-galactosidase in the femtogram range: application to quantitation of β-galactosidase in lacZ-transfected cells. Anal Biochem 1991;199:119-24.
3. Bronstein I, Martin CS, Fortin JJ, Olesen CEM, Voyta JC. Chemiluminescence: sensitive detection technology for reporter gene assays. Clin Chem 1996;42:1542-6.
4. Bronstein I, Martin CS, Olesen CEM, Voyta JC. Combined luminescent assays for multiple enzymes. In Hastings JW, Kricka LJ, Stanley PE, editors. Bioluminescence and Chemiluminescence: Molecular Reporting with Photons. Chichester, UK: John Wiley, 1997;in press.
5. O'Connor KL, Culp LA. Quantitation of two histochemical markers in the same extract using chemiluminescent substrates. BioTechniques 1994;17:502-9.
6. Hu M, Bigger CB, Gardner PD. A novel regulatory element of a nicotinic acetylcholine receptor gene interacts with a DNA binding activity enriched in rat brain. J Biol Chem 1995;270:4497-502.
7. Ruess FU, Coffin JM. Stimulation of mouse mammary tumor virus superantigen expression by an intragenic enhancer. Proc Natl Acad Sci U S A 1995;92:9293-7.
8. Shaper N, Harduin-Lepers A, Shaper JH. Male germ cell expression of murine β4-galactosyltransferase. A 796-base pair genomic region, containing two cAMP-responsive element (CRE)-like elements, mediates male germ cell-specific expression in transgenic mice. J Biol Chem 1994;269:25165-71.
9. Feldman LJ, Steg PG, Zheng LP, Chen D, Kearney M, McGarr SE, et al. Low-efficiency of percutaneous adenovirus-mediated arterial genetransfer in the atherosclerotic rabbit. J Clin Invest 1995;95:2662-71.
10. Wolf SS, Roder K, Schweizer M. Construction of a reporter plasmid that allows expression libraries to be exploited for the one-hybrid system. BioTechniques 1996;20:568-74.
11. Van Poucke SO, Nelis HJ. Development of a sensitive chemiluminometric assay for the detection of beta-galactosidase in permeabilized coliform bacteria and comparison with fluorometry and colorimetry. Appl Environ Microbiol 1995;61:4505-9.
12. Bourne Y, Watson MH, Hickey MJ, Holmes W, Rocque W, Reed SI, Tainer JA. Crystal structure and mutational analysis of the human CDK2 kinase complex with cell cycle-regulatory protein CksHs1. Cell 1996;84:863-74.

CHEMILUMINESCENT DETECTION OF GLYCOSIDIC ENZYMES WITH 1,2-DIOXETANE SUBSTRATES

JC Voyta, JY Lee, CEM Olesen, R-R Juo, B Edwards and I Bronstein[1]
Tropix, Inc., 47 Wiggins Avenue, Bedford, MA 01730 USA

Introduction

The benefits of increased detection sensitivity provided by chemiluminescent 1,2-dioxetane enzyme substrates compared to colorimetric and fluorometric substrates has been widely demonstrated (1, 2). Alkaline phosphatase substrates such as AMPPD®, CSPD®, and CDP-Star™ have been used in a variety of applications, including immunoassays (3), probe hybridization assays and immunoblot detection (4) and reporter gene assays (5). 1,2-Dioxetane substrates for β-galactosidase (β-gal), β-glucuronidase (GUS) and alkaline phosphatase have been incorporated into reagent systems utilized in quantitative ultrasensitive detection of the reporter enzymes β-gal, GUS and placental alkaline phosphatase (6). Recently, a chemiluminescent 1,2-dioxetane substrate for β-glucosidase has been developed. 1,2-Dioxetane substrates for other hydrolytic enzymes, including esterases, phospholipases and phosphotriesterases have also been described.

The majority of applications for 1,2-dioxetane substrates have been designed for the detection of enzyme labels and reporter enzymes. Recently, however, chemiluminescent 1,2-dioxetane substrates for β-gal have been successfully used in the rapid determination of coliform contamination of drinking water (2). The presence of one coliform cell in 100 mL of water is detected following a 6 hour culture with chemiluminescent Galacton™. This approach offers many advantages compared to conventional coliform tests which require a 24 hour or longer culture.

Quantitation of glycosidic enzymes has widespread application, including environmental testing, biomedical research, clinical evaluation, toxicology and pharmaceutical screening. β-Gal detection generates a measure of the total coliform population, while GUS activity provides a specific marker for *E. coli* within total coliform contamination. Measurement of GUS activity has been employed in a bioassay with *E. coli* to evaluate toxicity of metal ions in environmental samples (7). In addition, mammalian β-glucuronidase activity has been measured in assays for mast cell degranulation (8) and to monitor periodontal disease (9). A physiologic test strategy, including assays for β-glucuronidase and β-glucosidase has been used for the rapid identification and differentiation of enterococci and streptococci (10). Microbial β-glycosidases in the intestinal tract generate toxicants, and accurate analysis of these activities is critical in toxicology studies (11).

Chemiluminescent 1,2-dioxetane substrates enable highly sensitive detection of several important enzymes and are used in both solution and membrane-based assay formats. These substrates provide versatile, highly sensitive and rapid detection alternatives to current methodologies for glycosidic enzymes.

Materials and Methods

β-Galactosidase (G-5635, EC 3.2.1.23), β-glucuronidase (G-7896, EC 3.2.1.31) and β-glucosidase (G-6906, EC 3.2.1.21) and bovine serum albumin (BSA) were from Sigma (St. Louis, MO). Galacto-Light™ and GUS-Light™ reporter gene assay systems (Tropix, Bedford, MA) which include the substrates Galacton™ and Glucuron™ were used for the detection of β-gal and β-glucuronidase, as described in the assay protocols using purified enzyme. For β-glucosidase, reaction buffer consisted of 0.1 mol/L sodium phosphate, pH 6.5. Serial enzyme dilutions were

[1] Author of correspondence

prepared in reaction buffer containing 10 g/L BSA. Reaction buffer containing substrate (180 μL of Galacton, 0.025 mmol/L; Glucuron or Glucon, 0.05 mmol/L) was added to 20 μL enzyme aliquots and incubated for 30 min at room temperature. Accelerator solution was then added and the luminescent signal was integrated for 5 s in a Berthold LB952T luminometer (Berthold; Wallac Inc., Gaithersburg, MD). The average of triplicates is plotted.

Results and Discussion

The three chemiluminescent 1,2-dioxetane glycosidase substrates described differ in the enzyme cleavable sugar group:

Galacton (R = β-D-galactose); Glucuron (R = β-D-glucuronic acid); and Glucon (R = β-D-glucose). Enzyme were assayed with each substrate, to determine sensitivity and specificity. With β-gal, 10,000-fold higher signal intensity is obtained with Galacton compared to Glucuron and Glucon (Figure 1). β-Glucuronidase demonstrates no cleavage of Galacton and only very low levels of cleavage of Glucon (Figure 2).

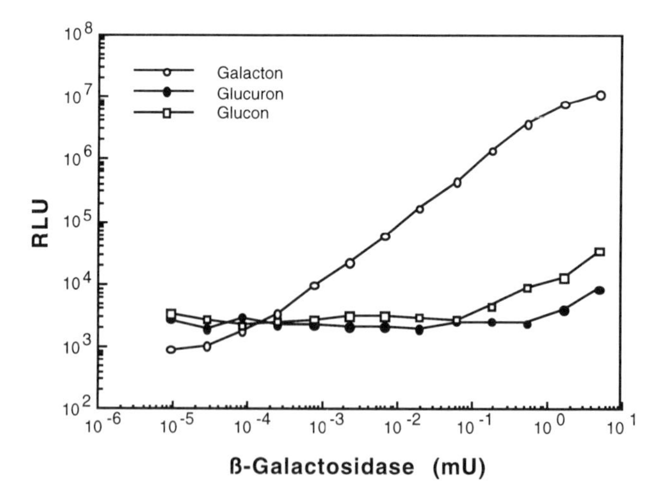

Figure 1. Chemiluminescent detection of β-galactosidase.

β-Glucosidase shows cleavage of both Glucon and Galacton, although an approximately 40-fold higher signal intensity is obtained with Glucon (Figure 3).

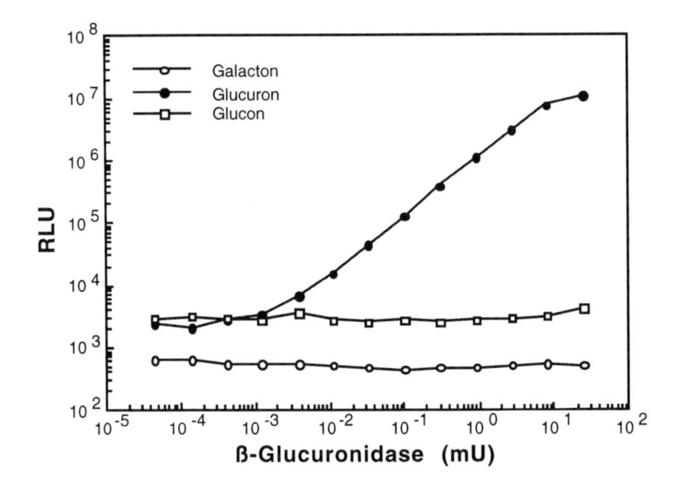

Figure 2. Chemiluminescent detection of β-glucuronidase.

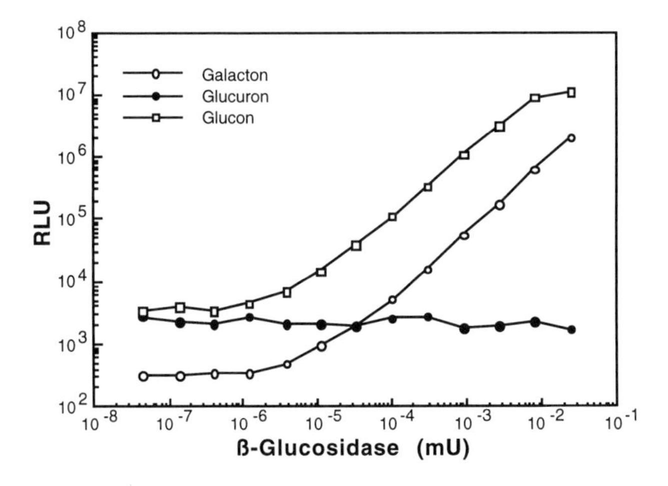

Figure 3. Chemiluminescent detection of β-glucosidase.

These results demonstrate specificities of β-gal, β-glucuronidase and β-glucosidase for these 1,2-dioxetane substrates. β-Glucuronidase is highly specific for Glucuron substrate, and β-gal is specific for Galacton, except for a low level of cleavage of Glucon at high enzyme concentrations. β-Glucosidase demonstrates some cleavage of the Galacton substrate. Additional 1,2-dioxetane substrates for β-gal, Galacton-Plus and Galacton-*Star*, may provide higher specificity for detection of β-gal.

Sensitive reporter gene assay systems have been previously described for β-gal and GUS (5). Vectors are also available for reporter gene constructs which contain the sequence for β-glucosidase (12). The chemiluminescent substrate described here will also be used in the development of an assay for the ultrasensitive detection of this reporter gene product. Thermostable β-glycosidic enzymes have been isolated and cloned (12), and their use as reporter genes in combination with chemiluminescent 1,2-dioxetane detection will provide extremely sensitive, robust reporter assay systems that will eliminate background from endogenous enzyme activity. Finally, 1,2-dioxetane substrates will be incorporated into a wide variety of assays for sensitive, quantitative glycosidase detection for clinical, environmental, research and pharmaceutical screening applications.

References

1. Jain VK, Magrath IT. A chemiluminescent assay for quantitation of β-galactosidase in the femtogram range: application to quantitation of β-galactosidase in *lacZ*-transfected cells. Anal Biochem 1991;199:119-24.
2. Van Poucke SO, Nelis HJ. Development of a sensitive chemiluminometric assay for the detection of β-galactosidase in permeabilized coliform bacteria and comparison with fluorometry and colorimetry. Appl Environ Micobiol 1995;61:4505-9.
3. Bronstein I, Sparks A. Sensitive enzyme immunoassays with chemiluminescent detection. In: Nakamura RM, Kasahara Y, Rechnitz GA, editors. Immunochemical Assays and Biosensor Technology for the 1990s. Washington, DC: American Society for Microbiology, 1992:229-50.
4. Bronstein I, Olesen CEM, Martin CS, Schneider G, Edwards B, Sparks A, Voyta JC. Chemiluminescent detection of DNA and protein with CDP™ and CDP-*Star*™ 1,2-dioxetane enzyme substrates. In Campbell AK, Kricka LJ, Stanley PE, editors. Bioluminescence and Chemiluminescence: Fundamentals and Applied Aspects. Chichester, UK: John Wiley, 1994:269-72.
5. Bronstein I, Martin CS, Fortin JJ, Olesen CEM, Voyta JC. Chemiluminescence: sensitive detection technology for reporter gene assays. Clin Chem 1996;42:1542-6.
6. Bronstein I, Fortin J, Voyta JC, Olesen CEM, Kricka LJ. Chemiluminescent reporter gene assays for β-galactosidase, β-glucuronidase and secreted alkaline phosphatase. In Campbell AK, Kricka LJ, Stanley PE, editors. Bioluminescence and Chemiluminescence: Fundamentals and Applied Aspects. Chichester, UK: John Wiley, 1994:20-3.
7. Mariscal A, García A, Carnero M, Gómez E, Fernández-Crehuet J. New toxicity detemination method that uses fluorescent assay of *Escherichia coli.* BioTechniques 1994;16:888-92.
8. Dreskin SC, Probluda VS, Metzger H. IgE receptor-mediated hydrolysis of phosphoinositides by cytoplasts from rat basophilic leukemia cells. J Immunol 1989;142:4407-15.
9. Lamster IB, Holmes LG, Gross KBW, Oshrain RL, Cohen DW, Rose LF, Peters LM, Pope MR. The relationship of β-glucuronidase activity in crevicular fluid to clinical parameters of periodontal disease. J Clin Periodontol 1994;21:118-27.
10. Kirby R, Ruoff KL. Cost-effective, clinically relevant method for rapid identification of beta-hemolytic streptococci and enterococci. J Clin Microbiol 1995;33:1154-57.
11. Chadwick RW, Allison JC, Talley DL, George SE. Possible errors in assay for β-glycosidase activity. Appl Environ Microbiol 1995;61:820-2.
12. Gabelsberger J, Liebl W, Schleifer K-H. Purification and properties of recombinant β-glucosidase of the hyperthermophilic bacterium *Thermotoga maritima.* Appl Microbiol Technol 1993;40:44-52.

LERS LIGHT
A LUMINESCENCE ENHANCED REAGENT SYSTEM FOR THE DETECTION OF HORSERADISH PEROXIDASE (HRP)

Thomas Schlederer[2], Gottfried Himmler[1]
[1] Institut of Applied Microbiology, Nußdorfer Lände 11, University of Agriculture, A-1190 Vienna, Austria/Europe
[2] MEDIATORS Diagnostika GmbH, Dragonerweg 21, 1220 Vienna, Austria/Europe

Introduction

Super sensitive Immunoassays based on HRP as detector enzymes can be developed using highly sensitive chemiluminescent detection methods (1,2). The advantages of the new detection methods are higher sensitivity and broader dynamic range compared to the photometric assays widely used in research and diagnostic industry. However, these super-sensitive substrates give slow-increasing and long-lasting (glowing) light signals which demand the use of white or black microtiterplates in order to avoid cross-talk.

Therefore, we have developed a reagent system which can be triggered to a flash reaction. In addition, the reagent system can be read photometrically to exclude high values in the luminescent detection system thus enabling the use of transparent microtiterplates. Moreover, the reagent system does not exhibit substrate consumption to produce light and therefore considerably high HRP amounts can be detected without depletion of signal.

LERS Light is based on chemically initiated electron exchange luminescence (CIEEL) using TCPO/H_2O_2 and a fluorescent activator. The intensity of light emission linearly relates to the concentration of the added fluorescent activator. Employing Dihydrorhodamine 6G that sets red light free, when oxidized by HRP via an enhancer to rhodamine 6G and triggered by TCPO/H_2O_2, allows the sensitive detection of horse radish peroxidase. This chemiluminescent detection system was successfully applied to the COULTER™ HIV-1 p24 Antigen Assay creating an ultrasensitive HIV-1 p24 Antigen Assay. The assay involves a tuned protocol in which the substrate TMB is substituted by LERS substrate solution.

Materials and Methods

Instrumentation: Light emission was measured using a filter equipped Anthos lucy1 microtiterplate luminometer (Anthos labtec, Austria). Optical Density was measured using an Anthos htIII microtiterplate photometer (Anthos labtec, Austria).
Reagents: LERS Light Reagent was supplied by Mediators Diagnostika (Vienna, Austria). Human Immunodeficiency Virus p24 Antigen Assay was purchased from Coulter Corp. (Miami, USA). HRP-conjugate was obtained from the p24 assay. Clinical specimens were provided by the Austrian Red Cross (Vienna, Austria). Microtiterplates were purchased from Nunc (Roskilde, Denmark). TMB substrate was generously provided by Organon Teknika (Boxtel, Netherlands).

Procedures: Dilution series experiments: HRP-conjugate was diluted in PBS/BSA at room temperature and 10 μL of the corresponding sample was mixed with 100 μL of LERS Light Substrate Solution in a transparent microtiter plate in 8-fold repeats. After incubation in the dark for 29 min the optical density was measured (520/620 nm), the plate being transferred to the luminometer integrating the light signal for 0.6 sec starting 0.2 sec after addition of 50 μL of Trigger Solution with the high pass filter at 570nm. For kinetic studies of the flash reaction the same protocol was applied, with the exception that light emission was recorded 50 times for 0.6 sec in the same well. For comparison of sensitivity the samples (10 μL) were diluted with 100 μL TMB substrate, incubated for 30 min, stopped with 100 μL of 1 mol/L H_2SO_4 and measured photometrically at 450 nm.

Elisa: The Human Immunodeficiency Virus p24 Antigen Assay was performed according to the manufacturer instructions except that the 200 μL of LERS Light Substrate Solution were added by the luminometer from dispenser 1 instead of TMB solution. After incubation in the dark for 30 min the light signal was integrated for 0.6 sec starting 0.2 sec after addition of 100 μL of Trigger Solution from dispenser 2. The standard curve was performed in duplicate and clinical specimens were analysed as singletons.

Results and Discussion

In order to make transparent microtiterplates usable for chemiluminescent detection of HRP we have developed a new reagent system. The system is performed in two steps. In the first step Dihydrorhodamine 6G is oxidized from H_2O_2 via catalytic action of HRP and an enhancer (4) at pH 5,0. At very high HRP concentrations all the Dihydrorhodamine 6G is consumed leading to a plateau rather than to a reduction of signal. This effect can be easily observed in substrates that directly "glow" upon addition of HRP. This feature is important in screening assays (e.g. HBsAg) where highly positive samples are expected. In the second step, TCPO in an organic solvent is added, forming an intermediate with H_2O_2 due to the action of a catalyst, both being already present from the foregoing step. This intermediate reacts with the fluorophor chemically exciting the latter (CIEEL). On relaxation to the ground state the fluorophor sets free with very high quantum efficiency light (while integrating on the PMT each fluorophor molecule can be excited a few time because TCPO and H_2O_2 are in surplus) making possible the use of transparent microtiterplates. While this system was successfully applied to detect oxidase activities (3) on their property to form H_2O_2 with a fluorophor added, the LERS system detects the fluorophor rhodamine 6G formed. "Filtering" the luminescent light by employing a high pass filter to reduce the intrinsic background light produced by the TCPO/H_2O_2 - system increases the signal-to-noise ratio. Although the reagent system produces a light-flash, a high light-emitting well could interfere with one emitting low intensity light when this second well is situated next to the first one and processed shortly after it. To enable the use of transparent microtiterplates very high samples can be detected photometrically and postponed prior to the chemiluminescent detection. To evaluate sensitivity we have performed dilution series experiments with HRP- conjugates which are used in Elisa techniques and have a reactivity 2-3 times less than pure HRP.

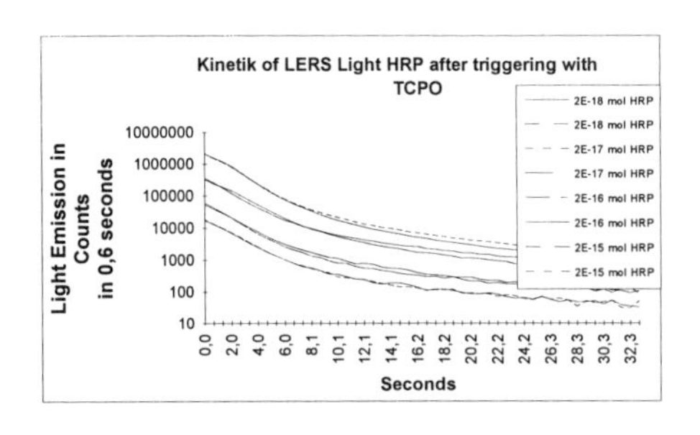

Figure 1. Kinetic of light emission

For comparison of sensitivity we use the concept of ES value where $ES_X = (X-B)/SD_B$ with X-signal for the respective sample and B-average signal value for the blank. This ratio corresponds to the sample reactivity relative to the corrected background and allows to compare the performance of different procedures.

The flash reaction exhibits a half life time of 1.5 sec. In order to avoid cross talk from high light emitting samples while performing detection with with LERS Light, two strategies can be employed. The first strategy is to program a "template" on the luminometer where the wells are pipetted in the following scheme: A1, A7, E1, E7, A2, A8, E2... Since it takes the luminometer 3 sec to process one well (for the addition of Trigger Solution, measuring and steering to the next well) and cross talk from one well to the next one in a transparent plate is always below 20%, this strategy allows to measure over a dynamic range of 2 orders of magnitude. In the second strategy the wells are photometrically measured (520/620 nm) prior to chemiluminescent detection. Wells that exhibit optical density greater than 50 mOD compared to the blanks are postponed and chemiluminometrically measured after the photometrically low values have been processed. This feature can be programmed on the anthos microtiterplate luminometer since photometric measurement is also available on this instrument. Of course a combination of the two strategies avoiding cross talk gives the best results while the dynamic range can be expanded to more than three orders of magnitude.

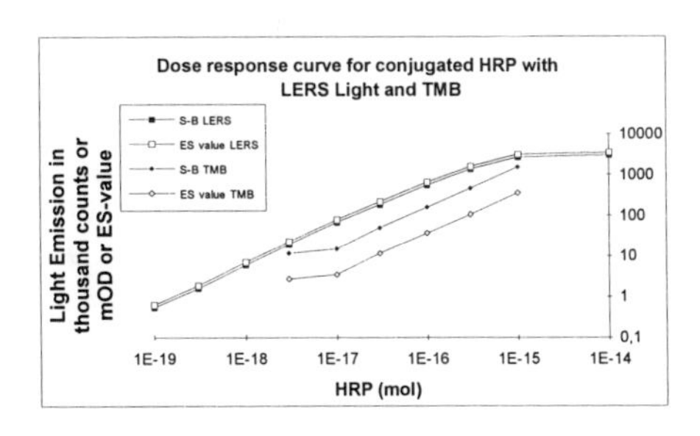

Figure 2. Quantitation of HRP

Figure 2 depicts the dose response curve for HRP conjugate detection. The comparison was performed with a highly sensitive TMB substrate. Assuming a significant ES - value of 3 (Signal-Blank)/ Standard deviation, it was shown that LERS Light HRP is 10 - fold more sensitive than TMB substrate, with the detection limit being less than 1 amol of HRP-conjugate.

LERS Light was successfully applied to the COULTER™ HIV-1 p24 Antigen Assay. Employing the tuned protocol, 28 negative sera were used in the assay with. Based on these samples a preliminary cut off was set which results in an assay sensitivity of 0.6 pg/mL of HIV-1 p24 antigen. In Figure 3 the standard curve and the mean and mean+3s of negative sera are depicted. The mean+3s of negative sera was used as a preliminary cut-off. LERS Light is an ultrasensitive detection system and the sensitivity of the ELISA only depends on the unspecific binding of the conjugates to the solid phase. This "conjugate blank" is raised when reagents are altered but within the manufacturers validation parameters; thus, diminishing the detection limit.

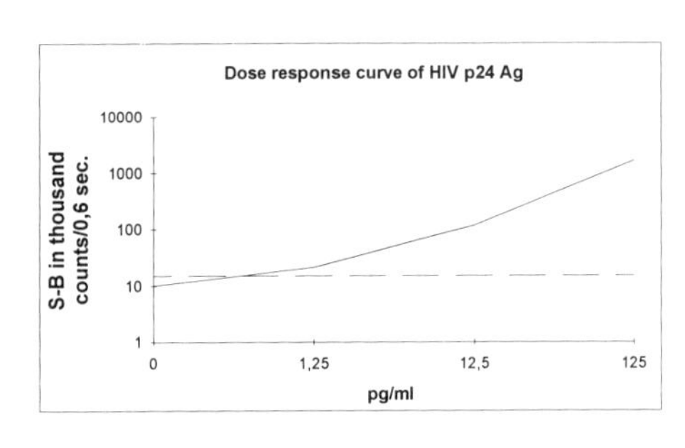

Figure 3. Detection of HIV-1 p24 Ag

Light emission is recorded immediately after addition of Trigger Solution and integrated for 0.6 sec. The flash kinetic along with high quantum efficiency allow the usage of transparent microtiterplates avoiding cross-talk, increasing the dynamic range to more than three orders of magnitude when detecting HRP.

Acknowledgements

We thank Peter G. Fritz, President of Anthos labtec instruments for his valuable suggestions and the support provided which enabled this project. We thank Ms. Amelia Thirring for her critical reading of this manuscript.

References

1. Kricka LJ, Ji X, Super-sensitive enzyme immunoassay for thyroid stimulating hormone using a new synergistic enhanced chemiluminescent endpoint. J Biolumin Chemilumin 1996; 11: 137-47.
2. Akhavan-Tafti H, Arghavani Z, DeSilva R, Schoenfelner BA, Schaap AP, Applications of Lumigen PS-1 Chemiluminescent Substrate to Immunoassays. In: Campbell AK, Kricka LJ and Stanley PE, editors. Bioluminescence and Chemiluminescence. Cambridge: Wiley, 1994:309-12.
3. Nakashima K, Kuroda N, Kawaguchi S, Wada M, Akiyama S, Peroxyoxalate chemiluminescent assay for oxidase activities based on detecting enzymatically formed hydrogen peroxide. J Biolumin Chemilumin 1995; 10: 185-91.
4. Nozaki O, Ji X, Kricka LJ, New enhancers for the chemiluminescent peroxidase catalysed chemiluminescent oxidation of pyrogallol and purpurogallin J Biolumin Chemilumin 1995; 10: 151-6.

PART 12

Instrumentation and Devices

THE NIGHT OWL MOLECULAR LIGHT IMAGER - A LOW-LIGHT IMAGING SYSTEM FOR BIO- AND CHEMILUMINESCENCE AND FLUORESCENCE

B Möckel, J Grand, R Ochs
EG&G Berthold, Calmbacher Str. 22, 75323 Bad Wildbad, Germany

Introduction

The sensitivity and spatial resolution of CCD cameras has resulted in a variety of new applications in the life sciences. Further improvements in camera technology and the development of a new generation of cooled CCD cameras allows the detection of very faint signals. Modern systems are based on peltier-air cooled CCD cameras. These cameras offer improved sensitivity, dynamic range, excellent image quality and robustness.

The Night OWL Molecular Light Imager is a complete system designed for macro imaging of gene expression in tissues or whole organisms, southern, northern blots or western blots, and the simultaneous measurement of multiple samples in microplates. At the micro level, reporter gene expression is visualized in single cells on the microscope. Two-dimensional photon detection is a versatile tool for quantitative and qualitative evaluation of luminescence signals. Measurements can be performed in vivo with highest sensitivity comparable to that of luminometers.

Materials and Methods

Instrumentation: The major requirement for low light imaging is a highly sensitive and versatile camera. The Night OWL slow scan CCD camera is a peltier-air cooled camera (-73°C). Exposure times can be selected from milliseconds to hours. For quantitation of results, the CCD shows excellent linearity and uniformity (over a full 16-bit range). The resolution of the camera is defined by 385 x 578 pixels. Sensitivity may be further improved by the ability to bin pixels together.

Reagents: Luciferase was purchased from Boehringer Mannheim (Mannheim, Germany). The Promega Madison, WI) Luciferase Assay System was used for detection of luciferase. The RAD-FREE Southern kit, NYTRAN-Plus membranes, and Lumi-Phos 530 substrate sheets were supplied by Schleicher & Schuell Inc. (Dassel, Germany). λ-DNA and Restriction enzymes were purchased from Boehringer Mannheim (Mannheim, Germany).

Procedures: For Luciferase detection dilutions of recombinant luciferase were prepared in cell culture lysis reagent, 1mg/mL BSA. 10µL of enzyme dilutions were added to the wells of a white microplate. Reactions were started by pipetting 50µL of Luciferase Assay Reagent to the samples and light emission was recorded for various time periods.

For Southern Blot detection DNA from H5 rat hepatoma cells was restricted with *Eco*R I and spiked with decreasing amounts of λ-DNA cut with *Hin*d III and *Eco*R I. The mixtures were separated in a 1% agarose gel and blotted onto NYTRAN-Plus membranes. The blot was allowed to air-dry and DNA was fixed to the membrane by UV-crosslinking. λ-DNA cut with *Hin*d III and *Eco*R I was used as a hybridization probe and was labeled with psoralen biotin using the RAD-FREE Southern kit. Hybridization and detection was performed according to the standard protocol. The concentration of labeled λ-DNA probe was 60ng/mL.

Results and Discussion

The high sensitivity of the Night OWL imaging system allows detection and quantification of even faint luminescence signals. The analytical performance of the Night OWL system for southern blot detection was evaluated. For detection of single copy genes in human genome DNA the RAD-FREE Southern kit [1] and Lumi-Phos 530 substrate sheets were used. The blot membrane was first exposed to x-ray film at room temperature and then light emission was recorded with the Night OWL Molecular Light Imager.

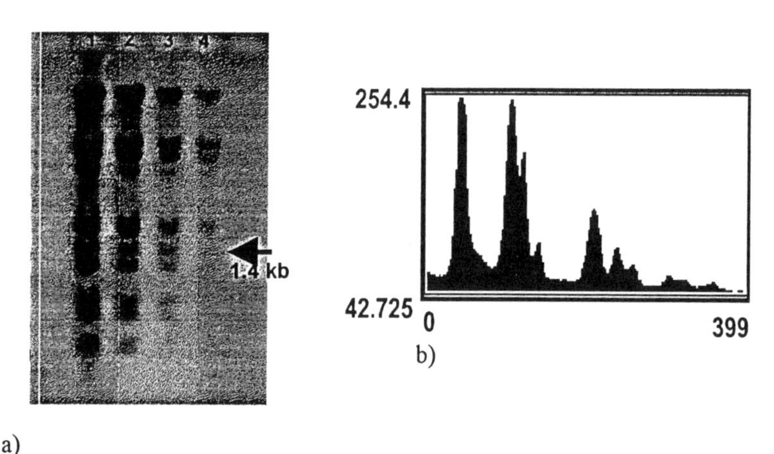

Figure 1
Visualization and quantification of DNA in southern blot analysis using the Schleicher & Schuell RAD-FREE Southern kit. DNA from H5 rat hepatoma cells restricted with *Eco*R I and spiked with 12.5ng, 1.25ng, 125pg or 12.5pg *Hin*d III / *Eco*R I cut λ-DNA (from left to right) using labeled *Hin*d III - *Eco*R I cut λ-DNA as a probe. a) Image of light emission integrated for 5 minutes using the NightOWL imaging system. b) Line plot of lane 2.

The 1.4kb fragment of 12.5pg λ-DNA cut with *Hin*d III and *Eco*R I has a mass of 0.35pg. This fragment can be easily detected in the large excess of 5μg genomic rat DNA. The rat genome size ($3x10^9$bp) is very similar to the human genome size. The detection of 0.35pg target DNA in 5μg of rat genomic DNA is thus equivalent to the single copy gene detection in 5μg of human genomic DNA. While the total probe concentration was 60ng/mL the actual concentration for the 1.4kb fragment specific probe was only 1.7ng/mL. Thus, even at such a low probe concentration the single copy gene detection is possible. Detection of hybridization events has been performed using x-ray films (data not shown) and the Night OWL CCD camera. The sensitivity of the CCD camera is equivalent to x-ray film or even better. Furthermore, when using substrates such as Lumi-Phos 530, shorter exposure times (for example 16h exposure to x-ray film versus 1h exposure to Night OWL CCD camera chip) are necessary to detect very faint signals. The dynamic range of the cooled CCD camera based Night OWL system is about 100 times higher than the x-ray film. Therefore, only a single exposure is necessary to detect very weak and strong signals on one blot membrane.

Imaging systems have also been proved to be versatile tools for detection and quantification of luminescence signals in microplates [2] such as reporter gene expression. A variety of substrates for reporter gene assays are now available that emit light which persists at a constant level for longer time periods and can be used for reporter gene studies with imaging systems [3]. Therefore, the simultaneous detection of large sample numbers for highest sample throughput is possible. Furthermore, with the available luciferase assays, light intensity is nearly constant for measurements of up to several minutes [4]. As an example, we describe quantification of firefly luciferase with the imaging system Night OWL.

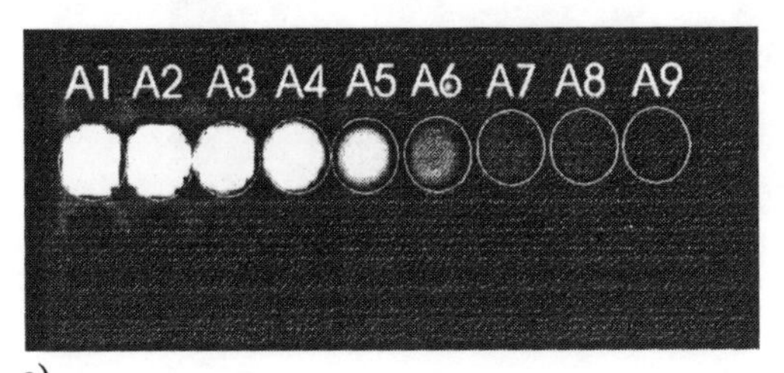

a)

sample	luciferase [pg]	sum grey [pW]
A1	100	50.566
A2	100	48.008
A3	20	10.695
A4	10	5.5150
A5	2	1.2063
A6	1	0.55545
A7	0.1	0.075689
A8	0.02	0.017296
Background		0.010387

b)

Figure 2
Quantification of firefly luciferase in microplates. Image of light emission of 8 dilution steps ranging from 100pg to 0.02pg luciferase. The image was acquired with the following instrument settings: exposure time 60 sec; pixel size 5 x 5; background subtraction; camera readout slow. (a) Visualization of luminescent signal in microplates. (b) Measurement report. Signals were measured in pW.

Exposure time and all system parameters are under comfortable interactive user interface control. Image acquisition parameters can be stored in a file which allows easy automation of these procedures. The system also allows the exposure repeated automatically for acquisition of image sequences. Areas of interest were defined for calculating the intensity of the luminescent signal either in counts, watts or photons.

For micro imaging the camera can easily be mounted to a microscope with a standard C-mount adapter. Imaging of luciferase expression in single cells is a powerful tool for quantification and localization of gene expression. Localization of luminescent signals in a sample is made by superposition of the luminescence signal in pseudo-colors to the light-image. A variety of new applications of imaging systems in molecular biology have been developed in the last few years [5]. The possibility to detect luminescent signals and to quantitate gene expression *in vivo* is one of the most fascinating new methods in molecular biology.

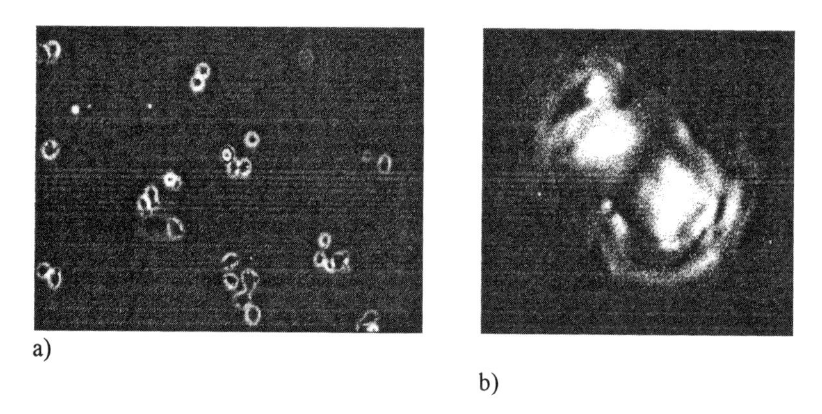

a)

b)

Figure 3
Visualization of luciferase expression. (a) in mammalian cells (microscope) and (b) in tissue sections of a brain of a transgenic luciferase expressing mouse (macro imaging).

References

1 Levenson C, Watson R, Sheldon E L. Biotinylated psoralen derivative for labeling nucleic acid hybridization probes. Method Enzymol 1990;184: 577-83.

2 Roda A, Pasini P, Musiani M, Girotti S, Baraldini M, Carrea G, Suozzi A. Chemiluminescent low-light imaging of biospecific reactions on macro- and microsamples using a videocamera-based luminograph. Anal Chem 1996;68: 1073-80.

3 Bronstein I, Fortin J, Stanley P E, Stewart G S A B, Kricka L J. Chemiluminescent and bioluminescent reporter gene assays. Anal Biochem 1994; 219:169-81.

4 Wood K V. Recent advances and prospects for use of beetle luciferase as genetic reporter. In: Stanley P, Kricka L. Bioluminescence & Chemiluminescence: Current status. John Wiley and Sons Ltd. Chichester. 1991.

5 Nicholas J. C. Applications of low-light imaging to life sciences. J Biolumin Chemilumin 1994;9: 139-144.

THE LUMI-IMAGER™, A SENSITIVE AND VERSATILE SYSTEM FOR IMAGING, ANALYSIS AND QUANTITATION OF CHEMILUMINESCENCE ON BLOTS AND IN MICROTITERPLATES

M Gutekunst[1], M Jahreis[2], R Rein[1] and HJ Hoeltke[1]
[1]Boehringer Mannheim GmbH, Mannheim / Penzberg / Tutzing, Germany
[2]Jahreis Photoengineering, Tutzing, Germany

Introduction

The most sensitive nonradioactive detection methods for proteins and nucleic acids employ chemiluminescent substrates. The use of chemiluminescence for all types of blotting applications (Southern, Northern, Western blots etc.) is rapidly increasing and provides an attractive alternative to radioactivity. A full replacement of radioactive techniques requires the availability of adequate instruments for imaging, documentation, and evaluation of chemiluminescent results. Traditionally, chemiluminescent signals on blots are detected with X-ray films. X-ray films have a limited dynamic range of two orders of magnitude resulting in a saturation of high signals, therefore multiple exposures are usually required. Real quantitation and image analysis is not possible with X-ray films. CCD cameras now provide the technical basis for imaging and analysis of chemiluminescent signals. But suitable instruments that combine good sensitivity, high resolution, reliable quantitation, and a user-friendly software were commercially not available.

Results and Discussion

We have developed an instrument for imaging of chemiluminescent signals on blots and in microtiter plates. The system consists of a Peltier cooled CCD camera and a specially developed lens system that allows image acquisition of samples up to 24 x 30 cm^2 or 4 microtiterplates (Fig. 1).

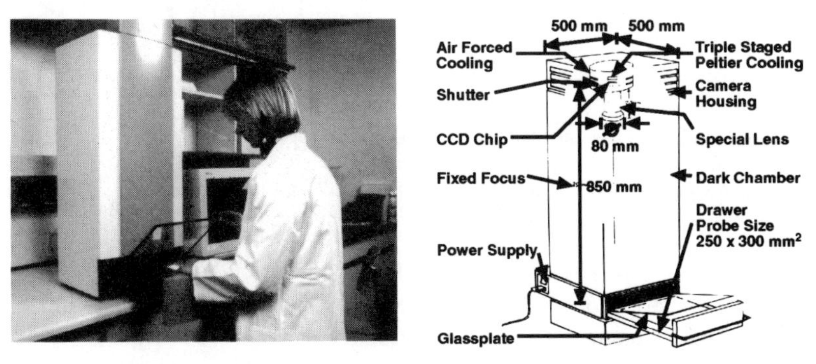

Figure 1: Photograph and schematic drawing of the Lumi-Imager™ instrument.

All blotting applications (Southern, Northern, dot blots, Western blots) and various microtiter plate assays can be documented in publication-ready quality with the Lumi-Imager™, and an appropriate software allows analysis and quantification of chemiluminescent signals of the 'glow' type, e.g., catalyzed by alkaline phosphatase or horseradish peroxidase.

The sensitivity and the image quality regarding resolution and contrast are absolutely comparable to X-ray film (Fig. 2). In addition, the Lumi-Imager™ provides an option to use a more sensitive 'binning' mode which reduces the exposure time by a factor of up to four compared to X-ray film. By 'binning' the resolution is reduced twofold but is still comparable to the resolution of most phosphoimagers.. The instrument has a dynamic range of 1:10,000, allowing to image signals in a single exposure and perform real quantitative analysis.

Because of the special optics and software for flatfield correction plus detector specific corrections, accurate quantitation of chemiluminescent signals over the entire sample is possible. For microtiterplate measurements, the Lumi-Imager™ demonstrates excellent linearity (< 1 %) over the whole dynamic range and compares very favourable to dedicated luminescence microtiter plate readers.

The high resolution of the detector (1280 x 1024 pixels) and the special designed optics achieve a resolution of two lines per millimeter over the whole sample area of 300 x 240 mm^2 and permit a fixed optical alignment. The fixed optical alignment and a specially designed focus stabilisation for temperature variations and long term effects avoid repeated, tedious focussing.

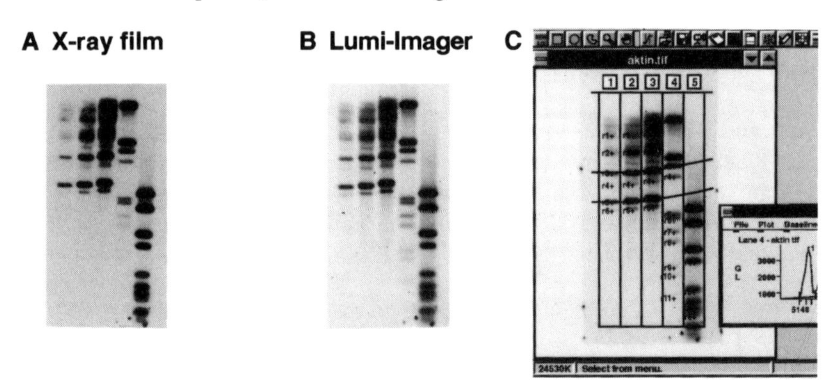

Figure 2: A human genomic Southern blot was hybridized with a DIG-labeled actin probe and detected with CDP-*Star*™. The signals were recorded by a 10 min exposure to (A) X-ray film or (B) 10 min image acquisition with the Lumi-Imager™. Some software features (automatic lane finding, automatic band finding, slant adjustement) are shown in (C).

The camera driver is integrated in the Windows™-based software and performs all system-corrections automatically before analyzing the data. The software enables fast and accurate molecular weight determination, quantification of bands and dots as well as microtiterplates including calibration. Experimental data can be stored together with the images in an integrated databank. Data can be directly transferred to Excel™ for further analysis and generation of graphics. The TIFF image files can be imported in other programs for documentation or publication purposes.

A NEW GENERAL, UNIFYING APPROACH TO THE THEORY AND PRACTICE OF LUMINOMETRIC METHODS OF ANALYSIS

DH Leaback
Biolink Technology Ltd., 5 Links Drive, Radlett, Herts, WD7 8BD, England

Introduction

The attractions of luminometric methods of analysis stem partly from the ease with which it is now possible to determine low levels of emitted light, but also from the potential simplicity of the procedures and instrumentation concerned. Unlike absorptiometric methods of analysis, the corresponding luminometric techniques suffer the disadvantage that the actual measurements are arbitrary (1). The latter have to be calibrated using typically tedious-to-make-up, unstable, standards and reagents. In 1993 Leaback (2) described calibrated, essentially stable, easy-to-use, light standards which make it possible to readily test luminometer performances and calibrations. Such standards also offer the possibility of the direct comparison of the analytical results from widely different luminometers.

How this may be done, can be illustrated by reference to the schematic below of a typical luminometer. Thus, if we consider a given volume of a solution of concentration c, of a chemiluminescent species with a quantum yield Q, the rate of total luminescent emission would then be:-

$L_T = c.Q$....................I

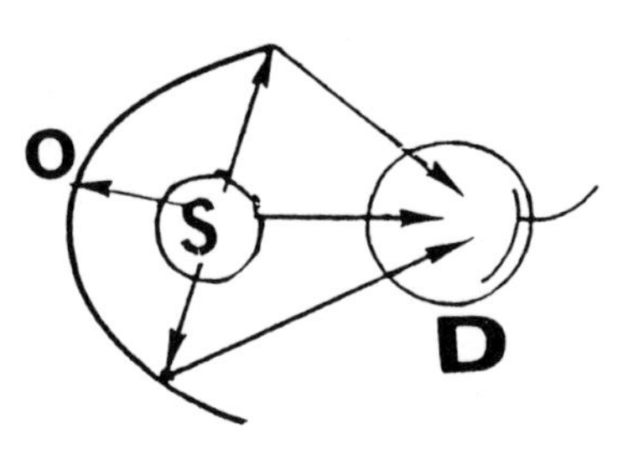

Schematic diagram of a typical luminometer comprising, a sample S, from which emitted light is transmitted to a detector D, via a mirror O - or other optics O.

Of course, no real luminometer can detect the total light emitted from a given luminescent sample. Accordingly, the above equation is modified to form the following experimentally - useful relationship, where the light detected (at D in the schematic) is:-

$L_D = c.d.Q.P_O.P_D$.....II

and where d is the 'depth' of solution concerned, P_O and P_D are the proportions of light transmitted by the optics, (eg. O in the schematic), and detected (at D in the schematic), respectively. The product $P_O.P_D$, the photon detection efficiency, (PDE) is a measure of the overall effectiveness of the particular luminometric system.

This publication will present examples illustrative of this approach.

Materials and Methods

ATP estimations were carried out using the firefly luciferase Lumit reagents (Lumac Ltd., Batley, Yorks.,UK), and performed as recommended by the manufacturers, but read out in the luminometers specified in the text. Calibrated Biolink light standards (Biolink Technology Ltd., Radlett, Herts., UK) were as described earlier (2). The luminometers employed here included, the tube instruments; M2010 Biocounter (Lumac Ltd., Batley, Yorks.,UK); Optocomp II, (MGM Instruments Inc., Hamden, CT, USA), and the BioOrbit 1251 & 1253 (BioOrbit, Turku, Finland). Also, the plate-reading Luminoskan (Labsystems, Helsinki, Finland). The imaging instruments were the Imaging Photon Detector (IPD) system described earlier (3); the Charge Couple Detector (CCD) system reported elsewhere (3,5); a Microphot microscope (Nikon Ltd., Telford, Salop, UK) fitted with an Isis Detector (Photonic Sciences, Robertsbridge, Sussex, UK), and an Argus system (Hamamatsu, Enfield, Middx., UK).

Results and Discussion

In Fig. 1 are shown logarithmic plots of the results of calibrating three typical, photon-counting, tube-luminometers, using appropriately calibrated Biolink light standards. Note that all three luminometers show about 4 orders of magnitude of linear, unit-slope photometric response, culminating in overload discontinuities of photon fluxes of around 1E4/sec. Applying equation II, gives PDE values of up to 4% (Table 1). Analogous results were obtained on a non-photon counting (BioOrbit 1251) tube-luminometer (Table 1), though calculation of the corresponding PDE value requires anciliary electronic data.

In Fig. 2 can be seen illustrative logarithmic plots of the results of the calibration of a non-photon-counting, plate-reading luminometer. This demonstrates about 4 orders of magnitude of linear response with unit slope, before cut-off by incorporated overload protection (Table 1).

Since the introduction of astronomical imaging detectors to biochemical purposes during the 1980s (1,2,3,4,), such detectors have been applied to a wide variety of studies involving luminometry. Such detectors generally involve either image intensifiers (It) amplifiers, or charge couple devices for the determination of quantitative images of emitted photons. Table 1 also lists the results of logarithmic plots of calibrations of a variety of imaging luminometers, based upon these two types of imaging detector.

Analogous calibrations on these imaging luminometers, using Biolink light standards, showed (Table 1) generally poorer dynamic ranges of linear, unit slope, calibration curves for the image-intensifier-based instruments, when compared with that of the CCD instrument.

All the instruments were ultimately limited by photon flux overload at the tops of their ranges and, to varying extents, by background or statistical exigencies at the lower ranges of photon flux. The latter limitation is obviously exacerbated by low values of PDE. In this context, the relative values for the IPD and CCD instruments listed in Table 1, can be attributed to their respective use of low and high demagnification optics used in the two instances (3,4,5).

It is instructive to enquire how these new insights into luminometer design and performance can be valuable in interpreting and improving the results of luminogenic assays. In Fig. 3 are logarithmic plots of the results of luminogenic

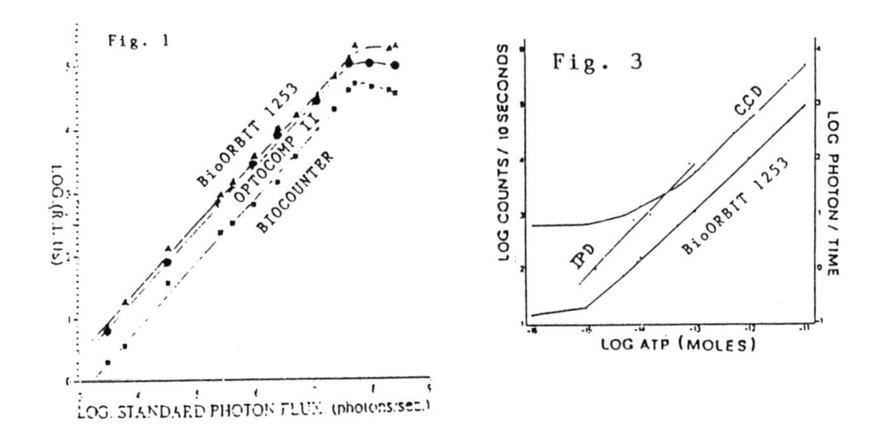

Fig. 1 Photometric characteristics of three, photon-counting, tube luminometers.

Fig. 2 Photometric characteristics of a (non-photon-counting) plate-reading, luminometer.

Fig. 3 Responses of one, tube and two types of imaging luminometers, to decreasing levels of ATP, determined using the firefly luciferasē reaction.

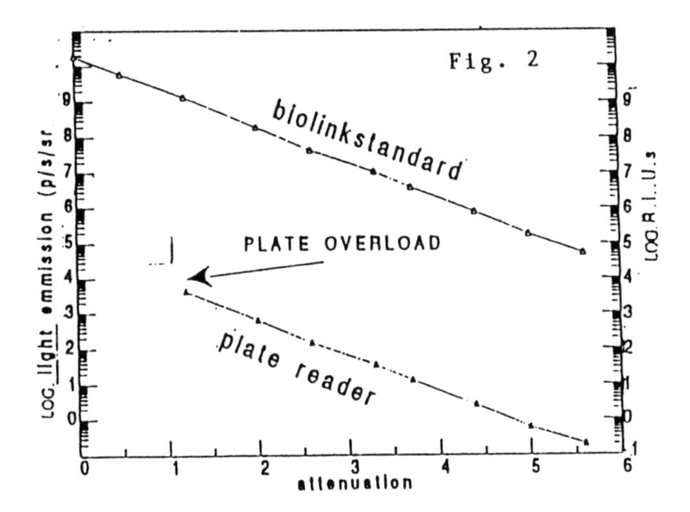

Luminometer	Type	Linear region (orders of mag.)	Limiting discontinuity	PDE %	Reference
Biocounter	T.P.	4	overload	1	Fig. 1
BioOrbit 1251	T.C.	4	overload	-	text
BioOrbit 1253	T.P.	4	overload	4	Fig. 1
Optocomp II	T.P.	4	overload	3	Fig. 1
Luminoskan	Pl.C.	4	overload	-	Fig. 2
I.P.D.	i.P.It.	3	overload/backgrnd	0.5	(3)
Isis.	i.P.It.	3	overload/backgrnd	-	text
Argus	i.P.It.	3	overload/backgrnd	-	text
C.C.D.	i.P.	4.5	overload/backgrnd	0.05	(4;5)

Table 1. Photometric characteristics of the nine luminometers concerned here. They include tube- (T), plate-reading (Pl), and imaging (i), luminometers operating with either photon counting (P), or current (c) read-out. Image intensifier-based systems are indicated thus- It.

ATP assays carried out using three widely-different luminometers with PDE values of 1, 0.5 and 0.05%. It can be seen (Fig. 3) that the respective instruments yielded about 4,3 and 3 orders of magnitude of linear, unit slope regions of ATP calibrationn curves. While these results are in accord with the data for the BioOrbit 1253 and IPD in Table 1, the departure from the ideal at the lower end of the CCD curve must be attributed to the low PDE value, consequent to the large demagnification involved in a set-up for imaging 96-well micro-test plates (3,4).
This approach manifestly yields new insight into luminometer design and performance and permits the selection of optimal conditions for particular analytical applications. Provided values of Q and PDE are available, this approach also makes possible, not only direct comparisons of results from divers instruments, but also direct measurements of parameters at the molecular level. It is difficult to imagine scientific aspects of this symposium on which this work does not impinge.

Acknowledgements
I am greatly indebted to RA Ladds, JA Leaback and PD Leaback for assistance, and to the various manufacturers, for access to their instruments.

References

1. Leaback DH. Anal Chim Acta 1989;227;1-9.
2. Leaback DH. Bioluminescence and Chemiluminescence: Status Report. Szalay AA, Kricka LJ, Stanley P, editors. Chichester. John Wiley and Sons Ltd. 1993:33-7.
3. Leaback DH, Hooper CE. Bioluminescence and Chemiluinescence: New Perspectives. Schölmerich J, Andreesen R, Kapp A, Ernst M,Woods WG, editors. Chichester. John Wiley and Sons Ltd. 1987, 439-42.
4. Leaback DH, Haggart R. J Biolumin Chemilumin 1989;4;512-22.
5. Haggart R, Leaback DH. Bioluminescence and Chemiluminescence: Current Status. Stanley PE, Kricka LJ, editors. Chichester. John Wiley and Sons Ltd. 1991: 365-8.

HYDROBIOPHYSICAL DEVICE 'SALPA' OF THE INSTITUTE OF BIOLOGY OF THE SOUTHERN SEAS USED FOR BIOLUMINESCENT INVESTIGATION OF THE UPPER LAYERS OF THE OCEAN

VI Vasilenko, EP Bitykov, BG Sokolov, YuN Tokarev
Institute of Biology of the Southern Seas, Sevastopol 335011, Ukraine

Introduction

Modern methods of bioluminescence field studies have been remarkably upgraded in the last decade, from the simple bathyphotometers (1-3) to a family of specialized hydrobiophysical instruments (4-8). The Laboratory of Biophysical Ecology of IBSS, one of the founders of bioluminescence studies in the Former Soviet Union, has implemented the development of several measuring and registering instruments in this domain. The first bathyphotometers designed in the laboratory had a cylindrical form, 65cm length and 15cm in diameter (9,10). The illuminator for the bioluminescent sensor was placed on its end plate. To calibrate the sensor a luminofor, etalons, activated by C-14 isotope, was used. This etalon has a permanent and stable illumination of 500nm.

Bathyphotometers, with a space angle of 3 steradians, were equipped with a light-protecting unit. The disk-like units of the bathyphotometers of the first generation reduced the astronomic background light by 70 times. The light protectors used in the latest modules, with rotor-type impellers reduce the astronomic background light by 106 times.

The necessity to study the spatial-temporal variability of the bioluminescent fields on the scale 1-100mm, together with the physical environment, has predetermined the development of the new generation of bathyphotometers of the last decade.

The aim of this paper is to give the basic technical characteristics of bathyphotometer **SALPA**.

Methods

The complex **SALPA** was designed for vertical profiling of the bioluminescent potential, temperature, conductivity and the hydrostatic pressure in the upper 200m layer of the ocean, when a vessel was drifting.

The bathyphotometer has an onboard and a submersible unit. Temperature, conductivity and hydrostatic pressure sensors are mounted on the upper cover of the titanium body. The rotor impeller is situated on the bottom cover. The rotation of rotor vanes induces the mechanical stimulation of the bioluminescent planktonic organisms. Their luminescence is detected by the sensor and transferred via the wave beam to the photocathode FEY-71 (6).

A signal, from the output of the submersible device is transferred via a cable to the onboard unit, in a form of bipolar binary code. The onboard device governs the mode of vertical profiling and restores the output signals from the submersible device.

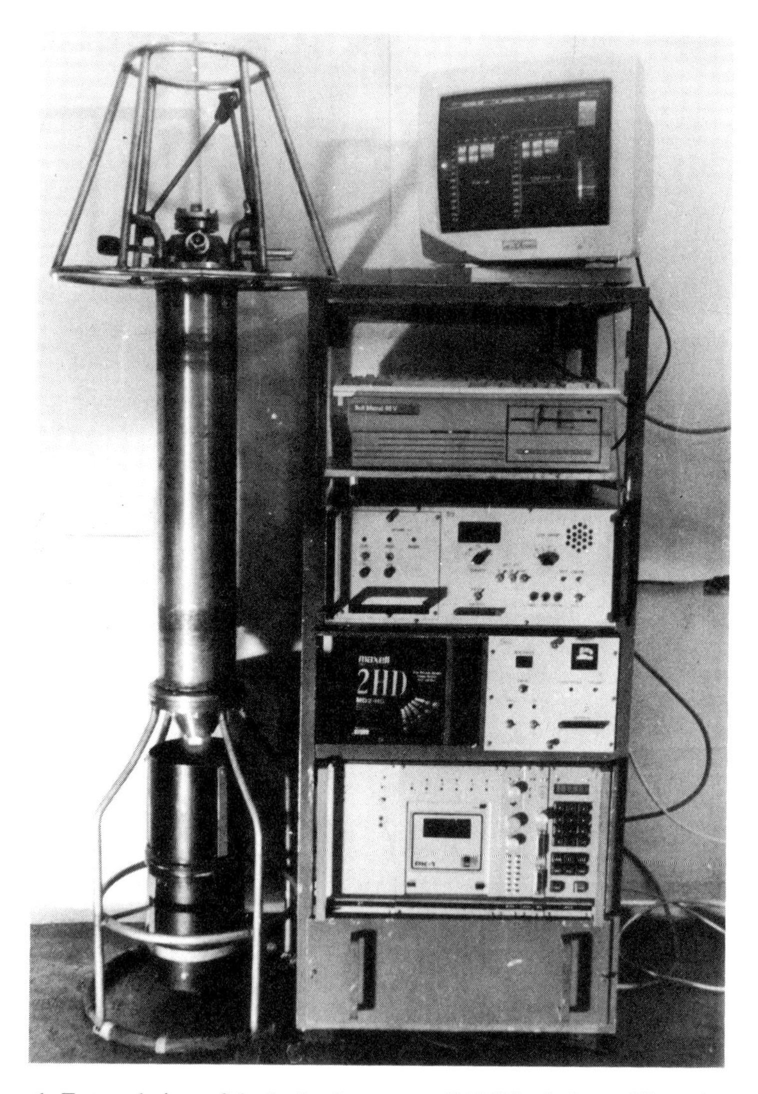

Figure 1. External view of the bathyphotometer **SALPA** (submersible and on deck units)

Results

Bathyphotometer **SALPA** has the following major characteristics:
-band of bioluminescence measurements, 10^{-13} - 10^{-8} Wcm^{-2};
-temperature band, -2 - 35°C;
-conductivity band, 1.5 - 6.5 cm m^{-1};

-pressure band, 0 - 2 Mpa;
-commutation frequency band, 6 - 100 Hz;
-the velocity of vertical profiling, 1.2 ms^{-1};
-spatial resolution of bioluminescence measurement in the vertical profiling mode, 0.2, 0.5, 1m.

To calibrate the bioluminescence sensor, the radioactive C-14 source (etalon) is used. It has the irradiation diameter 1.3mm and the absolute intensity of irradiation= 8.6 10^{-10} W, with the maximum of irradiation at 520nm.

The sensitivity of the bioluminescence channel (K) can be determined as the follows: $K = Fs/ Đ(h_2 + r_1^2 + r_2^2)N$, where Fs- is the irradiance flux of the etalon, h_2- the distance between photocathode and etalon, r_1^2 - the diameter of the irradiant surface of the etalon, r_2^2- the cathode diameter of photoelectric multiplier, N- value of the information code on the output of the bioluminescence channel.

The software designed for the operation of **SALPA** enables the graphic presentation of vertical profiles, the assessment of the routine statistical parameters, the spectral analysis etc.

Our experience of **SALPA** application in field oceanic research has shown that it is a reliable instrument which has enabled us to obtain long time series of biological and physical parameters in various regions of the world's oceans.

Acknowledgments
This work was part of an International cooperation program between IBSS and Plymouth Marine Laboratory and has been funded by ONR grant No. N00014-95-1-0089

References

1. Boden BP, Kampa EM. Records of bioluminescence in the ocean. Pacif Sci 1957;2229-35.
2. Clarke GL, Wertheim GK. Measurements of illumination at great depth and at night in the Atlantic Ocean by means of a new bathyphotometer. Deep-Sea Res 1956;4:189-205.
3. Clarke GL, Backus RH. Measurements of light penetration in relation to vertical migration and records of luminescence of deep sea animals. Deep-Sea Res 1956;4:1-14.
4. Arneson AC, Benyamin S, Janes B, Schmidt GW. A charge-coupled device (CCD) spectrophotometer for measuring marine bioluminescence. Mar Ecol 1988;43:277-83.
5. Batchelder HP, Swift E. Estimated near-surface mesopelagic bioluminescence in Western North Atlantic during July 1986. Limnol Oceanogr 1989;31:113-28.
6. Gitelson II, Levin LA, Utushev RN, Cherepanov OA, Chugunov YuN. Ocean bioluminescence. S.-Petersburg, Nauka, 1992 (in Russian).
7. Kiefer DA, Ondercin D. Mapping and modeling of bioluminescence in north Atlantic and Pacific oceans. In: Bioluminescence Symposium. Westin Maui, Kaanapali Beach, Hawaii.1993:82.

8. Lapota D, Paden S, Duckworth D, Rosenberg DE, Case JF. Coastal and oceanic bioluminescence trends in the Southern California Bight using MOOREX bathyphotometers. In: Campbell AK, Kricka LJ, Stanley PE, editors. Bioluminescence and Chemiluminescence: Fundamentals and Applied Aspects. Chichester: John Wiley & Sons, 1994:127-30.
9. Bityukov EP, Ribasov VP, Shaida VG. Annual variations of the bioluminescence field intensity in the neritic zone of the Black Sea. Oceanology (Okeanologiya) 1967;7:1089-99 (in Russian).
10. Bityukov EP, Vasilenko VI, Tokarev YuN, Shaida VG. Bathyphotometer with distance-switched sensitivity for estimating intensity of bioluminescent field. Hydrobiol J (Hidrobiologicheskii Jurnal) 1969;5;82-6 (in Russian).

COMPARISON OF REAL LIFE LUMINOMETER SENSITIVITY AS DETERMINED WITH ACTUAL LUMINESCENCE ASSAYS

PS Fullam and BS Fullam
NVE Inc. Chimayo, NM, 87522 USA

Introduction

Bioluminescence and chemiluminescence assays have been demonstrated to be extremely sensitive. A key to reproducibility for these assays is the standardization and calibration of the measurement instruments. Unfortunately the current methodology for the calibration relies on light sources which in most cases do not match the assay of interest. As has been pointed out previously (1) the peak spectral output of the most commonly used light source (tritium) is significantly different from the ATP assay.

The spectral output of luminescent reactions is not at single wavelength, but actually a distribution over a range of wavelengths typified as a skewed bell shaped curve with a peak output at the referenced wavelength. Various factors such as pH and contaminates can alter the curve shape as well as the peak wavelength.(2)

The photomultiplier tube (PMT) used in luminometers is characterized by a quantum efficiency verses wavelength curve. Instrument designers chose the PMT based on typical quantum efficient curves supplied by the PMT manufacturer. Individual PMTs can demonstrate significant variation from the published typical curve. This can result in final instruments which can produce disparate results when measurements are made at the limits of the sensitivity curve.

Materials and Methods

<u>Instrumentation:</u> The following luminometers were used:
Model BG-P (MGM Instruments, Hamden, CT, USA)
Model 1251 (BioOrbit Oy, Turku, Finland)
Flyte 400 (Cardinal Assoc., Santa Fe, NM USA)
PicoLite 6100 (Packard Inst., Downers Grove, IL, USA)
PicoLite 6500 (Packard Inst., Downers Grove, IL, USA)
Model 20 e (Turner Designs, Mountain View, CA, USA)
Wave 180 and Wave 250 (Coral Biotechnology, San Marcos CA, USA)
<u>Reagents:</u> Luciferin/Luciferase (NVE, Inc., Chimayo, NM, USA), Luminol, ATP (Sigma Chemical Co., St. Louis, MO, USA), Tritium/toluene (ARC Inc., St. Louis, MO, USA)
<u>Procedures:</u> Assays were performed per the instrument manufactures operating instructions, with data collected as sets of five replicates. All measurements were taken as ten second integrated values, with five second delay after reagent additions. In the case of the tritium sources the timing was the same, though no reagents were added.

ATP measurements were made by placing 25 uL of known concentration ATP solution (from $1x10^{-7}$ mol/L down to $1x10^{-15}$ mol/L) in a polystyrene tube, and then adding 100 uL of the Luciferin/Luciferase cocktail. The initial dilution of ATP was spectrophotometrically assayed to verify its concentration, and dilutions were made using reagent grade water.

The luminol assay was performed by placing 25uL of diluted H_2O_2 (from 0.03 mol/L to $1x10^{-11}$ mol/L) into polystyrene tubes, and then adding 100uL of $1x10^{-4}$ mol/L Luminol (initially dissolved in DMSO and 0.01 mol/L NaOH). The Luminol was chosen due to its emission spectrum which is close to that of the tritium light sources. (3)

The tritium sources were made using 0.06uCi to 5 uCi. The volume was sealed in a glass ampule, which was then sealed in a polystyrene tube.

Reagents were added automatically for the Model 6500, Model 20 e, Flyte 400, Model 1251, Wave 180 and Wave 250. Manual additions were required for the Model BG-P and the Model 6100. Careful timing was observed in order to obtain the best reproducibility.

Results and Discussion

For each instrument identical sets of data were collected for the ATP concentrations ranging from $1x10^{-15}$ mol/L to $1x10^{-7}$ mol/L, tritium sources, and luminol acting on hydrogen peroxide. The results were averaged and plotted on figures 1, 2, and 3. The relative light unit (RLU) values were normalized in order to use the same scale.

The repeatability of the data varies with the light producing system. The best comparative data was obtained with the ATP/Lucuferin/luciferase system, followed by the tritium sources, with the luminol system having the worst reproducibility between instruments.

Figure 1 displays the data from the ATP measurement demonstrating a close agreement between luminometers within the measurement range of $1x10^{-11}$ mol/L to $1x10^{-7}$ mol/L. This implies a simple relationship between RLU measured and ATP concentration and suggests that luminometer data could be translated from one model to another.

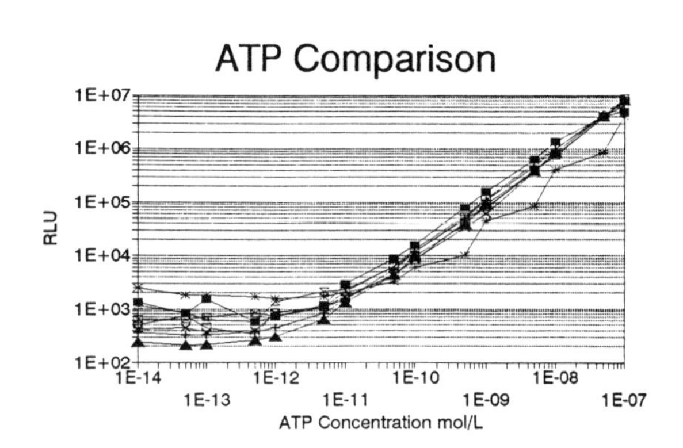

Model 6500 — Wave 180 — Model 1251
Model BP-G — Model 20 e — Wave 250
Model 6100 — Flyte`400

Figure 2 is for the Tritium sources. This data is also correlated between luminometers, but requires different scaling values than the ATP measurements. This counters the above suggestion of a simple relationship between instruments.

Figure 3 is for Luminol. This data is not as tightly correlated between instruments. It is probable that the kinetics of the reaction account for an increased experimental error. Due to the nature of the luminol reaction, this data suggests that reaction rates may have been effected by a lack of catalyst.

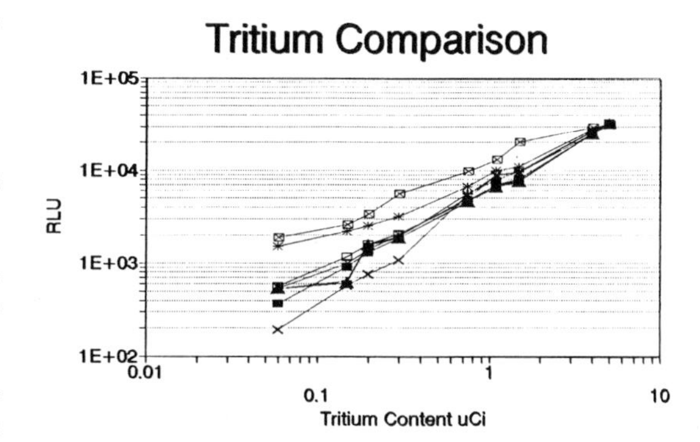

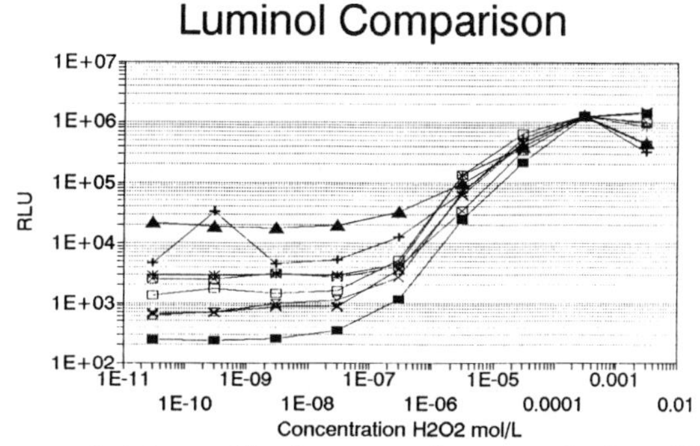

Each set of data was scaled (normalized) to produce equivalent RLU values. Table one lists the scale values used for each luminometer and reagent system. Both the ATP system and the tritium sources produced relatively constant light output, while the luminol system produces a high peak with rapid decline. (4)

	ATP	Tritium	Luminol
Model 6500	0.53	0.39	0.14
Model 6100	3.68	5.24	15.71
Flyte 400	0.31	0.78	1.85
Model 20 e	820.79	1826.25	4639.51
Model BP-G	0.47	0.47	3.77
Model 1251	70.97	152.41	358.72
Wave 180	3.00	1.31	6.90
Wave 250	5.08	0.92	13.13

TABLE 1 Scale factors used to normilize raw data based on regression coeficients.

While there appears to be general agreement between luminometers on relative measurements of various chemiluminescence reactions, there is significant inconsistency between reactions. If each luminometer was accurately measuring photons (either by count or current averaging) a simple equation should be able to be developed to convert any reading to another instrument. Yet the current data does not yield such an equation.

This is of significance when calibrating different luminometers or switching data collection from on instrument to another through the use of a tritium light source. Such changes in instruments can produce distortions in the data collected, and the test conclusions. Such calibration methodology could result in significant errors and misleading results.

The implication is that some other mechanism may be influencing the data, such as the spectral distribution of the light or kinetic response of the reaction. The data demonstrates a strong correlation for all instruments and the shape of the curves for each of the chemiluminescence systems measured. The presumption of a simple scale factor which would normalize all readings between luminometers was proved false. There appears to be different scaling depending on specific chemiluminescence reaction.

These reaction specific scaling factors raise a question as to what the luminometers are measuring. It can be argued that the luminol / H_2O_2 reaction involves rapidly changing (high peak followed by an exponential decline) light output as well as the effect of a catalyst limited reaction (4) which can help explain some of the differences. The Luciferin/Luciferase reaction and tritium source produce relatively constant light output.

The reaction specific response of the luminometers raises concern on development of calibration and standardization methodologies.

Acknowledgements
This research was funded by NVE, Inc. of Chimayo, N.M. and facilitated by Cardinal Associates of Santa Fe, N.M. who generously provided use of their laboratory space. Additionally George Kyrala,PhD of Los Alamos National Laboratory provided significant understanding necessary for light emission and measurement.

References

1. Turner GK. Measurement of light from chemical or biochemical reactions. In: Van Dyke K, editor. Bioluminescence and Chemiluminescence: Instruments and Applications Vol 1. Boca Raton: CRC Press, 1985:43-76
2. DeLuca M, McElroy WD. Purification and properties of firefly luciferase. In: DeLuca MA, editor. Methods in Enzymology Vol LVII. New York: Academic Press, 1978:3-15
3. Murphy ME, Sies H. Visible-range low-level chemiluminescence in biological systems. In: Packer L, Glazer AN, editors. Methods in Enzymology, Vol 186. New York: Academic Press, 1990:595-610
4. Nieman TA, Detection based on solution-phase chemiluminescence systems. In: Birks JW, editor. Chemiluminescence and Photochemical Reaction Detection in Chromatography. New York: VCH Publishers, Inc., 1989:99-123